Clinical Ophthalmic Genetics and Ge[

Clinical Ophthalmic Genetics and Genomics

Edited by

Graeme C.M. Black
Professor of Genetics and Ophthalmology, University of Manchester, Manchester, UK

Jane L. Ashworth
Consultant Ophthalmic Surgeon and Paediatric Ophthalmologist, Manchester Royal Eye Hospital, Manchester, UK

Panagiotis I. Sergouniotis
Senior Lecturer and Consultant Ophthalmologist, University of Manchester, Manchester, UK

Academic Press is an imprint of Elsevier
125 London Wall, London EC2Y 5AS, United Kingdom
525 B Street, Suite 1650, San Diego, CA 92101, United States
50 Hampshire Street, 5th Floor, Cambridge, MA 02139, United States
The Boulevard, Langford Lane, Kidlington, Oxford OX5 1GB, United Kingdom

Copyright © 2022 Elsevier Inc. All rights reserved.

No part of this publication may be reproduced or transmitted in any form or by any means, electronic or mechanical, including photocopying, recording, or any information storage and retrieval system, without permission in writing from the publisher. Details on how to seek permission, further information about the Publisher's permissions policies and our arrangements with organizations such as the Copyright Clearance Center and the Copyright Licensing Agency, can be found at our website: www.elsevier.com/permissions.

This book and the individual contributions contained in it are protected under copyright by the Publisher (other than as may be noted herein).

Notices
Knowledge and best practice in this field are constantly changing. As new research and experience broaden our understanding, changes in research methods, professional practices, or medical treatment may become necessary.

Practitioners and researchers must always rely on their own experience and knowledge in evaluating and using any information, methods, compounds, or experiments described herein. In using such information or methods they should be mindful of their own safety and the safety of others, including parties for whom they have a professional responsibility.

To the fullest extent of the law, neither the Publisher nor the authors, contributors, or editors, assume any liability for any injury and/or damage to persons or property as a matter of products liability, negligence or otherwise, or from any use or operation of any methods, products, instructions, or ideas contained in the material herein.

Library of Congress Cataloging-in-Publication Data
A catalog record for this book is available from the Library of Congress

British Library Cataloguing-in-Publication Data
A catalogue record for this book is available from the British Library

ISBN 978-0-12-813944-8

For information on all Academic Press publications
visit our website at https://www.elsevier.com/books-and-journals

Publisher: Andre G. Wolff
Acquisitions Editor: Peter B. Linsley
Editorial Project Manager: Megan Ashdown
Production Project Manager: Debasish Ghosh
Cover Designer: Greg Harris

Typeset by STRAIVE, India

Printed in the United States of America

Last digit is the print number: 9 8 7 6 5 4 3

Contents

Contributors xi
Preface xv

Section I
Genomics and the eye

1. **Genetic disorders and genetic variants** 1

   Sofia Douzgou Houge, Graeme C.M. Black, and Panagiotis I. Sergouniotis

2. **Genetic testing techniques** 7

   Caroline Van Cauwenbergh, Kristof Van Schil, Miriam Bauwens, and Elfride De Baere

3. **Genetic variant interpretation** 13

   Simon C. Ramsden

4. **Genetic counselling and family support** 21

   Georgina Hall, Avril Daly, Christina Fasser, and Jane L. Ashworth

5. **Syndromic conditions and the eye** 25

   Jill Clayton-Smith

6. **Ophthalmic phenotyping: Electrophysiology** 33

   Neil R.A. Parry and Panagiotis I. Sergouniotis

7. **Ophthalmic phenotyping: Imaging** 53

   Johannes Birtel, Martin Gliem, Wolf M. Harmening, and Frank G. Holz

8. **Gene therapy and treatment trials** 63

   Robert E. MacLaren and Jasmina Cehajic-Kapetanovic

Section II
Genetic disorders affecting the anterior segment

9. **Genetic disorders affecting the cornea** 67

   9.1 **Primary megalocornea** 68

   Graeme C.M. Black

   9.2 **Brittle cornea syndrome** 70

   Graeme C.M. Black, Emma Burkitt-Wright, and Susmito Biswas

   9.3 **Corneal dystrophies** 73

   Petra Liskova, Susmito Biswas, Graeme C.M. Black, and Francis L. Munier

Meesmann epithelial corneal dystrophy	74
Epithelial–stromal *TGFBI* corneal dystrophies	77
Corneal endothelial dystrophies	81
Miscellaneous corneal dystrophies	83

   9.4 **Keratopathy in inborn errors of metabolism** 85

   Susmito Biswas and Jane L. Ashworth

Mucopolysaccharidoses	85
Mucolipidoses	85
Cystinosis	86
Tyrosinaemia type II	88
Wilson disease	89
Fabry disease	89
LCAT-related metabolic disease	89
Tangier disease	91

   9.5 **Keratopathy in ectodermal dysplasias** 93

   Colin E. Willoughby and Michele Callea

Hypohidrotic ectodermal dysplasia	93
Ectodermal dysplasia with cleft lip/palate	94
Keratitis-ichthyosis-deafness syndrome	95

10. Anterior segment developmental disorders — 97

Arif O. Khan, Panagiotis I. Sergouniotis, and Graeme C.M. Black

10.1 Primary congenital glaucoma — 98
10.2 Primary juvenile glaucoma — 102
10.3 Axenfeld-Rieger spectrum — 105
10.4 Peters anomaly — 109

11. Cataract — 113

I. Christopher Lloyd, Sofia Douzgou Houge, Rachel L. Taylor, Jill Clayton-Smith, and Graeme C.M. Black

11.1 Non-syndromic congenital cataract — 114
11.2 Syndromic congenital cataract — 118
  Cockayne syndrome — 119
  Warburg Micro syndrome — 121
  Oculofaciocardiodental syndrome — 125
  Nance–Horan syndrome — 127
  Cataract in inborn errors of metabolism — 129
  Lowe oculocerebrorenal syndrome — 133
  Hyperferritinemia-cataract syndrome — 137
  Galactosaemia — 140
  Cerebrotendinous xanthomatosis — 144

12. Ectopia lentis — 147

Aman Chandra, Panagiotis I. Sergouniotis, Jane L. Ashworth, and Elias I. Traboulsi

12.1 *ADAMTSL4*-related disorders, including isolated ectopia lentis — 149
12.2 Marfan syndrome — 151
12.3 Weill–Marchesani syndrome — 155
12.4 Homocystinuria — 157

Section III
Genetic disorders affecting the posterior segment

13. Genetic disorders affecting the retina, choroid and RPE — 159

13A. Genetic disorders causing non-syndromic retinopathy — 161

13A.1 Non-syndromic retinitis pigmentosa — 162

Mays Talib, Caroline Van Cauwenbergh, and Camiel J.F. Boon

13A.2 Choroideremia — 172

Maria I. Patrício, Kanmin Xue, Miguel C. Seabra, and Robert E. MacLaren

13A.3 Enhanced S-cone syndrome and *NR2E3*-associated disorders — 176

Pascal Escher, Kaspar Schuerch, Martin Zinkernagel, Viet H. Tran, and Francis L. Munier

13A.4 Congenital stationary night-blindness — 181

Anthony G. Robson, Eva Lenassi, Panagiotis I. Sergouniotis, Isabelle Audo, and Christina Zeitz

13A.5 Leber congenital amaurosis and severe early childhood onset retinal dystrophies — 189

Bart P. Leroy and Panagiotis I. Sergouniotis

13A.6 Cone dysfunction disorders — 194

Nashila Hirji, Michalis Georgiou, and Michel Michaelides

13A.7 Cone/cone-rod dystrophies — 200

Alberta A.H.J. Thiadens and Caroline C.W. Klaver

13A.8 *ABCA4*-related disorders — 207

Eva Lenassi

13A.9 *BEST1*-related disorders (bestrophinopathies) — 217

Ine Strubbe, Panagiotis I. Sergouniotis, and Bart P. Leroy

| 13A.10 | Pattern dystrophies | 225 |

Omar A. Mahroo

| 13A.11 | X-linked retinoschisis | 236 |

Laryssa A. Huryn, Catherine A. Cukras, and Paul A. Sieving

| 13A.12 | Occult macular dystrophy | 241 |

Yu Fujinami-Yokokawa, Anthony G. Robson, Panagiotis I. Sergouniotis, and Kaoru Fujinami

| 13A.13 | North Carolina macular dystrophy | 246 |

Kent W. Small, Jingyan Yang, and Fadi Shaya

| 13A.14 | Genetic disorders mimicking age-related macular disease | 250 |

Veronika Vaclavik, Panagiotis I. Sergouniotis, Graeme C.M. Black, and Francis L. Munier

| 13A.15 | Genetic architecture of age-related macular degeneration | 261 |

Johanna M. Colijn and Caroline C.W. Klaver

13B.	Syndromic retinal disease	267
13B.1	Ciliopathies	268
13B.1.1	Bardet-Biedl syndrome	270

Hélène Dollfus

| 13B.1.2 | Joubert syndrome | 274 |

Frauke Coppieters, Elise Héon, and Monika K. Grudzinska Pechhacker

| 13B.1.3 | Alström syndrome | 277 |

Isabelle Meunier, Hélène Dollfus, Vasiliki Kalatzis, Béatrice Bocquet, and Catherine Blanchet

| 13B.1.4 | Usher syndrome | 281 |

Francesco Testa and Francesca Simonelli

| 13B.2 | Retinopathy in inborn errors of metabolism | 285 |
| 13B.2.1 | Gyrate atrophy of the choroid and retina | 288 |

Karolina M. Stepien, Panagiotis I. Sergouniotis, and Graeme C.M. Black

| 13B.2.2 | Bietti corneoretinal crystalline dystrophy | 292 |

Erin C. O'Neil and Tomas S. Aleman

| 13B.2.3 | Pseudoxanthoma elasticum | 295 |

Julie De Zaeytijd

| 13B.2.4 | MIDD (maternally inherited diabetes, deafness and maculopathy) | 302 |

Neruban Kumaran and Michel Michaelides

| 13B.2.5 | Long-chain L-3 hydroxyacyl-CoA dehydrogenase (LCHAD) deficiency | 306 |

Kristina Teär Fahnehjelm, Anna Nordenström, and Jane L. Ashworth

| 13B.2.6 | Neuronal ceroid lipofuscinosis (Batten disease) | 309 |

Dipak Ram and Jane L. Ashworth

| 13B.2.7 | Cobalamin C deficiency | 313 |

Tomas S. Aleman, Bart P. Leroy, and Giacomo M. Bacci

| 13B.2.8 | Cohen syndrome | 317 |

Kate E. Chandler

| 14. | Familial vitreoretinopathies | 323 |
| 14.1 | Familial exudative vitreoretinopathy spectrum | 324 |

Johane Robitaille and Graeme C.M. Black

Familial exudative vitreoretinopathy	324
Norrie disease	329
KIF11-related disorders	331

14.2 Incontinentia pigmenti — 334

Arundhati Dev Borman and Robert H. Henderson

14.3 Stickler syndrome and allied collagen vitreoretinopathies — 339

Martin P. Snead

14.4 Wagner disease — 347

Cyril Burin des Roziers, Pierre-Raphaël Rothschild, Antoine Brézin, and Sophie Valleix

14.5 Knobloch syndrome — 351

Irina Balikova

15. Genetic disorders affecting the optic nerve — 355

15.1 Leber hereditary optic neuropathy — 356

Ungsoo S. Kim and Patrick Yu-Wai-Man

15.2 Autosomal dominant optic neuropathy — 362

Guy Lenaers, Patrizia Amati-Bonneau, Alvaro J. Mejia-Vergara, and Alfredo A. Sadun

15.3 Autosomal recessive optic neuropathy — 368

Valerio Carelli, Chiara La Morgia, and Piero Barboni

15.4 Wolfram syndrome spectrum — 371

Nicole Balducci, Chiara La Morgia, Michele Carbonelli, Piero Barboni, and Valerio Carelli

Section IV
Genetic disorders affecting both the anterior and posterior segment

16. Developmental eye disorders — 377

16.1 Microphthalmia–anophthalmia–coloboma spectrum — 378

Mariya Moosajee and Graeme C.M. Black

16.2 Nanophthalmia and posterior microphthalmia — 385

Arif O. Khan

17. Aniridia — 389

Graeme C.M. Black and Mariya Moosajee

18. Albinism — 393

Eulalie Lasseaux, Magella M. Neveu, Mathieu Fiore, Fanny Morice-Picard, and Benoît Arveiler

Section V
Genetic disorders affecting ocular motility

19. Infantile nystagmus — 403

Jay E. Self and Helena Lee

20. Congenital cranial dysinnervation disorders — 407

20.1 Congenital fibrosis of the extraocular muscles — 408

Mary C. Whitman and Elizabeth C. Engle

20.2 Duane retraction syndrome — 412

Arif O. Khan

20.3 Horizontal gaze palsy and progressive scoliosis — 418

Arif O. Khan

20.4 Moebius syndrome — 422

Mary C. Whitman and Elizabeth C. Engle

21. Progressive external ophthalmoplegia — 425

Ungsoo S. Kim, Graeme C.M. Black, and Patrick Yu-Wai-Man

Section VI
Tumour predisposition syndromes

22. Phakomatoses 429

22.1 Neurofibromatosis type 1 430
D. Gareth Evans

22.2 Neurofibromatosis type 2 434
D. Gareth Evans

22.3 Von Hippel–Lindau disease 439
Anthony T. Moore

22.4 Tuberous sclerosis complex 446
Graeme C.M. Black and Elizabeth A. Jones

23. Naevoid basal cell carcinoma syndrome 449
D. Gareth Evans

24. Congenital hypertrophy of retinal pigment epithelium (CHRPE) 453
Fiona Lalloo and Graeme C.M. Black

25. Retinoblastoma 457
Manoj V. Parulekar and Brenda L. Gallie

Index 465

Contributors

Numbers in parenthesis indicate the pages on which the authors' contributions begin.

Tomas S. Aleman (292, 313), Scheie Eye Institute at The Perelman Center for Advanced Medicine and The Children's Hospital of Philadelphia, University of Pennsylvania, Philadelphia, PA, United States

Patrizia Amati-Bonneau (362), CHU Angers, Angers, France

Benoît Arveiler (393), Université de Bordeaux and CHU de Bordeaux, Bordeaux, France

Jane L. Ashworth (21, 85, 151, 306, 309), Manchester Royal Eye Hospital and University of Manchester, Manchester, United Kingdom

Isabelle Audo (181), Sorbonne Université and CHNO des Quinze-Vingts, Paris, France

Giacomo M. Bacci (313), Meyer Children's University Hospital, Florence, Italy

Nicole Balducci (371), Sant'Orsola-Malpighi Hospital and Studio Oculistico d'Azeglio, Bologna, Italy

Irina Balikova (351), Leuven University Hospital, Leuven, Belgium

Miriam Bauwens (7), Ghent University and Ghent University Hospital, Ghent, Belgium

Piero Barboni (368, 371), IRCCS Istituto San Raffaele di Milano, Milan, Italy

Johannes Birtel (53), University Hospital Bonn, Bonn, Germany; Oxford Eye Hospital and University of Oxford, Oxford, United Kingdom

Susmito Biswas (70, 73, 85), Manchester Royal Eye Hospital and University of Manchester, Manchester, United Kingdom

Graeme C.M. Black (1, 68, 70, 73, 98, 102, 105, 109, 118, 125, 137, 250, 288, 329, 331, 378, 389, 425, 446, 453), University of Manchester and Manchester Centre for Genomic Medicine, Manchester, United Kingdom

Catherine Blanchet (277), Université de Montpellier and CHRU de Montpellier, Montpellier, France

Béatrice Bocquet (277), Université de Montpellier and CHRU de Montpellier, Montpellier, France

Camiel J.F. Boon (162), Leiden University and Leiden University Medical Center, Leiden; Amsterdam University and Amsterdam University Medical Centers, Amsterdam, The Netherlands

Antoine Brézin (347), Hôpital Cochin and Université de Paris, Paris, France

Cyril Burin des Roziers (347), Hôpital Cochin and Université de Paris, Paris, France

Emma Burkitt-Wright (70), Manchester Centre for Genomic Medicine, St Mary's Hospital, Manchester, United Kingdom

Michele Callea (93), Meyer Children's University Hospital, Florence, Italy

Michele Carbonelli (371), Università di Bologna, Bologna, Italy

Valerio Carelli (368, 371), IRCCS Istituto delle Scienze Neurologiche di Bologna; Università di Bologna, Bologna, Italy

Jasmina Cehajic-Kapetanovic (63), University of Oxford and Oxford Eye Hospital, Oxford, United Kingdom

Kate E. Chandler (317), Manchester Centre for Genomic Medicine and University of Manchester, Manchester, United Kingdom

Aman Chandra (149, 155, 157), Southend University Hospital, Southend-on-Sea; Anglia Ruskin University, Cambridge, United Kingdom

Jill Clayton-Smith (25, 118, 119), Manchester Centre for Genomic Medicine and University of Manchester, Manchester, United Kingdom

Johanna M. Colijn (261), Erasmus MC, Rotterdam, The Netherlands

Frauke Coppieters (274), Ghent University, Ghent, Belgium

Catherine A. Cukras (236), National Eye Institute, NIH, Bethesda, MD, United States

Avril Daly (21), Retina International, Dublin, Ireland

Elfride De Baere (7), Ghent University and Ghent University Hospital, Ghent, Belgium

Julie De Zaeytijd (295), Ghent University Hospital, Ghent, Belgium

Arundhati Dev Borman (334), Bristol Eye Hospital, Bristol, United Kingdom

Hélène Dollfus (270, 277), Université de Strasbourg and CHU de Strasbourg, Strasbourg, France

Sofia Douzgou Houge (1, 118, 121, 127), Haukeland University Hospital, Bergen, Norway; University of Manchester, Manchester, United Kingdom

Elizabeth C. Engle (408, 422), Harvard Medical School and Howard Hughes Medical Institute, Bethesda, MD; Boston Children's Hospital, Boston, MA, United States

Pascal Escher (176), Inselspital, Bern University Hospital, Bern, Switzerland

D. Gareth Evans (430, 434, 449), Manchester Centre for Genomic Medicine and University of Manchester, Manchester, United Kingdom

Kristina Teär Fahnehjelm (306), St. Erik Eye Hospital and Karolinska Institutet, Stockholm, Sweden

Christina Fasser (21), Retina International, Zurich, Switzerland

Mathieu Fiore (393), CHU de Bordeaux, Bordeaux, France

Kaoru Fujinami (241), UCL Institute of Ophthalmology and Moorfields Eye Hospital, London, United Kingdom; National Institute of Sensory Organs, Tokyo Medical Center, Tokyo, Japan

Yu Fujinami-Yokokawa (241), National Institute of Sensory Organs, Tokyo Medical Center; Keio University, Tokyo, Japan

Brenda L. Gallie (457), Hospital for Sick Children; Princess Margaret Cancer Center; University of Toronto, Toronto, ON, Canada

Michalis Georgiou (194), UCL Institute of Ophthalmology and Moorfields Eye Hospital, London, United Kingdom; UAMS Harvey and Bernice Jones Eye Institute, Little Rock, AR, United States

Martin Gliem (53), Boehringer Ingelheim, Ingelheim am Rhein, Germany

Monika K. Grudzinska Pechhacker (274), St. Erik Eye Hospital and Karolinska Institutet, Stockholm, Sweden

Georgina Hall (21), Manchester Centre for Genomic Medicine and University of Manchester, Manchester, United Kingdom

Wolf M. Harmening (53), University Hospital Bonn, Bonn, Germany

Robert H. Henderson (334), Great Ormond Street Hospital and Moorfields Eye Hospital, London, United Kingdom

Elise Héon (274), Hospital for Sick Children and University of Toronto, Toronto, ON, Canada

Nashila Hirji (194), UCL Institute of Ophthalmology and Moorfields Eye Hospital, London, United Kingdom

Frank G. Holz (53), University of Bonn and University Hospital Bonn, Bonn, Germany

Laryssa A. Huryn (236), National Eye Institute, NIH, Bethesda, MD, United States

Elizabeth A. Jones (446), Manchester Centre for Genomic Medicine and University of Manchester, Manchester, United Kingdom

Vasiliki Kalatzis (277), Université de Montpellier, Montpellier, France

Arif O. Khan (98, 102, 105, 109, 385, 412, 418), Cleveland Clinic Abu Dhabi, Abu Dhabi, United Arab Emirates; Cleveland Clinic Lerner College of Medicine of Case Western University, Cleveland, OH, United States

Ungsoo S. Kim (356, 425), Kim's Eye Hospital, Seoul, South Korea

Caroline C.W. Klaver (200, 261), Erasmus MC, Rotterdam; Radboud University Medical Center, Nijmegen, The Netherlands; Institute of Molecular and Clinical Ophthalmology, Basel, Switzerland

Neruban Kumaran (302), UCL Institute of Ophthalmology and Guy's and St Thomas' Hospitals, London, United Kingdom

Chiara La Morgia (368, 371), IRCCS Istituto delle Scienze Neurologiche di Bologna, Bologna, Italy

Fiona Lalloo (453), Manchester Centre for Genomic Medicine, St Mary's Hospital, Manchester, United Kingdom

Eulalie Lasseaux (393), CHU de Bordeaux, Bordeaux, France

Helena Lee (403), University of Southampton and Southampton General Hospital, Southampton, United Kingdom

Guy Lenaers (362), Université d'Angers and CHU Angers, Angers, France

Eva Lenassi (181, 207), Manchester Centre for Genomic Medicine, Manchester Royal Eye Hospital, and University of Manchester, Manchester, United Kingdom

Bart P. Leroy (189, 217, 313), Ghent University and Ghent University Hospital, Ghent, Belgium; The Children's Hospital of Philadelphia, Philadelphia, PA, United States

Petra Liskova (73), Charles University and General University Hospital in Prague, Prague, Czech Republic

I. Christopher Lloyd (114, 118, 133, 140), Great Ormond Street Hospital, London; University of Manchester, Manchester, United Kingdom

Robert E. MacLaren (63, 172), University of Oxford and Oxford Eye Hospital, Oxford, United Kingdom

Omar A. Mahroo (225), UCL Institute of Ophthalmology and Moorfields Eye Hospital, London, United Kingdom

Alvaro J. Mejia-Vergara (362), Oftalmo-Sanitas Eye Institute, Ophthalmology Department, Sanitas University, Bogotá, Colombia

Isabelle Meunier (277), Université de Montpellier and CHRU de Montpellier, Montpellier, France

Michel Michaelides (194, 302), UCL Institute of Ophthalmology and Moorfields Eye Hospital, London, United Kingdom

Anthony T. Moore (439), University of California San Francisco, San Francisco, CA, United States; UCL Institute of Ophthalmology, London, United Kingdom

Mariya Moosajee (378, 389), UCL Institute of Ophthalmology and Moorfields Eye Hospital, London, United Kingdom

Fanny Morice-Picard (393), CHU de Bordeaux, Bordeaux, France

Francis L. Munier (73, 176, 250), Jules-Gonin Eye Hospital and University of Lausanne, Lausanne, Switzerland

Magella M. Neveu (393), Moorfields Eye Hospital and UCL Institute of Ophthalmology, London, United Kingdom

Erin C. O'Neil (292), Scheie Eye Institute at The Perelman Center for Advanced Medicine and The Children's Hospital of Philadelphia, University of Pennsylvania, Philadelphia, PA, United States

Anna Nordenström (306), Astrid Lindgren Children Hospital and Karolinska Institutet, Stockholm, Sweden

Neil R.A. Parry (33), Manchester Royal Eye Hospital and University of Manchester, Manchester; University of Bradford, Bradford, United Kingdom

Maria I. Patrício (172), University of Oxford and Oxford Eye Hospital, Oxford, United Kingdom

Manoj V. Parulekar (457), Birmingham Children's Hospital, Birmingham; Oxford Eye Hospital, Oxford, United Kingdom

Dipak Ram (309), Royal Manchester Children's Hospital, Manchester, United Kingdom

Simon C. Ramsden (13), Northwest Genomic Laboratory Hub, Manchester Centre for Genomic Medicine, Manchester, United Kingdom

Johane Robitaille (324, 329, 331), Dalhousie University and IWK Health Centre, Halifax, NS, Canada

Anthony G. Robson (181, 241), Moorfields Eye Hospital and UCL Institute of Ophthalmology, London, United Kingdom

Pierre-Raphaël Rothschild (347), Hôpital Cochin and Université de Paris, Paris, France

Alfredo A. Sadun (362), Doheny Eye Institute and UCLA, Los Angeles, CA, United States

Kaspar Schuerch (176), Inselspital, Bern University Hospital, Bern, Switzerland

Miguel C. Seabra (172), UCL Institute of Ophthalmology, London, United Kingdom; NOVA Medical School, Lisbon, Portugal

Jay E. Self (403), University of Southampton and Southampton General Hospital, Southampton, United Kingdom

Panagiotis I. Sergouniotis (1, 33, 98, 102, 105, 109, 151, 181, 189, 217, 241, 250, 288), University of Manchester, Manchester Royal Eye Hospital and Manchester Centre for Genomic Medicine, Manchester, United Kingdom

Fadi Shaya (246), Molecular Insight Research Foundation, Glendale, CA, United States

Paul A. Sieving (236), University of California Davis, Sacramento, CA, United States

Ine Strubbe (217), Ghent University and Ghent University Hospital, Ghent, Belgium

Francesca Simonelli (281), University of Campania Luigi Vanvitelli, Naples, Italy

Kent W. Small (246), Molecular Insight Research Foundation, Glendale, CA, United States

Martin P. Snead (339), University of Cambridge and Addenbrooke's Hospital, Cambridge, United Kingdom

Karolina M. Stepien (288), Salford Royal Hospital, Manchester, United Kingdom

Mays Talib (162), Leiden University and Leiden University Medical Center, Leiden, The Netherlands

Rachel L. Taylor (114, 118, 129, 144), University of Manchester and Manchester Centre for Genomic Medicine, Manchester, United Kingdom

Francesco Testa (281), University of Campania Luigi Vanvitelli, Naples, Italy

Alberta A.H.J. Thiadens (200), Erasmus MC, Rotterdam, The Netherlands

Elias I. Traboulsi (149, 155, 157), Cole Eye Institute, Cleveland Clinic, Cleveland, OH, United States

Viet H. Tran (176), Jules-Gonin Eye Hospital, Lausanne, Switzerland

Veronika Vaclavik (250), Jules-Gonin Eye Hospital, Lausanne, Switzerland

Sophie Valleix (347), Hôpital Cochin and Université de Paris, Paris, France

Caroline Van Cauwenbergh (7, 162), Ghent University and Ghent University Hospital, Ghent, Belgium

Kristof Van Schil (7), Ghent University Hospital, Ghent; Antwerp University Hospital, Antwerp, Belgium

Mary C. Whitman (408, 422), Harvard Medical School and Boston Children's Hospital, Boston, MA, United States

Colin E. Willoughby (93), Biomedical Sciences Research Institute, Ulster University, Coleraine; Belfast Health and Social Care Trust, Belfast, Northern Ireland, United Kingdom

Kanmin Xue (172), University of Oxford and Oxford Eye Hospital, Oxford, United Kingdom

Jingyan Yang (246), Molecular Insight Research Foundation, Glendale, CA, United States

Patrick Yu-Wai-Man (356, 425), Addenbrooke's Hospital and University of Cambridge, Cambridge; Moorfields Eye Hospital and UCL Institute of Ophthalmology, London, United Kingdom

Christina Zeitz (181), Sorbonne Université, INSERM, CNRS, Institut de la Vision, Paris, France

Martin Zinkernagel (176), Inselspital, Bern University Hospital, Bern, Switzerland

Preface

The management of individuals and families with ophthalmic genetic disorders is challenging.

The overwhelming majority of these conditions are rare and highly variable. Although genetic testing has transformed the diagnostic process for these diseases, it remains imperfectly integrated into clinical pathways. Notably, a deep knowledge of the underlying genes has historically been held by only a small group of people with long-standing specialist interest. Consequently, for general clinicians, who infrequently see such ophthalmic genetic disorders, the prospect that they may be missing diagnoses, and hence opportunities to treat or prevent complications, is a significant concern. The paradox, of course, is that it is these same clinicians who will be the first point of contact for most affected families.

The purpose of this book is to provide clinical ophthalmologists, geneticists, and genetic counsellors with a framework for approaching what is, collectively, a surprisingly large group of conditions. The book is divided into two parts:

- The first part, encompassed by Chapters 1–8, is designed to outline a structured clinical process that may be employed in the diagnosis of ophthalmic genetic disorders. These chapters describe key aspects of the commonest techniques and investigations including genomic sequencing (Chapters 2 and 3), clinical imaging (Chapter 7), and visual electrophysiology (Chapter 6). The aim of these introductory sections is to describe an effective, multidisciplinary diagnostic strategy that is both patient-focused (Chapter 4) and broadly relevant.
- The second part of this book, encompassed within Chapters 9–25, comprises over 80 sub-chapters. These are intentionally highly formulaic to allow those in the consulting room to gain a thorough and, hopefully, easily digestible understanding (literally, a working knowledge) of the clinical characteristics, molecular pathology, and clinical management of the greatest majority of ophthalmic genetic disorders. These sections cover all parts of the eye, discuss conditions of both children and adults, and have been written by clinical and scientific experts from across the globe.

There are a number of reasons why we think this book is timely.

First, while ophthalmology as a specialty has experienced, over a number of decades, unprecedented progress in terms of identifying the genes that underlie ophthalmic genetic disorders, it is now likely that the majority of these genes are known. Next, thanks to the methodical approach of clinicians and scientists worldwide, we are now gaining a deep understanding of how genetic variation can impact upon phenotypic expression and disease progression. Lastly, the introduction into the clinical arena of large-scale genomic sequencing approaches (including exome and whole genome sequencing), now offers a fast and cost-effective frontline diagnostic tool that is valuable in directing clinical decision-making.

Treatment and management options for ophthalmic genetic disorders are expanding rapidly. Consequently, it has become an imperative to demystify and simplify the field, and to begin to open it up to a much wider group of clinicians, both ophthalmic and genetic. We hope that this book can contribute to that process and in so doing, the ultimate aim must be to extend better care to more affected families in all regions of the world.

Graeme C.M. Black
Jane L. Ashworth
Panagiotis I. Sergouniotis

Section I

Genomics and the eye

Chapter 1

Genetic disorders and genetic variants

Sofia Douzgou Houge, Graeme C.M. Black, and Panagiotis I. Sergouniotis

Genetic disorders

Individual disease susceptibility is typically influenced by the complex interplay between genetic, environmental, lifestyle, and stochastic factors. The relative contribution of genetic alterations to disease predisposition is variable, and genetic disorders form a continuum. On the one end of the spectrum are common conditions (keratoconus, age-related macular degeneration, etc.) that are influenced by multiple factors to the degree that the direct effect of individual genetic changes is blurred; the term multifactorial is used to describe this group of conditions. On the other end of the spectrum, there are conditions like albinism and retinitis pigmentosa, for which the identification of defects in a single gene can predict disease development with relatively high accuracy; the term monogenic or Mendelian is used to describe this group of disorders (Fig. 1.1). Notably, Mendelian conditions can be recognised by the characteristic pedigree patterns that they give.

Mendelian inheritance

The human genome (i.e. the totality of a person's DNA) is divided among 46 physically distinct chromosomes: 22 pairs of autosomes (non-sex chromosomes) plus two sex chromosomes (two X chromosomes in females, one X and one Y chromosome in males). Mendelian disorders may be associated with genes on autosomes (chromosomes 1–22) or the X and Y chromosomes. Autosomal characters in both genders and X-linked characters in females can be dominant or recessive. This section will focus on the three main Mendelian pedigree patterns: autosomal dominant, autosomal recessive, and X-linked recessive (Fig. 1.2A–D). Mitochondrial inheritance, an atypical mode of inheritance that can be viewed as an expansion of Mendelian concepts, is also described (Fig. 1.2E). A more detailed discussion on pedigree patterns, including X-linked dominant, Y-linked and digenic inheritance, can be found in. Ref. 3.

Autosomal dominant inheritance

In autosomal dominant disorders, affected individuals carry one normal and one mutated copy of a gene. This means that a single copy of the mutation is sufficient to cause disease (i.e. the condition is expressed in the heterozygous state). Typically, an affected person has at least one affected parent and there are affected individuals in multiple generations. A person with an autosomal dominant disorder has a one in two chance of passing the mutated gene to an offspring (i.e. of having an affected child). This basic Mendelian pattern can be disguised by phenomena such as incomplete penetrance (Fig. 1.2B), variable expressivity and new/*de novo* mutations (discussed later in this chapter).

Autosomal recessive inheritance

In autosomal recessive disorders, affected individuals carry alterations in both copies of a given gene (biallelic defects). They can be either homozygote, if both copies carry the same mutation, or compound heterozygote, if each copy carries a different mutation. Typically, affected individuals are born to asymptomatic carrier parents; this means that the parents carry one normal and one mutant gene copy but are not affected as one normal copy is sufficient for normal cell function and structure. After the birth of a child with an

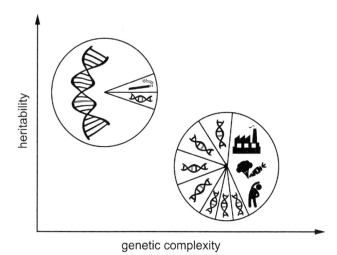

FIG. 1.1 The spectrum of genetic disorders on the basis of heritability and genetic complexity. Mendelian disorders like aniridia are highly heritable and have a simple genetic aetiology; in these conditions, alterations in a single gene are responsible for most of the disease risk (with possible minor contributions of modifier genes or environmental factors). In multifactorial disorders like diabetic retinopathy, multiple variants, each with a relatively small effect, contribute to disease risk along with environmental and lifestyle factors. *(Adapted from. Ref. 1.)*

autosomal recessive condition, each subsequent child has a one in four chance of being affected.

Autosomal recessive disorders can appear as 'sporadic' in a family where all parents and siblings are healthy. This is particularly common in small families. In real life, predicting autosomal recessive inheritance can be challenging and maybe inferred based on unaffected parents and the exclusion of X-linked inheritance.

A major risk factor for autosomal recessive disorders is parental consanguinity. Consanguineous marriage is in general defined as a union between a couple related as second cousins or closer. Although these unions increase the likelihood that spouses carry an identical gene mutation, in many ethnic groups, they are an important part of family culture. Personalised genetic counselling should be offered to consanguineous couples but this must be done with sensitivity and appreciation of cultural issues.

X-linked recessive inheritance

An X-linked disorder is caused by a mutation in a gene carried on the X chromosome. Males have only one X chromosome and, consequently, such a mutation will be manifest. Females have two copies of each X chromosome gene, but in individual cells of the female body, only one of these two copies tends to predominantly be active. However, in the context of an X-linked recessive condition, the presence of a normal copy is enough for normal development and organ function.

The essential features of X-linked inheritance are the presence of affected males (of greater severity than females) and the lack of father-to-son (male-to-male) transmission. The mothers of affected males are obligate carriers and have a one in two chance of passing the mutation to their offspring. Consequently, each son has a one in two chance of being affected and each daughter has a one in two chance of being a carrier.

Classically, it is assumed that X-linked recessive conditions only affect males. This is true for disorders such as X-linked retinoschisis and Norrie disease. However, for others such as X-linked retinitis pigmentosa, there may be phenotypic manifestations in heterozygous females, although these will usually be milder and of later-onset than in males.[4] In these cases, it might be difficult to distinguish X-linked from autosomal dominant inheritance, especially in the absence of detailed family history. It is worth highlighting that the phenotypic manifestations of X-linked disorders in females are highly variable; this is mainly due to a genetic phenomenon termed X chromosome inactivation or Lyonisation. It is of interest that in several X-linked conditions including choroideremia and Lowe syndrome, females may show characteristic ocular signs despite the absence of symptoms. These signs can help the identification of apparently unaffected female carriers.[5,6]

Complications to the basic Mendelian pedigree patterns

Various clinical phenomena can often disguise an underlying Mendelian inheritance pattern. An example of such a complication is incomplete penetrance.[7] Incomplete penetrance is common in dominant disorders where heterozygous mutation carriers are classically expected to be affected. However, in reality, the probability that a mutation carrier will develop symptoms is rarely 100%; in other words, mutations often show incomplete penetrance. In practice, this means that, for many conditions including certain subtypes of juvenile glaucoma and retinitis pigmentosa, individuals may carry a mutation but have no signs of the disorder. An extreme example is an asymptomatic individual with an affected parent and child; such a pedigree structure would suggest that the proband certainly carries the mutation but there is incomplete penetrance (Fig. 1.2B).

Another genetic phenomenon with similarities to incomplete penetrance is variable expressivity. A mutation is said to demonstrate variable expressivity if, within a family, affected individuals carry the same genetic defect but the manifestations of the condition vary widely. The causes of variable expressivity are generally expected to be the same as for incomplete penetrance (e.g. other genetic changes or environmental factors). Notable examples of variable expressivity include Stickler syndrome and

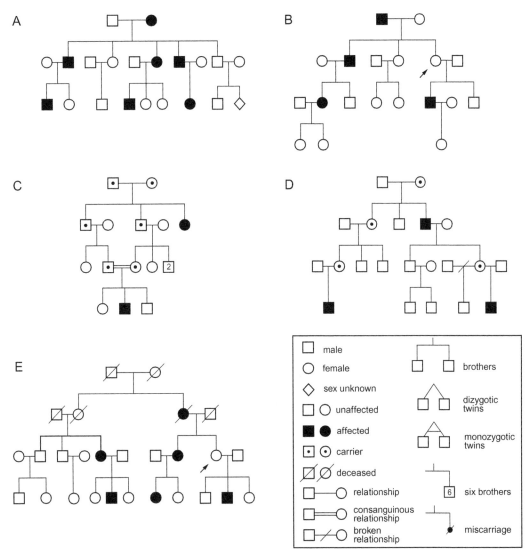

FIG. 1.2 Pedigree patterns in autosomal dominant (A, B), autosomal recessive (C), X-linked recessive (D), and mitochondrial (E) conditions. *Arrows* highlight non-penetrant carriers of an autosomal dominant (B) and a mitochondrial (E) disorder. *(Adapted from. Ref. 2.)*

albinism in which ophthalmic and systemic manifestations vary among those who carry identical mutations. In these cases, disease severity in one individual may give little or no indication of the likely disease severity for siblings or offspring. Incomplete penetrance and variable expressivity are important reasons for examining unaffected parents of children with conditions such as cataracts or anterior segment dysgenesis.

Many cases of the severe dominant disease are the result of new mutations. These so-called *de novo* mutational events arise as a result of a copying error from a parent's DNA and manifest without warning in families with no previous history of the disorder. This phenomenon is observed in many individuals with aniridia or retinoblastoma.[8,9] In such cases, the recurrence risk for future siblings is much lower than one in two. However, the figure will not be zero due to the risk of gonadal mosaicism (i.e. one parent carrying the mutation only in a proportion of his sperm or her eggs).

Mitochondrial inheritance

Mitochondria are cytoplasmic cellular organelles that contain a small genome (~16,500 bases) that is distinct from the nuclear genome (~3,200,000,000 bases). Similar to mutations carried on the nuclear chromosomes, mutations in the mitochondrial DNA can be associated with disease; notable examples include Leber hereditary optic neuropathy and Kearnes–Sayre syndrome.

Mitochondria are inherited exclusively from the ovum and a mitochondrial DNA mutation can only be passed on from a mother to her child. Consequently, while mitochondrially inherited disorders can affect both genders, they are transmitted only by mothers who carry the mitochondrial DNA defect. Although, at first sight, the pedigree pattern may appear dominant, it should be noted that affected males never transmits the condition.

Maternally inherited disorders are typically highly variable, making visual prognosis estimation challenging. In many cases where individuals carry mitochondrial DNA mutations in all of their cells, as is seen for most patients with Leber hereditary optic neuropathy, the basis for this variability is poorly understood. However, with mitochondrial myopathies such as Kearnes–Sayre syndrome, only a proportion of mitochondria carry mutations and, since each cell has many mitochondria, variability can arise. This state is termed heteroplasmy and is associated with a variable ratio of mutant to normal mitochondria between the cells/tissues of a single individual or between different individuals of the same family.

Heterogeneity in genetic disorders

There is very rarely a one-to-one correspondence between genes and genetic disorders and this has implications for diagnosis, counselling, and genetic testing. Two types of heterogeneity are commonly observed.

- Locus heterogeneity occurs when abnormalities in more than one gene can cause indistinguishable phenotypic manifestations. Examples include congenital cataracts and Bardet–Biedl syndrome; individuals with these disorders do not all have mutations in the same gene.
- Clinical heterogeneity is when mutations in the same gene produce two or more distinct disorders in different individuals (e.g. *SOX2* mutations can cause microphthalmia, anophthalmia, or coloboma[10]). As the clinical outcome of a mutation can be influenced by its position, type, and effect on the encoded protein, clinical heterogeneity may be associated with different effects of distinct mutations.

Multifactorial disorders

Although collectively Mendelian ophthalmic disorders are common, individually they tend to be rare. Notably, many prevalent ophthalmic conditions (e.g. Fuchs endothelial dystrophy, glaucoma, or age-related macular degeneration) appear to show familial clustering and are likely to be caused by a complex interplay of multiple genes and the environment. In these multifactorial disorders, the presence of a disease-implicated DNA change is not highly predictive of the development of disease. It is clear that there is a continuum of disorders from fully penetrant Mendelian to multifactorial conditions; compartmentalising genetic disorders into 'Mendelian, simple and rare' and 'multifactorial, complex and common' is an oversimplification (Fig. 1.1).

Genetic variants and mutations

Although the term 'mutation' can be used in different senses, it has developed a negative connotation and is typically used to indicate a disease-causing DNA change; that is a change that contributes mechanistically to disease but may not be sufficient in isolation to cause disease (e.g. in cases with incomplete penetrance).[11] In contrast, the term 'variant' is used to indicate any difference in a DNA sequence in comparison with the normal reference sequence. A variant may or may not affect the function of a gene/protein. Variants for which a functional effect is unknown are called 'variants of unknown significance'.[12] Other ways to classify variants include common versus rare and small-scale versus large-scale.

Human genetic variation ranges from single base/nucleotide changes (known as SNVs, single nucleotide variants) through to structural variants and gains or losses of whole chromosomes. Single nucleotide substitutions are the most numerous variants. These small-scale DNA sequence alterations may result in the substitution of one encoded amino-acid for another (missense changes) or convert an internal amino-acid of a protein into a termination codon (nonsense changes). Small insertions/deletions (indels), in which less than 50 bases/nucleotides are inserted and/or deleted, are another common subtype of small-scale DNA variation. Importantly, if the overall loss or gain of bases is not a multiple of three, the reading frame of the gene will be altered and a premature termination codon will be probably introduced (frameshift change).

Some changes can alter transcription (i.e. the process of copying a particular segment of DNA into mRNA) and in particular, the stage during which the immature mRNA molecule (consisting of introns and exons) is modified to form mature mRNA (consisting of joined exons only). This process is called splicing and requires the interaction of a large protein complex, the spliceosome, with mRNA. Many mutations, in particular those that lie at or close to the junctions between introns and exons, cause disease by interrupting splicing. These splicing mutations lead to reduced transcription of a gene and therefore reduced gene expression.

Any sequence change can be described using the conventions set out on the website of the Human Genome Variation Society (https://varnomen.hgvs.org/); common prefixes include 'c.' (coding DNA level) and 'p.' (protein level). Useful variant databases include the Human Gene Mutation Database (http://www.hgmd.cf.ac.uk/) and ClinVar (http://www.ncbi.nlm.nih.gov/clinvar/), two online repositories of disease-implicated variants, as well as the Genome Aggregation Database (gnomAD) (https://gnomad.broadinstitute.org/) which includes variant data from over 140,000 unrelated individuals.

Genetic testing

Genetic testing for ophthalmic disorders aims to provide an accurate diagnosis and to enable appropriate counselling to the patient/family; in some cases, guiding management and determining eligibility for gene-based interventions is also possible.[13] Testing is typically performed on DNA extracted from a tissue sample. Peripheral blood is the most commonly used and reliable source but mouthwashes or buccal scrapes can be used when sampling has to be non-invasive. The various available genetic testing strategies are discussed in Chapter 2 and a framework for interpreting genetic test results is discussed in Chapter 3. Importantly, testing can have significant and far-reaching consequences for an individual and their family. Informed consent and effective pre- and post-test genetic counselling are essential and can maximise the benefits and minimise the risks associated with genetic investigations (see also Chapter 4).

The role of the ophthalmologist

Ophthalmologists looking after people with suspected genetic ophthalmic disorders have a multifaceted role. Among others, they create a roadmap for diagnosis (including requesting appropriate genetic investigations), help interpret the genetic test results (as part of a multidisciplinary team) and implement personalised care pathways based on the findings. In the first instance though, two key questions should be considered:

- does the patient have features that suggest a Mendelian disorder (e.g. presence of a relevant family history)?
- are the findings associated with an isolated (non-syndromic) ophthalmic condition or are they part of a multisystemic (syndromic) disorder?

To answer these questions, a focused medical/family history, a targeted clinical examination and appropriate investigations are required. Phenotypic features that could point to a syndromic diagnosis should be sought (see Chapter 5) and the visual function and ocular structure should be carefully evaluated (see Chapters 6 and 7).

References

1. Sergouniotis PI. Inherited retinal disorders: using evidence as a driver for implementation. *Ophthalmologica* 2019;**242**(4):187–94.
2. Sergouniotis PI, Black GC. Genetics and paediatric ophthalmology. In: Lambert S, Lyons C, editors. *Taylor and Hoyt's pediatric ophthalmology and strabismus*. Elsevier; 2016. p. 94.
3. Strachan T, Read AP. *Human molecular genetics 5*. London: Garland Science; 2018.
4. De Silva SR, Arno G, Robson AG, Fakin A, Pontikos N, Mohamed MD, et al. The X-linked retinopathies: Physiological insights, pathogenic mechanisms, phenotypic features and novel therapies. *Prog Retin Eye Res*. 2021;**82**:100898. https://doi.org/10.1016/j.preteyeres.2020.100898.
5. Cibis GW, Waeltermann JM, Whitcraft CT, Tripathi RC, Harris DJ. Lenticular opacities in carriers of Lowe's syndrome. *Ophthalmology* 1986;**93**(8):1041–5.
6. Morgan JI, Han G, Klinman E, Maguire WM, Chung DC, Maguire AM, et al. High-resolution adaptive optics retinal imaging of cellular structure in choroideremia. *Invest Ophthalmol Vis Sci* 2014;**55**(10):6381–97.
7. Cooper DN, Krawczak M, Polychronakos C, Tyler-Smith C, Kehrer-Sawatzki H. Where genotype is not predictive of phenotype: towards an understanding of the molecular basis of reduced penetrance in human inherited disease. *Hum Genet* 2013;**132**(10):1077–130.
8. Hall HN, Williamson KA, FitzPatrick DR. The genetic architecture of aniridia and Gillespie syndrome. *Hum Genet* 2019;**138**(8–9):881–98.
9. Munier FL, Thonney F, Girardet A, Balmer A, Claustre M, Pellestor F, et al. Evidence of somatic and germinal mosaicism in pseudo-low-penetrant hereditary retinoblastoma, by constitutional and single-sperm mutation analysis. *Am J Hum Genet* 1998;**63**(6):1903–8.
10. Williamson KA, Yates TM, FitzPatrick DR. SOX2 disorder. In: Adam MP, Ardinger HH, Pagon RA, et al., editors. *GeneReviews((R))*. Seattle, WA: University of Washington; 2020.
11. Jarvik GP, Evans JP. Mastering genomic terminology. *Genet Med* 2017;**19**(5):491–2.
12. Richards S, Aziz N, Bale S, Bick D, Das S, Gastier-Foster J, et al. Standards and guidelines for the interpretation of sequence variants: a joint consensus recommendation of the American College of Medical Genetics and Genomics and the Association for Molecular Pathology. *Genet Med* 2015;**17**(5):405–23.
13. Lenassi E, Clayton-Smith J, Douzgou S, Ramsden SC, Ingram S, Hall G, et al. Clinical utility of genetic testing in 201 preschool children with inherited eye disorders. *Genet Med* 2020;**22**(4):745–51.

Chapter 2

Genetic testing techniques

Caroline Van Cauwenbergh, Kristof Van Schil, Miriam Bauwens, and Elfride De Baere

Over the past two decades, technological advances have transformed molecular genetic testing for ophthalmic disorders. There has been a shift from targeted testing of known mutations (using arrayed primer extension (APEX) technology) and serial analysis of a small number of genes (using Sanger sequencing) to testing comprehensive gene panels (using massively parallel sequencing, also known as high-throughput sequencing or next-generation sequencing). Cost-effective interrogation of all protein-coding genes at once is now possible (using exome sequencing) and soon, routine genetic testing is likely to involve in-depth analysis of both the protein and the non-protein-coding parts of the genome (using genome sequencing). In this chapter, we discuss technical aspects of the main genetic testing strategies relevant to ophthalmic practice.

Sanger sequencing

Since the early 1990s, testing of small sets of disease-implicated genes has been carried out using Sanger sequencing, also known as the dideoxy or chain termination method. This approach determines the sequence of an amplified DNA segment (up to approximately 1000 base-pair long) with per-base accuracy as high as 99.999%. These DNA templates are 'cycle sequenced' involving several rounds of DNA template denaturation, primer annealing, and extension. Combined use of regular deoxynucleotides (dNTPs) and a lower amount of terminating fluorescently labelled dideoxynucleotides (ddNTPs, chemically modified dNTPs) results in the generation of a mixture of termination fragments with different lengths. Indeed, the elongation of each DNA strand is stopped at a different cycle when a ddNTP is incorporated in the growing nucleotide chain. After size separation of these fragments by capillary electrophoresis, the fluorescent labels of the ddNTPs are detected by laser excitation, resulting in an electropherogram as output (Fig. 2.1). Analysing the sequencing results against a reference sequence can reveal the presence of DNA sequence changes (variants) in the studied sample. Although Sanger sequencing is very accurate, genetic testing using this approach has been limited to a rather small number of indications and clinical scenarios, due to low throughput and a high cost per base.[1]

Arrayed primer extension

The APEX microarray technology is a historical technique that can be used for parallel testing of specific DNA changes (e.g. in genes implicated in ophthalmic disorders). The surface of each predesigned APEX chip is covered with 25 base-pair long DNA segments (primers). Each of these immobilised primers is associated with one of the known genetic variants to be tested. The single-base primer extension reaction is performed by a specific DNA polymerase that extends the primer with one of the four different fluorescently labelled terminators, after which the fluorescence is measured. Multiple such reactions occur in parallel. Based on the observed signals, the corresponding genotypes can be deduced for each of the variants included in the APEX chip design.

Commercially available APEX microarray chips (Asper Ophthalmics) used to be a commonly utilised technique for genetic testing in ophthalmic diseases. APEX chips were developed for several retinal conditions, including Leber congenital amaurosis,[2] retinitis pigmentosa, and Usher syndrome.

Massively parallel sequencing (next-generation sequencing)

Since the early 2010s, testing strategies based on massively parallel sequencing (also known as next-generation sequencing or NGS) approaches have been implemented. These methods allow the processing of millions of DNA sequence fragments ('reads') in parallel. This significantly reduces the per-base cost of sequencing. Although several NGS technologies are commercially available, the sequencing-by-synthesis, short-read approach utilised in Illumina platforms has played the most prominent role both

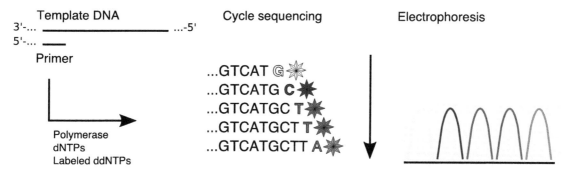

FIG. 2.1 Sanger sequencing. (Left) Sanger sequencing starts following denaturation of the target DNA fragment to obtain single-stranded DNA molecules. Subsequently, a short oligonucleotide (primer) anneals and the synthesis of a new DNA strand that is complementary to the template starts. This elongation reaction is catalysed by a DNA polymerase enzyme and requires a mixture of four deoxyribonucleotide triphosphates (dNTPs: dATP, dCTP, dGTP, and dTTP) and small amounts of four dideoxynucleotide triphosphates (ddNTPs), each labelled with a different fluorophore. (Middle) The incorporation of a ddNTP results in termination, leading to a set of molecules with different lengths. During electrophoresis, labelled molecules are size separated and move past a fluorescent detector, which can discriminate the attached labels. (Right) This is an example of a DNA sequence trace (electropherogram), where the sequence is represented as a series of peaks, one for each nucleotide position.

in clinical genetic testing and in gene discovery.[1,3] It is therefore the main focus of this section.

During the sample (library) preparation, DNA is randomly cut into fragments of 100–300 base pairs in length. This is followed by the ligation of specific adaptors onto both ends of each DNA fragment. The tagged DNA is then PCR amplified. Occasionally, multiplexing is used (analysing different samples in one sequencing experiment) in which case the utilised primers contain additional index sequences; this helps to track the sample from which a specific sequence originates. Denatured single-stranded DNA is added to a glass flow cell coated with two different oligonucleotide primers complementary to the adapters added to the sequences. The adaptor-containing sequences hybridise with the oligonucleotides on the flow cell, followed by a PCR reaction that synthesises the second strand for each fragment. The resulting double-stranded DNA is denatured, and the original DNA strand is washed away. The newly synthesised strand bends and binds to a complementary oligonucleotide on the flow cell, resulting in a bridge formation. Another PCR reaction is carried out starting from this new oligonucleotide, creating again a double-stranded DNA molecule. After denaturation, both the original and newly synthesised strands are attached to the flow cell nearby. When performing several rounds of this bridge amplification, 100–200 million clonal clusters are generated each one originating from one specific DNA fragment. Finally, sequencing of these clusters is performed according to a cyclic reversible termination approach using chemically modified nucleotides. In every cycle, a mixture of the four fluorescently labelled dNTPs are added. The elongation of every sequence stops after the incorporation of one nucleotide and the surface is then imaged to identify which dNTP was incorporated at each cluster. Removal of the fluorophore and blocking group initiates a new cycle. This process is repeated, ultimately revealing the full sequence of the DNA fragments in each cluster (Fig. 2.2). After sequencing, the read-outs from all clusters are mapped to the genome reference sequence using alignment software tools.[1,3]

Two NGS-based approaches that are frequently used for genetic testing in a clinical setting include targeted sequencing of gene panels and exome sequencing.

Targeted sequencing of gene panels

In 2011 the Illumina MiSeq sequencer was released. This low throughput and fast turnaround time instrument paved the way to implement NGS in a clinical setting. It could be used for targeted sequencing of a specific region of interest or sequencing gene panels. This approach was most suitable for testing small gene panels (e.g. up to 10 genes) flexibly (as different sets of genes can easily be added and combined). Subsequently, Illumina developed platforms with much higher throughput. These have enabled testing of comprehensive, customised gene panels for several indications including retinopathy, albinism, or cataract.[4]

Panel-based testing is a reliable tool to detect known and novel mutations in individuals with genetic ophthalmic disorders. The detection rate depends on the underlying condition, the specific gene panel and the studied population.[5]

Exome sequencing

Testing of a gene panel using targeted sequencing is cost- and time-effective in a diagnostic setting. However, this only allows screening for variants within a selected set of genes (which may not contain the disease-causing gene). In contrast, exome sequencing captures virtually all protein-coding genes (i.e. the exome, comprising 1% of the genome).

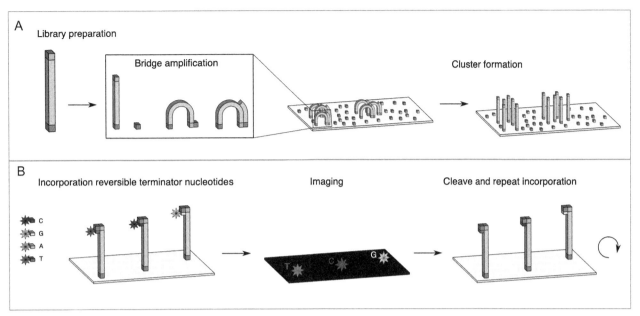

FIG. 2.2 Next-generation sequencing (Illumina's sequencing-by-synthesis technology). (A) Solid-phase amplification. Left: During sample (library preparation, fragmented double-stranded DNA is flanked by known adapter sequences (indicated in *blue* and *orange*) that are complementary to the two oligonucleotides that are coated on the flow cell surface. Middle: Cluster generation proceeds when denatured DNA libraries hybridise to the oligonucleotides, generating a template for extension, which is catalysed by a polymerase. The original strand is then denatured and washed away. A complimentary copy of the newly synthesised strand is generated through bridge amplification by which the strand bends and hybridises to an adjacent complementary oligonucleotide, allowing the extension by a polymerase. Right: Bridge amplification is repeated multiple times to produce clusters of DNA clones, each containing up to 1000 copies of a template molecule. (B) Four-colour cyclic reversible termination method. Left: During this sequencing-by-synthesis reaction, in each round, a fluorescently labelled and a reversibly terminated nucleotide is incorporated. Middle and right: Following imaging, a cleavage step removes the fluorophore and reverses the terminator, allowing the incorporation of the next base. This process is repeated until the predetermined sequence length is reached.

Exome capture occurs before sequencing and involves the hybridisation of fragmented genomic DNA to oligonucleotide probes that are complementary to the exonic sequences. Magnetic beads can selectively pull down and enrich the baits that bind the targeted regions; the non-targeted portion of the genome is subsequently washed away. The target DNA is then amplified, producing a sequence ready 'library'. It is worth highlighting that several approaches are available and, over the years, updates have been made including around capture technologies and probe design.

In the past decade, targeted testing of gene panels has been used as the primary diagnostic tool for conditions like congenital cataracts or retinal dystrophies.[6–8] However, exome sequencing often outperforms these approaches in a clinical setting, generally lowering costs and improving sensitivity and specificity.

Genome sequencing

Despite the implementation of gene panel testing and exome sequencing in the clinical setting, many individuals with genetic ophthalmic disorders remain without a molecular diagnosis.[9] This could be partially attributed to technical limitations inherent to the sequencing methods (e.g. low coverage in guanine-cytosine content (GC) rich regions), to the presence of disease-causing structural variants (such as copy number variants) or the presence of genetic defects in non-coding regions (such as deep intronic variants affecting splicing or variants impacting on gene regulatory regions).[8–10] The latter two are often missed by the standard targeted sequencing approaches.

Studies on cohorts of patients with retinal disorders or ocular anterior segment dysgenesis have compared genome sequencing to exome sequencing for the detection of structural variants and variants in non-coding regulatory regions.[8–11] Examples of structural variants identified by genome sequencing included a small deletion of *USH2A* and an inverted duplication in *EYS*, which would have likely been missed by exome sequencing.[8] Striking examples of deep intronic mutations causing retinal disorders included the recurrent deep intronic mutation c.2991+1655A>G in *CEP290*, accounting for 15% of congenital blindness in Europe,[12] the c.7595-2144A>G variant in *USH2A*,[13] and several deep intronic mutations in *ABCA4* in patients with Stargardt disease.[14] Conversely, only a few mutations affecting *cis*-regulatory regions have been identified in retinal disorders so far with mutations dysregulating the *PRDM13* gene in individuals with North Carolina macular dystrophy being the most striking example.[15] Identification and interpretation of these genetic variants can be challenging and requires high-throughput evaluation tools.

Structural variant analysis

Detecting structural variation, such as copy number variants, can be challenging, especially when NGS strategies that include a target enrichment step are used (i.e. gene panel and exome sequencing tests). A wide spectrum of methods has been utilised instead including multiplex ligation-dependent probe amplification (MLPA), quantitative polymerase chain reaction (qPCR) and array comparative genomic hybridisation (aCGH); computational tools analysing NGS data are also now commonly used for this purpose.

Multiplex ligation-dependent probe amplification

MLPA is a PCR-based semi-quantitative method that relies on sequence-specific probes that hybridise with genomic DNA. Each of the, on average, 40 probe sets consists of two oligonucleotides complementary to the DNA region of interest. Each probe has a tail complementary to universal amplification primers. One of both oligonucleotides contains a stuffer sequence. The stuffer sequence has a different length for all 40 probes. Only if both oligonucleotides hybridise adjacent to a DNA template, the probes can be joined using a DNA ligase. These ligated probes form amplifiable templates that are size separable. Only a limited number of probes[4] can be separated at once and the usability of probe sets depends on the availability of commercial kits (MRC-Holland). Such kits currently only cover a fraction of the genes that are associated with ophthalmic disease.

Molecular karyotyping

Molecular karyotyping typically involves using oligonucleotide microarray-based comparative genomic hybridisation (custom array CGH, also called microarray). This is a powerful tool that allows simultaneous analysis of a large number of targets. In brief, equal amounts of test and control genomic DNA are differentially labelled with fluorescent tags (usually Cy5 and Cy3), denatured and hybridised on a microarray slide containing immobilised DNA probes. The fluorescence intensities are measured with a laser scanner and the fluorescence ratio (Cy3:Cy5) is determined, potentially revealing copy number differences between the control and patient samples[16] (Fig. 2.3).

Several studies have already shown the power of custom high-resolution array CGH for detecting exonic copy number variants in heterogeneous ophthalmic disorders.[8,17,18]

Computational tools for analysing sequencing data

Several computational tools have been developed for structural variant detection in NGS data.[19] Read-depth based approaches that evaluate the number of reads mapped to each genomic region are often used. These analyses can be performed using NGS data on the condition that a sufficient sequence read depth (coverage) is obtained.[20–23] However, coverage can be variable when target enrichment methods such as gene panel testing or exome sequencing are used. Also, read-depth based approaches fail to detect neutral structural variants (e.g. inversions, balanced translocations) and cannot delineate the precise breakpoints. Compared to gene panel and exome sequencing, genome sequencing has a greater power to detect structural variation.[9–11] This is because the PCR-free library preparation that is used in genome sequencing results in more reliable sequence coverage. Moreover, genome sequencing

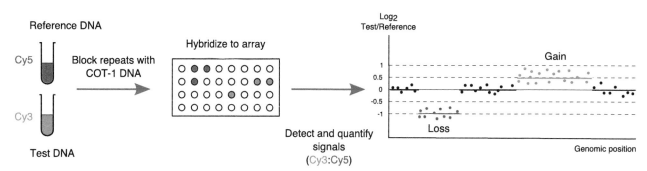

FIG. 2.3 Array comparative genomic hybridisation (array CGH). (Left) Equal amounts of test and reference DNA are differentially labelled with fluorochromes (e.g. Cyanine 5 (Cy5) and Cy3), mixed with Cot-1 to block repetitive sequences. (Middle) Following denaturation, the mixture is hybridised on a microarray slide that contains immobilised DNA probes. After scanning, the fluorescent intensities are quantified. Read out of the fluorescent signal for each of the oligonucleotide probes shows whether sample and reference DNA are present in equal amounts (equal emission of *red* and *green*, leading to *yellow*), or if there is a gain (duplication, more emission of *green*) or loss (deletion, more emission of *red*) of patient DNA. (Right) The signal intensity ratio (i.e. Log_2[Cy3:Cy5] ratios) are plotted corresponding to the genomic position and represent the relative DNA copy number. A deletion (in *red*) gives a theoretical log_2 ratio of -1 (log_2 1/2), whereas a duplication (in *green*) gives a theoretical log_2 ratio of 0.58 (log_2 3/2).

covers the entire genome, making it feasible to detect a wide variety of structural changes, including translocations, deletions, tandem duplications, insertions, and inversions.

Several alternatives to read-depth based approaches have been described. These include strategies based on: discordantly mapped paired reads (paired-end mapping methods); the incomplete mapping of a read from a read pair (split read based methods); *de novo* assembly; or a combination of those. Split read based methods even allow the detection of breakpoints at base-pair resolution.

Concluding remarks

In this chapter, we discussed the principles underlying various genetic testing approaches. For each of these tests, the analytical validity (i.e. the ability to accurately identify variants of interest; associated with the technical performance of the test), the clinical validity (i.e. the ability to detect the clinically defined disorder of interest; associated with the diagnostic yield of the test) and the clinical utility (i.e. the ability to improve patient management and healthcare outcomes) should be considered for different test indications/scenaria.[5] Another parameter to consider is the cost-effectiveness of each approach. This can be only evaluated at the site level. Importantly, although the overall cost of the sequencing process is expected to continue to drop, the cost of clinical interpretation is often not trivial and should be factored in.

References

1. Shendure J, Ji H. Next-generation DNA sequencing. *Nat Biotechnol* 2008;**26**(10):1135–45. Available from: http://www.nature.com/doifinder/10.1038/nbt1486.
2. Yzer S, Leroy BP, De Baere E, de Ravel TJ, Zonneveld MN, Voesenek K, et al. Microarray-based mutation detection and phenotypic characterization of patients with Leber congenital amaurosis. *Invest Ophthalmol Vis Sci* 2006;**47**(3):1167–76. https://doi.org/10.1167/iovs.05-0848. 16505055.
3. Chaitankar V, Karakülah G, Ratnapriya R, Giuste FO, Brooks MJ, Swaroop A. Next generation sequencing technology and genomewide data analysis: perspectives for retinal research. *Prog Retin Eye Res* 2016;**55**:1–31. https://doi.org/10.1016/j.preteyeres.2016.06.001.
4. Schouten JP, Schouten JP, McElgunn CJ, McElgunn CJ, Waaijer R, Waaijer R, et al. Relative quantification of 40 nucleic acid sequences by multiplex ligation-dependent probe amplification. *Nucleic Acids Res* 2002;**30**(12), e57.
5. Lenassi E, Clayton-Smith J, Douzgou S, Ramsden SC, Ingram S, Hall G, et al. Clinical utility of genetic testing in 201 preschool children with inherited eye disorders. *Genet Med* 2020;**22**:745–51. https://doi.org/10.1038/s41436-019-0722-8. 31848469.
6. Stone EM, Andorf JL, Whitmore SS, DeLuca AP, Giacalone JC, Streb LM, et al. Clinically focused molecular investigation of 1000 consecutive families with inherited retinal disease. *Ophthalmology* 2017;**124**(9):1314–31. https://doi.org/10.1016/j.ophtha.2017.04.008.
7. Haer-Wigman L, van Zelst-Stams WA, Pfundt R, van den Born LI, Klaver CC, Verheij JB, et al. Diagnostic exome sequencing in 266 Dutch patients with visual impairment. *Eur J Hum Genet* 2017;**25**(5):591–9. https://doi.org/10.1038/ejhg.2017.9.
8. Ma A, Yousoof S, Grigg JR, Flaherty M, Minoche AE, Cowley MJ, et al. Revealing hidden genetic diagnoses in the ocular anterior segment disorders. *Genet Med* 2020;**22**(10):1623–32. https://doi.org/10.1038/s41436-020-0854-x.
9. Ellingford JM, Barton S, Bhaskar S, Williams SG, Sergouniotis PI, O'Sullivan J, et al. Whole genome sequencing increases molecular diagnostic yield compared with current diagnostic testing for inherited retinal disease. *Ophthalmology* 2016;**123**(5):1143–50. 26872967.
10. Nishiguchi KM, Tearle RG, Liu YP, Oh EC, Miyake N, Benaglio P, et al. Whole genome sequencing in patients with retinitis pigmentosa reveals pathogenic DNA structural changes and NEK2 as a new disease gene. *Proc Natl Acad Sci* 2013;**110**(40):16139–44. 24043777. Available from: http://www.pnas.org/cgi/doi/10.1073/pnas.1308243110.
11. Carss KJ, Arno G, Erwood M, Stephens J, Sanchis-Juan A, Hull S, et al. Comprehensive rare variant analysis via whole-genome sequencing to determine the molecular pathology of inherited retinal disease. *Am J Hum Genet* 2017;**100**(1):75–90. 28041643.
12. Coppieters F, Lefever S, Leroy BP, De Baere E. CEP290, a gene with many faces: mutation overview and presentation of CEP290base. *Hum Mutat* 2010;**31**(10):1097–108. https://doi.org/10.1002/humu.21337. 20690115.
13. Liquori A, Vaché C, Baux D, Blanchet C, Hamel C, Malcolm S, et al. Whole USH2A gene sequencing identifies several new deep intronic mutations. *Hum Mutat* 2016;**37**(2):184–93. https://doi.org/10.1002/humu.22926.
14. Khan M, Cornelis SS, Pozo-Valero MD, Whelan L, Runhart EH, Mishra K, et al. Resolving the dark matter of ABCA4 for 1054 Stargardt disease probands through integrated genomics and transcriptomics. *Genet Med* 2020;**22**(7):1235–46. https://doi.org/10.1038/s41436-020-0787-4.
15. Small KW, DeLuca AP, Whitmore SS, Rosenberg T, Silva-Garcia R, Udar N, et al. North Carolina macular dystrophy is caused by dysregulation of the retinal transcription factor PRDM13. *Ophthalmology* 2016;**123**(1):9–18. 26507665.
16. Menten B, Pattyn F, de Preter K, Robbrecht P, Michels E, Buysse K, et al. arrayCGHbase: an analysis platform for comparative genomic hybridization microarrays. *BMC Bioinform* 2005;**6**(1):124. Available from: http://www.biomedcentral.com/1471-2105/6/124%5Cnpapers3://publication/doi/10.1186/1471-2105-6-124.
17. Van Cauwenbergh C, Van Schil K, Cannoodt R, Bauwens M, Van Laethem T, De Jaegere S, et al. arrEYE: a customized platform for high-resolution copy number analysis of coding and noncoding regions of known and candidate retinal dystrophy genes and retinal noncoding RNAs [Internet]. *Genet Med* 2016. https://doi.org/10.1038/gim.2016.119.
18. Mayer AK, Van Cauwenbergh C, Rother C, Baumann B, Reuter P, De Baere E, et al. CNGB3 mutation spectrum including copy number variations in 552 achromatopsia patients. *Hum Mutat* 2017;**38**(11):1579–91.
19. Zhao M, Wang Q, Wang Q, Jia P, Zhao Z. Computational tools for copy number variation (CNV) detection using next-generation sequencing data: features and perspectives. *BMC Bioinform* 2013;**14 Suppl 1**(Suppl 11):S1. 24564169. Available from: http://bmcbioinformatics.biomedcentral.com/articles/10.1186/1471-2105-14-S11-S1.
20. Khateb S, Hanany M, Khalaileh A, Beryozkin A, Meyer S, Abu-Diab A, et al. Identification of genomic deletions causing inherited retinal degenerations by coverage analysis of whole exome sequencing data. *J Med Genet* 2016;**53**(9):600–7. 27208209.

21. Ellingford JM, Horn B, Campbell C, Arno G, Barton S, Tate C, et al. Assessment of the incorporation of CNV surveillance into gene panel next-generation sequencing testing for inherited retinal diseases. *J Med Genet* 2018;**55**(2):114–21. https://doi.org/10.1136/jmedgenet-2017-104791. 29074561.
22. Daiger SP, Sullivan LS, Bowne SJ, Cadena ED, Koboldt D, Bujakowska KM, et al. Detection of large structural variants causing inherited retinal diseases. *Adv Exp Med Biol* 2019;**1185**:197–202. 31884611. Available from: http://www.ncbi.nlm.nih.gov/pubmed/31884611.
23. Zampaglione E, Kinde B, Place EM, Navarro-Gomez D, Maher M, Jamshidi F, et al. Copy-number variation contributes 9% of pathogenicity in the inherited retinal degenerations. *Genet Med* 2020;**55**(2). Jmedgenet-2017-104791. 32037395. Available from: http://www.ncbi.nlm.nih.gov/pubmed/29074561.

Chapter 3

Genetic variant interpretation

Simon C. Ramsden

In routine clinical practice, a genetic test referral may be made to a local, national or international public health testing laboratory or a private testing company. It is now common for clinicians to request genetic testing on patients with ophthalmic disease. This can help establish a precise diagnosis and determine the recurrence risk in family members. Often, genomic results can offer predictive information, both around ophthalmic disease progression and around the occurrence of non-ophthalmic manifestations.[1] The clinician will hope to receive a clear unambiguous result, establishing the cause of the clinical diagnosis and explaining the family history.

Primary and secondary findings

The primary role of genomic analysis is to identify the variants in an individual's genome that differ from a reference sequence. Subsequent analysis of these changes will highlight variants of proven or potential clinical significance. Broadly speaking such variants can be divided into two types, primary and secondary.

Primary findings: These are findings that address the presenting phenotype. At the time of reporting these may or may not unequivocally explain the clinical presentation. Typical primary findings are as follows:

(A) Diagnosis confirmed: variant(s) have been found that are in all likelihood (>90%) pathogenic and are consistent with the clinical presentation.
(B) Variant(s) have been found that appear as good candidates, however, there is not, at the time of reporting, sufficient evidence to unequivocally attribute pathogenicity. This includes autosomal recessive disease where one variant has been identified that is of proven pathogenicity and the second candidate variant has been identified without proven pathogenicity.
(C) Autosomal recessive disease where only a single pathogenic allele is identified.

Undoubtedly it is scenarios B and C that are the most challenging to the referring clinician. In these situations, the clinical team may wish to revisit the clinical presentation in light of the genetic findings. This can help confirm a likely clinical association with the reported genetic variants before acting on the findings. A close working relationship between the lab and the clinician is optimal and, in many healthcare systems, this assessment is completed in a multidisciplinary team (MDT) meeting attended by both scientists and clinicians. Where referring clinicians are remote from the testing laboratory virtual multidisciplinary team meetings may be possible.

Secondary findings: Secondary findings are genetic test results that provide information unrelated to the primary purpose for testing. Secondary findings are a common feature of many pathology or imaging tests and are not unique to genomics. For example, a routine X-ray to investigate a bone fracture may also identify a potential malignancy. However, genomics is perhaps more unusual in that secondary findings are not only obtained incidentally (i.e. unsolicited) but may also be specifically requested by the clinician.

In 2013 (and again in 2017 and 2021), the American College of Medical Genetics and Genomics (ACMG) recommended that all labs performing exome and genome sequencing tests should include the reporting of secondary findings, in addition to any primary findings.[2,3,4] A list of 73 genes that are associated with a variety of conditions, from cancer to heart disease, was proposed. These 73 genes were chosen because they are associated with conditions that have: a definable set of clinical features; the possibility of early diagnosis; a reliable clinical genetic test; effective intervention or treatment. The goal of reporting these secondary findings to an individual is to provide medical benefits by preventing (or better managing) health conditions. Only variants that are known to cause disease are expected to be reported. Variants of unknown significance (see below), whose involvement in disease at the current time is unclear, should not be reported. However, as with any type of medical diagnosis, the news of an unexpected potential health problem may lead to additional health costs and stress for individuals and their families. Thus, practice differs across the world in this regard and it is not currently routine practice to look for secondary findings in Europe.

Which test to choose?

There is a wide range of genomic tests available to a referring ophthalmologist. In most cases it is considered most prudent to focus a test on a narrow genomic region to maximise the chances of identifying causative variants whilst, at the same time, minimise the chances of detecting secondary findings. For some presentations, testing can be relatively targeted:

- **Single variant analysis**: For several well-defined phenotypes, a single causal variant has been described. For example, a single *EFEMP1* mutation c.1033C>T (p.Arg345Trp) is associated with autosomal dominant drusen (Doyne honeycomb retinal dystrophy).[5]
- **Single exon analysis**: Sorsby fundus dystrophy is an autosomal dominant disease caused by mutations in exon 5 of the *TIMP3* gene.[6]
- **Single gene analysis**: There are several routine presentations where there will be a single strong candidate gene; examples include choroideraemia[7] and X-linked retinoschisis.[8]

However, many Mendelian ophthalmic conditions display considerable genetic (locus) heterogeneity, i.e. heterozygous (autosomal dominant), hemizygous (X-linked recessive) or biallelic (autosomal recessive) pathogenic variants in several different genes can cause overlapping phenotypes. The ability to simultaneously analyse multiple genes at a cheaper cost per base makes current sequencing approaches an attractive solution for diagnostic testing of many Mendelian ophthalmic disorders. In these cases, a broader genomic test may be required.

- **Gene panel:** Targeted analysis of a set of genes, any one of which might explain the phenotype
- **Exome analysis:** Looking at the protein-coding sequences of all genes (approximately 20,500).
- **Genome analysis:** Sequencing all bases in a genome (3 million base-pairs per genome).

Confusingly, 'virtual' gene panels can be applied to both exome and genome data. The laboratory will offer a range of diagnostic gene panels and the clinician will decide the most appropriate given the presenting symptoms. For example, with a diagnosis such as Usher syndrome, the clinician may restrict testing to just the genes known to cause that specific indication. However, with a less specific diagnosis, such as retinitis pigmentosa, a broader panel incorporating both syndromic as well as non-syndromic presentations may be indicated. A point worth noting is that, for a typical gene panel for retinal disorders (approximately 100 genes), experience shows that approximately 1 in 5 people are carriers of a recessive disease-implicated variant that is unrelated to the primary reason for referral. There is ongoing debate as to whether these variants should even be reported in routine practice. However, there is clear clinical utility in knowing this information in many families (e.g. in the presence of consanguinity).

Variant analysis

The starting point of any genomic analysis is to identify genomic events that differ from a 'normal' reference sequence. In reality, however, there is no such as a single normal reference. Instead, there is a single composite genomic sequence (in fact there are two currently used; either GRCh37 released in February 2009 or GRCh38, released in December 2013).

Deviations from the reference sequence may be in the form of structural changes (including copy number variation) or smaller scale DNA sequence variants (insertions, deletions, single base-pair changes). Traditionally, different testing platforms have been used to optimally identify the full range of sequence variation (see Chapter 2). Here, we focus primarily on the interpretation of small scale genomic variants. However, evolving technologies will soon allow all the different sequence changes to be comprehensively interrogated in a single test.

The data generated through sequencing is often highly complex and requires several bioinformatics steps before they can be interpreted. These steps include:

- **Alignment (or mapping)**: The DNA sequences generated by the sequencer do not come with position information, that is, we do not know what part of the genome they came from; we need to use the sequence of the read itself to find the corresponding region in the reference sequence.
- **Variant calling:** Once the DNA sequence fragments have been mapped, we can look for differences between the patient sequence and the reference sequence. Any detected alteration may potentially be a disease-causing change. Critically appraising these variants for evidence of causation is therefore key.

Automated variant interpretation

Sequencing typically identifies large numbers of positions in the tested genome that differ from the reference sequence. Any one of these can be plausibly pathogenic. Therefore, we must first try to exclude as many of these as possible using automated filtering. There are a variety of automated filtering steps we may choose to apply to reduce the number of plausibly pathogenic variants for further consideration; two of these are discussed below:

- **Variant effect predictors:** Given the position and nature of a variant relative to a gene of interest we can predict its most likely effect. For example, a variant 2 base-pairs away from an intron/exon boundary might be *predicted* to affect splicing; a variant within an exon that affects the third base of a codon might be *predicted* to be a synonymous change; a variant changing an amino-acid to a stop codon might be *predicted* to prematurely terminate the protein during translation. We can choose to exclude some of these categories of variants from further consideration to simplify our analysis (for example on the presumption that synonymous changes are less likely to be pathogenic). However, it must be stressed that such findings are merely predictions based upon broad generalisations and that many exceptions exist. Most diagnostic testing pipelines for example will not automatically exclude synonymous changes as they may be exerting an effect (for example through aberrant splicing) independent of any effect on the amino-acid sequence.

- **Allele frequency:** We now have accurate population allele frequencies across the entire genome for several groups of people across the world. This knowledge can be used to screen out common variants (on the assumption that a common change is unlikely to be the cause of a Mendelian ophthalmic condition, given that these disorders are individually relatively rare). An arbitrary allele frequency cut-off of 5% has been generally agreed for this purpose. However even this somewhat conservative cut-off needs critical application.[9] In practice, most clinical genetic labs will apply a more stringent filter (e.g. 1%) to arrive at a smaller and more manageable filtered variant list. This approach has underestimated the significance of several common yet pathogenic 'hypomorphic' recessive alleles within ophthalmic disease-implicated genes such as *TYR*[10] and *ABCA4*.[11]

Manual variant interpretation

Following automated variant analysis and to identify putative causal changes, clinical genetic laboratories undertake a more labour intensive and stringent analysis of each variant that has passed through the automated process. This process relies heavily on the critical skills of trained practitioners. However, until recently, this process was not standardised. This has led to variability in the classification of variants which, in rare cases, has resulted in discrepant interpretations of the same variants even within the same families.

The process of variant interpretation took a significant step forward in 2015 with the publication of a set of rules issued jointly by the American College of Medical Genetics and the Association of Molecular Pathology.[12] This framework aimed to standardise the interpretation of single nucleotide high impact variants in rare disease patients. These guidelines have been accepted either formally or informally as the standard of practice in many countries across the world and have been subsequently adapted multiple times to different disease paradigms.

Richards et al.[12] provided a set of criteria that can be used in the interpretation of all single nucleotide variants agnostic of pathogenic mechanism, disease model or inheritance pattern. They also provided a scoring framework enabling evidence to be weighted according to importance. In doing so we place a variant into one of the following categories; (1) benign, (2) likely benign (>90% certainty if a variant being benign), (3) uncertain significance, (4) likely pathogenic (>90% certainty of a variant being disease-causing), (5) pathogenic (>99% certainty of a variant being disease-causing). A likely pathogenic or pathogenic variant is generally regarded as an actionable finding, i.e. it can be used for the clinical management of the patient and family. Variants of uncertain significance may be tested in family members to help aid in their interpretation. However, clinicians must practice extreme caution before using these variants in the clinical management of the patient.

Table 3.1 shows the list of lines of evidence can be used to establish pathogenicity and Table 3.2 shows the lines of evidence that can be used to establish the benign nature of a variant. The rules for collating this evidence into a final interpretation are given in Richards et al.[12] It is beyond the scope of this chapter to detail each line of evidence individually, however, it is useful to consider some of these further within the context of ophthalmic disease.

PVS1: A "null" variant is one predicted to destroy the function of the protein encoded by that allele. Due to the complex and varied biological possibilities encompassed by this type of variation, further guidance has been drawn up to assist in this decision making.[13] Many ophthalmic genes are sensitive to loss of function effects and potential functionally null mutations should usually be regarded as potentially pathogenic. However, there are notable exceptions, for example, *RHO* where the gain of function is the main recorded disease mechanism in dominant disease. For other genes such as *PROM1* null variants tend to be recessive, whereas missense mutations tend to be dominant.

PS1 and **PM5:** These lines of evidence relied upon the legitimacy of existing published mutation reports. One should treat published reports of pathogenicity with caution; these can only be applied after a complete and critical appraisal of all the evidence. Often published evidence for pathogenicity is weak and, until recently, a large emphasis has been incorrectly placed on the absence of a variant in a relatively small cohort of controls as evidence of pathogenicity. More recent publications use larger publicly available datasets to establish allele frequency (see PM2).

PM2: A major advance in recent years has been the availability of large cohorts of control populations (collected largely as part of studies on common disease and hence generally perceived to be free of, or at least not enriched for, highly penetrant Mendelian disorders). Most notable is the Genome Aggregation Database (gnomAD) provided by the Broad Institute, US (http://gnomad.

TABLE 3.1 Criteria for classifying pathogenic variants.

Very strong	PVS1	Null variant (nonsense, frameshift, canonical ±1 or 2 splice sites, initiation codon, single or multi-exon deletion) in a gene where loss of function is a known disease mechanism	*Caveats*: Use caution • if loss of function is not a known disease mechanism. • Interpreting loss of function variants at the extreme 3′ end of the gene. • with splice variants that are predicted to lead to exon skipping but leave the remainder of the protein intact. • in the presence of multiple transcripts.
Strong	PS1	Same amino-acid change as previously established as a pathogenic variant regardless of nucleotide change.	*Example*: Val→Leu caused by either G>C or G>T in the same codon. *Caveat*: Beware of changes that impact splicing rather than at the amino-acid/protein level.
	PS2	*De novo* (both maternally and paternally confirmed) in a patient with the disease and no family history	*Note*: confirmation of paternity only is insufficient. Egg donation, surrogate motherhood, errors in embryo transfer etc., can contribute to non-maternity.
	PS3	Well established *in vitro* or *in vivo* functional studies supportive of a damaging effect on the gene or gene product	*Note*: Functional studies that have been validated and shown to be reproducible and robust in a clinical diagnostic setting are considered the most well established.
	PS4	The prevalence of the variant in affected individuals is significantly increased compared with the prevalence in controls	*Note*: In instances of very rare variants where case–control studies may not reach statistical significance, the prior observation of the variant in multiple unrelated patients with the same phenotype, and its absence in controls may be used as a moderate level of evidence.
Moderate	PM1	Located in a mutational hotspot and/or critical and well-established functional domain (e.g. active site of an enzyme) without benign variation.	
	PM2	Absent from controls (or at extremely low frequency if recessive) e.g. in the Genome Aggregation Database (gnomAD)	*Caveat*: Population data for insertions and deletions may be poorly called by next-generation sequencing.
	PM3	For recessive disorders, detected in trans with a pathogenic variant	*Caveat*: This requires testing of parents (or offspring) to determine the phase
	PM4	Protein length changes as a result of in-frame deletions or insertions in a non-repeat region or stop-loss variants	
	PM5	Novel missense change at an amino-acid residue where a different missense change determined to be pathogenic has been seen before	*Example*: Arg156His is pathogenic; now you observe Arg156Cys *Caveat*: Beware of changes that impact splicing rather than at the amino-acid/protein level.
	PM6	Assumed *de novo*, but without confirmation of maternity and paternity	
Supporting	PP1	Co-segregation with the disease in multiple affected family members in a gene known to cause the disease	*Note*: May be used as stronger evidence with increasing segregation data.
	PP2	Missense variant in a gene that has a low rate of benign missense variation and in which missense variants are a common disease mechanism.	
	PP3	Multiple lines of computational evidence support a deleterious effect on the gene or gene product (conservation, evolutionary splicing impact etc.).	*Caveat*: Because many *in silico* algorithms use the same or very similar input for their predictions, each algorithm should not be counted as an independent criterion. PP3 can be used only once in any evaluation of a variant.
	PP4	The patient phenotype & family history is highly specific for a disease with a single genetic etiology.	
	PP5	A reputable source recently reports variant as pathogenic, but the evidence is not available to the laboratory to perform an independent evaluation	

Adapted from Ref. 12.

TABLE 3.2 Criteria for classifying benign variants.

Stand alone	BA1	Allele frequency is >5% in the Genome Aggregation Database (gnomAD)	
Strong	BS1	Allele frequency is greater than expected for the disorder	
	BS2	Observed in a healthy adult individual for a recessive (homozygous), dominant (heterozygous), or X-linked (hemizygous) disorder, with full penetrance expected at an early age	
	BS3	Well established *in vitro* or *in vivo* functional studies show no damaging effect on protein functioning or splicing	
	BS4	Lack of segregation in affected members of a family	*Caveat*: The presence of phenocopies for common phenotypes (i.e. age-related macular degeneration) can mimic a lack of segregation among affected individuals. Also, families may have more than one pathogenic variant contributing to an autosomal dominant disorder, further confounding an apparent lack of segregation.
Supporting	BP1	Missense variant in a gene for which primarily truncating variants are known to cause disease	
	BP2	Observed in trans with a pathogenic variant for a fully penetrant dominant gene/disorder or observed in cis with a pathogenic variant in any inheritance pattern.	
	BP3	In-frame deletions/insertions in a repetitive region without a known function.	
	BP4	Multiple lines of computational evidence suggest no impact on gene or gene product (conservation, evolutionary, splicing impact, etc.).	*Caveat*: Because many *in silico* algorithms use the same or very similar input for their predictions, each algorithm cannot be counted as an independent criterion. BP4 can be used only once in any evaluation of a variant.
	BP5	The variant found in a case with an alternative molecular basis for disease.	
	BP6	A reputable source recently reports variant as benign, but the evidence is not available to the laboratory to perform an independent evaluation.	
	BP7	A synonymous (silent) variant for which splicing prediction algorithms predict no impact to the splice consensus sequence not the creation of a new splice site AND the nucleotide is not highly conserved.	

Adapted from Ref. 12.

broadinstitute.org/). This database allows us to assess the frequency of variants in unaffected people. This can be then used as a standalone piece of evidence (the 5% cut off mentioned above can be critically applied here: see[9]).

PS4: It is possible to critically compare the prevalence of a variant in affected individuals against the prevalence in controls. This has proved to be an extremely powerful means to determine pathogenicity in disciplines such as Cardiology where large variant datasets in carefully phenotyped patient cohorts exist. However, such cohorts are not yet widely available for ophthalmic conditions.

PP3: We can use computational algorithms to predict the consequences of a missense change where one amino-acid is replaced by another. These predictions rely on comparing the physiochemical properties of one amino-acid to its replacement, as well as the evolutionary conservation of that position in the protein through time. Such studies have a questionable predictive value and the guidelines rightly relegate this level of evidence to 'supporting' rather than 'high'. Similarly, we can use computational tools to predict the effect of any variant on the splicing patterns of a gene. However, the predictive value of these approaches may also be suboptimal, especially for intronic locations distal from the intron-exon junction.

How can the clinician facilitate effective variant interpretation?

Firstly, the clinician must be able to choose the correct panel of genes to test or interrogate. There is a wide range of panels available and it is often difficult to get the balance right—the more genes you test, the more variants you will receive back that may be difficult to interpret.

It is vital to provide an accurate phenotype and family history to help the laboratory focus on the correct candidate gene(s). If a patient's phenotype or family history is highly specific for a disease with a single aetiology then this information can be used when interpreting the variants. It is therefore particularly important that this information is available to the clinical genetic laboratory.

There are several follow-up studies that will provide essential information when trying to interpret the pathogenicity of a variant.

- If two potentially pathogenic variants are reported in an autosomal recessive gene you may suspect compound heterozygosity. However, it is important to confirm this suspicion by performing segregation analysis testing in family members (usually parents).
- If an unclassified variant is identified and there is a high level of suspicion then testing other family members (especially affected ones) can add to the weight of evidence. However, the clinician must be careful to avoid unsolicited predictive testing by investigating an unaffected individual for a variant potentially associated with the later onset or incomplete penetrance.
- If a potentially pathogenic variant is identified in a dominant gene in a patient with no family history of the disease, testing parents may help establish whether this is a *de novo* event. Such a finding will have a maximal effect on variant interpretation if the lab can also confirm paternity.

Expect the unexpected

Historically individuals with sporadic disorders have often been counselled as 'most likely' having recessive disease (and concomitant low risk of passing the disease to offspring). Such advice is now considered premature in the absence of a molecular confirmation. Pseudo-dominance is not uncommon, especially in the more prevalent ophthalmic presentations, such as Stargardt disease, due to the high population frequencies of pathogenic variants.

Many genes with ophthalmic manifestations can harbour both dominant and recessive pathogenic variants. New biological models are emerging to explain some of these complicated situations. Falkenberg et al.[14] for example describe several individuals who had heterozygous mutations in *PEX6* (a typically recessive gene) with ophthalmic presentations consistent with the Zellweger syndrome spectrum, but no second mutations in the same gene; this observation was attributed to an allelic imbalance of expression.

Be aware of atypical presentations

Many genes associated with the ophthalmic disease tend to present with a wide range of phenotypes that are not always restricted to the eye. Ophthalmologists should therefore be versed in the complications of genotype-phenotype variability. However, unusual genomic events can add to this variability. Burin-des-Roziers et al.[15] for example decribed a somatic mosaic *VCAN* mutation in a patient with an unusually mild asymmetrical phenotype. It can also be useful to consider the possibility of multiple genetic diagnoses in a single patient and risk factors for this should be born in minds such as founder mutations, common chromosomal disorders (Down, Turner, Klinefelter), family history (e.g. microduplication/deletion) and consanguinity.[16]

Be alert to what you might be missing

Hopefully, your analysis has led you to the identification of one (dominant and X-linked disease) or two (autosomal recessive disease) causal variants in your patient. However, if this has not been possible, it is only very rarely that we can be certain that genetic analysis has fully excluded all pathogenic variants.

As highlighted previously, there are many ways to deliver genetic testing and even more methodologies to interpret the sequencing data. There is currently no single accepted best practice in this area and different bioinformatics methods have their strengths. Notably, the gradual adoption of genome sequencing has highlighted the importance of disease-associated non-coding variation deep in the intron,[1] in gene regulatory elements[17,18] and the gene untranslated region.[19] It is tempting to consider whether we should be moving wholesale to genome sequencing to capture this additional heritability. However, the added cost of sequencing and storing the data is currently a large obstacle to the more widespread adoption of genome sequencing in clinical genetic diagnostics. In addition, even if we could identify this non-coding variation the tools at our disposal to interpret its clinical significance is limited.

It is worth considering whether or not your genetic test includes copy number variation analysis. Although this is now routine in many bioinformatics pipelines, it is not universal. Ellingford et al.[20] estimated that at least 7% of individuals referred for diagnostic testing on a retinal disease gene panel had a copy number of variants within genes relevant to their clinical diagnosis.

Concluding remarks

Genomic medicine is increasingly used in mainstream medicine and ophthalmologists have been among the earlier adopters of these approaches due to the high burden of genetic ophthalmic disease. It is important to realise that all tests are not the same; it is worth checking whether a test detects copy number variants and inquiring as to the levels of mosaicism that can be reliably detected. Notably, variant analysis is a new skill to many and can seem a little daunting as the consequences of getting it wrong can be severe. Remember that a test revealing a 'negative' result may need repeating at later date as technology changes.

References

1. Cascella R, Strafella C, Caputo V, Errichiello V, Zampatti S, Milano F, et al. Towards the application of precision medicine in age-related macular degeneration. *Prog Retin Eye Res* 2018;**63**:132–46.
2. Kalia SS, Adelman K, Bale SJ, Chung WK, Eng C, Evans JP, et al. Recommendations for reporting of secondary findings in clinical exome and genome sequencing, 2016 update (ACMG SF v2.0): a policy statement of the American College of Medical Genetics and Genomics. *Genet Med* 2017;**19**:249–55.
3. Green RC, Berg JS, Grody WW, Kalia SS, Korf BR, Martin CL, et al. ACMG recommendations for reporting of incidental findings in clinical exome and genome sequencing. American College of Medical Genetics and Genomics. *Genet Med* 2013;**15**:565–74.
4. Miller DT, Lee K, Chung WK, Gordon AS, Herman GE, Klein TE, et al. ACMG Secondary Findings Working Group. *Genet Med.* 2021;**23**(8): 1381–1390. https://doi.org/10.1038/s41436-021-01172-3.
5. Stone EM, Lotery AJ, Munier FL, Héon E, Piguet B, Guymer RH, et al. A single EFEMP1 mutation associated with both Malattia Leventinese and Doyne honeycombretinal dystrophy. *Nat Genet* 1999;**22**:199–202.
6. Gliem M, Müller PL, Mangold E, Holz FG, Bolz HJ, Stöhr H, et al. Sorsby fundus dystrophy: novel mutations, novel phenotypic characteristics, and treatment outcomes. *Invest Ophthalmol Vis Sci* 2015;**56**:2664–76.
7. Ramsden SC, O'Grady A, Fletcher T, O'Sullivan J, Hart-Holden N, Barton SJ, et al. A clinical molecular genetic service for United Kingdom families with choroideraemia. *Eur J Med Genet* 2013;**56**: 432–8.
8. Strupaitė R, Ambrozaitytė L, Cimbalistienė L, Ašoklis R, Utkus A. X-linked juvenile retinoschisis: phenotypic and genetic characterization. *Int J Ophthalmol* 2018;**11**:1875–8.
9. Ghosh R, Harrison SM, Rehm HL, Plon SE, Biesecker LG, ClinGen Sequence Variant Interpretation Working Group. Updated recommendation for the benign stand-alone ACMG/AMP criterion. *Hum Mutat* 2018;**39**:1525–30.
10. Monfermé S, Lasseaux E, Duncombe-Poulet C, Hamel C, Defoort-Dhellemmes S, Drumare I, et al. Mild form of oculocutaneous albinism type 1: phenotypic analysis of compound heterozygous patients with the R402Q variant of the *TYR* gene. *Br J Ophthalmol* 2018;**103**(9): 1239–47. https://doi.org/10.1136/bjophthalmol-2018-312729.
11. Zernant J, Lee W, Collison FT, Fishman GA, Sergeev YV, Schuerch K, et al. Frequent hypomorphic alleles account for a significant fraction of ABCA4 disease and distinguish it from age-related macular degeneration. *J Med Genet* 2017;**54**:404–12.
12. Richards S, Aziz N, Bale S, Bick D, Das S, Gastier-Foster J, et al. Standards and guidelines for the interpretation of sequence variants: a joint consensus recommendation of the American College of Medical Genetics and Genomics and the Association for Molecular Pathology. *Genet Med* 2015;**17**:405–24.
13. Abou Tayoun AN, Pesaran T, DiStefano MT, Oza A, Rehm HL, Biesecker LG, et al. Recommendations for interpreting the loss of function PVS1 ACMG/AMP variant criterion. *Hum Mutat* 2018;**39**:1517–24.
14. Falkenberg KD, Braverman NE, Moser AB, Steinberg SJ, Klouwer FCC, Schlüter A, et al. Allelic expression imbalance promoting a mutant PEX6 allele causes Zellweger spectrum disorder. *Am J Hum Genet* 2017;**101**:965–76.
15. Burin-des-Roziers C, Rothschild PR, Layet V, Chen JM, Ghiotti T, Leroux C, et al. Deletions overlapping VCAN exon 8 are new molecular defects for Wagner disease. *Hum Mutat* 2017;**38**:43–7.
16. Kurolap A, Orenstein N, Kedar I, Weisz Hubshman M, Tiosano D, Mory A, et al. Is one diagnosis the whole story? Patients with double diagnoses. *Am J Med Genet A* 2016;**170**:2338–48.
17. Radziwon A, Arno G, Wheaton DK, EM MD, Baple EL, Webb-Jones K, et al. Single-base substitutions in the CHM promoter as a cause of choroideremia. *Hum Mutat* 2017;**38**:704–15.
18. Small KW, DeLuca AP, Whitmore SS, Rosenberg T, Silva-Garcia R, Udar N, et al. North Carolina macular dystrophy is caused by dysregulation of the retinal transcription Factor PRDM13. *Ophthalmology* 2016;**123**:9–18.
19. Devanna P, van de Vorst M, Pfundt R, Gilissen C, Vernes SC. Genome-wide investigation of an ID cohort reveals de novo 3'UTR variants affecting gene expression. *Hum Genet* 2018;**137**: 717–21.
20. Ellingford JM, Horn B, Campbell C, Arno G, Barton S, Tate C, et al. Assessment of the incorporation of CNV surveillance into gene panel next-generationsequencing testing for inherited retinal diseases. *J Med Genet* 2018;**55**:114–21.

Chapter 4

Genetic counselling and family support

Georgina Hall, Avril Daly, Christina Fasser, and Jane L. Ashworth

This chapter explores the impact of genetic testing for ophthalmic disorders on patients and their families. The role of the genetic counsellor is discussed and potential issues around consent for testing are highlighted.

Genetic testing—The patient and family perspective

The decision to have a genetic investigation requires consideration by the individual and their families (although ultimately it is the choice of the person having the test). Patients should have timely, equitable access to genetic testing.[1] Barriers may include limited funding, clinician knowledge and availability of specialist genetic ophthalmic services. Advances in genomic technologies, increasingly robust evidence on the clinical utility of testing and higher healthcare provider genomic competency are key for improving genetic testing for individuals with genetic disorders.[2,3]

In considering a genetic test, patients should initially be allowed to discuss the options with a clinician and to subsequently talk about the potential impact of a molecular diagnosis with a genetic counsellor. This process strongly influences the way that an individual deals with their diagnosis throughout their life.[4] Genetic counsellors and ophthalmologists with specialist genetic expertise should be well placed to provide clear and accurate information and relevant patient education material.

Genetic testing can bring benefits to those living with genetic ophthalmic disorders. A molecular diagnosis is helpful to affected individuals as it:

- refines the clinical diagnosis and gives a name to the disease.
- clarifies the mode of inheritance.
- provides insights into the prognosis for vision.

These pieces of information can help accept and deal with the disease.[5] Furthermore, for several genetic ophthalmic conditions, notably retinal dystrophies, a molecular diagnosis is a prerequisite for access to new and emerging therapies, participation in appropriate clinical trials, and inclusion in registries. It is worth noting though that a genetic diagnosis may raise several psychological issues in an individual and their families.[6] These may include:

- *Guilt*: Parents and grandparents very often develop a feeling of guilt, relating to having passed on or 'caused' the condition. This can have an impact on decisions, such as having children, or a parent's reaction to the diagnosis in their child. Counselling and support can help to acknowledge and normalise these feelings and support families in adjusting to the heritable nature of the condition.
- *Stigma:* Genetic diseases have often been stigmatised and this can reinforce the feeling of guilt. Individuals with genetic ophthalmic conditions may be coping with not only the burden of visual disability but also with the impact of having a rare disease that is often not understood by doctors, family or friends.
- *Denial:* Denial can be expressed by one or both parents, but even more frequently by the extended family of an affected person. This attitude is common and should be addressed. The roots of denial often lie with the fear of the unknown that is often concealed, and/or with concerns around the fact that the disease may also affect their family or children.
- *Anxiety while waiting for results:* Although patients are anxious to have access to genetic testing services, they often report that during the period waiting for, and ultimately receiving, the results can be stressful and this can often lead to a lot of emotion in families. With this in mind the need to talk through the possible timelines and outcomes in advance—be they positive or negative—cannot be overemphasised.
- *Impact on insurance:* Accessing insurance cover following a genetic test can be an issue in some countries. In the European Union, it is illegal to use this information to weigh a policy. However, the question may be asked in insurance forms. In general, genetic discrimination is illegal but precise details vary from country to country.

Genetic testing—Types of tests

It is important to consider the intended clinical application of a genetic test. Applications include diagnostic genetic testing, segregation analysis, carrier testing, predictive genetic testing, and reproductive genetic testing. An additional consideration is whether a test is performed in response to a specific clinical indication or as part of a screening process (e.g. newborn screening).

Diagnostic testing

Diagnostic tests aim to identify the underlying cause of a genetic condition. There are many approaches to diagnostic genetic testing including a single gene test, a gene panel test or exome/genome sequencing. The ophthalmologist or geneticist will need to understand which testing strategy is most appropriate for their patient. Informed consent will need to be appropriate for the selected approach.

Segregation testing

Segregation analysis involves testing family members (affected or obligate carriers) beyond the proband. The main aim is to confirm a diagnostic result by demonstrating that the genomic variant segregates in the family as predicted. This may involve testing parents for heterozygous (carrier) status to confirm that recessively-acting variants have been inherited on opposite copies of the gene. For variants of uncertain significance, segregation testing of other family members may provide evidence supporting the variant's causality in the family (if segregation of the variant matches segregation of the phenotype, e.g. if it is shared among affected family members). Extended family histories are often required for this approach.

Carrier testing

Carrier testing, like predictive and reproductive testing (see below), is a testing approach that can only be considered once the pathogenic variant has been identified in the family. In autosomal recessive and X-linked recessive conditions, where a heterozygous carrier has no clinical features of the disease but can pass the condition down to their children, carrier testing can be offered to inform reproductive decisions. Unlike diagnostic genetic testing which can be utilised in ophthalmology clinics, carrier testing is normally recommended within genetic counselling services.

Predictive testing

Predictive tests aim to identify genetic variants that increase a person's risk of developing a heritable disorder before signs or symptoms appear. This type of test is sometimes called presymptomatic test and is usually considered for adults in a family segregating an adult-onset genetic condition, where an individual can make an informed choice to find out about their own future risk of vision loss. Predictive testing in childhood is more challenging and has ethical implications around the autonomy of the child. It is recommended that predictive testing is undertaken within genetic counselling services for careful consideration of the implications within recommended guidelines.

Reproductive testing

Reproductive genetic tests offer the opportunity to identify people who are at increased risk for having a child who is affected with a genetic disease or to identify an affected embryo or foetus. It may involve carrier; prenatal or pre-implantation testing and it can help inform reproductive choices such as pre-implantation genetic diagnosis and invasive/non-invasive prenatal testing. Couples should be referred to genetic services if they wish to discuss these choices. Legal frameworks and availability of reproductive testing vary in different countries.

Research testing

Diagnostic testing is not always available or not always successful (due to limitations of genomic knowledge) and patients may be offered research testing. Difficulties have arisen for patients who may not fully understand the differences between genetic testing in a clinical setting as opposed to testing for research. There are many examples of patients not being aware that being part of a research study does not automatically mean they will receive a diagnosis. The difference between research and clinical-grade tests must be explicitly explained. Often this lack of understanding can be mitigated through the involvement of a genetic counsellor.

For more information about genetic testing and recommendations on genetic counselling, see:

- Eurogentest Recommendations for Genetic Counselling related to Genetic testing. http://www.eurogentest.org/index.php?id=674
- European Society of Human Genetics Genetic testing in asymptomatic minors: Recommendations of the European Society of Human Genetics.[8]

The family tree and the role of the genetic counsellor

Documentation of an accurate family tree is a key part of an ophthalmic genetic clinic. Detailed family history may reveal patterns that aid with diagnosis. For example, a patient with high myopia and a cleft palate may have a parent with a retinal detachment and a relative with hearing loss, indicating a possible family diagnosis of Stickler syndrome. A three-generation family tree is required with accurate recording of relationships and medical symptoms. Occasionally, obtaining and reviewing hospital records from family members (following appropriate consent) may be necessary.

Genetic conditions have an impact on families and a new diagnosis in a proband can highlight the risk to their relatives. For example, if a young boy presents with night-blindness and pigmentary retinopathy, making the diagnosis of X-linked retinitis pigmentosa will have implications for the proband's maternal aunt.

When drawing a family tree, the genetic counsellor will consider the possible pattern of inheritance and potential implications for the wider family. The genetic counsellor will also listen for cues around the family relationships, both for knowledge and accuracy in reporting medical symptoms in the family and also to prepare for a discussion about inheritance. For example, a patient may reveal a breakdown in communication with her sister: "*I haven't spoken to her since my mum died but I have heard that she has just had a baby boy.*" This may prompt the genetic counsellor to ask further questions and to explore the lines of communication that may be open in the family.

As well as facilitating communication, the family can have an important role in coping and adjusting to vision impairment. Some patients with a family history often identify a relative as a role model in the way in which they have coped with their vision impairment; "*My uncle lost his sight in his 40s but he never let it stop him, he continued to run his own business.*" Family adjustment or conversely negative associations with vision loss will impact how a patient may approach their condition, their resilience or determination to find ways to adjust.

The family impact is not only determined by the affected family members. Relationships between relatives, the reaction of family members and the way the family function around support, will all have an important influence on how an individual will adapt and adjust to their condition. A systemic approach to individuals with vision impairment should be extended beyond the family to include friends, teachers, work colleagues, and medical professionals. Family belief systems, and emotional responses such as guilt or denial, will also impact how a family functions around a genetic condition. Genetic counsellors may use family or narrative counselling approaches to support the family as a whole adjusting, making decisions and communicating with children (Box 4.1).

Consent for genetic testing

Clinicians who order genetic tests need to be familiar with the issues around testing. This will allow them to take

BOX 4.1 The role of genetic counsellors in ophthalmology

In the ophthalmic genetic clinic, ophthalmologists may work alongside genetic counsellors in a multidisciplinary team. In other settings, ophthalmologists should work in close liaison with genetic services and refer patients for genetic counselling particularly in these key areas:

- Discussion and further investigation following complex genomic results, including variants on uncertain significance, unexpected clinically significant results or incidental findings that may have reproductive implications (e.g. heterozygous carrier finding in consanguineous families).
- Counselling and cascade testing for relatives at risk, especially around carrier testing with reproductive risk or predictive testing.
- Family support and communication both within and beyond the family (employers, friends)
- Emotional support and signposting (Box 4.2).

BOX 4.2 The role of the Eye Clinic Liaison Officer

Research has demonstrated that patients with genetic ophthalmic conditions feel there is a significant lack of support around practical aspects of adjusting to vision loss.[5] They identify a complexity of services, a lack of joined-up working and poor signposting from their clinicians to the social care sector. This includes even basic support such as registration with sight impairment.

Eye clinic liaison officers (ECLO), are dedicated specialists working alongside clinicians. They can provide point of care support at the critical time of diagnosis, and follow-up support to find the most appropriate agencies/services within the health, social care and charity sectors. For affected children, the ECLO can work with qualified teachers for the visually impaired. For adults at work, the ECLO can signpost to support and services for employees/employers. More broadly the ECLO can highlight relevant activities and groups or refer for emotional counselling when needed.

appropriate consent from the patient and family, and to ensure that the implications of having the test are appreciated.

Issues that need to be discussed with the patient/family as part of the consent process include:

- The reasons for doing the test: to make a precise molecular diagnosis, to inform prognosis, to determine the need for systemic investigation and follow-up, to

enable family counselling etc. The reasons for offering a particular test should also be discussed (e.g. panel testing versus genome sequencing; see below).
- The possibility of not making a molecular diagnosis (i.e. of having a negative test result).
- The possibility of receiving an inconclusive test result ('variant of unknown significance').
- The possibility of incidental findings (e.g. a carrier of a recessively acting genetic variant that does not relate to the ophthalmic condition). These may not have direct clinical implications for the patient but could have implications for the wider family (e.g. in consanguineous families).
- The possibility of diagnosing a syndromic condition which may have implications for systemic health, or which may need further investigation.
- The possibility that the test result may have implications for the wider family including (future) children.
- The process for interpretation of the result (e.g. that the findings will be discussed in a multidisciplinary team meeting before delivery to the referring clinician and the family).
- The likely time it will take to receive the result, and how the result will be delivered to the family.
- A possible need for further genetic testing or clinical interpretation with emerging evidence around the genetic variant in the future as technology and genomic knowledge advances.

These points highlight that clinician should have an understanding of the testing approach, the time it will take and the possible outcomes. Patients should be aware that the results may reveal risks to other family members. While consent cannot cover every possible outcome (for example every syndromic gene on the gene panel), a broad consent approach can be used to ensure that the patient is sufficiently informed.[7]

Discussions around testing with the patient/family can take some time. It can therefore be difficult to achieve this in the context of a busy clinic, and so having ancillary clinical staff trained in this process can be beneficial. A standardised consent form can help with the consistency of this process. If this is not available, then appropriate documentation of the consent process should be made within the clinical records. In addition, information leaflets or web-based material should be available and the details of a contact person should be provided in case there will be further queries.

Referrals can be made to genetic services when the results are complex or family studies such as segregation (testing other family members to help interpreting results) are required. Patients/families who have more questions about genetic testing or the risks to the wider family could be referred for genetic counselling.

Concluding remarks

The diagnosis of a genetic ophthalmic condition impacts an individual beyond the management and treatment plan of the ophthalmologist. As illustrated in this section, the emotional impact, the additional burden of the implications to the wider family and increasingly, the complexity of genetic testing means that an integrated approach linking ophthalmology, genetics and the social and charity sector is critical in developing services. Patient-centred care requires time, listening, empathy, addressing patient's needs, a full and transparent discussion/explanation and working at the pace of each patient. Services must be developed to integrate care, work with families rather than individuals and adopt a long-term approach that responds to both emerging needs in the family and genomic developments in variant interpretation and personalised treatments.

References

1. Galvin O, Chi G, Brady L, Hippert C, Del Valle Rubido M, Daly A, et al. The impact of inherited retinal disease in the Republic of Ireland (ROI) and the United Kingdom (UK) from a cost-of-illness perspective. *Clin Ophthalmol* 2020;**14**:707–19. 32184557.
2. National Academies of Sciences, Engineering, and Medicine; Health and Medicine Division; Board on Health Care Services; Board on the Health of Select Populations; Committee on the Evidence Base for Genetic Testing. *An evidence framework for genetic testing.* Washington, DC: National Academies Press (US); 2017. 28418631.
3. Genomics Education Programme. *Facilitating genomic testing: a competency framework*. Genomics Education Programme, Health Education England; 2020. https://www.genomicseducation.hee.nhs.uk/consent-a-competency-framework/.
4. McVeigh E, Jones H, Black G, Hall G. The psychosocial and service delivery impact of genomic testing for inherited retinal dystrophies. *J Community Genet* 2019;**10**(3):425–34. https://doi.org/10.1007/s12687-019-00406-x.
5. Combs R, McAllister M, Payne K, Lowndes J, Devery S, Webster AR, et al. Understanding the impact of genetic testing for inherited retinal dystrophy. *Eur J Hum Genet* 2013;**21**(11):1209–13. 23403902.
6. Clarke A. *Harper's practical genetic counselling.* 8th ed. CRC Press; 2019.
7. RCP London. *Consent and confidentiality in genomic medicine.* UK Joint Committee on Genomics in Medicine Working Party Report; July 2019. https://www.rcplondon.ac.uk.
8. European Society of Human Genetics. Genetic testing in asymptomatic minors: Recommendations of the European Society of Human Genetics. *Eur J Hum Genet.* 2009;**17**(6):720–721. https://doi.org/10.1038/ejhg.2009.26.

Chapter 5

Syndromic conditions and the eye

Jill Clayton-Smith

Many ophthalmic conditions, particularly those affecting children, are developmental in origin and occur in association with other problems (including structural malformations and disorders of growth or development). Hundreds of syndromic (from the Greek syn and dromus, meaning 'running together') conditions have been described. Ophthalmic manifestations are often one of the first presenting features of these multisystemic disorders. Thus, ophthalmologists may be the first port of call for diagnosis. It is worth highlighting that syndromic features can occasionally be mild or clinically unsuspected. A good example of this is Bardet-Biedl Syndrome commonly detected nowadays on screening adults for genetic causes of retinal dystrophy.[1]

An ophthalmologist cannot be expected to recognise every single rare syndromic disorder. However, important skills to be mastered include being able to distinguish syndromic from non-syndromic conditions and being aware of the most common ophthalmic syndromes. Whilst the ideal situation is for an ophthalmologist to work jointly with a clinical geneticist in a multidisciplinary clinic, many patients will present first to the ophthalmology clinic and so recognition of a possible syndrome needs to occur at this stage. This chapter aims to provide a framework that ophthalmologists can adopt to improve their skills in recognising and making a syndrome diagnosis. It will also aim to provide some guidance about the utility of genetic testing for syndrome diagnosis as we move into an era where testing is being mainstreamed, with tests being ordered directly by members of the ophthalmology team.

The value of diagnosing an ophthalmic syndrome

Whilst some may believe that attaching a specific name to an ophthalmic disorder will make little difference to its management, it is now commonly acknowledged that making a syndrome diagnosis significantly benefits the patients concerned, their families and the professionals caring for them.[2] Stickler syndrome provides an excellent example.[3] Individuals with this condition occasionally present for the first time having lost vision due to retinal detachment. However, there may be important clues in the history and specific physical signs early in life that would suggest a diagnosis of Stickler syndrome and thereby lead to measures being put into place to reduce the risk of visual loss from the retinal detachment in both the patient and their affected relatives. Affected individuals may have a family history of retinal detachment, high myopia, cleft palate, hearing loss or premature osteoarthritis. Furthermore, infants with Stickler syndrome usually have micrognathia with or without a cleft palate; this is particularly notable in the neonatal period when they frequently suffer from airway obstruction and feeding difficulties and may be diagnosed as having Pierre Robin sequence. Early childhood onset myopia, variable hearing loss, joint laxity and arthritis may develop over time. Eliciting a history of any of these symptoms may lead to a suspicion of the syndrome diagnosis. The clinical examination may reveal differences in facial features including a small chin, flat nasal bridge, prominent eyes and a short nose. Joint hypermobility is common. Following diagnosis, appropriate surveillance can be offered for the hearing and joint problems and decisions can be made about monitoring and the need for prophylactic treatment for the risk of retinal detachment. Stickler syndrome is an autosomal dominant disorder with the ocular forms being caused by mutations in either the *COL2A1* gene or *COL11A1* gene. In many cases, there will be a family history with a parent and possibly other family members also being affected. Should the diagnosis be made in one individual, enquiry should be made about symptoms in other family members and investigations undertaken in those at prior risk of Stickler syndrome (as they may be at risk of future retinal detachment). Genetic testing and counselling can also then be offered to extended family members as appropriate.

Making an underlying syndrome diagnosis can provide answers for affected individuals or their parents who may have suspected that there was an overarching diagnosis for their own or their child's symptoms for some time. Having a diagnosis helps them to access appropriate information, services and benefits and may enable them to make contact with other families or sources of support. Finally, if a syndrome diagnosis is made then it can mean that individuals are then eligible to participate in relevant research studies, including clinical trials of any new treatments which are being developed.

When to suspect an ophthalmic syndrome

Syndrome diagnosis follows similar principles to those for clinical diagnosis in general. It is based on taking a thorough medical history and family history, followed by a general clinical examination as well as an ophthalmic examination and appropriate investigations. Though having time to undertake these steps comprehensively may be problematic for an ophthalmologist in a busy clinic, the key thing is to recognise that there may be a syndromic diagnosis to be made in a particular patient and then to refer on to the local genetic team or genetic ophthalmic clinic if a diagnosis cannot be made easily in the ophthalmology clinic.

Clues from medical and family history

An enquiry should be made about the patient's medical and developmental history. It is important to elicit whether there have been any other associated medical problems or whether the patient is currently being treated for other symptoms. In children particularly, a pregnancy history enquiring about teratogenic exposures, infections in pregnancy, any abnormalities seen on prenatal ultrasound scans and congenital anomalies noted after birth should be taken.

Several common teratogens can cause ophthalmic findings. Examples of this are foetal alcohol syndrome[4] where optic nerve hypoplasia is one of the commonest ophthalmic findings and foetal valproate syndrome in which there have now been several reports of coloboma and myopia. Though immunisation programs have reduced the incidence of congenital rubella, other infections during pregnancy such as Zika virus,[5] cytomegalovirus and toxoplasmosis[6] cause ophthalmic phenotypes. The increased rate of migration in recent years has led to ophthalmologists in some countries caring for diverse populations which include individuals who may have been born in countries without immunisation programmes or with high incidences of specific rare disorders. The presence of congenital malformations may also raise suspicion. A cleft palate might prompt consideration of Stickler syndrome as mentioned above. Short limbs or digit abnormalities seen in association with retinopathy might indicate the presence of one of the groups of disorders known as ciliopathies. Since many syndromic ophthalmic disorders affect development or learning, enquiry about developmental milestones, how a child is achieving at school, need for special education and whether an adult has any educational qualifications and their type of occupation can all be informative. Any previous investigations that have been undertaken and their results should be noted.

Ideally, a three-generation family tree should be recorded. With practice in systematically asking questions about family history, a clinician or clinic nurse will become efficient at eliciting a comprehensive history in a relatively short time. Guidance on this in the form of a video presentation can be found on the Health Education England Genomics Education website: https://www.genomicseducation.hee.nhs.uk/news/item/407-new-family-history-films-now-available/. In the case of ophthalmic syndromes, it is particularly important to ask about eyesight problems in other family members, but a general enquiry about other family health issues should also be made. The family history may indicate that one is dealing with a disorder that follows a particular Mendelian pattern of inheritance, e.g. autosomal dominant; this could narrow down the number of diagnoses to be considered. The presence of multiple early miscarriages suggests the possibility of a chromosome disorder, and an excess of affected females in a family raises the possibility of an X-linked dominant disorder with male lethality. There are some good examples of the latter in ophthalmology, including incontinentia pigmenti[7] which presents with early retinal detachment (the clue being the presence of skin blistering in the neonatal period); oculofaciocardiodental (OFCD) syndrome[8] caused by mutation or deletion of the BCOR gene and presenting with congenital cataracts, heart defects, 'hammer toe' deformity, characteristic facial features and abnormally long roots to the teeth is a further X-linked dominant disorder which is a likely under-diagnosed cause of congenital cataracts in females.

Clues from general examination

Although it may be difficult to carry out a full physical examination in the ophthalmology clinic, many general physical signs may be apparent or elicited quickly and easily. First of all, taking note of a child's muscle tone and posture may suggest a diagnosis. In Lowe syndrome for example, which usually presents with congenital cataracts,[9] there is profound infantile hypotonia. Noting the general build and body proportions of the patient and measuring standard growth parameters of height, weight and head circumference is useful. There are a group of syndromic diagnoses, for example, which present with obesity and retinal dystrophy. These include Bardet Biedl syndrome,[10] Alström syndrome[11] and Cohen syndrome,[12] all of which can be suspected from the patient's body habitus. A small head circumference (microcephaly), especially if this is not a familial feature, is an indicator of a syndromic diagnosis. One of the more recently recognised syndromic disorders is that caused by a mutation in the KIF11 gene.[13] The ophthalmic phenotype in this condition can be variable, with chorioretinopathy being the most common feature. However, a more consistent finding in this disorder is significant microcephaly which provides the main clue to the diagnosis.

Malformations and dysmorphic features

The presence of major congenital malformations should have been elicited from the history, but patients do not often mention more minor malformations. Minor malformations are those which do not affect the function and do not normally require any treatment. Good examples of minor malformations found in association with ophthalmic disorders are the presence of rudimentary postaxial digits in Bardet-Biedl Syndrome (often just a nubbin of tissue on the outside border of the hand), the conical teeth and redundant periumbilical skin seen in Rieger Syndrome, or the 'hammer toe' appearance associated with *BCOR* gene mutations (oculofaciocardiodental syndrome). A list of the most frequently encountered major and minor malformations in syndromic ophthalmic disorders is included in Table 5.1. Though minor malformations are common in the general population, the presence of multiple minor malformations in the same child should prompt one to search for an underlying syndromic diagnosis.

Dysmorphic facial features can provide an important clue to ophthalmic syndrome diagnosis. Whilst subtle facial dysmorphism may be difficult to discern, some syndromes are associated with striking facial features which are relatively easy to recognise. Cockayne syndrome[14] for example causes congenital cataracts and is associated

TABLE 5.1 Major and minor malformations associated with syndromic ophthalmic disorders.

Disorder	Malformations/dysmorphism	Ophthalmic phenotype
Stickler syndrome	Cleft palate, joint hypermobility, dislocated hip	High myopia, retinal detachment, vitreous abnormalities, prominent eyes
Bardet Biedl syndrome	Postaxial polydactyly, obesity, short fingers and toes, small genitalia, renal anomalies	Rod-cone dystrophy
Rubinstein Taybi syndrome	Broad medially deviated thumbs and great toes, microcephaly	Congenital glaucoma
Cockayne syndrome	Congenital contractures, deep-set eyes, microcephaly, carious teeth, sun sensitive skin	Congenital cataracts, pigmentary retinopathy
Rieger syndrome	Redundant periumbilical skin, small, misshapen or missing teeth	Anterior segment dysgenesis, secondary glaucoma
Nance Horan syndrome	'Screwdriver' incisors	Congenital cataracts
Oculofaciocardiodental syndrome	Cleft palate, bifid nasal tip, 'hammer toe', large dental roots	Congenital cataracts in females
Oculodentodigital syndrome	Syndactyly fingers 3,4,5, small. Misshapen or missing teeth	Microphthalmia, microcornea, glaucoma
Oculoauriculovertebral syndrome (Goldenhar syndrome)	Facial asymmetry, malformed auricles, accessory ear tags, macrostomia, vertebral abnormalities	Epibulbar dermoid
Kabuki syndrome	Cleft palate, preauricular pit, fingertip pads, renal agenesis	Microphthalmia, interrupted eyebrow, everted lower eyelid
Fanconi anaemia	Radial ray defects, renal agenesis, congenital heart defect, café au lait patches	Microphthalmia, anophthalmia
Goltz syndrome	Dermal hypoplasia in a linear or patchy distribution, cleft lip/palate, oligodontia, limb defects	Microphthalmia, anophthalmia, coloboma, aniridia, ectopia lentis
Incontinentia pigmenti	Blistering in linear distribution after birth healing with linear hyperpigmentation, dental anomalies, alopecia, breast aplasia	Microphthalmia, retinal detachment, retinal vascular defects, cataract
KIF11-associated syndrome	Microcephaly, lymphedema, upward-slanting palpebral fissures	Chorioretinopathy, retinal folds, cataract

with obvious enophthalmos due to lack of periorbital fat. The curvature of the spine and congenital contractures add to this recognisable phenotype. The flat midface and small jaw associated with Stickler syndrome are best appreciated in younger children when prominent eyes and short noses are also more remarkable. Some syndromes may be recognised based on a single feature, such as Fraser syndrome[15] where the presence of cryptophthalmos is the clue to the diagnosis. In Kabuki syndrome[16] where microphthalmia is a feature, the presence of an interrupted eyebrow can provide a clue. Requesting consent to take photographs of any facial dysmorphism and structural anomalies can provide an opportunity for later discussion with the clinical genetic team. Table 5.2 lists some syndromes with more recognisable facial features and Fig. 5.1 shows some recognisable clinical signs.

How should syndromic ophthalmic disorders be investigated?

Around 80% of ophthalmological syndromes will have a genetic basis. For those associated with developmental

TABLE 5.2 Selected ophthalmic syndromes with recognisable facial features.

Syndrome	Facial features; ophthalmic phenotype	Gene/aetiology
Stickler syndrome	Flat nasal bridge, short nose, megophthalmos; myopia, retinal detachment, vitreaous changes, cataract	COL2A1, COL11A1 are associated with ocular phenotypes
Kabuki syndrome	Arched, interrupted eyebrows, long palpebral fissures, everted lateral eyelids, large, unfolded ears, dimple below the lower lip	KMT2D, KDM6A
Rubinstein Taybi syndrome	Microcephaly, hirsute forehead, low columella, arched eyebrows	CREBBP, EP300, deletion 16p13.3
Cohen syndrome	Short philtrum, grimacing expression, wave-shaped palpebral fissures, bushy eyebrows; myopia, chorioretinal dystrophy	VPS13B
Oculodentodigital dysplasia	Microphthalmia, microcornea, narrow nasal alae, thin, anteverted nares, short palpebral fissures	GJA1
Fraser syndrome	Cryptophthalmos, bifid nasal tip, the tongue of hair extending from lateral eyebrow to temple, hypertelorism	FRAS1, FREM2, GRIP3
MIDAS syndrome	Linear skin defects on head and neck, cardiac defects, microphthalmia, sclerocornea	HCCS point mutation or deletion
Branchio-oculo-facial syndrome	Pseudoclefts of upper lip, skin lesions behind ears, upturned earlobes, premature greying of hair; microphthalmia, coloboma	TFAP2A
CHARGE syndrome	Facial palsy, choanal atresia, cupped ear; iris/retinal coloboma	CHD7
Foetal valproate syndrome	Broad nasal bridge, thin upper lip, small downturned mouth, neatly arched eyebrows, infraorbital groove; coloboma, myopia	Exposure to sodium valproate during pregnancy; dose-related effects
Foetal alcohol syndrome	Thin upper lip, flat philtrum, short, anteverted nose, microcephaly, optic nerve hypoplasia, retinal vascular anomalies	Exposure to alcohol during pregnancy

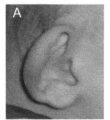 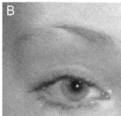 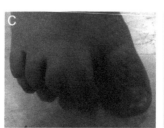 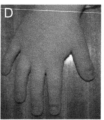 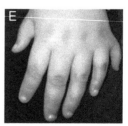

FIG. 5.1 Characteristic dysmorphic signs in paediatric ophthalmic syndromes. (A) cupped ear in CHARGE (coloboma, heart anomaly, choanal atresia, retardation, genital and ear anomalies) syndrome. (B) Interrupted eyebrow and everted lower lid in Kabuki syndrome. (C) Curved 'hammertoes' in oculofaciocardiodental syndrome. (D) Broad, medially deviated thumb in Rubinstein Taybi syndrome. (E) Postaxial polydactyly in Bardet Biedl syndrome.

problems, the paediatrician will usually have undertaken a battery of firstline diagnostic tests which will include routine biochemistry and haematology, urine amino and organic acids, creatine kinase, and chromosomal microarray analysis (array CGH, see Chapter 2). Of these, the microarray analysis is likely to have the highest diagnostic yield with the potential to reveal copy number variants (microdeletions and microduplications) that affect genes that have a role in ocular development or function. Where developmental problems are present, microarray analysis has a diagnostic yield of around 10%. Prime examples of chromosomal abnormalities affecting the eye are deletions of chromosome 6p25.3 region which harbours the *FOXC1* gene; deletion of the *BCOR* gene at Xp11.4; deletion of chromosome 4q25 which encompasses the *PITX2* gene and gives rise to Rieger syndrome. It should be noted that small microdeletions may involve just a single gene. If this gene has not been implicated in general development, then it is unlikely to be associated with developmental problems or intellectual disability. Sequencing of genes does not always reliably detect large structural variants. Therefore, consideration of microarray analysis or other techniques to assess chromosome dosage is important in circumstances where sequencing has proved negative but the phenotype remains typical for a particular gene. For example, approaches like MLPA (multiplex ligation-dependent probe amplification; see Chapter 2) may be required to exclude copy number variants as part of PAX6 gene analysis in patients with aniridia.[17]

Genetic testing has proven particularly fruitful in non-syndromic ophthalmic disorders. Whereas single gene analysis can be used in disorders that are clinically distinctive and known to be caused by mutations in a specific gene, sequencing of a larger panel of genes is particularly useful when investigating disorders that have overlapping phenotypes and are not clinically distinguishable. A gene panel may for example include all of the genes associated with a retinal disease or all those known to be associated with anterior segment dysgenesis. The inclusion of genes for both syndromic and non-syndromic ophthalmic disorders on the same panels has led to the realisation that some patients with the presumed isolated ophthalmic disease may have pathogenic variants in genes implicated in syndromic forms. In young patients, syndromic diagnoses will sometimes be made on gene panel testing before all the key features of the syndrome become apparent. These observations mean that it is very important that patients have consented for genetic testing appropriately so that any unexpected or incidental findings do not come as a complete surprise (see Chapter 4). Sequencing exomes (all the DNA encoding proteins) or genomes (both coding and non-coding DNA) provides the opportunity to identify diagnoses in individuals who have unique and distinctive ophthalmic phenotypes, and in those where screening panels of known genes have failed to find an answer.

Counselling before and following genomic testing

With the move from genetic to genomic testing (looking at the whole of an individual's genetic make-up rather than at individual genes) has come the requirement for ophthalmologists to understand more about the genomic tests they are ordering, to consent patients to testing in the most appropriate way. Microarray (array CGH) analysis looks for the number of copies of the genetic material across all of the chromosomes. It does not, however, look for changes in individual genes and the limitations of the test must be conveyed to the patients as well as the advantages. As microarray analysis involves examining a large amount of genetic material it should be explained to patients that the test may: provide a positive result; not show any abnormalities; detect a variant of unknown significance which might be difficult to interpret or identify an incidental finding that was unexpected but is significant. If patients understand these four possible outcomes, then, whatever the findings, they will be better prepared for a discussion of the results.[18] The same principles apply to families undergoing genomic testing of large gene panels, exomes or genomes. Here, the chance of finding variants that are difficult to interpret is higher, but the issues surrounding counselling are essentially the same. Interpretation of genomic findings and communicating these to patients can be complex. Prior discussion of these at a multidisciplinary team meeting can be of help in clarifying results and in refining the information that should be fed back to patients. Where the results are complex seeking the help of the clinical genetics team is useful, as there may be limited time in a busy ophthalmology clinic to devote the time needed to this. Also, supplementary information is perhaps less likely to be available.

Following the diagnosis of a rare ophthalmic syndrome, parents/patients will need to be given information about their condition and an explanation of what it will mean for them in the long term. For molecular diagnoses, they need to understand the implications of the particular genetic mechanism concerned for them and other family members. Extended family members at risk may need to be identified and offered testing. Where the relevant syndrome involves body systems other than the eyes, measures may have to be put in place to organise health surveillance. Summarising the results in a letter and signposting to appropriate sources of support (e.g. lay support groups) is of great help to the families concerned. The ophthalmologist may increasingly be able to take on some of these roles but referral to the genetics team for discussion of the implications for future pregnancies and extended family members is indicated in cases where risks are higher and there are significant implications for reproductive decisions or where specific multidisciplinary management advice is required.

Disorders that remain undiagnosed after genomic testing

It must be remembered that not all syndromic disorders have a genetic basis or can be diagnosed with the available genetic/genomic testing and other routine investigations. Although the percentage of diagnoses made after a thorough genetic investigation of non-syndromic ophthalmic disorders is high, diagnostic rates for syndromic disorders are lower as the genes concerned are often rare and may not be on established gene panels for ophthalmic genetic testing. For some rare syndromic disorders, e.g. Hallermann Streiff syndrome the genes remain unknown despite extensive research and some relatively common syndromes such as oculoauriculovertebral (Goldenhar) syndrome are considered to be multifactorial, with perhaps several genes and environmental factors combining to play a part.

Mitochondrial genes may not have been covered during testing, and some disorders will have environmental causes. One particular group of disorders that are difficult to diagnose are those caused by mosaicism. In this situation, an individual can have two or more cell lines, each having a different genetic pattern. Mosaicism is often difficult to detect if the abnormal cell type is present at a low level and sometimes it requires sampling of another tissue such as buccal cells or a skin biopsy. Clues to mosaicism are asymmetry and pigmentary skin changes following Blaschko's lines.[19] Unilateral ophthalmic findings may be present more often in mosaic disorders. Finally, a further reason for conditions that are difficult to diagnose is that of the 'blended phenotype'.[20] Several studies have now shown that up to 5% of patients have more than one disorder, making diagnosis more challenging.

Various strategies can be utilised to achieve a diagnosis in unsolved cases. First of all, if patient consent is granted, case sharing among colleagues at the local, national, or international levels may provide valuable input.[21] This is being facilitated now through the European Reference Networks and experts within the ERN-EYE can be accessed for opinions through the use of the European Union Clinical Patient Management System (CPMS). Syndrome identification sessions at ophthalmology meetings also provide a forum for similar discussions. There are several databases that can be searched to identify syndromic disorders. In addition to the OMIM (Online Mendelian Inheritance in Man) database where the clinical synopses provide good overviews, sites such as Phenomizer which can search for syndrome matches on specific combinations of clinical features might provide some clues. Where patients have unique phenotypes, they can be targeted for further research studies. Reporting of distinctive phenotypes in the literature can also help to identify similar patients. Where a diagnosis cannot be made immediately in an syndromic individual, patients should be kept under review as there is a chance that a diagnosis could be made at some stage in the future, possibly through reanalysis of existing genomic data,[22] or that new diagnostic techniques may become available. There is, for example, interest in face recognition technology[23] as an aid to clinical diagnosis and this is increasing as the software available becomes more sophisticated. Despite all efforts, some patients with the syndromic ophthalmic disease will remain undiagnosed currently. Though the number of undiagnosed patients will fall in the future the most important consideration is to make sure that the needs of undiagnosed patients are being met as well as possible. Contact with support groups for undiagnosed patients such as SWAN UK can also be of considerable benefit to families in these circumstances.

Concluding remarks

A significant proportion of genetic ophthalmic conditions are associated with extraocular signs and symptoms. For many of them, clinical history and examination provide important clues that can guide investigations. A key message is that findings like cataracts, ectopia lentis, retinal degeneration and optic neuropathy should be initially approached as phenotypic features rather than specific disease entities. Each of them can be a manifestation of many disorders and part of the diagnostic process should involve distinguishing multisystemic, nonsyndromic forms from isolated, non-syndromic ones.

References

1. Glöckle N, Kohl S, Mohr J, Scheurenbrand T, Sprecher A, Weisschuh N, et al. Panel-based next generation sequencing as a reliable and efficient technique to detect mutations in unselected patients with retinal dystrophies. *Eur J Hum Genet* 2014;**22**(1):99–104.
2. Strande NT, Berg JS. Defining the clinical value of a genomic diagnosis in the era of next-generation sequencing. *Annu Rev Genomics Hum Genet* 2016;**17**:303–32.
3. Snead MP, McNinch AM, Poulson AV, Bearcroft P, Silverman B, Gomersall P, et al. Stickler syndrome, ocular-only variants and a key diagnostic role for the ophthalmologist. *Eye* 2011;**25**(11): 1389–400.
4. Brennan D, Giles S. Ocular involvement in fetal alcohol spectrum disorder: a review. *Curr Pharm Des* 2014;**20**(34):5377–87.
5. Tsui I, Moreira MEL, Rossetto JD, Vasconcelos Z, Gaw SL, Neves LM, et al. Eye findings in infants with suspected or confirmed antenatal Zika virus exposure. *Pediatrics* 2018;**142**(4).
6. Kijlstra A, Petersen E. Epidemiology, pathophysiology, and the future of ocular toxoplasmosis. *Ocul Immunol Inflamm* 2014;**22**(2):138–47.

7. Holmström G, Thorén K. Ocular manifestations of incontinentia pigmenti. *Acta Ophthalmol Scand* 2000;**78**(3):348–53.
8. Ragge N, Isidor B, Bitoun P, Odent S, Giurgea I, Cogné B, et al. Expanding the phenotype of the X-linked BCOR microphthalmia syndromes. *Hum Genet* 2018;**138**:1051–69.
9. Couser NL, Masood MM, Aylsworth AS, Stevenson RE. Ocular manifestations in the X-linked intellectual disability syndromes. *Ophthalmic Genet* 2017;**38**(5):401–12.
10. Weihbrecht K, Goar WA, Pak T, Garrison JE, DeLuca AP, Stone EM, et al. Keeping an eye on Bardet-Biedl syndrome: a comprehensive review of the role of Bardet-Biedl syndrome genes in the eye. *Med Res Arch* 2017;**5**(9).
11. Álvarez-Satta M, Castro-Sánchez S, Valverde D. Alström syndrome: current perspectives. *Appl Clin Genet* 2015;**8**:171–9.
12. Taban M, Memoracion-Peralta DS, Wang H, Al-Gazali LI, Traboulsi EI. Cohen syndrome: report of nine cases and review of the literature, with emphasis on ophthalmic features. *J AAPOS* 2007;**11**(5):431–7.
13. Birtel J, Gliem M, Mangold E, Tebbe L, Spier I, Müller PL, et al. Novel insights into the phenotypical spectrum of KIF11-associated retinopathy, including a new form of retinal ciliopathy. *Invest Ophthalmol Vis Sci* 2017;**58**(10):3950–9.
14. Dollfus H, Porto F, Caussade P, Speeg-Schatz C, Sahel J, Grosshans E, et al. Ocular manifestations in the inherited DNA repair disorders. *Surv Ophthalmol* 2003;**48**(1):107–22.
15. Slavotinek AM, Tifft CJ. Fraser syndrome and cryptophthalmos: review of the diagnostic criteria and evidence for phenotypic modules in complex malformation syndromes. *J Med Genet* 2002;**39**(9):623–33.
16. McVeigh TP, Banka S, Reardon W. Kabuki syndrome: expanding the phenotype to include microphthalmia and anophthalmia. *Clin Dysmorphol* 2015;**24**(4):135–9.
17. Wawrocka A, Krawczynski MR. The genetics of aniridia – simple things become complicated. *J Appl Genet* 2018;**59**(2):151–9.
18. Patch C, Middleton A. Genetic counselling in the era of genomic medicine. *Br Med Bull* 2018;**126**(1):27–36.
19. Ruggieri M, Praticò AD. Mosaic neurocutaneous disorders and their causes. *Semin Pediatr Neurol* 2015;**22**(4):207–33.
20. Posey JE, Harel T, Liu P, Rosenfeld JA, James RA, Coban Akdemir ZH, et al. Resolution of disease phenotypes resulting from multilocus genomic variation. *N Engl J Med* 2017;**376**(1):21–31.
21. Douzgou S, Clayton-Smith J, Gardner S, Day R, Griffiths P, Strong K. Dysmorphology at a distance: results of a web-based diagnostic service. *Eur J Hum Genet* 2014;**22**(3):327–32.
22. Wright CF, McRae JF, Clayton S, Gallone G, Aitken S, FitzGerald TW, et al. Making new genetic diagnoses with old data: iterative reanalysis and reporting from genome-wide data in 1,133 families with developmental disorders. *Genet Med* 2018;**11**.
23. Dudding-Byth T, Baxter A, Holliday EG, Hackett A, O'Donnell S, White SM, et al. Computer face-matching technology using two-dimensional photographs accurately matches the facial gestalt of unrelated individuals with the same syndromic form of intellectual disability. *BMC Biotechnol* 2017;**17**(1):90.

Chapter 6

Ophthalmic phenotyping: Electrophysiology

Neil R.A. Parry and Panagiotis I. Sergouniotis

Genetic ophthalmic disorders exhibit significant clinical and genetic heterogeneity: defects in over 500 genes have been linked to various conditions affecting different levels of visual processing. A key diagnostic goal of clinicians managing genetic disorders is to identify the affected individual's molecular defect with sufficient accuracy. Advances in DNA sequencing technologies have led to the introduction of genetic tests that are larger in scope and sensitivity (e.g. exome sequencing; see Chapter 2). As a result, each test typically identifies many genetic changes that could potentially be relevant. Pinpointing the 'true positive' disease-causing DNA variant(s) among the many 'false positive' changes can be particularly challenging. To address this issue, it is very important to narrow the pre-genetic test hypothesis to the smallest number of genes possible and to efficiently communicate the clinical observations to the laboratory that performs the molecular investigation.[1]

Visual electrophysiology provides important mechanistic insights, improves diagnostic accuracy and helps guide the interpretation of genetic findings. Rarely, the electrodiagnostic results narrow the pre-genetic test hypothesis to a specific gene (e.g. *KCNV2*-retinopathy[2]). More commonly however, the findings suggest a range of possible molecular diagnoses.[3]

At every stage of the visual pathway, from the retina to the visual cortex, events are signalled by electrical changes which can be monitored non-invasively. By careful choice of stimulus and recording conditions, it is possible to dissect the visual system, both cross-sectionally (for example, separating the responses of rods and cones) and longitudinally (for example comparing retinal with cortical function). This chapter provides an overview of the visual pathway, highlighting the signals that are electrodiagnostically available. The various tests that are routinely employed are then described, concentrating mainly on the standard tests that apply to genetic disorders. The chapter ends with an overview of the role of electrophysiology in the diagnosis and management of genetic ophthalmic conditions, with examples.

Key electrical events along the visual pathway

Although at a macroscopic level there are eight main steps between the retina and the visual cortex, at each stage there is a complex cascade of events under the control of several hundred known molecules. These events are signalled by electrical changes, many of which can be recorded at electrodes some distance from the site of the activity. Some diseases have unambiguous electrodiagnostic signatures but many more (such as retinitis pigmentosa) have quite generic effects, mainly because the retina has a fairly limited repertoire of responses to insult or injury.

The photoreceptors respond to light with a graded response that is proportional to the strength of the stimulus; this is in contrast to most other neuronal cells which generally respond to stimuli through spiking, all-or-none action potentials. Increasing levels of light cause photoreceptors to become more hyperpolarised (i.e. more negatively charged). This is the result of a complex cascade of events that is triggered by the action of a single photon which ultimately results in the closure of an ion channel in the cell wall, reducing the influx of positively charged Na, K and Ca ions. The upshot of these mechanisms is that the release of the neurotransmitter glutamate is reduced under light stimulation.

Rod and cone photoreceptors have several key differences. Cones are spectrally tuned (i.e. sensitive to different wavelengths) thanks to the photopigments encoded by the *OPN1LW*, *OPN1MW* and *OPN1SW* genes; these molecules are found in the long-wavelength sensitive L-cones, the medium-wavelength sensitive M-cones and the short-wavelent sensitive S-cones respectively. Metabolically, cones are much faster that rods, for two main reasons. Unlike rods, cone outer segment discs are continuous with the cell membrane and have access to a rapid pathway for recycling bleached photopigment. Rods can only use the sluggish visual cycle to replenish rhodopsin, a process that takes many minutes. The visual cycle is managed by the RPE, a monolayer of cells in which the tips of the cone and rod outer segments are buried. It is important for photoreceptor maintenance, phagocytosing shed outer segment discs. Its many ion channels result in a standing electrical potential between outer and inner surfaces. This decreases during dark adaptation and a key component is the bestrophin-1 chloride channel, encoded by the BEST1 gene.

Photoreceptors synapse and provide input to retinal interneurons including bipolar cells. The bipolar cells either signal an increase (ON-bipolars) or a decrease in light (OFF-bipolars) and, like the photoreceptors, show graded responses. Critically, there are no identified rod OFF-bipolars and there are key structural and metabolic differences between the different types of bipolar cells. OFF-bipolars contain ionotropic glutamate receptors that are sign-conserving, i.e. light-induced changes in glutamate are preserved. ON-bipolars, on the other hand, respond to glutamate release via a more complex metabotropic cascade and are sign-inverting. These differences are reflected in the electrodiagnostic signature of certain diseases where glutaminergic receptors are affected, for example, congenital stationary night-blindness and melanoma-associated retinopathy. Aside from the ON–OFF distinction, L- and M-cone driven bipolar cells are broadly classified as a midget or diffuse whilst there is a single class of S-cone driven bipolar cells. Their connections are briefly discussed below.

The retinal ganglion cells are the neurons that make up the optic nerve. They can be broadly subdivided into three classes: parasol, midget and small bistratified. Parasol retinal ganglion cells take their input from diffuse bipolar cells and are thought to provide the main pathway for conveying luminance pattern information. Midget retinal ganglion cells connect to midget bipolars and have chromatically opponent (red vs green) centre-surround organisation. The small bistratified retinal ganglion cells are also chromatically opponent, deriving their excitatory input from S-cone bipolars whilst they are inhibited by L+M cone-derived input. These three classes of retinal ganglion cells are the starting points of three functionally distinct processing streams: the magnocellular (M), the parvocellular (P) and the koniocellular (K) pathways (named after the layers of the lateral geniculate nucleus in which each ganglion cell subtype terminates). These pathways subserve three relatively distinct aspects of vision: luminance/movement (M), red-green (P) and blue-yellow (K). The distinct temporal, chromatic and spatial properties of these pathways mean that they can be distinguished electrophysiologically by careful stimulus choice.

From retinal ganglion cells onwards, neurons in the visual pathway respond with action potentials, maintaining at rest a constant series of spikes and sending information along the optic nerve by changing firing rate. These fibres are myelinated, thus considerably enhancing the speed of transmission via saltatory conduction. Their calibre and thus their response properties vary according to the pathway they represent. The thickest fibres are found in the M pathway, whilst P fibres are thin and those of the K pathway even thinner. The upshot is that some diseases which might be thought to preferentially affect one or the other type of pathways might be characterised by P-specific (e.g. red-green colour vision), M-specific (e.g. movement sensitivity) or K-specific (blue-yellow colour vision) defects. One such is glaucoma, and many electrodiagnostic studies have employed stimuli designed to show this selectivity (e.g.[4]). In common with all mammals, approximately half the optic nerve fibres cross at the optic chiasm. This semi-decussation more or less respects the vertical midline: nasal fibres project to the contralateral visual cortex, whilst temporal fibres remain generally uncrossed and project to the ipsilateral cortex. This pattern is altered in conditions like albinism and achiasmia, giving a distinct electrodiagnostic signature.

Although the lateral geniculate nucleus provides a thalamic break-point, signals are transmitted more or less unchanged along the optic radiations, terminating in layer 4 of the striate visual cortex. The visual cortex is located deep in the calcarine fissure and varies widely in shape and precise location between individuals. This makes the location of the electrical signals received by occipital recording electrodes somewhat variable.

The retino-cortical pathway is only the beginning of the process of perception, but problems with higher visual areas are not simple to pin down so, for the most part, the role of electrodiagnostic testing stops here. Perhaps an exception is functional visual loss. This implies an intact visual pathway, so there is a role in establishing that this is the case.

Electrodiagnostic tests

There are two broad categories of electrodiagnostic tests which assess the integrity of retinal function, namely the electroretinogram (ERG) and the electrooculogram (EOG). Testing can also be performed at the level of the visual cortex, using the visual evoked potential (VEP). The basic components of an electrodiagnostic system, described in the next section, are common to all the tests.

Technology

Fig. 6.1 gives an overview of the essential components of an ophthalmic electrodiagnostic setup. Broadly this is divided into some form of visual stimulus, a physiological amplifier, and some means of analysing the signals (usually a computer). As the figure shows, one of the main differences between the categories of the test is the placement of the electrodes.

Stimulation

The key aim of any stimulus is to provide a change, whether this is a flash of light or the sudden reversal or appearance of

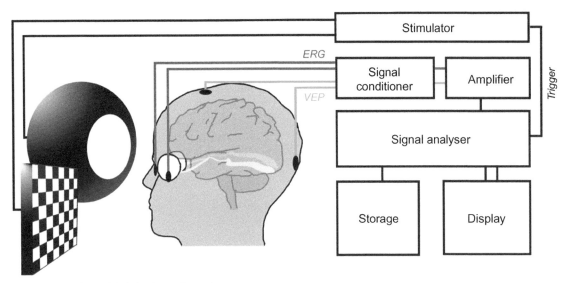

FIG. 6.1 Components of an ophthalmic electrodiagnostic system.

a pattern. Flashing lights are delivered using either a simple hand-held device or a device designed to stimulate as much of the retina a possible (a ganzfeld). The ganzfeld is an approximation of an integrating sphere, thus illuminating as much of the retina as possible, and as evenly as possible. It usually employs an array of LEDs for both test and background lights, although for stimulation there may be a Xenon discharge tube.

Patterned stimuli are usually computer-generated on a graphics monitor. Commonly a high-contrast black and white checkerboard is reversed periodically (black becoming white and vice versa) or is replaced periodically by a blank grey screen (on–off presentation). The other commonly used pattern is the multifocal array A vital property of any pattern stimulus is that there should be no overall change in mean luminance when the pattern is modulated, otherwise part of the response will be due to this artefactual component.

Amplification

Signals arriving at the receiving electrode are small, typically between 2 and 200 μV, and thus the first stage in processing the signals is an amplifier with a gain of around 50,000. The signals are first filtered to remove frequencies that are not of interest. The low pass filter removes high-frequency noise, and the high pass filter removes slow signals such as those generated by patient and electrode movement. Standard clinical systems generally have at least 4 or 5 amplifiers in a single unit, so that recordings can be made from multiple locations. This also enables simultaneous VEPs and ERGs or pattern ERGs to be recorded to the same stimulus.

Signal analysis

Once the signals are digitised, further analysis is provided by the host computer. One of the key forms of signal processing is averaging, which is described below. An important part of the analysis process is the ability to place cursors on the different waveforms to measure properties such as the size of the response and its timing characteristics. Although automated processes have been developed, cursoring is often done by the operator.

Averaging: Small electrophysiological signals are usually masked by large volumes of electrical noise. This is hardest to overcome with the VEP (because of the significant amount of electroencephalographic activity arising from non-visual brain events) and with the pattern ERG (because the signals themselves are very small and there is a lot of mechanical noise from eye movements). Although mathematically simple, averaging is a very effective way of improving signal-to-noise ratio. Fig. 6.2 shows a series of individual 250 ms samples of electroencephalographic activity, each triggered by the same event (in this case, the instantaneous reversal of a checkerboard pattern). These are summed and averaged to extract the non-varying signal change created by the event. Because the visual cortex is likely to respond similarly to each new reversal, and because the recording is time-locked to this event, one can expect that each new sweep contains, along with lots of noise, the same response. The process of averaging minimises any non-time-locked activity whilst allowing the time-locked response (in this case, the VEP) to remain. VEPs and pattern ERGs rely heavily on this process, but even the flash ERG, which usually has a much higher signal-to-noise ratio, can benefit from some averaging.

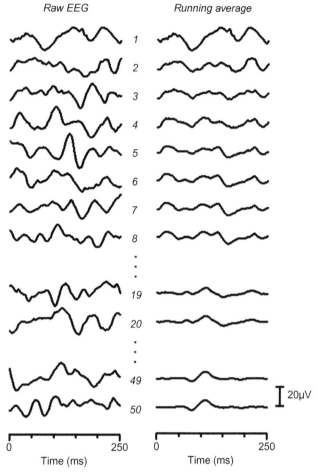

FIG. 6.2 Averaging a cortical signal to extract time-locked data and remove non-time-locked data. In the left column, selected raw samples of electroencephalographic (EEG) activity are shown from a run of 50 sweeps, each of 250 ms duration. The right column shows the running average. Thus row 2 is the average of sweeps 1 and 2, row 8 is the average of the first 8 sweeps, and row 50 is the final visual evoked potential, averaging all 50 sweeps.

Standardisation

In ophthalmology, there have been strenuous attempts to produce a set of standardised testing protocols to harmonise results of electrodiagnostic testing from one laboratory to another. This is overseen by the International Society for Clinical Electrophysiology of Vision (ISCEV), which regularly publishes and revises these standards (see www.iscev.org). One of the important features of standardisation is the use of consistent, unambiguous terminology. For example, one will often encounter a description of photopic and scotopic ERGs, when in fact a more proper description is light-adapted and dark-adapted ERG respectively. Here we have attempted to adhere to the more current terminology but, in the literature, these terms tend to be used interchangeably.

The flash electroretinogram

In its simplest variant, the ERG reflects electrical changes in the retina in response to flashes of light that illuminate large areas of the fundus. The flash ERG is largely derived from the photoreceptors and the bipolar cells. The earliest part of the response is a negative-going a-wave which is largely photoreceptor derived. As bipolar activity kicks in, the waveform takes a positive-going turn (the b-wave) and then returns towards the baseline.

Methods

Recording: Retinal signals are recorded from electrodes mounted on or close to the anterior pole of the eye. Corneal electrodes can be encased in contact lenses or simply be conductive silver/nylon fibres, silver wires or strips of gold foil mounted at the lower limbus. Since these are relatively invasive, an alternative approach, often used in young children and infants, is to employ a skin-mounted electrode as close to the lash line as possible. The signals are then sent to a differential amplifier; this requires a second, reference, electrode to be mounted elsewhere, usually a skin electrode at the outer canthus. Some contact lens electrodes use a bipolar montage in which the reference is built into the lens. Processing of the recorded signal usually involves some averaging to help reduce noise, and artefact rejection to remove unwanted signals, for example, those generated by blinks.

Stimulation: Where possible, flash ERGs are recorded using a ganzfeld stimulus. Some patients, particularly very young children, may find this hard to co-operate with; for them, a simpler, hand-held flash device is used. This is less intrusive and enables the operator to follow a particularly evasive patient's gaze. However, it is difficult to maintain standard adaptation in the light, and the dynamic range of the photostimulation is more limited. Thus, one is only able to record a subset of the ganzfeld ERGs, and results are inherently noisier and the signals are smaller. Nevertheless, the results are useful in establishing the presence or absence of individual components and this binary approach can go a long way towards providing a diagnosis.

After preparation, electrode montage and pupil dilatation, the patient sits close to the ganzfeld to ensure maximum retinal exposure. Testing occurs in two phases. After (usually) 20 min in full darkness, a series of dim flashes are shown every 2 s. These only signal rod system-mediated bipolar cell activity. As the flash intensity is increased, photoreceptor activity starts to contribute to the response, predominantly from rods, but with an increasing cone contribution at higher intensities. Thus, the ERG takes on a different shape, as shown in Fig. 6.3. At a minimum, 3 flash intensities are presented (0.01, 3 and between 10 and 30 cd.s.m.$^{-2}$), but intervening intensities

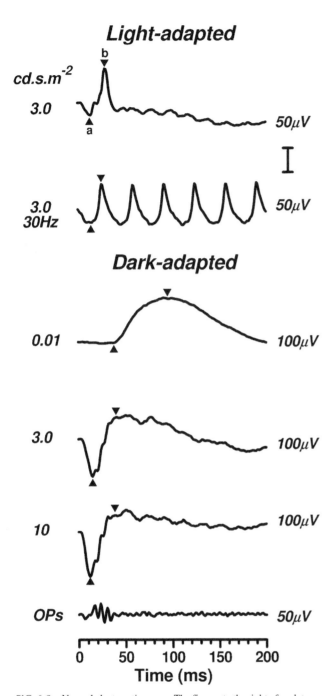

FIG. 6.3 Normal electroretinogram. The figures to the right of each trace show the value for the scale bar.

intensity of $3\,cd.s.m^{-2}$, one with a slow repetition rate (every 2s) to elicit a transient ERG, and one at a faster rate (30Hz).

Analysis

Typical light- and dark-adapted ganzfeld ERGs are shown in Fig. 6.3. The response to a standard flash of light in both light- and dark-adapted states is a negative-going a-wave which is largely formed by the descending portion of the photoreceptor-specific response. Both these stimuli also produce a prominent b-wave which, in the light-adapted retina, is either the ON-bipolar response (if the stimulus is an extended pulse of light) or a mixture of ON- and OFF-bipolar responses when, as is more common, the light is flashed for a shorter period. Notably, there is no direct OFF-bipolar response from the rods, so the dark-adapted b-wave is dominated by an unmixed ON-bipolar response. The response to a 30 Hz flickering light can be characterised by a succession of a- and b-waves. Although the presence of an adapting background light is intended to suppress rod activity, this is unlikely to be complete at this luminance. Thus, it is likely that the single flash response contains some activity from the rods. The flicker ERG is probably as close as it is possible to come to a 'pure' cone ERG using these standard stimuli. This is because rods have a very sluggish temporal response profile, and are unlikely to contribute significantly to the ERG at this high frequency.

In the dark-adapted state, the standard flash ERG is likely to contain a significant cone component, although it is dominated by the rods, now able to contribute because they have been released from bleaching by a period of dark adaptation. A 'rod-specific' ERG is obtained using lights that are much dimmer (by a factor of 300). Clearly, the a- and b-waves do not exist in isolation, as the ascending limb of one is the descending limb of the other. Nevertheless, it is convenient to attribute the a-wave primarily to photoreceptor function, and the b-wave to bipolar function. Significant ON-bipolar cell dysfunction produces an electronegative ERG, in which the dark-adapted b-wave fails to rise above the baseline. Thus, the ratio of b- to a-wave amplitude is less than one. This is usually only manifested in the dark-adapted ERG because, when flashes are brief, the OFF-bipolar cells still provide a cone-mediated response in the light.

Along with the prominent a- and b-waves, the standard dark-adapted flash response can be seen to contain a series of small fluctuations, so-called oscillatory potentials. These can be emphasised by filtering out frequencies below 70 Hz, as shown in Fig. 6.3. Their provenance is not well understood, but they are thought to reflect the operation of amacrine cells.

Other steps are routinely added to the assessment, even if they are not part of the standard protocol. The use of a dim

may be employed as well. As intensity increases, the inter-flash interval increases to 20s, so as not to bleach the rods. In the light-adapted phase, a steady white background light is switched on in the ganzfeld. It is common for the dark-adapted recordings to be made first so it is important to allow the patient to light adapt for 10 min before testing. Typically, two recordings are made at a standard

red flash in the dark-adapted eye produces a response in which cone and rod components are present, but separated in time. This is often used to aid the differentiation between rod- and cone-dominated retinal disease. Extending the duration of the stimulus to 150 ms enables the separation of ON- and OFF-ERGs. The use of multiprimary stimuli (employing four or more coloured LEDs rather than a simple white light) enables the user to isolate the responses of individual classes of photoreceptors utilising a technique called silent substitution[5] an example of which is shown in Fig. 6.10. It is worth noting that S-cone activity can also be measured using a short-wavelength flash on an L- and M-cone suppressing background.[6]

The normal ERG is affected by many non-disease-related factors, for example age, refraction, and cooperation. The relationship with age is complicated and for this reason, laboratories are recommended to have age-matched normative data. The amplitude of the response decreases with an axial length of the eye so that an individual with a refractive error of -5D might be expected to have an ERG approximately 80% as large as an emmetrope.[7] High myopia is often associated with retinopathy and a challenge is to differentiate the two, especially in cases of extreme myopia.

The amplitude of the response is related strongly to light flux and therefore pupil diameter is an important factor. Although the standard procedure is to dilate the pupil before testing, there is still considerable variation in light flux. An 8 mm pupil admits nearly twice as much light as a 6 mm pupil, and about 7 times as much as a 3 mm pupil. Co-operation can be a problem, especially if the stimulus induces blepharospasm. This is commonly encountered with the 30 Hz flicker stimulus, particularly in cases of photophobia. Most ganzfeld stimuli have an infrared camera incorporated so that the patient's compliance can be monitored.

Typical applications

The flash ERG is a measure of the integrity of the whole retina. Photoreceptor dystrophies often dramatically attenuate or extinguish all phases of the ERG; there are however more selective conditions, such as achromatopsia, which may be restricted to the cones and therefore mostly affects the light-adapted ERG. Some conditions affect only part of the ERG, such as some forms of congenital stationary night-blindness, which show specific b-wave abnormalities.

The ERG is not just used to diagnose retinal disease. It can also help determine the normality of the early visual pathway (i.e. the retina) in testing infants for cortical visual impairment with the VEP. Furthermore, it has a role in detecting and monitoring drug toxicity. Also, many acquired conditions affect the retina, and the ERG is an important part of their assessment (see Robson et al.[3] for an extensive list).

The multifocal electroretinogram

Multifocal electrophysiology permits a more localised approach. In this technique, different areas of the retina are stimulated using a mathematically designed sequence of stimuli that are encoded in the response[8,9].

Methods

The multifocal ERG stimulus is usually a patchwork of 61 or 103 hexagonal sectors which are scaled in size to produce ERGs of approximately equal amplitude across the stimulated field in the normal retina (see Fig. 6.4). The same pseudorandom m-sequence is run in each patch, asynchronously. Fig. 6.4 shows the result of such a sequence, using a 61-hexagon array, giving a total test time of just under 4 min, in 30-s sub-sequences (to allow the patient to rest, blink, etc.). Test time may be extended if significant artefacts (e.g. from blinks) are encountered. Good fixation is essential. As the patient may have a central field defect, the fixation cross is often extended. Monocular testing is preferred as a dilated patient may struggle to maintain a single image.

Analysis

Because the signals are small, the multifocal ERG is often re-analysed to average portions of the response. The most common approach is to use the ring analysis shown in Fig. 6.4E, often computing the ratio between each ring and the most peripheral ring.[10] As with the ganzfeld ERG, both amplitude and latency or implicit time are important parts of the analysis. Results often include a 3-dimensional topographical display (e.g. Fig. 6.4D). Whilst this has some uses and presents an attractive snapshot of the results, it is not useful on its own because, for example, it does not contain any timing information.

Typical applications

This is a response of the light-adapted eye to stimulation of the central retina. Therefore, it is most useful in detecting localised defects that might affect the cone-mediated response. It is common in a maculopathy for the ganzfeld ERG to be virtually normal, and thus the multifocal ERG provides a useful differentiation between macular disease and conditions affecting the more widespread retina. A useful example is Stargardt disease (*ABCA4*-retinopathy). As a measure of macular dysfunction, the multifocal ERG has much in common with the pattern ERG, described below.

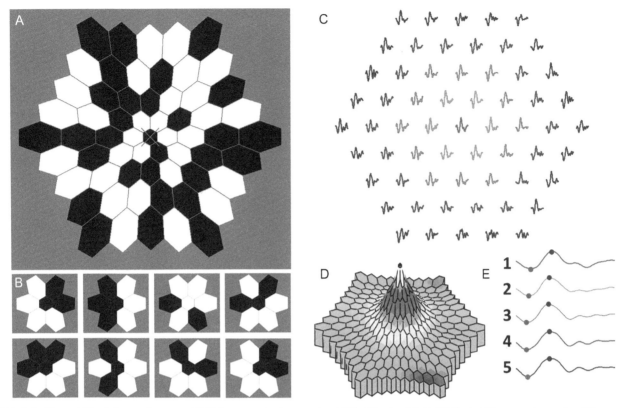

FIG. 6.4 The multifocal electroretinogram. (A) 61-hexagon stimulus covering ±30° of the visual field. (B) Central 7 hexagons in a series of 8 consecutive frames showing part of an m-series. (C) Normal multifocal ERG recorded to the 61-hexagon stimulus. (D) Density map in nV.deg^{-1}. (E) Ring analysis. Each row is the normalised root mean square average of each ring of the responses in (C), using the same colour code.

The pattern electroretinogram

Both ganzfeld and multifocal ERGs use diffuse stimuli or unpatterned patches to elicit retinal responses. These responses provide information about the early processing of the visual signal, mainly at the level of photoreceptors and bipolar cells. Using a patterned stimulus allows us to monitor more downstream activity, as the response is dominated by the retinal ganglion cells. Thus, the pattern ERG provides information that complements the flash response. As previously mentioned, the pattern ERG is also a useful indicator of macular dysfunction, since it is recorded from a relatively small area of the retina. In clinical practice, the pattern ERG is usually elicited by contrast reversal of a checkerboard and contains 2 important components which follow on from an initial negativity at around 35 ms post-trigger (N35). These are a positivity at about 50 ms (P50), and a negativity at around 95 ms (N95). Whilst the P50 is said to come at least in part from the retinal ganglion cells, the N95 is said to most fully represent ganglion cell response to contrast change.[11] Thus, N95 loss is a useful index of ganglion cell dysfunction (and, by extension, optic neuropathy), whereas generalised loss of both components is more akin to either a maculopathy or a generalised retinopathy. Thus, the test is best conducted as part of a battery of investigations including the ganzfeld ERG and the VEP (Fig. 6.5).

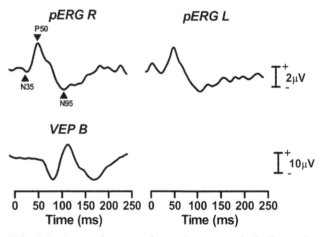

FIG. 6.5 A normal pattern electroretinogram, and simultaneously recorded binocular visual evoked potential.

Methods

Recording: Technically the recording side of pattern ERGs is identical to that used for the flash ERG. However, the signals recorded are comparatively small, of the order of

5 μV, whereas the ganzfeld ERG is perhaps 20–40 times larger. Because the low signal-to-noise ratio is a particular problem, extensive averaging is essential, incorporating as many as 200 sweeps in a single response. Furthermore, blink artefacts are significant and disrupt pattern ERG recording, so it is necessary to get the patient to minimise these whilst concentrating on the fixation target. Thus, recording is often broken down into more manageable parts.

Corneal electrodes are usually used. Contact lens variants, generally favoured for their high sensitivity, are not recommended because they degrade the visual image. As with the other ERGs, skin electrodes may be better tolerated by some (e.g. paediatric) patients, but the signal is attenuated by about 50% compared with corneal electrodes.

Stimulation: Usually, a high contrast (>80%) reversing black and white checkerboard is employed. For a standard pattern ERG, the recommendation is that checks should subtend 48′ along one side and that they reverse 4 times per second. The stimulus is approximately square and should subtend about 15°.

With a 2-channel amplifier, recording can be made from both eyes simultaneously; extra channels make it possible to simultaneously record a VEP. This can aid lesion localisation by recording from two different regions of the visual system in a single pass. If the pattern ERG is recorded with both eyes simultaneously, the VEP is a binocular recording. In cases of monocular functional visual loss (where the problem is non-organic), there is a great advantage in recording the pattern ERG from both eyes together, as a subject who is attempting to fool the system is less likely to be successful. Whilst binocular recording is also very time-efficient, there are some drawbacks, not least of which is that strabismic individuals cannot simultaneously fixate with each eye. In these cases, it is better to record the pattern ERG monocularly.

Analysis

The key features of the pattern ERG are the P50 and the N95 components. Effective analysis relies, as with all visual electrophysiology, on comparison with normative data. Because the normal stimulus extends only over the macula, the pattern ERG is essentially a measure of macular function. Thus, in a maculopathy, both components are likely to be absent. Of course, this may also be the case in the presence of widespread retinopathy, so this finding has to be interpreted as part of the wider picture. Selective attenuation may occur in some diseases, and more often than not it is the N95 that is affected. Since this component is known to be related to ganglion cell function, selective N95 loss is a feature of optic neuropathy. As well as the amplitude of the individual components, close attention must be paid to their time-to-peak as this can betray subtle maculopathy or optic neuropathy.

Typical applications

The relatively small (15°) stimulus size makes the pattern ERG sensitive to maculopathy, but it has a second role in detecting retinal ganglion cell dysfunction. In a genetic setting, it is useful for optic neuropathies such as Leber hereditary optic neuropathy (LHON), and maculopathies such as Stargardt disease (*ABCA4*-retinopathy).

The electrooculogram

From the patient's perspective, the EOG is probably the most arduous and longest test to perform and is generally the least commonly used. Nevertheless, it provides a useful adjunct to the other retinal tests in selected cases. In general, it provides a measure of the integrity of the RPE.

The standing potential of the RPE reflects the difference between membrane potentials of the basal and apical surfaces of this monolayer. In general, this results in a positive charge being measured at the front of the eye. The standing potential remains relatively constant if the state of adaptation of the eye is stable. As the normal visual system adapts to the dark and rods recover from bleaching, there is a slow change in the standing potential, typically reducing by at least a factor of 2. This takes of the order of 15 min.

Technically it is very hard to record such a slow change with standard electrophysiological amplifiers because the signals are filtered specifically to remove such slow changes. The solution was developed by Arden and colleagues in the 1960s, who used electronystagmography principles to show this slow change. If the electrodes are placed on either side of the eye and connected to a physiological amplifier, the recorded signal changes as the eye moves from one side to the other (e.g. from left to right). This causes one electrode (e.g. the right) to become more positively polarised and the other (e.g. the left) to become more negative; the polarity inverts when the eye moves to the other side (see Fig. 6.6). What is happening is that the electrical field associated with the standing potential is moving left and right as the eye rotates. Since, during dark adaptation, the size of the field reduces along with the standing potential, the deflection recorded to a standard saccade will reduce. The reverse happens when the eye is once more light-adapted.

Methods

The standard EOG is usually carried out using a ganzfeld dome containing lateral fixation lights 15° on either side of the centre; these can be alternated so that the patient can make 30° saccades. Eye movements are monitored by affixing electrodes close to the inner and outer canthi

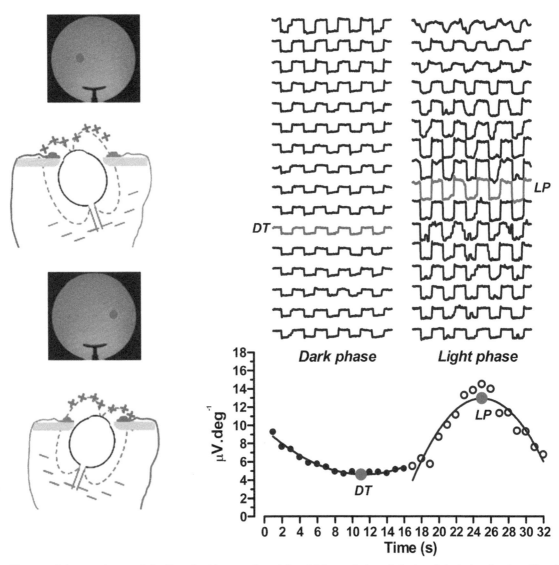

FIG. 6.6 The normal electrooculogram. *Left*: effect of making saccades to left and right on relative polarisation of electrodes placed on either side of the eye. *Right*: saccades recorded every minute in the dark and the light as the fixation lights changed every second. The graph plots the mean amplitude of each series of saccades and is fitted with a 2nd order polynomial curve. The *red markers* identify the dark trough (DT) and light peak (LP), the respective minimum and maximum of these curves, and the corresponding traces are drawn in *red*. Arden index is derived by LP:DT = 2.73.

of each eye. Eye movements are elicited for a few seconds every minute, the two fixation lights alternating every second. Testing is usually performed on the dilated eye in two phases. In the first phase, the test is done in darkness, repeating the sequence over at least 15 min. In the second phase, the background illumination is set to $100\,cd.m^2$ to commence light adaptation. Most patients find this transition quite uncomfortable, but recordings usually settle down within about a minute. Again, this phase lasts at least 15 min. It is common to incorporate a 2-min-long demonstration before commencing the test properly, with the background light on. This allows the patient to practice, assures the operator that the test is understood, and provides a quick pre-adaptation phase.

Analysis

Fig. 6.6 shows raw recordings made over the dark and light phases and identifies the dark trough (DT) and light peak (LP). The LP:DT ratio, also known as the Arden index, is computed and reported, along with the absolute DT and the time from light onset to the LP (if there is one).

Typical applications

Perhaps the most well-known application of the EOG is in the diagnosis of childhood-onset vitelliform macular degeneration (Best disease). Because this condition is classed as a maculopathy, it has little or no measurable effect on the ganzfeld, full-field ERG. However, the condition is associated with widespread Bestrophin protein abnormalities, typically resulting in a low LP:DT ratio. This dissociation between ERGs and EOGs in this condition is characteristic. Retinal disorders which significantly affect the ERGs (e.g. retinitis pigmentosa) will invariably also show reduced LP:DT ratios, but this extra information is of little use diagnostically. Adult-onset vitelliform maculopathy is often electrodiagnostically normal and does not show the EOG/ERG dissociation. Other conditions where the EOG might be helpful are the pattern dystrophies, which can show some reduction in LP:DT ratio in the presence of normal ERGs, though the dissociation is not as complete as in classic Best disease.

The visual evoked potential

Whilst changes in the VEP are well-correlated with the known properties of the retino-cortical visual system, the precise origin of the response is not well-understood. Generally, it represents the mass action of a large number of cortical units acting in concert. This activity results in a polarised electrical field, in which we place two or more electrodes. Because of the convoluted structure of the visual cortex, and the unpredictability of current flow across the scalp, localisation of the response is tricky. What we can be reasonably certain about is that the VEP is representative of activity in visual areas V1 and V2 (i.e. the primary visual cortex). Thus, it is a good measure of the quality of the primary visual pathway, before any cognitive activity.

The standard stimulus for recording a VEP is a reversing checkerboard pattern. Flash stimulation can also be used but has limited value on its own since it generates a very diffuse response in the visual cortex. Nevertheless, in cases where no pattern response can be recorded, it does give an idea of the overall patency of the visual pathway.

Methods

Recording: The simplest VEP is recorded from a single midline electrode placed on the scalp overlying the occiput. Standard electroencephalography (EEG) electrodes are typically used; these are domed discs made either of gold or chlorided silver and held in place using a conductive adhesive paste. A modification of the standard 10–20 electrode placement system[12] is used to locate this position, known as O_Z. This is 10% along a line measured from the inion (the external occipital protuberance) and the nasion (the indentation above the bridge of the nose) and is usually around 3.5 cm in an adult. The reference electrode is placed on a region of the scalp well away from the visually active occiput; the ISCEV standard position is mid-frontal (F_Z), which is 70% along the inion-nasion line. The earth electrode is usually placed on the vertex (C_Z) or the forehead. To ensure good electrical connections, the skin is usually gently abraded using a special paste.

If the scalp distribution of the cortical response is required, then two additional lateral electrodes are placed on either side of Oz, at locations O_1 (to the left of centre) and O_2 (to the right). These are formally defined as 10% of the inter-auricular (ear-to-ear) distance but are usually standardised as 3 cm. Additional electrodes may be placed more laterally.

Stimulation: For flash VEPs, the stimulator is usually a photic strobe. It is of limited value but can augment the more useful pattern response, especially if the latter is unrecordable. In clinical electrodiagnostic testing, the pattern is usually a high contrast checkerboard pattern, reversing between 1 and 3 times per second. At least two check sizes are used, whose sides subtend 15° and 60°, but often this is augmented with other sizes. Alternatively, it may be an onset-offset stimulus. This is particularly of benefit in patients with nystagmus and is part of the standard protocol for the albinism work-up. Averaging is always used, and a typical VEP will be the average of between 50 and 100 sweeps.

It is important with any pattern-based test (particularly the pattern ERG and the VEP) that a sharp retinal image is achieved. Therefore, the patient should be optimally refracted, especially if there is significant presbyopia. Good fixation is essential, and this needs to be carefully monitored should the patient be suspected of malingering; voluntary defocusing and misdirected gaze can produce artefactually attenuated VEPs. When testing children, it helps to provide additional encouragement such as interleaving the stimulus with a video or waving small unobtrusive toys in front of the screen.

Analysis

Fig. 6.7A shows normal VEP responses recorded at a midline electrode to reversal of four pattern sizes and a flash stimulus. A typical pattern response is a positive-going wave with a latency of around 100–120 ms, the P100, flanked by two negative waves, N75 and N135. The key measures are the amplitude of the wave (measured from the trough of N75 to the peak of P100), and the latency of P100. The normal latency is to some extent dependent on the stimulus itself and how the system is triggered, stressing the importance of lab norms. There are three key factors in the analysis. Firstly, whether it is possible to record a response to all check sizes, secondly whether responses from the two eyes are symmetrical, and thirdly whether there is any delay, either relative (between the two eyes) or absolute (affecting both eyes). It is not uncommon to encounter subtle interocular delays of the order of 10 ms; these are unlikely to be of great significance. The flash VEP usually comprises three positive peaks, P1, P2 and P3, with latencies around 80, 120 and

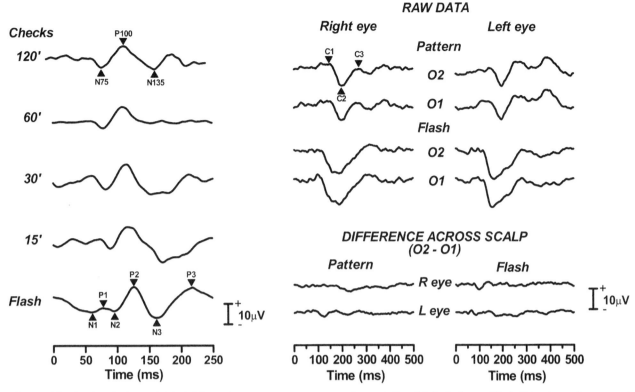

FIG. 6.7 The visual evoked potential. *Left*: Monocular response to pattern reversal presentation of 4 check sizes and flash. Triangles mark cursor positions. *Right*: Raw data are lateral scalp recordings to monocular on–off stimulation with 30° checks and flash stimuli. Difference plots are the result of subtracting the left (O1) from the right (O2) electrode recordings, showing, in this case, that the scalp distribution was symmetrical.

200 ms. Each is preceded by a negative wave, respectively N1, N2 and N3.

Fig. 6.7B shows the results of recording from two lateral electrodes, straddling the midline at positions O1 and O2. In this test, it is important to establish whether there is any asymmetry recorded across the scalp and, more importantly, whether this asymmetry flips when right and left eyes are stimulated. The purpose of this is to explore the extent of chiasmal decussation. In general, chiasmal semi-decussation means that signals from each nasal hemiretina cross at the chiasm, to be transmitted to the contralateral cortex, whilst temporal fibres remain uncrossed. Thus, one would normally see no change in the responses recorded at the lateral electrodes when changing from the right eye to the left. In each case, the left visual hemifield would be signalled by the right visual cortex and vice versa. This is easily seen by looking at the difference between the two electrodes, as demonstrated in the lower portion of Fig. 6.7, which shows more or less symmetrical results. It is not uncommon to have considerable asymmetry between right and left scalp recordings, and in this case, the difference plots would also contain significant signals. But since this is the result of anatomical asymmetry of the occipital cortex, there is no reason why the symmetry would flip when comparing the two eyes.

Typical applications

The VEP is an essential part of the investigation of the integrity of the entire visual pathway, particularly when visual acuity is under investigation, or when visual cortex assessment is required (e.g. in infants with suspected cortical visual impairment). In neuro-ophthalmology, it is useful to establish pathway delays, for example in optic neuritis. Furthermore, it can help diagnose albinism and other conditions with abnormalities at the chiasm.

Other conditions that affect the optic nerve, such as optic nerve hypoplasia and optic atrophy, can attenuate and ultimately extinguish the VEP. Importantly, the VEP helps establish functional visual loss, especially if the patient claims visual acuity of worse than 0.6 logMAR, and particularly if the claimed loss is asymmetrical. In the ensuing section, the use of the VEP will be restricted to common examples of genetic disorders affecting the optic nerve.

The role of electrophysiology in genetic ophthalmic practice

This section will provide key illustrations of the role of electrodiagnostic testing in the phenotypic work-up of patients with genetic disorders. The presented list is by no means comprehensive but we have attempted to include examples from all the main categories. To simplify this overview, an empiric electrophysiological classification system is introduced, using the following high-order diagnostic categories:

- macular dysfunction
- generalised retinal dysfunction—predominant cone system involvement
- generalised retinal dysfunction—predominant rod system involvement
- generalised retinal dysfunction—equal rod and cone system involvement
- optic nerve/visual pathway abnormalities

Macular dysfunction

An abnormal multifocal ERG or an abnormal P50 component of the pattern ERG suggest central retinal dysfunction. In conditions primarily affecting the macula, these abnormalities are combined with a normal full-field ERG in the dark- and light-adapted states. In the context of genetic disorders, the electrophysiological phenotype characterised by macular abnormalities without generalised retinal dysfunction is encountered in several conditions resulting from anatomically circumscribed primary defects in macular rods, cones and/or RPE.

Selected disorders/examples

The most common genetic cause of macular degeneration is mutations in *ABCA4*, a photoreceptor-specific gene. In *ABCA4*-maculopathy (Stargardt disease), affected individuals carry DNA variants that lead to reduced *ABCA4* activity. Often, this reduction is not sufficient to cause direct

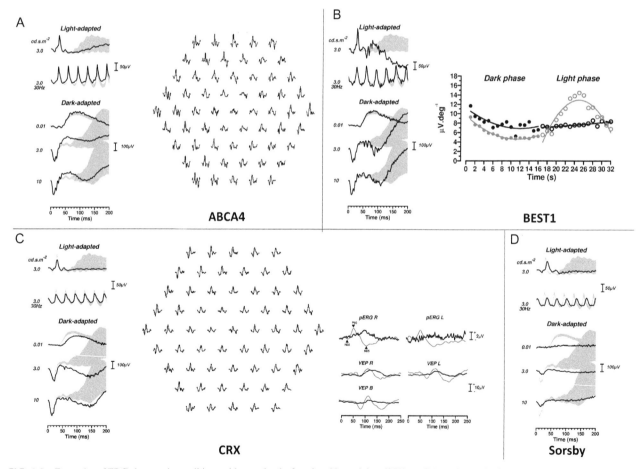

FIG. 6.8 Examples of ERGs in genetic conditions with macular dysfunction. Normal data (95% confidence intervals) for comparison are shown in grey. (A) *ABCA4*-maculopathy (Stargardt disease) showing normal full-field ERGs and centrally depressed multifocal ERG; (B) *BEST1*-maculopathy with normal full-field ERGs and grossly reduced light rise on EOG, with LP/DT ratio of 1.25 or 125%; (C) CRX maculopathy, showing normal full-field ERGs, attenuated central multifocal ERG, extinguished pattern ERG and delayed VEPs; (D) Sorsby fundus dystrophy showing attenuated dark-adapted ERGs with mild attenuation and delay of light-adapted responses. All ERGs and pattern ERGs were recorded with silver/nylon corneal fibre electrodes. We are grateful to Claire Delaney, Elisabeth Parry and Lindsi Williams for the collection of the clinical data shown in Figs. 6.8–6.14.

injury to the photoreceptors. As a result, there is usually little effect on global photoreceptor function and dysfunction is mainly due to the accumulation of toxic material in the macular RPE following outer segment phagocytosis. It is noteworthy that *ABCA4*-related disorders exhibit significant phenotypic heterogeneity and electrodiagnostic testing facilitates the distinction between the macular dystrophy and the cone/cone-rod dystrophy subtype. In the latter, there are additional full-field ERG abnormalities and this finding can have significant prognostic implications.

A group of dominant macular dystrophies (including dominant *CRX*-maculopathy and dominant *PROM1*-maculopathy) can present with localised macular dysfunction on electrodiagnostic testing and a 'bull's eye' pattern of perifoveal RPE atrophy on fundoscopy. It is worth highlighting that this fundus appearance is most commonly due to defects in *ABCA4* and that it is often associated with generalised retinal dysfunction.

Another important genetic cause of macular dysfunction is abnormalities in *BEST1*, a gene that is predominantly expressed in the RPE. In dominant *BEST1*-maculopathy (Best disease or Best vitelliform macular dystrophy), affected individuals carry variants that lead to structural abnormalities at the RPE-photoreceptor interface. This results in weakening of the RPE-neuroretina interactions and consequent formation of the vitelliform lesions that are characteristic of this condition. In general, the macular function can be normal at early disease stages and the ERG abnormalities are thought to add little additional information to clinical assessment. In contrast, EOG findings are characteristic although many experts no longer support the use of electrophysiology as a frontline test in suspected *BEST1*-related disorders and instead routinely order genetic testing in its place.[13]

In addition to *BEST1*-related disease, several other genetic conditions can present with localised macular dysfunction on electrodiagnostic testing and vitelliform macular lesions on fundoscopy. These include dominant *PRPH2*-maculopathy and dominant *IMPG1/2*-maculopathies.[14]

Other characteristic macular dystrophies include *RP1L1*-associated occult macular dystrophy and maternally inherited diabetes and deafness (MIDD) with maculopathy.

Generalised retinal dysfunction—Predominant cone system involvement

An abnormal light-adapted full-field ERG combined with a dark-adapted ERG that is either normal or less severely affected than in the light-adapted state suggest predominant cone system involvement. The pattern and multifocal ERGs are typically abnormal. In the context of genetic disorders, this electrophysiological phenotype is encountered in a variety of conditions that predominantly affect the photoreceptors including cone dysfunction disorders (e.g. achromatopsia) and cone/cone-rod dystrophies. Post-phototransduction defects can give a similar appearance; *CACNA1F*-retinopathy (also known as incomplete congenital stationary night-blindness, congenital stationary synaptic disorder or congenital cone-rod synaptic disorder) is a notable example of this.

Selected disorders/examples

Cone dysfunction disorders are a genetically heterogeneous group of relatively stationary conditions that include achromatopsia (also known as rod monochromacy) and blue cone monochromatism (also known as S-cone monochromacy). Achromatopsia is a recessively inherited disorder that has been associated with defects in several genes including *CNGA3*, *CNGB3*, *GNAT2*, *PDE6C*, *PDE6H* and *ATF6*. The first five of these genes are involved in the cone-specific phototransduction cascade and the biological function of the encoded proteins explains the classical electrodiagnostic findings of achromatopsia: undetectable light-adapted ERGs and normal dark-adapted ERGs (Fig. 6.9). *ATF6* is a ubiquitously expressed gene that has a role in endoplasmic reticulum homeostasis; it is unclear how the encoded protein causes cone dysfunction in isolation.[15] Blue cone monochromacy is an X-linked disorder associated with defects in the gene cluster that controls the expression of the L- and M-cone photoreceptor opsins. The expression of the opsin for the third cone subtype, the S-cones, and the rod pigment are not affected. The ERGs in blue cone monochromacy can resemble those in achromatopsia although the light-adapted response may be less affected (i.e. severely attenuated in blue cone monochromacy compared to undetectable in achromatopsia). Specialised testing may reveal a preserved S-cone ERG (Fig. 6.10).

Cone/cone-rod dystrophies are progressive disorders characterised by predominant cone photoreceptor dysfunction. They often present as diffuse retinopathy with significant macular involvement. Both non-syndromic and syndromic forms (e.g. Bardet-Biedl or Alström syndrome) have been described. Common molecular subtypes are due to defects in *ABCA4* (moderate/severe gene defects), *CDHR1*, *GUCY2D*, *PROM1*, *CRX*, *GUCA1A*, *RPGR* and *CERKL*; there is a significant genetic overlap between macular and cone/cone-rod dystrophies, as defects in many of these genes can cause both types of disorders. The encoded proteins perform a diverse range of functions in the photoreceptors, including ones relating to phototransduction and visual cycle (*GUCY2D*, *GUCA1A*, *ABCA4*), ciliary structure and transport (*RPGR*), outer segment structure (*CDHR1*, *PROM1*) and photoreceptor differentiation (*CRX*).[16] An early ERG finding in these conditions is a delay in the light-adapted 30 Hz flicker ERG implicit time. Dark-adapted ERGs are preserved early on but are usually affected at a later

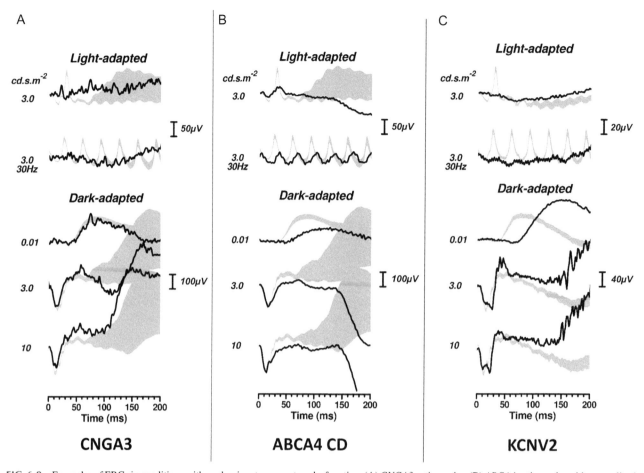

FIG. 6.9 Examples of ERGs in conditions with predominant cone system dysfunction. (A) *CNGA3*-retinopathy; (B) *ABCA4*-retinopathy with generalised photoreceptor dysfunction; (C) *KCNV2*-retinopathy: note that, in addition to the markedly attenuated light-adapted ERGs, the dark-adapted dim flash response (DA 0.01) is dramatically delayed and that the standard and bright flash responses (DA 3 and DA 10) have a characteristic broad bifid a-wave. (A) and (B) Recorded with silver/nylon corneal thread electrodes, (C) with skin electrodes.

stage. It is worth highlighting that, in one subtype, *KCNV2*-retinopathy, the ERG findings are considered pathognomonic[2] (Fig. 6.9C).

CACNA1F-retinopathy is a congenital, X-linked condition that is typically associated with a normal fundus and a non-progressive disease course. *CACNA1F* encodes a voltage-sensitive Ca^{2+} channel that is found in the rod and cone photoreceptor synaptic terminals. This channel plays a role downstream from the phototransduction cascade by transmitting signals to ON- and OFF-bipolar cells. The ERG findings in *CACNA1F*-retinopathy are characteristic: the light-adapted 30 Hz flicker ERG is markedly attenuated and delayed with most traces having a distinctive bifid peak (if they are recordable at all). The light-adapted single flash ERG is also markedly abnormal with a reduced b:a ratio (typically a- and b-wave are of similar size). The dark-adapted bright flash has a normal a-wave but the b:a ratio is reduced, giving an electronegative waveform. *CACNA1F*-retinopathy can be distinguished from the more rapidly progressive cone/cone-rod dystrophies as the latter do not classically display an electronegative ERG in the dark-adapted state and do not have the characteristic bifid peak in the light-adapted 30 Hz flicker response[17] (Fig. 6.11).

Generalised retinal dysfunction—Predominant rod system involvement

An abnormal dark-adapted full-field ERG combined with a light-adapted ERG that is either normal or less severely affected than in the dark-adapted state suggest predominant rod system involvement. The pattern and multifocal ERGs are variable but typically normal, at least at the early disease stages. In the context of genetic disorders, this electrophysiological phenotype is encountered in a variety of conditions that predominantly affect the photoreceptors (e.g. retinitis pigmentosa) or are associated with post-phototransduction defects, for example, complete congenital stationary night-blindness (CSNB) and X-linked retinoschisis.

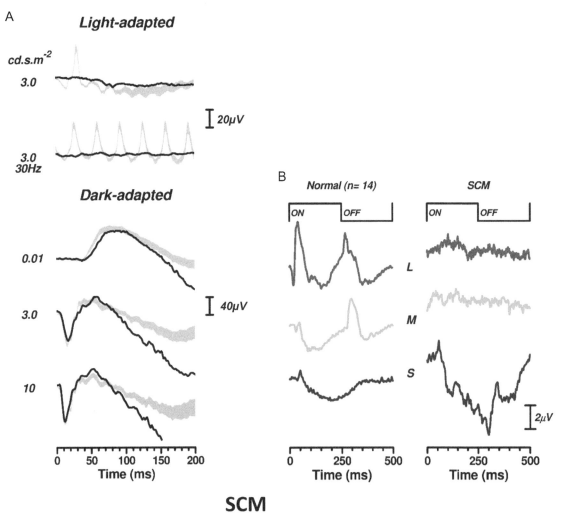

SCM

FIG. 6.10 Examples of ERGs in S-cone monochromacy. (A) Standard ISCEV full-field ERGs showing achromatopsia-like extinguished light-adapted ERGs. (B) L, M and S-cone isolated ERGs recorded using silent substitution in the same patient as (A), compared with average normal cone-isolated ERGs. Note the different morphology of the ERGs from the three cone classes, and that there is no recordable L and M cone ERG in the patient, whilst the S-cone stimulus produces a clear on and off response. *(Data courtesy of Prof D. McKeefry and Dr. J Maguire, University of Bradford, UK.)*

Selected disorders/examples

Retinitis pigmentosa (also known as rod-cone dystrophy) is a group of progressive disorders characterised by predominant rod photoreceptor dysfunction. The mid-peripheral retina is initially affected and the macular cones tend to be relatively preserved until late in the disease course. Both non-syndromic and syndromic forms (e.g. Usher syndrome or other ciliopathies and mucopolysaccharidoses) have been described. Common molecular subtypes are due to defects in *USH2A, RPGR, RHO, EYS, RP1, PRPF31, PRPH2* and *CRB1*. The encoded proteins perform a diverse range of functions in the photoreceptors, including ones relating to phototransduction and visual cycle *(RHO)*, ciliary structure and transport *(USH2A, RPGR)*, outer segment structure *(PRPH2, RP1)*, photoreceptor differentiation *(NR2E3)* and splicing *(PRPF31)*.[18]

An early ERG finding in these conditions is attenuation of the dark-adapted dim flash ERG. Light-adapted ERGs are preserved early on but are usually affected at a later stage. Notably, in one subtype, *NR2E3*-retinopathy (enhanced S-cone syndrome), the ERG findings are considered characteristic or pathognomonic[2] (see also Section 13A.3).

Chorioretinal dystrophies are progressive genetic conditions in which the primary site of dysfunction is at the level of the RPE and/or choriocapillaris. Two notable examples are choroideraemia and gyrate atrophy. In these conditions, the dark-adapted ERGs are typically affected earlier and to a greater extent than light-adapted responses. The macular function is usually preserved until late in the disease course.

Congenital stationary night-blindness (CSNB) is a genetically heterogeneous group of disorders that are

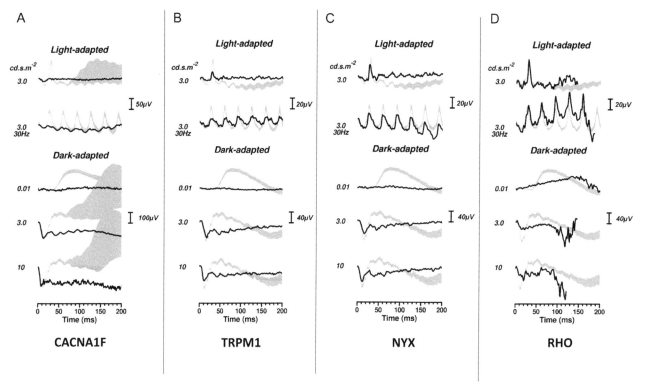

FIG. 6.11 ERGs in three individuals with different forms of congenital stationary night-blindness (CSNB) and one patient with RHO-associated retinitis pigmentosa. (A) *CACNA1F*-retinopathy (also known as X-linked incomplete CSNB): note the absence of a dark-adapted dim flash ERG (DA 0.01) and the electronegative dark-adapted standard and bright flash responses (DA 3 and DA 10). Light-adapted ERGs may be extinguished but, if present, show a bifid flicker response (LA 30Hz), as here; (B) *TRPM1*-retinopathy (autosomal recessive complete CSNB); (C) *NYX*-retinopathy (X-linked complete CSNB); (D) *RHO*-associated retinitis pigmentosa. (A) was recorded with corneal electrodes; (B), (C) and (D) with skin electrodes.

typically associated with a non-progressive disease course. The retinal examination is typically unremarkable in these conditions, although two rare subtypes, fundus albipunctatus and Oguchi disease, exhibit characteristic fundus abnormalities. It is noteworthy that, in these two subtypes, the standard dark-adapted ERG is abnormal but normalisation of this response can occur after extended periods (several hours) of dark adaptation. The so-called complete forms of CSNB are due to defects in molecules that are important for glutamate-induced signalling from the photoreceptors to the ON-bipolar cells (*NYX*, *TRPM1*, *GRM6*, *GPR179*, *LRIT3*). These forms are characterised by distinct ERG abnormalities suggestive of ON-bipolar cell dysfunction: the dark-adapted dim flash ERG is undetectable (thus the term 'complete') and there is an electronegative waveform with a normal a-wave in the dark-adapted bright flash ERG (pointing to dysfunction that occurs post-phototransduction). Light-adapted ERGs are grossly normal although there is a degree of variability.[17]

X-linked retinoschisis is an X-linked condition associated with foveal schisis and, in some cases, splitting of inner retinal layers in the peripheral retina. It is caused by defects in *RS1*, a gene that is expressed in the photoreceptor and bipolar cells. The encoded protein is secreted and it is thought to have a role in the maintenance of the photoreceptor/bipolar synapse. The classical electrophysiological findings in this condition include an electronegative waveform in the dark-adapted bright flash ERG and macular dysfunction on pattern and/or multifocal ERG. However, ERG findings tend to be variable and although electrophysiological testing was the major diagnostic tool for this condition for many years, OCT and genetic screening are now the primary diagnostic tests.

Generalised retinal dysfunction—Equal rod and cone system involvement

Occasionally, the dark- and light-adapted full-field ERGs are equally affected. This most commonly occurs in severe early childhood onset retinal dystrophies (including Leber congenital amaurosis). In these conditions, both the dark- and light-adapted ERGs are undetectable or severely attenuated from the first years of life. This pattern of dysfunction may also be encountered in advanced disease linked to conditions that may have initially caused either predominantly cone or predominantly rod system dysfunction (Fig. 6.12).

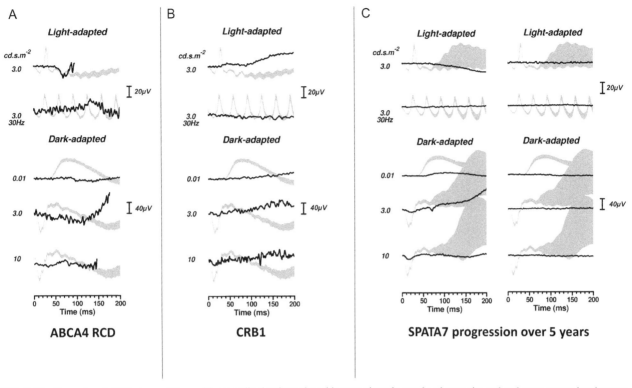

FIG. 6.12 Examples of ERGs in conditions with generalised retinopathy with approximately equal and extensive rod and cone system involvement. (A) Advanced *ABCA4*-retinopathy (RCD, rod-cone dystrophy); (B) *CRB1*-related early-onset retinal dystrophy; (C) progression of *SPATA7*-retinopathy over a 5-year period. (A) and (B) were recorded with skin electrodes, (C) with silver/nylon corneal thread electrodes.

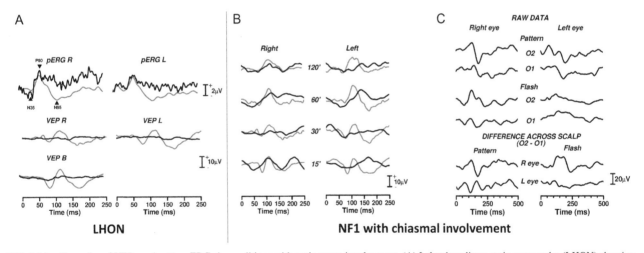

FIG. 6.13 Examples of VEPs and pattern ERGs in conditions with optic nerve involvement. (A) Leber hereditary optic neuropathy (LHON) showing attenuation of pattern ERG N95 and extinguished VEPs; (B) NF1-associated chiasmal glioma predominantly affecting the left optic nerve (note the delay in comparison to the right side); (C) Crossed asymmetry due to chiasmal involvement in the same patient.

Optic pathway abnormalities

Normal full-field ERGs and pattern ERG P50 component (or multifocal ERG) with abnormal VEPs suggest optic nerve dysfunction. In the context of genetic disorders, this electrophysiological phenotype is encountered in Leber hereditary optic neuropathy and dominant optic atrophy. In these hereditary optic neuropathies, there is primary ganglion cell dysfunction that typically leads to marked pattern ERG N95 component loss; this abnormality may occur before the VEP findings. The pattern ERG P50 component is preserved, at least at early disease stages, and pattern VEP is often markedly delayed and attenuated (Fig. 6.13).

A condition associated with a characteristic VEP phenotype is albinism, a clinically and genetically heterogeneous disorder characterised by ocular abnormalities and variable degrees of skin hypopigmentation. In both ocular and oculocutaneous forms of albinism, there is often an abnormally high percentage of decussating temporal optic nerve fibres that project to the contralateral occipital lobe. This 'contralateral predominance' is readily detected by VEP testing. The dominance of one visual cortex over the other is manifested as significant scalp asymmetry when lateral electrodes are used. When there is marked temporal decussation, the dominance can flip when changing from the right eye to the left, and thus the asymmetry will also flip. This results in what is known as crossed asymmetry, examples of which are shown in Fig. 6.14. The asymmetry can manifest as shorter latency, higher amplitude or both. It is most easily visualised by examining the difference plots (bottom of Fig. 6.14), subtracting one lateral signal from the other. In general, the electrophysiological features of albinism are said to be most readily demonstrated using a

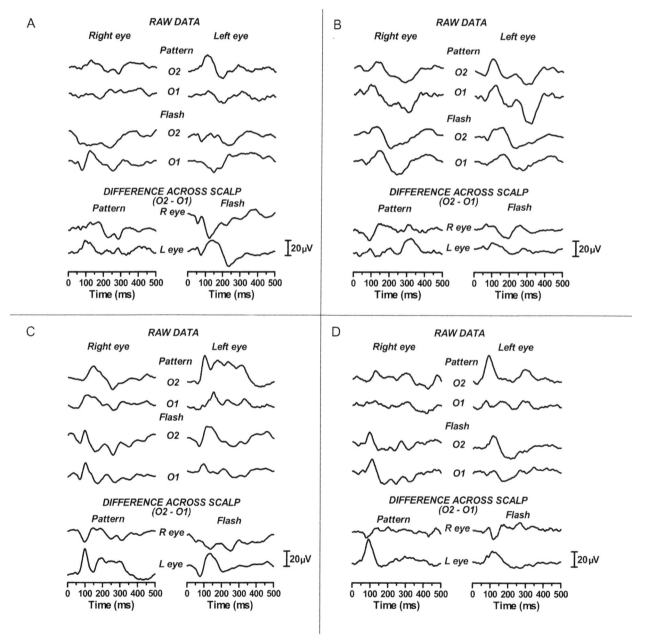

FIG. 6.14 Crossed asymmetry in 2 individuals with *TYR*-associated oculocutaneous albinism, both aged 4, but one showing crossed asymmetry with flash (A) and one with pattern (B) stimulation. (C) Patient with oculocutaneous albinism showing crossed asymmetry with both flash and pattern stimulation (D) *SLC38A8*-associated foveal hypoplasia.

flash stimulus in children under the age of 4 years and using a pattern stimulus in older patients. It is advisable, where possible, to use both flash and pattern stimulation, especially in children over the age of 5 years.[19]

Other conditions associated with chiasmal abnormalities and related VEP findings include achiasmia, *SLC38A8*-associated foveal hypoplasia (also known as foveal hypoplasia, optic-nerve-decussation defects, and anterior segment dysgenesis (FHONDA) syndrome; Fig. 6.14D) and space-occupying lesions leading to chiasmal compression (e.g. optic nerve gliomas associated with neurofibromatosis type 1 (NF1), Fig. 6.13C).

Concluding remarks

In this chapter, we have attempted to provide an overview of the utility of electrophysiology in genetic ophthalmic disorders. The key to these investigations is the context. For example, it is important to establish the integrity of retinal function before reaching any conclusions about higher visual function. Thus, tests are rarely conducted in isolation. However, an unplanned but extensive examination can take up to three hours, and planning needs to take into account the clinical question being posed. For example, conditions associated with nystagmus would require mainly VEPs to look for albinoid crossed asymmetry, and ERGs to detect either a cone or bipolar cell mediated retinopathy, as these conditions can all manifest nystagmus. This review by no means provides a fully extensive review of visual electrodiagnosis; a minute examination of every aspect of the general subject is provided by Heckenlively and Arden[20] and a useful overview by Robson et al.[3]

References

1. Stone EM, Andorf JL, Whitmore SS, DeLuca AP, Giacalone JC, Streb LM, et al. Clinically focused molecular investigation of 1000 consecutive families with inherited retinal disease. *Ophthalmology* 2017;**124**:1314–31.
2. Vincent A, Robson AG, Holder GE. Pathognomonic (diagnostic) ERGs. A review and update. *Retina* 2013;**33**:5–12.
3. Robson AG, Nilsson J, Li S, Jalali S, Fulton AB, Tormene AP, et al. ISCEV guide to visual electrodiagnostic procedures. *Doc Ophthalmol* 2018;**136**:1–26.
4. Horn FK, Bergua A, Junemann A, Korth M. Visual evoked potentials under luminance contrast and color contrast stimulation in glaucoma diagnosis. *J Glaucoma* 2000;**9**:428–37.
5. Maguire J, Parry NRA, Kremers J, Murray IJ, McKeefry D. Human S-cone electroretinograms obtained by silent substitution stimulation. *J Opt Soc Am A Opt Image Sci Vis* 2018;**35**:B11–8.
6. Arden G, Wolf J, Berninger T, Hogg CR, Tzekov R, Holder GE. S-cone ERGs elicited by a simple technique in normals and in tritanopes. *Vision Res* 1999;**39**(3):641–50.
7. Westall CA, Dhaliwal HS, Panton CM, Sigesmun D, Levin AV, Nischal KK, et al. Values of electroretinogram responses according to axial length. *Doc Ophthalmol* 2001;**102**:115–30.
8. Parks S, Keating D. The multifocal electroretinogram. In: *Handbook of Clinical Neurophysiology*. vol. 5; 2005. p. 99–115.
9. Müller PL, Meigen T. M-sequences in ophthalmic electrophysiology. *J Vis* 2016;**16**:15.
10. Marmor MF, Cabael L. Clinical display of mfERG data. *Doc Ophthalmol* 2018;**137**:63–70.
11. Holder GE. The pattern electroretinogram in anterior visual pathway dysfunction and its relationship to the pattern visual evoked potential: a personal clinical review of 743 eyes. *Eye* 1997;**11**:924–34.
12. Jasper HH. The ten-twenty system of the international federation. *Electroencephalogr Clin Neurophysiol* 1958;**10**:271–5.
13. Johnson AA, Guziewicz KE, Lee CJ, Kalathur RC, Pulido JS, Marmorstein LY, et al. Bestrophin 1 and retinal disease. *Prog Retin Eye Res* 2017;**58**:45–69.
14. Chowers I, Tiosano L, Audo I, Grunin M, Boon CJ. Adult-onset foveomacular vitelliform dystrophy: a fresh perspective. *Prog Retin Eye Res* 2015;**47**:64–85.
15. Hirji N, Aboshiha J, Georgiou M, Bainbridge J, Michaelides M. Achromatopsia: clinical features, molecular genetics, animal models and therapeutic options. *Ophthalmic Genet* 2018;**39**:149–57.
16. Gill JS, Georgiou M, Kalitzeos A, Moore AT, Michaelides M. Progressive cone and cone-rod dystrophies: clinical features, molecular genetics and prospects for therapy. *Br J Ophthalmol* 2019;**103**:711–20.
17. Zeitz C, Robson AG, Audo I. Congenital stationary night blindness: an analysis and update of genotype-phenotype correlations and pathogenic mechanisms. *Prog Retin Eye Res* 2015;**45**:58–110.
18. Verbakel SK, van Huet RAC, Boon CJF, den Hollander AI, Collin RWJ, Klaver CCW, et al. Non-syndromic retinitis pigmentosa. *Prog Retin Eye Res* 2018;**66**:157–86.
19. Neveu MM, Jeffery G, Burton LC, Sloper JJ, Holder GE. Age-related changes in the dynamics of human albino visual pathways. *Eur J Neurosci* 2003;**18**:1939–49.
20. Heckenlively JR, Arden GB. *Principles and practice of clinical electrophysiology of vision.* Cambridge, Mass: MIT Press; 2006.

Chapter 7

Ophthalmic phenotyping: Imaging

Johannes Birtel, Martin Gliem, Wolf M. Harmening, and Frank G. Holz

Genetic diseases affecting the retina are highly heterogeneous. Although fundoscopy remains a diagnostic mainstay, retinal imaging provides additional important information and has become an essential part of the management of these conditions. These *in vivo* imaging techniques have rapidly evolved over the past decades due to advances in electronics, optics and computer technology. Key strengths include their high resolution and the potential to document retinal pathologies that are barely visible or cannot be detected on fundus examination alone, as well as to accurately document findings/progression over time. Fundus imaging can also be used to consolidate the diagnosis of a specific retinal disorder; multimodal imaging allows in-depth analysis of phenotypic heterogeneity and differential diagnoses (e.g. distinguishing localised from panretinal diseases) and may even indicate mutations in particular genes. Imaging can also offer novel insights into the pathogenesis of retinal conditions and provides surrogate outcome measures to assess the efficacy of therapeutic interventions.[1-4]

This chapter focuses on imaging modalities that are key for phenotyping genetic disorders affecting the retina. This includes established imaging techniques like optical coherence tomography (OCT) and short-wavelength fundus autofluorescence (SW-AF), but also more recently developed imaging techniques such as near-infrared fundus autofluorescence (NIR-AF), quantitative fundus autofluorescence (qAF) and adaptive optics (AO).

Optical coherence tomography (OCT)

OCT has revolutionised the clinical practice of ophthalmology. This technology is often described as the optical analogue of ultrasound as it uses the time delay and magnitude of light echoes to generate images. It is based on the interference of a light beam split into two paths: one probe path directed onto the tissue, and a second reference path within the instrument. Interference occurs between the reflected portion of the probe and the reference.[5] As the optical properties of the latter are known, the interference characteristics carry tissue information.[5,6] Information derived from a single beam is summarised in an A-scan. By scanning multiple A-scans along an axis, a two-dimensional image is obtained (B-scan); this is reminiscent of a histological cross-sectional preparation of the tissue.[5] By scanning along a second orthogonal axis, volumetric information can be obtained. While early OCT imaging used time-domain detection, the development of high-resolution tools including superluminescent diodes and ultrashort pulsed lasers improved the imaging technique and now sub-micrometre resolution is possible.

Each retinal layer shows characteristic reflectivity. This gives the resulting OCT image a stratified, quasi-histologic appearance which enables the correlation of OCT layers to certain retinal structures. In general, the following different layers can be defined: retinal nerve fibre layer, ganglion cell layer, inner plexiform layer, inner nuclear layer, outer plexiform, inner plexiform, external limiting membrane, inner photoreceptor segment, ellipsoid zone, outer photoreceptor segment, interdigitation zone with RPE, RPE-Bruch's membrane complex and choroid.[6]

High-resolution OCT imaging enables early detection of subtle retinal changes that might be invisible on fundus examination or on other imaging modalities (e.g. early disease stages or in diseases such as occult macular dystrophy). Serial OCT imaging is a powerful tool for disease monitoring as it can visualise retinal changes over time. This allows disease progression estimation and provides important read-outs for natural history studies and interventional trials. Structural data derived from OCT are frequently combined with information from other imaging modalities such as SW-AF or NIR-AF imaging.[6]

Various OCT features that offer important diagnostic information have been described for genetic disorders affecting the retina. Analysis of the integrity of the RPE and of the photoreceptor layers (e.g. ellipsoid zone, outer nuclear layer) can be particularly useful. For instance, patients with retinitis pigmentosa show characteristic peripheral thinning of outer retinal layers and disruption

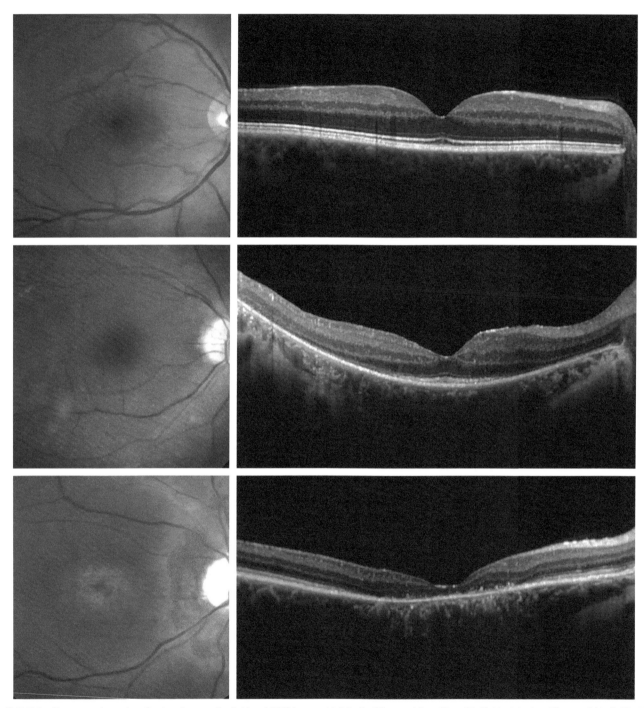

FIG. 7.1 Representative colour fundus photographs (left) and OCT images (right) of a 30-year-old unaffected individual (top), a 27-year-old individual with retinitis pigmentosa and a homozygous *EYS* mutation (middle), and a 30-year-old individual with cone/cone-rod dystrophy and a heterozygous *GUCA1A* mutation (bottom).

or disappearance of the ellipsoid zone due to a primary or predominant rod degeneration (Fig. 7.1).[7,8] Patients with cone/cone-rod dystrophies, on the contrary, show mainly atrophic changes of outer retinal layers within the central retina in keeping with predominant degeneration of cone photoreceptors. It is worth highlighting that these disease entities should be differentiated from stationary diseases (such as congenital stationary night-blindness or achromatopsia) which may present with comparable symptoms but without pronounced changes on OCT imaging.[1,2]

Short-wavelength fundus autofluorescence (SW-AF)

Fundus autofluorescence imaging of the retina with short-wavelength excitation light has been established as an important imaging modality for the diagnosis, follow-up and understanding of retinal diseases. This non-invasive imaging technique uses short-wavelength (SW) light in the blue or green spectrum to excite fluorophores at the fundus.[9] The main fluorophore reacting at this wavelength is lipofuscin; thus, SW-AF offers indirect information on the lipofuscin content of the RPE. Lipofuscin is composed of a mixture of lipids, proteins, and fluorescent compounds mainly derived from the visual cycle. It accumulates within the lysosomal compartment of the RPE throughout life.[11] The best characterised component is bisretinoid N-retinylidene-N-retinylethanolamine (A2E) that accumulates in RPE lysosomes and possesses toxic effects on RPE cells via multiple mechanisms.[11] In general, lipofuscin accumulation in the RPE is associated with a failure of clearance and/or with an abnormal metabolic load.[9]

SW-AF imaging may also provide information beyond that obtained by other imaging modalities. Retinal alterations may, for instance, be visible even in cases where fundoscopy does not detect obvious abnormalities and affected individuals have non-specific, minimal or no symptoms. The introduction of confocal scanning laser ophthalmoscopy using raster scanning of the retina has significantly improved image quality and thereby fostered the clinical application of SW-AF imaging. However, a few issues have been highlighted including glare during image acquisition and potential phototoxic effects.

A healthy fundus is typically characterised by a homogenous SW-AF signal. Retinal vessels and the optic disc appear darker due to light absorption by blood and the absence of fluorophores. The fovea typically shows reduced fluorescence due to masking by the luteal pigment, although this can be variable and there are differences between devices (Fig. 7.2). In general, increased signal is indicative of abnormal metabolite accumulation within the RPE whereas reduced signal is mostly related to RPE atrophy.[9]

Characteristic SW-AF imaging findings have been noted in several genetic disorders that affect the retina. In retinitis pigmentosa, for instance, many affected individuals demonstrate a ring of increased autofluorescence at the border of degenerated peripheral retina and areas of preserved central photoreceptors (Fig. 7.3). This ring, which is not visible on slit-lamp examination, correlates

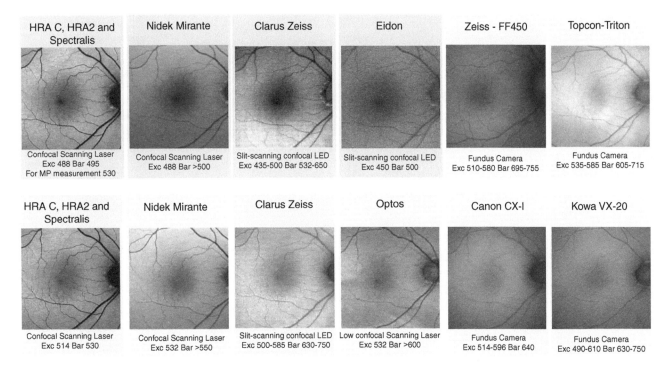

FIG. 7.2 Overview of fundus autofluorescence imaging of the same right normal eye with different devices. For each system, the excitation and emission range are shown. The four examples with *light blue* background colour operate with *blue* excitation light, the remaining with *green* background colour operate with excitation in the *green light* range. *(Adapted from Ref. 9.)*

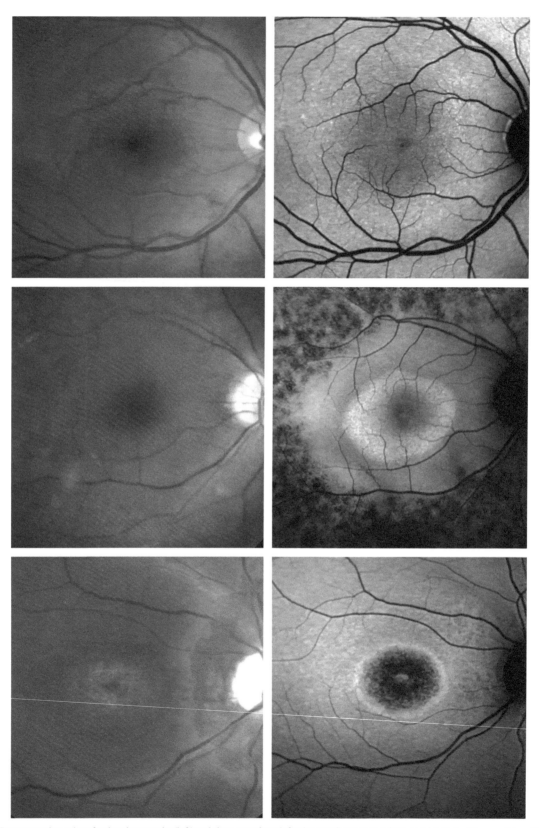

FIG. 7.3 Representative colour fundus photographs (left) and short-wavelength fundus autofluorescence images (right) of a 30-year-old unaffected individual (top), a 27-year-old individual with retinitis pigmentosa and a homozygous *EYS* mutation (middle), and a 30-year-old individual with cone/cone-rod dystrophy and a heterozygous *GUCA1A* mutation (bottom).

with visual field findings and can be an indicator of the remaining visual function.[14,15] The size of the ring and the rate of its constriction can be recorded with serial imaging and generally reflects disease progression.[16] However, rings of increased SW-AF are also found in other retinal dystrophies, including Leber congenital amaurosis and cone/cone-rod dystrophies. In the latter, a centrifugal expansion of the ring is typically seen.

Other characteristic autofluorescence findings include hyperautofluorescent flecks in *ABCA4*-retinopathy (Stargardt disease) and hyperautofluorescent vitelliform lesions in pattern dystrophies and autosomal-dominant Best disease. In carriers of X-linked diseases such as choroideremia or *RPGR*-associated retinitis pigmentosa, characteristic SW-AF findings may also facilitate a reliable diagnosis even before genetic testing.[17-19]

Near-infrared fundus autofluorescence (NIR-AF)

NIR-AF imaging was first described in 1996.[20] In contrast to SW-AF, the retina is excited with a longer wavelength (787 nm) using the indocyanine green (ICG) mode of a confocal scanning laser ophthalmoscope. The autofluorescence signal at this wavelength is less intense compared to SW-AF which has as yet hindered broader clinical applications. However, compared to SW-AF, NIR-AF imaging has several advantages: the excitation light is more comfortable for the patients (no glare), the image is less affected by cataract development, and the lower energy of the longer wavelength is less likely to be associated with retinal light toxicity.

In healthy subjects, the highest NIR-AF signal is observed in the central retina and not at about 10° eccentricity as in SW-AF images (Fig. 7.4). This is because there is no masking by the macular pigment in NIR-AF imaging; this allows the visualisation of subtle changes in the central retina, including abnormalities that might be invisible on SW-AF images.[21-23]

It is generally thought that the NIR-AF signal originates mainly from melanin.[24] Melanin is a pigment that is found at high concentrations in the RPE and choroid.[24-26] Apart from its light absorption properties, the exact function of melanin is unclear but may include scavenging of free radicals and other 'detoxifying' properties. Unlike lipofuscin, the fluorophore excited in SW-AF imaging, melanin is only synthesised *in utero* and perhaps in early childhood. Changes later in life may include the fusion of melanosomes and lysosomes to melanolysosomes, resulting in melanolipofuscin granules.

Several characteristic NIR-AF imaging findings have been described in genetic disorders that affect the retina. For instance, in *ABCA4*-retinopathy (Stargardt disease), changes on NIR-AF appear to precede those of SW-AF, suggesting superiority in detecting early disease progression.[27] Also, in choroideremia, it can be used to stratify affected individuals into groups with notable differences in terms of morphologic and functional alterations.[21] More broadly, there is emerging evidence suggesting that NIR-AF could have a role in early diagnostics and disease staging as well as for patient selection or as an outcome parameter in interventional trials.

Quantitative fundus autofluorescence (qAF)

As stated above, SW-AF imaging allows the visualisation of lipofuscin accumulation and individuals with retinal dystrophies frequently exhibit changes in autofluorescence signal levels. However, even with a standardised SW-AF imaging protocol, accurate comparison of the lipofuscin distribution can be complex. A technique that addresses these challenges is the qAF imaging approach that can be used as a quantitative surrogate marker for lipofuscin accumulation in the RPE.[28,29] This method encompasses a standard fluorescent reference that compensates for changes in laser power and detector gain, and allows the comparison of images obtained at different time points. The analysis compares the grey levels in the SW-AF image with the grey levels of the internal reference. It also includes corrections for magnification and optical media density based on normative data on lens transmission spectra.[28] When this approach is used together with a standardised image acquisition protocol, reproducible quantification of autofluorescence levels in individual patients, interpatient comparison, and monitoring of autofluorescence levels longitudinally are possible.[29] Comparison of data acquired on different devices and/or at different centres is also enabled.

As in conventional SW-AF images, qAF imaging shows the lowest values on retinal vessels, the optic disc and the fovea. Throughout the fundus, qAF values peak at an eccentricity of roughly 10° centered to the fovea with the highest values being supero-temporally. Consistent with age-related lipofuscin accumulation, qAF values increase with age.[28,29]

QAF can be a helpful tool for the diagnosis of genetic conditions affecting the retina. The prime example is *ABCA4*-retinopathy where qAF imaging can detect and quantify increased lipofuscin accumulation at the level of the RPE (Fig. 7.5).[30-34] This information can assist in establishing the diagnosis and in differentiating *ABCA4*-retinopathy from similar appearing disorders.[33] QAF imaging can also be used to monitor the

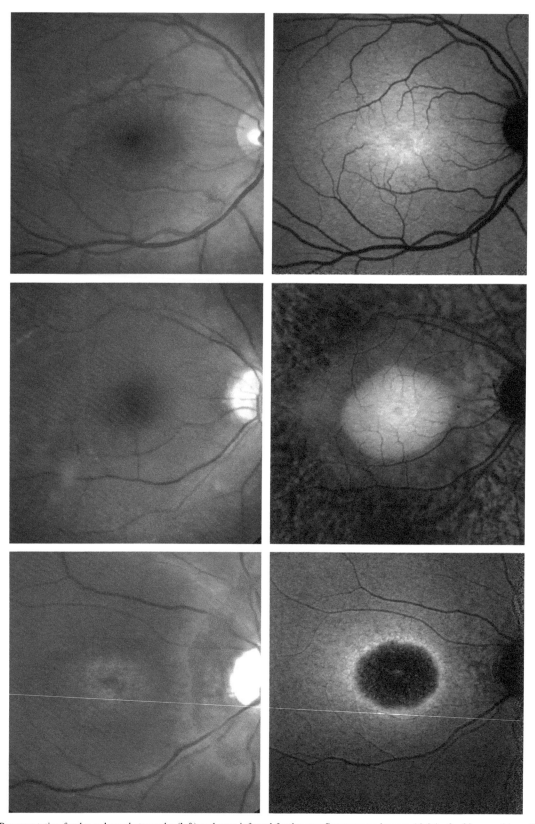

FIG. 7.4 Representative fundus colour photographs (left) and near-infrared fundus autofluorescence images (right) of a 30-year-old unaffected individual (top), a 27-year-old individual with retinitis pigmentosa and a homozygous *EYS* mutation (middle) and a 30-year-old individual with cone/cone-rod dystrophy and a heterozygous *GUCA1A* mutation (bottom).

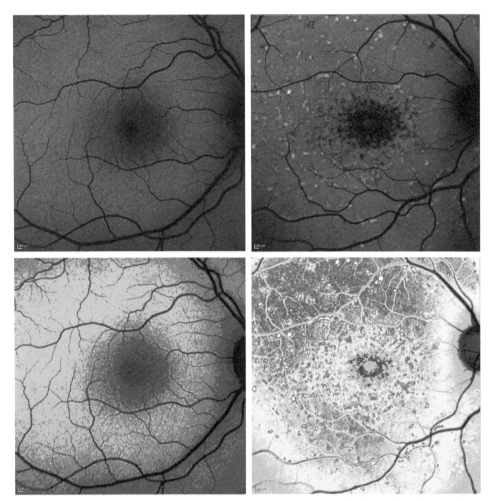

FIG. 7.5 Representative short-wavelength fundus autofluorescence (SW-AF; upper row) and quantitative autofluorescence (qAF; lower row) images of an unaffected individual (left) and a patient with *ABCA4*-retinopathy (Stargardt disease) (right), both 28 years old. QAF images can visualise and quantify the generalised lipofuscin accumulation in *ABCA4*-retinopathy.

accelerated lipofuscin accumulation over time. This could be helpful as a potential outcome measure for lipofuscin-lowering drugs which are currently under investigation.

QAF imaging can also provide insights into the pathophysiology of retinal diseases.[35] It has for example been shown that patients with *PRPH2*-related pattern dystrophy present with increased qAF values suggesting that lipofuscin accumulation is a key pathophysiological abnormality in this disorder.[32,33] In contrast, heterozygous carriers of *ABCA4* mutations have shown no evidence of increased lipofuscin accumulation confirming the autosomal recessive nature of the associated condition.[36,37]

Adaptive optics

Adaptive optics (AO) is a set of imaging tools that allow non-invasive retinal phenotyping on a microscopic scale.[38] The goal of an adaptive optics system is to reduce the contribution of the monochromatic aberrations of the ocular optics to achieve diffraction-limited lateral resolution at the largest possible physiological pupil size. Compensating ocular aberrations is achieved by a control loop consisting of a wavefront sensor and an adaptive optical element, often a deformable mirror.

Adaptive optics techniques are currently employed in three main imaging modalities: flood illumination ophthalmoscopy (AO fundus camera), scanning laser ophthalmoscopy (AO-SLO), and OCT (AO-OCT). These modalities have complementary advantages and allow visualisation of microscopic retinal structures such as the retinal nerve fibre layer; the lamina cribrosa; retinal ganglion cell bodies; blood flow with single leucocyte resolution in the smallest capillaries; individual cone and rod photoreceptors; the RPE mosaic. It is worth highlighting that imaging and stimulation can be spatially and temporally coupled in a scanning laser ophthalmoscope. As a result, AO-SLO systems can also produce stimulation lights with extremely high spatial precision. More specifically, with the combination of real-time eye motion compensation, high contrast light modulation and chromatic aberration correction, AO-SLO can be utilised as an optical platform to perform cellular-targeted visual function testing.[39–41]

Many genetic disorders affecting the retina are accompanied by structural changes of the photoreceptor mosaic. Adaptive optics platforms allow the visualisation of individual photoreceptor cells *in vivo* facilitating early disease detection and the development of precise imaging biomarkers (Fig. 7.6).[43] For instance, in patients with *ABCA4*-retinopathy (Stargardt disease), the photoreceptor mosaic is disrupted from early on, even in otherwise unremarkable regions of the retina.

Concluding remarks

Imaging facilitates precise visualisation and characterisation of retinal pathology including abnormalities that are not visible on fundoscopy. Multimodal imaging approaches are routinely used for diagnosis and follow-up, and are key for observational and interventional studies on genetic retinal disorders. It is expected that imaging-related applications of artificial intelligence will become

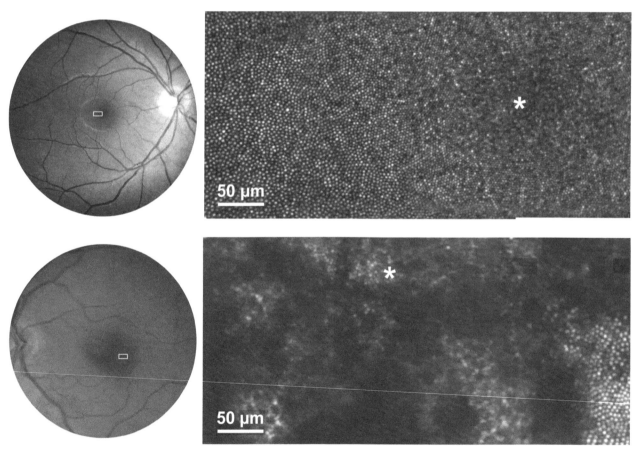

FIG. 7.6 Adaptive optics scanning laser ophthalmoscopy (AO-SLO) images of a healthy (top) and a diseased (bottom) retina close to the preferred retinal locus (*asterisk*) of the subject. In the healthy retina, the complete mosaic of the cone photoreceptors is visible, with naturally occurring differences in apparent cone brightness. In the diseased retina (here a case of idiopathic macular telangiectasia type II), only small patches of cones remain visible. The hexagonal mosaic of the underlying RPE cells is visible between larger patches of unresolved structure. Both images were recorded with 840 nm light and shown at a similar scale. *(Image material in lower part has been adapted from Ref. 42.)*

increasingly important in this group of conditions. Notable use-cases of artificial intelligence include: classification (including automated diagnosis and staging), segmentation (including delineation of anatomical structures or lesions) and prediction (including predicting visual function).

References

1. Birtel J, Eisenberger T, Gliem M, et al. Clinical and genetic characteristics of 251 consecutive patients with macular and cone/cone-rod dystrophy. *Sci Rep* 2018;**8**(1):4824.
2. Birtel J, Gliem M, Mangold E, et al. Next-generation sequencing identifies unexpected genotype-phenotype correlations in patients with retinitis pigmentosa. *PLoS One* 2018;**13**(12), e0207958.
3. Stone EM, Andorf JL, Whitmore SS, et al. Clinically focused molecular investigation of 1000 consecutive families with inherited retinal disease. *Ophthalmology* 2017;**124**(9):1314–31.
4. Birtel J, Gliem M, Holz FG, Herrmann P. Imaging and molecular genetic diagnostics for the characterization of retinal dystrophies. *Ophthalmologe* 2018;**115**(12):1021–7.
5. Huang D, Swanson EA, Lin CP, et al. Optical coherence tomography. *Science* 1991;**254**(5035):1178–81.
6. Staurenghi G, Sadda S, Chakravarthy U, Spaide RF. Proposed lexicon for anatomic landmarks in normal posterior segment spectral-domain optical coherence tomography: the IN*OCT consensus. *Ophthalmology* 2014;**121**(8):1572–8.
7. Hood DC, Lazow MA, Locke KG, Greenstein VC, Birch DG. The transition zone between healthy and diseased retina in patients with retinitis pigmentosa. *Invest Ophthalmol Vis Sci* 2011;**52**(1):101–8.
8. Jacobson SG, Aleman TS, Sumaroka A, et al. Disease boundaries in the retina of patients with usher syndrome caused by MYO7A gene mutations. *Invest Ophthalmol Vis Sci* 2009;**50**(4):1886–94.
9. Schmitz-Valckenberg S, Pfau M, Fleckenstein M, et al. Fundus autofluorescence imaging. *Prog Retin Eye Res* 2020;**100893**.
10. Delori FC, Staurenghi G, Arend O, Dorey CK, Goger DG, Weiter JJ. In vivo measurement of lipofuscin in Stargardt's disease—fundus flavimaculatus. *Invest Ophthalmol Vis Sci* 1995;**36**(11):2327–31.
11. Sparrow JR, Gregory-Roberts E, Yamamoto K, et al. The bisretinoids of retinal pigment epithelium. *Prog Retin Eye Res* 2012;**31**(2):121–35.
12. Cideciyan AV, Swider M, Aleman TS, et al. Reduced-illuminance autofluorescence imaging in ABCA4-associated retinal degenerations. *J Opt Soc Am A Opt Image Sci Vis* 2007;**24**(5):1457–67.
13. Cideciyan AV, Swider M, Jacobson SG. Autofluorescence imaging with near-infrared excitation:normalization by reflectance to reduce signal from choroidal fluorophores. *Invest Ophthalmol Vis Sci* 2015;**56**(5):3393–406.
14. Robson AG, Saihan Z, Jenkins SA, et al. Functional characterisation and serial imaging of abnormal fundus autofluorescence in patients with retinitis pigmentosa and normal visual acuity. *Br J Ophthalmol* 2006;**90**(4):472–9.
15. Robson AG, Egan CA, Luong VA, Bird AC, Holder GE, Fitzke FW. Comparison of fundus autofluorescence with photopic and scotopic fine-matrix mapping in patients with retinitis pigmentosa and normal visual acuity. *Invest Ophthalmol Vis Sci* 2004;**45**(11):4119–25.
16. Robson AG, Tufail A, Fitzke F, et al. Serial imaging and structure-function correlates of high-density rings of fundus autofluorescence in retinitis pigmentosa. *Retina* 2011;**31**(8):1670–9.
17. Birtel J, Yusuf IH, Priglinger C, Rudolph G, Charbel Issa P. Diagnosis of inherited retinal diseases. *Klin Monbl Augenheilkd* 2021;**238**(3):249–59. https://doi.org/10.1055/a-1388-7236. PMID: 33784788.
18. Wegscheider E, Preising MN, Lorenz B. Fundus autofluorescence in carriers of X-linked recessive retinitis pigmentosa associated with mutations in RPGR, and correlation with electrophysiological and psychophysical data. *Graefes Arch Clin Exp Ophthalmol* 2004;**242**:501–511.
19. Preising MN, Wegscheider E, Friedburg C, Poloschek CM, Wabbels BK, Lorenz B. Fundus autofluorescence in carriers of choroideremia and correlation with electrophysiologic and psychophysical data. *Ophthalmology* 2009;**116**(6):1201–9.e1–2. https://doi.org/10.1016/j.ophtha.2009.01.016. PMID: 19376587.
20. Piccolino FC, Borgia L, Zinicola E, Iester M, Torrielli S. Pre-injection fluorescence in indocyanine green angiography. *Ophthalmology* 1996;**103**(11):1837–45.
21. Birtel J, Salvetti AP, Jolly JK, et al. Near-infrared autofluorescence in Choroideremia: anatomic and functional correlations. *Am J Ophthalmol* 2019;**199**:19–27.
22. De Silva SR, Neffendorf JE, Birtel J, et al. Improved diagnosis of retinal laser injuries using near-infrared autofluorescence. *Am J Ophthalmol* 2019;**208**:87–93.
23. Birtel J, Hildebrand GD, Charbel IP. Laser pointer: a possible risk for the retina. *Klin Monatsbl Augenheilkd* 2020;**237**(10):1187–93.
24. Keilhauer CN, Delori FC. Near-infrared autofluorescence imaging of the fundus: visualization of ocular melanin. *Invest Ophthalmol Vis Sci* 2006;**47**(8):3556–64.
25. Huang Z, Zeng H, Hamzavi I, et al. Cutaneous melanin exhibiting fluorescence emission under near-infrared light excitation. *J Biomed Opt* 2006;**11**(3):34010.
26. Schmitz-Valckenberg S, Lara D, Nizari S, et al. Localisation and significance of in vivo near-infrared autofluorescent signal in retinal imaging. *Br J Ophthalmol* 2011;**95**(8):1134–9.
27. Müller PL, Birtel J, Herrmann P, Holz FG, Charbel Issa P, Gliem M. Functional relevance and structural correlates of near infrared and short wavelength fundus autofluorescence imaging in ABCA4-related retinopathy. *Transl Vis Sci Technol* 2019;**8**(6):46.
28. Delori F, Greenberg JP, Woods RL, et al. Quantitative measurements of autofluorescence with the scanning laser ophthalmoscope. *Invest Ophthalmol Vis Sci* 2011;**52**(13):9379–90.
29. Greenberg JP, Duncker T, Woods RL, Smith RT, Sparrow JR, Delori FC. Quantitative fundus autofluorescence in healthy eyes. *Invest Ophthalmol Vis Sci* 2013;**54**(8):5684–93.
30. Burke TR, Duncker T, Woods RL, et al. Quantitative fundus autofluorescence in recessive Stargardt disease. *Invest Ophthalmol Vis Sci* 2014;**55**(5):2841–52.
31. Duncker T, Tsang SH, Lee W, et al. Quantitative fundus autofluorescence distinguishes ABCA4-associated and non-ABCA4-associated bull's-eye maculopathy. *Ophthalmology* 2015;**122**(2):345–55.
32. Duncker T, Tsang SH, Woods RL, et al. Quantitative fundus autofluorescence and optical coherence tomography in PRPH2/RDS- and ABCA4-associated disease exhibiting phenotypic overlap. *Invest Ophthalmol Vis Sci* 2015;**56**(5):3159–70.

33. Gliem M, Müller PL, Birtel J, et al. Quantitative fundus autofluorescence and genetic associations in macular, cone, and cone-rod dystrophies. *Ophthalmol Retina* 2020;**4**(7):737–49.
34. Müller PL, Gliem M, McGuinnes M, Birtel J, Holz FG, Charbel IP. Quantitative fundus autofluorescence in ABCA4-related retinopathy—functional relevance and genotype-phenotype correlation. *Am J Ophthalmol*. 2021;**222**:340–50.
35. Gliem M, Müller PL, Finger RP, McGuinness MB, Holz FG, Charbel IP. Quantitative fundus autofluorescence in early and intermediate age-related macular degeneration. *JAMA Ophthalmol* 2016;**134**(7):817–24.
36. Müller PL, Gliem M, Mangold E, et al. Monoallelic ABCA4 mutations appear insufficient to cause retinopathy: a quantitative autofluorescence study. *Invest Ophthalmol Vis Sci* 2015;**56**(13):8179–86.
37. Duncker T, Stein GE, Lee W, et al. Quantitative fundus autofluorescence and optical coherence tomography in ABCA4 carriers. *Invest Ophthalmol Vis Sci* 2015;**56**(12):7274–85.
38. Williams DR. Imaging single cells in the living retina. *Vis Res* 2011;**51**(13):1379–96.
39. Harmening WM, Tiruveedhula P, Roorda A, Sincich LC. Measurement and correction of transverse chromatic offsets for multiwavelength retinal microscopy in the living eye. *Biomed Opt Express* 2012;**3**(9):2066–77.
40. Harmening WM, Tuten WS, Roorda A, Sincich LC. Mapping the perceptual grain of the human retina. *J Neurosci* 2014;**34**(16):5667–77.
41. Domdei N, Domdei L, Reiniger JL, et al. Ultra-high contrast retinal display system for single photoreceptor psychophysics. *Biomed Opt Express* 2018;**9**(1):157–72.
42. Wang Q, Tuten WS, Lujan BJ, et al. Adaptive optics microperimetry and OCT images show preserved function and recovery of cone visibility in macular telangiectasia type 2 retinal lesions. *Invest Ophthalmol Vis Sci* 2015;**56**(2):778–86.
43. Reiniger JL, Domdei N, Pfau M, Müller PL, Holz FG, Harmening WM. Potential of adaptive optics for the diagnostic evaluation of hereditary retinal diseases. *Klin Monbl Augenheilkd* 2017;**234**(3):311–9.

Chapter 8

Gene therapy and treatment trials

Robert E. MacLaren and Jasmina Cehajic-Kapetanovic

The complexity of the photoreceptor cell and the high energy environment required for phototransduction, make it particularly susceptible to cell death. This can be a result of a wide spectrum of mutations, some of which are not deleterious to any other cell type (e.g. changes in splicing factor genes like *PRPF31*). The surgical accessibility of the eye, the ability to image the retina at histological levels of accuracy and the capacity to perform complex visual function tests and compare them to a fellow 'control' eye make the retina an ideal target for assessing gene therapy. This chapter discusses retinal gene therapy and outlines an approach to clinical trials for genetic ophthalmic diseases.

Background to retinal gene therapy trials

The first retinal gene therapy trials around 2005 used adenoviral vectors to deliver DNA-encoding proteins that might have a therapeutic effect in disease states. These proteins included pigment epithelial growth factor (a molecule that inhibits blood vessel growth in age-related macular degeneration) and thymidine kinase (to ablate retinoblastoma tumour cells). Notably, these early trials used viral vector technology to deliver DNA that encoded a therapeutic protein rather than giving the protein endogenously, such as by injection.

Adenovirus however was found to be immunogenic even in the eye. Subsequent research efforts revealed the potential of adeno-associated viral (AAV) vectors. This small single-stranded DNA virus has no intrinsic replication ability and most importantly, no cell membrane. It was found to have considerable benefits in broadly being able to evade the immune system when given in therapeutic doses by subretinal injection (see Table 8.1 for comparisons of retinal gene therapy vectors). In 2007, after preclinical testing, most notably in the retina of the Briard dog, three clinical trials using AAV and aiming to replace the missing *RPE65* gene in human retinas were initiated. *RPE65* was chosen due to its critical role in the visual cycle and its role in Leber congenital amaurosis. The treated Briard dogs, which carried the same gene defect, had improved vision in dim light. This provided an instant and relatively straightforward endpoint that could be assessed in clinical trials. The US Food and Drug Administration (FDA) subsequently recognised the maze navigation test as another appropriate endpoint. Regulatory approval for the AAV gene therapy with *RPE65* was obtained in 2017 in the United States and 2018 in the European Union. The AAV *RPE65* gene therapy product voretigene neparvovec is now known commercially as Luxturna[1] and is currently licensed for use in many countries.

This represented the first FDA-approved gene replacement therapy, setting a precedent for other trials. It is therefore worth considering some novel aspects of this approval:

- The endpoint was an improvement in visual function. There was no obligation to demonstrate a slowing of retinal degeneration, which would have taken many years to validate and which might not be practical for a company to undertake. All the animal models to date have shown slowing of retinal degeneration when the AAV vector has been administered correctly and at a sufficient dose to ensure the majority of photoreceptors express the transgene.
- The company running the trial (Spark Therapeutics) was able to demonstrate to the FDA that another new test (the maze navigation test) could provide a meaningful clinically relevant endpoint for a phase III clinical trial, in place of traditional outcomes, such as visual acuity and visual field.
- The FDA approval also included recognition of the AAV vector production process. This has set the standard about biological properties such as genome titre, full-to-empty capsid ratio and efficacious dose for targeting the RPE (10^{10} to 10^{11} genome particles of AAV2).

Subsequently, retinal gene therapy clinical trials using AAV2 have started for choroideremia[2], X-linked retinitis pigmentosa replacing *RPGR*[3] and achromatopsia[4] replacing *CNGA3* and *CNGB3*. For an up-to-date list of ongoing trials, the reader is referred to https://clinicaltrials.gov/.

TABLE 8.1 Comparison of viral vectors used in subretinal gene therapy trials to date.

Vector and serotype	Transgene delivered	Cellular tropism	Advantages	Disadvantages
Adenovirus	Double-stranded DNA	Presumed any	Large (up to 50 kb) payload	Highly immunogenic
Lentivirus	RNA (with reverse transcriptase)	RPE	Up to 9 kb payload	Immunogenic, cannot easily diffuse across outer segments to transduce photoreceptors
AAV2	Single-stranded DNA	Rods and RPE	Low immunogenicity	Limited payload (up to 4.7 kb) and poor targeting of cones
AAV8	Single-stranded DNA	Rods, cones and RPE	Low immunogenicity	Limited payload
AAV2/8 capsid mutant	Single-stranded DNA	Rods, cones and RPE	Increased transduction efficiency	Limited payload and more immunogenic with capsid mutation

Ethical approval of a retinal gene therapy clinical trial

For any clinical trial, or indeed observational study, appropriate research ethics committee review and approval are required. For studies involving gene therapy, a key focus of this process should be patient consent. Trial participants should be fully informed about the risks of participating in the trial. Regulatory authority will also independently assess the safety of the viral vector production process. For example, in the United Kingdom this is the Medicines and Healthcare products Regulatory Authority (MHRA) and in the United States this would be the FDA. Whilst the standards for some processes differ between the European Medicines Agency (EMA) and the US FDA (e.g. how distilled water is prepared; strength of fans in tissue culture hoods), the processes can be set to standards such that vector production is aligned.

It has been repeatedly noted that the results of clinical trials in humans are unpredictable and differ from those in animal models. The ethics approval process is in place to support the need to justify risky procedures in human subjects. Furthermore, dose escalation trials are designed to enable novel therapeutics to move towards approval in a safe and measured fashion. Generally, phase I trials are primarily designed to demonstrate safety. Outcome measures and dose are assessed more closely in phase II studies (although some trials use a dose escalation approach and merge phases I and II). Efficacy is evaluated in phase III clinical trials using clinical endpoints accepted by the regulators and usually involving randomisation with blinded observers.

Clinical trial endpoints

Outcomes of clinical studies are reported in terms of clearly defined endpoints. These determine the success of a trial, and ultimately, the approval of a new treatment.[5] Choosing appropriate and reliable study endpoints is therefore critical. Endpoints, primary and secondary, are objective measures that must be clinically meaningful and relevant to be accepted by the regulatory agency. There can only be one primary endpoint, which determines the overall success of the study. Secondary endpoints assess additional trends that are supportive or can be correlated to the primary endpoint.

Historically, primary endpoints in vision research involve validated tests of visual function. Thus, the primary endpoint is usually defined as a change from baseline in best-corrected visual acuity in the treated eye. A significant change in vision is usually demonstrated when there is doubling (or halving) of the visual angle as demonstrated by a gain (or loss) of 15 letters on the Early Treatment Diabetic Retinopathy Study (ETDRS) chart. This can be particularly challenging for chronic conditions such as retinal dystrophies, which generally progress very slowly over several years. In addition, for those patients who enter the trial with normal visual acuity, any effect of gene therapy will need to rely on the treated eye preserving vision in comparison to the non-treated eye.

Regulatory authorities, recognising shortfalls of visual acuity testing, are increasingly accepting surrogate endpoints that correlate with visual function. Such endpoints include retinal sensitivity (central visual field as determined by microperimetry as well as traditional visual field tests), mobility tests (such as multi-luminance mobility test, MLMT) (Fig. 8.1), virtual street testing and structural endpoints.

- Mobility tests, such as MLMT, provide a measure of vision-related tasks in a setting that reflects conditions of daily living. Specifically, MLMT measures the speed and accuracy with which patients can navigate a mobility course under a range of lighting conditions. This is particularly appropriate for patients with a poor level of vision at the outset of a clinical trial. In these cases, improvements in ambulatory vision would signify a beneficial effect of a gene-based treatment.
- Structural endpoints, such as OCT and fundus autofluorescence imaging, give objective measures of retinal structure. They are both non-invasive and easily obtained in the clinic. These endpoints are particularly important in the context of slowly progressing retinal degenerations where anatomical changes can often precede functional loss.
- In addition to the currently accepted endpoints, it is increasingly recognised that patient-reported outcome measures from individuals taking part in clinical trials are key. Subjective data describing the quality of visual improvement after the delivery of treatment give considerable additional credibility to scientific findings.

Surgical considerations

The efficacy of retinal gene therapy is dependent on several factors including vector design and quality, appropriate timing of intervention and successful surgical delivery. The mode of gene delivery is critical in terms of both efficacy and safety and must be carefully considered when structuring a gene therapy trial. Two existing surgical techniques involve intravitreal and subretinal administration of viral vectors (Fig. 8.2).

- The intravitreal delivery is less technically challenging and has the potential for more widespread retinal gene

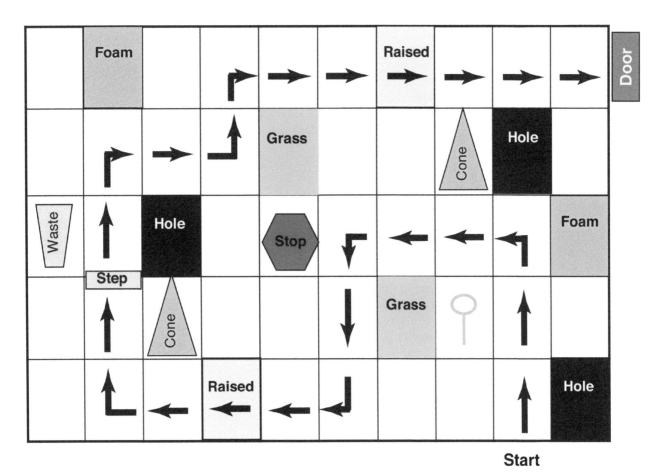

FIG. 8.1 Multi-luminance mobility test (MLMT) sample layout. This test was used as an efficacy endpoint in voretigene neparvovec trials. Patients were assessed according to their ability to quickly and accurately navigate a 5 ft. × 10 ft. course under increasing illumination. The custom-designed assessment was reconfigured before each run and standardised for each of 12 different configurations with a fixed number of turns and obstacles (including raised steps, elevated foam blocks, Styrofoam cones, a stop sign, a waist-high obstacle etc.). The assessment was validated for correlation with visual acuity and inter-rater agreement of successful completion. *(Adapted from Ref. 6.)*

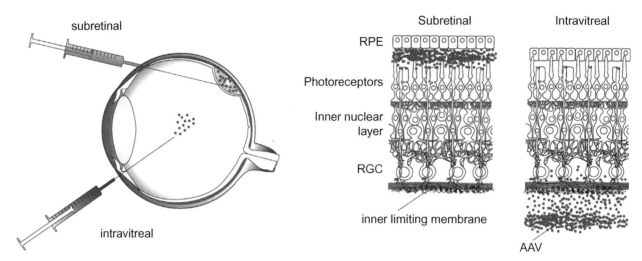

FIG. 8.2 Schematic drawing representing a subretinal and intravitreal injection of adeno-associated viral (AAV) vector particles into a human eye. *(Adapted from Ref. 7.)*

expression. It is currently the delivery of choice in several clinical trials including achromatopsia, X-linked retinoschisis, and Leber hereditary optic neuropathy. However, to be successful, the vector has to overcome the anatomical barriers at the vitreoretinal interface as well as an increased risk of ocular inflammation.

- The subretinal gene delivery was part of pioneering Leber congenital amaurosis studies and is currently used in clinical trials involving patients with other retinal disorders including choroideremia and X-linked retinitis pigmentosa. This approach delivers high viral loads to the target tissue and evades adaptive immune responses. Although generally well-tolerated, it necessitates a temporary retinal detachment which may be deleterious in the atrophic retinas of end-stage degenerations.[8] The technique of subretinal gene delivery involves pars plana vitrectomy followed by OCT guided two-step subretinal injection, first with normal saline to induce retinal detachment followed by slow infusion of a viral vector. It is imperative that, at preoperative planning, retinal structural integrity is demonstrated within the treatment zone (using OCT and autofluorescence imaging) to minimise effects of iatrogenic pressure and to allow for safe vector delivery.

The experience from early clinical trials has given us insight into the importance of the timing of gene therapy intervention. Early trials involved patients with advanced disease but emerging safety and efficacy data suggest that intervening at earlier stages may be beneficial to capture retinal cells before irreversible damage has occurred.

Concluding remarks

Over the past two decades, there has been tremendous progress in the field of gene therapy. It is now possible to target retinal disorders with molecular precision at the level of the individual genetic change. As a result, the dynamics in genetic retinal clinics have changed with the focus shifting towards establishing a precise and accurate genetic diagnosis, and towards enrolling participants into treatment trials and natural history studies. We expect that the success of future retinal gene therapy trials will depend on efforts to: improve clinical trial designs; validate and approve reliable trial endpoints; optimise viral vector design and surgical techniques.

References

1. Russell S, Bennett J, Wellman JA. Efficacy and safety of voretigene neparvovec (AAV2-hRPE65v2) in patients with RPE65-mediated inherited retinal dystrophy: a randomised, controlled, open-label, phase 3 trial. *Lancet* 2017;**390**(10097):849–60.
2. Xue K, et al. Beneficial effects on vision in patients undergoing retinal gene therapy for choroideremia. *Nat Med* 2018;**24**:1507–12.
3. Cehajic-Kapetanovic J, et al. Initial results from a first-in-human gene therapy trial on X-linked retinitis pigmentosa caused by mutations in RPGR. *Nat Med* 2020;**26**(3):354–9.
4. Fischer MD, et al. Safety and vision outcomes of subretinal gene therapy targeting cone photoreceptors in achromatopsia: a nonrandomized controlled trial. *JAMA Ophthalmol* 2020;**138**(6):643–51.
5. Csaky K, Ferris 3rd F, Chew EY, Nair P, Cheetham JK, Duncan JL. Report from the NEI/FDA endpoints workshop on age-related macular degeneration and inherited retinal diseases. *Invest Ophthalmol Vis Sci* 2017;**58**(9):3456–63.
6. Darrow JJ. Luxturna: FDA documents reveal the value of a costly gene therapy. *Drug Discov Today* 2019;**24**(4):949–54.
7. Dalkara D, Sahel JA. Gene therapy for inherited retinal degenerations. *C R Biol* 2014;**337**(3):185–92. https://doi.org/10.1016/j.crvi.2014.01.002.
8. Xue K, Groppe M, Salvetti AP, MacLaren RE. Technique of retinal gene therapy: delivery of viral vector into the subretinal space. *Eye (Lond)* 2017;**31**(9):1308–16.

Section II

Genetic disorders affecting the anterior segment

Chapter 9

Genetic disorders affecting the cornea

Chapter outline

9.1 Primary megalocornea 68
9.2 Brittle cornea syndrome 70
9.3 Corneal dystrophies 73
9.4 Keratopathy in inborn errors of metabolism 85
9.5 Keratopathy in ectodermal dysplasias 93

Corneal abnormalities are an important feature of a wide range of genetic disorders, reaching beyond the relatively small group of corneal dystrophies. The underlying molecular defect has been identified for many of these conditions and genetic testing can help obtain a timely diagnosis and inform management.

Assessment of corneal size may highlight microcornea (Fig. 9.1), often associated with congenital cataracts (Section 11.1) and/or microphthalmia (Section 16.1), or megalocornea (Fig. 9.2). The latter can be a manifestation of raised intraocular pressure in early life and hence a sign of congenital glaucoma (Section 10.1). An enlarged corneal diameter may also be noted in *LTBP2*-associated disease (megalocornea with zonular weakness and secondary lens-related glaucoma) (Sections 10.1 and 12.3) or, it can occasionally be a benign and isolated finding caused by mutations in the *CHRDL1* gene (Section 9.1).

The corneal dystrophies (Section 9.3) are a group of precisely delineated genetic conditions whose classification and recognition are strongly assisted by molecular analysis (Table 9.1). Genetic testing can inform prognosis and genetic counselling and can occasionally facilitate the diagnosis of gelsolin amyloidosis (also known as Meretoja syndrome or familial amyloidosis, Finnish type), a multisystemic disorder leading to corneal abnormalities that are virtually indistinguishable from those seen in lattice corneal dystrophy (Box 9.1).

Corneal clouding in childhood has a broad differential diagnosis including several genetic conditions. Among others, it can be caused by congenital glaucoma (Section 10.1), congenital hereditary corneal endothelial dystrophy (Section 9.3) or inborn errors of metabolism such as mucopolysaccharidoses, mucolipidoses or cystinosis (Section 9.4).

Other multisystemic disorders that affect the cornea include brittle cornea syndrome, a connective tissue disease associated with significant corneal thinning (Section 9.2), and ectodermal dysplasias, associated with limbal stem cell deficiency (Section 9.5). The latter can also be a feature of aniridia (Chapter 17).

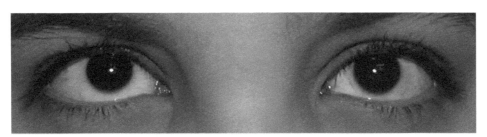

FIG. 9.1 External photograph of a 12-year-old male who has isolated microcornea and an otherwise unremarkable eye exam. The horizontal corneal diameter was 9.5 mm in each eye. *(Courtesy of Professor Arif O. Khan.)*

9.1

Primary megalocornea

Graeme C.M. Black

Primary megalocornea is a bilateral, non-progressive condition characterised by enlarged corneal diameter in the absence of raised intraocular pressure. It is typically non-syndromic and inherited as an X-linked recessive trait. As a result, most affected individuals are male but somewhat enlarged corneas may occasionally be noted in female carriers.

Clinical characteristics

Children with primary megalocornea are usually asymptomatic and have well-preserved corrected visual acuities. Occasionally there are parental concerns about the appearance of the eyes and/or there might be a relevant family history. However, it is not unusual for the condition to be overlooked early on.

Bilateral, symmetrical corneal enlargement is present from birth and the horizontal corneal diameter is often between 13 and 16.5mm (Fig. 9.2). The central corneal thickness is reduced (which may complicate the assessment of glaucoma risk) or within normal limits. The corneal endothelial cell density is usually normal. The anterior chambers are deep but the vitreous chamber length is often decreased; this results in a relatively normal overall axial length and a posteriorly positioned iris–lens diaphragm. Astigmatism and myopia are common and the intraocular pressures are within normal limits. In general, affected individuals are thought to have increased susceptibility to developing early cataracts, retinal detachment, and glaucoma.

It is important to differentiate primary megalocornea from globe enlargement due to congenital glaucoma (Section 10.1); the latter usually requires prompt intervention whereas megalocornea does not require treatment. The clinical history and main signs are different. Congenital glaucoma is typically symptomatic and often displays asymmetry; the intraocular pressures are raised and the increase in anterior chamber depth is less pronounced compared to megalocornea. A useful distinguishing feature is the presence of an anterior chamber depth to total axial length ratio below 0.19.[2]

Molecular pathology

Primary megalocornea is caused by pathogenic variants in the *CHRDL1* gene, which is located in the X chromosome. The encoded protein that has a role in the regulation of corneal and anterior segment growth, and retinal angiogenesis.[3] A range of mutations, presumed to result in loss of ventroptin function, have been described. The identification of a disease-causing variant in *CHRDL1* in an affected individual clarifies the diagnosis and

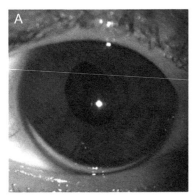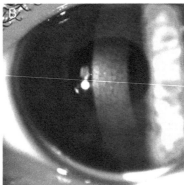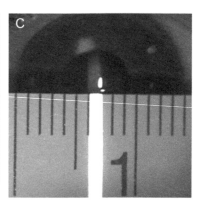

FIG. 9.2 Slit-lamp imaging from a 48-year-old visually asymptomatic male who has X-linked primary megalocornea and carries a loss of function variant in the *CHRDL1* gene in hemizygous state. Prominent arcus lipoides (A), crocodile shagreen (B), and horizontal corneal diameters of 14 mm (C) were noted. *(Adapted from Ref. 1.)*

inheritance pattern and facilitates estimation of visual prognosis (i.e. reducing glaucoma risk).

An intellectual disability syndrome with megalocornea has been described (Neuhauser syndrome). However, no disease-causing gene has been identified for this syndrome and, in one published example, it was demonstrated that the affected individual carried a mutation in *CHRDL1* and likely had two separate genetic conditions.[2]

Several children with megalocornea are found to have biallelic mutations in the *LTBP2* gene. In these cases, a wide range of developmental abnormalities is typically present including microspherophakia, zonular weakness and secondary lens-related glaucoma (Sections 10.1 and 12.3).[4]

Clinical management

Primary megalocornea is generally managed with routine refractive error correction to prevent amblyopia. Regular follow-up to monitor for cataracts and glaucoma should be considered. The surgical management of lens opacification in individuals with megalocornea is challenging due to zonular weakness. There is an increased risk of lens displacement and a modified technique should be considered (e.g. involving anchoring of the lens implant optic).

References

1. Liu S, Sergouniotis PI, Black GC. Primary X-linked megalocornea presenting in adulthood. *Am J Ophthalmol* 2021;**S0002-9394**(21): 0047907.
2. Davidson AE, Cheong SS, Hysi PG, Venturini C, Plagnol V, Ruddle JB, et al. Association of CHRDL1 mutations and variants with X-linked megalocornea, Neuhäuser syndrome and central corneal thickness. *PLoS One* 2014;**9**(8), e104163.
3. Webb TR, Matarin M, Gardner JC, Kelberman D, Hassan H, Ang W, et al. X-linked megalocornea caused by mutations in CHRDL1 identifies an essential role for ventroptin in anterior segment development. *Am J Hum Genet* 2012;**90**(2):247–59.
4. Khan AO, Aldahmesh MA, Alkuraya FS. Congenital megalocornea with zonular weakness and childhood lens-related secondary glaucoma—a distinct phenotype caused by recessive LTBP2 mutations. *Mol Vis* 2011;**17**:2570–9.

9.2

Brittle cornea syndrome

Graeme C.M. Black, Emma Burkitt-Wright, and Susmito Biswas

Brittle cornea syndrome is a multisystemic connective tissue disorder that is characterised by extreme corneal thinning and confers a high risk of globe rupture, often at an early age. Common extraocular features include deafness, skeletal abnormalities, joint hypermobility, and skin hyperelasticity. Early recognition can lead to optimal management including appropriate corneal protection.

Brittle cornea syndrome is inherited as an autosomal recessive trait and has been associated with biallelic mutations in the ZNF469 and PRDM5 genes.

Clinical characteristics

Brittle cornea syndrome should be suspected in patients with bilateral corneal thinning. Central corneal thickness under 400 µm was observed in all affected individuals in one series.[1] This is associated with a significant risk of perforation either spontaneously or from minor trauma (Fig. 9.3); such complication has been observed in children as young as 2 years of age.[3] As a result of the reduction in corneal thickness, patients are often highly myopic. Keratoconus and keratoglobus have both been described. Some patients have blue sclerae although this generally becomes less noticeable with age.[3,4]

Bilateral hearing impairment is a frequent association; it can be severe/profound and it may be present in early childhood.[3,4] There is a mixed picture with conductive and neurosensory components. Tympanometry may demonstrate hypermobile tympanic membranes. Skeletal features such as developmental dysplasia of the hip and scoliosis are also common. Repeated fractures have been reported in some patients. Small joint hypermobility is a significant problem for affected individuals; mild contractures of the small fingers, as well as hallux valgus (i.e. deformity of the joint connecting the big toe to the foot), have been noted. Patients often describe easy bruising and thin skin although skin healing and scarring are normal.

Brittle cornea syndrome was formerly classified as part of the kyphoscoliotic Ehlers Danlos syndrome (previously termed Ehlers Danlos syndrome type VI) spectrum of disorders although it is now thought of as a separate condition. Nonetheless, kyphoscopiotic Ehlers Danlos syndrome, a group of connective tissue diseases that are associated with musculoskeletal, vascular and ocular manifestations including blue sclerae, remains amongst the differential diagnoses. It is worth noting that, unlike Ehlers Danlos syndrome, brittle cornea syndrome has a normal urine lysyl pyridinoline: hyroxylysyl pyridinoline ratio. Also, the extraocular manifestations of brittle cornea syndrome are usually milder than those observed in kyphoscoliotic Ehlers Danlos syndrome, although significant cardiac valvular dysfunction (particularly frequently of the mitral valve) is known to occur.

Molecular pathology

Brittle cornea syndrome is caused by biallelic, presumed loss of function mutations of either ZNF469 or PRDM5.[2,5,6] These relatively large genes are often screened as part of a panel of genes implicated in corneal disease.

ZNF469 was the first gene to be associated with brittle cornea syndrome and encodes a zinc finger protein of unknown function.[5,6] The PRDM5 gene encodes a DNA-binding transcription factor involved in the regulation of collagen and extracellular matrix glycoprotein genes. Indeed, fibroblasts from patients with brittle cornea syndrome show significantly reduced expression of extracellular matrix components including collagen types I, III, and V.[2]

Clinical management

A management checklist has been published[3,4] and some key points are discussed below.

Early identification is essential and will allow advice of lifestyle that recognises the risk of even minor trauma (avoidance of sport/physical play). The use of protective spectacles is recommended. It is critical to be proactive about the examination and, where possible, genetic testing of family members at risk. Care is recommended particularly during manoeuvres that may result in increased intraocular pressure (e.g. coughing, including following extubation after general anaesthesia) as these may risk consequent globe rupture.

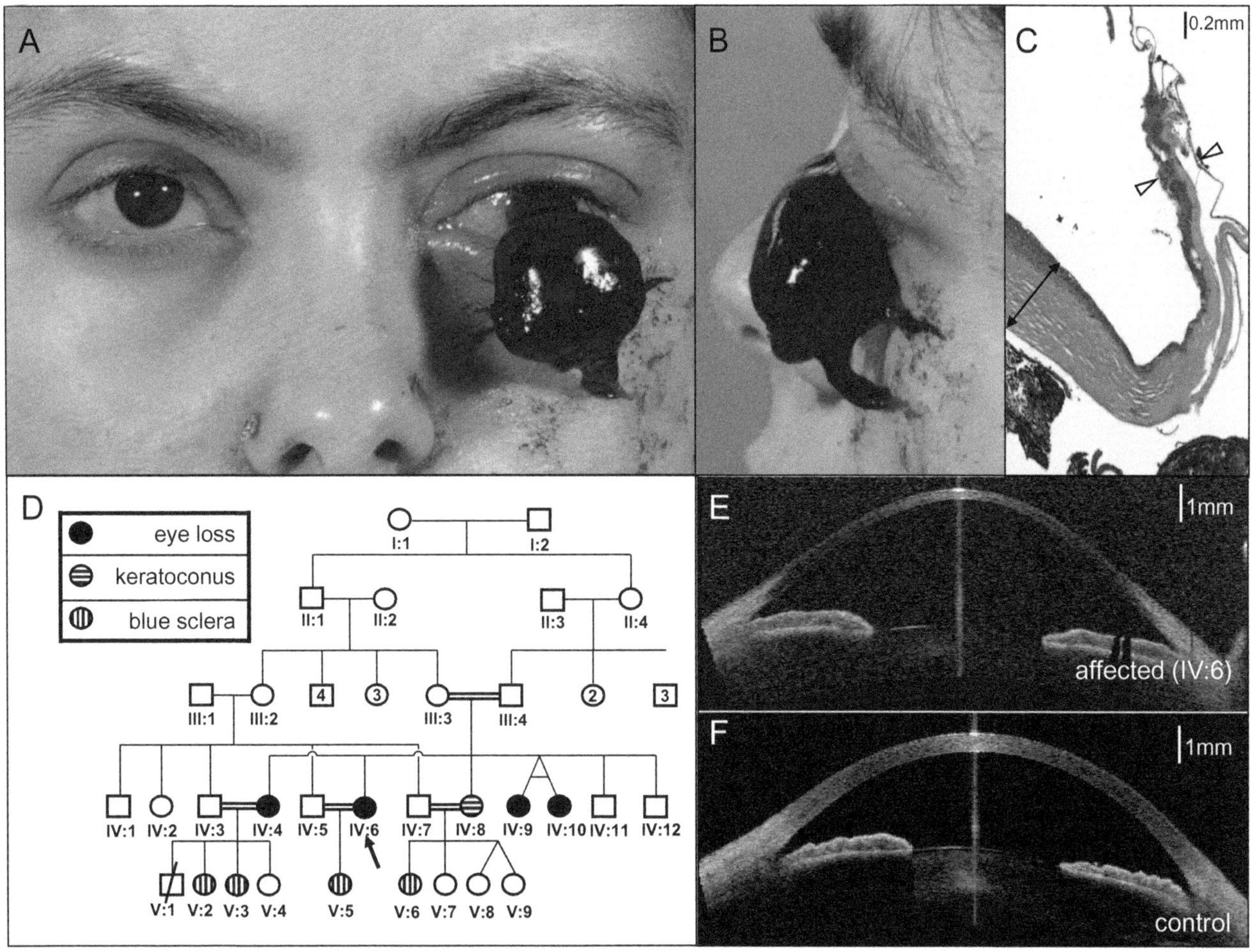

FIG. 9.3 Extreme corneal fragility in an individual with *PRDM5*-associated brittle cornea syndrome. (A and B) Expulsive haemorrhage affecting the left eye. Note also the blue-grey sclera in the right eye. (C) Histopathological *PRDM5*-associated findings (40 × magnification of haematoxylin and eosin-stained section) in the proband's eviscerated cornea. Note the relative preservation of peripheral corneal thickness (indicated by *double-headed arrow*) in contrast to extreme thinning of the central cornea in the region of rupture (indicated by *open arrowheads*). The stroma (stained *pink*) is almost absent in that location: much of the remaining corneal thickness derives from epithelial and endothelial layers. (D) The pedigree of the affected family. Filled shapes indicate affected individuals, as assessed by the history of the previous enucleation after minor trauma or phthisis. Blue sclerae and keratoconus in heterozygous individuals without corneal rupture are also indicated. The proband (IV:6) is indicated by an *arrow*. (E) Optical coherence tomography demonstrating the cross-sectional appearance of the right eye of the proband; note the extreme thinning of the central cornea compared to that in the control image (F). *(Adapted from Ref. 2.)*

Surgical management is challenging due to the extreme thinness of the cornea which can result in rupture simply following suture placement or rotation. Onlay corneoscleral grafting (epikeratoplasty) using a limbal sparing technique has been advocated (where a large diameter donor corneoscleral button is sutured over the top of the host's cornea).[7] Corneal cross-linking has been tried as a possible means of increasing corneal strength. Given the extent of corneal thinning in these patients, a modified protocol using low power ultraviolet light (UV-A) radiation for short intervals needs to be used (to reduce the risk of corneal endothelial damage). Importantly, the long-term efficacy of this treatment has not yet been established.[8]

While small corneal perforations may be potentially repaired with tissue adhesives and bandage contact lens, the repair of corneoscleral rupture in brittle cornea syndrome is complicated by cheese-wiring of the sutures and persistent leakage through suture tracks. A technique of using non-expansile perfluorocarbon gas tamponade and posturing postoperatively has been shown to prevent aqueous egress through corneal suture tracks enabling sufficient time for wound healing.[9]

As with all multisystemic conditions, a coordinated multidisciplinary approach is required. This may involve:

- Clinical genetics: for genetic counselling and molecular diagnosis.

- Paediatrics: mainly for musculoskeletal assessment (including monitoring for developmental hip problems and scoliosis).
- ENT/audiology: for the the assessment and management of hearing loss, and potentially dual sensory loss.
- Cardiology: assessment is advised as mitral valve dysfunction has been reported.
- Rheumatology: rheumatology or metabolic bone specialist input may be needed if problematic joints, osteopenia, or recurrent fractures are present.
- Obstetrics: obstetric and perinatal problems have been described including premature rupture of membranes.[3,4]

References

1. Eleiwa T, Raheem M, Patel NA, Berrocal AM, Grajewski A, Abou SM. Case series of brittle cornea syndrome. *Case Rep Ophthalmol Med* 2020;**2020**:4381273.
2. Burkitt Wright EMM, Spencer HL, Daly SB, Manson FDC, Zeef LAH, Urquhart J, et al. Mutations in PRDM5 in brittle cornea syndrome identify a pathway regulating extracellular matrix development and maintenance. *Am J Hum Genet* 2011;**88**(6):767–77.
3. Walkden A, Burkitt-Wright E, Au L. Brittle cornea syndrome: current perspectives. *Clin Ophthalmol* 2019;**13**:1511–6.
4. Burkitt Wright EM, Porter LF, Spencer HL, Clayton-Smith J, Au L, Munier FL, et al. Brittle cornea syndrome: recognition, molecular diagnosis and management. *Orphanet J Rare Dis* 2013;**8**:68.
5. Rohrbach M, Spencer HL, Porter LF, Burkitt-Wright EM, Bürer C, Janecke A, et al. ZNF469 frequently mutated in the brittle cornea syndrome (BCS) is a single exon gene possibly regulating the expression of several extracellular matrix components. *Mol Genet Metab* 2013;**109**(3):289–95.
6. Abu A, Frydman M, Marek D, Pras E, Nir U, Reznik-Wolf H, et al. Deleterious mutations in the zinc-finger 469 gene cause brittle cornea syndrome. *Am J Hum Genet* 2008;**82**(5):1217–22.
7. Muthusamy K, Tuft S. Use of onlay corneal lamellar graft for brittle cornea syndrome. *BMJ Case Rep* 2018;**2018**, bcr2017223824. Published 2018 Aug 16 https://doi.org/10.1136/bcr-2017-223824.
8. Kaufmann C, Schubiger G, Thiel MA. Corneal cross-linking for brittle cornea syndrome. *Cornea* 2015;**34**(10):1326–8.
9. Hussin HM, Biswas S, Majid M, Haynes R, Tole D. A novel technique to treat traumatic corneal perforation in a case of presumed brittle cornea syndrome. *Br J Ophthalmol* 2007;**91**(3):399.

9.3

Corneal dystrophies

Petra Liskova, Susmito Biswas, Graeme C.M. Black, and Francis L. Munier

Corneal dystrophies are a clinically and genetically heterogeneous group of hereditary disorders. They are generally considered to be bilateral, relatively symmetrical and not directly linked to environmental or systemic factors. Importantly, they exhibit marked variability in disease onset, progression and severity of visual impairment; the clinical presentation can be significantly different, even among affected individuals from the same family.

Numerous corneal dystrophy subtypes have been described with Fuchs endothelial corneal dystrophy being the most common. A classification system has been proposed by the International Committee for Classification of Corneal Dystrophies (IC3D). IC3D integrated clinical, pathological, and genetic information and provided a set of nomenclature recommendations that have been widely accepted.[1] At a high level, corneal dystrophies can be classified as:

- epithelial and subepithelial dystrophies,
- epithelial-stromal *TGFBI* (transforming growth factor β-induced) dystrophies,
- stromal dystrophies,
- endothelial dystrophies (Table 9.1).

In this chapter, we discuss the most common subtypes. A more comprehensive overview can be found at https://www.corneasociety.org/publications/ic3d.

Although patients with corneal dystrophies may initially be asymptomatic, the ophthalmic review is often prompted by the eventual development of visual disturbances and/or ocular surface irritation. The clinical diagnosis is made based on the characteristic features observed on slit-lamp examination (using direct beam and/or retro-illumination). Pachymetry and keratometry measurements may also provide valuable clues.[1-3] Examination of first-degree relatives is important and identification of other affected family members often confirms the genetic nature of the condition.

Genetic testing is particularly helpful in confirming the clinical diagnosis and excluding superficially similar conditions such as paraproteinemic keratopathy. In addition, it facilitates the differentiation between subtypes that may be indistinguishable upon clinical examination. For example, both congenital hereditary endothelial dystrophy and posterior polymorphous corneal dystrophy can manifest as corneal oedema detected at, or soon after, birth making a differential diagnosis based on examination alone challenging.

Management of corneal dystrophies at initial stages is typically conservative. Lamellar or penetrating keratoplasty may be indicated for advanced disease (e.g. if the corrected visual acuity is significantly decreased and/or if there is intractable pain or photophobia). Recurrence can occur post-procedure and the chances of this need to be considered individually for each condition. It is worth highlighting that in patients with *TGFBI*-associated corneal dystrophies, certain types of refractive surgery (including LASIK) may lead to keratocyte activation and rapid disease progression.[4]

TABLE 9.1 Corneal dystrophies with an established molecular pathology.

	Gene	Inheritance
Epithelial and subepithelial		
Meesmann epithelial corneal dystrophy, type 1	*KRT12*	AD
Meesmann epithelial corneal dystrophy, type 2	*KRT3*	AD
Epithelial recurrent erosion dystrophy	*COL17A1* (majority unknown)	AD
Epithelial basement membrane dystrophy	*TGFBI* (majority unknown)	Sporadic and AD
Lisch epithelial corneal dystrophy	Unknown	X-linked
Gelatinous drop-like corneal dystrophy	*TACSTD2*	AR
Epithelial-stromal		
Reis–Bücklers corneal dystrophy	*TGFBI*	AD
Thiel–Behnke corneal dystrophy	*TGFBI*	AD
Granular corneal dystrophy, type 1	*TGFBI*	AD
Granular corneal dystrophy, type 2	*TGFBI*	AD
Lattice corneal dystrophy, type 1 (classic) and variants	*TGFBI*	AD
Stromal		
Macular corneal dystrophy	*CHST6*	AR
Schnyder stromal corneal dystrophy	*UBIAD1*	AD
Congenital stromal corneal dystrophy	*DCN*	AD
Fleck corneal dystrophy	*PIKFYVE*	AD
Posterior amorphous corneal dystrophy	Deletion of *EPYC, KERA, LUM, DCN*	AD
Endothelial		
Fuchs endothelial corneal dystrophy, type 1	*COL8A2*	AD
Fuchs endothelial corneal dystrophy, type 3	*TCF4*	AD
Fuchs endothelial corneal dystrophy, type 4	*SLC4A11*	Not specified
Fuchs endothelial corneal dystrophy, type 6	*ZEB1*	AD
Fuchs endothelial corneal dystrophy, type 8	*AGBL1*	AD
Posterior polymorphous corneal dystrophy, type 1	*OVOL2*	AD
Posterior polymorphous corneal dystrophy, type 3	*ZEB1*	AD
Posterior polymorphous corneal dystrophy, type 4	*GRHL2*	AD
Congenital hereditary corneal endothelial dystrophy	*SLC4A11*	AR
X-linked endothelial corneal dystrophy	Unknown	X-linked

AD, autosomal dominant; *AR*, autosomal recessive.

Meesmann epithelial corneal dystrophy

The term Meesmann epithelial corneal dystrophy (MECD) has been used to describe an autosomal dominant disorder that predominantly affects the corneal epithelium.[1–3] Mutations in two keratin genes, *KRT3* and *KRT12*, have been implicated in this condition.

Clinical characteristics

Slit-lamp examination of affected individuals reveals multiple tiny intraepithelial vesicles surrounded by clear epithelium and affecting the entire cornea (Fig. 9.4A). These microcysts can be visible early in life (e.g. from 12 months

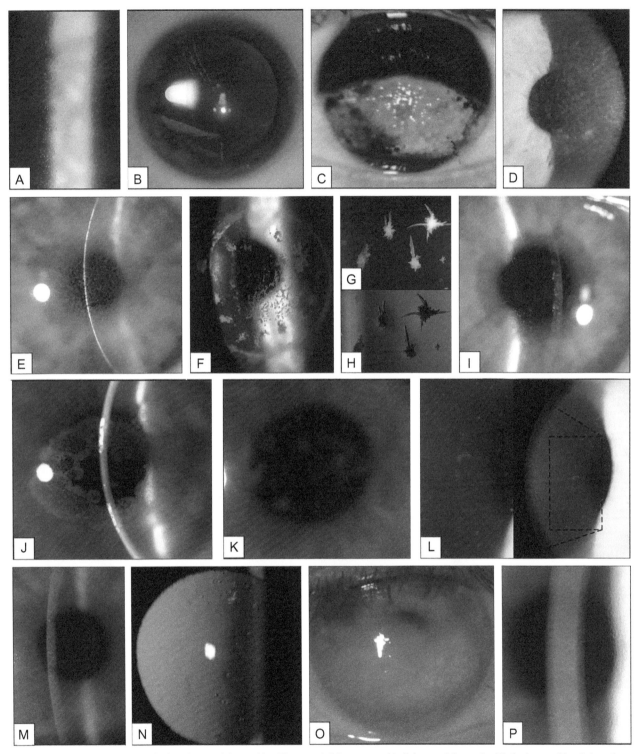

FIG. 9.4 Slit-lamp photographs of selected corneal dystrophies. Fine cysts in Meesmann epithelial corneal dystrophy (A); feathery and band-shaped grey opacity in Lisch epithelial corneal dystrophy (courtesy of C. Rapuano and W. Lisch) (B); subepithelial nodules in Gelatinous drop-like corneal dystrophy (C); honeycomb pattern in Thiel–Behnke corneal dystrophy (D); superficial geographical opacities in Reis–Bücklers corneal dystrophy (E); recurrence of stromal deposits in granular corneal dystrophy type I (F); granular dystrophy type II under diffuse light (G) and retro-illumination (H); fine translucent lattice lines and dots in a variant of lattice corneal dystrophy (I); central ring opacity with superficial crystalline deposits in Schnyder corneal dystrophy (J); central stromal opacities in macular corneal dystrophy (K); fine stromal opacities in fleck corneal dystrophy (L); milk stromal appearance in congenital hereditary endothelial dystrophy (M); vesicular and geographical lesions in retro-illumination (N) and corneal oedema with secondary subepithelial band keratopathy in posterior polymorphous corneal dystrophy (O); numerous guttae in Fuchs endothelial corneal dystrophy (P).

of age). However, patients are often asymptomatic throughout childhood. In adult life, the cysts rupture and may lead to symptoms resulting from an unstable corneal epithelium; these include recurrent erosions (pain, tearing) and scarring (photophobia, glare sensitivity, reduced vision). However, a completely asymptomatic clinical course is not uncommon. On light and electron microscopy the epithelium appears disordered with keratinocytes showing vacuolisation and the presence of dense intracytoplasmic inclusions that represent clumping of keratin filaments.[5]

Molecular pathology

MECD is caused by heterozygous mutations in either the KRT3 or the KRT12 gene encoding keratin K3 and K12 respectively.[6] Keratins are a heterogeneous group of proteins that form the structural framework of epithelial cells. Different combinations of keratins are found in different tissues. In each tissue, a type I keratin (acidic) pairs with a type II keratin (basic) to form heterodimers. These heterodimers interact with one another to form the intracytoplasmic 10 nm intermediate filament meshwork that provides tissue stability and resilience from mechanical stress. Several Mendelian blistering disorders, including the skin disorder epidermolysis bullosa simplex, are caused by mutations in keratins. Such mutations cause the basal epidermis to be prone to damage from mechanical stress, leading to increased skin fragility and blistering.

Corneal keratinocytes mostly express the keratin pair K3 and K12. This is the major keratin pair in all the layers of mature corneal epithelium and mutations in either of these cornea-specific molecules can lead to MECD. Each keratin protein comprises a rod domain flanked by highly conserved helix-initiation and helix-termination motifs that are required for heterodimerisation and keratin fibre assembly. These motifs are hotspots for missense mutations, which represent the majority of MECD-implicated variants. By altering multimerisation, these missense changes result in intracellular clumping of keratins and a failure to assemble normal intermediate filament structures. This points to the fact that mutations causing MECD act in a dominant negative fashion (i.e. the mutant proteins act antagonistically to the normal gene products). Generally, pathogenic variants in KRT12 are significantly commoner than those in KRT3.

Genetic testing of patients suspected of having MECD can help resolve diagnostic uncertainty and facilitates more accurate genetic counselling. Also, it might make a patient eligible for future gene-based therapeutic trials.[7,8]

Clinical management

MECD rarely requires treatment. Individuals affected by foreign body sensation, photophobia, and glare (noted when the epithelial inclusion cysts break onto the surface) may be treated palliatively with lubricants, or bandage contact lenses.[9]

Phototherapeutic keratectomy with or without mitomycin C has been tried but complications such as recurrence and subepithelial fibrosis are not uncommon.[10] Other interventions include lamellar and penetrating keratoplasty. These may be required for more severe phenotypes. However, recurrence is known to occur even within the grafts due to repopulation of the donor cornea with the host's defective epithelial cells (from their limbal stem cells).

Epithelial–stromal *TGFBI* corneal dystrophies

Historically, several autosomal dominant corneal disorders caused by mutations in the *TGBFI* gene had been recognised as distinct clinical entities. These include granular (types I, II, III), lattice (type I) and Thiel–Behnke corneal dystrophies.[1–3] The associated symptoms in this group of conditions depend on the initial site of pathology (predominantly subepithelial or predominantly stromal). Recurrence in corneal grafts (originating from the overlying epithelium) is common.

Clinical characteristics

Thiel–Behnke corneal dystrophy

The term Thiel–Behnke corneal dystrophy has been used to describe a superficial/epithelial *TGFBI*-related disorder, that is associated with recurrent erosions and may manifest in the first decades of life. In some cases, the clinical course may be mild or asymptomatic until there is a significant decrease in visual acuity. Slit-lamp examination reveals subepithelial reticular opacities producing a symmetric honeycomb appearance (Fig. 9.4D).

Reis–Bücklers corneal dystrophy

The term Reis–Bücklers corneal dystrophy (also known as granular corneal dystrophy type III) has been used to describe a superficial/epithelial *TGFBI*-related disorder, that is associated with recurrent erosions and manifests in early childhood. Slit-lamp examination shows confluent irregular subepithelial geographic-like opacities which emerge earlier than in Thiel–Behnke corneal dystrophy and lead to progressive visual deterioration (Figs. 9.4E and 9.5A).

Granular corneal dystrophy type I

The term granular corneal dystrophy type I has been used to describe a slowly progressive, predominantly stromal *TGFBI*-related disorder that may manifest as early as age 2 years with photophobia and pain accompanying recurrent erosions. The visual acuity usually remains good until the fifth decade of life. Clinically, a large number of stromal corneal opacities resembling bread crumbs with clear intervening stroma (Fig. 9.4F) can be observed; these increase in number and involve deeper layers of the stroma with age.

Granular corneal dystrophy type II

The term granular corneal dystrophy type II (also known as Avellino corneal dystrophy) has been used to describe a slowly progressive, predominantly stromal *TGFBI*-related disorder that may manifest during adolescence with photophobia, but can often remain largely asymptomatic. Granular corneal dystrophy type II is considerably milder than granular dystrophy type I and recurrent erosions occur only rarely. Vision decreases with age only if the central visual axis is affected. Slit-lamp examination typically reveals snowflake-like, ring-shaped or star-shaped lesions of fewer number than in granular corneal dystrophy type I (Figs. 9.4G, H and 9.5B).

Lattice corneal dystrophy type I and variants

The term lattice corneal dystrophy type I has been used to describe a slowly progressive, predominantly stromal *TGFBI*-related disorder that may become symptomatic in the first decade of life as a consequence of recurrent erosions. Superficial dots can be initially seen in the central cornea. These tend to accumulate forming radial branching lines. They generally start centrally and spread both centrifugally and from anterior to the posterior stroma. A diffuse stromal haze develops later causing visual loss, typically between the third and fourth decade of life (Figs. 9.4I and 9.5C).

Later onset lattice corneal dystrophy variants show lattice lines that may be either larger, thinner or almost absent; a deeper position within the stroma is common. Affected individuals have a significantly delayed disease onset and often do not develop corneal erosions.

Molecular pathology

This group of epithelial-stromal corneal dystrophies is caused by heterozygous mutations in the *TGFBI* gene. *TGFBI* is widely expressed and encodes an extracellular protein containing four FAS1 domains and, at the C (end) terminus, an RGD (Arg-Gly-Asp) motif that is capable of binding integrins (i.e. transmembrane receptors that facilitate cell–cell and cell–extracellular matrix adhesion).[11,12] TGFBI is the second most abundant protein in the cornea stroma where, in addition to integrins, it interacts with glycoproteins and collagens I, II, IV, VI, and XII. Although TGFBI is highly expressed in the cornea, it is not fully understood why, for a protein that is so ubiquitous, pathology is restricted to a single tissue. Furthermore, why different protein variants lead to different phenotypic manifestations is also incompletely understood. It is thought that missense changes causing granular, Thiel–Behnke and Reis–Bücklers corneal dystrophies alter protein solubility leading to aggregation of TGFBI while variants implicated in lattice corneal dystrophy the proteolytic processing of the TGFBI protein.

TGFBI-related dystrophies are all inherited in an autosomal dominant fashion and *de novo* (new) mutations are rare.[13–15] Over 70 pathogenic variants in *TGFBI* have been described. However, alterations of residues p.Arg124 and p.Arg555 (that are associated with multiple classical and recognisable phenotypes; Fig. 9.5) are by far the most

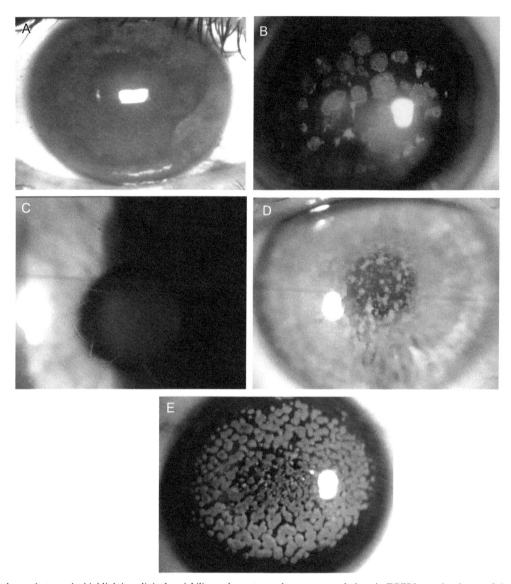

FIG. 9.5 Slit-lamp photographs highlighting clinical variability and genotype–phenotype correlations in *TGFBI*-associated corneal dystrophies. Four distinct clinical presentations result from different missense variants of the same residue: (A) Reis–Bücklers corneal dystrophy in a patient who is heterozygous for *TGFBI* c.371G > T (p.Arg124Leu); (B) granular dystrophy type II in a patient who is heterozygous for *TGFBI* c.371G > A (p.Arg124His); (C) lattice corneal dystrophy type I in a patient who is heterozygous for *TGFBI* c.370C > T (p.Arg124Cys); (D) granular dystrophy variant in a patient who is heterozygous for *TGFBI* c.370C > A (p.Arg124Ser). (E) Severe granular dystrophy type II in a patient who is homozygous for *TGFBI* c.371G>A (p.Arg124His).

common. Excluding variants in these two hotspots in the first instance is the preferred testing strategy in many diagnostic labs. It is worth noting though that additional consideration is required in cases with late onset subtypes of lattice dystrophy. This group of conditions is caused by a range of missense mutations (typically in the fourth FAS1 domain of *TGFBI*) and requires a different approach to testing that involves sequencing of a larger region of the gene.

Clinical management

Recurrent corneal erosions are common early in the course of many epithelial-stromal *TGFBI* dystrophies. Management of these can be undertaken using a step-wise approach starting with the regular use of ocular lubricants. Since waking up and rapid eye movements are felt to be associated with the onset of erosion, the use of lubricant ointment at night and drops first thing in the morning

BOX 9.1 Gelsolin amyloidosis— an important differential diagnosis of lattice corneal dystrophy

It is important to ensure that the autosomal dominant multisystemic disorder gelsolin amyloidosis (also known as Meretoja syndrome or familial amyloidosis, Finnish type) is not misdiagnosed as non-syndromic lattice corneal dystrophy.[16] Gelsolin amyloidosis is typically caused by a single recurrent missense mutation, c.640G > A (p.Asp187Tyr) in the gelsolin (GSN) gene. Clinically, it is associated with corneal amyloidosis that is present in a lattice pattern that is virtually indistinguishable from that seen in lattice corneal dystrophy; indeed, formerly the condition was known as lattice dystrophy type II. This disorder is associated with a large number of fine lattice lines in the corneal stroma (Fig. 9.6A). Notably, all patients with lattice corneal dystrophy who are found to lack a *TGFBI* mutation should be tested for *GSN*.

Gelsolin amyloidosis has neurological (peripheral neuropathy including cranial nerve VI and VII involvement; carpal tunnel syndrome), cardiac (amyloid cardiomyopathy), and renal (renal failure) features. Dermatological features (cutis laxa) and a characteristic 'mask-like' facial expression make the condition highly recognisable later in life (Fig. 9.6B). Identifying and treating some of these sequalae is crucial. The loss of corneal sensation is important to recognise, especially in those undergoing corneal transplantation.

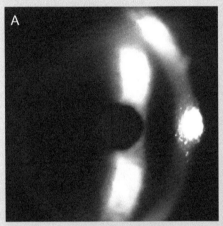
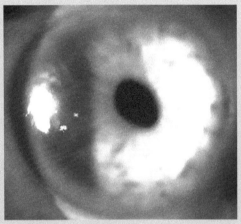
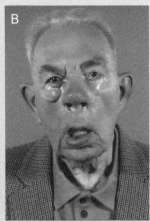

FIG. 9.6 Findings in gelsolin amyloidosis (Meretoja syndrome). (A) Slit-lamp photographs showing multiple fine lattice lines in the cornea of a 61-year-old with a longstanding history of cutis laxa at the elbows. Six years previously he was noted to have droopy eyelids that required corrective surgery. He also reported leaking food and drink when eating and had been found to have bilateral facial weakness. (B) Facial photograph of an 80-year-old affected individual illustrating smoothing of the brow, droopy face and lips.

may be helpful. Treatment of patients with an acute presentation of corneal erosion includes topical lubricants and topical antibiotic prophylaxis applied until the corneal epithelium has healed. Patching and cycloplegia have also been advocated. Therapeutic bandage contact lenses may provide quicker symptomatic relief, but ultimately, this approach is no more effective than the use of topical lubricants. Furthermore, the use of contact lenses may carry an additional risk of microbial keratitis. The practice of anterior stromal puncture may be considered unless the area involved is centrally located.[17]

Typically, the involvement of the central anterior stroma of the cornea results in visual symptoms as these diseases progress. Significant deterioration in vision may warrant intervention. Phototherapeutic keratectomy using excimer laser for example has become an established technique to treat corneal opacities associated with *TGFBI* mutations.[18] The advantages include the precision of the ablation and the ability to smoothen the surface of the cornea. The disadvantages include the expense of the equipment, the resultant thinning of the cornea with hypermetropic shift, and known recurrence of the condition which may necessitate further ablative treatments. Alcohol epitheliectomy with mechanical debridement has been advocated for relatively superficial deposits particularly when they are at the level of the Bowman layer; this results in good visual outcomes but without the thinning of the cornea and hyperopic shift associated with phototherapeutic keratectomy.[19] The disadvantage of this technique is that deeper anterior stromal lesions are not amenable to this approach.

For deeper stromal lesions extending further toward the posterior stroma, phototherapeutic keratectomy is not an option. Deep anterior lamellar keratoplasty is preferred over penetrating keratoplasty due to the preservation of the host's uninvolved Descemet membrane. Recurrence within the new graft is still a problem and tends to occur sooner in lattice corneal dystrophy compared to granular corneal dystrophies.[20,21] Delayed epithelial healing is also observed in lattice dystrophy. Recurrence of the abnormalities can be treated with phototherapeutic keratectomy or repeat corneal grafting. Repeat deep anterior lamellar keratoplasty is relatively straightforward compared to the primary procedure. This is another reason why lamellar keratoplasty may be considered a superior procedure for *TGFBI*-related dystrophies as the patient may require many repeat grafts over their lifetime.

The lack of animal models of *TGFBI*-related dystrophies has hampered the investigation of biological mechanisms underlying these conditions and hence the development of novel therapeutics. In the future, it is to be hoped that a better understanding of the mechanisms underlying *TGFBI* aggregation may highlight approaches to prevent or reverse such accumulations.[22]

Corneal endothelial dystrophies

The posterior corneal dystrophies result from defects of the corneal endothelium. This monolayer of endothelial cells maintains the corneal stroma in a relatively dehydrated state ensuring normal corneal thickness and clarity. Dysfunction of the active corneal endothelial pump leads to corneal oedema, pain, and reduced vision. The corneal endothelial dystrophies are a heterogeneous group of disorders that may present early in life (e.g. congenital hereditary endothelial dystrophy (CHED)) or late in adulthood (Fuchs endothelial corneal dystrophy (FECD)). Making a precise diagnosis in individuals presenting with these conditions is important as it can assist the identification of familial cases and highlight potential ophthalmic or systemic associations.[2,23,24]

Clinical characteristics

Congenital hereditary endothelial dystrophy (CHED)

CHED is an autosomal recessive condition that presents at, or soon after, birth and is an important cause of early corneal clouding. It is commonly associated with bilateral diffuse corneal opacification that ranges from a hazy to a milky appearance (Fig. 9.4M) and can result in nystagmus and amblyopia. Measurement of corneal thickness shows values double or triple to those seen in unaffected individuals.

Posterior polymorphous corneal dystrophy (PPCD)

Posterior polymorphous corneal dystrophy (PPCD) is an autosomal dominant condition associated with variable expressivity. Whilst corneal oedema may present at a young age, the majority of patients are asymptomatic. In most cases, PPCD is slowly progressive or static, while about one third develop corneal oedema (Fig. 9.4O) necessitating keratoplasty. Endothelial cells are of abnormal morphology and may produce a multilayered endothelium and migrate to the anterior chamber angle and over the iris. Deposition of abnormal basement membrane may result in iridocorneal adhesions and posterior–anterior synechiae. Secondary glaucoma may occur, for which monitoring is required. Slit-lamp examination detects a range of 'polymorphic' lesions (vesicular, band-shaped, or geographical) at the level of corneal endothelium and Descemet membrane, which is thickened due to abnormal collagenous deposition (Fig. 9.4N).

Fuchs endothelial corneal dystrophy (FECD)

FECD is a multifactorial condition characterised by progressive and premature corneal endothelial cell depletion. This can result in oedema, thickening and reduced clarity of the cornea which may lead to visual loss. FECD is usually seen as a late onset disorder with the corneal signs generally beginning to manifest in the fourth or fifth decade of life. Many years after the first detectable changes on slit-lamp examination, a subset of patients will become symptomatic. This typically involves developing impaired vision in the morning which becomes gradually permanent as the disease progresses. Advanced stages with bullous keratopathy can be painful.

On examination FECD is characterised by the presence of guttae, i.e. centrally located excrescences on the posterior corneal surface that over time coalesce (giving a beaten metal-like appearance) and spread peripherally (Fig. 9.4P). With gradual failing of the endothelial cell barrier function, corneal thickness increases due to stromal oedema, leading to epithelial bullae (bullous keratopathy) in advanced stages. In the chronic stage subepithelial fibrous scarring can be noted. Early-onset forms of FECD are more highly penetrant and considerably rarer; they are also more severe in terms of their impact on vision.

Molecular pathology

CHED is associated with biallelic pathogenic variants in the *SLC4A11* gene. *SLC4A11* encodes a transmembrane protein found on the basolateral membrane of the corneal endothelium where it modulates water transport. Pathogenic variants are hypothesised to cause loss of function although many are missense substitutions.[25]

Harboyan syndrome is a rare autosomal recessive condition in which CHED presents in individuals who also have neurosensory hearing loss. Importantly, both CHED and Harboyan syndrome can be caused by loss of function mutations in *SLC4A11*. Given this and the fact that individuals with apparently non-syndromic CHED have been shown to develop progressive hearing loss, it is likely that these two conditions form part of a spectrum.[26] Patients with CHED should therefore be monitored for progressive hearing loss which may develop in childhood.

In contrast to CHED, PPCD is genetically heterogeneous and typically inherited as an autosomal dominant trait. Three subtypes have been described PPCD1 (associated with the *OVOL2* gene), PPCD3 (associated with the *ZEB1* gene) and PPCD4 (associated with the *GRHL2* gene).[2,3,27] There are recognised genotype–phenotype correlations. Corneal steepening and astigmatism, which may be a significant cause of amblyopia if not managed adequately, are common with PPCD3 caused by heterozygous mutations in *ZEB1*. This subtype is thought to be associated with an increased risk for inguinal hernias.

Early-onset forms of FECD are caused by recurrent missense changes in the *COL8A2* gene. *COL8A2* encodes the α2 chain of collagen type 8, which is an important component of the Descemet membrane. Molecular analysis of later-onset forms has highlighted that a minority of cases are caused by rare, heterozygous, high effect variants. Presumed Mendelian forms include FECD4 (associated with the *SLC4A11* gene), FECD6 (associated with the *ZEB1* gene) and FECD8 (associated with the *AGBL1* gene). Importantly, the majority of the heritability that underlines late onset FECD results from variants in the CTG trinucleotide repeat expansion within the third intron of the gene encoding transcription factor 4 (*TCF4*). Around 80% of individuals with FECD carry copies with tandem repeat lengths of over 50, whereas normal individuals carry repeats of under 20. Some studies have shown that severity may be linked to repeat length.[28]

Clinical management

Management of corneal decompensation in the early stages involves the use of topical hyperosmotic agents such as 5% NaCl. In cases where conservative treatment fails or is unlikely to adequately clear the cornea, corneal transplantation is required. Where once penetrating keratoplasty was commonly performed, this has been superseded by selected replacement of the posterior corneal layers with Descemet Stripping Endothelial Keratoplasty (DSEK) or Descemet Stripping Automated Endothelial Keratoplasty (DSAEK). These offer the advantage of shorter recovery time, reduced postoperative steroid requirement, reduced risk of rejection and less intense follow-up. Another approach is Descemet Membrane Endothelial Keratoplasty (DMEK). This is technically more challenging than DSAEK or DSEK but has a faster recovery and better visual outcomes compared to either procedure. However, graft dislocation is more frequent.[29]

PPCD and CHED may present in early infancy or with congenital corneal opacity. Visual rehabilitation may necessitate early corneal transplantation. Penetrating keratoplasty is still performed in the majority of these cases although several reports have highlighted that DSAEK and DMEK can also be effective.[30] Posterior lamellar procedures may become more commonplace in children due to the attraction of maintaining a closed system and the faster recovery times.[31]

Miscellaneous corneal dystrophies

Lisch epithelial corneal dystrophy

Patients with this X-linked condition remain asymptomatic as long as the visual axis is uninvolved, which may take until the third decade of life. At that point, slight photophobia and ultimately blurred vision can occur. On slit-lamp examination the disease is characterised by the presence of well-delineated epithelial opacities that initially spare the central cornea and that adopt various patterns (including radial, band-shaped, and club-shaped) with clear corneal epithelium between them (Fig. 9.4B).

Epithelial recurrent erosion dystrophy

Patients with this autosomal dominant condition experience recurrent corneal erosions starting in childhood and, typically, ceasing by the third or fourth decade of life. On slit-lamp examination, subepithelial patchy grey-white scarring and nodules can be noticed, leading to impaired vision.

Gelatinous drop-like corneal dystrophy

Patients with this autosomal recessive condition typically experience irritation, redness and tearing within the first decade of life. This is associated with the appearance of subepithelial nodules and recurrent erosions (Fig. 9.4C). With age, these nodules increase in diameter and number, leading to significant visual loss as they coalesce and infiltrate the anterior stroma. This may give rise to the kumquat-like lesions seen in later life. Superficial neovascularisation may occur. Recurrence in corneal grafts is common.

Macular corneal dystrophy

Macular corneal dystrophy is an autosomal recessive disorder predominantly affecting the corneal stroma. Photophobia, impaired vision, and reduction of corneal sensitivity are the main symptoms. Recurrent erosions can also occur. Clinically, the disease is recognisable between the first and third decade by a superficial stromal haze initially affecting the central cornea then extending to the limbus. Later, poorly delineated whitish opacities (macules) appear within all stromal layers (Fig. 9.4K). The vision is significantly reduced usually by the third decade of life.

Macular corneal dystrophy is caused by biallelic variants in or close to the *CHST6* gene. *CHST6* encodes an enzyme that catalyses the transfer of a sulphate group to keratan GlcNAc residues. Keratan sulphate is key to the maintenance of corneal transparency and reduced CHST6 function leads to keratan chains that are less water-soluble, leading to reduced corneal clarity.

Schnyder stromal corneal dystrophy

Schnyder corneal dystrophy is an autosomal dominant disorder associated with corneal opacification resulting from abnormal deposition of cholesterol and phospholipids in the corneal stroma. Disease onset can be in early childhood, but most affected individuals are diagnosed in the second or third decade of life. The disease course is slowly progressive and most patients initially report problems glare. Visual acuity and corneal sensation decrease with age. Initially, the central cornea develops superficial haze (in the anterior stroma) and/or subepithelial crystals (Fig. 9.4J). It is worth highlighting though that only a subset of affected individuals will demonstrate the crystalline form. Later, an arcus lipoides is noted and this is often complicated by a mid-peripheral panstromal haze typically in the fifth decade of life. Schnyder corneal dystrophy is caused by heterozygous mutations in *UBIAD1*, a gene that has a role in lipid metabolism.

Congenital stromal dystrophy

The diagnosis of this autosomal dominant stromal dystrophy is made at birth in the presence of diffuse bilateral corneal clouding with flake-like, whitish opacities, equally distributed throughout the stroma. The corneal thickness may be increased. The condition is amblyogenic and may be associated with nystagmus. Otherwise, it is either non-progressive or slowly progressive.

Fleck stromal corneal dystrophy

This autosomal dominant, congenital, non-progressive, asymptomatic dystrophy is characterised by the presence of small discrete, discoid, flat opacities (Fig. 9.4L) equally scattered throughout the entire stroma from limbus to limbus. Disease expressivity is highly variable ranging from few to hundreds of lesions. Asymmetric distribution has been described in some cases.

Posterior amorphous corneal dystrophy

This autosomal dominant condition is associated with partial or complete posterior lamellar corneal opacification on slit-lamp examination. Affected individuals also demonstrate diffuse corneal thinning and flattening. Bilateral superior corneal scleralisation is common, and some individuals may show iris abnormalities, such as iridocorneal adhesions, correctopia, atrophy and coloboma. Subtle endothelial changes have been reported in some patients. Although most affected individuals retain good corrected vision throughout their lives, visually significant corneal opacification requiring corneal transplantation may occur.

X-linked endothelial corneal dystrophy

This X-linked condition may manifest in males as congenital corneal clouding leading to amblyopia and nystagmus. Some affected individuals only exhibit endothelial changes that resemble moon craters; decreased endothelial cell density and band-shaped keratopathy may also be present. Keratoplasty is necessary for a subset of male patients. In females, asymptomatic moon crater-like endothelial changes have been noted.

References

1. Weiss JS, Møller HU, Aldave AJ, Seitz B, Bredrup C, Kivelä T, et al. IC3D classification of corneal dystrophies—edition 2. *Cornea* 2015;**34**(2):117–59.
2. Lisch W, Weiss JS. Clinical and genetic update of corneal dystrophies. *Exp Eye Res* 2019;**186**:107715.
3. Soh YQ, Kocaba V, Weiss JS, Jurkunas UV, Kinoshita S, Aldave AJ, et al. Corneal dystrophies. *Nat Rev Dis Primers* 2020;**6**(1):46.
4. Chao-Shern C, Me R, DeDionisio LA, Ke BL, Nesbit MA, Marshall J, et al. Post-LASIK exacerbation of granular corneal dystrophy type 2 in members of a Chinese family. *Eye (Lond)* 2018;**32**(1):39–43.
5. Irvine AD, Coleman CM, Moore JE, Swensson O, Morgan SJ, McCarthy JH, et al. A novel mutation in KRT12 associated with Meesmann's epithelial corneal dystrophy. *Br J Ophthalmol* 2002;**86**(7):729–32.
6. Irvine AD, Corden LD, Swensson O, Swensson B, Moore JE, Frazer DG, et al. Mutations in cornea-specific keratin K3 or K12 genes cause Meesmann's corneal dystrophy. *Nat Genet* 1997;**16**(2):184–7.
7. Courtney DG, Atkinson SD, Allen EH, Moore JE, Walsh CP, Pedrioli DM, et al. siRNA silencing of the mutant keratin 12 allele in corneal limbal epithelial cells grown from patients with Meesmann's epithelial corneal dystrophy. *Invest Ophthalmol Vis Sci* 2014;**55**(5):3352–60.
8. Courtney DG, Moore JE, Atkinson SD, Maurizi E, Allen EH, Pedrioli DM, et al. CRISPR/Cas9 DNA cleavage at SNP-derived PAM enables both in vitro and in vivo KRT12 mutation-specific targeting. *Gene Ther* 2016;**23**(1):108–12. https://doi.org/10.1038/gt.2015.82.
9. Bourne WM. Soft contact lens wear decreases epithelial microcysts in Meesmann's corneal dystrophy. *Trans Am Ophthalmol Soc* 1986;**84**:170–82.
10. Ghanem RC, Piccinini AL, Ghanem VC. Photorefractive keratectomy with mitomycin C in Meesmann's epithelial corneal dystrophy. *J Refract Surg* 2017;**33**:53–5.
11. Nielsen NS, Poulsen ET, Lukassen MV, Chao Shern C, Mogensen EH, Weberskov CE, et al. Biochemical mechanisms of aggregation in TGFBI-linked corneal dystrophies. *Prog Retin Eye Res* 2020;**29**:100843.
12. Skonier J, Neubauer M, Madisen L, Bennett K, Plowman GD, Purchio AF. cDNA cloning and sequence analysis of beta ig-h3, a novel gene induced in a human adenocarcinoma cell line after treatment with transforming growth factor-beta DNA. *Cell Biol* 1992;**11**(7):511–22.
13. Munier FL, Korvatska E, Djemaï A, Le Paslier D, Zografos L, Pescia G, et al. Kerato-epithelin mutations in four 5q31-linked corneal dystrophies. *Nat Genet* 1997;**15**(3):247–51.
14. Korvatska E, Munier FL, Djemaï A, Wang MX, Frueh B, Chiou AG, et al. Mutation hot spots in 5q31-linked corneal dystrophies. *Am J Hum Genet* 1998;**62**(2):320–4.
15. Munier FL, Frueh BE, Othenin-Girard P, Uffer S, Cousin P, Wang MX, et al. BIGH3 mutation spectrum in corneal dystrophies. *Invest Ophthalmol Vis Sci* 2002;**43**(4):949–54.
16. Stewart HS, Parveen R, Ridgway AE, Bonshek R, Black GC. Late onset lattice corneal dystrophy with systemic familial amyloidosis, amyloidosis V, in an English family. *Br J Ophthalmol* 2000;**84**(4):390–4.
17. Lin SR, Aldave AJ, Chodosh J. Recurrent corneal erosion syndrome. *Br J Ophthalmol* 2019;**103**:1204–8.
18. Gruenauer-Kloevekorn C, Braeutigam S, Froster UG, Duncker GIW. Surgical outcome after phototherapeutic keratectomy in patients with TGFBI-linked corneal dystrophies in relation to molecular genetic findings. *Graefes Arch Clin Exp Ophthalmol* 2009;**247**:93–9.
19. Ashar JM, Latha M, Laddavalli PK. Phototherapeutic keratectomy versus alcohol epitheliectomy with mechanical debridement for superficial variant of granular dystrophy: a paired eye comparison. *Cont Lens Anterior Eye* 2012;**35**:236–9.
20. Unal M, Arslan OS, Atalay E, Mangan MS, Bilgin AB. Deep anterior lamellar keratoplasty for the treatment of stromal corneal dystrophies. *Cornea* 2013;**32**(3):301–5.
21. Marcon AS, Cohen EJ, Rapuano CJ, et al. Recurrence of corneal stromal dystrophies after penetrating keratoplasty. *Cornea* 2003;**22**:19–21.
22. Christie KA, Robertson LJ, Conway C, Blighe K, DeDionisio LA, Chao-Shern C, et al. Mutation-independent allele-specific editing by CRISPR-Cas9, a novel approach to treat autosomal dominant disease. *Mol Ther* 2020;(20):30236–7. S1525-0016.
23. Aldave AJ, Han J, Frausto RF. Genetics of the corneal endothelial dystrophies: an evidence-based review. *Clin Genet* 2013;**84**(2):109–19.
24. Matthaei M, Hribek A, Clahsen T, Bachmann B, Cursiefen C, Jun AS. Fuchs endothelial corneal dystrophy: clinical, genetic, pathophysiologic, and therapeutic aspects. *Annu Rev Vis Sci* 2019;**5**:151–75.
25. Alka K, Casey JR. Molecular phenotype of SLC4A11 missense mutants: setting the stage for personalized medicine in corneal dystrophies. *Hum Mutat* 2018;**39**(5):676–90.
26. Siddiqui S, Zenteno JC, Rice A, Chacón-Camacho O, Naylor SG, Rivera-de la Parra D, et al. Congenital hereditary endothelial dystrophy caused by SLC4A11 mutations progresses to Harboyan syndrome. *Cornea* 2014;**33**(3):247–51.
27. Davidson AE, Hafford-Tear NJ, Dudakova L, Sadan AN, Pontikos N, Hardcastle AJ, et al. CUGC for posterior polymorphous corneal dystrophy (PPCD). *Eur J Hum Genet* 2020;**28**(1):126–31.
28. Fautsch MP, Wieben ED, Baratz KH, Bhattacharyya N, Sadan AN, Hafford-Tear NJ, et al. TCF4-mediated Fuchs endothelial corneal dystrophy: insights into a common trinucleotide repeat associated disease. *Prog Retin Eye Res* 2020;**28**:100883.
29. Dapena I, Ham L, Melles GR. Endothelial keratoplasty: DSEK/DSAEK or DMEK—the thinner the better? *Curr Opin Ophthalmol* 2009;**20**(4):299–307.
30. Hermina Strungaru M, Ali A, Rootman D, Mireskandari K. Endothelial keratoplasty for posterior polymorphous corneal dystrophy in a 4-month-old infant. *Am J Ophthalmol Case Rep* 2017;**7**:23–6.
31. Madi S, Santorum P, Busin M. Descemet stripping automated endothelial keratoplasty in pediatric age group. *Saudi J Ophthalmol* 2012;**26**:309–13.

9.4

Keratopathy in inborn errors of metabolism

Susmito Biswas and Jane L. Ashworth

Corneal abnormalities are a feature of many inborn errors of metabolism (i.e. genetic disorders in which the impairment of a biochemical pathway is intrinsic to disease pathophysiology[1]). Most of these conditions present in childhood and are associated with a spectrum of clinical findings affecting multiple organ systems. Central nervous system involvement is common and developmental delay, hypotonia/spasticity, neuropathy, and/or epilepsy are often observed. Non-neurological findings can include failure to thrive, hepatomegaly, splenomegaly, and renal tubular acidosis. Some of these conditions can be detected on newborn screening but, on many occasions, the diagnosis is challenging since the clinical presentation is non-specific (especially in later-onset forms that are less severe, more variable and less easily recognisable).

Inborn errors of metabolism that are associated with keratopathy are discussed in this section. The main focus is on the corneal findings in these disorders and for further information on the systemic manifestations, it is recommended that the reader refers to relevant knowledgebases and textbooks (e.g. 2,3).

Mucopolysaccharidoses

Mucopolysaccharidoses (MPS) are a clinically and genetically heterogeneous group of recessive neurometabolic disorders caused by the absence or malfunctioning of lysosomal enzymes that are required for glycosaminoglycan degradation. Glycosaminoglycans (formerly known as mucopolysaccharides) are complex carbohydrate molecules that are a major component of the extracellular matrix. Accumulation of partially degraded glycosaminoglycans is a key feature of MPS and results in a multisystemic phenotype that may include skeletal abnormalities (e.g. dysplasia, decreased joint mobility), short stature, characteristic facial features (e.g. flat nasal bridge, thickening of the lips/tongue), cardiac abnormalities, hepatosplenomegaly, and/or neurological abnormalities.

More than 10 distinct MPS forms have been identified (Table 9.2). The associated signs, symptoms and disease severity vary significantly by subtype. Corneal involvement is common and notable corneal haze is often seen in the following MPS forms: MPS type I-Hurler, MPS type I-Hurler–Sheie, MPS type VI-Maroteaux–Lamy, and MPS type VII-Sly (Table 9.2; Fig. 9.7). Typically, slit-lamp examination reveals a diffuse, 'ground-glass' appearance throughout the cornea. This is often associated with progressive visual loss and photophobia. However, milder corneal haze can be compatible with good levels of vision.[4]

Histopathology of affected corneas reveals increased intra- and extracellular glycosaminoglycan staining, thinning of the corneal epithelial cell layers and alterations/discontinuity of Bowman layer. Electron microscopy reveals glycosaminoglycans within keratocytes, epithelium and corneal endothelium. Confocal microscopy of the cornea demonstrates increased reflectivity in basal epithelial cells, abnormally increased reflectivity of keratocytes (giving a reticular pattern) and a poorly distinguishable sub-basal nerve plexus.

Corneal clouding associated with MPS can be treated with penetrating keratoplasty. More recently, deep anterior lamellar keratoplasty is beginning to be favoured as endothelial function is often not severely affected. The final visual outcome may be compromised however by other MPS-associated ophthalmic morbidities including optic atrophy, retinal dysfunction, or cerebral visual impairment. Importantly, there are significant risks associated with surgery under general anaesthesia for patients with MPS.

Treatment with enzyme replacement therapy does not significantly improve corneal clouding in MPS, but there is some evidence suggesting that early treatment with haematopoietic stem cell transplantation may affect the corneal phenotype in patients with MPS type I-Hurler. Gene replacement therapy has shown promise in reducing glycosaminoglycans in tissues and may be considered as a treatment option for corneal clouding in the future, following further clinical trials.[5]

Mucolipidoses

Mucolipidosis type IV is an autosomal recessive, lysosomal storage disorder characterised by psychomotor retardation and ophthalmic manifestations. It is caused by biallelic mutations in the *MCOLN1* gene and it is mainly encountered in people of Ashkenazi Jewish ancestries (due to a geographic founder effect). *MCOLN1* encodes mucolipin 1, a cation channel. It has been proposed that abnormal endocytosis of mutant mucolipin 1 leads to irregular

TABLE 9.2 Genetic and ophthalmic characteristics of mucopolysaccharidoses.

MPS type	Enzyme deficiency	GAG accumulation	Inheritance	Associated gene	Corneal haze	Other ophthalmic features
MPS I-H (Hurler)	α-L-Iduronidase	DS, HS	AR	IDUA	+++	ONA (++), R (++), G(++)
MPS I-HS (Hurler–Sheie)	α-L-Iduronidase	DS, HS	AR	IDUA	++	ONA (++), R (++), G(++)
MPS I-S (Sheie)	α-L-Iduronidase	DS, HS	AR	IDUA	+	ONA (+), R(++), G(+)
MPS II (Hunter)	Iduronate-2-sulfatase	DS, HS	XLR	IDS	–	ONA(++), R(++), G(+)
MPS IIIA (Sanfilippo A)	Heparan sulfaminidase	HS	AR	SGSH	+/–	ONA(+), R(+++), G(+)
MPS IIIB (Sanfilippo B)	N-Acetyl-α-D-glucosaminidase	HS	AR	NAGLU	+/–	ONA(+), R(+++), G(+)
MPS IIIC (Sanfilippo C)	Acetyl-CoA: α glucosaminidase N-Acetyltransferase	HS	AR	HGSNAT	+/–	ONA(+), R(+++), G(+)
MPS IIID (Sanfilippo D)	N-Acetylglucosamine-6-sulfatase	HS	AR	GNS	+/–	ONA(+), R(+++), G(+)
MPS IV (Morquio)	Galactose-6-sulphate sulfatase	KS	AR	GALNS	+/–	ONA(+), R(++), G(+)
MPS VI (Maroteaux–Lamy)	N-Acetylgalactosamine-4-sulfatase	DS	AR	ARSB	+++	ONA(++), G (++)
MPS VII (Sly)	β-D-Glucuronidase	DS,HS,CS	AR	GUSB	++	ONA(++)
MPS IX	Hyaluronidase	CS	AR	HYAL1	Not known	Not known

DS, dermatan sulphate; HS, heparan sulphate; KS, keratan sulphate; CS, chondroitin sulphate; AR, autosomal recessive; XLR, X-linked recessive; ONA, optic nerve abnormalities; R, retinal abnormalities; G, glaucoma.

lysosomal processing of normal constituents of cellular membranes. This is thought to have an impact on the cell's ability to carry out the normal turnover of various molecules.

A key ophthalmic feature of mucolipidosis type IV is corneal opacification. This is typically present at birth and tends to be bilateral and broadly symmetrical. It primarily affects the central cornea and the corneal epithelium. Other frequently encountered ophthalmic anomalies include strabismus and nystagmus. Patients may also have pigmentary retinopathy, optic atrophy, and/or vascular attenuation. The ophthalmic abnormalities are likely to be progressive.[6]

Other relevant lysosomal storage disorders include mucolipidosis type II and mucolipidosis type III. Mucolipidosis type II and mucolipidosis type III α/β are caused by biallelic mutations in the GNPTAB gene; mucolipidosis type III γ is caused by changes in the GNPTG gene. Significant corneal opacification may be seen in mucolipidosis type II and less so in type III. In general, there is no definitive treatment for this group of disorders and affected individuals often have a shortened lifespan.

Cystinosis

Cystinosis is an autosomal recessive multisystemic disorder associated with mutations in the CTNS gene. CTNS encodes cystinosin, a protein that transports the disulphide amino-acid cystine out of lysosomes into the cytoplasm. Dysfunction results in intralysosomal cystine accumulation with crystal formation. Crystals accumulate in many

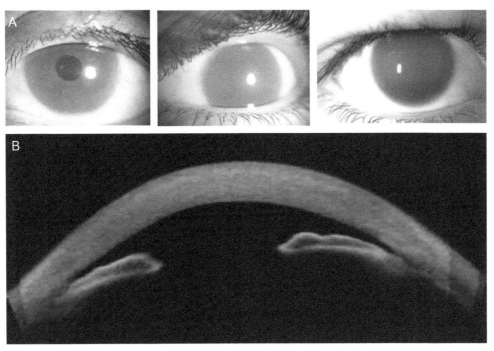

FIG. 9.7 Corneal findings in people with mucopolysaccharidoses (MPS). (A) Mild, moderate, and severe corneal clouding in affected individuals. (B) Anterior segment OCT showing increased corneal thickness in a patient with MPS type VI-Maroteaux–Lamy.

tissues, notably the kidney, eye, bone marrow, pancreas, muscles, and brain.

Nephropathic (infantile-onset) cystinosis is the most prevalent subtype and the commonest genetic cause of renal Fanconi syndrome (a syndrome associated with inadequate reabsorption in the proximal renal tubules) and growth failure between 6 and 12 months of age.[7] It typically results in renal failure within the first decade of life. There are also non-nephropathic (adult-onset) and intermediate (juvenile-onset) forms which are less frequently seen. In the intermediate form, renal failure occurs between the second and third decades. The adult-onset form has no renal involvement. In both the intermediate and adult-onset form the patients are typically compound heterozygotes for *CTNS* variants with one of the mutations being milder and retaining some residual function.

Corneal crystals appear within the first year of life, initially in the peripheral and anterior cornea, progressing centrally and posteriorly. They appear as numerous needle-shaped highly refractile opacities which are easily seen on slit-lamp examination (Fig. 9.8). By 16 months, all affected children have evident corneal crystals. By 6–7 years of age, crystals are found throughout the cornea, including the epithelium, stroma, and endothelium. Other ocular tissues affected include the conjunctiva, uveal tract, anterior lens capsule, retina, RPE, and optic nerve. Symptoms are predominantly photophobia and pain occurring in the first few years of life and occasionally resulting in debilitating blepharospasm. Photophobia may be due to light scattering caused by the corneal crystals, increased presence of inflammatory cells (dendritic cells) and/or corneal nerve alterations.[8]

Following successful renal transplantation, longer life expectancy results in the progression of ophthalmic manifestations. Ocular surface disease develops including superficial punctate keratopathy, filamentary keratopathy, peripheral corneal neovascularisation, and band keratopathy. Other ophthalmic features include retinopathy, maculopathy, papilledema, and angle-closure glaucoma. Together, these ophthalmic complications can result in severe visual impairment.

Treatment with oral cysteamine bitartrate has little impact on the corneal phenotype. However, treatment with topical cysteamine (mercaptamine) can reduce corneal crystal deposition but needs frequent instillation and there are issues include stinging (which often leads to poor compliance) and limited efficacy (due to unstable drug formulation of pharmacy compounded topical cysteamine). More stable pharmaceutical preparations are now available which, to an extent, overcome the shortfalls of previous topical treatments. Severe corneal involvement may necessitate corneal transplantation but recurrence is a feature with poor outcomes. Other management options include topical lubricants and management of photophobia with tinted glasses.

Slit-lamp diagnosis is relatively straightforward, but immunoglobulin deposition in the cornea associated with

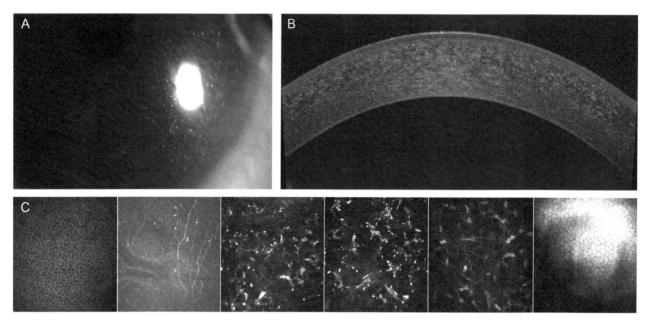

FIG. 9.8 Corneal findings in an 11-year-old girl with nephropathic cystinosis. Corneal crystals are noted on slit-lamp image (A), anterior segment OCT (B), and confocal microscopy (C). The visual acuity was normal in both eyes.

multiple myeloma can look similar. Other crystalline keratopathies, however, generally look significantly different.

The corneal crystals in cystinosis have characteristic confocal microscopy and anterior segment OCT features. On confocal microscopy, they can be visualised as needle-shaped hyper-reflective bodies. Crystals can be found in the epithelium, Bowman layer, through stroma and extending posteriorly to Descemet layer and endothelium; the density of crystals may vary within the stroma. In the conjunctiva, the crystals appear round or oval. OCT of the cornea reveals hyper-reflective punctate deposits predominantly in the anterior stroma. The depth of crystal deposition can be used as a measure of the disease severity and is greater in the peripheral cornea than in the centre.

Tyrosinaemia type II

Tyrosinaemia type II is an autosomal recessive disorder associated with dysfunction of tyrosine aminotransferase. Reduced activity of this enzyme results in a build-up of tyrosine crystals in tissue and serum. Painful hyperkeratotic palmar and plantar lesions are characteristic. Corneal features include dendritic and pseudo-dendritic lesions and ulcers that may be mistaken for herpes simplex keratitis (Fig. 9.9). These lead to eventual corneal clouding, scarring, and impaired vision.

It is noteworthy that similar corneal lesions appear as a side effect of nitisinone (NTBC), a medication used to alleviate the effects of tyrosinemia type I. In this condition, nitisinone is known to increase serum tyrosine.

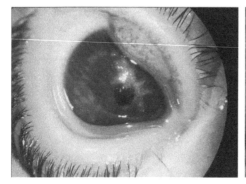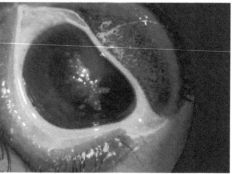

FIG. 9.9 Corneal findings in a 6-month-old child with tyrosinaemia type II. A dendritiform corneal lesion with an infero-central location is observed. Note the lack of terminal bulbs. Fluorescein staining reveals a poor staining pattern. *(Adapted from Ref. 9.)*

In vivo confocal microscopy in tyrosinaemia type II shows hyper-reflective linear crystal deposition in the superficial cornea (epithelium—superficial and middle layers). The basal epithelium and sub-basal layer are unaffected. The stroma and Bowman layer are not involved. The crystals are typically aligned in parallel with the epithelial cells, with occasional deeper extensions. A normal sub-basal nerve plexus is evident and the most likely source of crystal deposits is felt to be tear fluid.

Dietary restriction of protein to reduce the dietary source of phenylalanine and tyrosine results in a rapid reduction in the crystals found in the cornea. The dietary restriction also helps to resolve painful skin lesions.

Wilson disease

Wilson disease is an autosomal recessive disorder of copper metabolism associated with mutations in the *ATPB7* gene. It results in copper accumulation in the liver, kidney, brain, and cornea.

The liver disease typically occurs between the ages of 5 and 20 years. Neurological problems generally occur between the 3rd and 5th decade of life and, occasionally, in childhood, manifesting as dystonia, dysarthria, dysphagia, tremor, parkinsonism, and psychiatric problems.

A high proportion (95%) of affected individuals with neurological problems are found to have a Kayser–Fleischer ring at the Descemet membrane (i.e. brown-yellow-green peripheral pigmentation that starts inferiorly and has no clear interval at the limbus) (Fig. 9.10). Kayser–Fleischer rings are best identified by gonioscopy and may also be seen in pre-symptomatic patients (but in a much smaller proportion, ~60%).[11] It is worth highlighting that Kayser–Fleischer rings are not entirely pathognomonic of Wilson disease; fewer common causes include cholestasis and primary biliary cirrhosis.

Other ophthalmic findings in Wilson disease may include 'sunflower' cataracts, saccadic pursuit movement abnormalities and retinopathy. Treatment with copper chelating agents (d-penicillamine and/or zinc salts) result in regression of Kayser–Fleischer rings. Occasionally liver transplantation is required.

Fabry disease

Fabry disease is an X-linked sphingolipidosis caused by a deficiency of the enzyme α-galactosidase. It is associated with the widespread accumulation of glycosphingolipids (globotriaosylceramide), particularly in vascular endothelial cells, and results in vascular narrowing and insufficiency. Progressive multisystemic involvement can result in premature death. Males are often severely affected with no residual enzyme function and display classic features. Female obligate carriers may exhibit severe symptoms due to skewed X-chromosome inactivation (lyonisation).

Symptoms of Fabry disease include acroparasthesia, autonomic dysfunction leading to hyperhidrosis or hypohidrosis, flushing, gastrointestinal disturbance; neurological sequelae due to vascular insufficiency, including ischaemic strokes, transient ischaemic attack, cranial nerve palsies; skin lesions including classic angiokeratomas (macular non-blanching lesions in bathing trunk distribution and also present on palms); cardiac arrhythmias, coronary artery stenosis; and renal involvement resulting in renal failure.

Several ophthalmic features have been described. Corneal verticillata are superficial corneal opacities that do not affect vision. These abnormalities originate at the limbus where lipid is deposited in the basal epithelium. Other accompanying lesions include posterior subcapsular and propeller cataracts, which are usually visually insignificant. Conjunctival vessel abnormalities occur such as tortuosity, and aneurysmal dilation (Fig. 9.11). Retinal vascular tortuosity may also be present.

Treatment with enzyme replacement therapy-recombinant α-galactosidase (agalsidase α or agalsidase β) has been used with benefit for cardiac features (left ventricular hypertrophy stabilisation) but there was no significant effect in other organ systems.

LCAT-related metabolic disease

Lecithin-cholesterol acyltransferase is a lipoprotein-associated enzyme involved in the esterification of free cholesterol, in the formation of mature high density lipoprotein (HDL) and in the intravascular stage of reverse cholesterol transport.

LCAT deficiency is an autosomal recessive disorder associated with mutations in the LCAT gene. More than 80 disease-causing variants have been described. In general, mutations in this gene are associated with at least two syndromes—familial LCAT deficiency and fish-eye disease. Both these disorders lead to corneal haze in middle age associated with the formation of a lipoid arcus in the peripheral cornea.

Fish-eye disease (partial LCAT deficiency): In this familial disorder, there is partial LCAT enzyme function with loss of α-LCAT activity but the preservation of β-LCAT activity. This condition is characterised by severe HDL deficiency and extensive corneal opacification (Fig. 9.12). Diffuse corneal clouding occurs, which is denser in the periphery. The name of the disorder was coined for the density of the corneal opacity which resembled that of a boiled fish. The presentation can be initially of an arcus lipoides encroaching towards the pupil at an early age; in contrast to typical arcus lipoides, there is no

90 SECTION | II Genetic disorders affecting the anterior segment

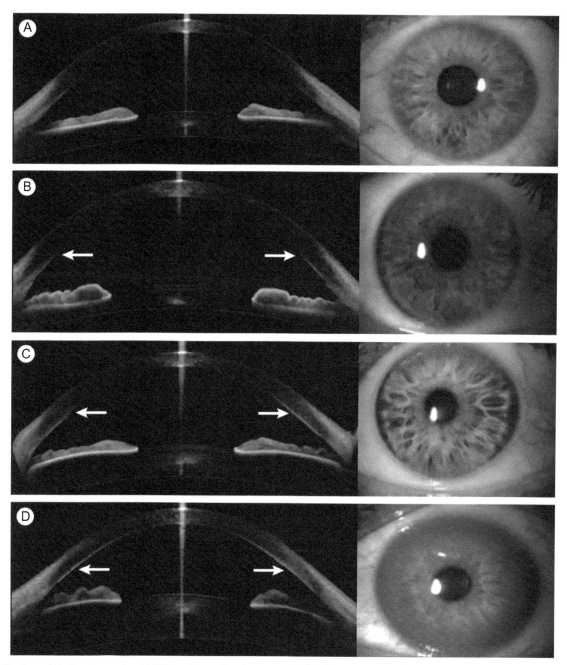

FIG. 9.10 Corneal findings in people with Wilson disease. Anterior segment OCT of the eye in a healthy individual and patients with Wilson disease. In some patients with Wison disease and healthy individuals, corneal copper deposits are not noted (A). However, in a substantial proportion of affected cases, copper deposits can be visualised by anterior segment OCT as hyper-reflective points that are discrete (B), on the superior and inferior part of the cornea (C) or in the formation of a complete Kayser–Fleischer ring (D). *(Adapted from Ref. 10.)*

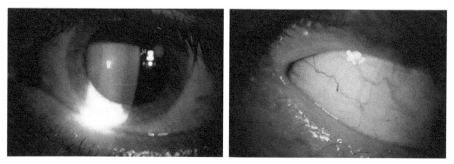

FIG. 9.11 Ocular surface findings in a patient with Fabry disease. Corneal verticillata (i.e. deposits at the level of the basal epithelium forming a faint golden-brown whorl pattern) and conjunctival vessel tortuosity are noted.

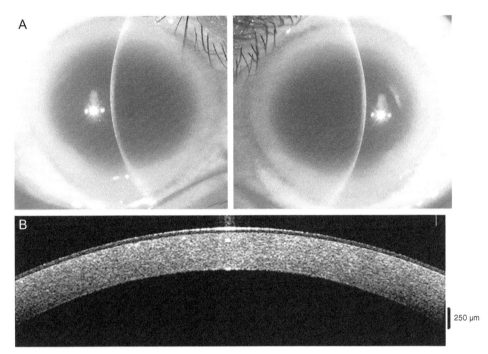

FIG. 9.12 Corneal findings in a 36-year-old patient with partial LCAT deficiency. The slit-lamp photograph shows bilateral, diffuse corneal clouding (A) and anterior segment OCT reveals a homogeneously hyper-reflective corneal stroma (B). The corneal thickness was within normal limits. *(Adapted from Ref. 12.)*

clear zone between the outer border of the opacity and the limbus. Progressive deposition occurs over time, until the entire cornea is involved, resulting in photophobia and visual loss. Corneal thickness may be normal. Histology reveals numerous small lipid-containing vacuoles distributed throughout the corneal stroma, but more prominent in the anterior layers. On electron microscopy, vacuoles are present within and between stromal lamellae. Involvement of the Bowman layer with vacuolisation has been described in some cases.

LCAT deficiency: In this disorder, there is a loss of function of both α- and β-LCAT activity. The *LCAT* protein may be present or absent but is inactive. There is an increase in unesterified cholesterol in all plasma lipoproteins. The corneal opacification commences in childhood with an early onset arcus. This dense, irregular, white arcus lacks the clear demarcated outer border and lucid space between its edge and the limbus; this feature can help differentiate LCAT deficiency associated arcus from the typical age-related arcus lipoides. Diffuse, fine, granular, stromal opacities progress to diffuse corneal haze. Systemic effects include haemolytic anaemia and renal disease progressing to end-stage renal failure, hypertension, and hypertriglyceridemia.

Histopathology reveals vacuoles in the corneal stroma with a compact layer of vacuoles just anterior to the Descemet membrane. These extracellular lesions are interspersed between collagen fibres. Electron microscopy reveals that these vacuoles contain myriad membranous deposits. Bowman layer shows small vacuoles that are finer than in the rest of the stroma. Descemet membrane, endothelium, epithelium, and keratocytes appear normal on electron microscopy. Despite elevated levels of unesterified cholesterol in the cornea, no crystals are found. OCT of the cornea performed on a patient with LCAT showed an overall increase in central and peripheral corneal thickness, thinned corneal epithelium and hyperreflectivity throughout the entire stroma with focal disruption of the Bowman layer. Confocal microscopy revealed focal oval spots of hyperreflectivity with multiple dark striae, and reduced and irregular keratocytes.[13]

Tangier disease

Tangier disease (also known as familial high-density lipoprotein deficiency) is an autosomal recessive disorder caused by mutations in the *ABCA1* gene. This transporter gene plays an important role in the efflux and transfer of free cholesterol from peripheral cells.

Tangier disease is characterised by severe reduction in HDL cholesterol and accumulation of cholesterol ester-rich lipids in various macrophage laden tissues of the body, particularly those at lower body temperatures such as the tonsils. Over time the tonsils enlarge and obtain a yellow/orange appearance. Other symptoms include anaemia, thrombocytopenia, peripheral neuropathy, and corneal opacification. There is a risk of cardiovascular disease. In general, disease expression tends to be highly variable even within the same family.

There is a progressive clouding of the cornea with enhancement of the opacity at the 3 and 9 o'clock quadrants. Histopathology demonstrates the accumulation of esterified cholesterol and phospholipids, with numerous vacuoles containing myelin-like membranous lamellar bodies. Some cases of reduced corneal sensation have been described, but this may be secondary to the chronic exposure keratopathy or to the lagophthalmos associated with the disease. On confocal microscopy, the epithelial and sub-basal nerve plexus appear normal. Diffuse light scattering is noted at the corneal stroma caused by lipid deposits. These deposits appear as small granular bodies in both the anterior and posterior parts of the stroma.

References

1. Ferreira CR, van Karnebeek CDM, Vockley J, Blau N. A proposed nosology of inborn errors of metabolism. *Genet Med* 2019;**21**(1):102–6.
2. Lee JJY, Wasserman WW, Hoffmann GF, van Karnebeek CDM, Blau N. Knowledge base and mini-expert platform for the diagnosis of inborn errors of metabolism. *Genet Med* 2018;**20**(1):151–8.
3. Saudubray J-M, Baumgartner M, Walter J, editors. *Inborn metabolic diseases diagnosis and treatment*. 6th ed. Springer; 2016. https://doi.org/10.1007/978-3-662-49771-5.
4. Ashworth JL, Biswas S, Wraith E, Lloyd IC. Mucopolysaccharidoses and the eye. *Surv Ophthalmol* 2006;**51**(1):1–17.
5. Belur LR, Temme A, Podetz-Pedersen KM, et al. Intranasal adeno-associated virus mediated gene delivery and expression of human iduronidase in the central nervous system: a noninvasive and effective approach for prevention of neurologic disease in mucopolysaccharidosis type I. *Hum Gene Ther* 2017;**28**(7):576–87. https://doi.org/10.1089/hum.2017.187.
6. Smith JA, Chi-Chao Chan C-C, Goldin E, et al. Noninvasive diagnosis and ophthalmic features of mucolipidosis type IV. *Ophthalmology* 2002;**109**:588–94.
7. Gahl WA. Cystinosis coming of age. *Adv Pediatr* 1986;**33**:95–126.
8. Liang H, Baudouin C, Hassani RTJ, Brignole-Baudouin F, Labbé A. Photophobia and corneal crystal density in nephropathic cystinosis: an in vivo confocal microscopy and anterior segment optical coherence tomography study. *Invest Ophthalmol Vis Sci* 2015;**56**:3218–25.
9. Macsai MS, Schwartz TL, Hinkle D, Hummel MB, Mulhern MG, Rootman D. Tyrosinemia type II: nine cases of ocular signs and symptoms. *Am J Ophthalmol* 2001;**132**(4):522–7.
10. Członkowska A, Litwin T, Dusek P, et al. Wilson disease. *Nat Rev Dis Primers* 2018;**4**(1):21.
11. Taly AB, Meenakshi-Sundaram S, Sinha S, Swamy HS, Arunodaya GR. Wilson disease: description of 282 patients evaluated over 3 decades. *Medicine (Baltimore)* 2007;**86**(2):112–21.
12. Kanai M, Koh S, Masuda D, Koseki M, Nishida K. Clinical features and visual function in a patient with fish-eye disease: quantitative measurements and optical coherence tomography. *Am J Ophthalmol Case Rep* 2018;**10**:137–41.
13. Palmiero P-M, Sbeity Z, Liebman J, Ritch R. In vivo imaging of the cornea in a patient with lecithin-cholesterol acetyltransferase deficiency. *Cornea* 2009;**28**:1061–4.

9.5

Keratopathy in ectodermal dysplasias

Colin E. Willoughby and Michele Callea

Ocular surface abnormalities may be seen in the context of ectodermal dysplasias, a heterogeneous group of developmental disorders characterised by alterations in two or more ectodermal structures (including the hair, teeth, nails, and/or sweat glands).[1] There are over 200 clinically defined subtypes and many of these have an identified genetic basis.[1] Ectodermal dysplasias can show several ophthalmic features that are characteristic and can be sight-threatening.[2-4] Fundamentally, this group of conditions can affect any ocular structure derived from the ectoderm: eyelids (including cilia, meibomian glands), conjunctiva, corneal epithelium, main and accessory lacrimal glands, lacrimal drainage system and lens.

The clinical assessment of an individual with ectodermal dysplasia requires a multidisciplinary approach. This can often be coordinated by a clinical geneticist who can instigate molecular genetic testing. A molecular diagnosis will define inheritance patterns, facilitate genetic counselling of patients and families and guide management.

Although ectodermal dysplasias are congenital disorders, features may not manifest until the emergence of the teeth or with hair growth. The earliest manifestations typically result from abnormalities in sweating resulting in unexplained fever, heat intolerance and febrile seizures; there is a reported mortality rate of up to 30% in the first year of life due to severe fever, respiratory infection and feeding difficulties.[5] Ectodermal dysplasias may also present with limb defects, cleft lip/palate, facial dysmorphism, and ophthalmic features. The early ophthalmic presentation can include lid adhesions (ankyloblepharon), nasolacrimal duct anomalies, recurrent conjunctivitis, severe corneal disease, and photophobia.[2,3] In this chapter, three prevalent ectodermal dysplasia subtypes are discussed.

Hypohidrotic ectodermal dysplasia

Hypohidrotic ectodermal dysplasia is the commonest ectodermal dysplasia subtype and mutations in four genes account for up to 90% of all cases. These include *EDA* (X-linked), *EDAR* (autosomal dominant or autosomal recessive), *EDARADD* (autosomal dominant or autosomal recessive) and *WNT10* (autosomal recessive). The clinical phenotypes that result from mutations in these different genes are often indistinguishable as the same developmental pathway is impacted at a molecular and cellular level.

Hypohidrotic ectodermal dysplasia is characterised by hypodontia, hypotrichosis, and hypohidrosis.[2] Sweating is either reduced (hypohidrotic) or absent (anhidrotic) and can lead to abnormal thermoregulation, unexplained infantile fevers and febrile seizures. The skin can be abnormal at birth and subsequently appears dry with patches of hyperkeratosis and/or eczematous type changes. Affected individuals have sparse body and scalp hair (hypotrichosis) which are often hypochromic, brittle and slow-growing. Often, the diagnosis is overlooked until dental anomalies present; these can include delayed eruption or hypodontia with teeth that are peg-shaped or conical in shape. Abnormalities in mucosal secretions lead to respiratory infections, chronic rhinitis, and gastrointestinal problems. ENT problems including otitis media and hearing loss are frequent.[6] In hypohidrotic ectodermal dysplasia, management is directed towards thermoregulation using environmental alterations and cooling vests. Early recognition and diagnosis are essential to prevent febrile seizures and treat respiratory infections appropriately given the high potential mortality rate. Dental and orthodontic input is vital for dental rehabilitation.[7]

Abnormalities in the meibomian glands are the commonest manifestation of hypohidrotic ectodermal dysplasia with total or partial absence of these structures (Fig. 9.13).[2,8] Most affected individuals show features of evaporative dry eye with a rapid tear film breakup time.[8] Corneal changes are usually mild and consist of an age-related pannus.[8] Infantile glaucoma is a rare association. In general, the ophthalmic management for hypohidrotic ectodermal dysplasia aims to support the ocular surface with topical lubrication.

Male patients with X-linked hypohidrotic ectodermal dysplasia (associated with changes in the *EDA* genes) have a characteristic facial appearance with periorbital hyperpigmentation and a depressed nasal bridge with flat cheekbones (malar hypoplasia). Examination of female family members can identify a female carrier state, for example in a male proband's mother or daughters. Due to random X-chromosome inactivation, female carriers may show abnormal dentition and patchy distribution of sweating or

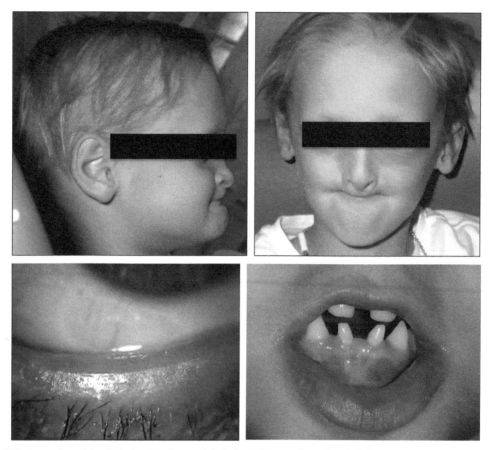

FIG. 9.13 Hypohidrotic ectodermal dysplasia showing characteristic facies with sparse, hypochromic hair, maxillary hypoplasia, and periorbital hyperpigmentation and dryness; absent meibomian glands; and hypodontia with conical incisors.

hair defects; an attenuated ophthalmic phenotype (i.e. patchy loss of meibomian glands) may be present.

Ectodermal dysplasia with cleft lip/palate

There are several ectodermal dysplasias associated with clefting of the lip and/or palate. Conditions associated with significant ocular surface abnormalities and corneal scarring include ectrodactyly-ectodermal dysplasia-cleft lip/palate (EEC) syndrome and ankyloblepharon-ectodermal dysplasia-cleft lip and palate (AEC) syndrome. Both these disorders are inherited as autosomal dominant traits and result from heterozygous mutations in the TP63 (p63) gene.[3]

Ectrodactyly-ectodermal dysplasia-cleft lip/palate (EEC) syndrome: EEC syndrome is the commonest p63-related ectodermal dysplasia. Patients show features of ectodermal dysplasia with cleft lip/palate and variable limb defects including soft-tissue and bony syndactyly and split hand foot malformation or ectrodactyly.[3,9] Hypotrichosis, a variety of dental abnormalities (hypodontia/anodontia/microdontia/enamel hypoplasia and peg-shaped incisors),

nail abnormalities (onychodysplasia) and, occasionally, hypohidrosis may also be present. Apart from the three cardinal features of EEC, other clinical characteristics that have been described include abnormalities of the lacrimal system, genitourinary anomalies, hearing defects, choanal atresia, mammary hypoplasia and facial dysmorphism.[9] EEC syndrome patients have characteristic facies with maxillary hypoplasia, a short philtrum and a broad nasal tip.

The commonest ophthalmic abnormality in EEC is the absence or a significant reduction in the number of meibomian glands with a resultant evaporative dry eye and reduced tear film breakup time (Fig. 9.14).[3] The aqueous component of the tear film resulting from lacrimal gland secretion is also commonly reduced. Lacrimal drainage system defects have been reported in most EEC syndrome cases. These include absence, stenosis or occlusion of the puncta and/or canaliculi; absence of the lacrimal sac; lacrimal fistula; nasolacrimal duct stenosis and obstruction including the complete absence of the membranous and bony nasolacrimal duct components.[3,9] These abnormalities can lead to epiphora from birth, bilateral dacryocystitis, lacrimal abscess and mucocele.[10] The major cause

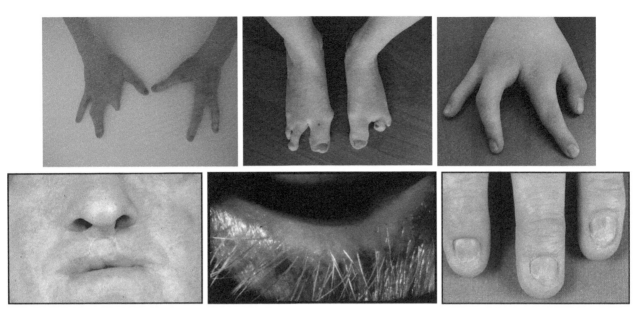

FIG. 9.14 Ectrodactyly-ectodermal dysplasia-cleft lip/palate (EEC) syndrome showing limb phenotype (syndactyly and ectrodactyly); repaired cleft lip; absent meibomian glands; and nail dysplasia.

of visual morbidity in EEC syndrome is limbal stem cell deficiency leading to corneal ulceration and subsequent, progressive corneal neovascularisation and/or scarring.[3] EEC patients with limbal stem cell deficiency have absent limbal palisades of Vogt and develop progressive 'conjunctivalisation' of the cornea. Limbal stem cell deficiency is age-related and more common in the 4th decade resulting in significant visual morbidity; there is no apparent relationship between limbal stem cell failure and the severity of the systemic phenotype in EEC syndrome. Conjunctival cicatrisation can also be seen. Despite its ectodermal origin, the crystalline lens has only been reported to be involved in a minority of EEC patients.[3]

Ankyloblepharon-ectodermal dysplasia-cleft lip/palate (AEC) syndrome: AEC syndrome (also known as Hay–Wells syndrome) combines ectodermal dysplasia with clefting and ankyloblepharon (fusion of the eyelids). Ankyloblepharon is rarely complete, and the lids are usually joined by a solid band at the lateral or medial margins, or by numerous epithelial strands (ankyloblepharon filiforme adnatum). The ectodermal dysplasia in AEC syndrome produces wiry scalp hair with complete or partial hair loss, hypohidrosis, severe nail dystrophy or absent nails, palmoplantar keratoderma, and dental anomalies (anodontia or hypodontia with pointed and spaced teeth). Associated abnormalities include maxillary hypoplasia, lacrimal duct anomalies, supernumerary nipples, syndactyly and auricular deformities. The ophthalmic phenotype in AEC syndrome resembles EEC syndrome with absence of the meibomian glands, ocular surface abnormalities and corneal scarring although the number of reported cases is small.[8]

The p63-related ectodermal dysplasias (EEC and AEC syndromes) can have a severe ophthalmic phenotype resulting in corneal scarring and vascularisation. Support of the ocular surface with preservative-free lubrication, staphylococcal decolonisation strategies and judicious use of topical steroids can retain corneal function. Persistent epithelial defects may require bandage contact lenses, serum eyedrops and tarsorrhaphy to promote resolution. The management of the corneal disease associated with EEC syndrome is challenging as treatments are not aimed at the underlying condition and the clinical effects are variable. Corneal transplantation without addressing the underlying limbal stem cell deficiency is likely to fail.

Keratitis-ichthyosis-deafness syndrome

Keratitis-ichthyosis-deafness (KID) syndrome is an autosomal dominant ectodermal dysplasia characterised by vascularising keratitis, sensorineural hearing loss and erythrokeratoderma. Heterozygous mutations in the gap junction protein 26 (or connexin 26) gene have been associated with this condition.[4]

The skin lesions do not represent true ichthyosis and are classified as an erythrokeratoderma forming symmetrical, well-circumscribed, erythematous and hyperkeratotic plaques on the face and extremities. The skin is thickened and often has a coarse-grained appearance. Patients usually develop follicular hyperkeratosis and palmoplantar palmar keratoderma. Congenital sensorineural hearing loss is normally severe and bilateral. Dystrophic hair and nails, heat intolerance and dental defects are other typical features.

There is commonly an increased susceptibility to mucocutaneous infection which can be fatal in the neonatal period. The development of squamous cell carcinoma of the skin and oral mucosa is a rare but life-threatening complication.

The ocular phenotype in KID syndrome can be severe with intensive superficial and deep neovascularisation, recurrent corneal epithelial defects and limbal stem cell deficiency.[4,11] Often there is hyperkeratotic thickened lid skin with rounding, vascularity and hyperkeratosis of the lid margin; madarosis of the upper and lower lashes may also be present. The meibomian glands are present but often stenosed and keratinised. Ophthalmic management in KID syndrome is equally challenging and some patients required living related or allogenic limbal grafts or keratoprosthesis. There is some evidence that systemic antifungal (chronic oral ketoconazole therapy) may improve the dermal and ophthalmic phenotype in KID syndrome and reduce the risk of squamous cell carcinoma.[12]

References

1. Pagnan NAB, Visinoni ÁF. Update on ectodermal dysplasias clinical classification. *Am J Med Genet A* 2014;**164**(10):2415–23. https://doi.org/10.1002/ajmg.a.36616.
2. Guazzarotti L, et al. Phenotypic heterogeneity and mutational spectrum in a cohort of 45 Italian males subjects with X-linked ectodermal dysplasia. *Clin Genet* 2015;**87**(4). https://doi.org/10.1111/cge.12404.
3. Di Iorio E, et al. Limbal stem cell deficiency and ocular phenotype in ectrodactyly-ectodermal dysplasia-clefting syndrome caused by p63 mutations. *Ophthalmology* 2012;**119**(1):74–83. https://doi.org/10.1016/j.ophtha.2011.06.044.
4. Richard G, et al. Missense mutations in GJB2 encoding connexin-26 cause the ectodermal dysplasia keratitis-ichthyosis-deafness syndrome. *Am J Hum Genet* 2002;**70**(5):1341–8. 2002/03/26 S0002-9297(07)62527-0 [pii]. https://doi.org/10.1086/339986.
5. Clarke A, et al. Clinical aspects of X-linked hypohidrotic ectodermal dysplasia. *Arch Dis Child* 1987;**62**(10):989–96.
6. Callea M, Teggi R, Yavuz I, Tadini G, Priolo M, Crovella S, et al. Ear nose throat manifestations in hypoidrotic ectodermal dysplasia. *Int J Pediatr Otorhinolaryngol*. 2013;**77**(11):1801–4.
7. Montanari M, Callea M, Battelli F, Piana G.Oral rehabilitation of children with ectodermal dysplasia. *BMJ Case Rep*. 2012 Jun 21;2012:bcr0120125652.
8. Kaercher T. Ocular symptoms and signs in patients with ectodermal dysplasia syndromes. *Graefes Arch Clin Exp Ophthalmol* 2004;**242**(6):495–500. https://doi.org/10.1007/s00417-004-0868-0.
9. Buss PW, Hughes HE, Clarke A. Twenty-four cases of the EEC syndrome: clinical presentation and management. *J Med Genet* 1995;**32**(9):716–23.
10. McNab AA, Potts MJ, Welham RA. The EEC syndrome and its ocular manifestations. *Br J Ophthalmol* 1989;**73**(4):261–4.
11. Messmer EM, et al. Ocular manifestations of keratitis–ichthyosis–deafness (KID) syndrome. *Ophthalmology* 2005;**112**(2):e1–6. https://doi.org/10.1016/j.ophtha.2004.07.034.
12. Hazen PG, et al. Keratitis, ichthyosis, and deafness (KID) syndrome: management with chronic oral ketoconazole therapy. *Int J Dermatol* 1992;**31**(1):58–9. https://doi.org/10.1111/j.1365-4362.1992.tb03524.x.

Chapter 10

Anterior segment developmental disorders

Chapter outline

10.1 Primary congenital glaucoma 98
10.2 Primary juvenile glaucoma 102
10.3 Axenfeld-Rieger spectrum 105
10.4 Peters anomaly 109

Ocular anterior segment developmental disorders represent a spectrum of developmental abnormalities that involve a single (e.g. trabecular meshwork in congenital glaucoma) or multiple (e.g. anterior segment dysgenesis) anterior segment structures. Conditions predominantly affecting the cornea or the lens can be included in this heterogeneous group of disorders but most of these diseases are discussed in separate sections of this book (Chapters 9, 11 and 12).

Many anterior segment developmental disorders involve the trabecular meshwork and often result in high intraocular pressure in the first year of life. Associated signs and symptoms including corneal haze, globe enlargement (buphthalmos), tearing, and photophobia are often what brings affected children to medical attention. Other presenting features may include obvious iris malformations, corneal opacification, nystagmus and sensory strabismus. Children with mild disease subtypes may not be picked up until school screening or adulthood.

Anterior segment developmental disorders may be unilateral or bilateral. In unilateral cases, genetic investigations very rarely yield a molecular diagnosis, particularly in the absence of systemic features or relevant family history (e.g. of neurofibromatosis). A positive genetic test result is much more likely in bilateral cases, including in cases that are initially thought to be unilateral but are ultimately found to be bilateral but asymmetrical following detailed examination.

A large number of genes have been implicated in this group of conditions (Table 10.1) and genetic analysis typically involves multi-gene panel testing or, exome/genome sequencing. It is worth highlighting though that, certain phenotypes are highly suggestive of mutations in specific genes. These include cornea plana (*KERA*), congenital hereditary endothelial dystrophy (*SLC4A11*) (Section 9.3), classic aniridia (*PAX6*) (Chapter 17), and megalocornea with zonular weakness and secondary lens-related glaucoma (*LTBP2*) (Section 12.3).

Anterior segment developmental abnormalities are occasionally the presenting feature of a multisystemic condition and genetic testing can highlight individuals who are at risk of developing significant extraocular manifestations (e.g. risk of Wilms tumour in infants presenting with sporadic aniridia; Chapter 17). Genetic analysis can also help reduce prognostic uncertainty and refine the risk to future pregnancies and family members.

TABLE 10.1 Anterior segment development disorder: Main phenotypes and underlying disease genes.

Phenotype	Main gene(s)	Other key gene(s)
Aniridia	*PAX6*	*PITX2, FOXC1, COL4A1, CPAMD8*
Primary congenital glaucoma	*CYP1B1*	*LTBP2, TEK*
Primary juvenile glaucoma	*MYOC*	*OPTN, TBK1, CYP1B1, LTBP2, CPAMD8*
Axenfeld-Rieger anomaly spectrum	*PITX2, FOXC1*	*PAX6, COL4A1, CYP1B1, CPAMD8*
Peters anomaly/severe corneal opacification	*PAX6, PITX2, FOXC1, PITX3, FOXE3*	*CYP1B1, COL4A1, PXDN, GJA8*

10.1

Primary congenital glaucoma

Panagiotis I. Sergouniotis, Arif O. Khan, and Graeme C.M. Black

Primary congenital glaucoma is a non-syndromic condition that becomes clinically apparent in the first few years of life and is characterised by developmental abnormalities of the trabecular meshwork and anterior chamber angle. These defects prevent adequate drainage of aqueous humour, leading to ocular hypertension early in life and often resulting in optic neuropathy and visual loss.[1]

Primary congenital glaucoma is usually bilateral although it can be asymmetrical. Its incidence varies geographically and, intriguingly, it appears to be somewhat more prevalent in males than females. Sporadic cases (i.e. without a family history) are common and both autosomal recessive and autosomal dominant forms of this condition have been described.[2,3]

Clinically diagnosing primary congenital glaucoma is often straightforward but can occasionally be difficult, especially when the affected child's cooperation is suboptimal. A particular challenge is around distinguishing primary congenital glaucoma from secondary forms of childhood glaucoma, including those associated with Axenfeld-Rieger spectrum disorders (Section 10.3), Peters anomaly (Section 10.4), ectopia lentis (Section 12.3), Lowe syndrome (Section 11.2.6), and neurofibromatosis (Section 22.1).[4] Genetic testing can help exclude these secondary causes and can inform genetic counselling.[2]

Clinical characteristics

Affected individuals are typically diagnosed in the first year of life. Based on the age at presentation, the following subtypes have been described:

- newborn-onset (most severe, clinically apparent between birth and age 1 month),
- infantile-onset (clinically apparent between age 1 month and age 2 years),
- late-onset (or late-recognised; clinically apparent after age 2 years but typically before age 4 years, see also Section 10.2 on juvenile open-angle glaucoma)
- spontaneously arrested (i.e. presence of Haab striae with or without corneal haze in the context of normal intraocular pressure).[5]

The symptoms are non-specific and include tearing, photophobia, eye rubbing, irritability and failure to thrive.

High intraocular pressure leads to globe/corneal enlargement (buphthalmos) (Fig. 10.1A–C), corneal haze that is associated with breaks in Descemet membrane (Haab striae) (Figs 10.1D and 10.2), optic nerve cupping and myopia. Gonioscopy shows a featureless angle, often with a characteristically flat iris insertion (Fig. 10.1E). Asymmetry between eyes is common and, by definition, there are no other ocular or extraocular developmental anomalies.

Molecular pathology

Primary congenital glaucoma is genetically heterogeneous and several genes have been implicated. These include:

- *CYP1B1*: Biallelic variants in the *CYP1B1* gene, which encodes the cytochrome *p*450 B1 enzyme, are the commonest cause of primary congenital glaucoma and account for a significant proportion of newborn-onset cases. A wide range of mutation types has been described including missense, nonsense, frameshift and splice site changes. No clear genotype–phenotype correlation has emerged. A few specific mutations account for most congenital glaucoma cases in selected populations including the c.1159G>A (p.Glu387Lys) variant in Slovakian Roma populations and the c.182G>A (p.Gly61Glu) variant in Saudi Arabian populations. In contrast, the c.1103G>A (p.Arg368His) mutation has been seen in affected individuals of Chinese, Iranian, Indian, and Pakistani ancestries.[3]
- *TEK*: Families lacking mutations in *CYP1B1* are occasionally found to have heterozygous loss of function variants in the *TEK* gene. These mutations result in a form of primary congenital glaucoma that is inherited as an autosomal dominant trait with reduced penetrance. *TEK* encodes the TIE2 receptor tyrosine kinase angiopoietin receptor, which has been shown to interact directly with *CYP1B1*.
- *LTBP2*: Biallelic variants in *LTBP2* have also been described in children who had been diagnosed with primary congenital glaucoma. However, such variants are more commonly seen in association with other ocular developmental abnormalities including megalocornea (which may be mistaken as secondary buphthalmos), microspherophakia, zonular weakness and secondary lens-related glaucoma (Section 12.3).

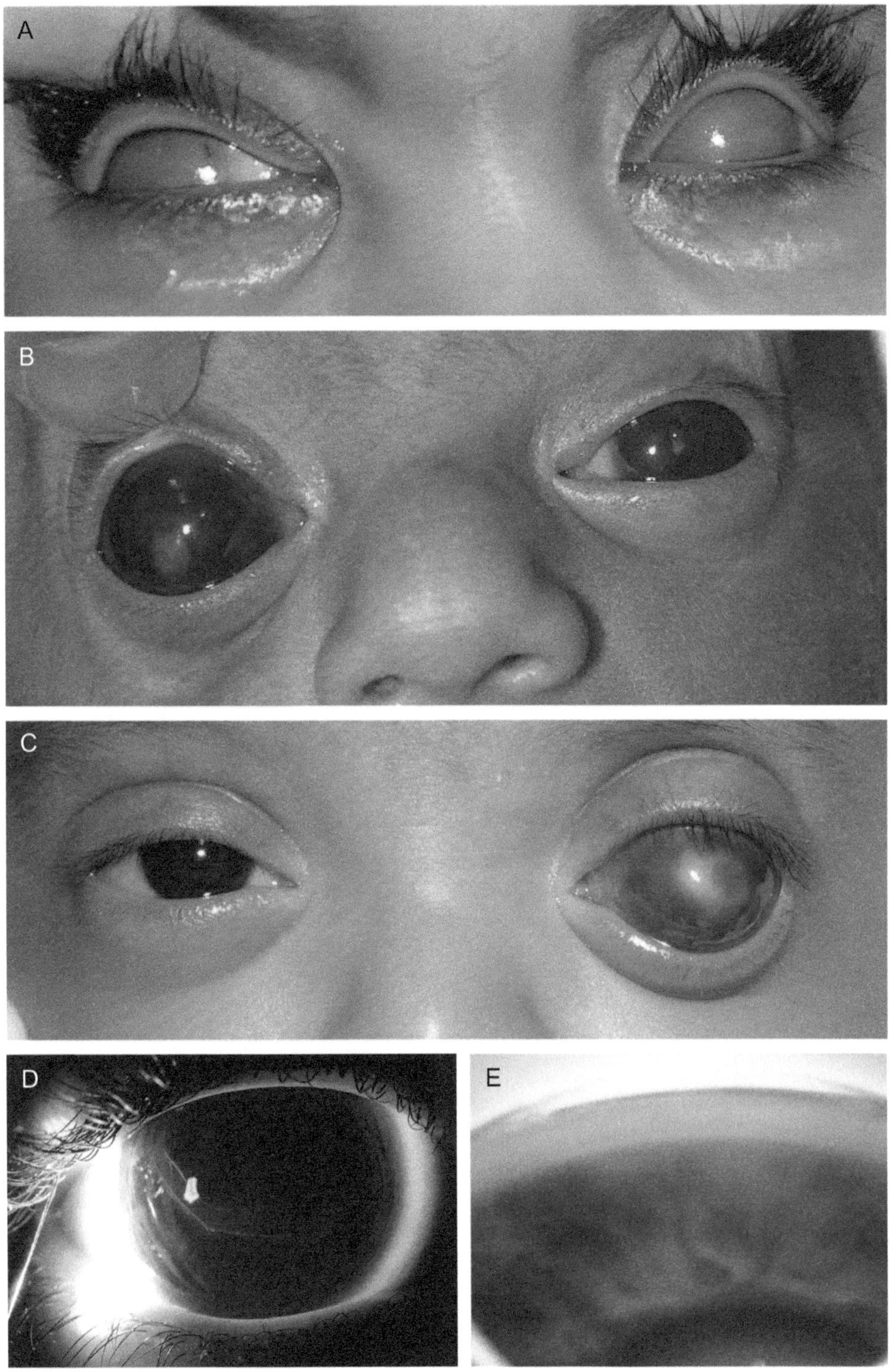

FIG. 10.1 Findings in primary congenital glaucoma. (A) External photograph showing bilateral globe enlargement and corneal opacification in an affected neonate. (B) External photograph showing right globe enlargement (asymmetrical buphthalmos) and corneal opacification in an affected neonate. (C) External photograph showing left globe enlargement (asymmetrical buphthalmos) and corneal opacification in an affected infant. (D) External photograph showing Haab striae with minimal associated corneal haze in a 7-year-old with spontaneously arrested primary congenital glaucoma (see also Fig. 10.2). (E) Gonioscopic image revealing a high/anterior iris insertion that masks the trabecular meshwork in an affected child.

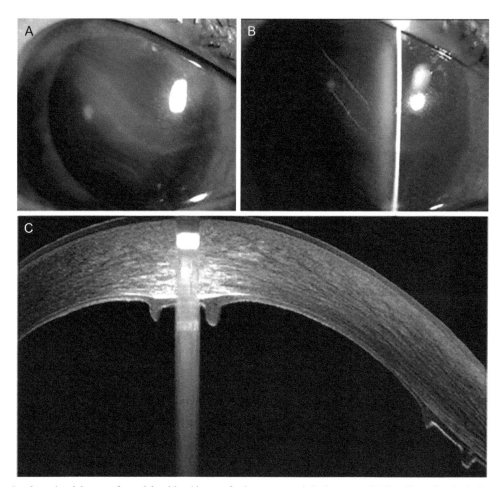

FIG. 10.2 Imaging from the right eye of an adult with a history of primary congenital glaucoma. (A) Curvilinear breaks in Descemet membrane (Haab striae). These breaks are typically oriented horizontally or are concentric to the limbus (in contrast to Descemet membrane tears resulting from birth trauma that are usually vertical or obliquely oriented). (B) Retro-illumination highlights the margins of the Descemet membrane defect. (C) Anterior segment optical coherence tomography showing the thickened edges of the Descemet membrane protruding into the anterior chamber. *(Adapted from Ref. 6.)*

- *PXDN:* A small number of families with congenital/developmental glaucoma have been found to carry biallelic missense mutations in *PXDN*. *PXDN* encodes peroxidasin and is also known to cause congenital cataracts, anterior segment dysgenesis, sclerocornea and microcornea.

Clinical management

Management of individuals with primary congenital glaucoma is generally directed towards optimising intraocular pressure control, correcting existing errors of refraction and treating secondary complications. It is generally surgical and the preferred approach depends on the severity of the phenotype. Options include goniotomy, trabeculotomy (conventional or microcatheter-assisted 360° approach), trabeculectomy, cycloablation (endoscopic or trans-scleral) and glaucoma drainage implant insertion (e.g. Baerveldt or Paul tube). Medical management is less effective although it can often temporise the need for surgery.[7] Topical β-blockers, carbonic anhydrase inhibitors and prostaglandins are the most commonly used agents; α-agonists should be avoided in young children.[8]

References

1. Chaudhary RS, Gupta A, Sharma A, Gupta S, Sofi RA, Sundar D, et al. Long-term functional outcomes of different subtypes of primary congenital glaucoma. *Br J Ophthalmol* 2019. https://doi.org/10.1136/bjophthalmol-2019-315131.
2. Ma AS, Grigg JR, Jamieson RV. Phenotype-genotype correlations and emerging pathways in ocular anterior segment dysgenesis. *Hum Genet* 2019;**138**(8–9):899–915.

3. Lewis CJ, Hedberg-Buenz A, DeLuca AP, Stone EM, Alward WLM, Fingert JH. Primary congenital and developmental glaucomas. *Hum Mol Genet* 2017;**26**(R1):R28–36.
4. Midha N, Sidhu T, Chaturvedi N, Sinha R, Shende DR, Dada T, et al. Systemic associations of childhood glaucoma: a review. *J Pediatr Ophthalmol Strabismus* 2018;**55**(6):397–402.
5. European Glaucoma Society. *Terminology and guidelines for glaucoma.* 5th ed; 2020. https://www.eugs.org/eng/guidelines.asp.
6. Mandal AK, Raghavachary C, Peguda HK. Haab's Striae. *Ophthalmology* 2017;**124**(1):11. https://doi.org/10.1016/j.ophtha.2016.07.002.
7. Gagrani M, Garg I, Ghate D. Surgical interventions for primary congenital glaucoma. *Cochrane Database Syst Rev* 2020;**8**, CD008213.
8. Freedman SGG, et al; 2020 https://eyewiki.aao.org/Primary_Congenital_Glaucoma.

10.2

Primary juvenile glaucoma

Graeme C.M. Black, Panagiotis I. Sergouniotis, and Arif O. Khan

Primary juvenile glaucoma (or juvenile open-angle glaucoma) is an early-onset form of glaucoma that is often passed down through families as an autosomal trait. It is most commonly diagnosed between age 4 and 40 years and it typically occurs in the context of normal-appearing, open-angle structures (Box 10.1). By definition, it is not associated with other ocular developmental anomalies or syndromic features.[1,2]

The molecular pathology of juvenile glaucoma is complex, and comprehensive genetic analysis points to a molecular diagnosis in only about one in four cases (most of whom have heterozygous variants in the *MYOC* gene). Despite this, genetic testing is recommended in this cohort, especially in affected individuals who have a strong family history or whose parents are related (consanguineous). Identifying a disease-causing variant enables ophthalmologists to plan closer monitoring and, often, more aggressive treatment regimens.[3]

Clinical characteristics

Individuals with juvenile glaucoma are generally asymptomatic until there is advanced field loss. Unlike primary congenital glaucoma, signs such as globe enlargement, corneal haze and Descemet membrane breaks (Haab striae) are absent. Compared to primary open-angle glaucoma of later onset, the juvenile form tends to be more rapidly progressive and to have more severely elevated intraocular pressures (often in the range of 40–50 mmHg). There is male preponderance and affected individuals often report a family history of glaucoma. Myopia is common and the iridocorneal angles appear open on gonioscopy; prominent iris processes may be noted. Similar to other forms of glaucoma, there is progressive degeneration of the optic nerve leading to visual field loss.[2]

BOX 10.1 Genetic testing for primary open-angle glaucoma.

Primary open-angle glaucoma can be classified by age of onset into juvenile forms (onset in general between 4 and 40 years) and typical forms (onset over 40 years). Both these subtypes have significant heritability that warrants clinical evaluation of first-degree relatives.

Juvenile glaucoma represents a small percentage of total cases and may be either an autosomal dominant or an autosomal recessive trait. In a significant proportion of cases (around 25%), the underlying genetic aetiology comprises either heterozygous (dominant) or biallelic (recessive) variants of large effect.[1] Genetic diagnostic testing is therefore justified in this group of patients since the identification of a highly penetrant pathogenic variant provides diagnostic certainty and facilitates directed family screening.

For open-angle glaucoma of later onset, current approaches for genetic analysis (e.g., multi-gene panels) have a low pick-up rate and are very seldom justified. However, different testing strategies utilising polygenic scores have shown considerable promise.

Polygenic scores: Primary open-angle glaucoma is a complex genetic disorder and its molecular basis is contributed to by a range of single nucleotide variants scattered throughout the genome. Although most of these genetic variants are present in the general population, cumulatively they are more common in individuals with, or at risk of, glaucoma. Using this information, it has been possible to develop a so-called 'polygenic score' or 'polygenic risk score' for glaucoma (Fig. 10.3)[9,10] (see also Section 13A.15). In a research environment, polygenic scores have been shown to:

- stratify populations by risk (i.e. identify individuals at high risk of developing glaucoma)
- predict individuals with glaucoma who are likely to show progressive disease or require surgical intervention.

Although currently a research tool, it is likely that, in the future, the use of polygenic scores as a screening tool will be a valuable adjunct to the management of primary open-angle glaucoma. It has the potential to facilitate a more rational allocation of resources through appropriate screening and timely treatment of high-risk patients, with reduced clinical monitoring burden in low-risk groups.[7] Ultimately, polygenic scores are likely to extend the use of genetic testing amongst general ophthalmologists.

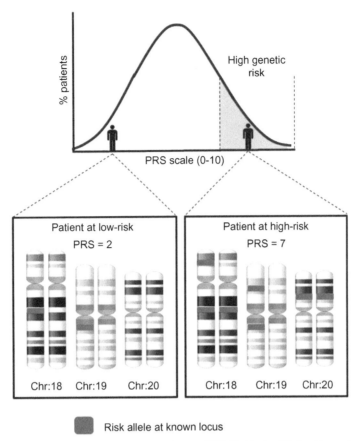

FIG. 10.3 Schematic explanation of the polygenic score (or polygenic risk score, PRS). In this simplified and imprecise example, differences in the DNA sequence of five genetic locations (loci) in three chromosomes are assumed to contribute to differences in the risk of glaucoma. The presence of risk alleles in these loci is assessed and the cumulative impact—the PRS—is calculated. *(Adapted from Ref. 8.)*

Molecular pathology

Mutations in several genes have been identified in people with juvenile glaucoma. These include:

- *MYOC*: Heterozygous *MYOC* mutations are frequent amongst individuals with juvenile glaucoma (around 20% of cases) and are also encountered in people with typical open-angle glaucoma (around 2% of cases). *MYOC* encodes myocilin, a 504 amino-acid secreted glycoprotein of broad expression and uncertain function. Over 150 mutations have been described, the majority of which are missense variants that lie within the olfactomedin domain of the protein. It has been shown that pathogenic variants prevent normal myocilin secretion from ocular tissues into the aqueous humour and lead to intracellular accumulation of mutant protein in the trabecular meshwork. Disease severity and penetrance vary between mutations. The c.1109C>T (p.Pro370Leu) and c.1430T>A (p.Ile477Asn) variants for example have over 80% penetrance at 25 years of age whereas the common c.1102C>T (p.Gln368Ter) variant has less than 5% penetrance at the same timepoint. At 75 years of age, the corresponding penetrance is around 100% and 50% respectively; the mean intraocular pressure is 40 mmHg and 30 mmHg respectively.[3] It is worth highlighting that a presumed loss of function variant, c.136C>T (p.Arg46Ter), is not associated with glaucoma in either a heterozygous or a homozygous state. This suggests that MYOC mutations cause disease via a gain of function (dominant negative) rather than a loss of function mechanism.[1,3]

- *CYP1B1*: Biallelic *CYP1B1* mutations are a common cause of primary congenital glaucoma (around 15% of cases) (Section 10.1) and have also been identified in cases with juvenile glaucoma (around 1.5% of cases). It is has been suggested that rare heterozygous *CYP1B1* variants may contribute to open-angle glaucoma.[1,4]

- *LTBP2*: Biallelic *LTBP2* mutations are associated with a wide range of developmental phenotypes (Sections 10.1 and 12.3). Occasional patients with juvenile glaucoma have been described as carrying either biallelic or heterozygous presumed pathogenic variants in *LTBP2*.[1,5,6]

- *CPAMD8*: Biallelic *CPAMD8* mutations are a known cause of anterior segment dysgenesis (Section 10.4). A small number of individuals with juvenile glaucoma have been found to carry biallelic, loss of function variants in *CPAMD8*.[7]

- *OPTN*: Heterozygous *OPTN* mutations have been identified in a small number of individuals with juvenile normal-tension glaucoma and strong family history. The *OPTN* gene is expressed in several tissues including the trabecular meshwork and ganglion cells. It encodes optineurin, a protein to which several biological functions have been attributed including a role in optic vesicle formation during embryogenesis and roles in autophagy and neuroprotection in the adult eye. The c.148G > A (p.Glu50Lys) variant has been identified in multiple affected individuals; it has been shown to cause ganglion cell apoptosis in animal models and is a key focus of optineurin glaucoma research.[3]
- *TBK1*: Heterozygous copy number variants affecting *TBK1* (e.g., whole gene duplications or triplications) have been identified in a small number of individuals with juvenile normal-tension glaucoma. Like *OPTN*, *TBK1* is not associated with high intraocular pressures. It is broadly expressed in human tissues and, in the eye, it is most abundant in ganglion cells. Like *OPTN*, it is thought to have a role in autophagy.[3]

Clinical management

Patient management involves lowering the intraocular pressure and regularly assessing the rate of disease progression (with fundus imaging and visual field testing). The main goal is to prevent further optic nerve damage and the principles are the same as in typical primary open-angle glaucoma. It is worth highlighting though that laser trabeculoplasty tends to be ineffective and is generally not recommended in this group of patients. Also, people with juvenile glaucoma are highly likely to require surgical intervention; options, include trabeculectomy, trabecular bypass stents, drainage implants, angle procedures (goniotomy, trabeculotomy), and cycloablative procedures.

References

1. Zhou T, Souzeau E, Siggs OM, Landers J, Mills R, Goldberg I, et al. Contribution of mutations in known mendelian hlaucoma henes to advanced early-onset primary open-angle glaucoma. *Invest Ophthalmol Vis Sci* 2017;**58**(3):1537–44.
2. Turalba AV, Chen TC. Clinical and genetic characteristics of primary juvenile-onset open-angle glaucoma (JOAG). *Semin Ophthalmol* 2008;**23**(1):19–25.
3. Sears NC, Boese EA, Miller MA, Fingert JH. Mendelian genes in primary open angle glaucoma. *Exp Eye Res* 2019;**186**:107702.
4. Lewis CJ, Hedberg-Buenz A, DeLuca AP, Stone EM, Alward WLM, Fingert JH. Primary congenital and developmental glaucomas. *Hum Mol Genet* 2017;**26**(R1):R28–36.
5. Saeedi O, Yousaf S, Tsai J, Palmer K, Riazuddin S, Ahmed ZM. Delineation of novel compound heterozygous variants in LTBP2 associated with juvenile open angle Glaucoma. *Genes (Basel)* 2018;**9**(11):527.
6. Khan AO, Aldahmesh MA, Alkuraya FS. Congenital megalocornea with zonular weakness and childhood lens-related secondary glaucoma - a distinct phenotype caused by recessive LTBP2 mutations. *Mol Vis* 2011;**17**:2570–9.
7. Siggs OM, Souzeau E, Taranath DA, Dubowsky A, Chappell A, Zhou T, et al. Biallelic CPAMD8 variants are a frequent cause of childhood and juvenile open-angle glaucoma. *Ophthalmology* 2020;**127**(6):758–66.
8. Hernández-Beeftink T, Guillen-Guio B, Villar J, Flores C. Genomics and the acute respiratory distress syndrome: current and future directions. *Int J Mol Sci* 2019;**20**(16):4004.
9. Craig JE, Han X, Qassim A, Hassall M, Cooke Bailey JN, Kinzy TG, et al. Multitrait analysis of glaucoma identifies new risk loci and enables polygenic prediction of disease susceptibility and progression. *Nat Genet* 2020;**52**(2):160–6.
10. Torkamani A, Wineinger NE, Topol EJ. The personal and clinical utility of polygenic risk scores. *Nat Rev Genet* 2018;**19**(9):581–90.

10.3

Axenfeld-Rieger spectrum

Arif O. Khan, Panagiotis I. Sergouniotis, and Graeme C.M. Black

Axenfeld and Rieger anomalies are eponymous names used to describe slightly different developmental abnormalities of the anterior segment. It is now understood that these form part of a spectrum; they share an identical mechanistic aetiology and can therefore be considered together.[1] Here, they are described under the grouping of 'Axenfeld-Rieger spectrum'.

Axenfeld-Rieger spectrum is a manifestation of a heterogeneous collection of disorders with ophthalmic and, often, systemic features. The associated ophthalmic features tend to be bilateral, congenital, and are thought to be caused by abnormal differentiation and migration of neural crest cells during the formation of anterior ocular structures.[2,3] Systemic features may include abnormal dentition, redundant periumbilical skin, hypospadias, hearing impairment, cardiac and pituitary abnormalities. Several genes have been implicated including *PITX2*, *FOXC1* (both associated with the dominant disease potentially with syndromic features), and *COL4A1* (associated with dominant disease potentially with angiopathy involving the brain, kidney and/or muscles).

Clinical characteristics

Ophthalmic features: Most affected individuals have a white line on the posterior peripheral cornea (posterior embryotoxon) and a range of iris abnormalities (Fig. 10.4). Posterior embryotoxon represents a thickened and anteriorly displaced Schwalbe line (i.e. the peripheral termination of Descemet membrane). In isolation, this feature is thought to be present in a significant proportion of the healthy population (quoted in some studies as 8%–15%).[2] However, the presence of posterior embryotoxon is most significant in association with additional features, including peripheral broad bands of iris that project across the iridocorneal angle and up onto the cornea (iridocorneal adhesions, peripheral anterior synechiae).

Iris abnormalities are variable and include corectopia (decentered pupil), polycoria (additional holes in the iris) and generalised hypoplasia (Fig. 10.4). A high iris insertion may restrict aqueous outflow through the trabecular meshwork and cause high intraocular pressure. Microscopic abnormalities of the trabecular meshwork may also impair aqueous drainage. As a result, elevated intraocular pressure and glaucoma develop in about 50% of Axenfeld-Rieger spectrum patients.[4,5]

In some cases, features of the Axenfeld-Rieger spectrum are combined with central corneal opacification due to defects in the posterior corneal epithelium; iridocorneal and/or lenticulocorneal adhesions may also be present. The term Peters anomaly has been used to describe this constellation of findings (see Section 10.4). Peters anomaly can represent the most severe end of the Axenfeld-Rieger spectrum and generally cannot—from an aetiological viewpoint—be fully separated from this group of conditions.

Axenfeld-Rieger spectrum shows significant variability even amongst individuals with identical mutations. Also, there is often considerable asymmetry between the eyes of the same patient; affected individuals occasionally have

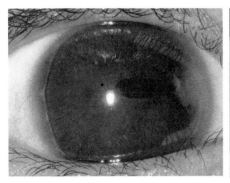 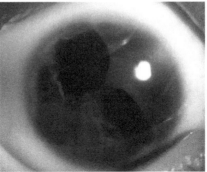 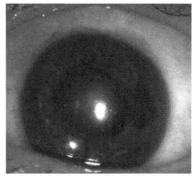

FIG. 10.4 External photographs from three unrelated children with Axenfeld-Rieger spectrum. Features shown include posterior embryotoxon, corectopia, polycoria and iris hypoplasia.

different phenotypes in each eye, e.g., Peters anomaly in one eye and Axenfeld-Rieger anomaly in the contralateral eye. This underlines the notion that this group of conditions form a spectrum.

Reduced vision in Axenfeld-Rieger spectrum generally results from visually significant corneal opacification or congenital/developmental glaucoma. Cataract is rarely a major part of this phenotypic spectrum and, apart from a small number of cases resulting from *PAX6* gene defects, foveal development tends to be normal.

Systemic manifestations: The two commonest genes mutated in patients with Axenfeld-Rieger spectrum are *PITX2* and *FOXC1*. Affected individuals carrying mutations in *PITX2* often have systemic features including a characteristic facial appearance (broad nasal root, maxillary hypoplasia), dental anomalies (hypodontia, partial anodontia) and an unusual umbilical morphology with redundant periumbilical skin (Fig. 10.5). A range of other abnormalities including cardiac, pituitary and gastrointestinal (e.g., anal stenosis) have also been described. Individuals with mutations in *FOXC1* are more likely to present with an isolated ophthalmic phenotype although hearing impairment is common and patients with cardiac abnormalities and various other systemic manifestations have been reported.[3]

A number of syndromic disorders in which the Axenfeld-Rieger spectrum is a commonly encountered aspect of the overall phenotype have been described. These include:

- *COL4A1-related disorders*, a group of autosomal dominant conditions that are caused by mutations in a type IV collagen gene. They are generally characterised by the presence of cerebrovascular disease with variable ophthalmic, renal, and muscular involvement. Ophthalmic abnormalities are variably observed and may include microphthalmia, cataract, Axenfeld-Rieger spectrum, Peters anomaly, glaucoma and retinal arterial tortuosity.[2,7] Cerebrovascular abnormalities occur from foetal life onward and their severity may range from subclinical small vessel brain disease to fatal cerebral haemorrhage. In some cases, recurrent perinatal or childhood cerebrovascular episodes leading to childhood hemi/quadriplegia, leukoencephalopathy or porencephaly (i.e. cysts or cavities within the cerebral hemispheres) may occur.

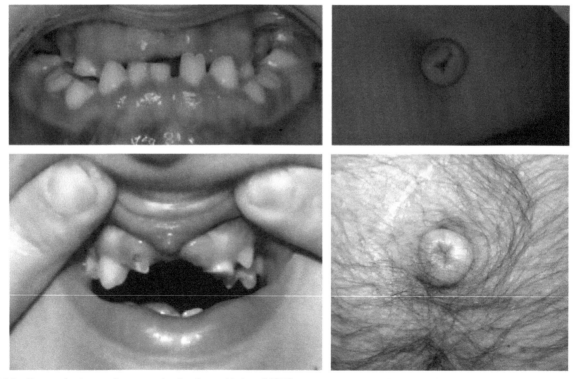

FIG. 10.5 Extraocular features in two unrelated patients with Axenfeld-Rieger spectrum. External photographs showing absent permanent upper and some lower incisors (left-hand side images) and redundant periumbilical skin (right-hand side images). *(The images in the upper panel have been adapted from Ref. 6.)*

- *Alagille syndrome*, an autosomal dominant disorder primarily affecting the liver (leading to jaundice secondary to abnormal bile duct development) and the heart (causing pulmonary stenosis, ventricular septal defect and/or tetralogy of Fallot). It is caused by mutations in either *JAG1* or *NOTCH2*, genes that encode proteins that are crucial to the notch gene signalling cascade. Anterior segment dysgenesis (typically posterior embryotoxon) and optic disc drusen are frequent features of this condition.
- *SHORT syndrome*, an autosomal dominant disorder associated with short stature, joint hyperextensibility, facial dysmorphism (ocular depression), Rieger anomaly and delayed tooth eruption. It is caused by heterozygous variants in *PIK3R1*. There is a high risk of diabetes mellitus and lipoatrophy; hearing loss may also be present.

Molecular pathology

The development of the anterior segment of the eye is dependent on the migration of cells that are largely derived from the neural crest associated periocular mesenchyme. Several transcription factors are critical to this process including *PITX2*, *FOXC1* and *PAX6*. These developmental genes are expressed early in embryonic development within the periocular mesenchyme and often work in concert. *PITX2* and *FOXC1* for example interact physically with one another with the former being a negative regulator of the latter.

Although the genetic basis of anterior segment dysgenesis is incompletely understood, studies have shown that comprehensive genetic testing can identify a molecular defect in over 50% of affected individuals, including sporadic cases.[2] *PITX2* and *FOXC1* are the main genes involved in the Axenfeld-Rieger spectrum. Both are associated with the autosomal dominant disease with high penetrance but variable expressivity.

For both *PITX2* and *FOXC1*, a range of mutations have been described including missense (in particular within their DNA binding domains) and, less frequently, presumed loss of function (nonsense, frameshift, splicing) changes. Whole gene deletions (for *PITX2* and *FOXC1*) and duplications (for *FOXC1*) have also been reported highlighting the importance of incorporating copy number variation analysis in genetic tests for these conditions.

Other genes that underly anterior segment dysgenesis and cause predominantly ophthalmic phenotypes include:

- *PXDN*, encoding peroxidasin, a protein that plays a role in collagen IV basement membrane cross-linking. Biallelic variants in *PXDN* can cause a complex array of anterior segment developmental phenotypes including microcornea, cataract and congenital glaucoma.
- *CPAMD8*, a gene that is expressed in the periocular mesenchyme and encodes a member of the alpha-2-macroglobulin/complement 3 protein family. Biallelic variants in *CPAMD8* can also cause an array of ophthalmic phenotypes including Axenfeld-Rieger spectrum, lens dislocation, cataract and congenital/developmental glaucoma.

Genes underlying syndromic forms of anterior segment dysgenesis include:

- *COL4A1*, a gene encoding the α1 chain of type IV collagen. Heterozygous, frequently *de novo*, pathogenic variants in *COL4A1* have been described. These mutations are often missense changes within the molecule's triple-helical domain (usually at the Gly of the Gly-X-Y motif).
- *ADAMTS17*, a gene also implicated in Weill Marchesani syndrome (Chapter 12.3). Biallelic mutations in *ADAMTS17* can infrequently cause Axenfeld-Rieger spectrum with short stature and skeletal abnormalities.
- *JAG1*, a gene associated with Alagille syndrome, an autosomal dominant condition. Most disease-causing variants in this single-pass, transmembrane protein are loss of function changes.

Clinical management

Axenfeld-Rieger spectrum represents the ophthalmic manifestation of a range of genetic conditions. As such, a combined ophthalmic and genetic approach that involves early genomic testing is valuable in facilitating precise diagnosis. This allows the timely identification of individuals who are at risk of multisystemic complications which, in turn, enables appropriate, multidisciplinary, patient-specific care. A range of inheritance patterns are recognised and examination of family members can clarify segregation. Proactive follow-up of relatives at risk of developing glaucoma is recommended.

Ophthalmic management should be individualised as the clinical presentation of patients with Axenfeld-Rieger spectrum is highly variable. More than half of affected individuals will develop glaucoma. Although this may present in early infancy, most cases occur during adolescence or early adulthood. Interestingly, individuals with *FOXC1* mutations display an earlier age of glaucoma onset than patients with *PITX2* mutations.[3] It is also worth noting that the extent of iris defects and iris stands in the angle do not correlate well with the severity of glaucoma. In general, appropriate lifelong monitoring should be considered for patients with the Axenfeld-Rieger spectrum.

Glaucoma management in this cohort can be challenging. As in congenital glaucoma, surgical intervention is more efficacious than medical management. Surgical options include, include goniotomy, trabeculotomy, trabeculectomy, cycloablation and glaucoma drainage implant insertion.

References

1. Seifi M, Walter MA. Axenfeld-Rieger syndrome. *Clin Genet* 2018;**93**(6):1123–30.
2. Ma A, Yousoof S, Grigg JR, Flaherty M, Minoche AE, Cowley MJ, et al. Revealing hidden genetic diagnoses in the ocular anterior segment disorders. *Genet Med* 2020. https://doi.org/10.1038/s41436-020-0854-x.
3. Gauthier AC, Wiggs JL. Childhood glaucoma genes and phenotypes: focus on FOXC1 mutations causing anterior segment dysgenesis and hearing loss. *Exp Eye Res* 2020;**190**:107893.
4. Lewis CJ, Hedberg-Buenz A, DeLuca AP, Stone EM, Alward WLM, Fingert JH. Primary congenital and developmental glaucomas. *Hum Mol Genet* 2017;**26**(R1):R28–36.
5. Souzeau E, Siggs OM, Zhou T, Galanopoulos A, Hodson T, Taranath D, et al. Glaucoma spectrum and age-related prevalence of individuals with FOXC1 and PITX2 variants. *Eur J Hum Genet* 2017;**25**(7):839–47.
6. Bloch-Zupan A, Sedano HO, Scully C. 2 - Missing Teeth (Hypodontia and Oligodontia). In: Bloch-Zupan A, Sedano HO, Scully C, editors. *Dento/oro/craniofacial anomalies and genetics*. Elsevier; 2012. p. 9–74. ISBN:9780124160385. https://doi.org/10.1016/B978-0-12-416038-5.00002-0.
7. Meuwissen ME, Halley DJ, Smit LS, Lequin MH, Cobben JM, de Coo R, et al. The expanding phenotype of COL4A1 and COL4A2 mutations: clinical data on 13 newly identified families and a review of the literature. *Genet Med* 2015;**17**(11):843–53.

10.4

Peters anomaly

Arif O. Khan, Panagiotis I. Sergouniotis, and Graeme C.M. Black

Peters anomaly is a term used to describe a complex spectrum of kerato-iridolenticular dysgeneses. It is thought to develop due to failure of separation of the lens from the surface ectoderm leading to subsequent abnormal adhesions between the lens, iris and cornea. It can be unilateral or bilateral and it is usually sporadic.

Clinical characteristics

Peters anomaly usually presents at birth and is characterised by a congenital central corneal opacity (associated with a posterior corneal defect) (Fig. 10.6). Attachments between the posterior cornea and the lens and/or iris may also be present. Based on clinical examination, Peters anomaly has been often classified into three groups:

- posterior corneal defect with corneal opacity alone
- posterior corneal defect with corneal opacity and adherent iris strands
- posterior corneal defect with corneal opacity, adherent iris strands, and keratolenticular contact or cataract (Fig. 10.7)

Like Axenfeld-Rieger syndrome, Peters anomaly exhibits significant intra- and inter-familial variability and is commonly associated with congenital/developmental glaucoma.

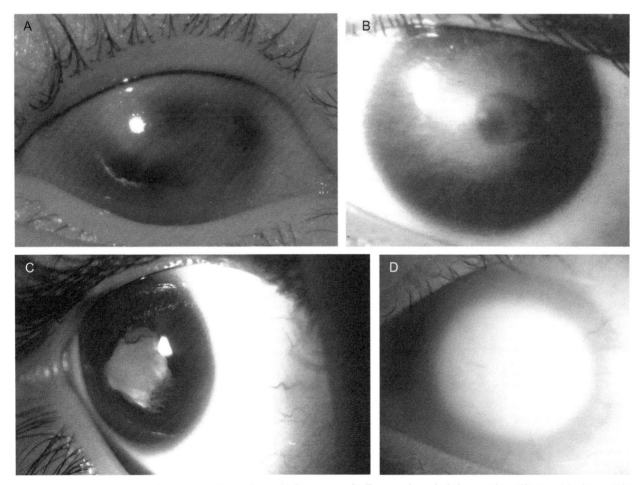

FIG. 10.6 External photographs from four unrelated infants with Peters anomaly. Features shown include corneal opacification, iris abnormalities, cataracts, and corneolenticular adhesions.

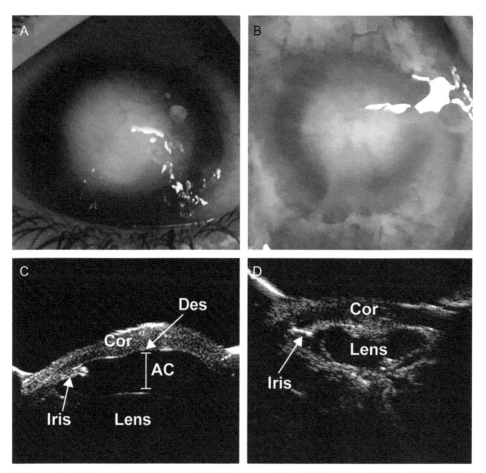

FIG. 10.7 External photograph and ultrasound biomicroscopy (UBM) from two patients with Peters anomaly. The individual shown in (A) and (B) has central corneal opacification and ultrasonographic evidence of iridocorneal adhesions, a thickened cornea (Cor), and an abnormal Descemet (Des) membrane. The central anterior chamber (AC) is formed and the lens is in an anatomic position. The individual shown in (C) and (D) also has central corneal opacification; however, the UBM shows corneolenticular adhesions with loss of the central anterior chamber. *(Adapted from Ref. 1.)*

Several systemic associations have been described, mostly in the context of Peters plus syndrome, an autosomal recessive multisystemic disorder associated with mutations in *B3GALTL*. Systemic features of this condition include growth retardation, short stature, and rhizomelic limb shortening, brain/heart/urogenital malformations, cleft lip/palate and hearing loss.

Molecular pathology

It is possible to identify a molecular diagnosis only in a subset of Peters anomaly cases. There is a clear association with periocular mesenchyme developmental defects and mutations in *PAX6*, *FOXC1* and *PITX2* have been described in some patients. Other affected individuals carry mutations in lens-specific genes that are also implicated in childhood cataracts; these include *PITX3*, *FOXE3*, *GJA8* and *MAF*. Lastly, a small subset of patients with the Peters anomaly carries mutations in genes implicated in congenital glaucoma, including *CYP1B1*.[2]

Clinical management

For the Peters anomaly, a key aim is to clear the central visual axis to allow for visual maturation. Surgical options include full thickness penetrating keratoplasty, surgical iridectomy and, in some cases with significant lens involvement, cataract extraction. The overall visual prognosis in this group of patients is however guarded.

Monitoring and treatment for glaucoma are also required. Controlling the intraocular pressure can be very challenging and surgical intervention is frequently required.

In cases with syndromic features, a multidisciplinary approach to management is required. Notably, there should be a low threshold for testing for hearing loss, especially in individuals who carry mutations in *FOXC1*.[3]

References

1. Dolezal KA, Besirli CG, Mian SI, Sugar A, Moroi SE, Bohnsack BL. Glaucoma and cornea surgery outcomes in Peters anomaly. *Am J Ophthalmol* 2019;**208**:367–75.
2. Ma A, Yousoof S, Grigg JR, Flaherty M, Minoche AE, Cowley MJ, et al. Revealing hidden genetic diagnoses in the ocular anterior segment disorders. *Genet Med* 2020. https://doi.org/10.1038/s41436-020-0854-x.
3. Gauthier AC, Wiggs JL. Childhood glaucoma genes and phenotypes: focus on FOXC1 mutations causing anterior segment dysgenesis and hearing loss. *Exp Eye Res* 2020;**190**:107893.

Chapter 11

Cataract

Chapter outline

11.1 Non-syndromic congenital cataract 114
11.2 Syndromic congenital cataract 118
    11.2.1 Cockayne syndrome 119
    11.2.2 Warburg Micro syndrome 121
    11.2.3 Oculofaciocardiodental syndrome 125
    11.2.4 Nance–Horan syndrome 127
    11.2.5 Cataract in inborn errors of metabolism 129
    11.2.6 Lowe oculocerebrorenal syndrome 133
    11.2.7 Hyperferritinemia-cataract syndrome 137
    11.2.8 Galactosaemia 140
    11.2.9 Cerebrotendinous xanthomatosis 144

Congenital/childhood cataract is an important cause of lifelong visual impairment. It occasionally arises from trauma, maternal infection and intrauterine exposure to drugs or radiation. Most cases however have a strong genetic basis and over 100 genes have been associated with cataract formation.

A key initial aim of clinicians managing children with cataracts is to determine the precise underlying cause of lens opacification (including distinguishing non-syndromic from syndromic forms). Genomic analysis such as gene panel testing can be particularly helpful, especially in children with bilateral cataracts. In this group of patients, genetic testing can streamline care pathways, inform genetic counselling and facilitate early and more precise interventions.

Cataracts caused by genetic abnormalities can be broken down into three main categories:

- *Non-syndromic (isolated) congenital cataract*: lens opacities without other ophthalmic or syndromic features. This group of conditions is discussed in Section 11.1.

- *Cataract with other ophthalmic abnormalities*: lens opacities combined with other eye phenotypes such as anterior segment dysgenesis (Chapter 10), microphthalmia (Chapter 16), ectopia lentis (Chapter 12) or aniridia (Chapter 17). Mutations in several genes including some key transcription factors (e.g. *FOXE3, PITX3, MAF, PAX6*) are detected in a subset of these cases. Persistent fetal vasculature, an important cause of unilateral cataract, can be included in this category although it is only very rarely found to be associated with Mendelian mutations.

- *Cataracts associated with multisystemic disorders*: lens opacities combined with various metabolic and syndromic disorders; these can be inherited as X-linked (e.g. oculofaciocardiodental syndrome), autosomal recessive (e.g. Warburg Micro syndrome) or autosomal dominant (e.g. Stickler syndrome) traits. This group of conditions is discussed in Section 11.2.

11.1

Non-syndromic congenital cataract

I. Christopher Lloyd and Rachel L. Taylor

Non-syndromic (isolated) cataracts account for the majority of congenital and childhood-onset cases.[1] This group of conditions has been associated with over 30 genes (Table 11.1).[2]

Clinical characteristics

Cataracts vary widely in morphology and degree of opacification. At the one end of the spectrum, there are small white dots on the lens that can only be detected through careful slit-lamp examination. At the other end, there are cases of total lens opacification that is readily apparent to the patient's family. The age at presentation also varies and the rate of progression is difficult to predict.

Affected children are often noted to have an abnormal fundal (red) reflex on ophthalmoscopy or photography. Children with unilateral lens opacities can present with sensory strabismus and/or dense amblyopia. The visual behaviour is typically unaffected in these unilateral cases and the parents may not be aware of any vision problems. In contrast, dense bilateral cataracts are usually associated with abnormal visual behaviour including, in severe cases, failure to develop social smiles and to track objects visually.

Children with mild opacities may not be picked up until adolescence or early adulthood. Many of these affected individuals have a family history of early-onset cataracts and it is this that prompts an ophthalmic assessment.

Cataracts can be classified according to their morphological appearance and location. In general, certain types (including anterior polar, lamellar or sutural) have a better prognosis than others (e.g. posterior polar or nuclear). However, using the morphology and location of a lens opacity to pinpoint the underlying genetic diagnosis is generally futile (Fig. 11.1).

Molecular pathology

Non-syndromic congenital cataract is typically inherited as an autosomal dominant trait. Autosomal recessive forms are less prevalent and occur more frequently in populations where parental consanguinity is common.[2]

Genes associated with non-syndromic cataracts commonly encode major lens structural proteins.[3] Mutations in crystallin genes for example account for at least half of the cases.[4] Crystallins are water-soluble intracellular molecules that constitute 90% of the total lens protein. There are two major types of crystallins. The first group is α-crystallins and includes CRYAA and CRYAB. These are not only major structural proteins but also have a role in facilitating protein folding (i.e. have chaperone-like properties). The second major group is the βγ-crystallins which are purely structural molecules. The β-crystallins comprise acidic proteins (CRYBA1, CRYBA2, CRYBA3, CRYBA4) and basic proteins (CRYBB1, CRYBB2, CRYBB3). The γ-crystallins include CRYGA, CRYGB, CRYGC, CRYGD. CRYAB is the only crystallin expressed in significant levels outside the lens; expression of this gene has been found in the retina, heart, brain, and within cardiac and striated muscle. Notably, *CRYAB* mutations can be associated with cardiomyopathy and myofibrillar myopathy as well as congenital cataract.[5]

A less frequent genetic cause of cataract in children is mutations in *GJA3* or *GJA8*, two lens-specific connexin genes. These genes encode two transmembrane gap junction proteins that have a role in cell-to-cell communication. Abnormalities in these molecules are occasionally reported in association with nuclear cataracts that have calcium accumulation within the centre of the lens.[6]

Cytoskeletal (intermediate filament) proteins are another group of molecules associated with non-syndromic cataracts. These include the lens-specific beaded filament proteins BFSP1 (filensin) and BFSP2 (phakinin).

A list of the major genes associated with non-syndromic lens opacification can be found in Table 11.1.

Clinical management

Most unilateral cases arise sporadically in otherwise well children and rarely need to be investigated. In contrast, bilateral (and, rarely, unilateral) cases may be associated with multisystemic disorders and warrant exploration of the underlying cause. Genomic testing should therefore be requested in this group of patients.[7]

The identification of mutations associated with non-syndromic cataracts is an important step that allows reassurance of the affected family that no systemic sequelae are expected. This may be obvious to large families with

TABLE 11.1 Selected genes implicated in childhood cataract without syndromic features.

Gene	Protein function	Mode of inheritance	Examples of observed ophthalmic phenotype(s)
CRYAA	Structural lens protein; molecular chaperone	AD/AR	Lamellar, nuclear, sutural, anterior/posterior polar, membranous cataract
CRYAB	Structural lens protein; molecular chaperone; roles in retina, skeletal muscle and cardiac muscle	AD	Nuclear, post polar, lamellar cataract; myofibrillar myopathy, cardiomyopathy
CRYBA1	Structural lens protein	AD	Nuclear, lamellar, pulverulent, sutural cataract
CRYBA2	Structural lens protein	AD	Multifocal lens opacities; myopia; glaucoma; eccentric pupil
CRYBA4	Structural lens protein	AD	Nuclear, lamellar cataract; microcornea; microphthalmia
CRYBB1	Structural lens protein	AD/AR	Nuclear, pulverulent, cortical cataract; microcornea; glaucoma
CRYBB2	Structural lens protein	AD/AR	Total, subcapsular, cerulean, nuclear, membranous, lamellar, progressive cataract; microcornea
CRYBB3	Structural lens protein	AD/AR	Nuclear, cortical cataract; microcornea
CRYGB	Structural lens protein	AD	Anterior polar, lamellar cataract
CRYGC	Structural lens protein	AD	Pulverulent, zonular, nuclear, lamellar cataract; microphthalmia; microcornea; glaucoma; iris atrophy; corneal opacification; myopia
CRYGD	Structural lens protein	AD	Nuclear, punctate, coralliform, lamellar, cerulean, anterior and posterior polar cataract; microcornea
CRYGS	Structural lens protein	AD	Nuclear, cortical, progressive, lamellar cataract
GJA3	Gap junction protein; molecule/ion transporter	AD	Nuclear, pulverulent, posterior polar, total, cortical, lamellar, coralliform, punctate cataract
GJA8	Gap junction protein; molecule/ion transporter	AD/AR	Total, nuclear, pulverulent, Y-sutural, posterior subcapsular, diffuse, lamellar cataract; microcornea; glaucoma; myopia
FOXE3	Transcription factor	AD/AR	Cerulean cataract; microphthalmia; sclerocornea; microcornea; optic disc coloboma; dysplastic irides; Peters anomaly; glaucoma; aphakia
EPHA2	Signaling receptor	AD/AR	Nuclear, cortical, posterior polar, zonular cataract; persistent fetal vasculature
PXDN	Peroxidase	AR	Microphthalmia; microcornea; sclerocornea; developmental glaucoma; anterior segment dysgenesis
FYCO1	Transport adaptor	AR	Nuclear cataract
BFSP2	Beaded filament protein	AD/AR	Cortical, nuclear, Y-sutural, lamellar cataract; myopia
PITX3	Transcription factor	AD/AR	Posterior polar, posterior subcapsular cataract; anterior segment dysgenesis; microphthalmia; microcornea; sclerocornea; Peters anomaly
PAX6	Transcription factor	AD	Anterior and posterior polar; cortical cataract; aniridia, iris and foveal hypoplasia; corneal abnormalities; glaucoma
MIP	Lens fibre membrane channel protein	AD	Cerulean, lamellar, punctate, Y-sutural, total, posterior polar cataract; myopia
VIM	Intermediate filament protein	AD	Pulverulent cataract
MiR184	microRNA; post-transcriptional regulator of gene expression	AD	Anterior polar cataract; corneal endothelial dystrophy; keratoconus; iris hypoplasia
HSF4	Transcription factor	AD/AR	Cortical, total, lamellar, sutural, nuclear cataract

AD, autosomal dominant; AR, autosomal recessive.

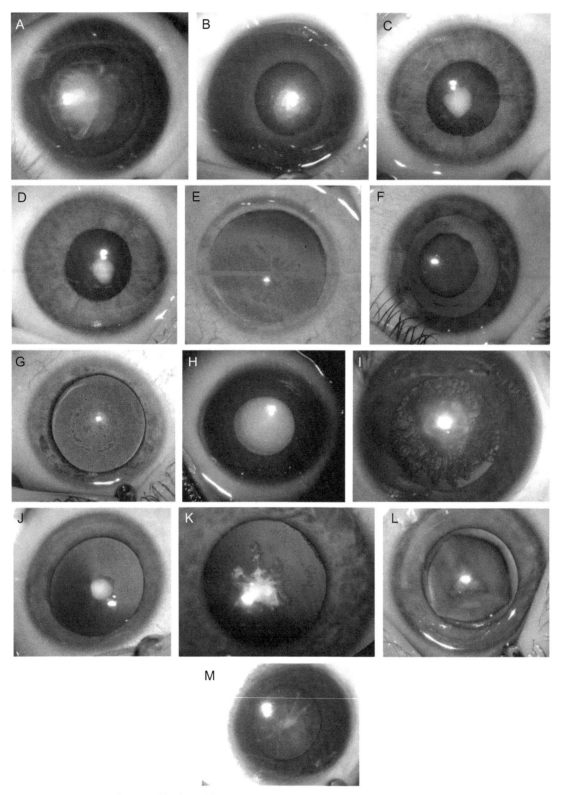

FIG. 11.1 Slit-lamp photographs of lens opacities from children with cataracts. (A and B) Right and left eyes of an infant with dense nuclear cataracts and asymmetric microcornea and microphthalmia. (C and D) Right and left eyes of an infant with dense nuclear cataracts, cortical opacities and irido-lenticular adhesions. (E and F) Examples of congenital lamellar cataract. (G) Congenital cerulean (blue dot) cataract. (H) Congenital total cataract. (I) Congenital nuclear cataract with extensive cortical opacification. (J) Anterior polar congenital cataract. (K) Coralliform congenital cataract. (L) Congenital sutural cataract. (M) Congenital spoke-like cataract.

multiple affected individuals and a clear autosomal dominant inheritance pattern. However, it is crucial for children with sporadic bilateral cataracts. In these cases, genetic testing can also determine the mode of inheritance of the condition, leading to more precise genetic counselling.

Treatment depends on the type and severity of cataracts. In general, though, many children will have visually significant lens opacities and will require surgical intervention (followed by optical correction and amblyopia management). Timely removal is a prerequisite for a good outcome. It is noted that the ophthalmic management of cataracts is common to all conditions that feature lens opacification and a detailed discussion on paediatric cataract management can be found in Ref. 7.

References

1. Haargaard B, Wohlfahrt J, Fledelius HC, Rosenberg T, Melbye M. A nationwide Danish study of 1027 cases of congenital/infantile cataracts: etiological and clinical classifications. *Ophthalmology* 2004;**111**(12):2292–8.
2. Reis LM, Semina EV. Genetic landscape of isolated pediatric cataracts: extreme heterogeneity and variable inheritance patterns within genes. *Hum Genet* 2019;**138**(8–9):847–63.
3. Shiels A, Hejtmancik JF. Biology of inherited cataracts and opportunities for treatment. *Annu Rev Vis Sci* 2019;**5**:123–49.
4. Hejtmancik JF. Congenital cataracts and their molecular genetics. *Semin Cell Dev Biol* 2008;**19**(2):134–49.
5. Vicart P, Caron A, Guicheney P, Li Z, Prevost MC, Faure A, et al. A missense mutation in the alphaB-crystallin chaperone gene causes a desmin-related myopathy. *Nat Genet* 1998;**20**(1):92–5.
6. Gao J, Sun X, Martinez-Wittinghan FJ, Gong X, White TW, Mathias RT. Connections between connexins, calcium, and cataracts in the lens. *J Gen Physiol* 2004;**124**(4):289–300.
7. Self JE, Taylor R, Solebo AL, Biswas S, Parulekar M, Dev Borman A, et al. Cataract management in children: a review of the literature and current practice across five large UK centres. *Eye (Lond)* 2020. https://doi.org/10.1038/s41433-020-1115-6.

11.2

Syndromic congenital cataract
Jill Clayton-Smith, Sofia Douzgou Houge, Rachel L. Taylor, I. Christopher Lloyd, and Graeme C.M. Black

Bilateral (and rarely unilateral) lens opacities can be a manifestation of a multisystemic condition. These disorders are thought to account for approximately 15% of all congenital and childhood cataract cases (although this varies in different populations). Examples of relevant conditions include oculocerebrorenal (Lowe) syndrome, in which cataract presents alongside hypotonia, intellectual disability, and renal abnormalities (Section 11.2.6); Nance–Horan syndrome, which features characteristic dental anomalies together with cataract (Section 11.2.4); and Cockayne syndrome, a premature aging syndrome in which cutaneous photosensitivity and sensorineural hearing loss manifest in conjunction with lens opacities (Section 11.2.1). Notably, cataract can occasionally be the presenting feature of some of these disorders in otherwise well infants. Furthermore, in a small subset of cases, metabolic diseases that are amenable to treatment may be highlighted. Examples include galactosaemia (Section 11.2.8) and cerebrotendinous xanthomatosis (Section 11.2.9).

11.2.1

Cockayne syndrome

Jill Clayton-Smith

Cockayne syndrome, first described in 1936, is a rare, autosomal recessive, multisystemic disorder characterised by growth failure, microcephaly, and progressive physical and neurological deterioration that leads to reduced lifespan.[1,2] Vision and hearing impairment are common features. Although Cockayne syndrome has been divided into three types, classic (type I), severe (type II) and milder (type III)[3], these represent a continuum with regards to the severity of phenotype. Cockayne syndrome arises due to defective DNA repair and the two genes involved, ERCC6 and ERCC8, play a critical role in this process.

Clinical characteristics

The main clinical characteristics of Cockayne syndrome include impaired growth and development, microcephaly, joint contractures, scoliosis, lack of subcutaneous fat, sensorineural hearing loss, visual impairment due to several different ophthalmological abnormalities, skin photosensitivity, and brain MRI abnormalities. These manifestations can vary in severity but ultimately culminate in the same clinical picture. At the severe end of the spectrum, life expectancy may be as little as 5 years, whereas individuals with milder features have been diagnosed in adulthood[4] and some have survived over 30 years. The natural history of this disorder has been reported through the study of several large cohorts.[1,5]

Many infants with Cockayne syndrome have a normal birth weight, but fail to thrive and have progressive growth failure. Microcephaly is often congenital and severe, with occipitofrontal circumference being around 6 standard deviations below the expected value. From early on, congenital contractures of the large joints and spinal curvature develop, giving rise to a characteristic posture with a curved back. There is a general lack of adipose tissue and this contributes to the classical appearance of enophthalmos and to coldness of the extremities. It is this presentation, together with a history of skin photosensitivity which often prompts the consideration of Cockayne syndrome as a diagnosis.

Development in infants with Cockayne syndrome can initially appear normal but within the first year of life, delay becomes apparent. In classic and severe cases there is increasing spasticity of the limbs and inability to gain ambulation. Peripheral neuropathy and tremor frequently develop and older children have intellectual disability. Brain imaging by MRI demonstrates calcification, often in the region of the basal ganglia.[4] There may be white matter changes and a smaller cerebellum.

Progressive sensory impairment in Cockayne syndrome is due to sensorineural hearing loss and several visual abnormalities.[6] Around 50% of Cockayne syndrome patients have cataracts.[5] In some cases, these are congenital but the majority of affected individuals develop lens opacification by the age of 5 years. The cataract type may vary. Generally, infants with congenital cataracts have a worse prognosis. In virtually all patients, retinal degeneration with associated electroretinographic abnormalities and a 'salt and pepper' pigmentary appearance develops over time.[7] Despite this, useful vision is typically retained. Nystagmus and light hypersensitivity are commonly seen, and many patients benefit from tinted lenses.

Over time, general health and mobility deteriorate. Abnormal liver function may be detected and frequent chest infections are common. A prematurely aged facial appearance becomes apparent (Fig. 11.2) and there are associated changes in the hair, which are sparse, and the teeth, which are decayed. Renal impairment and pneumonia are two major causes of death. Intriguingly, even though this condition arises due to defective DNA repair, there does not appear to be an increased risk for the development of cancer.

Molecular pathology

Cockayne syndrome is one of the groups of disorders termed DNA repair defects. Fibroblasts from Cockayne syndrome patients have a reduced ability to resume RNA synthesis following exposure to ultraviolet light which induces DNA damage.[9] This assay was used to confirm the diagnosis of Cockaye syndrome over many years until the genes responsible were identified.

Cockayne syndrome is caused by biallelic mutations in ERCC6 or ERCC8, two genes encoding proteins involved in a process called transcription-coupled nucleotide excision repair. ERCC6 encodes a DNA dependent ATPase and ERCC8 a component protein of an E3 ubiquitin ligase complex. In general, around 75% of patients have variants within ERCC6 (Cockayne syndrome B, CSB) and 25%

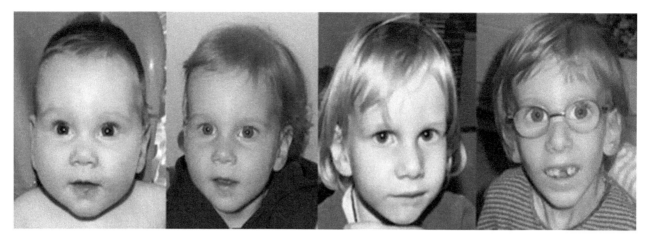

FIG. 11.2 Evolution of the facial appearance in a child with Cockayne syndrome over time. Appearance at age 8 months, 18 months, 3 years and 6 years is shown highlighting that the facial features often only develop with time and may not be easily recognisable in the early stages of the disease. *(Adapted from Ref. 8.)*

within *ERCC8* (Cockayne syndrome A, CSA). Variants of all types have been reported including frameshift, splice site, nonsense, in-frame deletions and missense changes.[10] Some of the variants are recurrent and cluster at hotspots. There is also a suggestion that there may be founder effects to account for recurrent changes. There appears to be very little genotype–phenotype correlation; the phenotypes seen in association with both genes are broadly similar but patients with *ERCC6* variants are reported to have slightly higher incidences of cataracts, microphthalmia and low birth weight.[10]

Clinical management

Clinical management of Cockayne syndrome has been reviewed in several publications in the biomedical literature.[5,11] This multisystemic disorder requires the input of several different health professionals. Ophthalmic assessment should be performed and tinted lenses are often prescribed. Feeding and growth are particular problems and gastrostomy may be required. The majority of patients require treatment for oesophageal reflux. Growth should be monitored regularly but there is no recognised benefit from growth hormone treatment. Regular medical checks should include monitoring of liver function and blood pressure. Hearing and vision require regular testing and many Cockayne syndrome individuals benefit from hearing aids. Cochlear transplantation has been carried out in some patients. Regular dental check-ups to monitor dental caries should be undertaken and high factor sunscreen is needed because of photosensitivity. Development and educational progress should be monitored and educational input should be tailored accordingly (with an education and healthcare plan in place). There should be access to speech, language, and occupational therapies. Regular physiotherapy may help to prevent the progression of contractures, maintain mobility and encourage good posture. As this is an autosomal recessive disorder, genetic counselling should be offered to parents following diagnosis.

References

1. Nance MA, Berry SA. Cockayne syndrome: review of 140 cases. *Am J Med Genet* 1992;**42**:68–84.
2. Vessoni AT, Guerra CCC, Kajitani GS, Nascimento LLS, Garcia CCM. Cockayne syndrome: the many challenges and approaches to understand a multifaceted disease. *Genet Mol Biol* 2020;**43**(1 Suppl. 1):e20190085 [Published 2020 May 20] https://doi.org/10.1590/1678-4685-GMB-2019-0085.
3. Lowry RB. Early onset of Cockayne syndrome. *Am J Med Genet* 1982;**13**:209–10.
4. Cocco A, Calandrella D, Carecchio M, Garavaglia B, Albanese A. Adult diagnosis of Cockayne syndrome. *Neurology* 2019;**93**(19):854–5.
5. Wilson B, Stark Z, Sutton R, et al. The Cockayne Syndrome Natural History (CoSyNH) study: clinical findings in 102 individuals and recommendations for care. *Genet Med* 2016;**18**:483–93.
6. Traboulsi EI, De Becker I, Maumenee IH. Ocular findings in Cockayne syndrome. *Am J Ophthalmol* 1992;**114**(5):579–83.
7. Figueras-Roca M, Budi V, Morató M, Camós-Carreras A, Muñoz JE, Sánchez-Dalmau B. Cockayne syndrome in adults: complete retinal dysfunction exploration of two case reports. *Doc Ophthalmol* 2019;**138**(3):241–6.
8. Laugel V. Cockayne syndrome: the expanding clinical and mutational spectrum. *Mech Ageing Dev* 2013;**134**(5–6):161–70.
9. Nakazawa Y, Yamashita S, Lehmann AR, Ogi T. A semi-automated non-radioactive system for measuring recovery of RNA synthesis and unscheduled DNA synthesis using ethynyluracil derivatives. *DNA Repair (Amst)* 2010;**9**(5):506–16.
10. Calmels N, Botta E, Jia N, et al. Functional and clinical relevance of novel mutations in a large cohort of patients with Cockayne syndrome. *J Med Genet* 2018;**55**(5):329–43.
11. Laugel V. Cockayne syndrome. In: Adam MP, Ardinger HH, Pagon RA, et al., editors. *GeneReviews®*. Seattle, WA: University of Washington, Seattle; December 28, 2000 [updated August 2019].

11.2.2

Warburg Micro syndrome

Sofia Douzgou Houge

Warburg Micro syndrome is a rare, autosomal recessive disorder associated with mutations in genes that cause Rab-18 protein deficiency. It usually presents to the ophthalmologist and characteristic features include congenital cataract, microphthalmia with microcornea, microcephaly and hypogonadism (Table 11.2). Although making a definitive diagnosis in the eye clinic is challenging, the role of the ophthalmologist is critical in: considering the possibility of Warburg Micro syndrome in the differential diagnosis, ordering appropriate genetic testing, and initiating referral to a clinical geneticist and other specialists.

Clinical characteristics

Warburg Micro syndrome is a developmental disorder characterised by pan-ocular morbidity that includes congenital bilateral cataracts, microphthalmia with microcornea, progressive optic atrophy and severe cortical visual impairment. Characteristically, affected patients have small atonic pupils that do not react to light or mydriatic agents, and the vision remains poor despite early cataract surgery.

Systemically, affected individuals present with neurodevelopmental and endocrine or growth symptomatology. The vast majority of patients have brain abnormalities that are invariably associated with postnatal progressive microcephaly (-4 to -6 standard deviations) and, variably, with epilepsy. Brain MRI evidences polymicrogyria of the frontal and parietal lobes, hypogenesis of the corpus callosum, increased subdural spaces, and cerebellar vermis hypoplasia. Most patients develop severe developmental impairment with no progress beyond the 4-month milestones (e.g. smiling). The neurological clinical presentation includes congenital hypotonia which evolves to ascending spasticity (involving the lower limbs from about 8–12 months and spastic quadriplegia later in life) leading to contractures. Affected individuals frequently have significant feeding difficulties (chewing and swallowing and/or dysphagia) and require gastrostomy tube placement to achieve appropriate nutritional status. They are also at risk of aspiration and pulmonary infection. A few individuals also manifest clinical hypothalamic hypogonadism (micropenis and/or cryptorchidism in males and hypoplastic labia minora, clitoral hypoplasia and small introitus in females) and/or short stature. In rare cases, the systemic morbidity of the condition also includes infantile-onset, dilated cardiomyopathy.

Facially, the patients have soft dysmorphic features that can be partly assigned to microphthalmia (small, deep-set eyes) and progressive microcephaly (low anterior hairline, bitemporal narrowing, flat occiput and prominent ears)

TABLE 11.2 Red flags that should highlight the possibility of Warburg Micro syndrome in infants and young children with congenital cataracts.

Finding	Recommendation
Consanguinity	Enquire about similarly affected individuals amongst first cousins
Microphthalmia	Clinical suspicion arises from a corneal diameter less than 10 mm at birth; an axial length <16 mm at birth (or <19 mm at 12 months of age or < 21 mm in an adult) substantiates the diagnosis of microphthalmia
Microcephaly and developmental delay	Measure and chart the occipitofrontal circumference; request microarray genetic analysis
Hypogenitalism	Remember that the hypogenitalism associated with Warburg Micro syndrome is a useful diagnostic clue in boys (micropenis and cryptorchidism) but may be absent or easily overlooked in girls
Visual and neurodevelopmental impairment in a neonate/infant following bilateral, congenital cataract surgery	Refer for or request brain MRI

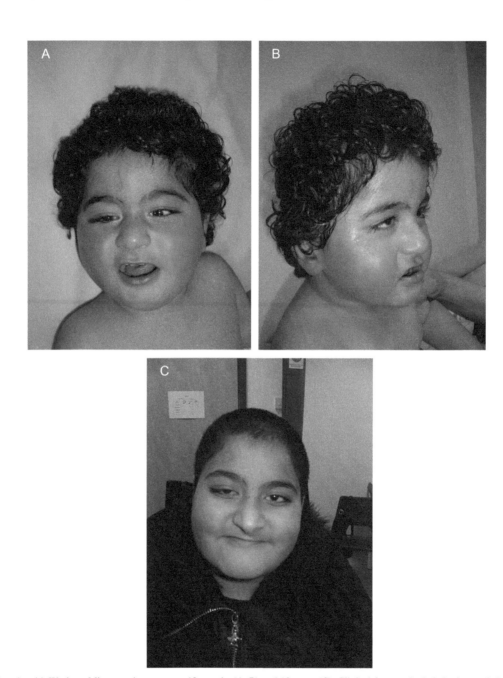

FIG. 11.3 Female with Warburg Micro syndrome at age 18 months (A, B) and 12 years (C). Clinical features included microcephaly, developmental delay, microphthalmia, bilateral aphakia (following congenital cataract surgery) and bilateral optic nerve hypoplasia. Genome sequencing revealed a homozygous duplication involving exons 4 to 8 of the *RAB3GAP1* gene. *(Courtesy of Dr. Kate Chandler.)*

(Fig. 11.3). These features, however, do not always form a recognisable pattern.

A small subgroup of patients has a milder clinical presentation known as Martsolf syndrome. They present with mild to moderate intellectual disability, preserved cortical structure with polymicrogyria confined to the frontal lobes, and infantile hypotonia which is followed by slowly progressive spasticity that remains limited to the lower limbs and allows independent ambulation.

Molecular pathology

Warburg Micro syndrome is a genetically heterogeneous condition; it is known to be caused by biallelic loss of function variants in at least 5 genes: *RAB3GAP1* (type 1),[1] *RAB3GAP2* (type 2), *RAB18* (type 3), *TBC1D20* (type 4)[2] and *ITPA*. Additional genetic loci may exist so the absence of causative variants in these genes does not necessarily exclude a clinical diagnosis of Warburg Micro syndrome.

TABLE 11.3 Differential diagnosis of Warburg Micro syndrome.

	Disorder	Inheritance	How to distinguish from Warburg Micro syndrome
Teratogen	TORCH	Not applicable	Perform TORCH screen (toxoplasmosis, rubella cytomegalovirus, herpes simplex virus, HIV); sensorineural hearing loss, congenital heart defect, hepatosplenomegaly may be present
	Zika virus	Not applicable	Check travel history; microphthalmia has been described in a few cases; microcephaly tends to be more severe; brain calcifications may be present
Chromosomal	Monosomy 1p36	*De novo*	Request genetic analysis; cardiac, skeletal and renal defects/malformations may be present
	Monosomy 1q21	*De novo*/autosomal dominant	Request genetic analysis; only a minority affected with severe developmental delay; congenital heart defects may be present
Mendelian	Cerebro-oculo-facial-skeletal syndrome	Autosomal recessive	Include relevant genes (*ERCC1, ERCC2, ERCC5, ERCC6, ERCC8*) in gene panel test; pigmentary retinopathy, contractures, cutaneous photosensitivity, brain calcifications may be present
	Smith–Lemli–Opitz syndrome	Autosomal recessive	Include relevant gene (*DHCR7*) in the panel; 2–3 toe syndactyly, cardiac, skeletal and renal defects/malformations may be present

Affected individuals have indistinguishable clinical presentations, irrespective of the disease-associated gene. Most of the genes that cause Warburg Micro syndrome encode members of a group of small G proteins (Rabs) that regulate vesicular trafficking in cells. The molecular pathology of the condition results from the physical (abolishing) or functional (dysregulation) absence of the Rab-18 protein.[3,4]

The condition has been mostly described in the offspring of consanguineous couples (homozygous mutations) but also, albeit more rarely, in the children of non-related parents (compound heterozygous mutations). Possible common (founder) pathogenic variants have been described in families of Pakistani, Turkish and Danish ancestries.[5,6]

Frameshift, nonsense, splice site, intragenic deletions and, very rarely, missense changes have been identified in patients with Warburg Micro syndrome. These variants typically abolish the function of the encoded protein. Mutations that diminish but do not abolish the gene and protein function cause the milder end of the phenotypic continuum (Martsolf syndrome).[7]

Given the genetic heterogeneity of Warburg Micro syndrome, the gold standard genetic analysis involves the use of a multi-gene panel test that includes *RAB3GAP1*, *RAB3GAP2*, *RAB18*, *TBC1D20*, *ITPA* and other genes of interest (Table 11.3). It is worth taking into consideration that the genes included in each panel and the diagnostic sensitivity of testing vary by laboratory and are likely to change over time.

Consent should be obtained before genetic analysis. The genetic testing strategy should be explained to the family sensitively to assist them in making informed medical decisions that best reflect their values and intentions.

Clinical management

There is no established treatment for the underlying disorder and affected individuals are supported and reviewed based on their symptoms. The condition is best approached by a multidisciplinary group of specialists which includes

- Ophthalmology: surgical removal of cataracts and follow-up for the risk of glaucoma.
- Neurology: epilepsy management and review.
- Paediatrics: neurodevelopmental assessment.
- Feeding/nutrition specialist: safe swallow assessment and feeding review.
- Endocrinology: possible surgical correction of cryptorchidism; management (HRT) of hypogonadism.
- Audiology: hearing review.
- Physiotherapy: support for the motor dysfunction.

References

1. Aligianis IA, Johnson CA, Gissen P, Chen D, Hampshire D, Hoffmann K, et al. Mutations of the catalytic subunit of RAB3GAP cause Warburg Micro syndrome. *Nat Genet* 2005;**37**:221–3.
2. Liegel RP, Handley MT, Ronchetti A, Brown S, Langemeyer L, Linford A, et al. Loss-of-function mutations in TBC1D20 cause cataracts and male infertility in blind sterile mice and Warburg micro syndrome in humans. *Am J Hum Genet* 2013;**93**(6):1001–14.
3. Handley MT, Carpanini SM, Mali GR, Sidjanin DJ, Aligianis IA, Jackson IJ, et al. Warburg Micro syndrome is caused by RAB18 deficiency or dysregulation. *Open Biol* 2015;**5**:150047.
4. Mark Handley PD, Eamonn Sheridan MD. RAB18 deficiency. In: Adam MP, Ardinger HH, Pagon RA, et al., editors. *Gene reviews*. Seattle, WA: University of Washington, Seattle; 1993–2019.

5. Morris-Rosendahl DJ, Segel R, Born AP, Conrad C, Loeys B, Brooks SS, et al. New RAB3GAP1 mutations in patients with Warburg Micro Syndrome from different ethnic backgrounds and a possible founder effect in the Danish. *Eur J Hum Genet* 2010;**18**:1100–6.
6. Ainsworth JR, Morton JE, Good P, Woods CG, George ND, Shield JP, et al. Micro syndrome in Muslim Pakistan children. *Ophthalmology* 2001;**108**:491–7.
7. Handley MT, Reddy K, Wills J, Rosser E, Kamath A, Halachev M, et al. ITPase deficiency causes a Martsolf-like syndrome with a lethal infantile dilated cardiomyopathy. *PLoS Genet* 2019;**15**: e1007605.

11.2.3

Oculofaciocardiodental syndrome

Graeme C.M. Black

Oculofaciocardiodental syndrome (OFCD) is an X-linked dominant condition that affects only females and is caused by heterozygous mutations in the *BCOR* gene. Characteristic ophthalmic abnormalities include congenital cataracts and, occasionally, microphthalmia.

OFCD remains under-recognised and represents an important cause of congenital cataracts in females. Genetic testing studies have shown that a significant number of female patients (between 5% and 10% of girls presenting with cataracts in childhood) carry *BCOR* mutations; most of these affected individuals have been subsequently shown to have non-ophthalmic features of OFCD[1].

Clinical characteristics

The main ophthalmic feature of OFCD is that of congenital or childhood-onset cataracts, which may be asymmetrical (Fig. 11.4A) or, occasionally, unilateral. Microphthalmia that is usually mild and may be associated with microcornea, has also been described.

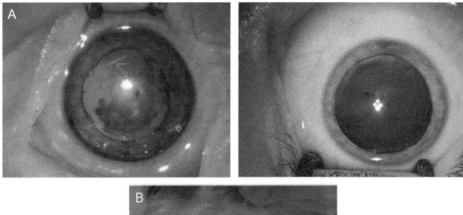

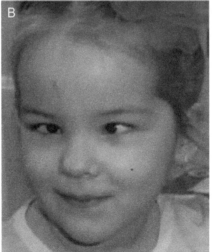

FIG. 11.4 Clinical features of oculofaciocardiodental syndrome (OFCD). Asymmetrical lens opacities (A) and facial features (B) in affected female patients; a high anterior hairline and a broad nasal tip are noted. *(B: Adapted from Ref. 1.)*

OFCD is associated with a wide range of non-ophthalmic sequelae. Defects of laterality (e.g. heart and other viscera[2]) have been described and other non-ophthalmic findings include

- Dysmorphic facial features: long and thin face, broad or bifid nasal tip (Fig. 11.4B), long philtrum (i.e. vertical indentation in the middle area of the upper lip).
- Cleft palate.
- Congenital cardiac defects: atrial/ventricular septal defects, complex heart defects.
- Dental abnormalities: canine radiculomegaly (i.e. presence of extremely large roots in the canine teeth), persistent primary dentition with delayed eruption, hypodontia, dental fusions.
- Skeletal anomalies: 2–3 toe syndactyly, radioulnar synostosis (i.e. abnormal connection between the radius and ulna bones of the forearm), hammertoe deformities (i.e upward bending at the middle joint of the toe and downward angling of the end of the toe).

Molecular pathology

OFCD is an X-linked dominant condition caused by loss of function mutations in *BCOR*.[3] This gene encodes the BCL6 co-repressor protein, i.e. a protein that does not bind to DNA by itself but interacts with the DNA-binding protein produced by the *BCL6* gene to suppress the activity of other genes. BCL6 plays critical role in specific immunological processes involving B and T cells. However, *BCOR* is expressed not just in the immune system but throughout the body and is thought to have an important role in early embryonic development.

In 2004, a single missense *BCOR* mutation was found to be causing a syndromic form of microphthalmia (Lenz microphthalmia syndrome). At the same time, *BCOR* variants were found in patients with OFCD including nonsense, frameshift, insertion/deletion and splice site mutations. Whole gene and partial gene deletions represent a significant (around 10%) number of cases of OFCD syndrome, occasionally encompassing wider genetic regions. Individuals with somatic mosaicism (i.e. presence of two genetically distinct populations of cells within an individual) have been recognised, providing one explanation for the observed phenotypic variability and the presence of more mildly affected individuals. Germline mosaicism (i.e. presence of a genetic change only in a subset of the testicular or ovarian cells of an individual) has also been seen on more than one occasion and this should be taken into account when attempting family counselling (e.g. the recurrence risks cannot be dismissed as negligible after one affected child in cases where a pathogenic *BCOR* variant has not been identified in the blood of the mother).

Clinical management

There are no syndrome-specific implications for ophthalmic management. However, it is notable that females with unilateral cataracts have been described with *BCOR* mutations. Whilst genomic testing of all children with unilateral lens opacities is not indicated this observation suggests that testing of children with unilateral cataract plus either a family history or non-ophthalmic developmental abnormalities is appropriate.

In female children with either congenital cataracts or microphthalmia and a heterozygous *BCOR* mutation, a systematic extraocular examination is indicated. In particular, this should focus on screening for cardiac abnormalities (septal defects); dental abnormalities (radiculomegaly, delayed loss of primary dentition, elongated tooth roots) and skeletal abnormalities (hammertoe deformities, radio-ulnar synostosis). Whilst developmental delay has been described, this has generally been noticed early on and may resolve after infancy; the majority of patients develop normally.

Although *BCOR* is thought to have a biological role in tumorigenesis, individuals with OFCD do not appear to have predisposition to cancer. As a result, cancer screening is not considered necessary at the current time.

Regarding family counselling, OFCD is considered an X-linked dominant (male-lethal) condition. Thus, all individuals with OFCD are female, with many incidences of mother-daughter transmission. For each affected person, 50% of their daughters are likely to be affected, whilst the condition will not affect their sons. There is a described excess of female liveborn children. However, perhaps surprisingly, there is no strong evidence for increased miscarriage rates. This may suggest that male pregnancies carrying loss of function mutations in BCOR are lost early, before pregnancy is recognised.

References

1. Redwood A, Douzgou S, Waller S, Ramsden S, Roberts A, Bonin H, et al. Congenital cataracts in females caused by BCOR mutations; report of six further families demonstrating clinical variability and diverse genetic mechanisms. *Eur J Med Genet* 2020;**63**(2):103658.
2. Hilton EN, Manson FD, Urquhart JE, Johnston JJ, Slavotinek AM, Hedera P, et al. Left-sided embryonic expression of the BCL-6 corepressor, BCOR, is required for vertebrate laterality determination. *Hum Mol Genet* 2007;**16**(14):1773–82.
3. Ng D, Thakker N, Corcoran CM, Donnai D, Perveen R, Schneider A, et al. Oculofaciocardiodental and lenz microphthalmia syndromes result from distinct classes of mutations in BCOR. *Nat Genet* 2004;**36**(4):411–6.

11.2.4

Nance–Horan syndrome

Sofia Douzgou Houge

Nance–Horan syndrome is an under-diagnosed, multisystemic disorder caused by mutations in the *NHS* gene. It often presents to the ophthalmologist and is primarily characterised by congenital cataracts and dental anomalies. It is inherited as an X-linked dominant trait with heterozygous carrier females presenting less severe features than affected males. Notably, the presence of lens opacities centred around the posterior Y-suture in females is thought to be a characteristic sign. Families with X-linked cataracts should be carefully examined for both ophthalmic and non-ophthalmic features to exclude Nance–Horan syndrome.

Clinical characteristics

Affected males have congenital cataracts, dental anomalies, facial dysmorphic features, and, occasionally, behavioural and learning difficulties.

Ophthalmic findings in male patients include bilateral severe congenital cataracts involving the fetal nucleus and posterior Y suture, with variable zonular extensions into the posterior cortex. These usually lead to significant visual loss and require surgery. Microcornea, nystagmus, and microphthalmia have also been reported.[1]

Dental abnormalities include Hutchinsonian incisors (screwdriver-shaped incisors), supernumerary maxillary incisors (mesiodens) and widely spaced teeth (diastema)[2] (Figs 11.5 and 11.6). Often, affected individuals have shortened lateral metacarpals and approximately 30% of male patients manifest behavioural and learning difficulties. Facially, the patients have soft dysmorphic features that include prominent nose and nasal bridge, long face and large ears with anteverted (i.e. forward-inclined) pinnae (Fig. 11.5). Notably, these may not always form a recognisable pattern.

Carrier females typically display congenital, posterior, Y-sutural lens opacities (Table 11.4); the dental and facial anomalies of the syndrome may be observed, but with a milder expression.[4]

Molecular pathology

Nance–Horan syndrome is caused by loss of function variants in the *NHS* gene or by complex rearrangements (copy number variants) of the chromosomal region Xp22.13.[1,5] Gene panels should identify the former but the latter is typically detected by microarray analysis. Genomic analysis of the region should aim to test for both.

Clinical management

There is generally no established treatment for the underlying defect and affected individuals are managed and reviewed based on their symptoms. The condition is best approached by relevant specialists:

- Ophthalmology: surgical removal of cataracts.
- Paediatrics: one-off neurodevelopmental assessment and ad hoc follow-up if necessary.
- Dentistry: orthodontic review if necessary.

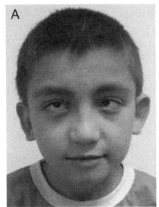  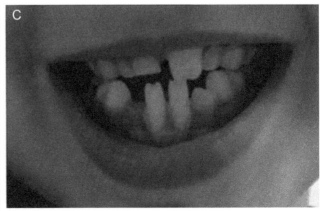

FIG. 11.5 Facial photos from two siblings with Nance–Horan syndrome. Characteristic facial features include large ears with anteverted pinnae and a prominent nose (A and B). Screwdriver shaped incisors and widely spaced teeth are noted (C). *(Adapted from Ref. 3.)*

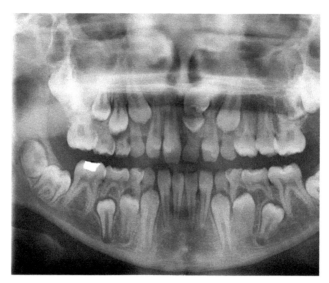

FIG. 11.6 X-ray orthopantomogram from an 11-year-old male with Nance–Horan syndrome, showing unerupted and supernumerary teeth, molars with misshapen cusps, and unusually shaped incisors.

TABLE 11.4 Red flags that should highlight the possibility of Nance–Horan syndrome in patients with cataracts.

Finding	Recommendation
Y-suture opacities in otherwise asymptomatic females	Examine the oral cavity or enquire about dental anomalies/extractions
Presence of encephalopathic symptoms or learning/behavioural difficulties in a male	Request microarray genetic analysis

References

1. Burdon KP, McKay JD, Sale MM, Russell-Eggitt IM, Mackey DA, Wirth MG, et al. Mutations in a novel gene, NHS, cause the pleiotropic effects of Nance-Horan syndrome, including severe congenital cataract, dental anomalies, and mental retardation. *Am J Hum Genet* 2003;**73**:1120–30.
2. Walpole IR, Hockey A, Nicoll A. The Nance-Horan syndrome. *J Med Genet* 1990;**27**:632–4.
3. Tug E, Dilek NF, Javadiyan S, Burdon KP, Percin FE. A Turkish family with Nance-Horan syndrome due to a novel mutation. *Gene* 2013;**525**(1):141–5.
4. Bixler D, Higgins M, Hartsfield Jr J. The Nance-Horan syndrome: a rare X-linked ocular–dental trait with expression in heterozygous females. *Clin Genet* 1984;**26**:30–5.
5. Coccia M, Brooks SP, Webb TR, Christodoulou K, Wozniak IO, Murday V, et al. X-linked cataract and Nance-Horan syndrome are allelic disorders. *Hum Mol Genet* 2009;**18**:2643–55.

11.2.5

Cataract in inborn errors of metabolism

Rachel L. Taylor

Cataract is a feature of many inborn errors of metabolism. This heterogeneous group of Mendelian conditions is associated with disruption of the activity of enzymes, and thus of specific biochemical pathways. Although metabolic diseases featuring cataracts are individually rare, collectively they are likely to account for a significant proportion of cases. A study of 50 childhood cataract cases investigated using gene panel testing found that around 15% of them carried mutations in genes that cause metabolic disorders.[1]

The early clinical signs of metabolic disorders can be subtle or ambiguous. Whilst newborn screening programmes play an important role in the early detection of a few selected inborn errors of metabolism,[2] timely diagnosis is challenging for most affected individuals. Importantly, children with cataracts represent a patient subgroup where rapid diagnosis is both possible and particularly impactful as many of these conditions are amenable to disease-modifying treatment or dietary intervention.

The traditional diagnostic approach involves sequential biochemical investigations aiming to confirm the presence of an inborn error of metabolism. However, this strategy is reactive and relies heavily on the recognition of specific disease features that guide each test.[3] Since 2010, strategies utilising high-throughput DNA sequencing technologies (including focussed gene panel testing, exome or genome sequencing) have emerged, enabling timely and cost-effective diagnostics in individuals with cataracts caused by metabolic disease (e.g. *GLUT1* and *PEX11B* mutations, Fig. 11.7)[4].

In this section, inborn errors of metabolism are first discussed as a group but four characteristic conditions (Lowe oculocelebrorenal syndrome, hyperferritinemia-cataract syndrome, galactosaemia, and cerebrotendinous xanthomatosis) are also highlighted in Sections 11.2.6 to 11.2.9.

Clinical characteristics

Inborn errors of metabolism are clinically highly heterogeneous. When cataracts feature as part of metabolic disturbance, they are usually bilateral and progressive and can occasionally be the primary presenting feature. They may be present at birth but often develop in infancy, childhood or adolescence. Age at presentation and cataract morphology depends largely on the type and severity of the metabolic disturbance. For example:

- Cataracts associated with peroxisomal disorders such as Zellweger syndrome and rhizomelic chondroplasia punctata are usually congenital and tend to form as dense, anterior or anterior polar opacities that can have a severe impact on vision.
- Cataracts associated with disorders of the cholesterol biosynthesis pathway such as cerebrotendinous xanthomatosis (described in detail in Section 11.2.9) and lathosterolosis, are usually childhood or juvenile-onset and develop as nuclear fleck or cortical opacities in the early stages of disease; they later progress to form dense posterior subcapsular opacities that are visually disruptive.

Multisystemic complications are a significant aspect of inborn errors of metabolism. These systemic manifestations are broad in range and there is a substantial degree of clinical overlap between certain disorders. It is worth noting that the majority of newborns with these conditions are healthy at birth; one study reported that only 25% of inborn errors of metabolisms have manifestations in the neonatal period.[9] Such early manifestations generally develop hours or days after birth and are usually non-specific. Common early clinical signs include low birth weight, failure to thrive, feeding difficulties, hypotonia, gastrointestinal problems, jaundice, organomegaly, respiratory distress and unexplained seizures; these can often be mistaken for symptoms of other conditions such as cardiopulmonary disorders or sepsis. Conditions that result in higher residual enzyme activity tend to present later in childhood, adolescence or early adulthood. Later onset manifestations occur during a progressive, chronic disease course and can occasionally be associated with fasting, fever or recurrent bouts elicited by excessive dietary intake of a particular substrate. The associated complications can be serious and life-threatening. They include neurological sequelae (developmental delay, spasticity, seizures, psychiatric or behavioural problems), skeletal problems, cardiac failure (e.g. cardiomyopathy), genitourinary abnormalities and liver failure (e.g. hepatosplenomegaly).

Molecular pathology

Inborn errors of metabolism are Mendelian conditions that result from the deficiency or abnormality of an enzyme or transporter molecule. The subset of these conditions that feature congenital or paediatric cataracts is genetically heterogeneous. Most of these disorders are inherited in an autosomal recessive manner. However, several autosomal dominant or X-linked subtypes have been described (Table 11.5). The vast majority of mutations causing

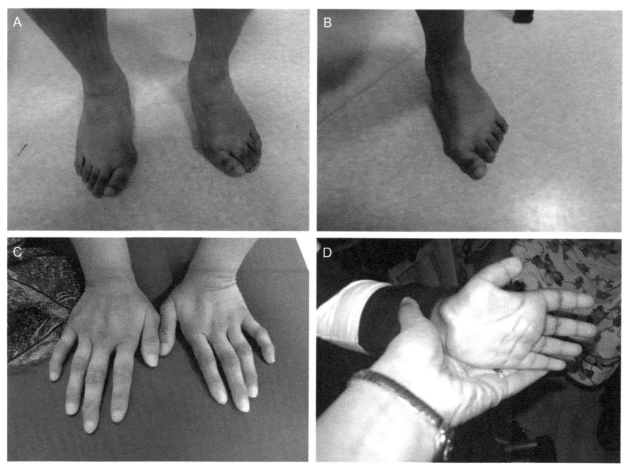

FIG. 11.7 Abnormalities of the hands and feet in two siblings with *PEX11B*-related peroxisomal biogenesis disorder. (A) and (B) show broad feet and hallux valgus in this sibling pair; (C) shows hands with abnormal metacarpal and phalangeal joints, and proximal implanting of the thumb; (D) shows hands with abnormally short distal phalanges and proximal implanting of the thumb. *(Adapted from Ref. 5.)*

BOX 11.1 Early diagnosis and treatment of stomatin-deficient cryohydrocytosis following genetic testing

A male proband, born to non-consanguineous parents, presented with bilateral congenital cataracts and tonic–clonic seizures in the first few months of life.

Genomic testing using a panel of 115 genes associated with congenital and paediatric cataracts was requested. This revealed a pathogenic missense change, *SLC2A1* c.226C>T p.(Gly76Ser), in heterozygous state. Parental testing found that neither parent carried this variant suggesting it was a *de novo* mutation, thereby reducing the recurrence risk for future pregnancies. The *SLC2A1* gene encodes a protein called solute carrier 2 (GLUT1). GLUT1 functions as a membrane-bound glucose transporter that facilitates the transmission of glucose across the blood–brain barrier; this is the main source of energy for brain growth and development. Mutations in *SLC2A1* have been associated with a complex and heterogeneous group of autosomal dominant GLUT1-deficiency disorders, including classical GLUT1-deficiency, paroxysmal exertion-induced dyskinesia and stomatin-deficient cryohydrocytosis. The latter of these three conditions has been associated with a phenotype that includes cataracts[6,7].

Immediately following diagnosis, the patient was commenced on a ketogenic diet which provides an alternative source of energy for the brain and has been shown to effectively control seizures[8].

This case highlights that gene panel testing offers an efficient means of diagnosis for inborn errors of metabolism complicated by lens opacities. Using these genomic testing approaches as a frontline means of diagnosis for cataract-related disease can streamline patient care, direct management, and facilitate personalised therapeutic intervention by removing the need to rely on accurate clinical phenotyping and numerous biochemical tests to establish the underlying cause of disease. Identification of the precise underlying cause of the condition permitted segregation analysis and the provision of accurate recurrence risk information for future pregnancies.

TABLE 11.5 Examples of inborn errors of metabolism featuring cataract listed by age at cataract onset alongside corresponding gene name, mode of inheritance and treatment options.

Appearance of cataract (age range)	Disorder	Gene	Inheritance	Treatment
Congenital (birth to 5 days)	Lowe oculocerebrorenal syndrome	OCRL	XL	Alkali supplements, adjusted sodium intake, vitamin D, drugs for the treatment of neurological manifestations; hormonal treatment of cryptorchidism
	Zellweger syndrome	PEX1	AR	–
	Rhizomelic chondroplasia punctata	PEX7	AR	–
	Stomatin-deficient cryohydrocytosis	SLC2A1	AD	Ketogenic diet
Neonatal (5 days to 4 weeks)	Galactosaemia	GALT	AR	Galactose and lactose restricted diet, supplementation of vitamins C, D and K, and calcium
	Hyperglycinuria	SLC36A2, SLC6A19, SLC6A20	AD	(Symptomatic treatment only) Sodium benzoate, dextromethorphan, anticonvulsants to control seizures
Infantile (4 weeks +)	Galactokinase deficiency	GALK	AR	Galactose restricted diet and calcium supplementation
	Peroxisome biogenesis disorder 14B	PEX11B	AR	Dietary supplementation of vitamin K, cholic and chenodeoxycholic, and docosahexaenoic acid. Dietary restriction of phytanic acid
	Cytochrome P450 cholesterol biosynthesis disorder	CYP51A1	AR	–
	α-mannosidosis	MAN2B1	AR	Enzyme replacement therapy in development
Childhood	Lathosterolosis	SC5D	AR	Simvastatin may normalise lathosterol levels
	Wilson disease	ATP7B	AR	Chelating agents (e.g. D-penicillamine or trientine dihydrochloride) and zinc salts
	Menke disease	ATP7A	XLR	D-penicillamine with vitamin C supplementation
	Alport syndrome	COL4A5	XLD	ACE inhibitor or ARB medicines may help to slow the progression of kidney disease
	Sialidosis type II	NEU1	AR	–
	Fabry disease	GLA	XL	Enzyme replacement therapy. Gene therapy trial underway
Juvenile/early adulthood	Cerebrotendinous xanthomatosis	CYP27A1	AR	Chenodeoxcholic acid and HMG-CoA-reductase inhibitors
	Female obligate carriers of Lowe syndrome	OCRL	XL	–
	Gyrate atrophy	OAT	AR	Dietary restriction of arginine and vitamin B_6 supplementation

AD, autosomal dominant; AR, autosomal recessive; XL, X-linked; XLR, X-linked recessive; XLD, X-linked dominant.

metabolic disease are coding single nucleotide variants although larger whole exon/gene deletions or duplications have also been reported.

Clinical management

Inborn errors of metabolism that feature cataracts are progressive and multisystemic. Care is lifelong, complex and requires the input of different specialties such as ophthalmology (including for surgical removal of cataracts where appropriate), paediatrics, neurology, cardiology, nephrology and genetics; input from specialist nurses, specialist pharmacists, dieticians and social workers is typically required.

Management and treatment are often needed both for acute disease manifestations and chronic complications. To optimise clinical outcomes, disease-specific surveillance following an early molecular diagnosis is key. Affected individuals are often closely monitored, requiring numerous and regular (sometimes weekly) hospital visits to control and/or optimise their metabolism. The diagnosis of an inborn error of metabolism and the prospect of witnessing their child undergoing intensive clinical care and disease management can be extremely overwhelming for a family and leave them faced with a significant disease burden.

Genetic counsellors play a unique part in supporting affected individuals and their families. Immediately following the diagnosis, the counsellor will play a role in managing the shock and distress that the family will inevitably experience. Once adjusted to the diagnosis, the family will likely have questions about the condition and their new situation. Genetic counsellors are well placed to educate the family about the nature and anticipated long-term complications and to help them adapt to their new situation. They will also educate the family on recurrence risk and ensure that they understand all options open to them in terms of treatment and reproductive choices. Providing the family with a sense of control over their circumstances can be of huge benefit to their wellbeing. For inborn errors of metabolism, genetic counselling is uniquely longitudinal, allowing the family to build a strong relationship with their counsellor and helping them overcome the overwhelming prospect of dealing with a potentially devastating disease.[10]

References

1. Gillespie RL, Urquhart J, Anderson B, Williams S, Waller S, Ashworth J, et al. Next-generation sequencing in the diagnosis of metabolic disease marked by pediatric cataract. *Ophthalmology* 2016;**123**(1):217–20.
2. Therrell BL, Padilla CD, Loeber JG, Kneisser I, Saadallah A, Borrajo GJ, et al. Current status of newborn screening worldwide. *Semin Perinatol* 2015;**39**(3):171–87.
3. Musleh M, Ashworth J, Black G, Hall G. Improving diagnosis for congenital cataract by introducing NGS genetic testing. *BMJ Qual Improv Rep* 2016;**5**(1). u211094.w4602.
4. Gillespie RL, O'Sullivan J, Ashworth J, Bhaskar S, Williams S, Biswas S, et al. Personalized diagnosis and management of congenital cataract by next-generation sequencing. *Ophthalmology* 2014;**121**(11):2124–37.
5. Taylor RL, Handley MT, Waller S, Campbell C, Urquhart J, Meynert AM, et al. Novel PEX11B mutations extend the peroxisome biogenesis disorder 14B phenotypic spectrum and underscore congenital cataract as an early feature. *Invest Ophthalmol Vis Sci* 2017;**58**(1):594–603.
6. Flatt JF, Guizouarn H, Burton NM, Borgese F, Tomlinson RJ, Forsyth RJ, et al. Stomatin-deficient cryohydrocytosis results from mutations in SLC2A1: a novel form of GLUT1 deficiency syndrome. *Blood* 2011;**118**(19):5267–77.
7. Bawazir WM, Gevers EF, Flatt JF, Ang AL, Jacobs B, Oren C, et al. An infant with pseudohyperkalemia, hemolysis, and seizures: cation-leaky GLUT1-deficiency syndrome due to a SLC2A1 mutation. *J Clin Endocrinol Metab* 2012;**97**(6):E987–93.
8. Klepper J, Akman C, Armeno M, Auvin S, Cervenka M, Cross HJ, et al. Glut1 deficiency syndrome (Glut1DS): state of the art in 2020 and recommendations of the international Glut1DS study group. *Epilepsia Open* 2020;**5**(3):354–65.
9. El-Hattab AW. Inborn errors of metabolism. *Clin Perinatol* 2015;**42**(2):413–39.
10. Hartley JN, Greenberg CR, Mhanni AA. Genetic counseling in a busy pediatric metabolic practice. *J Genet Couns* 2011;**20**(1):20–2.

11.2.6

Lowe oculocerebrorenal syndrome

I. Christopher Lloyd

In 1952 Dr. Charles Lowe of Massachusetts General Hospital described a unique syndrome: 'organic aciduria, decreased renal ammonia production, hydrophthalmos and mental retardation'.[1] By 1954, an associated renal Fanconi syndrome (proximal renal tubular transport dysfunction) was reported and in 1965 it was recognised that inheritance was via a recessive X-linked pattern.

Clinical characteristics

Lowe syndrome should be suspected in newborn males with bilateral dense cataracts, hypotonia and developmental delay; proteinuria due to proximal renal tubular transport dysfunction may also be noted.[2] The diagnosis is usually reached soon after birth through a combination of clinical assessment, genetic testing (*OCRL* gene screening), biochemical tests and ophthalmic examination of the parents and maternal family members.

Bilateral dense neonatal cataracts are present in nearly all affected individuals (Figs 11.8 and 11.9). Glaucoma has been described in up to 50% of cases; this has been associated with corneal enlargement (buphthalmos) and clouding occurring usually in the first year of life. Corneal and conjunctival scarring (keloids) have been reported in up to 25% of patients. Infantile onset nystagmus is common and is probably caused by both sensory deprivation amblyopia (associated with cataract-related visual axis opacification) and disruption to neurological development.

Proximal renal tubular transport dysfunction of the Fanconi type often develops with age and may not be symptomatic at birth. Notably, low molecular weight proteinuria can be seen early in life even in the absence of clinically significant aminoaciduria or other renal tubular abnormalities. It is therefore the most sensitive early marker of renal dysfunction in this disorder.[3] Other consequences of the renal abnormalities include aminoaciduria, proximal renal tubular acidosis and renal phosphate wasting that can lead to renal rickets, osteomalacia and pathological fractures. Hypercalciuria causing nephrocalcinosis and nephrolithiasis can also complicate management. Hypokalaemia due to secondary hyperaldosteronism may also be present. Renal bicarbonate, salt, and water loss lead to failure to thrive, and chronic renal failure eventually ensues in most individuals by the teenage years.[2] Renal transplantation has a variable success rate.

Neurological features include severe neonatal hypotonia and absent tendon reflexes, motor development delay and associated significant delay in walking. There may be poor feeding and neonatal respiratory problems. Cognition is usually severely affected but in approximately 10% of cases, there are only mild cognitive problems. The majority of patients exhibit behavioural problems including aggressiveness, irritability and stereotypic behaviour. Obsessive–compulsive behaviour is also often seen. There is increased susceptibility to febrile seizures and more than 50% of affected adults are epileptic. Mild ventriculomegaly may be apparent on neuroimaging and periventricular cystic lesions, which stabilise with time, have been described. Facial dysmorphism is common and includes frontal bossing, deep-set eyes and full cheeks (Fig. 11.10). Premature death usually occurs in the third to a fourth decade because of chronic renal disease, infections, respiratory illness, epilepsy or cardio-respiratory arrest.[2]

Mothers of affected children often have characteristic carrier signs: most heterozygous females older than age 15 years exhibit multiple radially arrayed, punctate grey–white opacities (snowflakes) in the peripheral layers of the lenticular cortex (Fig. 11.11). Slit-lamp examination

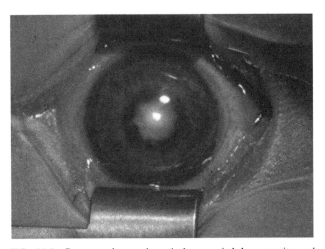

FIG. 11.8 Dense nuclear and cortical congenital lens opacity and irregular pupillary margin in an infant with Lowe oculocerebrorenal syndrome.

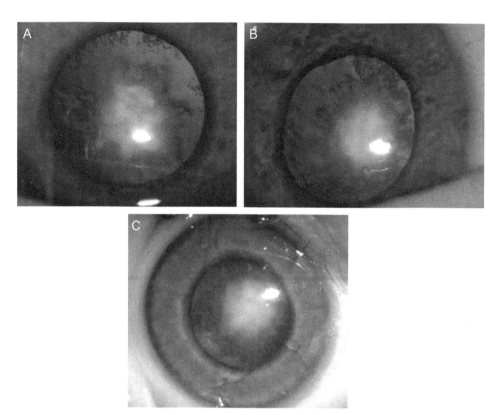

FIG. 11.9 Clinical pictures from a male infant with Lowe oculocerebrorenal syndrome. The patient was found to have dense congenital cataracts soon after birth [right eye (A), left eye (B)] and was noted to be hypotonic and irritable. Aminoaciduria was found on urinalysis and the diagnosis of Lowe syndrome was suspected. Notably, his mother's sister had given birth to another male infant who had bilateral dense congenital cataracts 1 month before [right eye (C)]. Both mothers were found to exhibit lenticular carrier features for Lowe syndrome and the diagnosis was subsequently confirmed by molecular genetic analysis (revealing a deletion of exons 1 to 3 of the *OCRL* gene). Both boys underwent bilateral lensectomy procedures before 12 weeks of age. Both went on to exhibit the classical renal and neurodevelopmental features of the condition.

after pupillary dilation is needed to accurately detect this characteristic sign.[5] Central posterior subcapsular lens opacities may rarely occur in carrier females (Fig. 11.11).

Molecular pathology

Lowe oculocerebrorenal syndrome is an X-linked disorder associated with mutations in the *OCRL* gene. *OCRL* encodes an enzyme, inositol polyphosphate 5 phosphatase, that is present in the trans-Golgi network and the endosomal and lysosomal compartments of a variety of cell types. Reduced enzymatic activity results in elevated intracellular levels of phosphatidylinositol-4,5-bisphosphate. High levels of this substrate have multiple effects on several cellular processes including protein trafficking, primary ciliary function and endocytosis.[2]

Most *OCRL* mutations occur in either the phosphatase domain or the C-terminal RhoGAP domain and result in loss of protein function. Over 300 disease-implicated variants have been described including nonsense, frameshift, splice site and missense changes. Missense variants are relatively uncommon (compared to truncating mutations) and can occur in or outside the catalytic domain of the enzyme. Large deletions and gross inversions have also been described.

Notably, mutations in *OCRL* are also found in a different disorder, Dent disease.[6,7] This condition is generally milder than Lowe syndrome and affects the kidney alone. Frameshift and nonsense *OCRL* variants associated with Dent disease are typically located in exons that precede those mutated in Lowe syndrome (exons 1–7 for Dent disease, exons 8–23 for Lowe syndrome). However, other types of *OCRL* variants (e.g. missense) that cause these two disorders do not map exclusively to specific gene regions.[8]

It is noted that around one-third of cases are due to new (*de novo*) mutations. Germline mosaicism (i.e. some sperm or eggs having a mutation that may not be present in other tissues of the body) has also been described. Thus, mothers of affected children should not necessarily be expected to have a mutation in *OCRL* or to exhibit any carrier features.[2]

On the rare occasion when the full features of Lowe syndrome are seen in a female, genetic testing may reveal two (biallelic) mutations (one on each X chromosome) or an X-autosome chromosomal translocation/rearrangement or

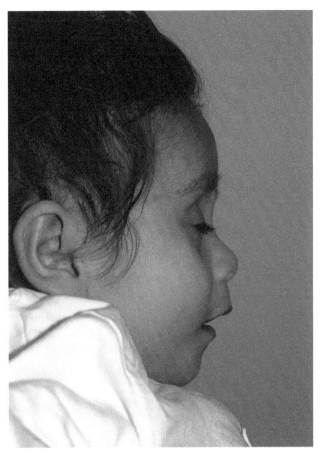

FIG. 11.10 Side view of the face of a child with Lowe oculocerebrorenal syndrome revealing a prominent forehead (frontal bossing), deep-set eyes and full cheeks. Protruding ears and sparse hair and can also be features of the condition. *(Courtesy of Professor Arif O. Khan.)*

skewed inactivation of the X chromosome carrying the normal copy (unfavourable lionisation).[2,9,10]

Clinical management

Infants with Lowe syndrome presenting with dense bilateral cataracts should, like other infants with congenital cataracts, undergo lensectomy surgery within the critical period of visual development (before 10 weeks of age) to minimise deprivation amblyopia. Contact lens correction of the subsequent aphakic refractive error may be challenging because of behavioural difficulties as well as concurrent glaucoma and corneal problems. Thus, many Lowe syndrome children are managed with aphakic spectacles. Strabismus surgery may be required in later childhood. Glaucoma is common[6] and should be managed rigorously.

Initially, reaching a diagnosis can be challenging and an examination under anaesthetic is often required to gather adequate information relating to intraocular pressure, corneal diameter, axial length, corneal thickness and optic disc appearance. Corneal haziness is not uncommon in the early postoperative period and may clear spontaneously. Glaucoma management may eventually require surgical intervention including the use of a seton drainage device such as a Baerveldt or a Paul tube.

Management of the extraocular manifestations of Lowe syndrome requires a multidisciplinary team approach involving different specialities including nephrologists and neurologists.

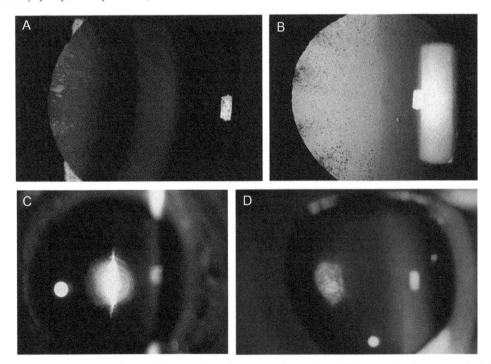

FIG. 11.11 Lens findings in mothers of patients with Lowe oculocerebrorenal syndrome. (A and B) Equatorial and anterior cortical clusters of smooth opacities of various sizes and shapes distributed in radial clusters are noted; retro-illumination emphasises the wedges of the clustered opacities. (C and D) Dense posterior lens opacities with smaller cortical dots are observed. *(A, B and C: Adapted from Ref. 4.)*

References

1. Lowe CU, Terrey M, MacLachan EA. Organic aciduria, decreased renal ammonia production, hydrophthalmos and mental retardation. *Am J Dis Child* 1952;**83**:164–84.
2. Lewis RA, Nussbaum RL, Brewer ED. Lowe syndrome. In: Adam MP, Ardinger HH, Pagon RA, et al., editors. *GeneReviews® [internet]*. Seattle, WA: University of Washington, Seattle; 1993–2020. 2001 July 24 [updated 2019 April 18]. Available from: https://www.ncbi.nlm.nih.gov/books/NBK1480/.
3. Laube GF, Russell-Eggitt IM, van't Hoff WG. Early proximal tubular dysfunction in Lowe's syndrome. *Arch Dis Child* 2004;**89**:479–80.
4. Lin T, Lewis RA, Nussbaum RL. Molecular confirmation of carriers for Lowe syndrome. *Ophthalmology* 1999;**106**(1):119–22.
5. Gardner RJM, Brown N. Lowe's syndrome: identification of carriers by lens examination. *J Med Genet* 1976;**13**:449–54.
6. Bokenkamp A, Bockenhauer D, Cheong HI, Hoppe B, Tasic V, Unwin R, et al. Dent-2 disease: a mild variant of Lowe syndrome. *J Pediatr* 2009;**155**:94–9.
7. Recker F, Reutter H, Ludwig M. Lowe syndrome/dent-2 disease: a comprehensive review of known and novel aspects. *J Pediatr Genet* 2013;**2**:53–68.
8. Song E, Luo N, Alvarado JA, Lim M, Walnuss C, Neely D, et al. Ocular pathology of oculocerebrorenal syndrome of Lowe: novel mutations and genotype-phenotype analysis. *Sci Rep* 2017;**7**:1442.
9. Hodgson SV, Heckmatt JZ, Hughes E, Crolla JA, Dubowitz V, Bobrow M. A balanced de-novo X/autosome translocation in a girl with manifestation of Lowe syndrome. *Am J Med Genet* 1986;**23**:837–47.
10. Mueller OY, Hartsfield Jr JK, Gallardo LA, Essig Y-P, Miller KL, Papemhausen PR, et al. Lowe oculocerebrorenal syndrome in a female with a balanced X;20 translocation: mapping of the X chromosome breakpoint. *Am J Hum Genet* 1991;**49**:804–11.

11.2.7

Hyperferritinemia-cataract syndrome

Graeme C.M. Black

Hyperferritinemia-cataract syndrome is an uncommon autosomal dominant condition characterised by early-onset cataracts associated with high serum ferritin levels, in the presence of normal serum iron. It can be misdiagnosed as a sign of hereditary hemochromatosis; this can lead to repeated venesection which may result in iron deficiency anaemia.

Clinical characteristics

Hyperferritinemia-cataract syndrome is reported to have cataracts as its only clinical manifestation. Cataracts arise early in infancy and childhood, i.e. they are not truly congenital. Lens changes are variable; many individuals have few symptoms other than photophobia and a mild impact on visual acuity. Cataract extraction is seldom indicated in childhood and is performed in early or mid-adulthood.

Lens changes are nuclear and cortical, with white or bluish, fleck-like or punctate opacities that are sometimes in a 'sunflower' distribution (Fig. 11.12). Where lens opacities have been followed longitudinally, they show age-related progression, which is generally slow.[1]

Affected individuals are found to have high ferritin levels, often 10–20 × normal values.[2] However, serum iron levels and transferrin saturation are typically within normal limits. Hyperferritinemia-cataract syndrome is therefore part of the differential diagnosis of individuals with hyperferritinemia without evidence of iron overload, a pattern seen in a wide range of non-genetic situations (including several liver disorders).[3]

In some cases, patients are suspected of having hereditary hemochromatosis and undergo repeated venesection with resultant iron deficiency anaemia. Notably, in hereditary hemochromatosis, iron levels, transferrin saturation

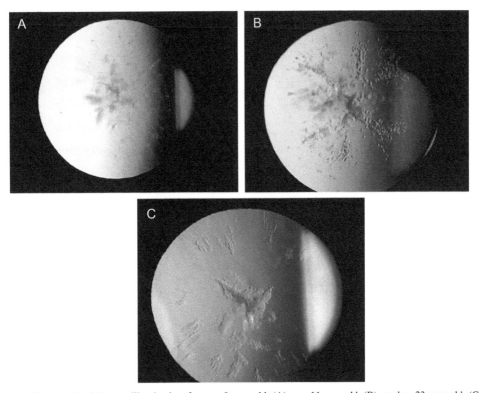

FIG. 11.12 Slit-lamp photograph with retro-illumination from a 9-year-old (A), an 11-year-old (B), and a 23-year-old (C) individual with hyperferritinemia-cataract syndrome. Cortical and nuclear opacities in a stellate pattern are evident.

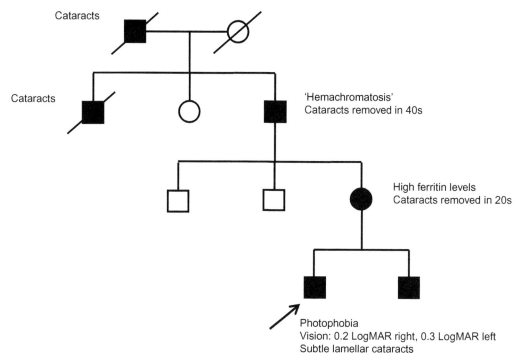

FIG. 11.13 Pedigree from a family with the hyperferritinemia-cataract syndrome. The proband was referred at 7 years of age for bilateral cataracts. Although these were having a subtle effect on his vision, they were not felt to require surgery. On direct questioning, the proband's mother who had childhood-onset cataracts removed in her 20s reported that she has been investigated for hyperferritinemia of unknown cause. Her father had been labelled with a diagnosis of hemochromatosis for which he was managed by regular venesection. Genomic testing was requested revealing a heterozygous c.-160A > G variant in the 5′ untranslated region of the *FTL* gene. This is a recurrent and previously reported disease-causing variant.

and liver tests are outside the normal range. Misdiagnosing hyperferritinemia-cataract syndrome for hemochromatosis has in the past led to unnecessary invasive investigations including liver biopsies, where histology shows normal iron staining (Fig. 11.13).

Molecular pathology

Ferritin is a universal intracellular protein that stores iron. It is essentially a protein complex consisting of 24 protein subunits. These subunits are of two types: heavy (encoded by the *FTH1* gene) and light (encoded by the *FTL* gene). The amount of expression of ferritin genes is sensitive to alterations in cytoplasmic iron concentration and is modulated by the binding of iron regulatory proteins to iron-responsive elements (IREs). These IREs are located within the 5′ untranslated region (i.e. the region of the mRNA that is directly upstream from the protein-coding sequence) of the *FTL* and *FTH1* genes. When the levels of cytoplasmic iron are high, structural changes of iron regulatory proteins reduce binding to IREs, facilitating gene translation and production of ferritin polypeptides. Conversely, in situations where there are low levels of intracytoplasmic iron, iron regulatory protein binding to IREs is unaffected and ferritin synthesis is repressed.

Hyperferritinemia-cataract syndrome is caused by non-coding mutations altering the IREs that are located in the 5′ untranslated region of the *FTL* gene. This region normally forms a three-dimensional stem-loop structure. These pathogenic variants alter this conformation such that it is unable to bind iron regulatory proteins, leading to unregulated translation and, consequently, raised levels of circulating ferritin.[4–6]

Over 30 disease-associated variants in this IRE stem-loop structure have been described; these are sometimes variably classified, e.g. by their position in the IRE or the geographical location of the laboratory where they were first described. The condition is inherited as an autosomal dominant trait. Interestingly, variants have an identical phenotypic impact (severity) whether heterozygous and homozygous.[7]

Since serum ferritin measurements are not part of the routine investigation of individuals with early-onset cataracts, a genomic approach that identifies patients with hyperferritinemia-cataract syndrome is required. Given that this condition results from non-coding mutations, genetic tests that focus only on coding regions will fail to highlight disease-causing variants. Also, this mutation may be overlooked by traditional variant prioritisation approaches used in standard sequencing data analyses.

Clinical management

Hyperferritinemia in this disorder does not usually cause any health problems other than cataracts and treatment is generally focused on improving vision (with glasses, contact lenses, and/or cataract surgery).

References

1. Cosentino I, Zeri F, Swann PG, Majore S, Radio FC, Palumbo P, et al. Hyperferritinemia-cataract syndrome: long-term ophthalmic observations in an Italian family. *Ophthalmic Genet* 2016;**37**(3):318–22.
2. Ismail AR, Lachlan KL, Mumford AD, Temple IK, Hodgkins PR. Hereditary hyperferritinemia cataract syndrome: ocular, genetic, and biochemical findings. *Eur J Ophthalmol* 2006;**16**(1):153–60.
3. Ferrante M, Geubel AP, Fevery J, Marogy G, Horsmans Y, Nevens F. Hereditary hyperferritinaemia-cataract syndrome: a challenging diagnosis for the hepatogastroenterologist. *Eur J Gastroenterol Hepatol* 2005;**17**(11):1247–53.
4. Beaumont C, Leneuve P, Devaux I, Scoazec JY, Berthier M, Loiseau MN, et al. Mutation in the iron responsive element of the L ferritin mRNA in a family with dominant hyperferritinaemia and cataract. *Nat Genet* 1995;**11**(4):444–6.
5. Girelli D, Corrocher R, Bisceglia L, Olivieri O, Zelante L, Panozzo G, et al. Hereditary hyperferritinemia-cataract syndrome caused by a 29-base pair deletion in the iron responsive element of ferritin L-subunit gene. *Blood* 1997;**90**(5):2084–8.
6. Halvorsen M, Martin JS, Broadaway S, Laederach A. Disease-associated mutations that alter the RNA structural ensemble. *PLoS Genet* 2010;**6**(8):e1001074.
7. Giansily-Blaizot M, Cunat S, Moulis G, Schved JF, Aguilar-Martinez P. Homozygous mutation of the 5'UTR region of the L-ferritin gene in the hereditary hyperferritinemia cataract syndrome and its impact on the phenotype. *Haematologica* 2013;**98**(4):e42–3.

11.2.8

Galactosaemia

I. Christopher Lloyd

Galactosaemia is a clinically and genetically heterogeneous group of autosomal recessive disorders associated with impaired degradation of galactose. It is caused by deficiency in four principal enzymes involved in the galactose metabolic pathway (Leloir pathway; Fig. 11.14): galactose-1-phosphate uridyltransferase (GALT), galactokinase (GALK), uridine diphosphogalactose-4-epimerase (GALE) or galactose mutarotase (GALM). Activation of alternative pathways of galactose metabolism is triggered in affected individuals leading to the accumulation of metabolites such as galactitol. Excessive galactitol is thought to cause direct damage to several tissues and its accumulation in the lens leads to cataracts.[2–6]

The most common and most severe form of galactosaemia is GALT deficiency (also known as type I or classic galactosaemia). In situations where infants affected with this condition are not treated promptly with a low-galactose diet, life-threatening complications appear within a few days after birth.[7] Other forms of galactosaemia tend to be more insidious and less severe; one of the few consistent clinical signs in GALK deficiency and GALM deficiency is early-onset cataract whilst GALE deficiency is variable but can occasionally be largely asymptomatic.[3,6]

Newborn screening for galactosaemia is recommended in many countries worldwide. The relevant biochemical assays are typically designed to detect both GALT

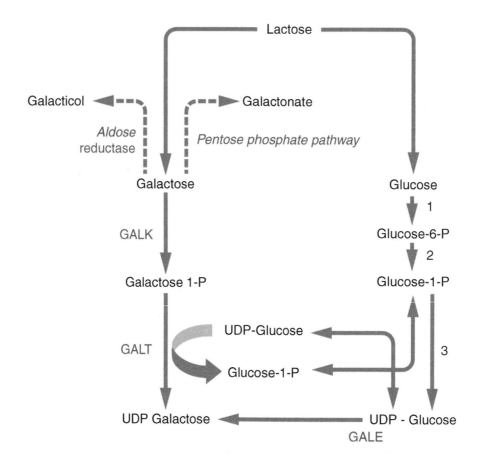

FIG. 11.14 Metabolism of galactose: Leloir pathway. 1 is hexokinase, 2 is phosphoglucomutase, and 3 is UDP-glucose pyrophosphorylase. *(Adapted from Ref. 1.)*

deficiency and other forms. However, cases that have less marked hypergalactosaemia may be missed.[2,7,8] Genetic testing confirms the diagnosis and/or allows the identification of milder cases that present with cataracts.[9] In general, the standard of care consists of a galactose-restricted diet which, if implemented in a timely manner, can prevent or resolve cataract formation.[3]

Clinical characteristics

Galactosaemia may be suspected based on suggestive clinical findings, a positive newborn screening test result and/or relevant laboratory findings (e.g. increased levels of galactose metabolites).

Individuals with GALT deficiency (type I galactosaemia) typically present in the neonatal period with vomiting, diarrhoea, feeding difficulties, lethargy, hypotonia, failure to thrive, jaundice and haemorrhagic diathesis. Dietary restriction prevents/resolves these neonatal complications but if the condition is left untreated, liver failure, sepsis and shock may ensue. Notably, diet is ineffective in preventing long-term complications of GALT deficiency. Even with early and adequate therapy, affected individuals develop cataracts, neurological sequelae (delay in language acquisition, speech problems, poor growth, poor intellectual function, ataxia) and premature ovarian insufficiency (leading to fertility problems in female patients).[3,7,10–12]

The main ophthalmic finding in classic galactosaemia is lens opacification. Due to increased osmolarity in the crystalline lens, an 'oil drop' refractive change may be noted initially (Fig. 11.15A). Early dietary treatment may enable reversal of these lenticular changes (Fig. 11.15B) but, in untreated individuals, permanent lens opacification ensues (Figs 11.16 and 11.17).[3,12]

GALK deficiency (type II galactosaemia), GALE deficiency (type III galactosaemia) and GALM deficiency (type IV galactosaemia) are less prevalent and present with different clinical signs and symptoms. GALK deficiency causes fewer serious medical problems than GALT deficiency has been associated with 'idiopathic' intracranial hypertension. Affected children develop cataracts, often in the first few years of life but are usually otherwise well.[3,13] GALE deficiency varies from mild peripheral epimerase deficiency (which can be asymptomatic or manifest with cataracts) through an intermediate form to severe generalised epimerase deficiency where dietary treatment is necessary to prevent acute clinical symptoms similar to those seen in GALT deficiency.[10] GALM deficiency is a rare subtype that has similarities to GALK deficiency. It can cause cataract formation in infancy but does not appear to lead to liver dysfunction or growth/developmental issues.[6,14]

Molecular pathology

Galactosaemia is associated with mutations in the *GALT*, *GALK1*, *GALE* and *GALM* genes. These genes encode enzymes that are essential for the processing of dietary galactose and its breakdown into glucose and other molecules that can be stored or used for energy.[2,3] More than 350 mutations have been identified in the *GALT* gene (i.e. the locus implicated in type I galactosaemia). Most of these changes eliminate enzyme activity, preventing normal processing of galactose, and resulting in severe disease.[12] In contrast, a relatively common GALT deficiency associated variant, c.940A>G (p.Asn314Asp), has been shown to lead to sufficient enzyme activity for the affected individual to remain asymptomatic. This form of galactosaemia is called Duarte or biochemical variant galactosaemia and patients are mainly identified due to abnormal newborn screening.[3] Another common *GALT* mutation that has been associated with a milder phenotype is the c.404C>T (p.Ser135Leu) variant, a change that appears to be particularly prevalent in individuals of African American and South African ancestries.[3,13]

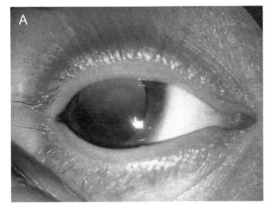 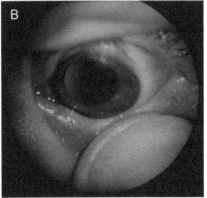

FIG. 11.15 'Oil drop' central lenticular refractive change in an infant with GALT deficiency (type I galactosaemia) (A). This resolved after appropriate dietary treatment was initiated (B). *(Courtesy of Professor David Taylor.)*

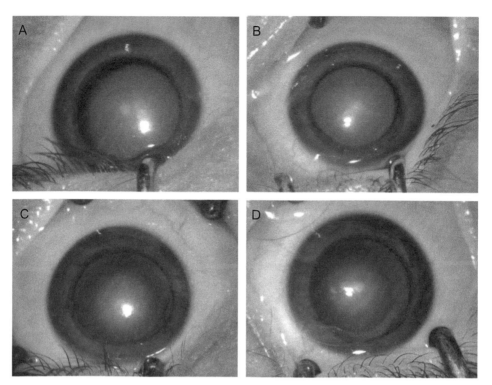

FIG. 11.16 Lens findings in identical twin brothers with GALT deficiency (type I galactosaemia). (A and B) Dense nuclear cataracts in the right and left eyes of a 6-month-old male infant twin. (C and D) Less marked galactosaemia-related cataracts in the right and left eyes of the identical twin brother of the above proband.

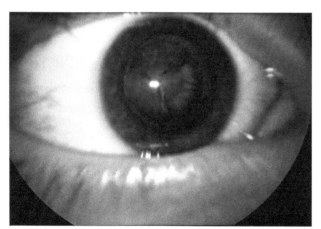

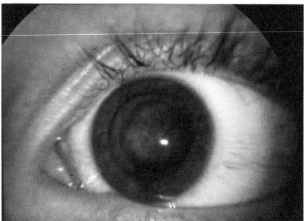

FIG. 11.17 Lamellar, sutural and nuclear lens opacification in both eyes of a child with late initiation of dietary treatment for GALT deficiency (type I galactosaemia). *(Courtesy of Professor David Taylor.)*

At least 50 *GALK1* mutations have been associated with GALK deficiency in the biomedical literature. Also, over 30 *GALE* variants have been described in individuals with GALE deficiency and at least 5 changes in *GALM* have been implicated in GALM deficiency.[7,15]

Clinical management

Newborn screening, if performed in the first few days of life, provides an opportunity for diagnosis either before or soon after an infant presents with symptoms. Early diagnosis of GALT deficiency (type I galactosaemia) allows a change to a soy-based formula and thus reduces the risk of liver failure and other complications. The patient's parents should be educated about the lifelong need for dietary restriction of cow's milk and dairy products. It should be highlighted that although the diet can reverse the neonatal clinical picture, it does not prevent the appearance of long-term complications. Careful monitoring by a multidisciplinary team of specialists is therefore required, including a metabolic physician, a developmental paediatrician, a speech therapist and an ophthalmologist. Cataract surgery is warranted in many affected children, although this is rarely needed in infancy.[3]

It is generally recommended that a galactose-restricted diet should be considered in all people affected by other types of galactosaemia (GALK, GALE and GALM deficiency). Management of childhood or presenile cataracts is likely to be needed in a proportion of patients with these diagnoses.

References

1. Broomfield A, Brain C, Grunewald S. Galactosaemia: diagnosis, management and long-term outcome. *Paediatr Child Health* 2015;**25**(3):113–8. ISSN 1751-7222.
2. Pasquali M, Yu C, Coffee B, et al. Laboratory diagnosis of galactosemia: a technical standard and guideline of the American College of Medical Genetics and Genomics (ACMG). *Genet Med* 2018;**20**:3–11.
3. Demirbas D, Coelho AI, Rubio-Gozalbo ME, Berry GT. Hereditary galactosemia. *Metabolism* 2018;**83**:188–96. https://doi.org/10.1016/j.metabol.2018.01.025.
4. Bosch AM. Classical galactosaemia revisited. *J Inherit Metab Dis* 2006;**29**:516–25.
5. Timson DJ. The molecular basis of galactosemia—past, present and future. *Gene* 2016;**589**(2):133–41.
6. Wada Y, Kikuchi A, Arai-Ichinoi N, Sakamoto O, Takezawa Y, Iwasawa S, et al. Biallelic GALM pathogenic variants cause a novel type of galactosemia. *Genet Med* 2019;**21**:1286–94.
7. Berry GT. Classic galactosemia and clinical variant galactosemia. In: Adam MP, Ardinger HH, Pagon RA, Wallace SE, Bean LJH, Stephens K, Amemiya A, editors. *GeneReviews® [internet]*. Seattle, WA: University of Washington, Seattle; 1993–2020. 2000 February 4 [updated 2020 July 2].
8. Lak R, Yazdizadeh B, Davari M, Nouhi M, Kelishadi R. Newborn screening for galactosaemia. *Cochrane Database Syst Rev* 2020;**6**(6):CD012272.
9. Gillespie RL, Urquhart J, Anderson B, Williams S, Waller S, Ashworth J, et al. Next-generation sequencing in the diagnosis of metabolic disease marked by pediatric cataract. *Ophthalmology* 2016;**123**(1):217–20. https://doi.org/10.1016/j.ophtha.2015.06.035.
10. Karadag N, Zenciroglu A, Eminoglu FT, Dilli D, Karagol BS, Kundak A, et al. Literature review and outcome of classic galactosemia diagnosed in the neonatal period. *Clin Lab* 2013;**59**(9–10):1139–46.
11. Fridovich-Keil JL, Gubbels CS, Spencer JB, Sanders RD, Land JA, Rubio-Gozalbo E. Ovarian function in girls and women with GALT-deficiency galactosemia. *J Inherit Metab Dis* 2011;**34**(2):357–66.
12. Beigi B, O'Keefe M, Bowell R, Naughten E, Badawi N, Lanigan B. Ophthalmic findings in classical galactosaemia—prospective study. *Br J Ophthalmol* 1993;**77**(3):162–4.
13. Bosch AM, Bakker HD, van Gennip AH, van Kempen JV, Wanders RJA, Wijburg FA. Clinical features of galactokinase deficiency: a review of the literature. *J Inherit Metab Dis* 2002;**25**:629–34.
14. Iwasawa S, Kikuchi A, Wada Y, Arai-Ichinoi N, Sakamoto O, Tamiya G, et al. The prevalence of GALM mutations that cause galactosemia: a database of functionally evaluated variants. *Mol Genet Metab* 2019;**126**:362–7.
15. Fridovich-Keil J, Bean L, He M, Schroer R. Epimerase deficiency galactosemia. In: Adam MP, Ardinger HH, Pagon RA, Wallace SE, Bean LJH, Stephens K, Amemiya A, editors. *GeneReviews® [internet]*. Seattle, WA: University of Washington, Seattle; 1993–2020 [updated 2016 June 16].

11.2.9

Cerebrotendinous xanthomatosis

Rachel L. Taylor

Cerebrotendinous xanthomatosis (CTX) is a rare inborn error of bile acid metabolism caused by biallelic mutations in the *CYP27A1* gene. This multisystemic disorder is characterised by chronic diarrhoea, bilateral cataracts, tendon xanthomas, behavioural issues, and cognitive and other neurologic impairments. Onset and presentation are variable and the disease course is slowly progressive. Early manifestations can be ambiguous making timely recognition and prompt diagnosis challenging. Despite early clinical signs, the average age at diagnosis has been reported as 32 years with a diagnostic delay of up to 25 years, in some cases.[1,2] Ophthalmologists are in a unique position to identify and diagnose CTX at a time when significant disability can be prevented.

Clinical characteristics

Infantile-onset chronic diarrhoea is often the earliest presenting feature, and childhood-onset bilateral cataract is a common early sign. Some patients will display cognitive impairment from infancy or slight intellectual disability up to puberty. Many experience a decline in their intellectual abilities or develop dementia in the third decade of life. Other adult-onset features of the disease include cardiovascular disease, spasticity, ataxia, seizures, peripheral neuropathy, osteopenia, and premature ageing.

A recent survey of 55 individuals with CTX found that 89% had developed cataracts; tendon xanthomas (appearing later in the disease process and often occurring on the Achilles tendon; Fig. 11.18) were present in 78%, osteoporosis in 67% and diarrhoea in 40%. Prevalent neurological features include peripheral neuropathy (70%), intellectual disability (60%) and psychiatric disturbance (44%).[2]

Ophthalmic: Childhood or juvenile-onset bilateral cataracts are the initial clinical finding in many CTX patients. These reflect the accumulation of cholestanol and cholesterol in the lens. In almost 50% of cases, cataracts develop before the age of 10 years, but 25% of patients will develop lens opacification after the age of 40 years.[3] CTX-associated cataracts often appear as multiple small or fleck-like opacities within the nucleus and/or cortex of the lens (Fig. 11.19).[4,5] They may be visually insignificant, but are often progressive and can develop to form dense posterior sub-capsular cataracts which may markedly reduce vision.[4,5] Most patients will require cataract surgery by age 20 years. Other ophthalmic signs include pale optic discs, retinal vessel sclerosis and the deposition of cholesterol-like deposits along the vascular arcades.[6]

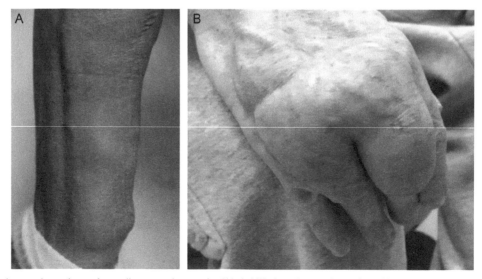

FIG. 11.18 Tendon xanthoma in cerebrotendinous xanthomatosis. (A) Achilles' tendon xanthoma in a 64-year-old individual with cerebrotendinous xanthomatosis; (B) tuberous xanthoma of the hand in a 38-year-old individual with cerebrotendinous xanthomatosis. *(Adapted from Ref. 1.)*

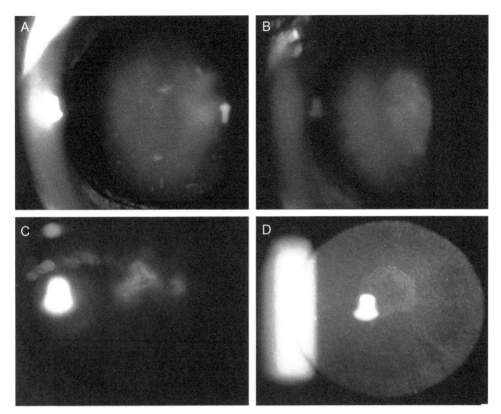

FIG. 11.19 Lens opacification associated with cerebrotendinous xanthomatosis. (A and B) Fleck-like opacities are apparent upon slit-lamp examination of a proband with cerebrotendinous xanthomatosis. (C and D) The lens of the proband's sibling shows barely visible inferonasal fleck opacities when retro-illumination is not used to examine the lens (C); multiple flecks with cortical changes are noted when the lens is examined with retro-illumination (D). *(Adapted from Ref. 4.)*

Cardiovascular: Patients with CTX are at increased risk of severe premature atherosclerosis and/or coronary artery disease. Most affected individuals will have very high 27-hydroxycholesterol and low high-density lipoprotein cholesterol levels.[7]

Skeletal: Premature osteoporosis and bone fractures are common.

Psychiatric: The most prevalent psychiatric manifestation is a behavioural or personality disorder.[8] Behavioural problems are particularly common in patients under 30 years of age. The psychiatric pathology of CTX tends to be more complex in people over 30 years and often includes features pre-dementia/dementia.[9]

Neurological: In childhood, neurological features may present as developmental delay, intellectual disability or learning difficulties.[10] Intellectual function progressively worsens with time. Pyramidal signs (i.e. spasticity, extensor plantar responses, hyperreflexia), cerebellar signs (e.g. dysarthria, nystagmus, and ataxia), or general symptoms of neurological dysfunction (i.e. gait disturbance and intellectual disability) usually become evident at around 20–30 years of age and worsen over time.

Enterohepatic: Enterohepatic system involvement is common and chronic infantile-onset diarrhoea is often the earliest manifestation of the disease.[3] Neonatal cholestatic jaundice and gallstones have also been described.

Molecular pathology

CTX is inherited as an autosomal recessive trait and is caused by mutations in the *CYP27A1* gene. This gene encodes sterol-27 hydroxylase (CYP27), a ubiquitously expressed mitochondrial enzyme that works in a pathway that breaks down cholesterol into bile acids that are necessary for the body to digest fats. Dysfunction of this enzyme results in the accumulation of cholesterol, cholestanol, and toxic bile acid intermediates in plasma and almost every tissue of the body.

Over 120 different *CYP27A1* mutations have been reported as the cause of CTX. Most of these are missense changes that result in severely reduced enzymatic function.[11] Mutation carriers (i.e. individuals who harbour a single heterozygous mutation in *CYP27A1*) can have 50% reduced sterol-27 hydroxylase function but normal to slightly increased levels of cholestanol and circulatory bile alcohols. Carriers are usually asymptomatic, although a few examples of symptomatic individuals have been described,[12,13] including carrier parents with Y-sutural fleck opacities of the lens.[4]

Clinical management

Diagnosis of CTX may be established through clinical assessment, biochemical profiling (5α-cholestanol blood plasma measurement) and neuroimaging (MRI). However, in younger patients, genomic analysis (including screening of the *CYP27A1* gene) is the most effective path to diagnosis. This approach can greatly facilitate timely treatment of the condition, before the development of more serious adult-onset manifestations.[14,15]

The input of a multidisciplinary team is necessary, usually involving paediatric metabolic medicine, ophthalmology, neurology, cardiology, clinical genetics, and pharmacy, as well as nursing, social workers, and genetic counsellors.

Treatment for CTX is available in the form of oral chenodeoxycholic acid (CDCA). Successful treatment should see plasma cholestanol levels decline gradually and normal plasma cholestanol to cholesterol ratio restored. Patient outcomes are improved with the early commencement of CDCA and early treatment is critical for preventing irreversible neurological damage or deterioration. For example, encephalopathy due to deposition of cholesterol and cholestanol in nervous tissues is potentially reversible with timely CDCA treatment. This is only possible if therapy is commenced before accumulated stanols and sterols reach a critical level at which permanent damage to the nervous system occurs.[16] CDCA treatment is also effective for non-neurological symptoms.[17]

Regular surveillance is recommended. It generally requires annual measurement of plasma cholestanol concentration, neurological and neuropsychological assessment, and brain MRI.

References

1. Duell PB, Salen G, Eichler FS, Debarber AE, Connor SL, Casaday L, et al. Diagnosis, treatment, and clinical outcomes in 43 cases with cerebrotendinous xanthomatosis. *J Clin Lipidol* 2018;**12**:1169–78.
2. Mignarri A, Gallus GN, Dotti MT, Federico A. A suspicion index for early diagnosis and treatment of cerebrotendinous xanthomatosis. *J Inherit Metab Dis* 2014;**37**:421–9.
3. Federico A, Dotti MT, Gallus GN. *Cerebrotendinous xanthomatosis.* In: GeneReviews® [Internet]. Seattle (WA): University of Washington, Seattle; 1993–2021. 2003 Jul 16 [updated 2016 Apr 14].
4. Khan AO, Aldahmesh MA, Mohamed JY, Alkuraya FS. Juvenile cataract morphology in 3 siblings not yet diagnosed with cerebrotendinous xanthomatosis. *Ophthalmology* 2013;**120**:956–60.
5. Cruysberg JR, Wevers RA, Van Engelen BG, Pinckers A, Van Spreeken A, Tolboom JJ. Ocular and systemic manifestations of cerebrotendinous xanthomatosis. *Am J Ophthalmol* 1995;**120**: 597–604.
6. Dotti MT, Rufa A, Federico A. Cerebrotendinous xanthomatosis: heterogeneity of clinical phenotype with evidence of previously undescribed ophthalmological findings. *J Inherit Metab Dis* 2001;**24**:696–706.
7. Weingartner O, Laufs U, Bohm M, Lutjohann D. An alternative pathway of reverse cholesterol transport: the oxysterol 27-hydroxycholesterol. *Atherosclerosis* 2010;**209**:39–41.
8. Nia S. Psychiatric signs and symptoms in treatable inborn errors of metabolism. *J Neurol* 2014;**261**(Suppl. 2):S559–68.
9. Moghadasian MH, Salen G, Frohlich JJ, Scudamore CH. Cerebrotendinous xanthomatosis: a rare disease with diverse manifestations. *Arch Neurol* 2002;**59**:527–9.
10. Larson A, Weisfeld-Adams JD, Benke TA, Bonnen PE. Cerebrotendinous xanthomatosis presenting with infantile spasms and intellectual disability. *JIMD Rep* 2017;**35**:1–5.
11. Gallus GN, Dotti MT, Federico A. Clinical and molecular diagnosis of cerebrotendinous xanthomatosis with a review of the mutations in the CYP27A1 gene. *Neurol Sci* 2006;**27**:143–9.
12. Hansson M, Olin M, Floren CH, Von Bahr S, Van'T Hooft F, Meaney S, et al. Unique patient with cerebrotendinous xanthomatosis. Evidence for presence of a defect in a gene that is not identical to sterol 27-hydroxylase. *J Intern Med* 2007;**261**:504–10.
13. Sugama S, Kimura A, Chen W, Kubota S, Seyama Y, Taira N, et al. Frontal lobe dementia with abnormal cholesterol metabolism and heterozygous mutation in sterol 27-hydroxylase gene (CYP27). *J Inherit Metab Dis* 2001;**24**:379–92.
14. Gillespie RL, Urquhart J, Anderson B, Williams S, Waller S, Ashworth J, et al. Next-generation sequencing in the diagnosis of metabolic disease marked by pediatric cataract. *Ophthalmology* 2016;**123**:217–20.
15. Nicholls Z, Hobson E, Martindale J, Shaw PJ. Diagnosis of spinal xanthomatosis by next-generation sequencing: identifying a rare, treatable mimic of hereditary spastic paraparesis. *Pract Neurol* 2015;**15**:280–3.
16. Berginer VM, Gross B, Morad K, Kfir N, Morkos S, Aaref S, et al. Chronic diarrhea and juvenile cataracts: think cerebrotendinous xanthomatosis and treat. *Pediatrics* 2009;**123**:143–7.
17. Koopman BJ, Wolthers BG, Van Der Molen JC, Waterreus RJ. Bile acid therapies applied to patients suffering from cerebrotendinous xanthomatosis. *Clin Chim Acta* 1985;**152**:115–22.

Chapter 12

Ectopia lentis

Chapter outline

12.1 *ADAMTSL4*-related disorders, including isolated ectopia lentis 149
12.2 Marfan syndrome 151
12.3 Weill–Marchesani syndrome 155
12.4 Homocystinuria 157

Ectopia lentis describes the displacement or dislocation of the crystalline lens. It is associated with defects of the zonular fibres that hold the lens in its natural position and can range from subtle sub-luxation to total dislocation. A lens is considered subluxed when it is partially displaced but remains in the retro-pupillary plane (Fig. 12.1). In contrast, it is considered dislocated (luxated) when it lies outside the lens space, either in the anterior chamber or in the vitreous cavity.

Ectopia lentis may be identified incidentally, a common scenario in children (e.g. Fig. 12.2), or, more frequently through its effects on vision due to induced refractive errors or distortion of the visual axis by the edge of the displaced lens. Notably, ectopic lenses are prone to becoming prematurely opacified, and vision may become affected if a significant cataract develops. If the lens acutely and completely dislocates posteriorly, sudden loss of vision occurs. Dislocation of the lens anteriorly may cause elevation of intraocular pressure (pupillary block mechanism) often leading to corneal decompensation.

Ectopia lentis can occur due to trauma, ophthalmic disorders (e.g. Section 12.1) or syndromic disease (e.g. Sections 12.2, 12.3 and 12.4). Enquiring about recent trauma and performing a targeted ophthalmic and systemic evaluation is therefore particularly important. As ectopia lentis may be the first sign of a serious multisystemic condition (e.g. Marfan syndrome or homocystinuria; Table 12.1), exploration of an underlying cause is warranted. Genomic analysis such as gene panel testing can be particularly helpful, especially in cases without a history of major trauma. These genetic tests enable the provision of a timely and definitive diagnosis and facilitate the development of personalised care pathways.

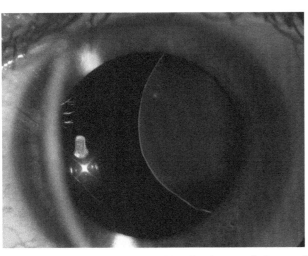

FIG. 12.1 Nasal lens sub-luxation. Note the almost total absence of zonular fibres.

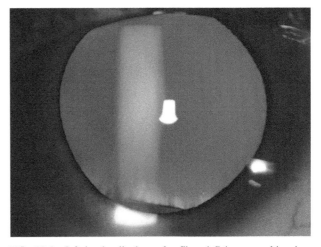

FIG. 12.2 Inferior localised zonular fibre deficiency resulting in a limited 'notch' of the lens in an asymptomatic patient; the zonular fibres appear elongated, a frequent feature in individuals with Marfan syndrome. In general, ectopia lentis is most reliably diagnosed by slit-lamp examination after maximal pupillary dilatation and with the patient looking in different directions; subtle iridodonesis may alert the clinician to the presence of mild sub-luxation.

TABLE 12.1 Selected genes implicated in ectopia lentis.

Category	Gene	Mode of inheritance	Associated condition(s)
Ophthalmic disorders	ADAMTSL4	Autosomal recessive	Isolated ectopia lentis
			Ectopia lentis et pupillae
	CPAMD8		Anterior segment dysgenesis, myopia, childhood-onset glaucoma
	PAX6	Autosomal dominant	Aniridia
Multi-systemic disorders	FBN1	Autosomal dominant	Marfan syndrome
			Dominant Weill–Marchesani syndrome
	CBS	Autosomal recessive	Homocystinuria
	SUOX		Sulphite oxidase deficiency
	AASS		Hyperlysinemia
	LTBP2		Microspherophakia
			Weill–Marchesani syndrome
	ADAMTS10		Weill–Marchesani syndrome
	ADAMTS17		
	ADAMTS18		MMCAT (microcornea, myopic chorioretinal atrophy, and telecanthus)
	ASPH		Traboulsi syndrome (ectopia lentis with facial dysmorphism, anterior segment abnormalities and spontaneous filtering blebs)
	COL18A1		Knobloch syndrome
	COL5A1		Ehlers-Danlos syndrome
	COL5A2		
	VPS13B		Cohen syndrome

12.1

ADAMTSL4-related disorders, including isolated ectopia lentis

Aman Chandra and Elias I. Traboulsi

ADAMTSL4-related disorders are a group of autosomal recessive conditions that predominantly affect the eye. They form a continuum that encompasses a range of phenotypes including autosomal recessive isolated ectopia lentis and ectopia lentis et pupillae.[1] Together with changes in *FBN1*, the gene implicated in Marfan syndrome, mutations in *ADAMTSL4* are the most common genetic causes of ectopia lentis. Although patients with *ADAMTSL4* variants generally have no extraocular features, they have been shown to have a more severe ophthalmic phenotype with a greater axial length and an earlier onset of visual symptoms compared to individuals with *FBN1* mutations.[2]

Clinical characteristics

Patients with *ADAMTSL4*-associated ectopia lentis typically present in the first decade of life. When the lens displacement is accompanied by significant iris abnormalities (e.g. in individuals with ectopia lentis et pupillae), the diagnosis is usually made soon after birth. Conversely, cases with mild ectopia lentis and normal pupils may remain undiagnosed until adulthood.[1]

Variability in the ophthalmic findings is observed not only among patients with identical mutations but also, occasionally, between the eyes of the same individual.[3] Lens displacement, for example, is usually bilateral but can be asymmetrical. Other ophthalmic features of *ADAMTSL4*-related disease include pupils that dilate poorly and may be oval-shaped or ectopic; a significant refractive error that may lead to amblyopia; early-onset lens opacification and an increased risk of retinal detachment and glaucoma. Remnants of the pupillary membrane are frequently found and can help differentiate this condition from other causes of ectopia lentis.

Although this group of conditions is generally not accompanied by significant extraocular features, a few affected individuals have been reported to have skeletal abnormalities.[1]

Molecular pathology

ADAMTSL4-related disorders are inherited as an autosomal recessive trait. They are caused by loss of function variants in *ADAMTSL4*, a gene encoding an ADAMTS-like protein. This protein belongs to a family of molecules that shares significant similarities with ADAMTS metalloproteases. However, ADAMTS-like proteins lack the protease domain of the ADAMTS family and are therefore thought to be catalytically inactive.

ADAMTSL4 is expressed in many tissues including the iris, ciliary body, and retinal pigment epithelium. Co-localisation with *FBN1* has been reported and, although the exact protein function remains unclear, a role in anchoring the zonular fibres to the lens has been suggested.[1]

Over 25 *ADAMTSL4* mutations have been reported. Many of them introduce a premature stop codon while disease-associated missense changes often impact critical parts of the protein. A 20 base-pair deletion, c.767_786del, appears to be the most common mutation in several populations.

Clinical management

The specific ophthalmic management of ectopia lentis is common for all conditions that feature lens displacement/dislocation (including *ADAMTSL4*-ophthalmopathy, Marfan syndrome and homocystinuria). This involves regular visual acuity checks and refraction; these are particularly important in children who are at risk of amblyopia. In general, appropriate contact lens or spectacle correction provide an adequate, stable vision for most affected individuals. Thus, if ectopia lentis is mild and corrected visual acuity is good, monitoring is sufficient (Fig. 12.3). If the lens is subluxated beyond the visual axis, a conservative approach to visual rehabilitation can also be undertaken with, for example, contact lenses refracting through the aphakic pupillary space. This is also true if complete posterior dislocation has occurred. The lens within its capsule is, for the most part, not uveogenic and can be left in situ.

Lens surgery is indicated in a subset of patients who either (i) have significantly reduced corrected visual acuity

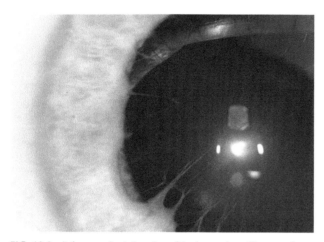

FIG. 12.3 Infero-nasal sub-luxation of the lens and pupillary membrane remnants in a patient with ectopia lentis et pupillae and biallelic mutations in *ADAMTSL4*. The patient also had lens opacities in both eyes from very early childhood. She was managed without lens removal and maintains good vision in both eyes at the age of 21 years. *(Adapted from Ref. 4.)*

due to lens displacement or cataracts or (ii) have anterior dislocation with associated secondary angle closure or corneal endothelial compromise.

Favourable outcomes have been reported with either a limbal or a pars plana approach lensectomy with subsequent aphakia. Various options for lens replacement following lensectomy have been described including iris-fixated, anterior chamber or scleral-fixated lens implants. Although a variety of techniques are gaining popularity, there is a lack of large series with long-term follow-up and decisions must be made on an individual basis.[5,6] It is worth highlighting that some authors recommend leaving patients aphakic because of the higher risk for retinal detachment that might be associated with lens implantation.[6]

It is common practice to recommend avoiding activities that risk collisions and exertion at near maximal capacity (although patients should be encouraged to remain active with aerobic activities performed in moderation). Also, affected individuals should be made aware of the potentially increased risk of retinal detachment and parents of children with ectopia lentis should maintain a degree of awareness for any loss of vision that the child may experience.[1,5]

References

1. Rødahl E, Mellgren AEC, Boonstra NE, et al. ADAMTSL4-related eye disorders. 2012 Feb 16 [updated 2020 Jul 9]. In: Adam MP, Ardinger HH, Pagon RA, et al., editors. *GeneReviews® [Internet]*. Seattle, WA: University of Washington, Seattle; 1993–2020. Available from: https://www.ncbi.nlm.nih.gov/books/NBK84111/.
2. Chandra A, Aragon-Martin JA, Hughes K, Gati S, Reddy MA, Deshpande C, et al. A genotype-phenotype comparison of ADAMTSL4 and FBN1 in isolated ectopia lentis. *Invest Ophthalmol Vis Sci* 2012;**53**(8):4889–96.
3. Morkin M. *Ectopia lentis*. American Academy of Ophthalmology, EyeWiki; 2020. August 29 https://eyewiki.aao.org/Ectopia_Lentis. Accessed 29 August 2020.
4. Salen G, Steiner RD. Epidemiology, diagnosis, and treatment of cerebrotendinous xanthomatosis (CTX). *J Inherit Metab Dis* 2017;**40**:771–81. https://doi.org/10.1007/s10545-017-0093-8.
5. Chandra A, Charteris D. Molecular pathogenesis and management strategies of ectopia lentis. *Eye (Lond)* 2014;**28**(2):162–8.
6. Miraldi Utz V, Coussa RG, Traboulsi EI. Surgical management of lens subluxation in Marfan syndrome. *J AAPOS* 2014;**18**(2):140–6.

12.2

Marfan syndrome

Panagiotis I. Sergouniotis and Jane L. Ashworth

Marfan syndrome is an autosomal dominant connective tissue disorder associated with mutations in the *FBN1* gene, encoding fibrillin-1. Ectopia lentis is a prominent feature of this multisystemic disorder and, although not universal, it is often the presenting sign. Consequently, Marfan syndrome should be suspected in all individuals with non-traumatic ectopia lentis especially if there is a relevant family history or suggestive clinical findings such as tall stature or cardiac abnormalities.

The diagnosis of Marfan syndrome relies on defined clinical criteria (Ghent 2010 criteria; Table 12.2). These comprise a set of major and minor manifestations in different organ systems. The two major clinical manifestations are ectopia lentis and aortic root aneurysm/dissection. Other clinical features are less influential in the diagnostic evaluation of affected individuals; they only contribute to a multisystemic score that guides diagnosis when the aortic disease is present but ectopia lentis is not. Genetic testing has a prominent role. Regardless of relevant family history, the diagnosis of Marfan syndrome is made when any two of the following three features are present:

- a disease-associated variant in the *FBN1* gene;
- ectopia lentis;
- aortic root enlargement.[1]

It is worth highlighting that a diagnosis of ectopia lentis and a previously described *FBN1* mutation, are sufficient to diagnose Marfan syndrome, regardless of aortic root dilatation or skeletal features. This may lead to re-diagnosing patients previously described as 'ectopia lentis syndrome' or 'simple ectopia lentis' as Marfan syndrome.

Clinical characteristics

Marfan syndrome exhibits high inter- and intra-familial variability. Its broad phenotypic spectrum ranges from rapidly progressive neonatal multisystemic disease to mild isolated ectopia lentis diagnosed in adulthood.[2] The median age at diagnosis is 19 years (range: 0–74 years).[3]

Ophthalmic features of Marfan syndrome include ectopia lentis, myopia, corneal flattening, early-onset cataracts and an increased risk of retinal detachment and glaucoma. Ectopia lentis is a hallmark feature although it is only present in ~60% of affected individuals. Lens displacement can range from a subtle posterior tilt (often with corresponding iridodonesis) to any vertical or horizontal displacement. It is typically bilateral and symmetrical although unilateral cases have been described.[2] Ectopia lentis is considered an early sign of Marfan syndrome with most dislocations occurring in childhood. However, there is a small risk of developing lens displacement in adulthood.[4] In general though, significant progression is unusual beyond the second decade of life and surgical intervention for lens sub-luxation is rarely required in adults with Marfan syndrome.

Myopia is the most prevalent ocular feature and tends to progress rapidly during childhood. Generally, the refractive state of the eye will depend on the interaction of several parameters: a longer axial length will induce myopia, a flatter cornea will result in a hypermetropic shift while significant displacement and spherical reshaping of the lens may cause high myopia and astigmatism.

The gene underlying Marfan syndrome, *FBN1*, is expressed in various tissues and cell types and, consequently, the condition affects many organ systems.[5] Cardiovascular and skeletal abnormalities are the most characteristic extraocular manifestations. Cardiovascular complications are primarily responsible for the attenuation of lifespan seen in the condition; they include aortic root aneurysm (i.e. enlargement of the aortic root that predisposes to aortic tear and rupture) and mitral valve prolapse. Aortic dilatation is present in ~80% of children with Marfan syndrome.[2] This tends to be progressive and regular monitoring by a cardiologist is required.

Individuals with Marfan syndrome tend to be taller than predicted for their family and commonly have a long, slim built.[2,5] The condition is otherwise associated with a remarkably wide range of skeletal findings including arachnodactyly (i.e. long and slender fingers and toes), chest wall deformities and scoliosis (Fig. 12.4). As the prevalence of skeletal features varies with age, their diagnostic value also varies. For example, joint hypermobility and the thumb sign (Table 12.2) are the most discriminating clinical parameters in children younger than 10 years of age; for

TABLE 12.2 Ghent 2010 criteria for Marfan syndrome.

The diagnosis of Marfan syndrome can be established if the following criteria apply[a]	
In the absence of a family history • Aortic root enlargement (Z-score ≥2.0) or dissection AND ectopia lentis • Aortic root enlargement (Z-score ≥2.0) or dissection AND *FBN1* mutation • Aortic root enlargement (Z-score ≥2.0) or dissection AND systemic score ≥7 • Ectopia lentis AND FBN1 mutation known to be associated with aortopathy[b]	
In the presence of a family history of Marfan syndrome (as defined above) • Family history of Marfan syndrome AND ectopia lentis • Family history of Marfan syndrome AND systemic score ≥7 • Family history of Marfan syndrome AND aortic root enlargement or dissection	
Multisystemic score	
Wrist and thumb sign[b]	3
Wrist or thumb sign	1
Pectus carinatum	2
Pectus excavatum or chest asymmetry	1
Hindfoot deformity	2
Pes planus	1
Pneumothorax	2
Dural ectasia	2
Protrusio acetabuli	2
Reduced upper segment/lower segment ratio AND increased arm/height ratio AND no severe scoliosis	1
Scoliosis or thoracolumbar kyphosis	1
Reduced elbow extension	1
Facial features (≥3 of dolichocephaly, enophthalmos, downslanting palpebral fissures, malar hypoplasia, retrognathia)	1
Skin striae	1
Myopia >3 dioptres[c]	1
Mitral valve prolapses (all types)	1

[a] Without discriminating features of Loeys–Dietz, Ehlers–Danlos, or Shprintzen–Goldberg syndrome.
[b] The thumb sign is elicited by asking the person to flex the thumb as far as possible and then close the fingers over it. A positive thumb sign is when there is significant protrusion of the thumb from the clenched fist (the distal phalanx of the adducted thumb extends beyond the ulnar border of the palm). The wrist sign is elicited by asking the person to curl the thumb and fingers of one hand around the other wrist. A positive wrist sign is when the little finger and the thumb overlap.
[c] According to the Ghent 2010 criteria, individuals who are over 20 years old and have ectopia lentis and an FBN1 mutation that is not known to be associated with aortopathy should receive the diagnosis of 'ectopia lentis syndrome'. However, this diagnosis remains controversial and many experts consider it potentially overly reassuring. The same cardiovascular assessment and follow-up regime is recommended in this group of patients as in individuals with Marfan syndrome.
Adapted from Ref. 1.

children older than 10 years the wrist sign (Table 12.2) and the presence of flat feet or a chest wall deformity have the greatest sensitivity.[6]

Facial features associated with Marfan syndrome include a long, narrow face with deep-set eyes, downslanting palpebral fissures, flat cheekbones and a small chin.[2] However, these features are less specific and, therefore less influential in the diagnostic evaluation of patients.[1]

Molecular pathology

Marfan syndrome is associated with fully penetrant, heterozygous variants in *FBN1*, a 65-exon gene encoding fibrillin-1. Fibrillin-1 is an extracellular matrix protein and a major constituent of microfibrils[7]. It contributes to the integrity and function of various connective tissues throughout the body including skin, vasculature, cartilage, tendon, cornea and lens zonules.[2]

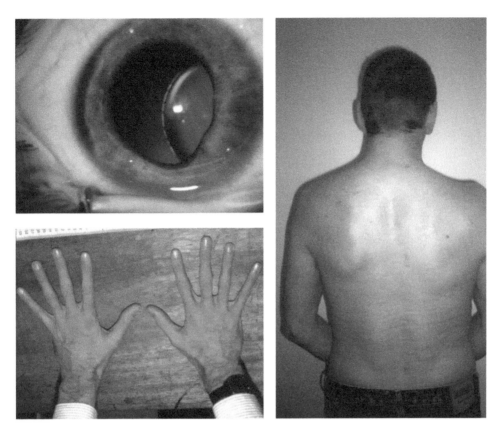

FIG. 12.4 Images from an adult with Marfan syndrome demonstrating ectopia lentis, arachnodactyly and scoliosis. *(Courtesy of Professor Graeme C.M. Black.)*

The first *FBN1* mutation associated with Marfan syndrome was identified in 1991. Since then, over 3000 mutations have been described; the majority of these variants are missense or protein-truncating (including nonsense, splice-site and frameshift). Large deletions/insertions affecting *FBN1* are not uncommon and analyses to identify such changes should be incorporated in routine diagnostic tests for Marfan syndrome.

Identifying reproducible genotype–phenotype correlations has proved challenging. In general, mutations in exons 25–33 of *FBN1* are associated with a severe, early-onset, rapidly progressive phenotype ('neonatal Marfan syndrome'). Furthermore, individuals carrying truncating variants or missense changes leading to loss of a cysteine residue have a higher risk of aortic events. In contrast, individuals with other missense variants tend to have milder aortic phenotypes.[8,9] Overall, no clear pattern has emerged as to which genotypes preferentially affect the eye, although the c. 1948C>T (p.Arg650Cys) variant has been reported multiple times in association primarily with ectopia lentis without major skeletal involvement and with only a very low risk of serious aortic complications.[10]

Genomic testing identifies disease-associated variants in *FBN1* in >90% of patients with typical Marfan syndrome.[2] However, these tests should not be reserved only for cases with the classical presentation. There is a high degree of overlap between Marfan syndrome and other connective tissue disorders such as. Loeys–Dietz syndrome, Stickler syndrome (Section 14.3), homocystinuria (Section 12.3), and Ehlers–Danlos syndrome.[2] Also, *ADAMTSL4*-associated ectopia lentis (Section 12.1) is an important differential in young children presenting with lens dislocation.

Clinical management

Establishing the extent of the disease is an important first step in the management of individuals with Marfan syndrome. A multidisciplinary team approach is required and there should be co-ordinated input from several health professionals (including an ophthalmologist, a clinical geneticist, and a cardiologist).

A detailed discussion on the management of extraocular manifestations of Marfan syndrome can be found in recent review articles.[2,5,11] Notably, general anaesthesia and pregnancy in patients with Marfan syndrome deserve special consideration as they present challenges due to the increased incidence of complications.

Agents that stimulate the cardiovascular system, such as caffeine, are relatively contraindicated. Medications that reduce haemodynamic stress on the aortic wall, such as β-blockers or angiotensin receptor inhibitors, have been shown to retard aortic enlargement and are routinely prescribed, often regardless of the dimensions of the aortic root.[2,11] However, it is the evolution of prophylactic surgical repair that has led to a life expectancy approaching that of the general population (at least in affected individuals who are diagnosed early and treated promptly).

Given the autosomal dominant inheritance pattern of Marfan syndrome, the risk to family members should be highlighted. Given that the children of a person with Marfan syndrome are at a 50% risk of being affected, some families may want to discuss options like prenatal or preimplantation genetic testing.

Approximately 75% of patients with Marfan syndrome have an affected parent but it is not unusual for the family history to be recorded as unremarkable because of failure to recognise the disorder in family members. Up to 25% of affected individuals are thought to have the disorder as a result of a *de novo* mutation, a finding that has implications for genetic counselling.[2]

References

1. Loeys BL, Dietz HC, Braverman AC, Callewaert BL, De Backer J, Devereux RB, et al. The revised Ghent nosology for the Marfan syndrome. *J Med Genet* 2010;**47**(7):476–85. https://doi.org/10.1136/jmg.2009.072785.
2. Dietz H. Marfan syndrome. 2001 Apr 18 [Updated 2017 Oct 12]. In: Adam MP, Ardinger HH, Pagon RA, et al., editors. *GeneReviews®* [Internet]. Seattle, WA: University of Washington, Seattle; 1993–2020. Available from: https://www.ncbi.nlm.nih.gov/books/NBK1335/.
3. Groth KA, Hove H, Kyhl K, Folkestad L, Gaustadnes M, Vejlstrup N, et al. Prevalence, incidence, and age at diagnosis in Marfan Syndrome. *Orphanet J Rare Dis* 2015;**10**:153.
4. Sandvik GF, Vanem TT, Rand-Hendriksen S, Cholidis S, Saethre M, Drolsum L. Ten-year reinvestigation of ocular manifestations in Marfan syndrome. *Clin Exp Ophthalmol* 2018;**47**(2):212–8. https://doi.org/10.1111/ceo.13408.
5. Child AH. Non-cardiac manifestations of Marfan syndrome. *Ann Cardiothorac Surg* 2017;**6**(6):599–609.
6. Stheneur C, Tubach F, Jouneaux M, Roy C, Benoist G, Chevallier B, et al. Study of phenotype evolution during childhood in Marfan syndrome to improve clinical recognition. *Genet Med* 2014;**16**(3):246–50.
7. Schrenk S, Cenzi C, Bertalot T, Conconi MT, Di Liddo R. Structural and functional failure of fibrillin-1 in human diseases (Review). *Int J Mol Med* 2018;**41**(3):1213–23.
8. Faivre L, Collod-Beroud G, Loeys BL, Child A, Binquet C, Gautier E, et al. Effect of mutation type and location on clinical outcome in 1,013 probands with Marfan syndrome or related phenotypes and FBN1 mutations: an international study. *Am J Hum Genet* 2007;**81**(3):454–66.
9. Aubart M, Gazal S, Arnaud P, Benarroch L, Gross MS, Buratti J, et al. Association of modifiers and other genetic factors explain Marfan syndrome clinical variability. *Eur J Hum Genet* 2018;**26**(12):1759–72.
10. Vatti L, Fitzgerald-Butt SM, McBride KL. A cohort study of multiple families with FBN1 p.R650C variant, ectopia lentis, and low but not absent risk for aortopathy. *Am J Med Genet A* 2017;**173**(11):2995–3002.
11. Pyeritz RE. Marfan syndrome: improved clinical history results in expanded natural history. *Genet Med* 2019;**21**(8):1683–90. https://doi.org/10.1038/s41436-018-0399-4.

12.3

Weill–Marchesani syndrome

Aman Chandra and Elias I. Traboulsi

Weill–Marchesani syndrome is a group of connective tissue disorders characterised by microspherophakia (small, spherical lens) in association with skeletal and cardiac defects. Autosomal dominant and autosomal recessive forms have been described.

Clinical characteristics

Weill–Marchesani syndrome usually presents in childhood with short stature and/or ophthalmic issues. Ophthalmic manifestations are typically recognised in the first decade of life and include microspherophakia, lenticular myopia, and lens-related secondary glaucoma. Microspherophakic lenses are mobile and sometimes dislocate. This can lead to high intraocular pressure resulting from a pupil block mechanism. This can be associated with either forward movement of the lens or with lens dislocation into the anterior chamber. In general, visual loss tends to occur earlier and to be more severe in Weill–Marchesani syndrome compared to other causes of ectopia lentis.[1]

The skeletal abnormalities are variable and include proportionate short stature (below 3rd centile; this feature is considered an essential part of the syndrome), brachydactyly (i.e. unusually short fingers and toes) and associated joint stiffness (Fig. 12.5). Cardiac abnormalities have been reported including pulmonary valve stenosis and mitral valve prolapse.[1]

Molecular pathology

Weill–Marchesani syndrome is genetically heterogeneous and both autosomal dominant and recessive forms have been described.

- Weill–Marchesani syndrome type 1: This autosomal recessive disorder is caused by loss of function mutations in the *ADAMTS10* gene.
- Weill–Marchesani syndrome type 2: This autosomal dominant disorder is indistinguishable phenotypically from type 1 and is associated with heterozygous mutations in *FBN1*. While a large number of *FBN1* variants have been implicated in Marfan syndrome, only a small number cause Weill–Marchesani syndrome including in-frame deletions and two missense changes in a specific cysteine domain. Such patients may be predisposed to aortic dissection.
- Weill–Marchesani syndrome type 3: In a small number of families, missense variants in *LTBP2* have been associated with autosomal recessive Weill–Marchesani syndrome. Notably, biallelic loss of function changes in this gene have been described in patients with isolated microspherophakia and individuals with microspherophakia and/or megalocornea with ectopia lents (with or without secondary glaucoma). Extraocular findings in these *LTBP2*-related disorders may include a high-arched palate, tall or short stature and cardiac anomalies.[2]

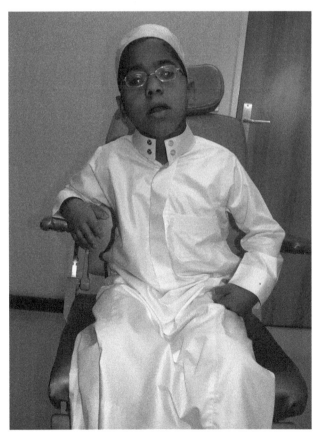

FIG. 12.5 Photograph of a child with Weill–Marchesani syndrome. Unusually short fingers (brachydactyly), relatively short stature and glasses to correct high myopia are noted. *(Courtesy of Professor Arif O. Khan.)*

- Weill–Marchesani syndrome type 4: This autosomal recessive disorder is caused by loss of function mutations in the *ADAMTS17* gene.

Clinical management

To establish the extent of the disease it is recommended that affected individuals have an ophthalmic assessment, an examination by a clinical geneticist and an echocardiogram. Early diagnosis and management of glaucoma and ectopia lentis are key. Medical management with anti-glaucoma drops is typically ineffective. A peripheral iridectomy can prevent or relieve pupillary block although lens extraction with or without glaucoma surgery is often needed. It is noteworthy that airway management during anaesthesia can be challenging in Weill–Marchesani syndrome.[1]

References

1. Tsilou E, MacDonald IM. Weill-Marchesani syndrome. 2007 Nov 1 [updated 2013 Feb 14]. In: Adam MP, Ardinger HH, Pagon RA, et al., editors. *GeneReviews® [Internet]*. Seattle, WA: University of Washington, Seattle; 1993–2020. Available from: https://www.ncbi.nlm.nih.gov/books/NBK1114/.
2. Morlino S, Alesi V, Calì F, Lepri FR, Secinaro A, Grammatico P, et al. LTBP2-related "Marfan-like" phenotype in two Roma/Gypsy subjects with the LTBP2 homozygous p.R299X variant. *Am J Med Genet A* 2019;**179**(1):104–12.

12.4

Homocystinuria

Aman Chandra and Elias I. Traboulsi

Homocystinuria is an autosomal recessive disorder of sulphur metabolism. It is caused by a deficiency of cystathionine β-synthase, a key enzyme of organic sulphur metabolism that is encoded by the *CBS* gene. The condition is rare but likely under-diagnosed.

Clinical characteristics

Diagnosis may be made by newborn screening. However, such programmes are not universally available and indeed individuals may be seen who were born after the programme was introduced in their country of birth. In such circumstances, where a patient is suspected of having this condition, diagnosis may be made via genetic testing (screening of the *CBS* gene) and/or metabolic screening (measurement of homocysteine and methionine levels in plasma or urine).

The metabolic disturbance in homocystinuria results in a condition that classically affects four organ systems.

- Eye: myopia and ectopia lentis.
- Skeleton/connective tissue: arachnodactyly and scoliosis; brittle thin skin and fine blond hair; low body mass index; osteoporosis.
- Blood vessels: increased risk of thromboembolism and stroke.
- Brain: intellectual disability; seizures.

When untreated, the phenotype of patients with homocystinuria is reminiscent of that of Marfan syndrome. Ectopia lentis is one of the commonest manifestations; it is associated with myopia and it is one of the early specific symptoms of the condition (Fig. 12.6). It is often noted in the first decade of life and it is present in around 90% of untreated affected individuals by the middle of the third decade. Patients tend to be thin, to have unusually tall stature and to have long fingers and toes. Affected children can be slow to attain milestones (sitting, standing walking) and will often develop intellectual disability which may be associated with seizures. Adult patients have osteoporosis or decreased bone mineral density. Untreated homocystinuria may lead to premature death from thromboembolism and stroke.[1]

Molecular pathology

Homocystinuria is caused by biallelic mutations in the *CBS* gene. *CBS* encodes cystathionine β-synthase, an enzyme involved in the conversion of the amino-acid methionine to the amino-acid cysteine. Deficiency of this enzyme results in significant accumulation of both methionine and the toxic compound homocysteine in tissues, plasma and urine. This leads to the characteristic signs and symptoms of homocystinuria.

Over 200 mutations have been described in the *CBS* gene, including loss of function variants as well as a significant number of missense changes.

Clinical management

Children with homocystinuria are typically normal at birth, but, when they are untreated, they gradually develop signs of the disease. Prompt diagnosis and therapeutic intervention are therefore particularly important in preventing thrombosis and reducing morbidity.

Treatment aims to reduce the levels of plasma homocysteine and to manage complications (e.g. surgery for ectopia lentis). The former relies on dietary restriction—through a low-protein, low-methionine diet— which has significant compliance issues. In a small subset of patients who have

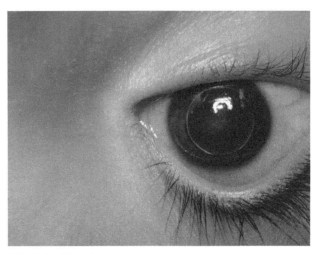

FIG. 12.6 Dislocation of the lens into the anterior chamber in a child with homocystinuria. *(Courtesy of Professor Arif O. Khan.)*

'pyridoxine-responsive', milder forms of the disease, pyridoxine (i.e. vitamin B6, a co-factor of cystathionine β-synthase that can boost its residual activity) can help reduce homocysteine levels. Betaine is another dietary supplement that is usually added to the therapeutic regimen. Precautions need to be taken during anaesthesia of patients with homocystinuria to prevent thromboembolic events and avoid hypoglycaemia.[1]

Reference

1. Sacharow SJ, Picker JD, Levy HL. Homocystinuria caused by cystathionine beta-synthase deficiency. 2004 Jan 15 [Updated 2017 May 18]. In: Adam MP, Ardinger HH, Pagon RA, et al., editors. *GeneReviews® [Internet]*. Seattle, WA: University of Washington, Seattle; 1993–2020. Available from: https://www.ncbi.nlm.nih.gov/books/NBK1524/.

Section III

Genetic disorders affecting the posterior segment

Chapter 13

Genetic disorders affecting the retina, choroid and RPE

The retina provides remarkably sensitive vision that relies on the integrity of a uniquely vulnerable cell, the photoreceptor. Dysfunction/degeneration of retinal photoreceptors is a major cause of visual impairment in the adult population.[1] This can occur as a primary or a secondary event and it is the shared feature of a heterogeneous group of conditions. Genetic factors play a role in many of these disorders, sometimes in the form of relatively rare Mendelian diseases, sometimes in the form of more common conditions caused by the interplay of multiple genes and the environment.[2] This section focuses on the former group.

Incremental advances in the field of retinal genetics have transformed our understanding of Mendelian retinal disorders and have led to the development of powerful diagnostic tests and promising gene-based therapies.[3] Also, large-scale cohort studies have provided important insights into disease prognosis (especially for the most common subtypes like *ABCA4*-retinopathy [Section 13A.8]), and have enhanced our ability to recognize gene-specific phenotypes (for example choroideremia [Section 13A.2] or enhanced S-cone syndrome [Section 13A.3]). For the paediatric population, genetic testing has removed diagnostic bottlenecks allowing early and timely differentiation between stationary and progressive disorders (e.g. distinguishing congenital stationary night-blindness [Section 13A.4] from early childhood onset retinal dystrophies [Section 13A.5]).

Although the majority of retinal conditions affect only the eye, photoreceptor degeneration can be one of the first presenting features of a syndromic condition such as a ciliopathy (Section 13B.1) or an inborn error of metabolism (Section 13B.2). In these cases, prompt diagnosis and implementation of multidisciplinary care have significant benefits for improving outcomes.[4]

Affected families often ask questions about the risk of disease recurrence in future pregnancies or other family members. Answering these questions is a key goal of genetic counselling (Chapter 4). Clarifying inheritance patterns and providing individualised, rather than empirical, risk estimates have a profound ability to empower patient decision making and to inform reproductive and life planning. Notably, a major development in this area has been the more widespread availability of preimplantation genetic testing including for severe Mendelian retinal disorders.

Most retinal dystrophies remain untreatable. It is therefore crucial not to miss the rare subtypes for which therapeutic options exist. This includes metabolic disorders amenable to dietary manipulation, such as gyrate atrophy and LCHAD deficiency (Section 13B.2.5). Importantly, the successful use of gene therapy for certain retinal dystrophies (including *RPE65*-retinopathy; Chapter 8 and Section 13A.5) has allowed considerable optimism that an increasing number of patients will be able to access treatments or clinical trials soon.

References

1. Wright AF, Chakarova CF, Abd El-Aziz MM, Bhattacharya SS. Photoreceptor degeneration: genetic and mechanistic dissection of a complex trait. *Nat Rev Genet* 2010;**11**(4):273–84.
2. Sheffield VC, Stone EM. Genomics and the eye. *N Engl J Med* 2011;**364**(20):1932–42.
3. Sergouniotis PI. Inherited retinal disorders: using evidence as a driver for implementation. *Ophthalmologica* 2019;**242**(4):187–94. https://doi.org/10.1159/000500574.
4. Lenassi E, Clayton-Smith J, Douzgou S, Ramsden SC, Ingram S, Hall G, et al. Clinical utility of genetic testing in 201 preschool children with inherited eye disorders. *Genet Med* 2020;**22**(4):745–51.

Chapter 13A

Genetic disorders causing non-syndromic retinopathy

Chapter outline

13A.1	Non-syndromic retinitis pigmentosa	162
13A.2	Choroideremia	172
13A.3	Enhanced S-cone syndrome and *NR2E3*-associated disorders	176
13A.4	Congenital stationary night-blindness	181
13A.5	Leber congenital amaurosis and severe early childhood onset retinal dystrophies	189
13A.6	Cone dysfunction disorders	194
13A.7	Cone/cone-rod dystrophies	200
13A.8	*ABCA4*-related disorders	207
13A.9	*BEST1*-related disorders (bestrophinopathies)	217
13A.10	Pattern dystrophies	225
13A.11	*X*-linked retinoschisis	236
13A.12	Occult macular dystrophy	241
13A.13	North Carolina macular dystrophy	246
13A.14	Genetic disorders mimicking age-related macular disease	250
13A.15	Genetic architecture of age-related macular degeneration	261

13A.1

Non-syndromic retinitis pigmentosa

Mays Talib, Caroline Van Cauwenbergh, and Camiel J.F. Boon

Retinitis pigmentosa (RP) or rod-cone dystrophy is a group of genetic conditions associated with progressive rod photoreceptor degeneration that is followed by cone photoreceptor involvement. Its worldwide incidence is estimated at 1:3500-1:4000,[1] but depending on the geographic location, incidence reports have varied between 1:750 and 1:9000.[2,3] Although at least two-thirds of RP cases are non-syndromic, some affected individuals display extraocular abnormalities. More than 30 different RP-associated syndromes have been described and some of these are discussed in Section 13B.[1,4]

Non-syndromic RP is a clinically and genetically heterogeneous disease entity. Over 100 genes have been implicated and mutations in these genes account for up to 80% of cases.[5] It is worth highlighting that each gene may be associated with several distinct forms of retinal disease including RP, Leber congenital amaurosis (Section 13A.5), and cone/cone-rod dystrophy (Section 13A.7). These conditions are generally thought to form a continuum with partially overlapping clinical and genetic findings; this overlap can complicate the diagnostic process in people with the retinal disease.[6]

Clinical characteristics

The age at which symptoms present and the speed at which they evolve are variable and depend, at least partially, on the involved gene and mode of inheritance. The autosomal dominant forms of RP (ADRP) are usually the mildest, with many affected individuals first presenting in mid-adulthood (although cases presenting in early childhood have also been described).[6,7] Autosomal recessive and X-linked RP (ARRP and XLRP, respectively) generally have a more severe disease course with an earlier onset, often within the first or second decade of life.[8] Typically, the initial symptom noticed by patients and/or their parents is a disturbance in night vision (nyctalopia). However, in today's artificially lit environments, it may take years for patients to notice night vision problems. Moreover, these symptoms may be subtle, and patients may recognise them only when comparing their night vision to others'. In general, most affected individuals report first noticing nyctalopia during adolescence.[1] Loss of (mid-)peripheral visual field usually becomes apparent in adolescence or young adulthood, and patients may compensate for visual field loss by scanning their environment. In later stages, central visual loss, colour vision disturbances, and light aversion may be present. Patients may also experience photopsias, which can manifest as static or moving flashes, luminous images (phosphenes) or shapes.[9] Depending on the location and frequency of these photopsias, the remaining vision may be disturbed.

Fundus examination in the very early stages of the disease can be unremarkable, as changes may be subtle or not yet present. Typical fundoscopic features in mid-stage RP include: optic disc pallor, which may be limited to the temporal optic disc; vascular attenuation; intraretinal bone spicule shaped pigment migration; variable degrees of retinal and RPE alterations/atrophy that typically start in the mid-periphery and progress in a centripetal fashion (Fig. 13A.1). As the disease advances, the posterior pole becomes involved although the central macula is often spared until the late disease stages. In ADRP, the retinal degenerative changes may be sectoral, usually affecting the lower quadrants. Additional fundus features that may be present include cystoid macular oedema (up to 50% of all RP cases),[10] epiretinal membrane (up to 35% of cases)[11] and optic disc drusen (up to 10% of cases).[12] Female carriers of XLRP may exhibit a characteristic 'tapetoretinal reflex', a golden-metallic sheen at the posterior pole. Furthermore, a proportion of female XLRP carriers develop symptoms and RP-associated fundus changes with advancing age (see Box 13A.1; Fig. 13A.2).

Non-retinal features that may be present include presenile posterior subcapsular cataracts,[13] and refractive error which, depending largely on the gene involved, could be myopia or hypermetropia.[14] Vitreous abnormalities may be present, such as dust-like particles. Although nystagmus has been described, this is more closely associated with severe early-onset retinal dystrophies, such as Leber congenital amaurosis, rather than RP.[15]

Measures of central visual function, such as best-corrected visual acuity and colour vision, are usually normal in early disease stages. With the progression of central cone photoreceptor degeneration, visual acuity will eventually deteriorate. As mentioned above, the age at which decline to severe visual impairment is expected depends partially on the involved gene and mode of inheritance. However, variability between individuals carrying

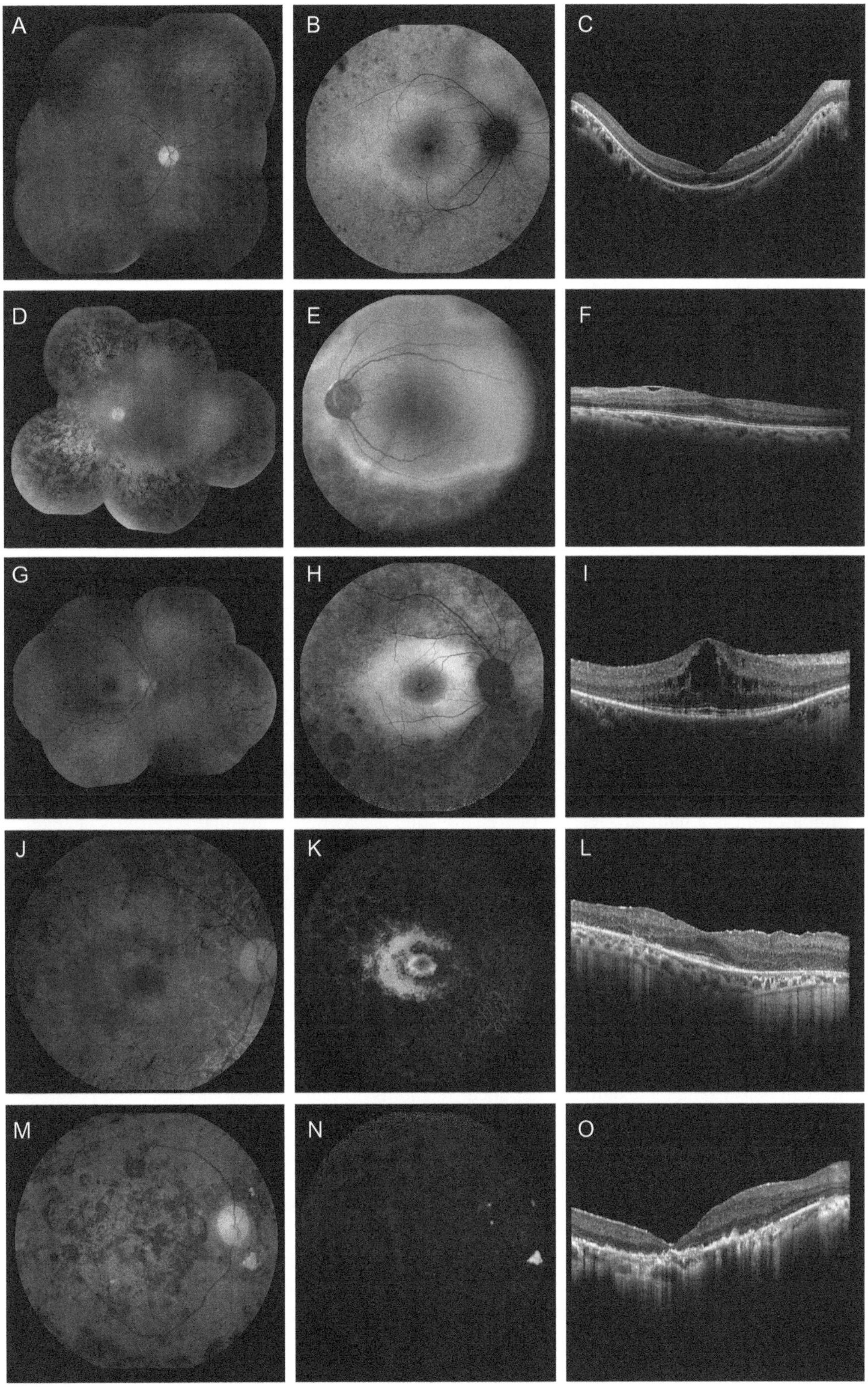

Genetic disorders causing non-syndromic retinopathy **Chapter | 13A** 163

FIG. 13A.1 See figure legend on next page.

FIG. 13A.1—CONT'D Multimodal imaging in non-syndromic retinitis pigmentosa (RP). (A–C). Findings in a 20-year-old female with RP due to a homozygous *IDH3A* mutation, c.524C>T (p.Ala175Val). Her nyctalopia and visual field symptoms started at the age of 14 years. Fundus photography (A) revealed attenuated arteries, and bone spicule pigmentation and mottled RPE in the periphery. Fundus autofluorescence imaging (B) revealed mottled hypoautofluorescence in the mid-periphery, with sparing of the posterior pole. OCT (C) revealed relative foveal sparing of the outer nuclear layer, the external limiting membrane and the ellipsoid zone at the foveola although disruption and/or thinning of these layers was noted in the parafovea and perifovea. Corresponding best-corrected visual acuity was 0.0 LogMAR in the right and 0.1 LogMAR in the left eye. (D–F) Findings in a 60-year-old male with RP and no conclusive molecular diagnosis (following gene panel testing). Fundus photography (D) revealed vascular attenuation, a relatively spared posterior pole, and atrophy and dense bone spicule-like pigmentation in the periphery. Fundus autofluorescence imaging (E) revealed abnormalities in regions that appeared relatively preserved on fundoscopy; hypoautofluorescent lesions with a hyperautofluorescent border that encroach upon the posterior pole from the inferior retina are noted. OCT (F) revealed an epiretinal membrane and relative preservation of the outer nuclear layer, the external limiting membrane and the ellipsoid zone; a degree of interdigitation zone disorganisation is also noted. The patient had a visually significant posterior subcapsular cataract in both eyes and the best-corrected visual acuity was 0.1 LogMAR in the right and 0.2 LogMAR in the left eye. After cataract surgery 2 years later, his best-corrected visual acuity improved to 0.0 LogMAR in the right and 0.1 LogMAR in the left eye. (G–I) Findings in a 38-year-old female with RP due to a heterozygous *RHO* mutation, c.541G>A (p.Glu181Lys). Fundus photography (G) revealed the typical RP-associated changes, which were more abundant in the inferior hemisphere. The posterior pole and the superior retina were relatively spared. Fundus autofluorescence imaging (H) revealed mottled hypoautofluorescent changes in the central macula and the mid-periphery extending to the posterior pole. In the inferior mid-periphery, sharply circumscribed patches of absent autofluorescence are visible. A perimacular hyperautofluorescent ring is visible. OCT (I) revealed relative preservation of the outer retina, with cystoid macular oedema mostly in the inner nuclear layer, and a few cystoid spaces in the outer nuclear and ganglion cell layers. (J–L) Findings in a 60-year-old male with RP due to a homozygous *FAM161A* mutation, c.1138T>C (p.Arg380Ter). He reported experiencing night vision problems since the age of 35–40 years. Fundus examination (J) revealed vascular attenuation, bone spicule pigmentation extending into the posterior pole and generalised retinal atrophy with relative sparing in the central macula and the temporal posterior pole. The latter is also visualised on fundus autofluorescence imaging (K). Due to the relative foveal sparing of the outer nuclear layer, the external limiting membrane and the ellipsoid zone, as seen on OCT (L), the corresponding best-corrected visual acuity was 0.0 LogMAR in each eye. (M–O) A 37-year-old male with RP and a heterozygous pathogenic mutation in *NRL*, c.654del (p.Cys219Valfs*4). There was no family history and this variant appeared to have arisen *de novo*. Fundus examination (M) revealed end-stage RP, with atrophy and dense intraretinal pigment migration in the posterior pole; several hamartomas were noted around the optic disc and in the posterior pole. Fundus autofluorescence imaging (N) revealed minimal remaining autofluorescence (apart from the hyperautofluorescence of the peripapillary hamartomas). OCT (O) revealed significant atrophy of the outer nuclear layer and loss/disorganisation of the hyper-reflective outer retinal bands. The corresponding best-corrected visual acuity was 1.3 LogMAR in the right and counting fingers in the left eye.

identical mutations has also been described. ADRP generally has the best visual prognosis, with some affected individuals maintaining good vision well into the 6th and 7th decade of life.[7,8]

Visual field impairment is a hallmark feature of RP. It typically begins with patches of sensitivity loss in the mid-periphery, which gradually form a ring scotoma. This ring scotoma then extends centripetally towards the far periphery and centre. In advanced disease, a central island and/or a peripheral wedge of vision are often seen. Other visual field loss patterns have been described, such as progressive concentric visual field constriction without a preceding ring scotoma, or a predilection for visual field loss in the superior quadrants.[16] These abnormalities and their progression over time can be periodically examined and documented with kinetic perimetry. In kinetic perimetry, visual stimuli of fixed size and intensity are moved by an operator from non-seeing areas into seeing areas of the visual field. The Goldmann perimeter was commonly used in the past but in recent years semi-automated perimeters such as the Octopus 900 (Haag-Streit International, Switzerland) are also being utilised.

Central visual field measurements map the sensitivity of the central macula, typically using stationary stimuli at different locations within the central 10° or 30° radius of the macula. The stimulus intensity is adjusted based on the patient's response. In fundus controlled perimetry (also known as microperimetry or fundus-related perimetry), fixation stability is assessed and retinal sensitivity is precisely correlated to the examined macular location. This type of perimetry testing is particularly advantageous over other forms of static perimetry in patients with fixation loss and it may detect macular sensitivity changes before visual acuity changes occur.[17]

Full-field electroretinography is an objective tool to measure the functional integrity of the inner and outer retina and to quantify rod and cone photoreceptor function (see also Chapter 6). This set of tests measures dark- and light-adapted responses to light flashes at different intensities, from which rod and cone electrical responses can be respectively deduced. In patients with RP, ERGs are usually performed for diagnostic purposes, and less commonly to clinically monitor disease progression. Reduced and delayed dark-adapted (rod) responses are noted from the early disease stages when light-adapted (cone) responses can still be within normal limits. As the disease advances, ERG amplitudes further diminish and the implicit times are further delayed, but the attenuation is more marked for dark-adapted responses, i.e. there is a rod-cone dysfunction pattern (Fig. 13A.3). In advanced RP, the ERG becomes extinguished, i.e. the residual responses have become too small to be detectable. In sectoral RP, the full-field ERG can remain within the normal range, as a smaller retinal area is affected.

The multifocal ERG measures retinal function across the central 40°-50° of the macula, mapping electrical

> **BOX 13A.1 X-linked retinitis pigmentosa (XLRP)**
>
> XLRP is a form of rod-cone degeneration that affects both males and females and is associated with mutations in the *RPGR* and *RP2* genes.
>
> **Clinical characteristics**
> Affected males are generally myopic and present in the first decade of life with night vision problems. The clinical presentation and disease natural history in this group are generally towards the severe end of the retinitis pigmentosa spectrum.
>
> Female carriers are also often myopic and, compared to affected males, generally experience considerably milder disease (although there is significant variability in clinical presentation). Fundoscopic findings in this group broadly correlate with the reported symptoms and adult-onset pigmentary retinopathy, often with a focal or patchy nature, is common. A significant proportion of young female carriers display a highly characteristic 'tapetoretinal reflex'—a golden-metallic sheen at the posterior pole or, less commonly, in the mid-peripheral retina. Notably, fundus autofluorescence imaging of carrier females may show a radial pattern, presumably reflecting X chromosome inactivation patterns within the retina (Fig. 13A.2).
>
> **Molecular pathology**
> Pathogenic variants in two genes are found in 80%–90% of XLRP cases.
>
> Mutations in *RP2* are detected in 10%–20% of families with XLRP. Affected males tend to have early macular involvement and thus worse central vision compared to age-matched individuals with mutations in *RPGR*. The *RP2* gene encodes a ubiquitously expressed protein with a role in ciliary trafficking. The majority of mutations are protein-truncating variants likely to result in loss of function. Missense variants represent around a third of disease-causing changes and cluster in the N-terminal region (i.e. towards the start) of the protein.
>
> Mutations in *RPGR* are by far the commonest cause of XLRP and account for 70%–80% of cases. A wide range of disease-causing variants (>300) has been described. Around 60% of these changes are nonsense or frameshift variants altering a highly repetitive, retina-enriched exon termed ORF15. *RPGR* is expressed in several tissues and, in the photoreceptor, the encoded protein (that includes ORF15 at its terminus) has a key role in protein trafficking through the connecting cilium.
>
> There are some loose genotype–phenotype correlations for mutations in *RPGR*.
> - while mutations in *RPGR* exons 2–14 are generally associated with XLRP, a small number of variants that sit within the RCC1 domain are implicated in a syndromic ciliopathy phenotype. This may include hearing loss, bronchiectasis and increased susceptibility to respiratory infections.
> - although the majority of nonsense and frameshift mutations in the *RPGR* ORF15 exon cause XLRP, as these move towards the C-terminal region (i.e. towards the end) of the protein, the phenotype is more likely to be an X-linked cone/cone-rod dystrophy.
>
> **Clinical management**
> Given the severity of the phenotype and its X-linked context, the diagnosis of XLRP is an important one to make; this is all the more pertinent since *RPGR*-associated XLRP has been the subject of promising clinical trials. Several challenges make identifying individuals with XLPR far from straightforward:
> - the phenotype in males, while severe, is not particularly distinctive
> - females can manifest symptoms and may thus hide (to some degree) X-linked pedigrees.
> - most current genetic diagnostic technologies are based on short-read sequencing and preclude analysis of *RPGR* ORF15; additional, targeted testing for this repetitive exon should be considered in all males with retinal dystrophy who are negative on conventional genetic tests.
>
> It is important to ensure that any clinical clues that may help identify families with XLRP are sought after:
> - females are affected later or progressing more slowly than males
> - affected males are affected in childhood and are myopic.
> - there is a 'tapetoretinal reflex'/radial autofluorescence pattern in carrier females.
>
> Families with X-linked conditions require intensive genetic counselling given that females who may not only be asymptomatic but may also have grown up amongst males with a severe visual disability have a high risk of transmitting the condition.
>
> Prenatal and pre-implantation genetic testing have both been described in families with XLRP and, where requested, will need a multidisciplinary liaison.
>
> In families with known XLRP-associated mutations, approaching the diagnosis of very young males who are born to carrier females is challenging. Pre-symptomatic diagnosis is discussed in Pre-symptomatic diagnosis Chapter 4. Briefly, it is important to ensure that parents: understand how either clinical examination or genetic testing may result in presymptomatic diagnosis; are fully prepared for such testing and understand in particular the psychosocial consequences of a positive test both for themselves and their children.
>
> Carrier testing in young girls is an area that is also challenging, and often requires time and input that is not available in busy outpatient clinics.

responses to specific regions and requiring stable fixation (in contrast to the full-field approach). In general, the multifocal ERG is less useful as a diagnostic tool for RP as the cone function in the central macula is often spared. It can however be utilised as a complementary test to document disease progression and macular function. The pattern ERG is another approach that can be used to evaluate macular function (see also Chapter 6).

OCT imaging enables near-histological evaluation of the retinal architecture and the integrity of the photoreceptor

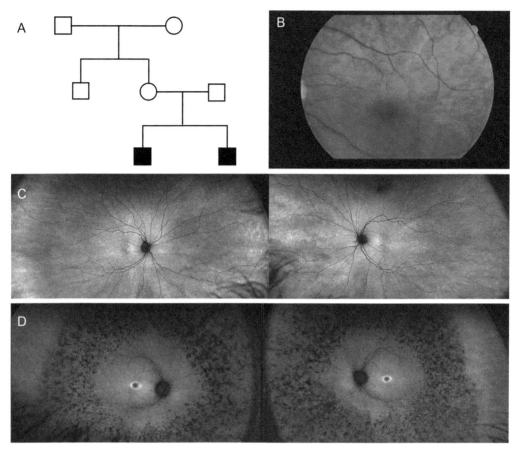

FIG. 13A.2 Pedigree and multimodal imaging in a family with XLRP segregating a disease-causing variant in the *RPGR* gene. In the family tree (A) two affected male siblings are noted; their parents and grandparents appear to be unaffected. Fundus photography (B) showing an irregular, tapetal-like reflex, a characteristic and frequently encountered fundoscopic finding in female carriers of X-linked RP. Fundus autofluorescence imaging (C) from the carrier mother reveals a radial pattern of differential autofluorescence signal. Fundus autofluorescence imaging from one of the two affected males (D) reveals irregular autofluorescence signal in the periphery as well as a concentric ring of hyperautofluorescence.

structures at the macula (see also Chapter 7). The photoreceptor nuclei are located in the outer nuclear layer. The external limiting membrane, which is the first hyper-reflective outer retinal band, is thought to represent the adherens junctions between the Müller cells and the outer part of the photoreceptors.[18] The second hyper-reflective outer retinal band, the so-called ellipsoid zone, represents the mitochondria-rich photoreceptor inner segments.[19] The photoreceptor outer segments co-localise with the interdigitation zone, the third hyper-reflective outer retinal band. As photoreceptors degenerate, disorganisation of these hyper-reflective outer retinal bands occurs. This results in visual acuity loss and is accompanied by thinning of the outer nuclear layer. These changes are observed in the peripheral macula first and gradually encroach upon the fovea. A 'transition zone' between degenerated and relatively spared outer retina is noted and retinal sensitivity has been shown to decline faster in this zone compared to other regions of the retina.

Short-wavelength (blue light) fundus autofluorescence imaging provides information on the integrity and function of the RPE through the visualisation of lipofuscin (Fig. 13A.1) (see also Chapter 7). Lipofuscin is material derived from degraded photoreceptor outer segments that have been shed from photoreceptors and phagocytosed by the RPE as part of a physiological outer segment turnover. Short excitation wavelengths (488 nm) of blue light will cause lipofuscin to autofluoresce, and light emitted at wavelengths between 500 and 800 nm is captured on the image. Areas of lipofuscin accumulation will appear hyperautofluorescent, while areas of atrophic RPE will appear hypoautofluorescent. In RP, a perimacular concentric hyperautofluorescent ring or arc is seen in many patients. This is considered to represent the transition zone between degenerated and relatively preserved retina.[20] When overlaying a fundus autofluorescence image with an OCT scan, the hyperautofluorescent ring indeed colocalizes with an area of outer nuclear layer thinning and ellipsoid zone attenuation. Relative sparing of the photoreceptor structures is noted internally to the ring and this ring tends to constrict progressively over time.[20] Double concentric hyperautofluorescent rings have also been described.[21] It is worth highlighting that hypoautofluorescent changes that

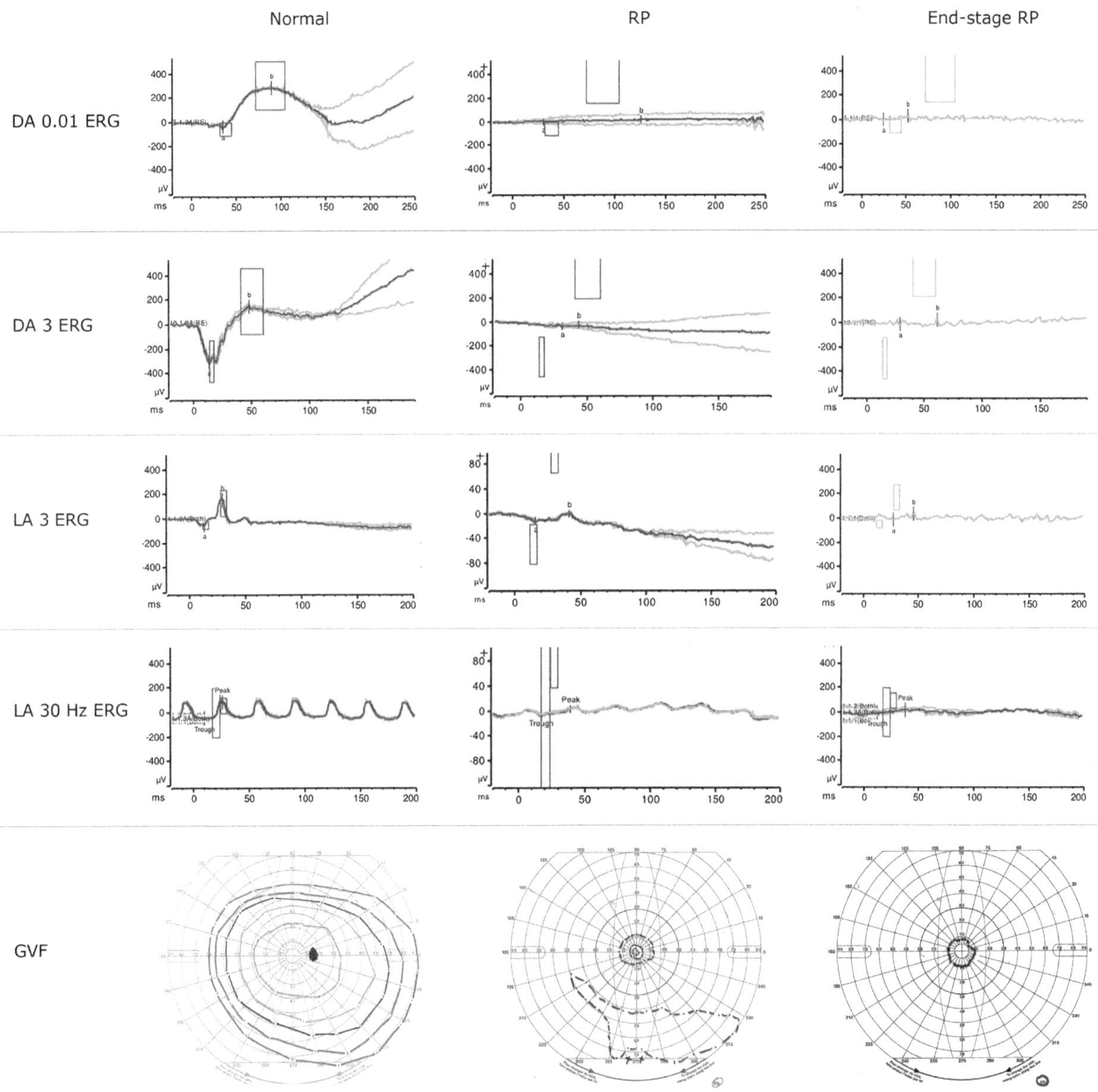

FIG. 13A.3 Overview of representative electrophysiological and visual field test results in retinitis pigmentosa (RP). Each row represents the stimulus used in electroretinogram (ERG) testing based on recommendations by the International Society for Clinical Electrophysiology of Vision (ISCEV). A dark-adapted dim flash (DA 0.01) elicits a rod-driven response from ON-bipolar cells; a dark-adapted standard flash (DA 3) elicits a mixed rod- and cone-driven (albeit rod-dominated) response originating from both photoreceptors and bipolar cells; a light-adapted standard flash (LA 3) elicits a cone-driven response from ON- and OFF-bipolar cells; a light-adapted 30 Hz flicker (LA 30 Hz) elicit a cone-driven response. Rod-driven responses are often undetectable in early disease stages, while a light-adapted 30 Hz flicker response progressively diminishes with advancing disease, i.e. there is a rod-cone pattern of dysfunction, as seen in the 2nd and 3rd columns. Goldmann visual field (GVF) testing may show various patterns of (mid)peripheral field loss, eventually resulting in a small central remnant of vision.

can be granular, mottled, or bone-spicule-shaped may also be seen in people with RP.

Molecular pathology

Non-syndromic RP can be inherited in an autosomal dominant, autosomal recessive or, less commonly, X-linked fashion.[4] About one-third of cases cannot be easily classified without genetic testing; patients in this group typically have no affected relatives and are denoted as sporadic or simplex RP. Although the majority of simplex cases are predicted to be ARRP, approximately 15% of male simplex patients have a mutation in an XLRP-associated gene (*RPGR* or *RP2*).[22] De novo ADRP mutations account for at least 1%–2% of simplex cases.[23]

In XLRP, mild to severe phenotypic expression can occur in female carriers, probably due to non-random or skewed X chromosome inactivation.[24] Consequently, some

XLRP pedigrees may be misclassified as ADRP.[25] Approximately 8% of families initially considered to display ADRP are caused by mutations in XLRP genes (see also Box 13A.1).[26] A small proportion of cases have been reported to result from non-Mendelian inheritance, such as digenic RP.[25]

Over 100 genes have been implicated in non-syndromic RP. The encoded proteins have diverse functions and the principal pathways affected include ciliary structure and transport, the phototransduction cascade and the visual cycle. Commonly mutated genes include *RHO*, *USH2A* and *RPGR* (Table 13A.1).[5] There is striking allelic heterogeneity with at least 3000 different mutations reported to be causing non-syndromic RP (including loss of function and gain of function variants). Notably, within the same gene, different mutations can occasionally result in a different phenotypic outcome.[27] Also, in some ADRP families, individual mutation carriers may not exhibit clinical signs of RP. Such incomplete or reduced penetrance has been reported for dominant mutations in several genes including *PRPF31*, *PRPF8* and *SNRNP200*.[28,29]

Choosing the right molecular test for a patient with RP can be challenging, especially with the plethora of disease genes and the different tests that are provided by molecular laboratories. Several factors need to be taken into account, including clinical findings, family history, presence or absence of consanguinity, previous molecular testing, the reimbursement options by the patient's insurance, and the expertise of the laboratory (see also Chapters 2 and 3). An overview of the available genetic tests provided by accredited laboratories can be found on the orpha.net and the National Institute of Health (NIH) Genetic Testing Registry portals.

At present, gene panel or exome sequencing-based approaches are used in most clinical diagnostic laboratories. However, despite screening the protein-coding sequence of all known RP genes, many cases remain unsolved. In these cases, several considerations should be taken into account:

- Some gene panel tests only target a limited set of genes (e.g. the most prevalent RP genes). A molecular report should provide information on the targeted regions.
- Some regions might not be adequately covered by the currently available short-read sequencing approaches that underly exome sequencing or gene panel tests. GC-rich and highly repetitive sequences are particularly problematic. The *RPGR* open reading frame (ORF) 15, a mutation hotspot for XLRP, is such an example (see also Box 13A.1). For XLRP it is therefore recommended to start with alternative testing approaches that provide sufficient coverage of the entire *RPGR* ORF15 sequence, especially since approximately 60% of the disease-causing *RPGR* mutations are located in this region.[30,31]
- Inherent to the testing approach used, some specific genetic alterations might be missed, such as deep intronic and/or structural variants.[32,33] Genome sequencing has proven to be a relatively sensitive method for the

TABLE 13A.1 Common clinical characteristics of selected genetic subtypes of non-syndromic retinitis pigmentosa.

Clinical feature	Associated gene
Age of onset <5 years	BBS1, C8orf37, CRB1, CNGA1, DHX38, IDH3A, IFT140, LRAT, MERTK, NR2E3, NRL, OFD1, PCARE, PDE6G, PRPF3, PRPF31, RBP3, RDH12, RP2, RPE65, RPGR, RPGRIP1, SNRNP200, SPATA7, TTC8, TULP1
Age of onset <10 years	ABCA4, AGBL5, BBS2, BEST1, CLN3, CNGB1, CWC27, HGSNAT, IFT172, IMPDH1, IMGP2, PDE6A, PDE6B, POMGNT1, PRCD, PROM1, PRPF8, REEP6, RHO, RLBP1, SLC7A14, ZNF513
Age of onset >50 years	CRX, MAK, RBP3, HGSNAT
Early macular involvement (including bull's eye maculopathy)	C8orf37, CDHR1, CERKL, CRB1, CRX, DHX38, FSCN2, GUCA1B, HK1, IDH3A, IFT140, IMPG2, MERTK, NRL, PCARE, PRCD, PROM1, PRPF6, RDH12, RP2, RPGR, RPGRIP1, SAG, SPATA7, TTC8, ZNF513
Dense pigment migration (including presence of 'nummular' pigmentation)	CDHR1, CRB1, EYS, IFT140, KCNJ13, PDE6A, PDE6B, PRPF8, RDH12, SNRNP200 (NR2E3, BEST1)
Absence/scarcity of retinal hyperpigmentation	CDHR1, CLN3, FAM161A, HGSNAT, LRAT, NRL, OFD1, RLBP1, RP1, RPE65, RPGRIP1, TTC8, USH2A
Sectoral distribution	CDH3, EYS, IMPDH1, MYO7A, PRPS1, RHO, RPGR, RP1, USH1C
Pericentral pigmentary atrophy	CERKL, CNGA1, CNGB1, CRX, DHDDS, HGSNAT, HK1, NR2E3, PDE6B, PRPF31, PROM1, PRPH2, RHO, TOPORS, TULP1, USH2A

Adapted from Ref. 6.

screening of these mutations and is expected to play a significant role as a standard diagnostic test.[34,35]

- Variants of uncertain significance (VUS; also known as ACMG class 3 mutations) can be reported. The evidence of pathogenicity of these variants can evolve over time based on emerging biological evidence or the presence of these changes in multiple individuals with RP. For partially solved cases, segregation analysis in affected and unaffected family members may provide insights into the pathogenicity of these variants of unknown significance[36] (see also Chapter 3).

Clinical management

For most forms of RP, no effective, commercially approved treatment is available. Patient management should therefore consist mostly of elaborate counselling on disease background, heredity and prognosis, low vision aids, and management of additional ophthalmic conditions. As avenues for gene-specific treatments are under active investigation, establishing a reliable molecular diagnosis has become increasingly important. A molecular diagnosis also allows screening of family members at risk, performing carrier testing in partners, and offering the option of preconception consultation or pre-implantation genetic diagnosis.

It is recommended the visual function and retinal structure are evaluated at regular intervals, preferably every 1–3 years, also taking into account the patient's needs and preferences. Some affected individuals may wish to be involved in patient groups or societies. Where possible, affected individuals should be referred to a tertiary centre, where the disease progression can be monitored, and where they can be kept up to date regarding ongoing and upcoming clinical trials.

Ophthalmic conditions associated with RP, such as refractive errors, macular oedema or cataract, should be monitored and managed. Cystoid macular oedema may be treated with systemic or topical carbonic anhydrase inhibitors (although the visual benefit of these treatments is uncertain). If the oedema is unresponsive to carbonic anhydrase inhibitors, other treatments, including topical, oral or intravitreal steroids, intravitreal anti-vascular endothelial growth factors, laser photocoagulation, or pars plana vitrectomy, have been tried, also with limited success.[37] Corrective glasses with specific colour filters may be helpful in patients with light aversion or decreased contrast sensitivity. Visually significant cataracts should be surgically treated as both outcomes and patient satisfaction are generally favourable.[38] However, in cases where macular atrophy is present preoperatively, it is important to highlight that only limited visual improvement should be expected. To reduce potential light-induced toxicity to the retina, the intraoperative microscope light intensity can be reduced and, between different surgical steps, the microscope light may be switched off to minimise light exposure.[38] Increased cataract surgery risks in the RP population include zonular weakness and the postoperative development of posterior capsule opacification and cystoid macular oedema.[38]

As for treatments aimed at halting the degenerative process, the approval of subretinal injections of gene therapy for *RPE65*-associated retinal diseases, including RP, in 2017 has been a milestone (see also Chapter 8).[39] Clinical gene augmentation therapy trials are ongoing, being prepared for or are in the preclinical phase for several ARRP and XLRP subtypes. For many forms of ADRP, gene augmentation alone is unlikely to be enough, and genome editing approaches are under investigation, for example, using CRISPR-Cas9 based systems in a gene replacement strategy. Most current approaches utilise viral vector-based delivery systems and are applied to patients with remaining functional photoreceptor cells. Certain early-onset and rapidly progressive forms of RP would ideally require therapeutic intervention in childhood or adolescence.[40] Expectations of the outcomes of such trials should be managed, and patients should be advised that these options are gene-specific or mutation-specific.

For more advanced disease stages, optogenetic and cell-based therapeutic options are investigated, which are not gene-specific. Examples of the latter include the intravitreal or subretinal administration of retinal progenitor cells or induced pluripotent stem cells. Retinal prostheses may also have a role in the management of legally blind patients with end-stage RP. Careful preoperative screening, expectation management and counselling are required before proceeding to this intervention. A comprehensive postoperative rehabilitation program at a specialised centre should be available.[41]

The role of vitamin A supplementation in RP remains debated. At age-adjusted dosages, vitamin A supplementation has been shown to slow the decline of cone photoreceptor amplitudes on full-field ERG.[24] As this effect has been demonstrated in genetically undifferentiated RP populations, it remains unclear whether vitamin A supplementation works for all RP subtypes. It is worth highlighting that a synthetic vitamin A derivative, 9-cis-retinyl acetate, has shown some efficacy in preserving visual field area and visual acuity in a clinical trial including patients with *RPE65*- and *LRAT*-associated retinal disorders, including RP.[42] However, care should be taken before prescribing oral supplements. Mouse studies have indicated that vitamin A supplementation should be avoided in *ABCA4*-associated retinal disorders with this observation raising caution in recommending vitamin A supplementation.[43] Moreover, hypervitaminosis A can lead to toxicity in several organ systems, and adverse reactions should be assessed at intervals. Notably, docosahexaenoic acid (fish oil) has shown no clear efficacy in slowing disease progression, and oral valproic acid has been shown to lead to a worse visual outcome than a placebo.[44,45]

References

1. Hartong DT, Berson EL, Dryja TP. Retinitis pigmentosa. *Lancet (London, England)* 2006;**368**(9549):1795–809.
2. Nangia V, Jonas JB, Khare A, Sinha A. Prevalence of retinitis pigmentosa in India: the Central India eye and medical study. *Acta Ophthalmol* 2012;**90**(8):e649–50.
3. Na K-H, Kim HJ, Kim KH, Han S, Kim P, Hann HJ, et al. Prevalence, age at diagnosis, mortality, and cause of death in retinitis Pigmentosa in Korea—a Nationwide population-based study. *Am J Ophthalmol* 2017;**176**:157–65.
4. Daiger SP, Bowne SJ, Sullivan LS. Perspective on genes and mutations causing retinitis pigmentosa. *Archiv Ophthalmol* 2007;**125**(2):151–8.
5. Stone EM, Andorf JL, Whitmore SS, DeLuca AP, Giacalone JC, Streb LM, et al. Clinically focused molecular investigation of 1000 consecutive families with inherited retinal disease. *Ophthalmology* 2017;**124**(9):1314–31.
6. Verbakel SK, van Huet RAC, Boon CJF, den Hollander AI, Collin RWJ, Klaver CCW, et al. Non-syndromic retinitis pigmentosa. *Prog Retin Eye Res* 2018;**66**:157–86.
7. Kemp CM, Jacobson SG, Faulkner DJ. Two types of visual dysfunction in autosomal dominant retinitis pigmentosa. *Invest Ophthalmol Vis Sci* 1988;**29**(8):1235–41.
8. Hamel C. Retinitis pigmentosa. *Orphanet J Rare Dis* 2006;**1**:40.
9. Bittner AK, Diener-West M, Dagnelie G. Characteristics and possible visual consequences of photopsias as vision measures are reduced in retinitis pigmentosa. *Invest Ophthalmol Vis Sci* 2011;**52**(9):6370–6.
10. Hajali M, Fishman GA, Anderson RJ. The prevalence of cystoid macular oedema in retinitis pigmentosa patients determined by optical coherence tomography. *Br J Ophthalmol* 2008;**92**(8):1065–8.
11. Testa F, Rossi S, Colucci R, Gallo B, Di Iorio V, Della Corte M, et al. Macular abnormalities in Italian patients with retinitis pigmentosa. *Br J Ophthalmol* 2014;**98**(7):946–50.
12. Grover S, Fishman GA, Brown Jr J. Frequency of optic disc or parapapillary nerve fiber layer drusen in retinitis pigmentosa. *Ophthalmology* 1997;**104**(2):295–8.
13. Fujiwara K, Ikeda Y, Murakami Y, Funatsu J, Nakatake S, Tachibana T, et al. Risk factors for posterior subcapsular cataract in retinitis pigmentosa. *Invest Ophthalmol Vis Sci* 2017;**58**(5):2534–7.
14. Hendriks M, Verhoeven VJM, Buitendijk GHS, Polling JR, Meester-Smoor MA, Hofman A, et al. Development of refractive errors-what can we learn from inherited retinal dystrophies? *Am J Ophthalmol* 2017;**182**:81–9.
15. Booij JC, Florijn RJ, ten Brink JB, Loves W, Meire F, van Schooneveld MJ, et al. Identification of mutations in the AIPL1, CRB1, GUCY2D, RPE65, and RPGRIP1 genes in patients with juvenile retinitis pigmentosa. *J Med Genet* 2005;**42**(11), e67.
16. Jacobson SG, McGuigan 3rd DB, Sumaroka A, Roman AJ, Gruzensky ML, Sheplock R, et al. Complexity of the class B phenotype in autosomal dominant retinitis pigmentosa due to rhodopsin mutations. *Invest Ophthalmol Vis Sci* 2016;**57**(11):4847–58.
17. Liu H, Bittencourt MG, Wang J, Sepah YJ, Ibrahim-Ahmed M, Rentiya Z, et al. Retinal sensitivity is a valuable complementary measurement to visual acuity – a microperimetry study in patients with maculopathies. *Graefe's Archiv Clin Exp Ophthalmol = Albrecht von Graefes Archiv fur klinische und experimentelle Ophthalmologie* 2015;**253**(12):2137–42.
18. Staurenghi G, Sadda S, Chakravarthy U, Spaide RF. Proposed lexicon for anatomic landmarks in normal posterior segment spectral-domain optical coherence tomography: the IN*OCT consensus. *Ophthalmology* 2014;**121**(8):1572–8.
19. Spaide RF, Curcio CA. Anatomical correlates to the bands seen in the outer retina by optical coherence tomography: literature review and model. *Retina* 2011;**31**(8):1609–19.
20. Lima LH, Burke T, Greenstein VC, Chou CL, Cella W, Yannuzzi LA, et al. Progressive constriction of the hyperautofluorescent ring in retinitis pigmentosa. *Am J Ophthalmol* 2012;**153**(4):718–27. 27.e1-2.
21. Escher P, Tran HV, Vaclavik V, Borruat FX, Schorderet DF, Munier FL. Double concentric autofluorescence ring in NR2E3-p.G56R-linked autosomal dominant retinitis pigmentosa. *Invest Ophthalmol Vis Sci* 2012;**53**(8):4754–64.
22. Branham K, Othman M, Brumm M, Karoukis AJ, Atmaca-Sonmez P, Yashar BM, et al. Mutations in RPGR and RP2 account for 15% of males with simplex retinal degenerative disease. *Invest Ophthalmol Vis Sci* 2012;**53**(13):8232–7.
23. Neveling K, Collin RW, Gilissen C, van Huet RA, Visser L, Kwint MP, et al. Next-generation genetic testing for retinitis pigmentosa. *Hum Mutat* 2012;**33**(6):963–72.
24. Berson EL, Weigel-DiFranco C, Rosner B, Gaudio AR, Sandberg MA. Association of vitamin a supplementation with disease course in children with retinitis pigmentosa. *JAMA Ophthalmol* 2018;**136**(5):490–5.
25. Sullivan LS, Bowne SJ, Birch DG, Hughbanks-Wheaton D, Heckenlively JR, Lewis RA, et al. Prevalence of disease-causing mutations in families with autosomal dominant retinitis pigmentosa: a screen of known genes in 200 families. *Invest Ophthalmol Vis Sci* 2006;**47**(7):3052–64.
26. Churchill JD, Bowne SJ, Sullivan LS, Lewis RA, Wheaton DK, Birch DG, et al. Mutations in the X-linked retinitis pigmentosa genes RPGR and RP2 found in 8.5% of families with a provisional diagnosis of autosomal dominant retinitis pigmentosa. *Invest Ophthalmol Vis Sci* 2013;**54**(2):1411–6.
27. Daiger SP, Sullivan LS, Bowne SJ. Genes and mutations causing retinitis pigmentosa. *Clin Genet* 2013;**84**(2):132–41.
28. Maubaret CG, Vaclavik V, Mukhopadhyay R, Waseem NH, Churchill A, Holder GE, et al. Autosomal dominant retinitis pigmentosa with intrafamilial variability and incomplete penetrance in two families carrying mutations in PRPF8. *Invest Ophthalmol Vis Sci* 2011;**52**(13):9304–9.
29. Rose AM, Bhattacharya SS. Variant haploinsufficiency and phenotypic non-penetrance in PRPF31-associated retinitis pigmentosa. *Clin Genet* 2016;**90**(2):118–26.
30. Talib M, van Schooneveld MJ, Thiadens AA, Fiocco M, Wijnholds J, Florijn RJ, et al. Clinical and genetic characteristics of male patients with RPGR-associated retinal dystrophies: a Long-Term Follow-up Study. *Retina* 2019;**39**(6):1186–99.
31. Vervoort R, Lennon A, Bird AC, Tulloch B, Axton R, Miano MG, et al. Mutational hot spot within a new RPGR exon in X-linked retinitis pigmentosa. *Nat Genet* 2000;**25**(4):462–6.

32. Van Cauwenbergh C, Van Schil K, Cannoodt R, Bauwens M, Van Laethem T, De Jaegere S, et al. arrEYE: a customized platform for high-resolution copy number analysis of coding and noncoding regions of known and candidate retinal dystrophy genes and retinal noncoding RNAs. *Genet Med* 2017;**19**(4):457–66.
33. Van Schil K, Naessens S, Van de Sompele S, Carron M, Aslanidis A, Van Cauwenbergh C, et al. Mapping the genomic landscape of inherited retinal disease genes prioritizes genes prone to coding and noncoding copy-number variations. *Genet Med* 2018;**20**(2):202–13.
34. Ellingford JM, Barton S, Bhaskar S, Williams SG, Sergouniotis PI, O'Sullivan J, et al. Whole genome sequencing increases molecular diagnostic yield compared with current diagnostic testing for inherited retinal disease. *Ophthalmology* 2016;**123**(5):1143–50.
35. Carss KJ, Arno G, Erwood M, Stephens J, Sanchis-Juan A, Hull S, et al. Comprehensive rare variant analysis via whole-genome sequencing to determine the molecular pathology of inherited retinal disease. *Am J Hum Genet* 2017;**100**(1):75–90.
36. Richards S, Aziz N, Bale S, Bick D, Das S, Gastier-Foster J, et al. Standards and guidelines for the interpretation of sequence variants: a joint consensus recommendation of the American College of Medical Genetics and Genomics and the Association for Molecular Pathology. *Genet Med* 2015;**17**(5):405–24.
37. Bakthavatchalam M, Lai FHP, Rong SS, Ng DS, Brelen ME. Treatment of cystoid macular edema secondary to retinitis pigmentosa: a systematic review. *Surv Ophthalmol* 2018;**63**(3):329–39.
38. Davies EC, Pineda R. Cataract surgery outcomes and complications in retinal dystrophy patients. *Can J Ophthalmol* 2017;**52**(6):543–7.
39. Voretigene neparvovec-rzyl (Luxturna) for inherited retinal dystrophy. *Med Lett Drugs Ther* 2018;**60**(1543):53–5.
40. Bennett J, Wellman J, Marshall KA, McCague S, Ashtari M, DiStefano-Pappas J, et al. Safety and durability of effect of contralateral-eye administration of AAV2 gene therapy in patients with childhood-onset blindness caused by RPE65 mutations: a follow-on phase 1 trial. *Lancet* 2016;**388**(10045):661–72.
41. Edwards TL, Cottriall CL, Xue K, Simunovic MP, Ramsden JD, Zrenner E, et al. Assessment of the electronic retinal implant alpha AMS in restoring vision to blind patients with end-stage retinitis pigmentosa. *Ophthalmology* 2018;**125**(3):432–43.
42. Koenekoop RK, Sui R, Sallum J, van den Born LI, Ajlan R, Khan A, et al. Oral 9-cis retinoid for childhood blindness due to Leber congenital amaurosis caused by RPE65 or LRAT mutations: an open-label phase 1b trial. *Lancet* 2014;**384**(9953):1513–20.
43. Radu RA, Yuan Q, Hu J, Peng JH, Lloyd M, Nusinowitz S, et al. Accelerated accumulation of lipofuscin pigments in the RPE of a mouse model for ABCA4-mediated retinal dystrophies following vitamin A supplementation. *Invest Ophthalmol Vis Sci* 2008;**49**(9):3821–9.
44. Hoffman DR, Hughbanks-Wheaton DK, Pearson NS, Fish GE, Spencer R, Takacs A, et al. Four-year placebo-controlled trial of docosahexaenoic acid in X-linked retinitis pigmentosa (DHAX trial): a randomized clinical trial. *JAMA Ophthalmol* 2014;**132**(7):866–73.
45. Birch DG, Bernstein PS, Iannacone A, Pennesi ME, Lam BL, Heckenlively J, et al. Effect of oral valproic acid vs placebo for vision loss in patients with autosomal dominant retinitis pigmentosa: a randomized phase 2 multicenter placebo-controlled clinical trial. *JAMA Ophthalmol* 2018;**136**(8):849–56.

13A.2

Choroideremia

Maria I. Patrício, Kanmin Xue, Miguel C. Seabra, and Robert E. MacLaren

Choroideremia is a degenerative disorder of the retina that is inherited as an X-linked trait and is caused by mutations in the *CHM* gene. First described by Mauthner in 1872, the name refers to the barren ('*eremos*' in Greek) atrophic appearance of the choroid; this is seen in advanced disease and leads to exposure of the underlying sclera. Choroideremia is characterised by a progressive and generally centripetal loss of RPE and photoreceptors, with choroidal atrophy occurring secondary to RPE loss.[1]

Clinical characteristics

Males with choroideremia typically develop night vision problems in the first decade of life. This is followed soon after by slowly progressive peripheral visual field loss/constriction. Due to the centripetal pattern of retinal degeneration in this condition, most patients retain central vision until at least the fifth or sixth decade of life (although there is a degree of variability).

On fundoscopy, an early finding is that of peripheral pigmentation at the level of the RPE. Subsequently, well-defined areas of chorioretinal atrophy are noted in the mid- and far-peripheral retina; there is sparing of the central macula until the later stages of the disease. The extent of degeneration is best appreciated on fundus autofluorescence imaging with normal levels of autofluorescence corresponding to areas of preserved RPE. These areas are closely associated with regions of surviving overlying photoreceptors whose healthy inner segments are indicated by the preservation of the ellipsoid zone on OCT (Fig. 13A.4). The full-field electroretinogram (ERG) is abnormal with a rod-cone pattern of dysfunction initially; these full-field responses typically become extinguished in early- to mid-adulthood.[1]

Some features differentiate choroideremia from retinitis pigmentosa. In the former, the bone-spicule type of pigmentation is uncommon, the areas of chorioretinal atrophy are sharply demarcated, the retinal vessels appear relatively preserved and there is often no optic disc pallor. All these abnormalities however are shared with several other genetic retinal disorders, with two notable differential diagnoses including gyrate atrophy and dominant *RPE65*-retinopathy.

Examination of family members, in particular the proband's mother, can help confirm X-linked inheritance and support the diagnosis of choroideremia. Assuming the causative variant did not arise *de novo*, the patient's mother would be predicted to be a carrier. A spectrum of phenotypes is seen in female carriers. Most tend to be asymptomatic and to have good visual acuity but a small minority have significant visual loss (most likely as a result of skewed X-chromosome inactivation during early retinal development). Fundus changes can range from patchy pigment clumping (including generalised RPE mottling) to extensive atrophy (similar to the male phenotype) (Fig. 13A.5).[2] ERG responses range from normal (despite RPE pigmentary changes) to moderate impairment. These electrophysiological abnormalities typically manifest at an older age than those in carriers of X-linked retinitis pigmentosa.[3]

Molecular pathology

Choroideremia is caused by mutations in the *CHM* gene, encoding the Rab escort protein-1 (REP1).[4] REP1 is a ubiquitously expressed 653 amino-acid protein involved in the lipid modification of Rab GTPases, a family of proteins that regulate membrane vesicle trafficking.[5] This lipid modification, known as prenylation, is crucial for the recruitment of Rab GTPases to their target membranes. REP1 participates in this process by escorting Rab GTPases to the Rab geranylgeranyl transferase (also known as Rab GGTase) where prenyl groups are added. Once prenylated, Rab GTPases are able to relocate to the target membrane. The tight regulation of Rab proteins in time and space is particularly important for the homeostasis of the RPE and photoreceptors, hence the finding of retinal degeneration in choroideremia patients.

While minor disturbances in blood lipid profile have been detected in some patients, the disease manifests almost exclusively in the retina. This is because of the presence of *CHML*, an intronless *CHM* gene copy (retrogene) that encodes REP2, a protein that is very similar to REP1 and provides functional redundancy. In the retina, however, this redundancy is not sufficient, most likely due to the critical

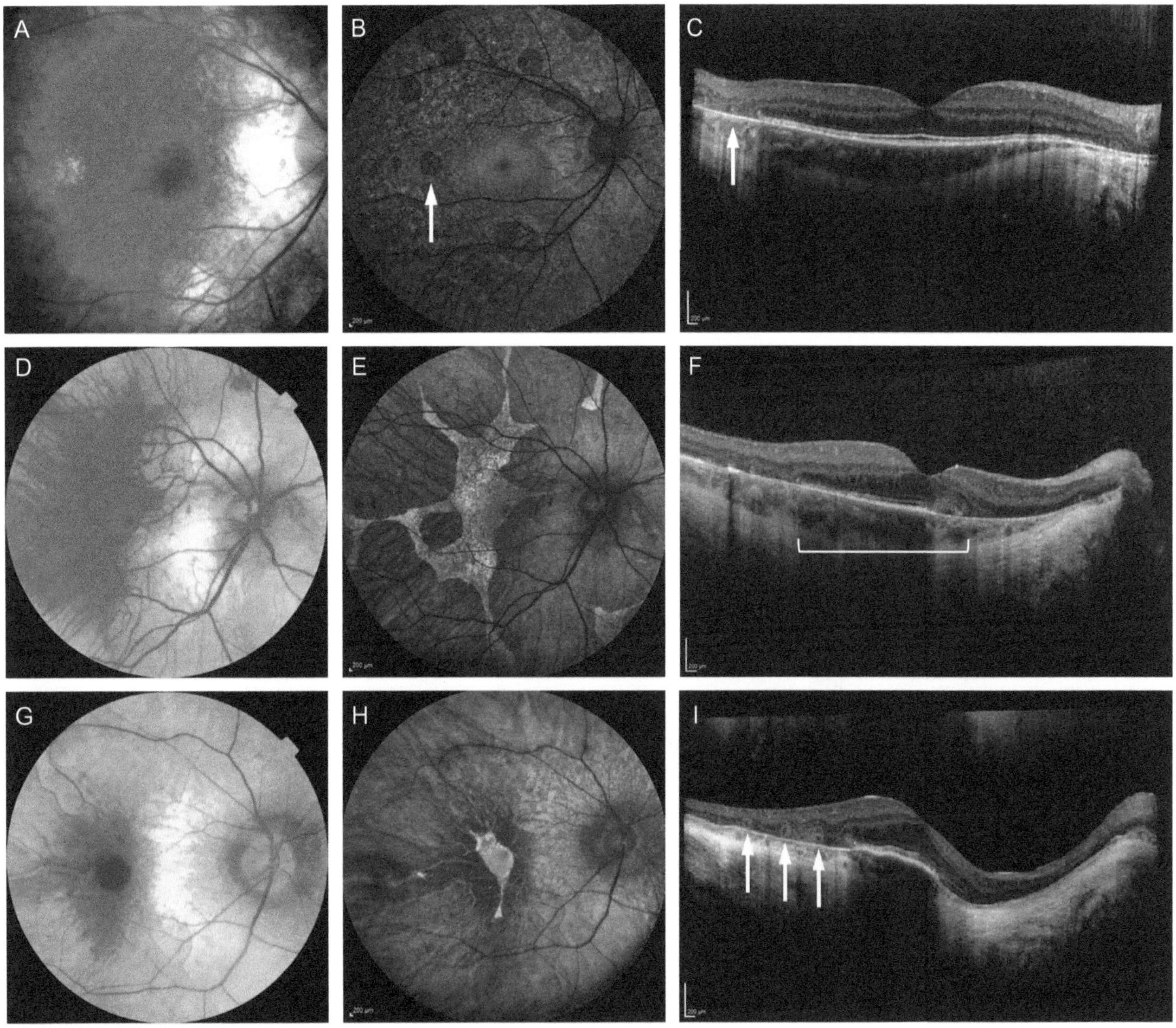

FIG. 13A.4 Progressive retinal degeneration in choroideremia. (A–C) Early choroideremia: fundus photograph (A), fundus autofluorescence imaging (B) and OCT (C) of the right eye of a 16-year-old male showing patches of outer retinal degeneration. The OCT ellipsoid zone is preserved over the central macula except temporally over a patch of RPE loss (corresponding arrows in B and C). (D–F) Moderately advanced choroideremia in the right eye of a 28-year-old male showing the emergence of a central island of the functional retina as peripheral degeneration becomes confluent (D, E). The edges of this central island correspond to the extent of preservation of the ellipsoid zone on OCT (F; bracket). (G–I) Advanced choroideremia in the right eye of a 46-year-old male showing a small residual island of the functional retina (H). Despite a severely restricted visual field, the fovea remains well preserved on OCT (I), thus providing normal visual acuity. Outer retinal tubulations *(arrows)* can be seen adjacent to the border of the surviving retina, which likely represents cross-sections of tubular rearrangement of photoreceptors following the loss of the underlying RPE support.

function that REP1 serves in retinal homeostasis, namely in the phagocytosis of shed photoreceptor outer segments.

More than 350 unique pathogenic variants have been described in *CHM*. The majority of these changes are loss of function variants (including nonsense, frameshift or splice-site mutations) which would be predicted to cause absent or inactive REP1. Other mutations include partial gene duplications, translocations, and occasionally missense variants. A disease-associated synonymous change, c.1359C>T, and variants within the promoter region (e.g. c.-98C>T) have also been described.[6] It is worth highlighting that a non-canonical splice site mutation, c.940+3delA, has been reported to be associated with a milder choroideremia phenotype. This change results in very low levels of correctly spliced *CHM* in peripheral blood (<1% to that seen in unaffected individuals) and this observation provides insights into the minimum dose required for gene therapy.[7] There are situations where mutations in the *CHM* gene are not detected in an individual with a clinical phenotype of choroideremia. In such cases, alternative biochemical tests may be considered, such as analysing the peripheral blood mononuclear cells for the absence of REP1.

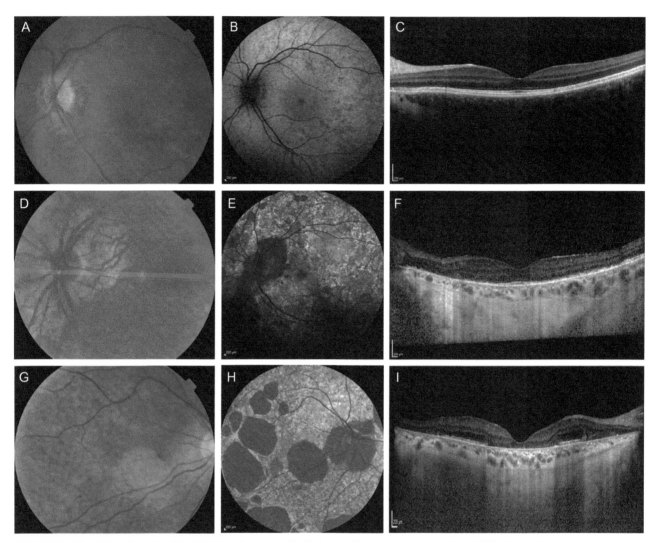

FIG. 13A.5 Choroideremia carriers. (A–C) Fundus photograph (A), fundus autofluorescence imaging (B) and OCT of the left eye of an asymptomatic 15-year-old choroideremia carrier female: fine patchy RPE mottling can be seen on fundal examination and autofluorescence imaging while the ellipsoid zone appears intact on the OCT. (D–F) The left eye of a mildly affected choroideremia carrier aged 51 years showing a peripapillary patch of outer retinal degeneration on a background of generalised granular disruption of the RPE (E) and ellipsoid zone (F). (G–I) The right eye of a severely affected choroideremia carrier aged 70 years showing a depigmented fundal appearance (G) and large patches of RPE loss (H), one of which has led to ellipsoid zone (photoreceptor) loss at the fovea (I).

Clinical management

Choroideremia is currently incurable. However, subretinal gene replacement therapy using recombinant adeno-associated viral (AAV) vectors carrying normal *CHM* copies holds promise for correcting the genetic defect and halting disease progression. Phase 1 and 2 clinical trials of choroideremia gene therapy have demonstrated the safety of the approach; there was a small gain in vision in treated eyes of a median of 4.5 letters, versus a loss of 1.5 letters in untreated eyes over 2 years. Importantly, choroideremia gene therapy has led to sustained preservation and, in some advanced cases, an increase of visual acuity in the treated eyes.[8] It remains to be seen, with longer-term follow-up, whether the peripheral retinal degeneration can be slowed or halted following gene therapy.

A prerequisite for retinal gene therapy or clinical trial involvement is the genetic confirmation of the underlying molecular defect, hence genetic testing should be offered to affected individuals. Genetic counselling should also be offered, including to daughters of affected choroideremia males of child-bearing age who would be obligate disease carriers with a 1 in 2 chance of having an affected male offspring.

It should be noted that patients with choroideremia, as with retinitis pigmentosa, tend to develop visually significant posterior subcapsular cataracts at an earlier age. This is likely the consequence of chronic low-grade vitreous inflammation as a result of retinal degeneration since a trace of vitreous cells can often be seen in this condition. Cataract extraction in choroideremia is generally uncomplicated although there is a higher incidence of zonular weakness, which could usually be overcome with the implantation of a capsular tension ring. In the early stages of the disease, retinal degeneration can rarely be complicated by choroidal neovascularisation. While this may be self-limiting, it is also highly sensitive to intravitreal anti-VEGF therapy, which helps limit the development of a central scotoma.

References

1. Simunovic MP, Jolly JK, Xue K, Edwards TL, Groppe M, Downes SM, et al. The spectrum of CHM gene mutations in choroideremia and their relationship to clinical phenotype. *Invest Ophthalmol Vis Sci* 2016;**57**(14):6033–9.
2. Edwards TL, Groppe M, Jolly JK, Downes SM, MacLaren RE. Correlation of retinal structure and function in choroideremia carriers. *Ophthalmology* 2015;**122**(6):1274–6.
3. De Silva SR, Arno G, Robson AG, Fakin A, Pontikos N, Mohamed MD, et al. The X-linked retinopathies: physiological insights, pathogenic mechanisms, phenotypic features and novel therapies. *Prog Retin Eye Res* 2020;**82**, 100898.
4. Cremers FP, van de Pol DJ, van Kerkhoff LP, Wieringa B, Ropers HH. Cloning of a gene that is rearranged in patients with choroideraemia. *Nature* 1990;**347**(6294):674–7.
5. Seabra MC, Brown MS, Goldstein JL. Retinal degeneration in choroideremia: deficiency of rab geranylgeranyl transferase. *Science* 1993;**259**(5093):377–81.
6. Radziwon A, Arno G, Wheaton D, McDonagh EM, Baple EL, Webb-Jones K, et al. Single-base substitutions in the CHM promoter as a cause of choroideremia. *Hum Mutat* 2017;**38**(6):704–15.
7. Fry LE, Patrício MI, Williams J, Aylward JW, Hewitt H, Clouston P, et al. Association of Messenger RNA level with phenotype in patients with choroideremia: potential implications for gene therapy dose. *JAMA Ophthalmol* 2020;**138**(2):128–35.
8. Xue K, Jolly JK, Barnard AR, Rudenko A, Salvetti AP, Patrício MI, et al. Beneficial effects on vision in patients undergoing retinal gene therapy for choroideremia. *Nat Med* 2018;**24**(10):1507–12.

13A.3

Enhanced S-cone syndrome and *NR2E3*-associated disorders

Pascal Escher, Kaspar Schuerch, Martin Zinkernagel, Viet H. Tran, and Francis L. Munier

The *NR2E3* gene encodes a photoreceptor-specific transcription factor with a role in rod photoreceptor development.[1] Biallelic variants in *NR2E3* cause a characteristic retinopathy called enhanced S-cone syndrome (also known as Goldmann-Favre syndrome). In this childhood-onset, slowly progressive condition, loss of *NR2E3* function leads to an excess of S-cone photoreceptors and to a lack of rod photoreceptors.[2] Notably, a unique, dominantly acting *NR2E3* missense variant, c.166G > A (p.Gly56Arg) causes autosomal dominant retinitis pigmentosa.[3]

Clinical characteristics

Enhanced S-cone syndrome

Individuals with enhanced S-cone syndrome typically present in the first decade of life. Night vision disturbance (with or without reduced visual acuity) is the most frequently reported initial complaint. Hypermetropia is common and visual function is highly variable ranging from normal to severely reduced. Poorer visual outcomes are associated with the presence of macular schisis, a characteristic finding present in approximately 40% of affected individuals. Importantly, the peripheral visual fields are relatively preserved and the visual acuity remains relatively stable over time in the majority of patients.[4]

In addition to macular schisis (which can accompany or precede macular atrophic changes), two other fundoscopic findings are observed in a large proportion of people with enhanced S-cone syndrome: yellow/white dots (corresponding to hyperautofluorescent spots at the level of the RPE) and 'nummular' pigmentation (i.e. round-shaped pigment clumps also at the level of the RPE, see also Fig. 13A.6 and Table 13A.4.1). Both these lesions are mainly found along the vascular arcades and in the mid-peripheral retina, especially temporally. They often co-exist but it is thought that the dots appear early in the disease process while pigmentary deposition occurs at a later stage.[5] These findings are not specific to enhanced S-cone syndrome but in the appropriate clinical context, they should raise the possibility of this condition.[4-9]

The full-field electroretinogram (ERG) findings are characteristic (Fig. 13A.7):

- rod function, assessed using a dim flash in dark-adapted conditions (DA 0.01 ERG), is typically undetectable;
- the reaction to a bright flash appears abnormal but identical in dark- and light-adapted conditions (i.e. the DA 3 ERG and the LA 3 ERG have similar simplified, delayed and usually reduced waveforms); this suggests that retinal responses are dominated by S-cone mechanisms.
- the light-adapted flicker response (LA 30 Hz flicker ERG) is of low amplitude and delayed.
- additional testing for S-cone activity, for example using a blue stimulus (~445 nm) on an orange background

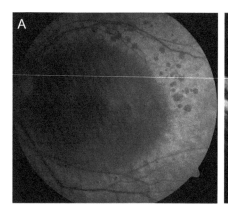
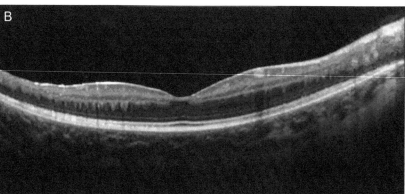

FIG. 13A.6 Fundus imaging findings in enhanced S-cone syndrome. (A) Fundus photography of a 10-year-old patient with typical nummular pigmentary deposits at the level of the RPE, along the vascular arcades. (B) OCT imaging of the same eye with the presence of retinal 'waves'. Macular schisis, cystoid maculopathy and disorganisation of retinal layers are common findings but were not present in this affected individual.

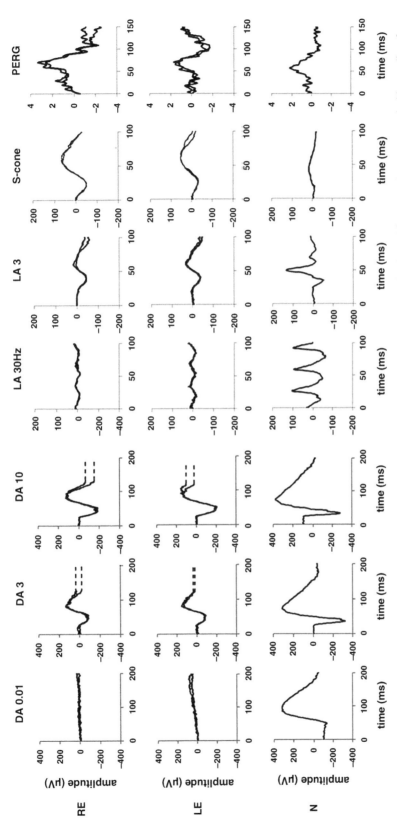

FIG. 13A.7 Full-field and pattern electroretinographic (ERG) recordings from the right eye (RE) and left eye (LE) of a patient with enhanced S-cone syndrome compared with recordings from a representative unaffected control individual (N). ERG recordings include dark-adapted (DA) responses (flash strengths, 0.01 and 10 cd.s/m^2; DA 0.01 and DA 10) and light-adapted (LA) responses for a flash strength of 3 cd.s/m^2 (LA 3; 30 Hz and 2 Hz). A 20-ms pre-stimulus delay in single-flash ERG recordings is present, except for the S-cone ERG response. Broken lines replace blink artefacts occurring after ERG b-waves, for clarity. Patient responses are superimposed to demonstrate reproducibility. In this patient, the pattern ERG (PERG) component is delayed but of normal amplitude. The rod-specific dim flash (DA 0.01) response is undetectable. The single-flash DA 3, DA 10, and LA 3 responses have similarly simplified and severely delayed waveforms, qualitatively comparable in shape with the S-cone ERG response and consistent with generation by the same (S-cone) mechanism. The S-cone ERG response is delayed and grossly enlarged. The LA 30-Hz response is smaller than the LA 3 a-wave, whereas, in the typical healthy participant, the LA 30-Hz ERG amplitude is between that of the LA 3 a- and b-waves. *(Adapted from Ref. 4.)*

(~620 nm), elicits a high amplitude response that is typically several times higher than normal; this is consistent with the increased number of S-cone photoreceptors.[4–8]

It is worth highlighting that a similar clinical and electrophysiological phenotype has been described in patients with biallelic mutations in *NRL*, a gene encoding a transcription factor that acts upstream of *NR2E3*.[10]

NR2E3-associated dominant retinitis pigmentosa

A missense variant in *NR2E3*, c.166G > A (p.Gly56Arg) has been found to underly autosomal dominant retinitis pigmentosa in multiple families.[3] Most affected individuals report night vision problems in the second or third decade of life; central vision is typically preserved until the fifth decade of life.[11] Similar to other retinitis pigmentosa subtypes, the progressive degeneration of rods leads eventually to bone spicule-like pigment deposition.[3,10] A concentric double ring of hyperautofluorescence is observed in many affected individuals.[12,13] While the disease progresses, both the innermost and outermost hyperautofluorescent rings expand centripetally toward the fovea (Fig. 13A.8).[12,13]

Molecular pathology

NR2E3, *CRX* and *NRL* encode major transcription factors (i.e. proteins that bind to DNA and stimulate or inhibit gene expression and protein synthesis) that are expressed in the fetal retina and regulate photoreceptor differentiation. During retinogenesis, a common population of pluripotent retinal progenitor cells gives rise to the various retinal cell types, including rod photoreceptors and L- (long-wavelength), M- (medium-wavelength) and S- (short-wavelength) cone photoreceptors. In general, cone precursors arise earlier than their rod counterparts. Also, S-cones appear before L- and M-cones and are therefore the default primordial cone cells (although they ultimately end up being the least prevalent photoreceptor cell type in the mature retina). *NR2E3* suppresses cone-specific genes, thus committing progenitor cells to a rod fate. When it is absent, rod photoreceptor cell development is hindered and S-cone population overexpansion occurs (Fig. 13A.9).[2,3,14] This cell fate misspecification is the cellular basis of the hallmarks of enhanced S-cone syndrome.

Over 60 pathogenic *NR2E3* variants have been reported in patients. A loss of function mechanism is likely for most of these changes. Several pathogenic missense variants have been described and it is thought that most of these

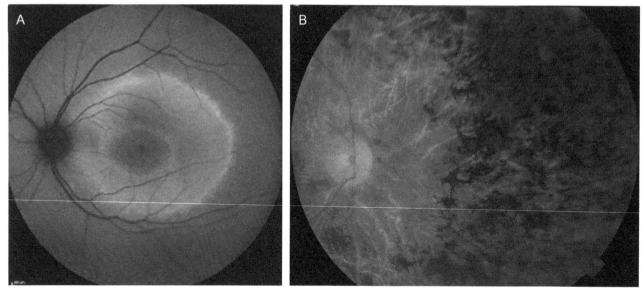

FIG. 13A.8 Retinal imaging findings in *NR2E3*-associated retinitis pigmentosa. (A) Fundus autofluorescence imaging reveals two hyperautofluorescent rings in a 16-year-old patient who has retinitis pigmentosa and carries the *NR2E3* c.166G > A (p.Gly56Arg) mutation in the heterozygous state. (B) Fundus photograph from a 46-year-old who has retinitis pigmentosa and carries the *NR2E3* c.166G > A (p.Gly56Arg) mutation. Bone spicule-like pigmentary deposits are noted. *(A: Adapted from Ref. 13.)*

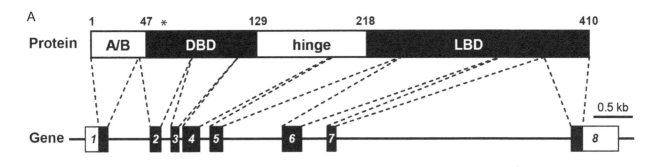

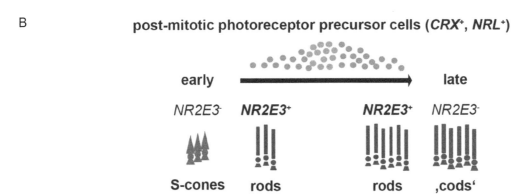

FIG. 13A.9 Molecular mechanisms of enhanced S-cone syndrome. (A) Schematic representation of the *NR2E3* gene *(bottom)* and protein *(top)*: The *NR2E3* gene includes 8 exons *(boxes)*. Exons 1 and 8 contain the non-coding 5'- and 3'-untranslated regions, respectively *(white boxes)*. The coding sequence *(black boxes)* is translated into a polypeptidic chain of 410 amino-acids. This protein comprises an N-terminal A/B domain (amino-acids 1–46), a highly conserved DNA-binding domain (DBD; amino-acids 47–128), a hinge domain (amino-acids 129–217) and a ligand-binding domain (LBD; amino-acids 218–410). The dominantly-acting c.166G>A (p.Gly56Arg) variant that causes retinis pigmentosa is located in the DNA-binding domain (*). Recessively-acting variants associated with enhanced S-cone syndrome cluster in the highly conserved DNA- and ligand-binding domains. (B) Disease mechanisms in enhanced S-cone syndrome: Rod photoreceptors are generated during embryonic retinal development over an extended period from photoreceptor precursor cells expressing the cone-rod homeobox transcription factor (CRX+) and the neural retina leucine zipper transcription factor (NRL+). The absence of *NR2E3* expression (NR2E3-) causes misspecification in cell fate toward S-cones in 'early-born' rod precursors, and the generation of non-functional hybrid photoreceptors expressing both rod- and cone-specific genes ('cods') in 'late-born' rod precursors.

either disrupt binding to DNA or impair dimerization of the *NR2E3* protein (Fig. 13A.9).[15] The c.166G>A (p.-Gly56Arg) change that has been associated with the dominant disease is located in the conserved DNA-binding domain. However, unlike other *NR2E3* variants, it is thought to influence the terminal differentiation and maintenance of rods rather than to have a major impact on retinogenesis.[3]

Clinical management

There is currently no established treatment for enhanced S-cone syndrome or *NR2E3*-associated retinitis pigmentosa. Macular schisis with acute-onset visual acuity loss may be treated with topical or oral carbonic anhydrase inhibitors although efficacy is variable.[4] Affected children should be monitored for refractive error and macular changes. The risk of choroidal neovascularisation (estimated to occur in up to 15% of cases) should be discussed.[16]

References

1. Bumsted O'Brien KM, Cheng H, Jiang Y, Schulte D, Swaroop A, Hendrickson AE. Expression of photoreceptor-specific nuclear receptor NR2E3 in rod photoreceptors of fetal human retina. *Invest Ophthalmol Vis Sci* 2004;**45**(8):2807–12.
2. Haider NB, Jacobson SG, Cideciyan AV, Swiderski R, Streb LM, Searby C, et al. Mutation of a nuclear receptor gene, NR2E3, causes enhanced S cone syndrome, a disorder of retinal cell fate. *Nat Genet* 2000;**24**(2):127–31.
3. Coppieters F, Leroy BP, Beysen D, Hellemans J, De Bosscher K, Haegeman G, et al. Recurrent mutation in the first zinc finger of the orphan nuclear receptor NR2E3 causes autosomal dominant retinitis pigmentosa. *Am J Hum Genet* 2007;**81**(1):147–57.

4. de Carvalho ER, Robson AG, Arno G, Boon C, Webster AA, Michaelides M. Enhanced S-cone syndrome: spectrum of clinical, imaging, electrophysiological and genetic findings in a retrospective case series of 56 patients. *Ophthalmol Retina* 2020;**5**(2):195–214.
5. Sharon D, Sandberg MA, Caruso RC, Berson EL, Dryja TP. Shared mutations in NR2E3 in enhanced S-cone syndrome, Goldmann-Favre syndrome, and many cases of clumped pigmentary retinal degeneration. *Arch Ophthalmol* 2003;**121**(9):1316–23.
6. Audo I, Michaelides M, Robson AG, Hawlina M, Vaclavik V, Sandbach JM, et al. Phenotypic variation in enhanced S-cone syndrome. *Invest Ophthalmol Vis Sci* 2008;**49**(5):2082–93.
7. Pachydaki SI, Klaver CC, Barbazetto IA, Roy MS, Gouras P, Allikmets R, et al. Phenotypic features of patients with NR2E3 mutations. *Arch Ophthalmol* 2009;**127**(1):71–5.
8. Hull S, Arno G, Sergouniotis PI, Tiffin P, Borman AD, Chandra A, et al. Clinical and molecular characterization of enhanced S-cone syndrome in children. *JAMA Ophthalmol* 2014;**132**(11):1341–9.
9. Ammar MJ, Scavelli KT, Uyhazi KE, Bedoukian EC, Serrano LW, Edelstein ID, et al. Enhanced S-cone syndrome: visual function, cross-sectional imaging and cellular structure with adaptive optics ophthalmoscopy. *Retin Cases Brief Rep* 2019. https://doi.org/10.1097/ICB.0000000000000891.
10. Littink KW, Stappers PTY, Riemslag FCC, Talsma HE, van Genderen MM, Cremers FPM, et al. Autosomal recessive NRL mutations in patients with enhanced S-cone syndrome. *Genes (Basel)* 2018;**9**(2):68.
11. Blanco-Kelly F, García Hoyos M, Lopez Martinez MA, Lopez-Molina MI, Riveiro-Alvarez R, Fernandez-San Jose P, et al. Dominant retinitis pigmentosa, p.Gly56Arg mutation in NR2E3: phenotype in a large cohort of 24 cases. *PLoS One* 2016;**11**(2):e0149473.
12. Escher P, Vaclavik V, Munier FL, Tran HV. Presence of a triple concentric autofluorescence ring in NR2E3-p.G56R-linked autosomal dominant retinitis pigmentosa (ADRP). *Invest Ophthalmol Vis Sci* 2016;**57**(4):2001–2.
13. Escher P, Tran HV, Vaclavik V, Borruat FX, Schorderet DF, Munier FL. Double concentric autofluorescence ring in NR2E3-p.G56R-linked autosomal dominant retinitis pigmentosa. *Invest Ophthalmol Vis Sci* 2012;**53**(8):4754–64.
14. Wang S, Cepko CL. Photoreceptor fate determination in the vertebrate retina. *Invest Ophthalmol Vis Sci* 2016;**57**(5), ORSFe1-6.
15. von Alpen D, Tran HV, Guex N, Venturini G, Munier FL, Schorderet DF, et al. Differential dimerization of variants linked to enhanced S-cone sensitivity syndrome (ESCS) located in the NR2E3 ligand-binding domain. *Hum Mutat* 2015;**36**(6):599–610.
16. Nowilaty SR, Alsalamah AK, Magliyah MS, Alabdullah AA, Ahmad K, Semidey VA, et al. Incidence and natural history of retinochoroidal neovascularization in enhanced S-cone syndrome. *Am J Ophthalmol* 2021;**222**:174–84.

13A.4

Congenital stationary night-blindness

Anthony G. Robson, Eva Lenassi, Panagiotis I. Sergouniotis, Isabelle Audo, and Christina Zeitz

Congenital stationary night-blindness (CSNB) is a group of genetically determined, largely non-progressive retinal disorders. These conditions exhibit significant clinical and genetic heterogeneity but all affected individuals share a common feature: rod system dysfunction. Full-field electroretinograms (ERG; see also Chapter 6) and genetic testing play a critical role in the diagnosis of CSNB and help distinguish different subtypes (Fig. 13A.10). Broadly, CSNB can be divided into:

- subtypes affecting post-phototransduction signalling ('Schubert-Bornshein type'). This group can be further subdivided into forms that affect the ON-bipolar cell pathway ('complete' CSNB) and forms that affect both the ON- and OFF-bipolar cell pathways ('incomplete' CSNB). It is noted that the majority of cones connect with at least two bipolar cells (one of which is ON and the other OFF) while rods directly interact mainly with ON-bipolar cells. This group of conditions is characterised by normal fundi (besides myopic changes) (Fig. 13A.11).
- subtypes affecting the rod phototransduction cascade or retinoid recycling ('Riggs type'). These disorders can be further subdivided into forms that are associated with a normal fundus and forms that are associated with fundus abnormalities; the latter group includes fundus albipunctatus (Fig. 13A.12) and Oguchi disease (Fig. 13A.13).

Phenotypic studies of patients with CSNB have highlighted that affected individuals are not always aware of problems with night vision. Some patients may present with significant light aversion and a small subset of cases show evidence of slowly progressive disease. These observations suggest that using the term CSNB can occasionally be confusing and imprecise.[1] An international initiative to refine and update CSNB nomenclature has been initiated.

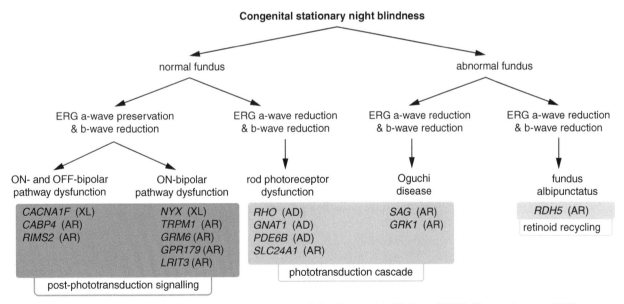

FIG. 13A.10 Categories and genotype–phenotype correlations in congenital stationary night-blindness (CSNB). Electroretinogram (ERG) a-wave and b-wave refer to dark-adapted strong flash (DA 10) ERG findings. *AR*, autosomal recessive; *AD*, autosomal dominant; *XL*, X-linked; *ERG*, electroretinogram. *(Adapted from Ref. 1.)*

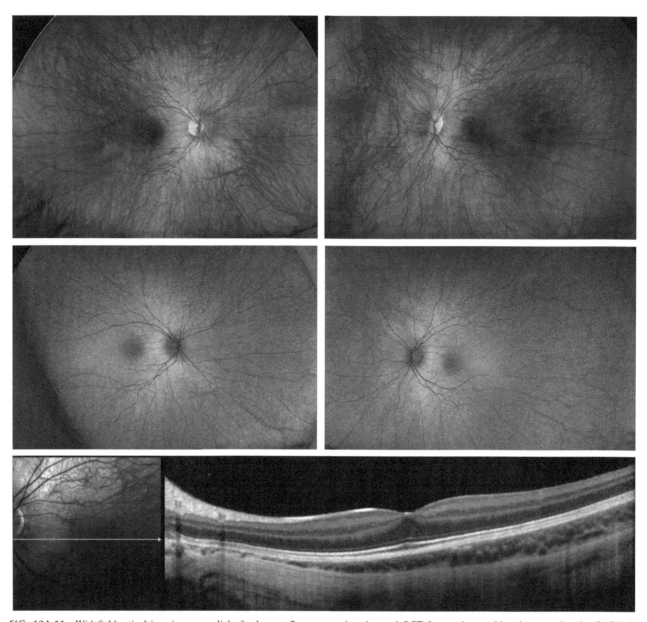

FIG. 13A.11 Widefield retinal imaging, green-light fundus autofluorescence imaging and OCT from a 4-year-old patient carrying the *CACNA1F* c.4093_3095delAAC, (p.Asn1365del) in the hemizygous state. Myopic fundi with slightly tilted optic discs are noted.

Clinical characteristics

CSNB subtypes affecting post-phototransduction signalling

This group of conditions is characterised clinically by an 'electronegative' ERG, i.e. the dark-adapted bright flash (e.g. DA 10) ERG a-wave is preserved (reflecting a dominant contribution from the rod photoreceptors and intact phototransduction) but the b-wave is attenuated and smaller than the a-wave (reflecting a locus of dysfunction that is post-phototransduction).[2] As mentioned above, CSNB with electronegative ERG can be further subdivided into complete and incomplete forms. In complete CSNB, there is no detectable response in the dark-adapted dim flash (DA 0.01) ERG (consistent with the term 'complete'). In contrast, incomplete CSNB is associated with a subnormal but detectable dark-adapted dim flash (DA 0.01) ERG (consistent with the term 'incomplete'). For further information on the electrodiagnostic findings in this group of conditions see Fig. 13A.14, Fig. 6.11 and Chapter 6.

Individuals with complete CSNB typically present in the first years of life. The most common presenting features are nystagmus and mild to moderate reduction in visual acuity; night vision problems, high myopia, and strabismus may also be present.[3,6] Fundus examination and imaging

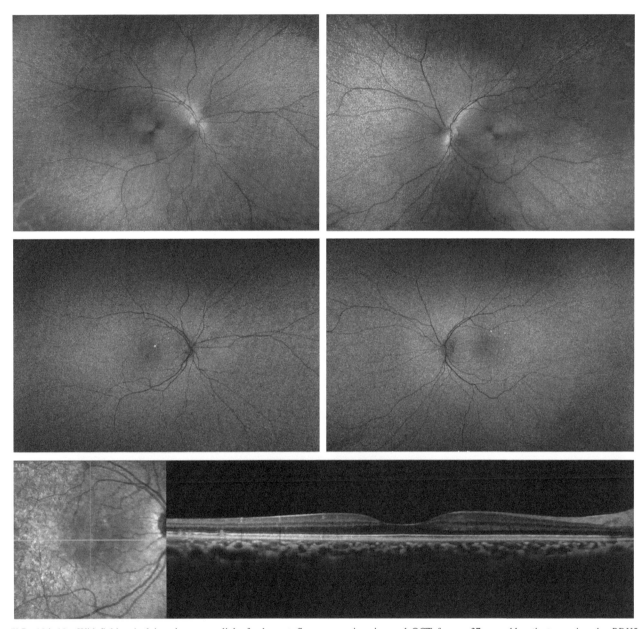

FIG. 13A.12 Widefield retinal imaging, green-light fundus autofluorescence imaging and OCT from a 37-year-old patient carrying the *RDH5* c.470G > A (p.Arg157Gln) variant in homozygous state. Multiple yellow-white spots are noted throughout the fundus with sparing of the central macula. Fundus autofluorescence reveals a grossly reduced background signal, reflecting life-long impairment of retinoid recycling and a lack of lipofuscin deposition (image enhanced to show lack of foci associated with the yellow dots). On OCT the spots correspond to hyper-reflective lesions extending from the RPE to the outer retina.

reveals no abnormalities other than myopic changes. Electrophysiological testing highlights that signal transmission from photoreceptors to ON-bipolar cells is defective; the function of the OFF-bipolar cell pathway is preserved (Fig. 13A.14).

Individuals with incomplete CSNB also typically present with nystagmus and/or abnormal central vision in the first years of life. Night vision problems have been reported to be present in just over half of cases. Notably, photophobia is common and there is a high incidence of colour vision impairment.[6] Myopia or hypermetropia may be present and fundus examination and imaging are generally within normal limits (Fig. 13A.11). Electrophysiological testing reveals that the light-adapted responses are more abnormal than in the complete form as both the ON- and OFF-bipolar cell pathways are affected (Fig. 13A.14). Consequently, individuals with incomplete CSNB tend to have more severe daylight symptoms and lower visual acuity on average compared to those with complete CSNB.[1,6]

184 SECTION | III Genetic disorders affecting the posterior segment

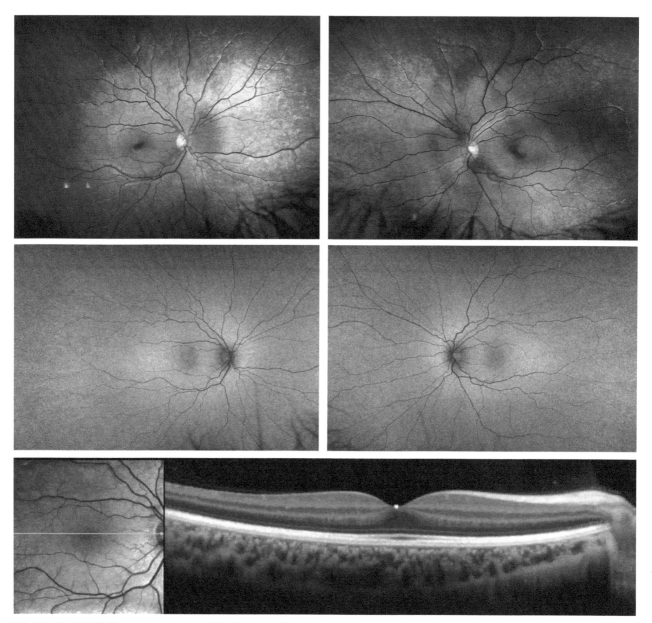

FIG. 13A.13 Widefield retinal imaging, green-light fundus autofluorescence imaging and OCT from a 6-year-old patient carrying the *SAG* c.916G>T (p.Glu306Ter) variant in the homozygous state. A golden fundus reflex is noted. Fundus autofluorescence imaging was normal. OCT was performed under light-adapted conditions and there was an evident reduction in the contrast of the hyporeflective photoreceptor outer segment band. This subtle finding is common in Oguchi disease and is likely to be associated with increased light scattering from the inner segment ellipsoid zone, the RPE layer, or both. Notably, the contrast of the hyporeflective outer segment band increased when imaging was performed following prolonged dark adaptation.

CSNB subtypes that affect rod phototransduction and are associated with a normal fundus

In these rare forms of 'Riggs type' CSNB, night vision problems are often the only visual symptom.[1] In some cases, a late manifestation of fundus abnormalities may be noted suggesting a slowly progressive disease course.[7]

Electrophysiological testing reveals normal light-adapted ERGs. The dark-adapted dim flash (DA 0.01) response is undetectable and the strong flash (DA 10) ERG a- and b-waves are subnormal. This is consistent with a primary defect affecting rod phototransduction. The typical dark-adapted ERG phenotype is indistinguishable from that seen in fundus albipunctatus and Oguchi disease but without recovery following prolonged dark-adaptation (see below and Fig. 6.11).

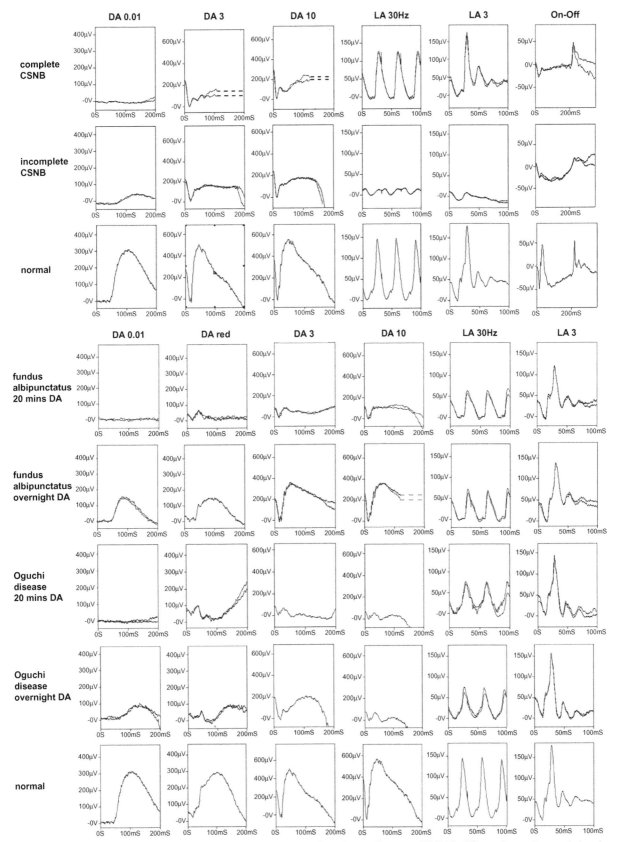

FIG. 13A.14 Examples of dark-adapted (DA) and light-adapted (LA) full-field electroretinograms (ERGs) in different forms of congenital stationary night-blindness (CSNB). International standard ERGs are shown for white flashes of 0.01, 3.0 and 10.0 cd.s.m.$^{-2}$ (DA 0.01, DA 3 and DA 10 respectively); light-adapted ERGs are shown for white flashes of 3.0 cd.s.m.$^{-2}$ (30 Hz and 2 Hz; LA 30 Hz and LA 3 respectively). Additional photopic ON–OFF ERGs and DA red flash ERGs are included. Where appropriate, traces from patients are superimposed to demonstrate reproducibility. Broken lines replace blink artefacts that occur soon after b-waves in some DA10 ERGs (for clarity). Recordings from a typical patient with complete CSNB are shown in the first row. The DA 0.01 ERG, arising in the rod ON-bipolar cells, is undetectable. There is an electronegative waveform in the DA 3 and DA 10 ERG with a preserved a-wave and a severely reduced b-wave (b:a amplitude ratio <1). The LA 30 Hz ERG is relatively preserved but shows a flattening of the trough between

Continued

FIG. 13A.14—CONT'D adjacent peaks; in some cases, peak times and amplitudes may also be mildly abnormal. The LA 3 (single flash) ERG is generally of normal or near-normal amplitude but there is a broad bifid a-wave and a sharply rising b-wave that lacks oscillatory potentials; in some cases, the b:a ratio can be mildly reduced. These ERG findings are pathognomonic for generalised rod and cone-mediated ON-bipolar cell dysfunction; this observation is consistent with findings in the light-adapted ON–OFF ERGs which reveals selective attenuation of the ON b-wave. It is worth highlighting that identical ERG abnormalities can occur in autoimmune and paraneoplastic retinopathies such as melanoma-associated retinopathy and interpretation of ERGs must be in a clinical context. Recordings from a typical patient with incomplete CSNB are shown in the second row. The DA 0.01 ERG is detectable but attenuated. The DA 3 and DA 10 ERG are electronegative with normal or largely preserved a-wave. The LA 30 Hz ERG is subnormal and there is a characteristic twin-peaked waveform. The LA 3 ERG is subnormal and simplified with a subnormal b:a ratio close to 1. There is a reduction in both the ON b-wave and OFF d-wave, consistent with the involvement of both the ON- and OFF- bipolar cell systems. Recordings in the case of fundus albipunctatus are shown in the fourth and fifth row (following 20 min and overnight dark adaptation respectively). The DA 0.01 ERG after a standard period of dark adaptation is undetectable. The DA red flash ERG shows preservation of the early cone-mediated x-wave but the rod system-mediated b-wave component is undetectable. The DA 3 and DA 10 ERG a-wave and b-wave are markedly subnormal in keeping with the absence of the normally dominant rod system contribution. It is worth noting that the residual DA 10 ERG represents the dark-adapted cone system contribution to a response that is normally a summation of dark-adapted rod and dark-adapted cone system activity. Light-adapted ERGs are within normal limits in this case. Following overnight dark adaptation, there is a recovery of rod-mediated ERG components. Importantly, a similar clinical and electrophysiological phenotype may occur in young patients with retinitis punctata albescens (Bothnia dystrophy), but a longer period of dark adaptation is required to recover rod function and patients develop a progressive rod-cone degeneration. Recordings in the case of Oguchi disease are shown in the sixth and seventh row (following 20 min and overnight dark adaptation respectively). The findings are generally similar to the fundus albipunctatus case. However, in fundus albipunctatus, after prolonged dark adaptation, the ERGs appear similar to successive flashes (interstimulus interval 20 s). In Oguchi disease, there is a significant recovery of the response to the first bright flash (DA 3 ERG) but responses to subsequent flashes resemble those obtained after 20 min of dark adaptation, despite an increased interstimulus interval (1 min). *(The case of complete CSNB is adapted from Ref. 3, the case of fundus albipunctatus is adapted from Ref. 4 and the case of Oguchi disease is adapted from Ref. 5).*

CSNB subtypes that affect rod phototransduction and are associated with an abnormal fundus

Two forms of CSNB are associated with abnormal fundi: fundus albipunctatus and Oguchi disease.

Individuals with fundus albipunctatus have multiple yellow-white spots that are scattered throughout the fundus but spare the central macula. These spots are present from early childhood and may decrease in number over time, making the diagnosis difficult in people presenting in adulthood (Fig. 13A.12). Affected individuals report night vision problems from the first decade of life and typically have normal visual acuity. Progressive macular/cone photoreceptor dysfunction can occur in older patients. The electrophysiological phenotype in this condition is characteristic; a key finding is that of markedly abnormal dark-adapted dim flash (DA 0.01) ERGs that completely or largely recover after prolonged dark adaptation[4,8] (see Fig. 13A.14 for more information). Light-adapted ERGs are normal or mildly abnormal in 50%–75% of cases.

Patients with Oguchi disease have a characteristic golden sheen that is most prominent in the mid-peripheral retina (Fig. 13A.13). The appearance is reversible and the sheen diminishes after prolonged dark adaptation (Mizuo-Nakamura phenomenon). Most affected individuals present with abnormal night vision in childhood; the visual acuity is typically preserved. A degree of progressive degeneration in adulthood has been described in some cases. The dark-adapted ERG findings are similar to fundus albipunctatus after a standard period of dark adaptation, but recovery after prolonged dark adaptation tends to be less striking. In Oguchi disease, a single strong flash after prolonged dark adaptation typically elicits a normal or near-normal response, but causes sustained desensitisation of rods, such that ERGs to a second and subsequent strong flashes resemble the baseline dark-adapted recordings. Further dark-adapted ERG recovery requires a second period of extended dark adaptation in keeping with abnormally slow rhodopsin regeneration. The light-adapted ERGs are typically normal (see Fig. 13A.14 for more information).[5,9]

Molecular pathology

At least 15 genes have been implicated in CSNB. These include molecules involved in the rod phototransduction cascade, in retinoid recycling and in signal transmission from the photoreceptors to the bipolar cells (Fig. 13A.10).

CSNB subtypes that affect post-phototransduction signalling (i.e. that have electronegative ERGs) are inherited as X-linked or autosomal recessive traits. The complete forms generally result from defects in genes that are expressed in the retinal ON-bipolar cells; common causes include hemizygous variants in the *NYX* gene and biallelic changes in the *TRPM1* gene; less commonly, biallelic variants in *GRM6*, *GPR179* or *LRIT3* are identified. Incomplete CSNB is inherited as an X-linked trait in most cases, resulting from hemizygous variants in the *CACNA1F* gene; rare cases with biallelic variants in *CABP4* have also been described. These two genes affect neurotransmitter release from the photoreceptor membrane and affect rod ON-bipolar cell and cone ON- and OFF-bipolar cell function.[1] A similar but more severe ERG phenotype has been documented in individuals with biallelic variants in *RIMS2*; these patients have a syndromic phenotype including pancreatic and neurodevelopmental disease.[10]

CSNB subtypes that affect rod phototransduction and are associated with normal fundi are inherited as autosomal dominant or, less commonly, as autosomal recessive traits. Specific heterozygous variants in *RHO* (a gene also associated with autosomal dominant retinitis pigmentosa) are the commonest cause of autosomal dominant CSNB. Other causes of stationary rod dysfunction include heterozygous variants in *GNAT1* and *PDE6B* and biallelic changes in *SLC24A1* and *CNGB1* (the latter associated with a specific variant, c.807G > C (p.Gln269His), when encountered in compound heterozygous state with a loss-of-function mutation).[1,7]

Most individuals with fundus albipunctatus have autosomal recessive mutations in the *RDH5* gene, encoding 11-cis retinol dehydrogenase. This enzyme converts 11-cis-retinol to 11-cis retinal in the RPE, before it is transported to the photoreceptors.[1,4,8] Loss of *RDH5* function can therefore result in delayed retinoid recycling.

Patients with Oguchi disease have autosomal recessive mutations in either *GRK1* or *SAG*. Rhodopsin kinase (encoded by *GRK1*) phosphorylates rhodopsin and forms a complex with arrestin (encoded by *SAG*) that ultimately stops downstream signalling. Pathogenic variants in either gene result in sustained rhodopsin activation causing continuous phototransduction and desensitisation of rod function.[1,9]

Targeted (single gene or gene panel) or comprehensive (exome or genome sequencing) genetic testing can be offered to families with CSNB. These approaches generally have a high diagnostic yield in this group of patients (over 90%). It is noted that disease-causing synonymous changes, intronic variants and large insertions/deletions have been previously described in CSNB-implicated genes.[1,6,11]

Clinical management

Presently there is no approved curative treatment for CSNB. Young patients should undergo regular eye examinations with refraction to monitor visual development. Coincident high myopia or hypermetropia should be managed with glasses or contact lenses. Amblyopia management is required in some cases, especially when there is an associated ocular misalignment (strabismus).

In the absence of obvious fundus abnormalities, the diagnosis of CSNB can be challenging and initial misdiagnosis or delayed diagnosis is common. Idiopathic nystagmus, myopia or retinal dystrophy may be suspected (Table 13A.2). Full-field ERGs can help establish or refine the diagnosis. Notably, there should be a low threshold for screening CSNB-implicated genes in infants with nystagmus and/or high myopia.

TABLE 13A.2 Main differential diagnoses for congenital stationary night-blindness (CSNB).

Conditions with infantile nystagmus
- severe early childhood onset retinal dystrophies (including Leber congenital amaurosis)
- cone dysfunction disorders (including achromatopsia)
- ocular or oculocutaneous albinism
- optic nerve hypoplasia
- infantile idiopathic nystagmus

Conditions with night vision abnormalities
- inherited: retinitis pigmentosa, choroideremia
- acquired: vitamin A deficiency, melanoma-associated retinopathy

Conditions with electronegative ERGs
- inherited: X-linked retinoschisis, juvenile Batten disease
- acquired: retinal ischaemia, melanoma-associated retinopathy, carcinoma-associated retinopathy (rarely), toxic retinopathy (quinine, methanol), birdshot retinochoroidopathy, posterior uveitis

Conditions resembling fundus albipunctatus
- retinitis punctata albescens (including Bothnia dystrophy)
- benign familial fleck retina
- vitamin A deficiency

Conditions resembling Oguchi disease
- X-linked retinoschisis
- carrier status of X-linked retinitis pigmentosa

Adapted from Ref. 1.

References

1. Zeitz C, Robson AG, Audo I. Congenital stationary night blindness: an analysis and update of genotype-phenotype correlations and pathogenic mechanisms. *Prog Retin Eye Res* 2015;**45**:58–110. https://doi.org/10.1016/j.preteyeres.2014.09.001.
2. Miyake Y, Yagasaki K, Horiguchi M, Kawase Y, Kanda T. Congenital stationary night blindness with negative electroretinogram. A new classification. *Archiv Ophthalmol* 1986;**104**(7):1013–20.
3. Sergouniotis PI, Robson AG, Li Z, Devery S, Holder GE, Moore AT, et al. A phenotypic study of congenital stationary night blindness (CSNB) associated with mutations in the GRM6 gene. *Acta Ophthalmol* 2012;**90**(3):e192–7.
4. Sergouniotis PI, Sohn EH, Li Z, McBain VM, Wright G, Moore AT, et al. Phenotypic variability in RDH5 retinopathy (Fundus albipunctatus). *Ophthalmology* 2011;**118**(8):1661–70.
5. Sergouniotis PI, Davidson AE, Sehmi K, Webster AR, Robson AG, Moore AT. Mizuo-Nakamura phenomenon in Oguchi disease due to a homozygous nonsense mutation in the SAG gene. *Eye* 2011;**8**:1098–101.
6. Bijveld MM, Florijn RJ, Bergen AA, van den Born LI, Kamermans M, Prick L, et al. Genotype and phenotype of 101 dutch patients with congenital stationary night blindness. *Ophthalmology* 2013;**120**(10):2072–81. https://doi.org/10.1016/j.ophtha.2013.03.002.

7. Ba-Abbad R, Holder GE, Robson AG, Neveu MM, Waseem N, Arno G, et al. Isolated rod dysfunction associated with a novel genotype of CNGB1. *Am J Ophthalmol Case Rep* 2019;**14**:83–6.

8. Katagiri S, Hayashi T, Nakamura M, Mizobuchi K, Gekka T, Komori S, et al. RDH5-related fundus albipunctatus in a large Japanese cohort. *Invest Ophthalmol Vis Sci* 2020;**61**(3):53. https://doi.org/10.1167/iovs.61.3.53.

9. Nishiguchi KM, Ikeda Y, Fujita K, Kunikata H, Akiho M, Hashimoto K, et al. Phenotypic features of Oguchi disease and retinitis pigmentosa in patients with S-antigen mutations: a long-term follow-up study. *Ophthalmology* 2019;**126**(11):1557–66. https://doi.org/10.1016/j.ophtha.2019.05.027.

10. Mechaussier S, Almoallem B, Zeitz C, Van Schil K, Jeddawi L, Van Dorpe J, et al. Loss of function of RIMS2 causes a syndromic congenital cone-rod synaptic disease with neurodevelopmental and pancreatic involvement. *Am J Hum Genet* 2020;**106**(6):859–71.

11. Zeitz C, Michiels C, Neuille M, Friedburg C, Condroyer C, Boyard F, et al. Where are the missing gene defects in inherited retinal disorders? Intronic and synonymous variants contribute at least to 4% of CACNA1F-mediated inherited retinal disorders. *Hum Mutat* 2019;**40**(6):765–87.

13A.5

Leber congenital amaurosis and severe early childhood onset retinal dystrophies

Bart P. Leroy and Panagiotis I. Sergouniotis

Leber congenital amaurosis (LCA) is a group of severe congenital retinal disorders, first described by Theodor Leber in 1869.[1] LCA is characterised by either no or very poor visual function from birth; involuntary eye movements and, in most instances, an undetectable electroretinogram (ERG) are noted typically in the first few months of life. The term severe early childhood onset retinal dystrophy (SECORD) is used to describe a group of related conditions characterised by onset of visual impairment before the age of 5 years. There is a significant overlap between the molecular causes of LCA and SECORD, with several genes associated with both clinical phenotypes.[2] Given these common mechanistic underpinnings, the term LCA/SECORD will be used thereafter to denote conditions in the most severe end of the retinal dystrophy spectrum.

In most cases of LCA/SECORD, disease is confined to the eye. However, syndromic subtypes have been described and early genetic testing can help identify children who should have baseline or ongoing systemic investigations.[2]

The differential diagnosis of nystagmus and visual impairment in infancy is broad and, in addition to LCA/SECORD, includes cone dysfunction disorders (achromatopsia or S-cone monochromacy; Section 13A.6), congenital stationary night-blindness (Section 13A.4), optic nerve hypoplasia, partial albinism (Chapter 18), ciliopathies (Section 13B.1), and neurometabolic disorders (Section 13B.2).

Clinical characteristics

Patients with LCA/SECORD present in the first year(s) of life. Initial clinical features include poor visual behaviour, sensory nystagmus or roving eye movements and a markedly reduced or undetectable ERG response to a full-field flash stimulus. Abnormal pupillary light responses and a natural tendency to push the eye (oculodigital sign), may also be observed; the latter may lead to deep-set eyes due to progressive atrophy of retro-ocular orbital adipose tissue and/or keratoconus later in life. Children with residual visual function, can either show photophobia (e.g. in *GUCY2D*-related LCA) or night-blindness (e.g. in *RPE65*-related LCA).[3]

The degree of visual impairment and the disease natural history vary significantly among individuals with LCA/SECORD. In general though, when affected children are old enough to be reliably tested, their vision is in the region of 1.3 LogMAR to light perception, often with a hypermetropic correction (although a subset of patients has no perception of light from birth). To a certain extent, the visual potential depends on the gene involved with some genotypes leading to more severe phenotypes than others. For example, mutations in *CEP290*, *GUCY2D* and *RDH12* tend to lead to more significant visual loss compared to mutations in *RPE65*.[2,4] Also, a fairly stable clinical course is common in *GUCY2D*-related LCA while progression of the disease over several decades has been observed in patients with mutations in *RDH12*, *RPE65* and *CRB1*.

Fundoscopy can be unremarkable at presentation or show a variety of retinal abnormalities such as pigmentary retinopathy (including RPE mottling or intraretinal pigmentation of the nummular and/or spicular type), yellow/white deposits at the level of the RPE, reduced retinal vascular calibre and macular atrophy (including central 'colobomatous' abnormalities or bull's eye macular lesions). Patients presenting with normal fundus often subsequently develop signs such as vascular attenuation, RPE mottling, and optic disc pallor. Characteristic retinal phenotypes are noted in some genetic subtypes (Table 13A.3, Fig. 13A.15, Fig. 13A.16, Fig. 13A.17).

Molecular pathology

Like other retinal dystrophies, the genetic background of LCA/SECORD is diverse. Mutations in over 25 genes account for up to 80% of non-syndromic cases. Most of these genes are expressed solely or predominantly in the retina or the RPE. The encoded proteins have diverse functions; the principal pathways affected include the phototransduction cascade (e.g. *GUCY2D*) the visual cycle (e.g. *RDH12*, *RPE65*), retinal patterning (e.g. *CRB1*) and ciliary structure and transport (e.g. *CEP290*).[2,5]

In populations of European ancestries, the most commonly mutated genes in LCA/SECORD patients include: *CEP290*, *GUCY2D*, *CRB1*, *RDH12*, and *RPE65*. Most

TABLE 13A.3 Characteristic retinal phenotypes in selected genetic subtypes of Leber congenital amaurosis and severe early childhood onset retinal dystrophy.

Associated gene	Clinical feature
CEP290	relatively normal fundus at presentation (Fig. 13A.15A); relatively preserved macular structure on OCT
GUCY2D	relatively normal fundus at presentation (Fig. 13A.15B); relatively preserved macular structure on OCT; significant photophobia
CRB1	peripheral nummular pigmentation at the level of the RPE; increased retinal thickness and abnormal retinal architecture on OCT (Fig. 13A.16); Coats-like vasculopathy or a later-onset maculopathy (including foveal schisis) in a minority of patients with CRB1-retinopathy (Fig. 13A.15C).
RDH12, NMNAT1	early progressive macular atrophy with macular excavation on OCT; dense intraretinal pigmentation developing over time (Fig. 13A.15D); milder phenotype in a minority of NMNAT1-retinopathy patients.
RPE65, LRAT	relatively normal fundus at presentation; relatively preserved macular structure on OCT; profound night-blindness; reduced or undetectable short-wavelength autofluorescence signal (Fig. 13A.17)

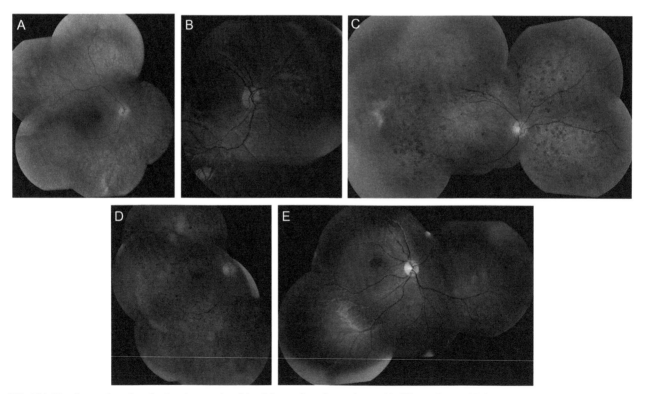

FIG. 13A.15 Composite colour fundus photographs of the right eye from five patients with different forms of Leber congenital amaurosis and severe early childhood onset retinal dystrophy (LCA/SECORD). (A) Patient with *CEP290*-associated LCA/SECORD at age 18 years. The macular area appears unaffected but mid-peripheral changes (that were not present in early childhood) have started to emerge. (B) Patient with *GUCY2D*-associated LCA/SECORD at age 9 years. The macula and nasal retina appear preserved except for some slight RPE mottling in the mid-periphery. (C) Patient with *CRB1*-associated LCA/SECORD at age 27 years. The characteristic features of this condition are noted including yellowish discolouration of the macula, clumped pigmentary changes with a nummular appearance (i.e. resembling the shape of a coin), periarteriolar sparing of retinal tissue, the greyish hue of outer retinal atrophy and intraretinal exudation due to a Coats-like vascular anomaly in the temporal periphery. See also Fig. 13A.16. (D) Patient with *RDH12*-associated LCA/SECORD at age 8 years. There is a yellowish discolouration of the macula in keeping with macular atrophy, a universal clinical finding in children with biallelic mutations in *RDH12*. Outer retinal atrophy with bone-spicule pigment migration in the mid-peripheral retina is also noted. The peripapillary retina appears relatively preserved. (E) Patient with *RPE65*-associated LCA/SECORD at age 9 years. The retina appears relatively preserved although a mild greyish hue and a few fine white spots are noted in the mid-periphery. See also Fig. 13A.17.

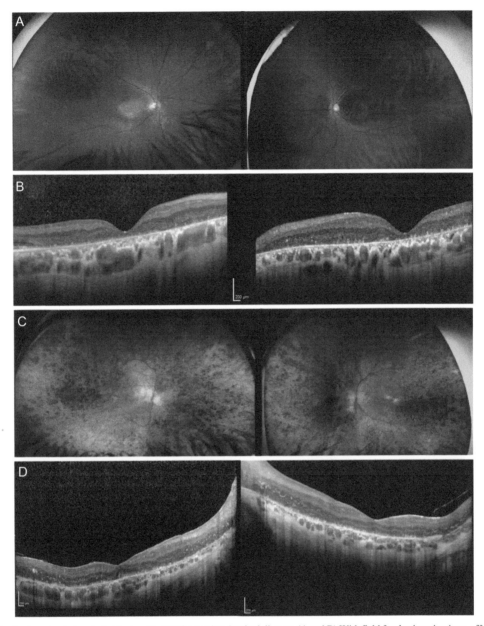

FIG. 13A.16 Fundus imaging in two individuals with *CRB1*-associated retinal disease. (A and B) Widefield fundus imaging in an affected child reveals macular abnormalities and an preserved peripheral retina. OCT highlights the loss of the outer retinal hyper-reflective bands; increased choroidal reflectance due to a 'window defect' effect caused by overlying RPE atrophy is also noted. The laminar structure of the retina appears disorganised and the inner retina appears thickened. (C and D) Widefield fundus imaging in an affected adult reveals nummular (round) intraretinal pigmentation. On OCT, there is evidence of significant photoreceptor and RPE loss; signs of inner retinal thickening and abnormal lamination are also noted. In general, early macular RPE alterations, nummular pigmentary changes and an abnormal retinal architecture on OCT are common features of *CRB1*-retinopathy.

types of LCA/SECORD are inherited as an autosomal recessive trait, although some are autosomal dominant (notably *CRX*- and *IMPDH1*-retinopathy).

Targeted (panel-based analysis) or comprehensive (exome or genome sequencing) genetic testing should be offered to individuals and families with LCA/SECORD. It is worth highlighting that several large insertions/deletions and disease-causing intronic variants have been described; an example of the latter is a deep intronic change in *CEP290* (c.2991+1655A>G) which is one of the most common LCA-causing mutations in certain populations.[6] These observations should be taken into account when requesting or designing genetic tests for patients with LCA/SECORD.

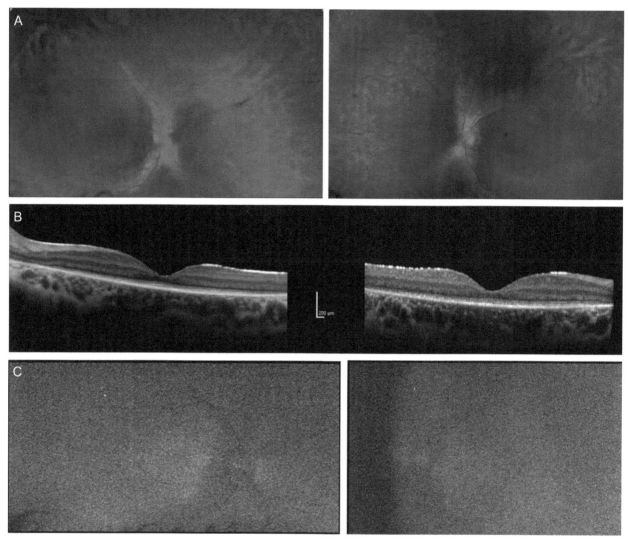

FIG. 13A.17 Fundus imaging in a child with *RPE65*-associated retinal disease. (A) Widefield fundus imaging reveals a relatively preserved retinal appearance with a few fine white spots nasally. (B) OCT reveals a relatively preserved central retinal structure although the hyper-reflective outer retinal bands appear irregular. (C) Fundus autofluorescence imaging highlights signal artefacts and points to severely diminished autofluorescence. This is observed as a consequence of the enzymatic blockade in the visual cycle that *RPE65* mutations can cause.

LCA/SECORD can be one of the first presenting features of a syndromic condition such as a ciliopathy (e.g. Joubert, Senior-Løken or Alström, syndrome; >30 genes implicated) or a neurometabolic disorder (e.g. neuronal ceroid lipofuscinoses or Refsum disease; >15 genes implicated). Genetic testing in children with LCA/SECORD can predict the development of extraocular involvement or point to mild, unrecognised, or previously disregarded syndromic features. This often leads to downstream clinical management that can have important implications for the child's health and development.[7,8]

A key driver for genetic testing in individuals with LCA/SECORD is the advent of gene-directed interventions including gene augmentation therapies and antisense oligonucleotide treatments. Clinical trials for multiple disease subtypes are in progress (see below) and genetic analysis is required to identify affected individuals for whom these treatments might be relevant.

Clinical management

Given the significant effect of congenital or early-onset blindness on quality of life, scientific efforts to better understand the disease pathogenesis in these conditions are numerous. *RPE65*-related LCA/EORD has been the subject of the development of the first successful ophthalmic gene therapy, which has, in turn, led to the first certified gene therapy available in the clinical setting (see also Chapter 8).[9] Ongoing research aims to develop treatments for *GUYC2D*-, *CEP290*- and *RDH12*-related LCA/EORD, using technologies such as gene augmentation, antisense oligonucleotides and CRISPR/Cas9 genome editing (up-to-date information on these can be found on the ClinicalTrials.gov platform).[10–18]

References

1. Leber T. Uber retinitis pigmentosa und angeborene amaurose. *Graefes Arch Klin Exp Ophthalmol* 1869;**15**:13–20.
2. Kumaran N, et al. Leber congenital amaurosis/early-onset severe retinal dystrophy: clinical features, molecular genetics and therapeutic interventions. *Br J Ophthalmol* 2017;**101**(9):1147–54.
3. Hanein S, et al. Leber congenital amaurosis: comprehensive survey of the genetic heterogeneity, refinement of the clinical definition, and genotype-phenotype correlations as a strategy for molecular diagnosis. *Hum Mutat* 2004;**23**(4):306–17.
4. Walia S, et al. Visual acuity in patients with Leber's congenital amaurosis and early childhood-onset retinitis pigmentosa. *Ophthalmology* 2010;**117**(6):1190–8.
5. den Hollander AI, et al. Leber congenital amaurosis: genes, proteins and disease mechanisms. *Prog Retin Eye Res* 2008;**27**(4):391–419.
6. den Hollander AI, et al. Mutations in the CEP290 (NPHP6) gene are a frequent cause of Leber congenital amaurosis. *Am J Hum Genet* 2006;**79**(3):556–61.
7. Ellingford JM, et al. Pinpointing clinical diagnosis through whole exome sequencing to direct patient care: a case of Senior-Loken syndrome. *Lancet* 2015;**385**(9980):1916.
8. Lenassi E, et al. Clinical utility of genetic testing in 201 preschool children with inherited eye disorders. *Genet Med* 2020;**22**(4):745–51.
9. Russell S, et al. Efficacy and safety of voretigene neparvovec (AAV2-hRPE65v2) in patients with RPE65-mediated inherited retinal dystrophy: a randomised, controlled, open-label, phase 3 trial. *Lancet* 2017;**390**(10097):849–60.
10. Cideciyan AV, et al. Effect of an intravitreal antisense oligonucleotide on vision in Leber congenital amaurosis due to a photoreceptor cilium defect. *Nat Med* 2019;**25**(2):225–8.
11. Cideciyan AV, Jacobson SG. Leber Congenital Amaurosis (LCA): potential for improvement of vision. *Invest Ophthalmol Vis Sci* 2019;**60**(5):1680–95.
12. McCullough KT, et al. Somatic gene editing of GUCY2D by AAV-CRISPR/Cas9 alters retinal structure and function in mouse and macaque. *Hum Gene Ther* 2019;**30**(5):571–89.
13. Maeder ML, et al. Development of a gene-editing approach to restore vision loss in Leber congenital amaurosis type 10. *Nat Med* 2019;**25**(2):229–33.
14. Garafalo AV, et al. Progress in treating inherited retinal diseases: early subretinal gene therapy clinical trials and candidates for future initiatives. *Prog Retin Eye Res* 2019;**77**:100827.
15. Lee JH, et al. Gene therapy for visual loss: opportunities and concerns. *Prog Retin Eye Res* 2019;**68**:31–53.
16. Burnight ER, et al. Using CRISPR-Cas9 to generate gene-corrected autologous iPSCs for the treatment of inherited retinal degeneration. *Mol Ther* 2017;**25**(9):1999–2013.
17. Ruan GX, et al. CRISPR/Cas9-mediated genome editing as a therapeutic approach for Leber Congenital Amaurosis 10. *Mol Ther* 2017;**25**(2):331–41.
18. Daich Varela M, Cabral de Guimaraes TA, Georgiou M, Michaelides M. Leber congenital amaurosis/early-onset severe retinal dystrophy: current management and clinical trials. *Br J Ophthalmol* 2021. https://doi.org/10.1136/bjophthalmol-2020-318483, bjophthalmol-2020-318483.

13A.6

Cone dysfunction disorders

Nashila Hirji, Michalis Georgiou, and Michel Michaelides

Cone dysfunction disorders are a group of clinically and genetically heterogeneous retinal conditions that are associated with significant visual impairment. Affected individuals typically present in early childhood with nystagmus, light hypersensitivity (photophobia), subnormal visual behaviour and impaired colour vision. These abnormalities result from the dysfunction of cone photoreceptor cells, which are mainly concentrated at the fovea. In contrast to cone/cone-rod dystrophies (Section 13A.7) and severe early childhood onset retinal dystrophies (SECORD; Section 13A.5) which are progressive, cone dysfunction disorders are considered to be primarily stationary. The clinical presentation of infants affected by these three groups of conditions can occasionally appear similar; early genetic testing can help obtain a timely diagnosis, inform genetic counselling and provide insights into prognosis.[1]

Autosomal recessive and X-linked subtypes of cone dysfunction disorders have been described and more than 9 genes have been implicated. Extraocular manifestations are typically not a feature of these conditions and the main disease subtypes are achromatopsia (also known as rod monochromacy), blue cone monochromatism (also known as S-cone monochromacy) and bradyopsia (also known as *RGS9/R9AP*-retinopathy).[1] These disorders share certain electroretinographic (ERG) findings: there is generalised reduction or absence of light-adapted (cone system-mediated) responses but the dark-adapted (rod system-mediated) responses are normal or well-preserved. Electrodiagnostic testing using extended protocols (beyond the standards set by the International Society for Clinical Electrophysiology of Vision, ISCEV)[2–4] is often recommended as it can support differential diagnosis and functional monitoring in this group of conditions (see also Chapter 6).

Clinical characteristics

Achromatopsia

Achromatopsia is the most common cone dysfunction disorder. It is inherited as an autosomal recessive trait and it is associated with lack of function of all three types of cone photoreceptors: red (L-, long-wavelength sensitive), green (M-, middle-wavelength sensitive) and blue (S-, short-wavelength sensitive) cones. Affected children typically present with nystagmus (often pendular, high frequency and low amplitude) and marked photophobia from birth or early infancy. Reduced visual acuity, markedly impaired or absent colour vision, and central scotomata are also present.[5]

There is a range of phenotypic severity and the terms 'complete' and 'incomplete' have been used to describe recognisable forms. Individuals with complete achromatopsia have little or no residual cone function; the visual acuity is typically 1.0 LogMAR or worse and there is total absence of colour perception. Paradoxical pupillary constriction to darkness is often present, i.e. there is pupillary constriction on transfer to darkness after 2–3 min of exposure to bright light. Incomplete achromatopsia is less common than the complete form. Affected individuals have residual cone function and therefore display a milder phenotype with visual acuity between 0.6 and 0.8 LogMAR and residual colour discrimination. Nystagmus and photophobia may or may not be present.

Achromatopsia is frequently associated with hypermetropic refractive errors; less commonly, myopia may be present. The visual acuity tends to be stable (hence the condition is considered functionally non-progressive), but nystagmus and photophobia may improve over time, with nystagmus often being absent in later adulthood.

On clinical examination, the fundus usually appears normal. However, some affected individuals may show subtle abnormalities at the macula, such as central pigment mottling, and/or loss of the foveal reflex. Less commonly, RPE atrophy may be evident at the fovea. Electrodiagnostic testing reveals non-recordable or markedly reduced light-adapted responses and normal or near-normal dark-adapted responses (Fig. 6.9).

Four distinct fundus autofluorescence phenotypes have been reported: normal appearance (Fig. 13A.18A), increased signal centrally (Fig. 13A.18B), reduced signal centrally (Fig. 13A.18C), and central area of decreased signal with a surrounding ring of hyperautofluorescence (Fig. 13A.18D).[1,5]

OCT imaging of the macula can be used to grade achromatopsia into five grades:

- continuous ellipsoid zone (Fig. 13A.18E),
- disrupted ellipsoid zone (Fig. 13A.18F),

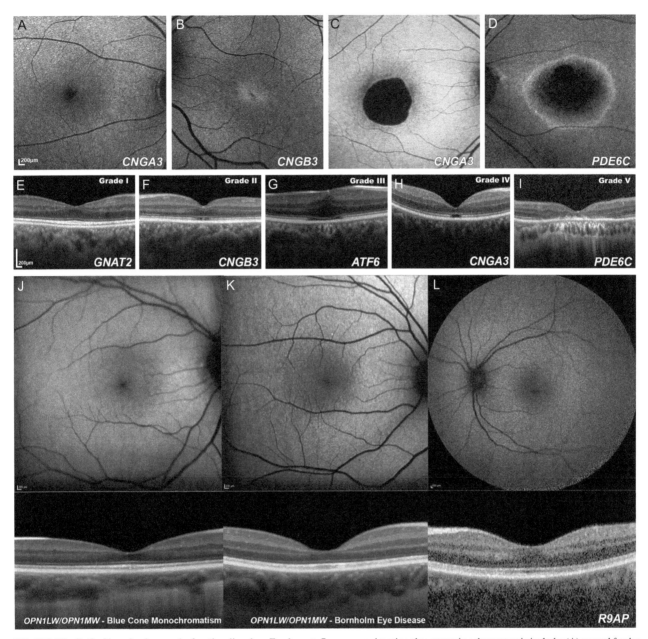

FIG. 13A.18 Retinal imaging in cone dysfunction disorders. Fundus autofluorescence imaging phenotypes in achromatopsia include: (A) normal fundus autofluorescence appearance, (B) central hyperautofluorescence, (C) reduced fundus autofluorescence signal centrally, and (D) a central area of decreased signal, with a surrounding ring of hyperautofluorescence. OCT grading in achromatopsia: (E) Grade I: continuous ellipsoid zone, (F) Grade II: disrupted ellipsoid zone, (G) Grade III: absent ellipsoid zone, (H) Grade IV: presence of a hyporeflective zone in the fovea, and (I) Grade V: outer retinal atrophy with RPE loss. Patients (F) and (G) have foveal hypoplasia, with retention of inner retinal layers at the fovea. Fundus autofluorescence imaging with corresponding horizontal transfoveal OCT: (J) Blue cone monochromatism, (K) Bornholm eye disease, (L) *RGS9/R9AP*-associated retinopathy ('bradyopsia').

- absent ellipsoid zone (Fig. 13A.18G),
- presence of a hyporeflective zone in the fovea (also described in the biomedical literature as 'foveal cavitation' or 'optical gap'[6]) (Fig. 13A.18H; Fig. 13A.19), and
- outer retinal atrophy with RPE loss (Fig. 13A.18I).

Preservation of the inner retinal layers over the fovea (foveal hypoplasia) is a common OCT finding (Fig. 13A.18G–F).[5] Notably, observations in a large cohort of achromatopsia patients with substantial follow-up, support the notion that the condition is predominantly stable in the vast majority of cases.[7,8]

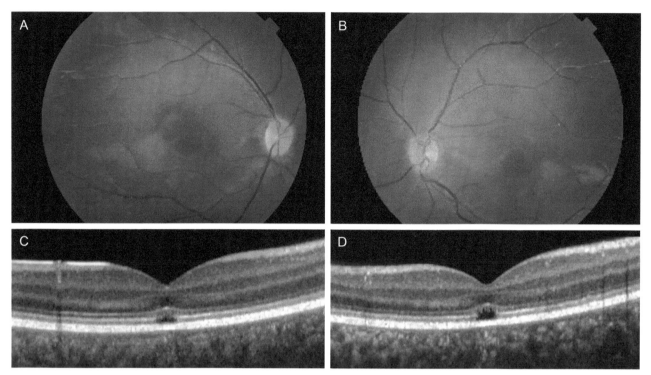

FIG. 13A.19 Imaging findings in a 13-year-old female with *CNGB3*-associated achromatopsia. (A and B) Colour fundus photographs demonstrating normal retinal appearances bilaterally. (C and D) OCT scans showing focal disruption of the foveal ellipsoid zone bilaterally. This corresponds to a loss of cone photoreceptors which are normally present at high density at the fovea.

Blue cone monochromatism

Blue cone monochromatism is an X-linked condition resulting from red (L-) and green (M-) cone photoreceptor dysfunction. This is due to genetic abnormalities in the corresponding opsin genes, both located on the X-chromosome. The function of blue (S-) cones (whose opsin is encoded by a gene on chromosome 7) and rods is retained. Affected individuals usually present at birth or early infancy with reduced visual acuity (0.6–0.8 LogMAR), photophobia and nystagmus. Myopia is very common. Fundus examination often reveals a normal myopic retina, or in some cases, macular RPE disturbance and atrophy. Unlike complete achromatopsia, individuals with blue cone monochromatism show significantly reduced, but typically detectable, light-adapted ERG responses. Dark-adapted responses are generally normal.[9,10] Electrodiagnostic testing beyond the ISCEV standard ERG should be considered in patients suspected to have blue cone monochromatism; assessing the function of the blue (S-) cone retinal pathway as part of an extended protocol[3] can help distinguish between this X-linked disorder and achromatopsia (Fig. 6.10).

Colour vision testing (e.g. using the H-R-R pseudo-isochromatic plates) reveals retained tritan (blue, yellow) function and absent protan (red) and deutan (green) colour discrimination. Fundus autofluorescence and OCT imaging can be similar to achromatopsia, with a variable degree of ellipsoid zone disruption (Fig. 13A.18J).[1,9,10]

Bornholm eye disease

Bornholm eye disease was initially reported in a family originating from the Danish island of Bornholm, and is best described as 'X-linked cone dysfunction with dichromacy'. Affected individuals demonstrate X-linked infantile myopia/astigmatism, and reduced visual acuity. Clinical examination can reveal signs of optic nerve hypoplasia and retinal pigmentary changes. ERG reveals moderately reduced light-adapted responses and normal dark-adapted responses. The condition is associated with deuteranopia or protanopia, and so colour vision testing probing these axes is necessary (e.g. using the Ishihara pseudo-isochromatic plates which are designed to screen for red-green colour deficiencies but not for tritan defects). Retinal thinning can be observed on OCT, and fundus autofluorescence imaging can appear normal (Fig. 13A.18K) but findings are variable.[1,11]

Bradyopsia (RGS9/R9AP-associated retinopathy)

Bradyopsia is a rare autosomal recessive retinopathy with a characteristic electrophysiological phenotype. Affected individuals demonstrate difficulty in adapting to sudden changes in luminance levels. Presentation is in early childhood with delayed dark-to-light and light-to-dark adaptation, mild photophobia, moderately reduced visual acuity and normal colour vision. The fundi appear normal on clinical examination and adaptive optics imaging reveals a normal cone photoreceptor mosaic.[1,12] Electrophysiological testing beyond the ISCEV standard enables the identification of the pathognomonic ERG findings of this condition. In general, ISCEV standard ERGs superficially resemble those of incomplete achromatopsia and/or blue cone monochromatism: there is a generalised reduction or absence of light-adapted responses with normal or mildly reduced dark-adapted responses. Extended protocols that should be considered include (i) a red flash dark-adapted (DA red) ERG[3] and (ii) a dark-adapted bright flash (DA 10.0) ERG using the increased interstimulus interval (2 min as opposed to the standard of 20 s). Individuals with bradyopsia have a normal DA red ERG (both the early cone and the later rod system components are preserved); this suggests that dark-adapted cone function is preserved despite the severe cone system abnormalities in light-adapted conditions. Furthermore, the DA 10.0 ERG may appear normal after the first flash but is then mildly attenuated in response to another flash occurring after a standard interstimulus interval (20 s). Increasing the interval to 2 min leads to normalisation of the ERG response. The overall picture is consistent with delayed photoreceptor recovery following each stimulus.[2,3]

Molecular pathology

Achromatopsia

Achromatopsia is inherited in an autosomal recessive manner. Disease-causing variants in six genes are currently thought to be responsible for over 90% of cases of achromatopsia. CNGA3 and CNGB3 mutations account for approximately 80% of cases. Variants in GNAT2, PDE6H, PDE6C and ATF6 are responsible for approximately 2% of cases each. CNGA3, CNGB3, GNAT2, PDE6H and PDE6C encode proteins involved in the cone-specific phototransduction cascade. ATF6 is implicated in endoplasmic reticulum homeostasis and is thought to have a role in normal foveal development.[1,13]

Most mutations in these achromatopsia-associated genes result in a complete loss of function of the respective protein. A small number of variants (particularly in CNGA3 and GNAT2) result in partial loss of function and may be associated with milder forms including incomplete achromatopsia. By far the most common achromatopsia-associated mutation in individuals of European ancestries is the c.1148delC (p.Thr383fs) variant in CNGB3. In contrast, CNGA3 variants have been found to be the major cause of achromatopsia in cohorts from China[14] and Israel.[15] It is noteworthy that most CNGB3 mutations are nonsense variants, frameshift insertions/deletions or splice site variants; only a small number of missense changes have been observed in this gene. Conversely, most CNGA3 mutations are missense variants.[13]

An intriguing case is the c.1208G>A (p.Arg403Gln) missense change in CNGB3. This variant is relatively prevalent in the general population (~2% allele frequency in people of South Asian ancestries) and has been reported in patients presenting with a rather variable retinal phenotype described as progressive macular dystrophy, macular degeneration, cone dystrophy or achromatopsia. Notably, there is some evidence to suggest that the effect of this variant is unlike that of classic Mendelian mutations.[16]

In general, individuals with mutations in GNAT2 typically have unremarkable OCT imaging with a continuous ellipsoid zone.[17] In contrast, mutations in PDE6C (Fig. 13A.18I) and ATF6 (Fig. 13A.18G) tend to be associated with more severe OCT phenotypes (with absent ellipsoid zone ± outer retinal atrophy). PDE6C-associated achromatopsia can also be slowly progressive. It is worth highlighting that certain PDE6C and GNAT2 genotypes show a significant phenotypic overlap with blue cone monochromatism, with a high incidence of myopia and relatively preserved blue (S-) cone ERGs.[1,17–19]

Blue cone monochromatism

Blue cone monochromatism is inherited in an X-linked recessive manner. It is caused by genetic variants in the red (L-) and green (M-) cone opsin gene array and affects the corresponding cone photoreceptor types. This gene cluster is located on the X chromosome (Xq28) and includes the OPN1LW gene (encoding red/L-cone opsin) and often multiple copies of the OPN1MW gene (encoding the green/M-cone opsin) arranged in a tandem array. In general, only the first two genes of the array appear to be expressed. There are three broad genetic mechanisms by which blue cone monochromatism can arise:

(i) deletions confined to the Locus Control Region (LCR), a specialised regulatory sequence that is located upstream of the OPN1LW and OPN1MW genes and is necessary for their expression;
(ii) a two-step process mechanism where there is recombination (i.e. exchange of DNA segments) between the OPN1LW and OPN1MW genes leading to the formation of a hybrid OPN1LW/OPN1MW gene. This is followed by a subsequent mutation in this hybrid gene (commonly the c.607T>C, (p.Cys203Arg) change or a large deletion[9]).

(iii) specific rare combinations of single nucleotide polymorphisms in exon 3 of the *OPN1LW* and *OPN1MW* genes ('L/M interchange haplotypes'); such variants cause irregular splicing of these genes with a variable degree of exon 3 skipping. The most common disease-associated variant combination is designated LIAVA, where L is the amino-acid leucine, V is valine, A is alanine and I is Isoleucine.[1,11]

The complexity of these mechanisms and the high degree of DNA sequence similarity between the *OPN1LW* and *OPN1MW* genes make genetic testing for this condition using standard approaches challenging.

Bornholm eye disease

Bornholm eye disease was initially mapped to the *OPN1LW* and *OPN1MW* gene cluster on the X chromosome in the original Danish family. Further investigation revealed rare genotypes in exon 3 of the *OPN1LW* gene (including the LIAVA variant combination discussed in the blue cone monochromatism section). It is worth highlighting that these genotypes have been described in isolated red-green dichromacy, i.e. the state in which only two of the three cone subtypes are functional, either red (L-) and blue (S-) cones (deuteranopia) or green (M-) and blue (S-) (protanopia). It is hypothesised that a skewed red (L-) to green (M-) cone ratio results in cone dysfunction if the most abundant cone subtype is inactivated by the above genetic changes. It has been argued that Bornholm eye disease differs from these common forms of dichromacy in which only red-green colour vision is affected but visual acuity is fully preserved.[1,11] However, its status as an independent clinical entity remains unclear.

Bradyopsia (RGS9/R9AP-associated retinopathy)

Bradyopsia is associated with biallelic variants in either the *RGS9* or the *R9AP* gene. The proteins encoded by these genes enable the return of the phototransduction cascade to its resting state quickly after light exposure. More specifically, *RGS9* encodes a signalling protein that is anchored to the photoreceptor outer segment membrane by the *R9AP* protein; the latter is thought to enhance the activity of the former by up to 70-fold. Generally, the prolonged recovery phase associated with dysfunction of these molecules is reflected on the electrophysiological findings of the condition (e.g. extended interstimulus intervals are required to obtain full recovery of the ERG following the previous flash).[1,12]

Clinical management

Although cone dysfunction disorders represent a significant cause of lifelong visual impairment, there are currently no curative treatments. At present, the mainstay of managing these conditions involves optimisation of residual vision, through correcting refractive errors, provision of low vision aids and educational support. This requires a multidisciplinary approach. Tinted glasses or contact lenses may help reduce the photophobia that affected individuals frequently experience, and improve comfort and vision in bright conditions. In achromatopsia, deep red tints aim to reduce rod saturation whilst maintaining residual cone function. Magenta tints in blue cone monochromatism preserve transmission of blue light, which activates the residual S-cone photoreceptors.

There has been a substantial amount of work aiming to identify treatments that will improve the quality of life of individuals with cone dysfunction disorders. Obtaining an accurate molecular diagnosis is of marked importance in guiding potential gene-based therapeutic approaches such as gene supplementation. Gene supplementation involves supplying affected cone photoreceptor cells with healthy copies of the mutated gene, in an attempt to restore normal function. Promising results for this approach have been shown in animal models of *CNGA3*-, *CNGB3*- and *GNAT2*-associated achromatopsia.[1] This has led to the initiation of translational studies investigating gene supplementation in affected humans.[20] The results of clinical trials evaluating this form of treatment may revolutionise the future of achromatopsia management. Gene therapy may similarly become applicable to other cone dysfunction disorders in future. Further research in this field, together with evaluation of other possible modes of intervention such as neuroprotection and pharmacological treatments, may ultimately lead to therapies which improve sight in affected individuals.

References

1. Aboshiha J, Dubis AM, Carroll J, Hardcastle AJ, Michaelides M. The cone dysfunction syndromes. *Br J Ophthalmol* 2016;**100**:115–21. https://doi.org/10.1136/bjophthalmol-2014-306505.
2. Robson AG, Nilsson J, Li S, Jalali S, Fulton AB, Tormene AP, et al. ISCEV guide to visual electrodiagnostic procedures. *Doc Ophthalmol* 2018;**136**(1):1–26. https://doi.org/10.1007/s10633-017-9621-y.
3. Perlman I, Kondo M, Chelva E, Robson AG, Holder GE. ISCEV extended protocol for the S-cone ERG. *Doc Ophthalmol* 2020;**140**(2):95–101.
4. Thompson DA, Fujinami K, Perlman I, Hamilton R, Robson AG. ISCEV extended protocol for the dark-adapted red flash ERG. *Doc Ophthalmol* 2018;**136**(3):191–7.
5. Hirji N, Aboshiha J, Georgiou M, Bainbridge J, Michaelides M. Achromatopsia: clinical features, molecular genetics, animal models and therapeutic options. *Ophthalmic Genet* 2018;**39**:149–57.
6. Oh JK, Ryu J, Lima de Carvalho Jr JR, Levi SR, Lee W, Tsamis E, et al. Optical gap biomarker in cone-dominant retinal dystrophy. *Am J Ophthalmol* 2020;**218**:40–53. https://doi.org/10.1016/j.ajo.2020.05.016.

7. Hirji N, Georgiou M, Kalitzeos A, Bainbridge JW, Kumaran N, Aboshiha J, et al. Longitudinal assessment of retinal structure in achromatopsia patients with long-term follow-up. *Invest Ophthalmol Vis Sci* 2018;**59**(15):5735–44.
8. Georgiou M, Singh N, Kane T, Zaman S, Hirji N, Aboshiha J, et al. Long-term investigation of retinal function in patients with achromatopsia. *Invest Ophthalmol Vis Sci* 2020;**61**(11):38.
9. Sumaroka A, Garafalo AV, Cideciyan AV, Charng J, Roman AJ, Choi W, et al. Blue cone Monochromacy caused by the C203R missense mutation or large deletion mutations. *Invest Ophthalmol Vis Sci* 2018;**59**(15):5762–72.
10. Luo X, Cideciyan AV, Iannaccone A, Roman AJ, Ditta LC, Jennings BJ, et al. Blue cone monochromacy: visual function and efficacy outcome measures for clinical trials. *PLoS One* 2015;**10**(4), e0125700.
11. De Silva SR, Arno G, Robson AG, Fakin A, Pontikos N, Mohamed MD, et al. The X-linked retinopathies: physiological insights, pathogenic mechanisms, phenotypic features and novel therapies. *Prog Retin Eye Res* 2020;**26**:100898.
12. Strauss RW, Dubis AM, Cooper RF, Ba-Abbad R, Moore AT, Webster AR, et al. Retinal architecture in RGS9- and R9AP-associated retinal dysfunction (Bradyopsia). *Am J Ophthalmol* 2015;**160**(6):1269–75.
13. Kohl S, Jägle H, Wissinger B, Zobor D. Achromatopsia. 2004 Jun 24 [updated 2018 Sep 20]. In: Adam MP, Ardinger HH, Pagon RA, Wallace SE, LJH B, Stephens K, Amemiya A, editors. *GeneReviews® [Internet]*. Seattle, WA: University of Washington, Seattle; 1993–2020. 20301591.
14. Sun W, Li S, Xiao X, Wang P, Zhang Q. Genotypes and phenotypes of genes associated with achromatopsia: a reference for clinical genetic testing. *Mol Vis* 2020;**26**:588–602.
15. Zelinger L, Cideciyan AV, Kohl S, Schwartz SB, Rosenmann A, Eli D, et al. Genetics and disease expression in the CNGA3 form of achromatopsia: steps on the path to gene therapy. *Ophthalmology* 2015;**122**(5):997–1007.
16. Burkard M, Kohl S, Krätzig T, Tanimoto N, Brennenstuhl C, Bausch AE, et al. Accessory heterozygous mutations in cone photoreceptor CNGA3 exacerbate CNG channel-associated retinopathy. *J Clin Invest* 2018;**128**(12):5663–75.
17. Georgiou M, Singh N, Kane T, Robson AG, Kalitzeos A, Hirji N, et al. Photoreceptor structure in GNAT2-associated achromatopsia. *Invest Ophthalmol Vis Sci* 2020;**61**(3):40.
18. Georgiou M, Robson AG, Singh N, Pontikos N, Kane T, Hirji N, et al. Deep phenotyping of PDE6C-associated achromatopsia. *Invest Ophthalmol Vis Sci* 2019;**60**(15):5112–23.
19. Daich Varela M, Ullah E, Yousaf S, Brooks BP, Hufnagel RB, Huryn LA. PDE6C: novel mutations, atypical phenotype, and differences among children and adults. *Invest Ophthalmol Vis Sci* 2020;**61**(12):1.
20. Fischer MD, Michalakis S, Wilhelm B, Zobor D, Muehlfriedel R, Kohl S, et al. Safety and vision outcomes of subretinal gene therapy targeting cone photoreceptors in achromatopsia: a nonrandomized controlled trial. *JAMA Ophthalmol* 2020;**138**(6):643–51.

13A.7

Cone/cone-rod dystrophies

Alberta A.H.J. Thiadens and Caroline C.W. Klaver

Cone/cone-rod dystrophies are a heterogeneous group of progressive genetic disorders characterised predominantly by cone photoreceptor degeneration.[1–4] A degree of rod involvement is present in most cases but this tends to occur at a later stage and to be less significant than the cone loss. Distinguishing cone dystrophies (i.e. progressive retinal disorders exclusively affecting the cones) from cone-rod dystrophies (i.e. progressive retinal disorders that affect the cones more than the rods) is often challenging. Although these groups of disorders were once thought to be distinct, they are now often considered to be parts of the same disease spectrum or to represent different stages of a single disease continuum. Given this and the significant overlap between the molecular causes of these two groups of conditions, the term cone/cone-rod dystrophy is preferred and will be used thereafter.

Autosomal dominant, autosomal recessive and X-linked subtypes have been described and more than 30 genes have been implicated in cone/cone-rod dystrophies. Confusingly, some of these genes are also associated with other retinal disorders such as retinitis pigmentosa (Section 13A.1), cone dysfunction disorders (Section 13A.6) and Leber congenital amaurosis (Section 13A.5).

In general, cone/cone-rod dystrophies often present as isolated disorders, but they can also be part of a syndrome as in Bardet–Biedl or Jalili syndromes or spinocerebellar ataxia. Some of these multisystemic disorders are discussed in Section 13B.

Clinical characteristics

Cone/cone-rod dystrophies are characterised by progressive central visual loss, photophobia, and colour vision abnormalities in childhood or early adulthood. Secondary rod system involvement can occur later in the disease process leading to night vision problems and peripheral visual field constriction. Nystagmus is not a typical feature at presentation, as cone function is generally not severely affected in the first few months of life.[1–4]

Fundus findings are variable and usually involve macular changes with or without visible peripheral retinal abnormalities. Common fundus patterns at diagnosis include: granular macular appearance (Fig. 13A.20), bull's eye maculopathy, macular atrophy and subtle/scarce pigment deposition in the peripheral retina. Peripheral RPE atrophy, intraretinal pigment clumping, bone spicule pigmentation and vascular attenuation are often observed in advanced disease.[4] Fundus flecks are occasionally noted; this is a feature suggestive of *ABCA4*-retinopathy, one of the commonest causes of cone/cone-rod dystrophy (discussed separately in Section 13A.8).

OCT and fundus autofluorescence imaging have greatly improved the characterisation of cone/cone-rod dystrophies. A range of abnormalities is seen on OCT imaging. Subtle changes at the level of the photoreceptor outer segments (ellipsoid zone) are often one of the first retinal findings to be recorded in patients who have cone/cone-rod

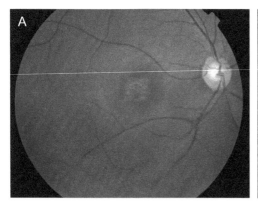
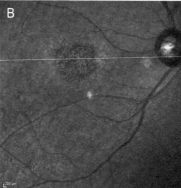
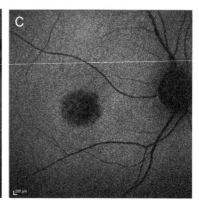

FIG. 13A.20 Fundus photography (A), infrared reflectance imaging (B) and fundus autofluorescence imaging (C) from the right eye of a 43-year-old patient who carries the *GUCY2D* c.2512C>T (p.Arg838Cys) variant in the heterozygous state. This change has been previously reported to be associated with autosomal dominant cone/cone-rod dystrophy and this patient's light-adapted electroretinographic responses were undetectable. Note the granular appearance at the macula. *(Courtesy of Professor Graeme C.M. Black.)*

dystrophy and an apparently normal fundus examination at presentation. A rare but characteristic finding is that of foveal cavitation (oval-shaped hyporeflective foveal lesion with associated ellipsoid zone disruption but with preservation of the external limiting membrane and RPE layers). This OCT feature is associated several other conditions including achromatopsia, S-cone monochromacy and macular dystrophy.[4]

Fundus autofluorescence imaging often highlights abnormalities corresponding to subtle changes on fundoscopy. Three frequently observed patterns include:

- a concentric macular hyperautofluorescent ring surrounding a central area of abnormal signal. This ring corresponds to the transition zone between preserved and degenerating retina. A similar ring is seen in retinitis pigmentosa (Section 13A.1) although in cone/cone-rod dystrophy degeneration occurs within (rather than outside) the ring. This abnormality is associated with several genetic defects including heterozygous mutations in *GUCY2D* (Fig. 13A.20 and Fig. 13A.21) and *CRX* (Fig. 13A.22), and biallelic mutations in *ABCA4*.
- heterogeneous background autofluorescence (in a granular pattern with alternating hyper- and hypoautofluorescent areas) affecting the macula with or without extension beyond the vascular arcades. This pattern is generally non-specific and is commonly seen in patients with *ABCA4*, *PRPH2* or *PCARE* mutations.
- sparing of the peripapillary autofluorescence. This sign is seen characteristically (but not exclusively) in *ABCA4*-retinopathy (see also Section 13A.8).[4]

Molecular pathology

Mutations in more than 30 genes have been shown to cause cone/cone-rod dystrophy. The encoded proteins are involved in a variety of processes including the cone phototransduction cascade (e.g. *GUCY2D*), the visual cycle (e.g. *ABCA4*; see also Fig. 13A.23), photoreceptor development (e.g. *CRX*), and ciliary structure and transport (e.g. *RPGR*) (Fig. 13A.24).[2]

Most individuals with cone/cone-rod dystrophy have autosomal recessive forms. Autosomal dominant inheritance is less common and only a few cases have the X-linked subtype. A significant proportion of cases (>30%) remain unsolved. Commonly mutated genes include *ABCA4* (autosomal recessive), *GUCY2D* (autosomal dominant), *PRPH2* (autosomal dominant), *CRX* (autosomal dominant), *KCNV2* (autosomal recessive) and *RPGR* (X-linked).[3] It is worth highlighting that:

- mutations in *ABCA4* and *PRPH2* are common causes not only of cone/cone-rod dystrophy but also of macular dystrophy, a distinct group of disorders in which there is only central (macular) cone involvement with peripheral cone function remaining within normal limits (see also Sections 13A.8 and 13A.10);

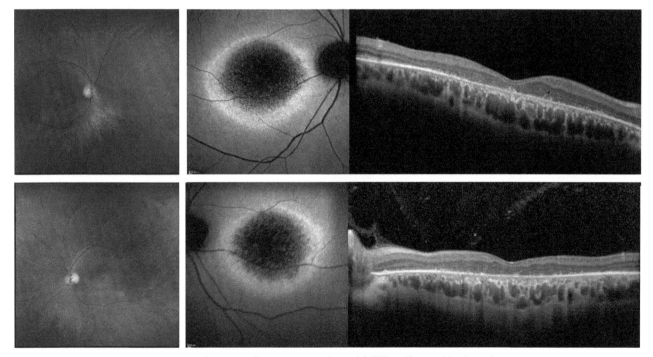

FIG. 13A.21 Multi-colour fundus imaging, fundus autofluorescence imaging and OCT in a 48-year-old patient who carries the *GUCY2D* c.2512C>T (p.Arg838Cys) variant in the heterozygous state. Note the hyperautofluorescent ring that surrounds a central area of abnormal signal. This ring corresponds to the transition zone between the abnormal and preserved retina. OCT highlights severe attenuation of the ellipsoid zone and outer nuclear layer thinning in both maculae. *(Courtesy of Professor Graeme C.M. Black.)*

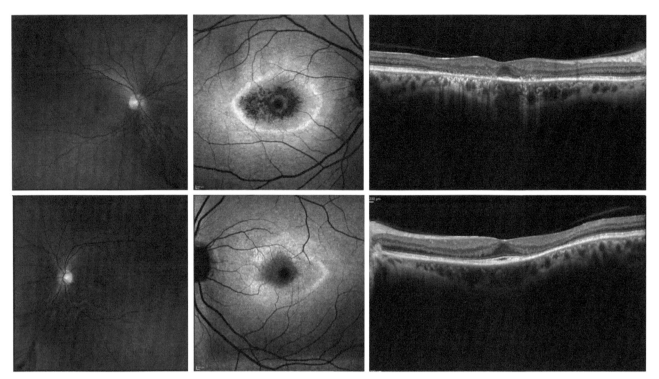

FIG. 13A.22 Multi-colour fundus imaging, fundus autofluorescence imaging and OCT in a 35-year-old patient who carries a nonsense variant in the *CRX* gene in heterozygous state. Note the hyperautofluorescent ring and the relative preservation of the ellipsoid zone at the fovea bilaterally. *(Courtesy of Dr. Panagiotis I. Sergouniotis.)*

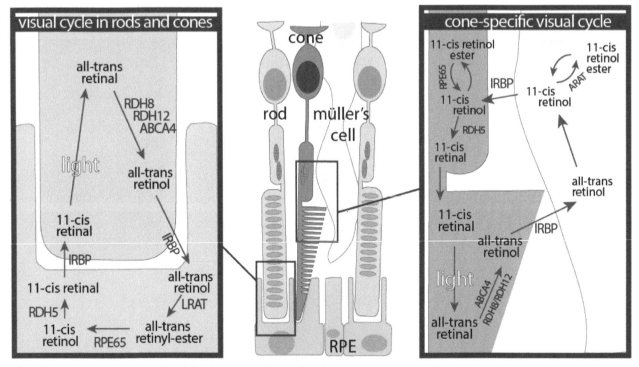

FIG. 13A.23 Schematic representation of the cone and rod visual cycle and outer segment structure. Cones share a visual cycle via the RPE with rods. However, as they function under light conditions, they require a more rapid supply of the light-sensing chromophore 11-cis retinal (vitamin A) and therefore need a second visual cycle via the Müller cells. 11-cis retinal is attached to opsin to form the visual pigment. Cones have three different opsin molecules which correspond to the red, green and blue cone photoreceptor subtypes; rods only have one, rhodopsin. Morphologically, the cone outer segment discs are continuous with each other and with the plasma membrane. In contrast, rod outer segments are physically separated from the cell membrane. This results in a larger cone outer segment surface area and more rapid exchange of substances between the cell interior and exterior.[5] *(Figure by Magda Meester.)*

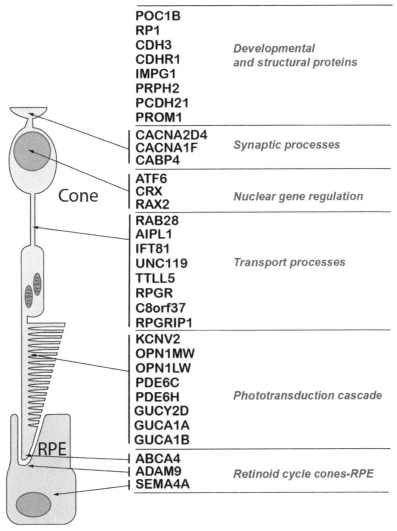

FIG. 13A.24 Schematic overview, function and localisation of a selection of proteins known to be associated with non-syndromic cone/cone-rod dystrophies. *(Figure by Magda Meester & Alberta Thiadens.)*

- biallelic mutations in *GUCY2D* are a common cause of Leber congenital amaurosis (see also Section 13A.5). However, a specific heterozygous missense mutations (mostly affecting the part of the protein around the p.Arg838 residue) are causing cone/cone-rod dystrophy[6] (Fig. 13A.20 and Fig. 13A.21).
- *CRX* is a transcription factor that is key for both the development and the survival of photoreceptors. Heterozygous mutations in this gene lead to a broad spectrum of clinical presentations ranging from Leber congenital amaurosis to adult-onset macular dystrophy (Fig. 13A.22).[7]
- mutations in *KCNV2*, a voltage-gated potassium channel, are associated with a characteristic cone/cone-rod dystrophy known as 'cone dystrophy with supernormal rod responses' (Fig. 13A.25). This childhood-onset, slowly progressive disorder has a characteristic/pathognomonic ERG signature. The light-adapted responses (LA 3 and LA 30 Hz) are reduced and delayed in keeping with generalised cone system dysfunction. The dark-adapted (DA) responses are unique. The rod-mediated dim flash (DA 0.01) ERG is severely delayed and typically of reduced amplitude; in contrast, the dark-adapted bright flash (DA 10) ERG a-wave has a normal amplitude (suggesting unaffected phototransduction) but with a broadened, 'squared' shape ('flattened trough') and it is followed by a b-wave of relatively high amplitude (that may be supernormal).[8]
- *RPGR* is a gene in the X chromosome that encodes two protein isoforms (versions). In the retina, the most common isoform is RPGR^{ORF15} which contains exons 1–14 plus an alternatively spliced version of exon 15, called exon ORF15 (which includes the regular exon

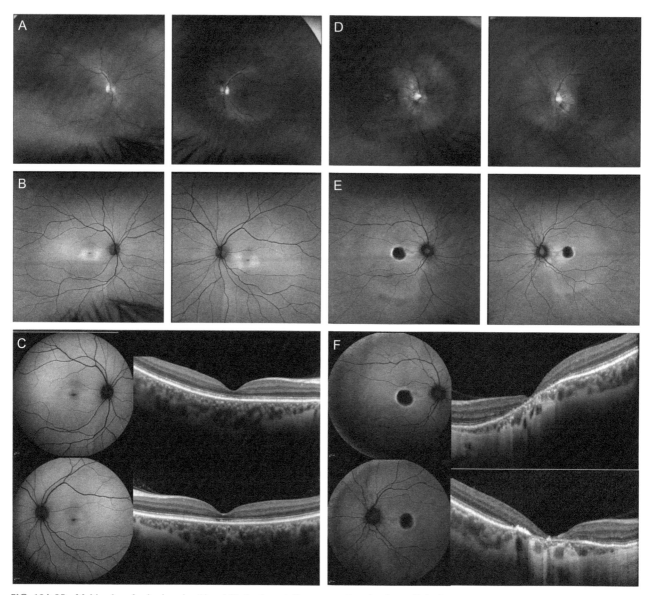

FIG. 13A.25 Multi-colour fundus imaging (A and D), fundus autofluorescence imaging [green light (B and E); blue light (C and F)], and OCT in two siblings who carry biallelic variants in *KCNV2*. Despite the patients having similar age, their phenotypes are different with the individual presented in (D), (E), and (F) having more extensive atrophy and a hyperautofluorescent ring. Note the differences in appearance between the green (B) and blue (C) light autofluorescence images. *(Courtesy of Dr. Eva Lenassi.)*

15 sequences plus part of what is otherwise intron 15). Exon ORF15 is highly repetitive and is a mutational hotspot. Most mutations in *RPGR* (including ORF15 variants) result in X-linked retinitis pigmentosa (see Box 13.1). However, a small number of sequence alterations at the C-terminal region (i.e. towards the end) of the protein lead to cone/cone-rod dystrophy (Fig. 13A.26). High myopia is a common finding in male patients with this disease subtype.

Other genes associated with non-syndromic cone/cone-rod dystrophy are shown in Fig. 13A.24. Several syndromic forms have also been described. Ophthalmic signs and symptoms may come after the systemic features and it is therefore recommended to incorporate a comprehensive ophthalmologic examination into the regular clinical work-up of patients with disorders such as Danon disease, Alström syndrome and Jalili syndrome. Danon disease is a rare X-linked condition consisting of muscle weakness, cardiomyopathy, mental impairment, and early mortality in males. Alström syndrome is an autosomal recessive disorder characterised by hearing loss, obesity, diabetes, short stature, cardiomyopathy, and progressive organ failure that often leads to premature

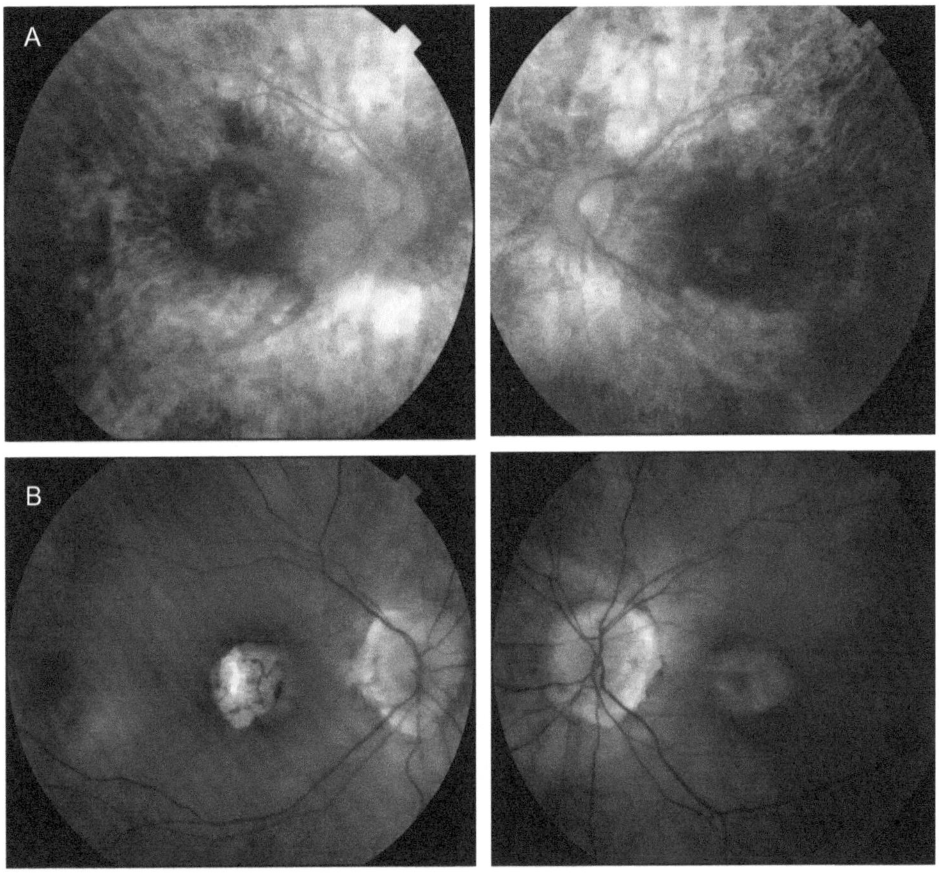

FIG. 13A.26 Colour fundus photography from an individual carrying the *RPGR* c.673_674delAG variant (A) and an individual carrying the *RPGR* c.1425_1426delGA variant (B). Both patients are male and in the fourth decade of their life. Both mutations are in a hemizygous state and alter the sequence of *RPGR* exon ORF15. The c.673_674delAG change results in a retinitis pigmentosa phenotype (like most changes in *RPGR*). The c.1425_1426delGA variant is closer to the C-terminal region of the gene and is associated with cone/cone-rod dystrophy with early macular involvement. *(Courtesy of Professor Graeme C.M. Black.)*

death (Section 13B.1.3). Jalili syndrome is an autosomal recessive condition associated with cone/cone-rod dystrophy and amelogenesis imperfecta (i.e. demineralisation of both primary and secondary dentition leading to teeth of abnormal colour).

Targeted (panel-based analysis) or comprehensive (exome or genome sequencing) genetic testing can be very helpful for differential diagnosis and can inform counselling around the prognosis and familial risk. Referrals for prenatal diagnostics or pre-implantation genetic testing should be considered for people whose offspring are at significant risk of having a cone/cone-rod dystrophy.[9]

Clinical management

There are no proven treatments for cone/cone-rod dystrophies, although research is getting closer to solutions. Current patient management mostly consists of conveying the diagnosis, prognosis, and inheritance pattern, and alleviating symptoms through referrals for filtered glasses or contact lenses and low vision aids. The frequency of visits depends on the patient's needs and will generally be more often at younger ages when significant cone photoreceptor loss is still ongoing.

Individuals with specific cone/cone-rod dystrophy subtypes can be advised to adopt strategies to try to slow degeneration, based on gene function or animal model studies. For example, some clinicians advise people with *GUCA1A*-retinopathy to sleep with the lights on; this aims to prevent accumulation of cGMP, which otherwise occurs at night and causes photoreceptor damage. In contrast, a degree of light avoidance using tinted glasses may be beneficial in people with *ABCA4*-retinopathy as it can reduce the production of toxic visual cycle by-products (including A2E). Vitamin A should also be avoided in *ABCA4*-retinopathy as it may enhance A2E production and, therefore, disease progression.[3]

There are ongoing efforts to develop novel treatments that aim to slow degeneration or improve visual function. Gene replacement therapy trials are already underway in people who have selected disease subtypes (including *ABCA4* and *RPGR*[3]).

References

1. Thiadens AA, Phan TM, Zekveld-Vroon RC, Leroy BP, van den Born LI, Hoyng CB, et al. Clinical course, genetic etiology, and visual outcome in cone and cone-rod dystrophy. *Ophthalmology* 2012;**119**(4):819–26.
2. Roosing S, Thiadens AA, Hoyng CB, Klaver CC, den Hollander AI, Cremers FPM. Causes and consequences of inherited cone disorders. *Prog Retin Eye Res* 2014;**42**:1–26.
3. Gill JS, Georgiou M, Kalitzeos A, Moore AT, Michaelides M. Progressive cone and cone-rod dystrophies: clinical features, molecular genetics and prospects for therapy. *Br J Ophthalmol* 2019;**103**(5):711–20.
4. Boulanger-Scemama E, Mohand-Saïd S, El Shamieh S, Démontant V, Condroyer C, Antonio A, et al. Phenotype analysis of retinal dystrophies in light of the underlying genetic defects: application to cone and cone-rod dystrophies. *Int J Mol Sci* 2019;**20**(19):4854.
5. Mustafi D, Engel AH, Palczewski K. Structure of cone photoreceptors. *Prog Retin Eye Res* 2009;**28**(4):289–302.
6. Sharon D, Wimberg H, Kinarty Y, Koch KW. Genotype-functional-phenotype correlations in photoreceptor guanylate cyclase (GC-E) encoded by GUCY2D. *Prog Retin Eye Res* 2018;**63**:69–91.
7. Fujinami-Yokokawa Y, Fujinami K, Kuniyoshi K, Hayashi T, Ueno S, Mizota A, et al. Clinical and genetic characteristics of 18 patients from 13 Japanese families with CRX-associated retinal disorder: identification of genotype-phenotype association. *Sci Rep* 2020;**10**(1):9531.
8. Guimaraes TAC, Georgiou M, Robson AG, Michaelides M. KCNV2 retinopathy: clinical features, molecular genetics and directions for future therapy. *Ophthalmic Genet* 2020;**41**(3):208–15.
9. Yahalom C, Macarov M, Lazer-Derbeko G, Altarescu G, Imbar T, Hyman JH, et al. Preimplantation genetic diagnosis as a strategy to prevent having a child born with an heritable eye disease. *Ophthalmic Genet* 2018;**39**(4):450–6.

13A.8

ABCA4-related disorders

Eva Lenassi

ABCA4-retinopathy is arguably the most common Mendelian retinal condition (estimated prevalence >1:10,000). It is caused by biallelic mutations in ABCA4 (a gene encoding a photoreceptor-specific ATP-binding protein) and has a broad phenotypic spectrum ranging from severe panretinal dystrophy presenting in early childhood to mild maculopathy with late onset or even an asymptomatic course. Several terms have been previously used to describe various manifestations of ABCA4-retinopathy including Stargardt disease, fundus flavimaculatus, pattern dystrophy, bull's eye maculopathy, macular dystrophy, cone/cone-rod dystrophy and retinitis pigmentosa. The universal phenotypic feature of the condition is progressive degeneration of the photoreceptors and the RPE at the macula ('macular affection'). Yellow-white fundus flecks are a characteristic but not pathognomonic finding; similar lesions are seen in other conditions including autosomal dominant PRPH2-retinopathy and autosomal dominant ELOVL4-retinopathy.[1,2]

Clinical characteristics

Most affected individuals present with bilateral central visual loss, although a minority of patients are asymptomatic and are coincidentally diagnosed during screening tests. Less commonly, patients present with central scotomas, dyschromatopsia or light hypersensitivity. Symptoms typically develop in the second or third decade of life although early- and late onset forms have been well recognised.

The diagnosis of ABCA4-retinopathy can be particularly challenging in children presenting in the first decade of life; up to 30% of these patients have no readily observable fundus abnormalities despite reduced visual acuity. To avoid misdiagnosis in this group, there should be a heightened awareness and a low threshold for requesting structural and functional tests (such as fundus autofluorescence imaging, OCT and electrodiagnostic testing).[3,4]

A small group of patients become symptomatic later in life (e.g. >45 years). These individuals often present with patchy parafoveal RPE atrophy, subtle flecks, and relatively preserved foveal structure and function (foveal sparing[5]). Rarely, this presentation can be misdiagnosed as age-related macular degeneration since fundus flecks may superficially resemble drusen. Drusen are round and show confluence towards the centre of the macula. In contrast, flecks are more irregularly shaped and tend to be less clustered in the fovea. Fundus autofluorescence imaging can be helpful in these cases as flecks show intense autofluorescence whereas drusen can be associated with mildly increased, normal, or decreased signal.

As the above paragraphs suggest, ABCA4-retinopathy exhibits significant clinical heterogeneity and there is a spectrum of associated clinical presentations, imaging findings and electrophysiological phenotypes. The fundus picture in most patients is characterised by the presence of flecks and RPE abnormalities that predominantly affect the posterior pole. During the course of the disease, RPE atrophy develops and most affected individuals have a central to peripheral expansion of the retinal degeneration over time. Intriguingly, the peripapillary retina and RPE tend to be spared from degeneration even in advanced disease. These three features—macular affection, fundus flecks and peripapillary sparing—are considered the diagnostic triad of ABCA4-retinopathy (i.e. when they occur together in a patient, they are highly indicative of this condition[1]).

A subset of patients with ABCA4-retinopathy has a 'bull's eye' pattern of macular atrophy with scarce or absent fundus flecks (Fig. 13A.27). Also, as mentioned above, it is not uncommon for affected individuals who become symptomatic in the first decade of life to present either with a normal fundus or with foveal RPE changes without flecks (Fig. 13A.28).[3,4] Patients who become symptomatic in late adulthood often have foveal sparing and therefore good visual acuity (Fig. 13A.29). In these cases, flecks might be confused with soft drusen (Fig. 13A.30) or pattern dystrophy type lesions (Fig. 13A.31). In contrast, individuals with severe, end-stage disease have widespread chorioretinal degeneration with pigment deposition and retinal vessel attenuation; the flecks disappear and that the fundus appearance may resemble advanced choroideremia or retinitis pigmentosa (Fig. 13A.32).

Fundus imaging has an important role in the diagnosis and monitoring of ABCA4-retinopathy. One of the hallmarks of the condition is the accelerated deposition of lipofuscin and other fluorescent compounds in the RPE monolayer due to the incomplete degradation of photoreceptor outer

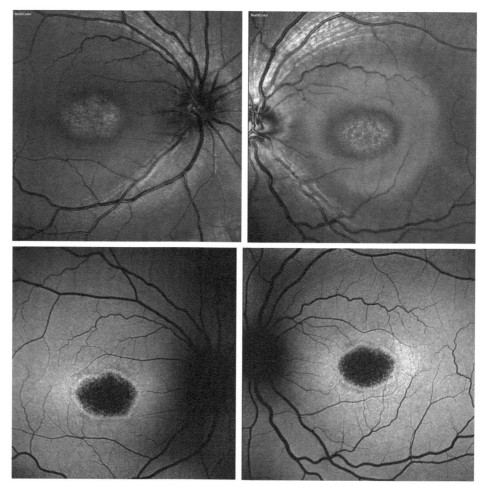

FIG. 13A.27 Multicolour fundus imaging and fundus autofluorescence imaging from a 14-year-old patient carrying two heterozygous variants in the *ABCA4* gene: c.2438A>C (p.Glu813Ala) and c.2588G>C (p.Gly863Ala). A 'bull's eye' pattern of macular atrophy with absent fundus flecks is noted.

segment discs. Fundus autofluorescence imaging allows topographic mapping of lipofuscin distribution in the RPE. Reduced signal reflects a reduction in RPE lipofuscin density (typically due to RPE loss), while increased signal reflects increased lipofuscin density (typically due to photoreceptor cell demise with augmented fluorophore formation). Given the underlying pathophysiology of *ABCA4*-retinopathy, fundus autofluorescence imaging abnormalities are often more extensive than the abnormalities seen on fundoscopy (Fig. 13A.33). Furthermore, early fundus changes such as small flecks or RPE atrophy can be detected on fundus autofluorescence imaging before they become clinically evident on slit-lamp examination (Fig. 13A.28). It is worth highlighting that the more rapidly progressive forms of *ABCA4*-retinopathy are linked to the presence of multiple areas of reduced signal at the posterior pole combined with heterogeneous background autofluorescence (Fig. 13A.34).

OCT imaging allows detection of foveal/parafoveal photoreceptor outer segment loss, a key finding at the early stages of *ABCA4*-retinopathy. Also, it enables detection of what has been shown to be one of the earliest structural changes in affected children: hyper-reflectivity at the base of the outer nuclear layer that gives the impression of external limiting membrane thickening (as the increased signal appears to be continuous with this membrane).[4]

Electrodiagnostic testing is valuable in *ABCA4*-retinopathy. Most patients have macular dysfunction on pattern or multifocal electroretinography. The full-field ERG is of particular interest as it helps classify affected individuals into three groups:

- group 1: dysfunction confined to the macula (i.e. full-field ERG is normal);
- group 2: macular and generalised cone system dysfunction present (i.e. abnormal light-adapted full-field ERG, normal or near-normal dark-adapted full-field ERG);
- group 3: macular and generalised cone and rod system dysfunction (i.e. abnormal light-adapted and dark-adapted full-field ERG).

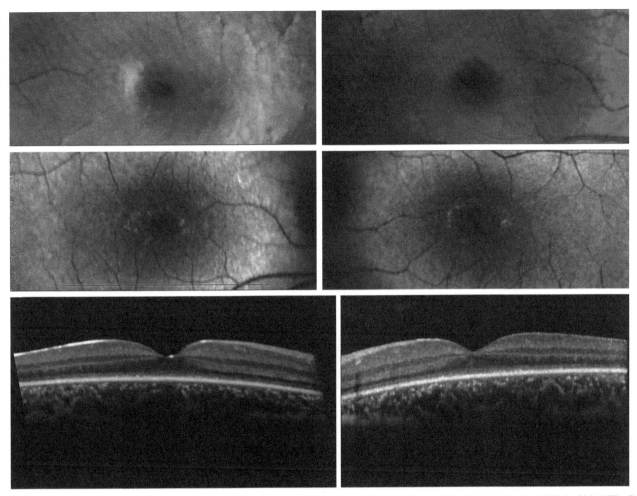

FIG. 13A.28 Multicolour fundus imaging, fundus autofluorescence imaging and OCT from a 6-year-old patient carrying the *ABCA4* c.5461-10T>C variant in the homozygous state. Only subtle abnormalities were noted on fundoscopy and the diagnosis of *ABCA4*-retinopathy was suspected after fundus autofluorescence imaging highlighted hyperautofluorescent lesions in the parafoveal region. OCT revealed an irregular photoreceptor ellipsoid zone bilaterally.

It has been suggested that these groups represent distinct disease subtypes and do not just reflect stages of the disease. Importantly, they appear to have significant prognostic implications: >95% of individuals with dark-adapted ERG involvement (group 3) at baseline have clinically significant electrophysiological deterioration over a mean time of 10.5 years; in contrast, only 20% of patients with normal full-field ERG (group 1) showed clinically significant progression in this time frame.[6]

Molecular pathology

ABCA4 (ATP-binding cassette transporter type A4) is a 50-exon gene encoding a transmembrane protein that is localised at the cone and rod photoreceptor outer segment discs. The role of this protein is to flip a retinoid intermediate of the visual cycle from the inner to the outer aspect of the disc membrane. Defects in *ABCA4* result in the intradiscal accumulation of this retinoid intermediate, which in turn leads to the formation of a toxic molecule known as A2E (a prominent component of lipofuscin).

ABCA4 was first linked to retinal disease in 1997 and since then >1500 disease-associated variants have been reported. Attempts to group these mutations based on the predicted severity of each change (null, severe, moderate, mild, hypomorphic) are ongoing.[1,7,8] A range of phenotypes have been linked to homozygous or compound heterozygous variants in *ABCA4* and a genotype–phenotype correlation model has been proposed to explain the significant clinical heterogeneity of the condition. According to the model, the residual activity of the mutant *ABCA4* protein determines the clinical outcome. *ABCA4* variants with a

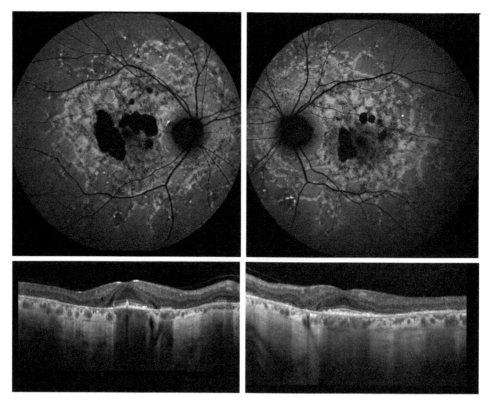

FIG. 13A.29 Fundus autofluorescence imaging and OCT from a 58-year-old patient carrying two heterozygous variants in *ABCA4*: c.3481C>A (p.Arg1161Ser) and *ABCA4* c.4519G>A (p.Gly1507Arg). There is sparing of the central fovea with preservation of the ellipsoid zone in the foveola bilaterally. The peripapillary retina and RPE are also spared from degeneration.

relatively mild combined effect (e.g. two moderate variants or a severe variant combined with a mild variant) generally result in A2E accumulation within and beneath the RPE and give rise to a mild, macular dystrophy phenotype. *ABCA4* variants with an intermediate combined effect (e.g. a severe variant combined with a moderate variant) generally result in a direct injury mainly to cone photoreceptors and lead to a cone/cone-rod dystrophy phenotype. Finally, *ABCA4* variants with the most detrimental combined effect (e.g. two severe variants, i.e. changes predicted to lead to loss of protein function, splicing, frameshifting or truncating variants) result in injury to both cone and rod photoreceptors and lead to early onset panretinal dysfunction with more rapid progression (Fig. 13A.35).[9]

A different but complementary model emphasises that the natural history of *ABCA4*-retinopathy consists of multiple trajectories that are determined by the patient's genotype. Assuming that disease severity is defined based on the spatial extent of the disorder, the following stages emerge: macular stage, extramacular stage and advanced stage. An individual carrying variants with a relatively mild combined effect will become symptomatic later in life and will only transition from macular to the extramacular stage in late adulthood; no further transition to the advanced stage should be expected. In contrast, a patient carrying variants with the most detrimental combined effect will become symptomatic in the first decade of life, the transition from the macular to the extramacular stage in their teenage years and transition from the extramacular to the advanced stage soon after. This highlights that an earlier onset of the disease is usually associated with a poorer prognosis.[1]

It is worth discussing three characteristic mild *ABCA4* alleles: c.5603A>T (p.Asn1868Ile), c.5882G>A (p.Gly1961Glu) and c.6089G>A (p.Arg2030Gln). The c.5603A>T variant was previously thought to be benign (as it is relatively common in the general population) but it has been now shown to account for at least 10% of all known *ABCA4*-retinopathy cases. It is phenotypically expressed only when it is paired with a severe variant and most affected individuals have foveal sparing and late onset of symptoms. Another common *ABCA4* disease-associated variant is c.5882G>A. This change is thought to result in a relatively mild phenotype with sparing

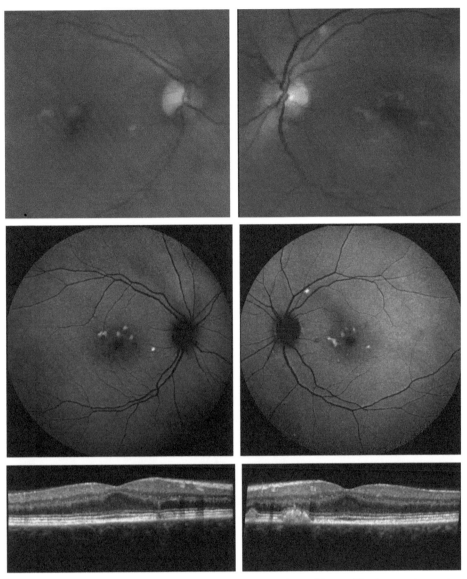

FIG. 13A.30 Multicolour fundus imaging, fundus autofluorescence imaging and OCT from a 59-year-old patient carrying two heterozygous variants in *ABCA4*: c.6079C>T (p.Leu2027Phe) and c.5603A>T (p.Asn1868Ile). Yellow-white macular lesions that superficially resemble soft drusen were noted on fundoscopy. However, the irregular shape and intense hyperautofluorescent nature of the lesions suggested that these are flecks and pointed to the diagnosis of *ABCA4*-retinopathy. OCT imaging in the left eye revealed hyper-reflective deposits traversing the photoreceptor-attributable bands.

of the peripheral retina in most cases; foveal cavitations are noted on OCT of many affected individuals that carry this mutation (i.e. there is focal loss of the photoreceptors in the foveola leading to an 'optical gap' appearance). The c.6089G>A variant is linked to a later age of onset and, like c.5603A>T it is commonly found in individuals with a foveal sparing phenotype.[1,2]

Molecular investigations should be considered in all individuals with suspected *ABCA4*-retinopathy as they improve the accuracy of diagnosis and make the patient eligible for trials of gene-based interventions. Targeted (single-gene or panel-based analysis) or comprehensive (e.g. genome sequencing) genetic testing could be offered. Presently over 90% of patients with presumed *ABCA4*-retinopathy receive a genetic diagnosis. Some important considerations when requesting a genetic test for *ABCA4*-retinopathy include: (i) *ABCA4* has a high allelic heterogeneity and a large number of mutations have been identified. Although the majority of these changes are located in the protein-coding area of the gene (exons), a significant number of intronic variants affecting splicing have been described. A genetic test should therefore include as

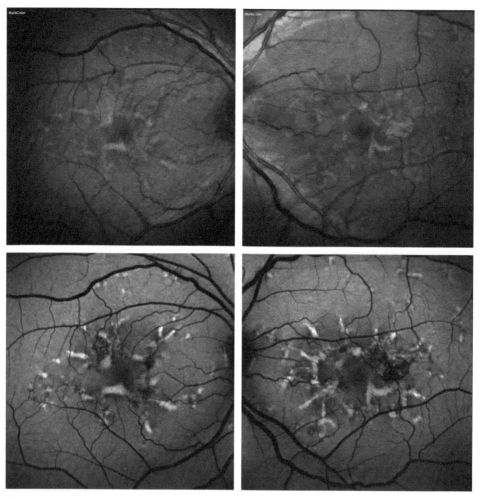

FIG. 13A.31 Multicolour fundus imaging and fundus autofluorescence imaging from a 51-year-old patient carrying two heterozygous variants in ABCA4: c.5196+1137G>A and c.3322C>T (p.Arg1108Cys). This retinal appearance resembles what some authors have described as butterfly-shaped pattern dystrophy.

many of these intronic variants as possible. (ii) Many previously unsolved cases have been explained by the presence of the relatively common c.5603A>T variant. This change is quite prevalent in the general population and may evade standard bioinformatics filtering strategies. Care should therefore be taken to ensure that this variant is not filtered out during data processing. (iii) Although large deletions/insertions affecting ABCA4 have been previously described, these are uncommon and represent only a small fraction of the variant spectrum.[1,2,7,8]

Testing for genes linked to phenotypes that may mimic ABCA4-retinopathy should be considered, especially when ABCA4 screening is negative or when there is a dominant family history. Some genes to consider testing include: PRPH2, ELOVL4, PROM1 and CRX; RPGR, GUCY2D, CTNNA1, BEST1, IMPG1, IMPG2, OTX2, KCNV2, TTLL5, DRAM2 and PROM1 should also be included in the differential in selected cases.

It is worth highlighting that the presence of affected individuals in more than one generation should not be considered incompatible with ABCA4-retinopathy; pseudo-dominant inheritance (in which a disease associated with autosomal recessive transmission appears in two successive generations simulating autosomal dominant inheritance) has been previously described and is well recognised in this disorder.

Clinical management

The finding that the retinal degeneration in an animal model of ABCA4-retinopathy was accelerated with

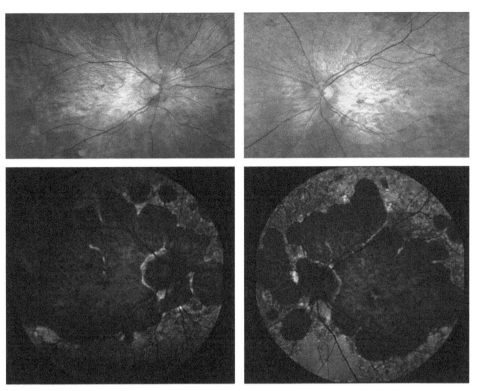

FIG. 13A.32 Multicolour fundus imaging and fundus autofluorescence imaging from a 60-year-old patient carrying two heterozygous variants in *ABCA4*: c.3481C>A (p.Arg1161Ser) and c.4519G>A (p.Gly1507Arg). Extensive RPE loss, intraretinal pigment deposition, scarcity of fundus flecks and relative preservation of the peripapillary RPE are noted.

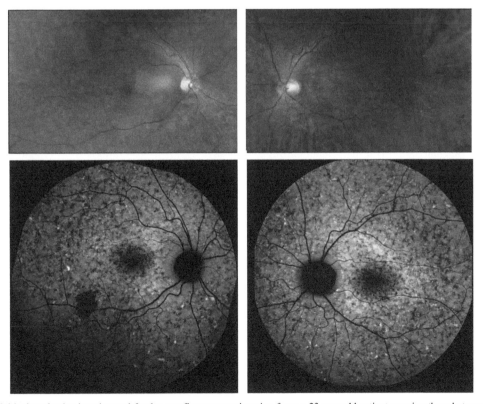

FIG. 13A.33 Multicolour fundus imaging and fundus autofluorescence imaging from a 23-year-old patient carrying three heterozygous variants in *ABCA4*: c.1622T>C (p.Leu541Pro), c.3113C>T (p.Ala1038Val) and c.5714+5G>A. There is heterogeneous background autofluorescence both within and outside the vascular arcades and fundus autofluorescence imaging abnormalities are more extensive than the abnormalities seen on fundoscopy.

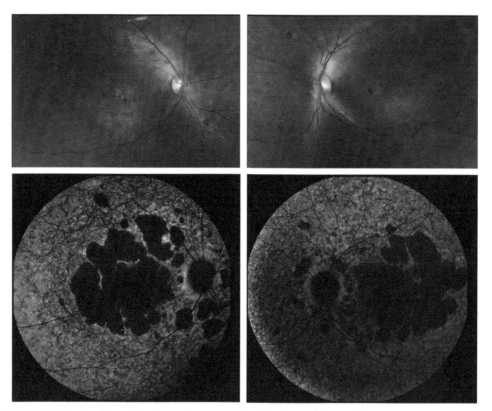

FIG. 13A.34 Multicolour fundus imaging and fundus autofluorescence imaging from a 36-year-old patient carrying two heterozygous variants in *ABCA4*: c.1937+1G>A and c.6112C>T (p.Arg2038Trp). Fundus autofluorescence imaging revealed multiple low signal areas at the posterior pole (corresponding to RPE atrophy) with a heterogeneous background. This appearance has been linked with more rapidly progressive disease.

exposure to significantly bright (ultraviolet) light and ingestion of large doses of vitamin A has led to patients with *ABCA4*-retinopathy being advised to avoid excessive exposure to bright sunlight and to avoid vitamin A supplementation.

Choroidal neovascularisation is a rare complication of *ABCA4*-retinopathy with only a few cases described in the literature. Previous reports of intravitreal anti-VEGF treatment for *ABCA4*-associated choroidal neovascularisation suggest variable efficacy.[2]

Presently there is no approved curative treatment for *ABCA4*-retinopathy. However, several therapeutic options are emerging including approaches using gene augmentation, small molecule drugs and stem cell therapy.[1] For an up-to-date list of ongoing trials, the reader is referred to https://clinicaltrials.gov/.

Family management and counselling are important components of the care of individuals affected with *ABCA4*-retinopathy. Given the autosomal recessive inheritance pattern of the condition, the risk to family members should be highlighted. The risk to future siblings of an individual with *ABCA4*-retinopathy is typically 25%. The risk to children of patients is low but challenging to precisely estimate. A key confounder is the fact that *ABCA4* mutation carrier frequency in the general population is relatively high: approximately one in every 25 people who have European ancestries is a carrier of a presumed pathogenic *ABCA4* variant (although it is unclear how penetrant some of the common pathogenic alleles are). At present, it is not routine practice to regularly offer testing to partners of individuals with *ABCA4*-retinopathy (to refine the risk to children).

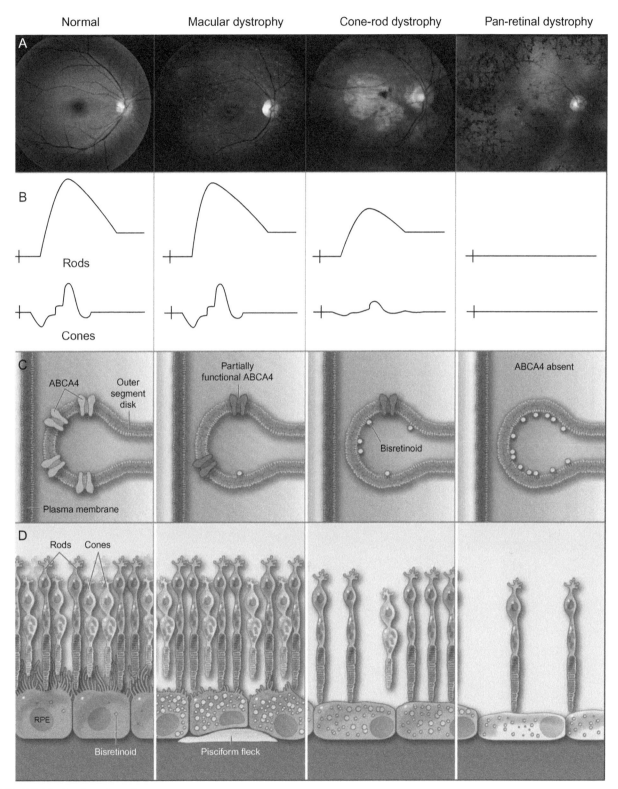

FIG. 13A.35 See figure legend on next page.

FIG. 13A.35—CONT'D Genotype–phenotype correlation model that has been proposed to explain the clinical heterogeneity of *ABCA4*-retinopathy. (A) Fundus images from individuals with progressively decreasing amounts of ABCA4 function *(left to right)*, ranging from a normal retina to macular dystrophy, cone-rod dystrophy, and panretinal dystrophy. (B) Effect of reduced ABCA4 function on dark-adapted (rod) and light-adapted (cone) single-flash full-field ERGs. Patients carrying variants that lead to a relatively mild reduction in ABCA4 activity are expected to have disease confined to the macula and there is, therefore, no effect on generalised retinal function. Patients with variants leading to moderate loss of ABCA4 function are expected to have cone-rod dystrophy and therefore photoreceptor dysfunction that is greater for cones than for rods. Patients with variants leading to complete loss of ABCA4 function are expected to have panretinal dystrophy with onset in early childhood. In these cases, there is extensive degeneration of both cone and rod photoreceptors resulting in undetectable full-field ERGs. (C) Effects of reduced ABCA4 function on the accumulation of bisretinoid (A2E) in photoreceptor outer segment discs. Mild reduction in ABCA4 activity (in macular dystrophy) is associated with A2E formation that is only slightly above normal; moderate loss of function (in cone-rod dystrophy) is associated with intermediate amounts of A2E accumulation; complete loss of function (in panretinal dystrophy) results in maximum accumulation. (D) Predicted histopathological effects of reduced ABCA4 activity. In patients with disease confined to the macula, the rate of A2E formation in the outer segments is relatively slow and the photoreceptors are not directly injured. Consequently, A2E is delivered to the secondary lysosomes of the RPE during the normal phagocytosis of photoreceptor outer segments. Some of this material accumulates beneath the RPE and forms lesions that are visible on ophthalmoscopy as flecks. In patients with cone-rod dystrophy, moderate loss of ABCA4 function results in sufficient accumulation of A2E in outer segments to cause some direct photoreceptor degeneration (with cones more severely affected than rods). In patients with panretinal dystrophy, complete loss of ABCA4 function causes extensive accumulation of A2E in the outer segments leading to a degeneration of both rod and cones. *(Adapted from Ref. 9.)*

References

1. Cremers FPM, Lee W, Collin RWJ, Allikmets R. Clinical spectrum, genetic complexity and therapeutic approaches for retinal disease caused by ABCA4 mutations. *Prog Retin Eye Res* 2020;**9**:100861.
2. Tanna P, Strauss RW, Fujinami K, Michaelides M. Stargardt disease: clinical features, molecular genetics, animal models and therapeutic options. *Br J Ophthalmol* 2017;**101**(1):25–30.
3. Bax NM, Lambertus S, FPM C, Klevering BJ, Hoyng CB. The absence of fundus abnormalities in Stargardt disease. *Graefes Arch Clin Exp Ophthalmol* 2019;**257**(6):1147–57.
4. Khan KN, Kasilian M, Mahroo OAR, Tanna P, Kalitzeos A, Robson AG, et al. Early patterns of macular degeneration in ABCA4-associated retinopathy. *Ophthalmology* 2018;**125**(5):735–46.
5. Lambertus S, Lindner M, Bax NM, Mauschitz MM, Nadal J, Schmid M, et al. Foveal sparing atrophy study team (FAST). Progression of late-onset Stargardt disease. *Invest Ophthalmol Vis Sci* 2016;**57**(13):5186–91.
6. Fujinami K, Lois N, Davidson AE, Mackay DS, Hogg CR, Stone EM, et al. A longitudinal study of Stargardt disease: clinical and electrophysiologic assessment, progression, and genotype correlations. *Am J Ophthalmol* 2013;**155**(6):1075–1088.e13.
7. Bauwens M, Garanto A, Sangermano R, Naessens S, Weisschuh N, De Zaeytijd J, et al. ABCA4-associated disease as a model for missing heritability in autosomal recessive disorders: novel noncoding splice, cis-regulatory, structural, and recurrent hypomorphic variants. *Genet Med* 2019;**21**(8):1761–71. https://doi.org/10.1038/s41436-018-0420-y.
8. Cornelis SS, Bax NM, Zernant J, Allikmets R, Fritsche LG, den Dunnen JT, et al. In silico functional meta-analysis of 5,962 ABCA4 variants in 3,928 retinal dystrophy cases. *Hum Mutat* 2017;**38**(4):400–8.
9. Sheffield VC, Stone EM. Genomics and the eye. *N Engl J Med* 2011;**364**(20):1932–42.

13A.9

BEST1-related disorders (bestrophinopathies)

Ine Strubbe, Panagiotis I. Sergouniotis, and Bart P. Leroy

Mutations in the *BEST1* gene result in several distinct retinal phenotypes collectively referred to as bestrophinopathies or *BEST1*-related disorders. These include Best vitelliform macular dystrophy, autosomal recessive bestrophinopathy and autosomal dominant vitreoretinochoroidopathy.

Best vitelliform macular dystrophy, also known as Best disease, is the most common subtype and was the first condition to be associated with genetic variants in *BEST1*.[1] This gene encodes bestrophin-1, a transmembrane protein that, within the eye, is uniquely expressed in the RPE where it predominantly localises to the basolateral surface.[2]

Clinical characteristics

Best vitelliform macular dystrophy (BVMD)

This condition was first described by J. Adams in 1883[1] but was later named after F. Best who reported an affected family in 1905.[2] It is one of the most common macular dystrophies and it is characterised by central, yellow-white macular lesions in both eyes, classically described as resembling an egg yolk (Fig. 13A.36).[3,4]

Central visual loss and, less commonly, distortion or metamorphopsia are the main presenting symptoms. Acute loss of vision may occur due to a haemorrhage from a sub-retinal neovascular membrane, which is a known complication (Fig. 13A.37). There is significant variability in age of onset. Although most affected individuals present in childhood or early adulthood, a subset of patients present later in life. Notably, it is not unusual to diagnose the condition in an asymptomatic individual through family work-up, triggered by a diagnosis in a family member. It is worth highlighting that the diagnosis of Best vitelliform macular dystrophy can be made in people who have neither fundoscopic signs nor symptoms (e.g. based on electro-oculography and/or genetic testing).

Best-corrected visual acuity progressively declines in affected eyes but generally remains fair in at least one eye until later in life. Patients are often hypermetropic.[5] The anterior chambers can be shallow and the incidence of angle-closure (narrow-angle) glaucoma is thought to be higher than that of the general population.[6] Fundoscopy reveals deposition of lipofuscin in central macular lesions, best highlighted on short (blue) wavelength autofluorescence imaging. On OCT, separation of photoreceptors from the underlying RPE is typically observed. There is often significant variability in the size and shape of the macular lesions both between eyes and between patients. The following stages of the disease have been described:

- stage 1 or pre-vitelliform or carrier stage, present in asymptomatic patients.
- stage 2 or vitelliform stage, with classic egg yolk-like lesions of different sizes (Fig. 13A.36).

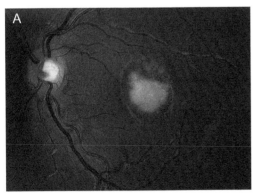
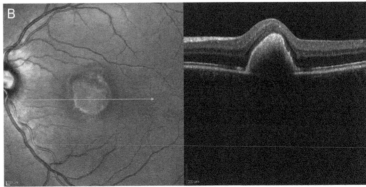

FIG. 13A.36 Colour fundus photograph (A), infrared reflectance image and foveal OCT scan (B) from the left eye of a 10-year-old child who carries a *BEST1* c.73C>T (p.Arg25Trp) variant in the heterozygous state. Note the egg-yolk-like, vitelliform macular lesion that corresponds to a conical, lipofuscin-containing neurosensory retinal detachment.

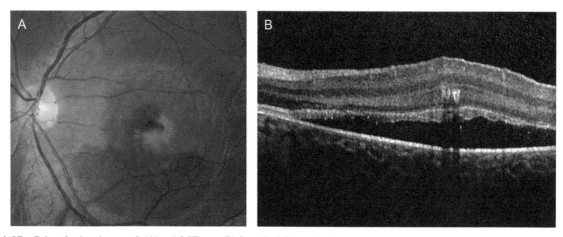

FIG. 13A.37 Colour fundus photograph (A) and OCT scan (B) from the left eye of a 5-year-old child who carries a *BEST1* c.652C>T (p.Arg218Cys) variant in the heterozygous state. Note the greyish hue corresponding to a subretinal neovascular membrane in the foveal region. An associated subretinal haemorrhage is present, mostly at the inferior aspect of the neurosensory detachment. A horizontal OCT scan of the superior half of the macular neurosensory detachment (above the level of the subretinal neovascular membrane) reveals thickened outer retinal layers and several 'stalactites' hanging off in the area of detachment (probably corresponding to photoreceptor outer segment tips). The identification of individual layers below the external limiting membrane is challenging.

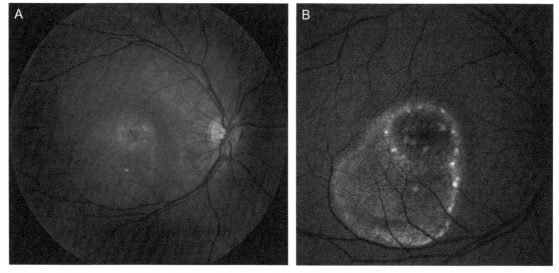

FIG. 13A.38 Colour fundus photograph (A) and fundus autofluorescence image (B) from the right eye of a 17-year-old individual who carries a *BEST1* c.431G>A (p.Ser144Asn) variant in the heterozygous state. Note the subtle 'pseudohypopyon' within the area of neurosensory retinal detachment and the hyperautofluorescence of deposited material, mostly in the watershed zone of the detachment.

- stage 3 or pseudo-hypopyon stage, where break-up of vitelliform lesions leads to a hypopyon-like image, the level of which changes depending on gravity (Fig. 13A.38).
- stage 4 or vitelliruptive stage ('scrambled egg'), with irregular distribution of subretinal lipofuscin, occasionally combined with subretinal fibrosis (Fig. 13A.39).
- stage 5 with atrophy/scarring with or without choroidal subretinal neovascularisation (Fig. 13A.40).

Changing from one stage to another is generally seen from the vitelliform stage onwards but this evolution may not be observed when the lesions are small. Although most people with Best vitelliform macular dystrophy have a single macular lesion, some cases present with multiple lesions in each eye. Furthermore, lesions are typically bilateral but a unilateral presentation has been reported in a small number of affected individuals.[7]

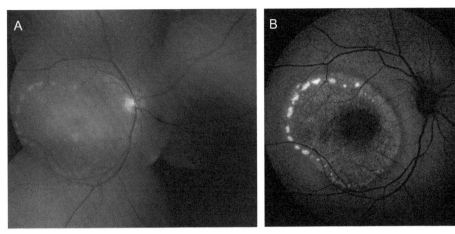

FIG. 13A.39 Composite colour fundus photograph (A) and fundus autofluorescence image (B) from the right eye of a 45-year-old individual who carries a *BEST1* c.92T>G (p.Leu31Arg) variant in the heterozygous state. Note the scattered deposits that are more pronounced at the borders of the area of macular neurosensory detachment, giving a scrambled egg appearance. These scattered lesions contain lipofuscin material and tend to hyperautofluorescence.

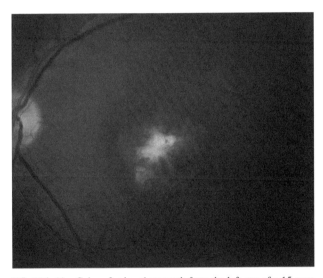

FIG. 13A.40 Colour fundus photograph from the left eye of a 15-year-old individual who carries a *BEST1* c.92T>G (p.Leu31Arg) variant in the heterozygous state. Note the subretinal fibrotic scar, a consequence of previous subretinal neovascularisation in the central macula.

The electrooculogram (EOG) light peak:dark trough (LP:DT) ratio, which tests RPE function (Section 6.2.6), is typically, but not universally, decreased in affected individuals. This ratio is normally above 1.7 and mostly 2.0.[8,9] In people with Best vitelliform macular dystrophy, it is usually below 1.3. Otherwise, full-field electroretinography (ERG) is within normal limits and pattern ERG responses may be normal or attenuated.

Atypical forms of the condition may look more like pattern dystrophy (Section 13A.10). This includes cases described as adult-onset vitelliform macular dystrophy, as well as cases who have a spoke-like pattern around the central macula without vitelliform lesions; LP:DT ratios are often near-normal or normal in these groups of patients.[10–14]

Autosomal recessive bestrophinopathy (ARB)

This condition was first described by Burgess and co-workers in 2008.[15] It is characterised by multiple yellow-white retinal lesions and it is associated with slowly progressive, central visual loss. This is typically first noted in late childhood or early adulthood, although cases presenting as late as in the sixth decade of life have been described. Night vision problems and, later on, photophobia may be reported as the disease progresses from a central to a more generalised retinopathy. A subset of cases develops acute/subacute symptoms either linked to the development of choroidal neovascularisation or associated with iridocorneal angle closure.[14–18]

Many patients with autosomal recessive bestrophinopathy are hypermetropic and have shallow anterior chambers that predispose them to develop angle-closure glaucoma.[14–17,19] Several cases that required surgical intervention to control the intraocular pressures have been described.[20]

Fundoscopy in autosomal recessive bestrophinopathy often reveals a grey-white diffuse irregularity of the RPE throughout much of the fundus. Shallow neurosensory detachments with dispersed, punctate, yellow-white flecks (mostly in the watershed zone at the periphery of these detachments) are noted (Fig. 13A.41). These lesions typically affect the macula but may also be seen beyond the vascular arcades and/or around the optic disc (with a predilection for the temporal or nasal side superior to the optic disc). In areas of detachment, 'stalactites' of photoreceptor outer segments can be seen hanging off the inner part of the photoreceptor layers on OCT. Apart from the neurosensory detachments, schisis of the retinal layers reminiscent of cystic macular oedema may be observed. These cysts are most prominently seen at the level of the inner nuclear layer on OCT. Thickening of the outer retinal layers is also commonly observed (Fig. 13A.42).

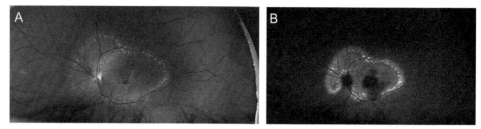

FIG. 13A.41 Widefield fundus image (A) and fundus autofluorescence image (B) of the left eye of 31-year-old individual who is compound heterozygote for *BEST1* c.172_173dup (p.Gln58HisfsTer4) and c.584C>T (p.Ala195Val). Note the shallow neurosensory detachment at the posterior pole, extending beyond the macula and including the area superonasal to of the optic disc. Dispersed, punctate, yellow-white flecks, mostly in the watershed zone at the periphery of detachment are noted. The visibility of these lipofuscin-containing lesions is increased when autofluorescence imaging is used.

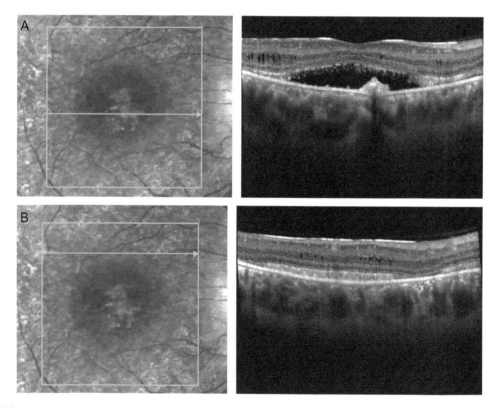

FIG. 13A.42 Infrared reflectance images and OCT scans from the right eye of a 17-year-old individual who is homozygous for *BEST1* c.948G>C (p.Gln316His). Note the thickening of the outer retinal layers (A); the loss of individualisation and irregularity of signal from the inner segment ellipsoid zone to the RPE layer (A); outer retinal patches with complete absence of photoreceptor layers that alternate with areas of loss of individualisation and irregularity of all layers, including those representing the photoreceptors (B).

In some patients, the initial presentation of the macular lesions is reminiscent of the vitelliruptive stage ('scrambled egg') of Best vitelliform macular dystrophy, with some extramacular involvement. These phenotypes may be described as 'multifocal Best disease', and are more likely to represent autosomal recessive bestrophinopathy than Best vitelliform macular dystrophy.

An integral part of the clinical phenotype is the reduced LP:DT ratio on EOG. This is noted before a generalised, progressive rod-cone system involvement is seen on full-field ERG. Indeed, there are no other conditions except for the generalised bestrophinopathies, namely autosomal dominant vitreoretinochoroidopathy (see below) and autosomal recessive bestrophinopathy, in which a decreased EOG light rise precedes progressive rod-cone dysfunction.[14,17,21,22]

Autosomal dominant vitreoretinochoroidopathy (ADVIRC)

This very rare disorder is characterised by a distinctive hyperpigmented 360° band around the retinal periphery. It was first described by Hermann and colleagues in 1958[23] but the name autosomal dominant vitreoretinochoroidopathy was introduced by Kaufman and co-workers in 1982.[24] The identification of heterozygous mutations in the *BEST1* gene was reported by Yardley and co-workers in 2004.[25]

Affected individuals often present with progressive night and/or peripheral vision problems. However, occasionally, patients present to the clinic with loss of central vision. This can be due to acute angle-closure glaucoma, cataracts, vitreous haemorrhage or macular oedema. Sometimes, patients unaware of their condition, are diagnosed through a family work-up, triggered by a diagnosis in a family member.

Up to the age of 40 to 50 years, most affected individuals will have a fair visual function, with a rapid decline to complete loss of vision occurring in at least 50% of patients after that. A significant majority of cases will have an abnormal LP:DT ratio on EOG long before any rod-cone dysfunction becomes apparent on ERG.

On ophthalmic examination, a small corneal diameter may be noted ('microcornea') and the iridocorneal angles are often narrow. Further ocular developmental anomalies that might be a feature of the condition include iris dysgenesis, posterior staphyloma and nanophthalmos.[26–28] Cataracts generally develop more often and earlier than in the general population.[22,27]

A circumferential, coarsely hyperpigmented band with a discrete border in the retinal periphery is the characteristic sign of this condition; this band extends between the equatorial region and the ora serrata and is typically present early, before any subjective visual complaint (Fig. 13A.43).[22,25,27,29] Other fundus abnormalities include macular oedema, superficial punctate yellow-white retinal lesions, vascular abnormalities (breakdown of the blood-retina barrier with or without retinal neovascularisation), vitreous fibrillar condensations and cells, and chorioretinal atrophy (at the later stages of disease) (Fig. 13A.44).[27,29]

The hyperpigmented band in the retinal periphery is certainly not always equally clear. It can differ in terms of intensity of pigmentation not only from one patient to the other but also within the same eye or between the two eyes of one patient. In this light, it is likely that individuals reported to have retinitis pigmentosa due to *BEST1* mutations, may actually represent cases with a late stage of autosomal dominant vitreoretinochoroidopathy when a widespread rod-cone degeneration and generalised pigmentary retinopathy has come to the fore.[30–32]

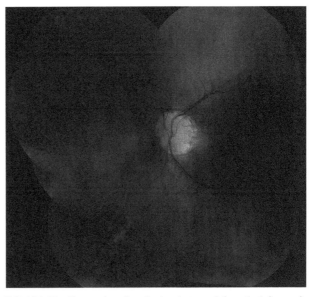

FIG. 13A.43 Composite colour fundus photograph from the left eye of a 32-year-old patient with autosomal dominant vitreoretinochoroidopathy (ADVIRC) who is heterozygous for *BEST1* c.256G > A (pVal86Met). Note the discrete hyperpigmented band in the retinal periphery, as well as the chorioretinal developmental and degenerate abnormalities in the peripapillary area.

FIG. 13A.44 Composite colour fundus photograph from the left eye of a 54-year-old patient with autosomal dominant vitreoretinochoroidopathy (ADVIRC) who is heterozygous for *BEST1* c.256G > A (pVal86Met). Note the widespread chorioretinal atrophy, the circumferential hyperpigmented band in the retinal periphery and the peripapillary chorioretinal maldevelopment and atrophy.

The absence of normal peripapillary RPE development seems to create specific abnormalities of the retina in the peripapillary area. These have been described as optic disc dysplasia,[27] although it is more likely that this manifestation is secondary to the RPE changes.

Molecular pathology

Best vitelliform macular dystrophy, autosomal recessive bestrophinopathy and autosomal dominant vitreoretinochoroidopathy are all associated with mutations in the *BEST1* gene. *BEST1* is primarily expressed in the RPE and brain.[33,34] The encoded protein, bestrophin-1, forms a homopentameric Ca^{2+}-activated Cl^- channel, located on the basolateral membrane of the RPE.[35] In general, bestrophin-1 plays a key role in the regulation of the ionic environment of the RPE and/or subretinal space. Abnormal protein function leads to a range of biochemical and structural abnormalities at the RPE-photoreceptor interface and can affect the adhesion between the retina and the RPE (with the macular region being particularly vulnerable given the abundance of shorter cone outer segments).[36]

Best vitelliform macular dystrophy is generally inherited as an autosomal dominant trait, although biallelic *BEST1* mutations have been reported in a small number of families.[37] A large proportion (>90%) of the variants implicated in this disorder are missense changes in the N-terminal half (i.e. front end) of the protein. Only a few nonsense, splice site, frameshift, and in-frame deletion mutations have been reported. Most pathogenic variants are private and there are at least 3 reports of *de novo BEST1* mutations. Among the most commonly encountered disease-implicated variants are c.16A>C (p.Thr6Pro); c.652C>T (p.Arg218Cys); c.653G>A (p.Arg218His) and c.898G>A (p.Glu300Lys). Another prevalent mutation is c.728C>T (p.Ala243Val); this change is associated with a recognisable pattern dystrophy-like phenotype with carriers developing symptoms at a later age and generally having more preserved EOG ratios.[11]

Autosomal recessive bestrophinopathy is caused by biallelic variants in *BEST1*, i.e. both gene copies must be mutated.[15] Less than 20% of known *BEST1* mutations are associated with this phenotype and the mutation spectrum is wide and includes missense, nonsense and other variant types. Parents of patients are heterozygous mutation carriers and are generally unaffected with normal EOG ratios. This condition is likely due to very low levels of functional protein (i.e. represents the bestrophin-1 complete loss of function or 'null' phenotype) but experimental evidence on this remains inconclusive.[36,38] Mutations commonly associated with autosomal recessive bestrophinopathy include: c.422G>A (p.Arg141His) and c.584C>T (p.Ala195Val).

Autosomal dominant vitreoretinochoroidopathy is caused by a small number of specific, heterozygous missense changes in *BEST1*. These variants may have an effect on splicing[25] and there is evidence supporting the hypothesis that they are associated with a gain-of-function effect (i.e. lead to increased anion (Cl^-) transport) [45]. Mutations associated with autosomal dominant vitreoretinochoroidopathy include c.256G>A (p.Val86Met); c.704T>C (p.Val235Ala); c.707A>G (p.Tyr236Cys) and c.715G>A (p.Val239Met).

A classification scheme for *BEST1* mutations has been proposed and further details on this can be found in Ref. 38.

Clinical management

At present, there are no therapies to reverse vision loss in people with bestrophinopathies, and so management is mainly supportive. Affected children should be followed up regularly to ensure that adequate refraction is performed and that appropriate glasses are prescribed to avoid amblyopia. A decrease in best-corrected vision should always be taken seriously and there should be a low threshold for examining with dilated fundoscopy, OCT and, if required, angiography (fluorescein and/or OCT-A) to exclude the presence of the subretinal neovascular membrane. Such membranes react very well to anti-VEGF intravitreal injections but the importance of early treatment cannot be overstated.

Careful assessment and regular monitoring for narrow iridocorneal angles and associated glaucoma is recommended in all patients with bestrophinopathies. This is particularly important in individuals with biallelic mutations as anterior chamber depth is shallower and angle-closure risk is greater compared to those with single heterozygous variants.[19] Anti-glaucoma medications and prophylactic/therapeutic YAG laser peripheral iridotomy should be considered in relevant cases. If the intraocular pressure remains suboptimally controlled, glaucoma surgery is indicated. Filtration procedures (including trabeculectomy) should be approached with caution as there is a high risk of aqueous misdirection; deep sclerectomy or pars plana vitrectomy with pars plana drainage tube insertion can be considered as a primary procedure.[20,39]

Regular (e.g. annual) follow-up to evaluate the progression of the retinopathy should be offered. Among others, this can help appropriately inform patients regarding abilities like driving, disability status and, if required, the prescription of low vision aids.

Cystoid oedema in patients with autosomal recessive bestrophinopathy can be treated with carbonic anhydrase inhibitors, which can reduce oedema, and improve visual acuity in some cases.[16] Presenile cataracts in autosomal dominant vitreoretinochoroidopathy may need removal if they are a major reason for decreased best-corrected visual acuity.

All patients should have their clinical diagnosis confirmed by genetic testing. Molecular diagnostic confirmation provides the patient and the family with appropriate reproductive advice. Counselling regarding autosomal dominant inheritance with a recurrence risk of 50% is essential in families with Best vitelliform macular dystrophy and autosomal dominant vitreoretinochoroidopathy. All first-degree relatives to such patients should be given the option to undergo a clinical, electrophysiological and molecular genetic evaluation if so desired. Patients with autosomal recessive bestrophinopathy require genetic counselling focusing on autosomal recessive inheritance. Segregation analysis in these cases helps confirm the biallelic character of the disease-associated variants.

Naturally occurring dog models with canine multifocal retinopathy due to biallelic *BEST1* mutations have been described. Gene therapy using subretinal injections of adeno-associated virus carrying a copy of the *BEST1* gene has been shown to be successful in this large animal model. It is expected that such treatments will become available to human patients with autosomal recessive bestrophinopathy in the coming years.[40,41] For Best vitelliform macular dystrophy studies using approaches based on gene augmentation, CRISPR-Cas9 editing or small molecule therapies have shown considerable promise in human induced pluripotent stem cell-derived RPE models.[42,43]

References

1. Adams JE. Case showing peculiar changes in the macula. *Trans Ophthalmol Soc UK* 1883;**3**:113.
2. Best F. Ueber eine hereditare Maculaaffektion: Beitrage zur Vererbungslehre. *Z Augenheilkd* 1905;**13**:199.
3. Nordstrom S. Hereditary macular degeneration – a population survey in the country of Vsterbotten, Sweden. *Hereditas* 1974;**78**(1):41–62.
4. Bitner H, et al. Frequency, genotype, and clinical spectrum of best vitelliform macular dystrophy: data from a national center in Denmark. *Am J Ophthalmol* 2012;**154**(2):403–12. e4.
5. Coussa RG, et al. Predominance of hyperopia in autosomal dominant Best vitelliform macular dystrophy. *Br J Ophthalmol* 2020. https://doi.org/10.1136/bjophthalmol-2020-317763, bjophthalmol-2020-317763.
6. Liu J, et al. Novel BEST1 mutations and special clinical features of best vitelliform macular dystrophy. *Ophthalmic Res* 2016;**56**(4):178–85.
7. Arora R, et al. Unilateral BEST1-associated retinopathy. *Am J Ophthalmol* 2016;**169**:24–32.
8. Constable PA, et al. Correction to: ISCEV standard for clinical electro-oculography (2017 update). *Doc Ophthalmol* 2018;**136**(2):155.
9. Constable PA, et al. ISCEV standard for clinical electro-oculography (2017 update). *Doc Ophthalmol* 2017;**134**(1):1–9.
10. Boon CJ, et al. Clinical and molecular genetic analysis of best vitelliform macular dystrophy. *Retina* 2009;**29**(6):835–47.
11. Khan KN, et al. The fundus phenotype associated with the p.Ala243Val BEST1 mutation. *Retina* 2018;**38**(3):606–13.
12. Yu K, Cui Y, Hartzell HC. The bestrophin mutation A243V, linked to adult-onset vitelliform macular dystrophy, impairs its chloride channel function. *Invest Ophthalmol Vis Sci* 2006;**47**(11):4956–61.
13. Yu K, et al. Chloride channel activity of bestrophin mutants associated with mild or late-onset macular degeneration. *Invest Ophthalmol Vis Sci* 2007;**48**(10):4694–705.
14. Boon CJF, Leroy BP. The bestrophinopathies. In: De Laey J-J, Puech B, Holder GE, editors. *Inherited chorioretinal dystrophies*. Berlin: Springer; 2014. p. 197–212.
15. Burgess R, et al. Biallelic mutation of BEST1 causes a distinct retinopathy in humans. *Am J Hum Genet* 2008;**82**(1):19–31.
16. Boon CJ, et al. Autosomal recessive bestrophinopathy: differential diagnosis and treatment options. *Ophthalmology* 2013;**120**(4):809–20.
17. Leroy BP. Bestrophinopathies. In: Traboulsi EI, editor. *Genetic diseases of the eye*. Oxford University Press; 2011. p. 426–36.
18. Casalino G, et al. Autosomal recessive bestrophinopathy: clinical features, natural history and genetic findings in preparation for clinical trials. *Ophthalmology* 2020;**128**(5):706–18.
19. Xuan Y, et al. The clinical features and genetic Spectrum of a large cohort of Chinese patients with vitelliform macular dystrophies. *Am J Ophthalmol* 2020;**216**:69–79.
20. Low S, et al. A new paradigm for delivering personalised care: integrating genetics with surgical interventions in BEST1 mutations. *Eye (Lond)* 2020;**34**(3):577–83.
21. Borman AD, et al. Childhood-onset autosomal recessive bestrophinopathy. *Archiv Ophthalmol* 2011;**129**(8):1088–93.
22. Lafaut BA, et al. Clinical and electrophysiological findings in autosomal dominant vitreoretinochoroidopathy: report of a new pedigree. *Graefes Arch Clin Exp Ophthalmol* 2001;**239**(8):575–82.
23. Hermann P. Associated microphthalmia: microphthalmia-pigmentary retinitis-glaucoma syndrome. *Bull Soc Ophtalmol Fr* 1958;**1**:42–5.
24. Kaufman SJ, et al. Autosomal dominant vitreoretinochoroidopathy. *Arch Ophthalmol* 1982;**100**(2):272–8.
25. Yardley J, et al. Mutations of VMD2 splicing regulators cause nanophthalmos and autosomal dominant vitreoretinochoroidopathy (ADVIRC). *Invest Ophthalmol Vis Sci* 2004;**45**(10):3683–9.
26. Reddy MA, et al. A clinical and molecular genetic study of a rare dominantly inherited syndrome (MRCS) comprising of microcornea, rod-cone dystrophy, cataract, and posterior staphyloma. *Br J Ophthalmol* 2003;**87**(2):197–202.
27. Vincent A, et al. BEST1-related autosomal dominant vitreoretinochoroidopathy: a degenerative disease with a range of developmental ocular anomalies. *Eye (Lond)* 2011;**25**(1):113–8.
28. Chen CJ, et al. Long-term macular changes in the first proband of autosomal dominant vitreoretinochoroidopathy (ADVIRC) due to a newly identified mutation in BEST1. *Ophthalmic Genet* 2016;**37**(1):102–8.
29. Blair NP, et al. Autosomal dominant vitreoretinochoroidopathy (ADVIRC). *Br J Ophthalmol* 1984;**68**(1):2–9.
30. Dalvin LA, et al. Retinitis pigmentosa associated with a mutation in BEST1. *Am J Ophthalmol Case Rep* 2016;**2**:11–7.
31. Davidson AE, et al. Missense mutations in a retinal pigment epithelium protein, bestrophin-1, cause retinitis pigmentosa. *Am J Hum Genet* 2009;**85**(5):581–92.

32. Shah M, et al. Association of clinical and genetic heterogeneity with BEST1 sequence variations. *JAMA Ophthalmol* 2020;**138**(5):544–51.
33. Petrukhin K, et al. Identification of the gene responsible for Best macular dystrophy. *Nat Genet* 1998;**19**(3):241–7.
34. Marmorstein AD, Cross HE, Peachey NS. Functional roles of bestrophins in ocular epithelia. *Prog Retin Eye Res* 2009;**28**(3):206–26.
35. Yang T, et al. Structure and selectivity in bestrophin ion channels. *Science* 2014;**346**(6207):355–9.
36. Guziewicz KE, et al. Bestrophinopathy: an RPE-photoreceptor interface disease. *Prog Retin Eye Res* 2017;**58**:70–88.
37. Bitner H, et al. A homozygous frameshift mutation in BEST1 causes the classical form of Best disease in an autosomal recessive mode. *Invest Ophthalmol Vis Sci* 2011;**52**(8):5332–8.
38. Nachtigal AL, et al. Mutation-dependent pathomechanisms determine the phenotype in the bestrophinopathies. *Int J Mol Sci* 2020;**21**(5):1597.
39. Shi Y, et al. Pathogenic role of the vitreous in angle-closure glaucoma with autosomal recessive bestrophinopathy: a case report. *BMC Ophthalmol* 2020;**20**(1):271.
40. Guziewicz KE, et al. BEST1 gene therapy corrects a diffuse retina-wide microdetachment modulated by light exposure. *Proc Natl Acad Sci U S A* 2018;**115**(12):E2839–48.
41. Guziewicz KE, et al. *AAV-mediated gene therapy for bestrophinopathies*. Berlin, Germany: International Society for Eye Research; 2012.
42. Liu J, et al. Small molecules restore Bestrophin 1 expression and function of both dominant and recessive bestrophinopathies in patient-derived retinal pigment epithelium. *Invest Ophthalmol Vis Sci* 2020;**61**(5):28.
43. Sinha D, et al. Human iPSC modeling reveals mutation-specific responses to gene therapy in a genotypically diverse dominant maculopathy. *Am J Hum Genet* 2020;**107**(2):278–92.

13A.10

Pattern dystrophies

Omar A. Mahroo

Pattern dystrophies of the RPE are a genetically and phenotypically diverse group of disorders, with variable effects on visual function. They usually, but not exclusively, present in adulthood and are characterised by macular abnormalities that are apparent on fundus examination. Pigmentary changes and yellow-white retinal lesions are common, and the findings are often more evident on fundus autofluorescence imaging than on clinical examination or colour fundus photography.

Certain forms of pattern dystrophy are Mendelian traits and are caused by variants in one of several genes, most commonly *PRPH2* (Figs 13A.45 and 13A.46), *BEST1* (Section 13A.9) and *ABCA4* (Section 13A.8). Multisystemic syndromes and toxicity to certain medications

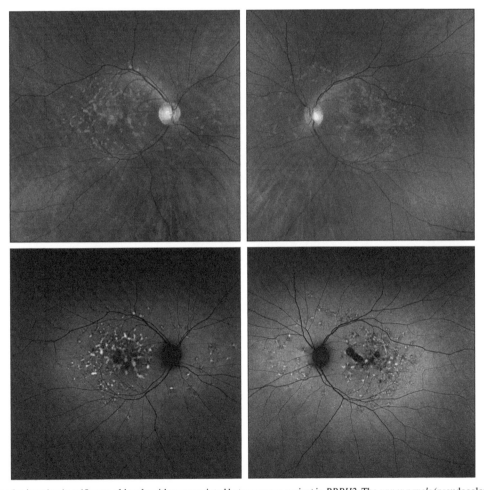

FIG. 13A.45 Fundus imaging in a 45-year-old male with an associated heterozygous variant in *PRPH2*. The *upper panels* (pseudocolour fundus images) show multiple yellow deposits and macular pigmentary abnormalities. The *lower panels* correspond to fundus autofluorescence images (532 nm) and demonstrate that the yellow lesions in the upper panels are hyperautofluorescent in the lower panels. The black areas in the lower panels represent zones of atrophy. It is worth noting that retinal changes can vary widely in association with variants in this gene, ranging from a few deposits, sometimes with minimal impact on visual acuity, to large zones of outer retinal atrophy. In general, the multifocal deposits seen in *PRPH2*-related pattern dystrophy have similarities to those seen in *ABCA4*-retinopathy (Stargardt disease). However, by this age, more atrophy and visual impairment are usually seen in *ABCA4*-retinopathy.

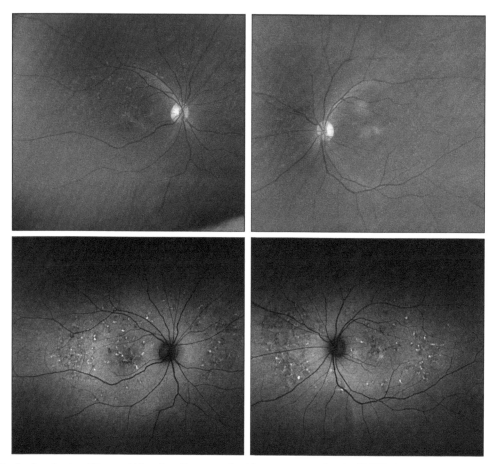

FIG. 13A.46 Fundus imaging in a 58-year-old female with an associated heterozygous variant in *PRPH2*. The *upper panels* (pseudocolour fundus images) show multiple yellow deposits and macular pigmentary abnormalities. The *lower panels* show autofluorescence images (532 nm), demonstrating that many of the yellow lesions show increased hyperautofluorescence. The left macula had choroidal neovascularisation (resulting in an altered macular reflex, sub-retinal fluid and impaired visual acuity), which was treated with intravitreal anti-VEGF injections.

TABLE 13A.4 Classification of pattern dystrophies by aetiology.

Category	Aetiology/associations
Non-syndromic	Idiopathic
	Mendelian (heterozygous variants in *PRPH2*, *BEST1*, *CTNNA1*, *IMPG1*, *IMPG2*, *ELOVL4*; biallelic variants in *ABCA4*)
Syndromic	MIDD/MELAS, pseudoxanthoma elasticum, myotonic dystrophy, McArdle disease, Kjellin syndrome
Medication-associated	Desferrioxamine, pentosan polysulfate[a]

[a]*Pentosan polysulfate, a drug used for bladder pain syndrome, has been associated with a pigmentary maculopathy; desferrioxamine, an iron-chelating agent, has been associated with macular changes, which can include vitelliform lesions.*
MIDD, maternally inherited diabetes and deafness; MELAS, mitochondrial myopathy, encephalopathy, lactic acidosis and stroke-like episodes.

(Table 13A.4) are implicated in a minority of pattern dystrophy cases. Hence, medical history, medication history, family history and examination of relatives are key. It is worth noting that probands often have no affected family members, even in cases where the associated genetic variant is dominantly inherited; this suggests that incomplete penetrance and variable expressivity are common in this group of conditions.[1,2]

Traditionally, pattern dystrophies were classified by clinical appearance into five categories: adult-onset foveomacular vitelliform dystrophy; butterfly-shaped pigment dystrophy; reticular dystrophy; multifocal pattern dystrophy simulating Stargardt disease; fundus pulverulentus. More recently, a genetic classification is becoming increasingly relevant, with the recognition that variants in some genes can give rise to a range of phenotypes.[1-3]

Clinical characteristics

In many cases, people with pattern dystrophies are asymptomatic and fundus abnormalities are identified as part of a routine optometric/ophthalmic examination. In other cases, affected individuals present with mildly impaired central or paracentral vision, which can include symptoms of distortion. At the most severe end of the spectrum, there is extensive outer retinal atrophy and central vision is severely impaired bilaterally. Subretinal or choroidal neovascularisation may develop in a subset of cases.[1,2]

The main sign on ophthalmic examination is macular pigmentary changes in both eyes; this is often associated with yellow-white outer retinal deposits. A central yellow-white subretinal lesion (vitelliform lesion) may be seen in some cases, classically in Best disease (a condition associated with heterozygous variants in *BEST1*; Section 13A.9).[3,4] Often the pigmentary abnormalities are semi-linear, sometimes radiating from the foveal centre, for example in a butterfly shape. Areas of atrophy may be seen, and these can be multifocal, with associated areas of visual loss. Short (blue; 488 nm) and medium (green; 532 nm) wavelength autofluorescence imaging may show particular patterns, including areas of hyper- and hypoautofluorescence that are multifocal, reticular or semi-linear. OCT imaging frequently shows abnormalities at the level of the outer retina and/or the RPE; these include hyper-reflective material deposition and areas of loss/disruption of outer retinal layers. These changes are evident on near-infrared reflectance imaging usually as areas of increased signal (Fig. 13A.47). It is worth highlighting that outer retinal abnormalities with similarities to those seen in pattern dystrophies may be seen in age-related macular degeneration and pachychoroid spectrum diseases (including central serous chorioretinopathy). These common conditions, therefore, represent key differential diagnoses. However, if the findings are reasonably symmetrical, particularly in young patients, and especially if other family members are affected (with similar appearances on imaging even if they are asymptomatic), then a Mendelian pattern dystrophy is more likely.[1,5,6]

A pattern dystrophy phenotype may be one of the many manifestations of a multisystemic disorder.[7-9] In the majority of cases, the diagnosis of a syndromic condition precedes the detection of retinal changes (although the retinal changes can aid diagnosis). In two syndromes, in particular, the retinal features may point to the underlying disorder, namely maternally inherited diabetes and deafness (MIDD) (Section 13B.2.4) and pseudoxanthoma elasticum PXE (Section 13B.2.3).

MIDD (associated with the m.3243A>G mitochondrial DNA variant) might be suspected in a patient with diabetes and parafoveal areas of atrophy associated with widespread irregular autofluorescence that usually extends over the macula and nasal to the optic disc (Fig. 13A.48). Patients may have a maternal history of diabetes and/or deafness and are typically reported to have more severe nephropathy, but less severe diabetic retinopathy, than diabetic patients in general.[10]

Pseudoxanthoma elasticum (associated with biallelic variants in the *ABCC6* gene) should be suspected in a patient with angioid streaks. These irregular, red/brown/grey lines that typically radiate from the area around the optic nerve may also be a feature of Paget disease (a chronic, slowly progressive skeletal condition), sickle cell disease, thalassaemia, or abetalipoproteinaemia. In pseudoxanthoma elasticum, there may also be mottling of the RPE (so-called 'peau d'orange' changes) and/or hyper-reflectivity of Bruch's membrane on OCT (Fig. 13A.49). Subretinal fluid that may or may not be associated with choroidal neovascularisation is observed in some cases.[7] Characteristic skin changes are typically noted and other systemic abnormalities may also be observed.

The key phenotypic features of the various genetic subtypes of pattern dystrophy are discussed in Table 13A.5 (non-syndromic conditions) and Table 13A.6 (syndromic disorders). Figs 13A.45–13A.47 (non-syndromic conditions) and Figs 13A.48–13A.51 (syndromic conditions) illustrate several retinal findings associated with these disorders.

Electrophysiological testing is helpful in pattern dystrophies. The pattern ERG (which evaluates macular cone-driven retinal function) is variably affected in these disorders: in cases with pigmentary changes only (such as butterfly-shaped pigment dystrophy), the pattern ERG is often normal; in patients with significant central retinal atrophy, the pattern ERG is usually subnormal and can be undetectable. The full-field ERG evaluates generalised rod system and cone system function. In macular dystrophies, including the majority of macular pattern dystrophies, the full-field ERG tends to be within normal limits. The electro-oculogram (EOG) light peak:dark trough (LP:DT) ratio, which evaluates the general health of the RPE, may be affected in some cases: in *BEST1*-associated vitelliform macular dystrophy, the LP:DT ratio is almost always profoundly subnormal even when fundal changes are minimal (Fig. 6.8); in *CTNNA1*-associated pattern dystrophy, the LP:DT ratio is often, but not always, subnormal.[1,11]

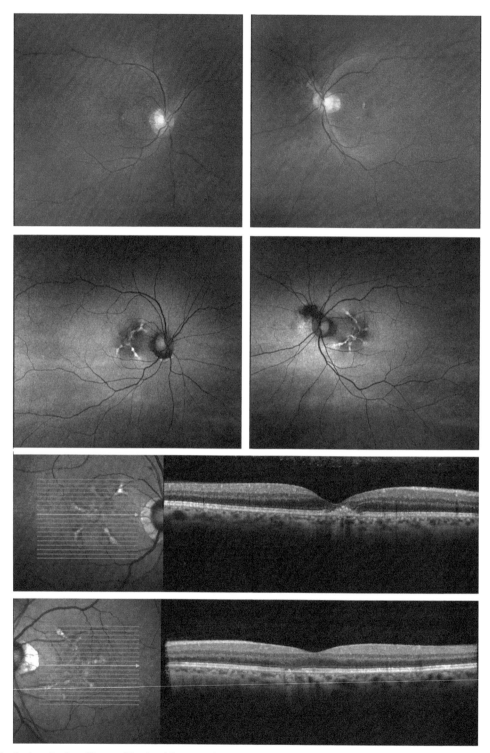

FIG. 13A.47 Fundus imaging in a 59-year-old male with an associated heterozygous variant in *CTNNA1*. The *upper panels* are pseudocolour fundus images, depicting linear pigmentary anomalies, and autofluorescence images (532 nm) showing linear areas of hyperautofluorescence. The *lower panels* are OCT scans through the fovea showing areas of the abnormal outer retina. On the *left* in the *lower panels* are near-infrared reflectance images; sometimes abnormalities are more evident on these images.

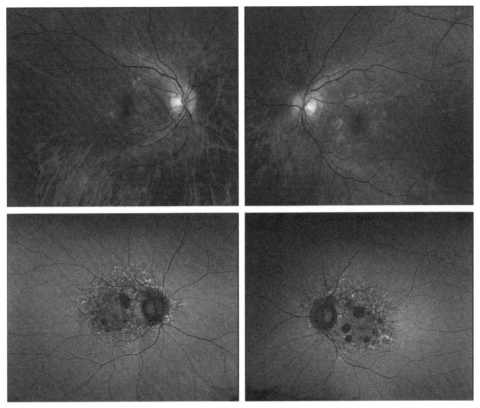

FIG. 13A.48 Fundus imaging from a 38-year-old female with the mitochondrial variant associated with maternally inherited diabetes and deafness (MIDD). The *upper panels* show pseudocolour fundus images and the *lower panels* show autofluorescence (532 nm) images. Over time, the atrophic areas enlarged leading to more severe impairment of vision.

Molecular pathology

Pattern dystrophies are genetically heterogeneous and more than 13 genes have been implicated (see Tables 13A.5 and 13A.6 for a list). *PRPH2*, *BEST1* and *ABCA4* are the most commonly mutated genes. The latter two are discussed in detail in Sections 13A.8 and 13A.9 but two points about *BEST1*-retinopathy are worth also highlighting here: that uniocular presentation may occur[12] and that certain variants, including c.728C>T (p.Ala243Val), are consistently associated with adult-onset pattern dystrophy phenotypes.[13]

PRPH2 encodes peripherin-2, a protein that plays a structural role in the outer segment discs in rod and cone photoreceptors. Mutations in this gene are associated with a wide variety of phenotypes including pattern dystrophy, central areolar choroidal dystrophy, cone-rod dystrophy and rod-cone dystrophy.[14] Many missenses, nonsense, indel and splice site mutations have been described and most of these variants cause disease in the heterozygous state. It has been speculated that variants leading to *PRPH2* haploinsufficiency are generally associated with rod-dominated conditions whereas variants with dominant negative effects are associated with cone-dominated pathologies including macular pattern dystrophy.[1]

A small subset of patients with pattern dystrophy has mutations in either the *IMPG1* or the *IMPG2* gene. These genes encode components of the interphotoreceptor matrix and have been associated with a range of adult-onset phenotypes including macular pattern dystrophy and rod-cone dystrophy. Many mutations in these genes cause disease in a heterozygous state but a subset of variants have been implicated in autosomal recessive disease.[13,15,16]

In addition to the genes listed in Tables 13A.5 and 13A.6, other molecules have been associated with pattern dystrophies in small numbers of cases. For example, a missense variant in *OTX2* has been implicated in autosomal dominant pattern dystrophy in at least two families.[17] Affected individuals variably had additional developmental anomalies; this is unsurprising as defects in this gene have been previously associated with severe developmental abnormalities[18] In addition, a pattern dystrophy phenotype has been reported in a female patient with a variant in *RP2*, a gene associated with X-linked retinitis pigmentosa.[19] Also, an association has been drawn between late

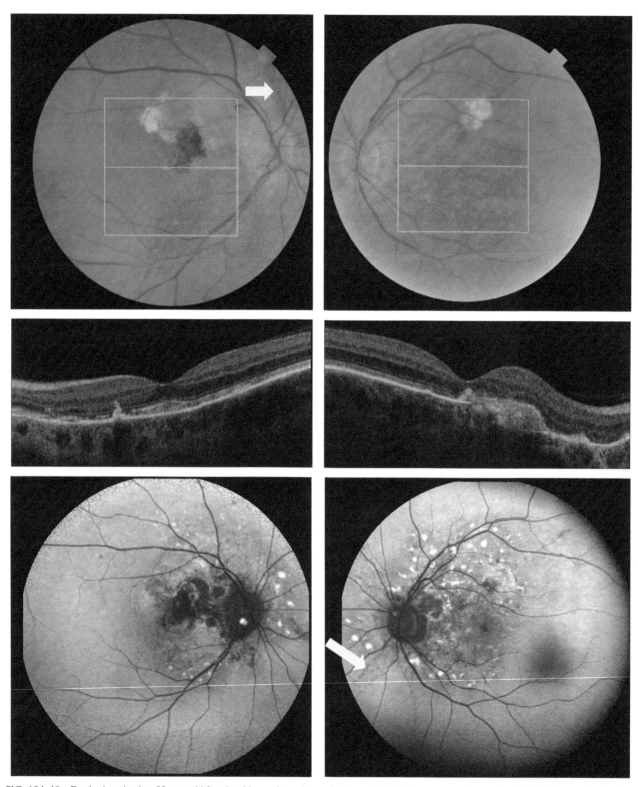

FIG. 13A.49 Fundus imaging in a 55-year-old female with pseudoxanthoma elasticum (PXE). The upper panels are colour fundus photographs showing angioid streaks radiating from the optic disc (one is highlighted by the *yellow arrow*). Other features include a mottled appearance and fibrosis. The *middle panels* are OCT scans showing increased reflectivity of Bruch's membrane and areas of outer retinal material deposition (including a lesion associated with choroidal neovascularisation in the left eye). The *lower panels* are autofluorescence images (488 nm), showing multiple areas of increased and decreased signal. Sometimes angioid streaks are more apparent on the autofluorescence images *(yellow arrow, bottom right panel)*. Optic disc drusen are occasionally associated with this condition (evident as an area of hyperautofluorescence within the right eye optic disc, *bottom left panel*).

TABLE 13A.5 Main genes associated with non-syndromic pattern dystrophy together with the associated phenotypic features.

Gene	Chromosomal location	Protein product	Mode of inheritance	Pattern dystrophy phenotypes reported
PRPH2	6p21.1	Peripherin 2	Autosomal dominant[a]	Vitelliform lesions; multifocal deposits and atrophy; butterfly-shaped pigment dystrophy
BEST1	11q12.3	Bestrophin 1	Autosomal dominant[a]	Vitelliform lesions; also multifocal deposits and atrophy
ABCA4	1p22.1	ATP-binding cassette A4	Autosomal recessive	Multifocal deposits and central atrophy; peripapillary sparing
ELOVL4	6q14.1	Elongation of very long fatty acids protein 4	Autosomal dominant	Multifocal deposits, areas of atrophy, Stargardt-like appearance
CTNNA1	5q31.2	Alpha catenin 1	Autosomal dominant	Butterfly-shaped pigment dystrophy
IMPG1	6q14.1	Interphotoreceptor matrix proteoglycan 1	Autosomal dominant[a]	Vitelliform lesions
IMPG2	3q12.3	Interphotoreceptor matrix proteoglycan 2	Autosomal dominant[a]	Vitelliform lesions

[a]Recessively inherited biallelic variants in PRPH2, BEST1, IMPG1 and IMPG2 give rise to a more widespread retinal dystrophy.
PRPH2 was also previously known as RDS. Variants in ABCA4 give rise to Stargardt disease (Section 13A.8), but features can resemble pattern dystrophy in some cases.

TABLE 13A.6 Selected genetic syndromes in which a pattern dystrophy phenotype might be seen.

Syndrome	Gene	Chromosomal location	Protein product	Main mode of inheritance	Ophthalmic features	Systemic features
MIDD/ MELAS	MTTL1	Mitochondrial	Mitochondrial transfer RNA for leucine	Mitochondrial	Parafoveal areas of atrophy; irregular autofluorescence in posterior pole; may have ptosis or ophthalmoplegia	MIDD: diabetes, deafness, variably might also have ptosis, hearing problems, cardiac conduction abnormalities, nephropathy. MELAS: headaches, vomiting, convulsions, deafness, myopathy, neuropathy
PXE	ABCC6	16p13.11	ATP-binding cassette C6	Autosomal recessive	Angioid streaks; irregular autofluorescence; CNV; subretinal drusenoid deposits; RPE mottling (peau d'orange); subretinal fluid; comet lesions; hyper-reflective Bruch's membrane on OCT	Abnormal mineralisation of skin and other organs; arterial calcification; coronary artery fibrosis; GI bleeds
Myotonic dystrophy	DMPK	19q13.32	Dystrophia myotonica protein kinase	Autosomal dominant	Butterfly-shaped pigment dystrophy; reticular pattern dystrophy; peripheral atrophic polygonal-shaped retinal changes; cataracts	Myotonia, muscular dystrophy, hypogonadism, cardiac conduction abnormalities

Continued

TABLE 13A.6 Selected genetic syndromes in which a pattern dystrophy phenotype might be seen—cont'd

Syndrome	Gene	Chromosomal location	Protein product	Main mode of inheritance	Ophthalmic features	Systemic features
McArdle disease	PYGM	11q13.1	Muscle isoform of glycogen phosphorylase	Autosomal recessive	Outer retinal deposits, areas of atrophy (in some cases)	Exercise intolerance, rhabdomyolysis
Kjellin syndrome	ZFYVE26	14q24.1	Zinc finger FYVE containing domain-containing protein 26	Autosomal recessive	Central retinal degeneration, flecks (carriers might have flecks without systemic features), pigmentary abnormalities and atrophy	Progressive spastic paraplegia and other features of neurologic dysfunction, thin corpus callosum
	SPG11[a]	15q21.1	Spatacsin			

[a]Variants in SPG11 are also associated with other diseases.
MIDD, maternally inherited diabetes and deafness; MELAS, mitochondrial myopathy, encephalopathy, lactic acidosis, and stroke-like episodes (see also Section 13B.2.4). PXE, pseudoxanthoma elasticum (see also Section 13B.2.3). CNV, choroidal neovascularisation.

adulthood onset pattern dystrophy and a single-nucleotide polymorphism in *HTRA1* (rs11200638).[20]

Most pattern dystrophies are dominantly inherited, and so examining and testing family members, particularly parents, can be informative. However, a large number of affected individuals have no relevant family history. This suggests that other factors, including polygenic predisposition, may play a role in disease pathogenesis. This model would explain the extensive intra- and interfamilial phenotypic variability observed in this group of conditions.[1] Importantly, this variability has implications for genetic counselling: although children of affected individuals have a 50% chance of inheriting a heterozygous genetic variant, given the variable mutation effects on visual function, there is a significantly lower chance of them developing severe visual impairment.

Genetic testing can help refine the risk of children/relatives being affected. It typically involves screening all known pattern dystrophy associated genes using panel-based tests or, increasingly, exome or genome sequencing. Notably, in cases with MIDD, the causative variant can be difficult to detect from peripheral blood samples using conventional methods, especially in elderly individuals (possibly due to selection against haematopoietic cells carrying the mutation). It is also worth highlighting that myotonic dystrophy, an important cause of syndromic pattern dystrophy, is associated with CTG trinucleotide repeats in the 3′ untranslated region of the *DMPK* gene; this type of genetic alteration requires targeted testing/analysis and has been shown to reduce the expression of the relevant gene.

Despite the discovery of multiple genes associated with pattern dystrophy, the diagnostic yield of genetic testing in this group of patients remains low.[1,15,16,21,22] However, the probability of a positive genetic test is increased when there is relevant family history and/or an earlier age of onset (e.g., <40 years).[22]

Clinical management

The visual acuity is often stable in many people with pattern dystrophies, especially when the fundus findings are incidentally detected. However, in some cases, vision can be severely affected, with the development of outer retinal atrophy. There are currently no medical interventions to reverse vision loss in this group of patients, and so management is mainly supportive, including optical aids and low vision assessments. Some pattern dystrophies can be complicated by choroidal neovascularisation. In these cases, intravitreal injections of anti-VEGF agents can help improve or preserve vision, especially if treatment is commenced early. Thus, patients should be advised to seek attention promptly if they notice the onset of an abrupt reduction in vision or distortion in the vision of either eye. As monocular pathology can go unnoticed, patients are advised to check each eye sequentially (by covering one eye at a time).

Unlike choroidal neovascularisation associated with age-related macular degeneration, there are no defined protocols for the number or frequency of intravitreal injections in pattern dystrophies complicated by neovascularisation. Treating physicians might undertake a trial of one or three

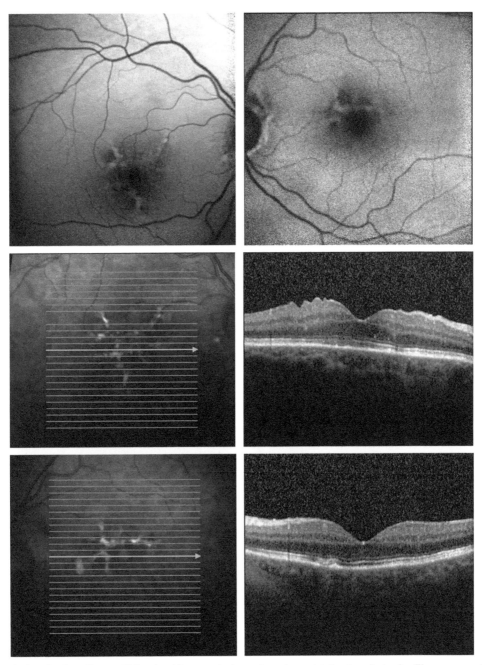

FIG. 13A.50 Fundus imaging in a 65-year-old female with myotonic dystrophy and associated pattern dystrophy. The *upper panels* show autofluorescence images (488 nm) of the macula in both eyes. The *middle panels* show near-infrared reflectance imaging and OCT from the right eye (the OCT shows an epiretinal membrane and outer retinal abnormalities). The *lower panels* show near-infrared reflectance imaging and OCT from the left eye. Generally, the abnormalities here appear more discernible on near-infrared reflectance than on autofluorescence images.

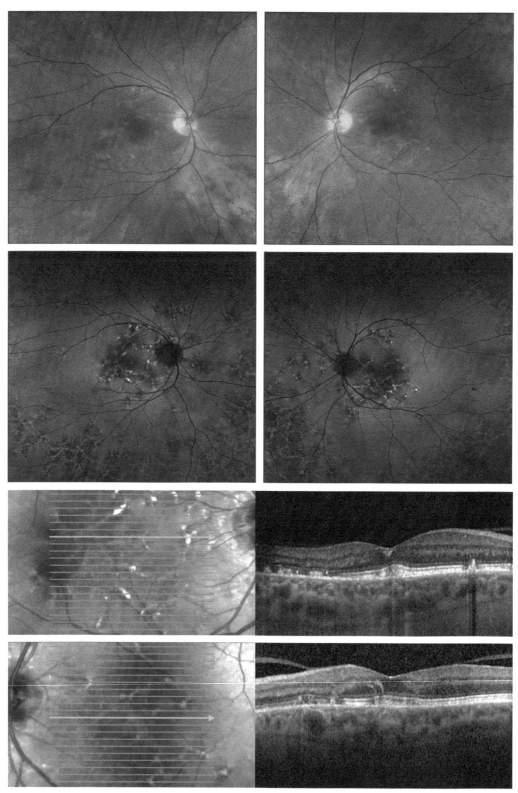

FIG. 13A.51 Fundus imaging in a 70-year-old male with McArdle disease (also known as glycogen storage disease type V), a metabolic muscle disorder associated with exercise intolerance and muscle pain. The proband was visually asymptomatic and was noted to have retinal changes on routine optometric examination. The *upper panels* show pseudocolour fundus images and the *middle panels* show fundus autofluorescence images (532 nm); there are pigmentary abnormalities at the macula and associated yellow-white linear lesions, many of which are associated with increased signal on autofluorescence imaging. The lower panels show near-infrared reflectance and OCT images; disruption and material deposition are noted at the level of the outer retina.

injections (each one month apart), and then evaluate response, before deciding on further treatment. It might be reasonable to continue monthly treatment until stability (in visual acuity or on retinal imaging), and then to observe (with treatment re-initiated if there are signs of activity), or to continue for a while, but with extended intervals between injections.

In the various syndromes listed in Table 13A.6, multidisciplinary specialist care is warranted, with referral to other specialities as appropriate.

Research is progressing in the search for therapies to prevent or reverse visual loss in non-exudative macular degeneration and some of these approaches (such as stem cell therapies) may be also relevant to pattern dystrophies. Gene replacement or editing approaches may also become relevant in the future. However, as visual loss is mild in many (but not all) cases, interventions with a substantial risk of complications may not be appropriate.

References

1. Chowers I, Tiosano L, Audo I, Grunin M, Boon CJ. Adult-onset foveomacular vitelliform dystrophy: a fresh perspective. *Prog Retin Eye Res* 2015;**47**:64–85.
2. Hanif AM, Yan J, Jain N. Pattern dystrophy: an imprecise diagnosis in the age of precision medicine. *Int Ophthalmol Clin* 2019;**59**(1):173–94. https://doi.org/10.1097/IIO.0000000000000262.
3. Rahman N, Georgiou M, Khan KN, Michaelides M. Macular dystrophies: clinical and imaging features, molecular genetics and therapeutic options. *Br J Ophthalmol* 2020;**104**(4):451–60.
4. Boon CJ, Klevering BJ, Leroy BP, Hoyng CB, Keunen JE, den Hollander AI. The spectrum of ocular phenotypes caused by mutations in the BEST1 gene. *Prog Retin Eye Res* 2009;**28**(3):187–205.
5. Khan KN, Mahroo OA, Khan RS, Mohamed MD, McKibbin M, Bird A, et al. Differentiating drusen: Drusen and drusen-like appearances associated with ageing, age-related macular degeneration, inherited eye disease and other pathological processes. *Prog Retin Eye Res* 2016;**53**:70–106. https://doi.org/10.1016/j.preteyeres.2016.04.008.
6. Saksens NT, Fleckenstein M, Schmitz-Valckenberg S, Holz FG, den Hollander AI, Keunen JE, et al. Macular dystrophies mimicking age-related macular degeneration. *Prog Retin Eye Res* 2014;**39**:23–57.
7. Agarwal A, Patel P, Adkins T, Gass JD. Spectrum of pattern dystrophy in pseudoxanthoma elasticum. *Arch Ophthalmol* 2005;**123**(7):923–8.
8. Kimizuka Y, Kiyosawa M, Tamai M, Takase S. Retinal changes in myotonic dystrophy. Clinical and follow-up evaluation. *Retina* 1993;**13**(2):129–35.
9. Mahroo OA, Khan KN, Wright G, Ockrim Z, Scalco RS, Robson AG, et al. Retinopathy associated with biallelic mutations in PYGM (McArdle disease). *Ophthalmology* 2019;**126**(2):320–2.
10. Massin P, Virally-Monod M, Vialettes B, Paques M, Gin H, Porokhov B, et al. Prevalence of macular pattern dystrophy in maternally inherited diabetes and deafness. GEDIAM Group. *Ophthalmology* 1999;**106**(9):1821–7.
11. Tanner A, Chan HW, Pulido JS, Arno G, Ba-Abbad R, Jurkute N, et al. Clinical and genetic findings in CTNNA1-associated macular pattern dystrophy. *Ophthalmology* 2020;**128**(6):952–5.
12. Arora R, Khan K, Kasilian ML, Strauss RW, Holder GE, Robson AG, et al. Unilateral BEST1-associated retinopathy. *Am J Ophthalmol* 2016;**169**:24–32.
13. Khan KN, Islam F, Moore AT, Michaelides M. The fundus phenotype associated with the p.Ala243Val BEST1 mutation. *Retina* 2018 Mar;**38**(3):606–13.
14. Boon CJ, den Hollander AI, Hoyng CB, Cremers FP, Klevering BJ, Keunen JE. The spectrum of retinal dystrophies caused by mutations in the peripherin/RDS gene. *Prog Retin Eye Res* 2008;**27**(2):213–35.
15. Meunier I, Manes G, Bocquet B, Marquette V, Baudoin C, Puech B, et al. Frequency and clinical pattern of vitelliform macular dystrophy caused by mutations of interphotoreceptor matrix IMPG1 and IMPG2 genes. *Ophthalmology* 2014;**121**(12):2406–14. https://doi.org/10.1016/j.ophtha.2014.06.028.
16. Guo J, Gao F, Tang W, Qi Y, Xuan Y, Liu W, et al. Novel BEST1 mutations detected by next-generation sequencing in a Chinese population with vitelliform macular dystrophy. *Retina* 2019;**39**(8):1613–22.
17. Vincent A, Forster N, Maynes JT, Paton TA, Billingsley G, Roslin NM, et al. OTX2 mutations cause autosomal dominant pattern dystrophy of the retinal pigment epithelium. *J Med Genet* 2014;**51**(12):797–805. https://doi.org/10.1136/jmedgenet-2014-102620.
18. Ragge NK, Brown AG, Poloschek CM, Lorenz B, Henderson RA, Clarke MP, et al. Heterozygous mutations of OTX2 cause severe ocular malformations. *Am J Hum Genet* 2005;**76**(6):1008–22. Erratum in: Am J Hum Genet. 2005 Aug;77(2):334.
19. Misky D, Guillaumie T, Baudoin C, Bocquet B, Beltran M, Kaplan J, et al. Pattern dystrophy in a female carrier of RP2 mutation. *Ophthalmic Genet* 2016;**37**(4):453–5. https://doi.org/10.3109/13816810.2015.1081253.
20. Jaouni T, Averbukh E, Burstyn-Cohen T, Grunin M, Banin E, Sharon D, et al. Association of pattern dystrophy with an HTRA1 single-nucleotide polymorphism. *Arch Ophthalmol* 2012;**130**(8):987–91. https://doi.org/10.1001/archophthalmol.2012.1483.
21. Meunier I, Sénéchal A, Dhaenens CM, Arndt C, Puech B, Defoort-Dhellemmes S, et al. Systematic screening of BEST1 and PRPH2 in juvenile and adult vitelliform macular dystrophies: a rationale for molecular analysis. *Ophthalmology* 2011;**118**(6):1130–6. https://doi.org/10.1016/j.ophtha.2010.10.010.
22. Grunin M, Tiosano L, Jaouni T, Averbukh E, Sharon D, Chowers I. Evaluation of the association of single nucleotide polymorphisms in the PRPH2 gene with adult-onset foveomacular vitelliform dystrophy. *Ophthalmic Genet* 2016;**37**(3):285–9.

13A.11

X-linked retinoschisis

Laryssa A. Huryn, Catherine A. Cukras, and Paul A. Sieving

X-linked retinoschisis is a non-syndromic congenital vitreoretinopathy that was first described in 1898. It is associated with mutations in the *RS1* gene, it is inherited as an X-linked recessive trait and female carriers are almost always asymptomatic. Affected males typically present with reduced vision in the first decade of life and the visual acuity remains relatively stable in the majority of cases.[1–4] Characteristic features include foveal schisis and an electronegative ERG (see below). About half of patients also have peripheral retinal schisis; these individuals are at higher risk of complications such as vitreous haemorrhage and retinal detachment.[5] The differential diagnosis of X-linked retinoschisis includes *CRB1*-maculopathy (an autosomal recessive condition associated with the presence of a specific hypomorphic mutation in the *CRB1* gene, c.498_506del (p.Ile167_Gly169del)[6]), enhanced S-cone syndrome (an autosomal recessive condition associated with mutations in the *NR2E3* gene, see Section 13A.3) and other genetic and inflammatory causes of macular oedema. Genetic testing is helpful to establish a precise diagnosis and to identify carriers at risk of passing on the condition.

Clinical characteristics

Clinical findings and disease severity in X-linked retinoschisis are highly variable, even within families. Affected males have reduced central visual acuity generally ranging from 0.30 to 0.80 LogMAR. This typically comes to parental and medical attention during early school-age years when children manifest difficulties with reading. Rarely, affected individuals present in infancy with sensory strabismus or nystagmus. Hypermetropia is a frequent finding and a family history indicating X-linked inheritance of visual abnormalities can help in the differential diagnosis. Colour vision is typically not compromised and patients do not complain of night vision problems.

Bilateral foveal schisis is present in virtually all patients at some point in their clinical history and may be the only sign of the disease. The foveal schisis appears in a classic 'spoke-wheel' pattern resulting from microcysts within the nerve fibre layer architecture (Figs 13A.52 and 13A.53). The schitic changes seen early in life typically collapse and develop into an atrophic stage, with areas of foveal thinning or macular atrophy observable on OCT. Peripheral findings are also noted in approximately half of the patients; these include bullous schisis involving the inferotemporal region and flat lamellar schisis sometimes containing dendritiform-appearing retinal vessels (Fig. 13A.54). Large bullous schitic areas may progress toward the central retina or flatten, leaving RPE abnormalities or retinal scars. Peripheral retinoschisis can cause absolute scotomas and evaluation of the peripheral visual field can help determine the extent of a patient's functional impairment.[1–5]

Vitreous changes may be present. These are often described as 'vitreous veils' and may be associated with delamination of partial-thickness retinal layers into the vitreous (Figs 13A.52 and 13A.55). Traction of superficial retinal vessels can occur and this may result in bleeding into the vitreous cavity. In the biomedical literature, the frequency of complications such as significant vitreous haemorrhage is reported to range from 3% to 21% while retinal detachment is reported to range from 5% to 40%.[2–5]

Affected individuals can have an unusual fundal sheen when the dark-adapted retina is exposed to light, termed the Mizuo phenomenon; this is also seen in Oguchi disease (Section 13A.4) and is believed to result from abnormal potassium processing in the retina. Other atypical presentations of X-linked retinoschisis include a macular hole, macular dragging with pigmentary changes, foveal ectopia, and retinal yellow-white flecks that may simulate fundus albipunctatus.[1,2,4]

OCT (including imaging obtained using hand-held devices) is very useful in detecting macular cysts. Even in areas where the fovea appears flat, OCT can reveal a shallow layer of lamellar schisis. Affected individuals often demonstrate a characteristic ERG response to a bright flash in dark-adapted conditions (DA 3 or DA 10 ERG): the b-wave amplitude is markedly and selectively reduced in the context of a preserved a-wave (Fig. 13A.52D). The term electronegative ERG has been used to denote that the ratio of b-wave amplitude to a-wave amplitude (b:a ratio) is less than 1. Electronegative ERGs suggest inner retina dysfunction; they are seen in several conditions[7] and are not pathognomonic of X-linked retinoschisis (although they are highly suggestive in the context of macular schisis).

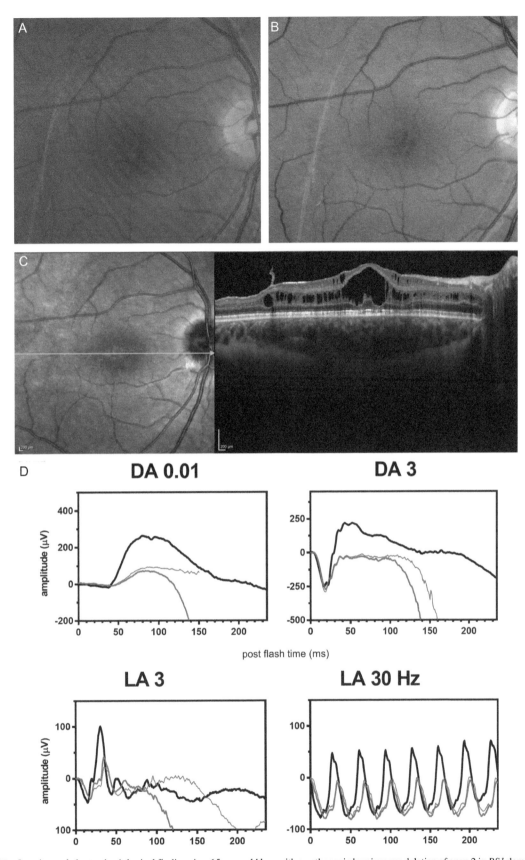

FIG. 13A.52 Imaging and electrophysiological findings in a 15-year-old boy with a pathogenic hemizygous deletion of exon 2 in *RS1* demonstrating the typical presentation of X-linked retinoschisis. (A) Colour fundus imaging showing foveal cysts and a temporal vitreous veil. (B) Red-free image highlighting the 'spoke-wheel' appearance of foveal cysts. (C) Macular OCT showing the classic retinal cystic changes at the fovea as well as the temporal attachment of the vitreous veil to the inner retina. (D) Full-field electroretinogram (ERG) in the dark-adapted state (DA 3) demonstrating the classic 'electronegative' waveform with a reduction of the b-wave in the presence of a preserved a-wave. Patient waveforms *(red)* are compared with a healthy individual *(black)*.

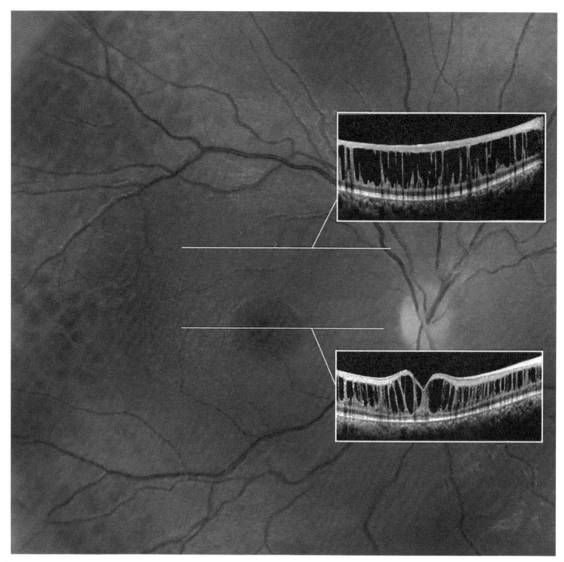

FIG. 13A.53 Classic OCT findings of X-linked retinoschisis in the macular region. A relatively flat, 'lamellar' schisis involving the posterior pole is noted.

Measures of acuity and ERG parameters tend to remain stable over time. This is in keeping with a relatively stationary natural history (unless complicated by retinal detachment or vitreous haemorrhage).[2]

Molecular pathology

Mutations in the *RS1* gene were first found to be the cause of X-linked retinoschisis in 1997. This gene is present across the animal kingdom, from slime mould to primates. In humans, it contains 6 exons and it encodes a 224-amino-acid protein called retinoschisin. Retinoschisin is produced and secreted by photoreceptor and bipolar cells and adheres to their cell surface. It is thought to function as an extracellular matrix and cell adhesion molecule and lack of *RS1* protein impair neural transmission between photoreceptors and bipolar cells.[1]

More than 300 mutations in *RS1* have been reported. Most variants are missense mutations that are clustered in the discoidin domain (amino-acids 62–219 in exons 4–6; known to facilitate cell adhesion) and lead to a misfolded protein. These changes are generally milder than variants encountered in exons 1–3 that typically lead to premature protein termination. It is worth noting that missense changes involving the addition or removal of a cysteine residue are associated with severe phenotypes, potentially through disrupting disulphide bonds important for tertiary folding.[8,9]

Genetic testing in X-linked retinoschisis has a high diagnostic yield and genotypic confirmation has a high degree of disease specificity.

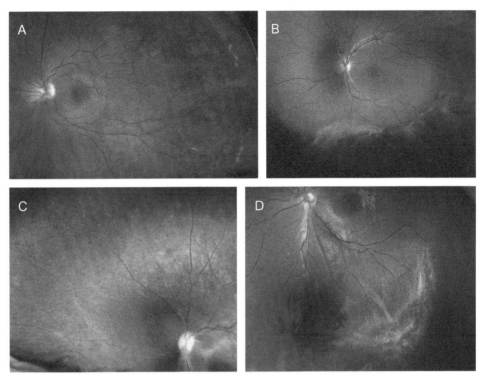

FIG. 13A.54 Peripheral retinal findings in patients with X-linked retinoschisis. (A) Peripheral schisis with inner retinal holes and vitreous veils. (B) Bullous schisis inferotemporal. (C) Peripheral lamellar schisis. (D) Dendritiform peripheral vessels.

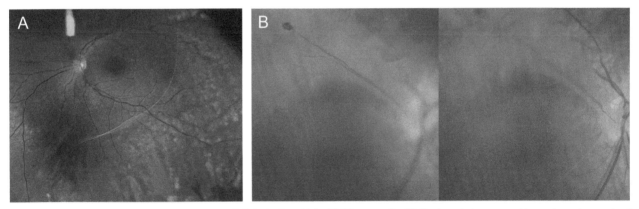

FIG. 13A.55 Vitreous abnormalities in patients with X-linked retinoschisis. (A) Vitreous veils. (B) Condensed vitreous attaching to the optic nerve head.

Clinical management

There are no approved curative treatments for X-linked retinoschisis at this time. However, various strategies have been explored in an attempt to improve retinal structure and function. Reports of the use of carbonic anhydrase inhibitors to decrease macular cysts have provided mixed results. The proposed therapeutic mechanism is an increase in the ability of the RPE to transport fluid from the apical side through the basement membrane with subsequent transport into the choroid. While variable degrees of diminished schisis cavity size have been reported, there is often little or no effect on visual acuity.[10] Although topical treatment (i.e. topical dorzolamide three times daily) is well tolerated and has a favourable safety profile, suitable studies with longer duration are needed to evaluate true efficacy. In our experience, this has little effect on functional outcomes.

Affected individuals are advised to refrain from contact sports and avoid activities with significant coup-contrecoup movement. Prophylactic laser or surgical treatment of peripheral schisis is not recommended. However, given the increased rates of vitreous haemorrhage and retinal detachment, surgical intervention is sometimes necessary. Surgery for rhegmatogenous retinal detachment can be challenging as there are often outer- and inner-wall holes in the area of schisis.

Hypermetropia and strabismus are common in X-linked retinoschisis. Any visual acuity disparity between eyes has the potential to contribute to amblyopia. It is often not possible to completely separate this from any biological disparity. Surveillance of affected individuals throughout the amblyogenic period of childhood should carefully evaluate and treat refractive errors, ocular misalignment, and amblyopia even when structural retinal changes are evident. Affected individuals often benefit from visual aids, including hand-held spectacle magnifiers and telescopes; referral to a low vision specialist should be therefore initiated when appropriate.

Genetic counselling is an important part of the evaluation and management of families with X-linked retinoschisis. A careful family history should be taken and women with an affected brother generally benefit from directed counselling, as they have a 50% chance of being a carrier. Female carriers have a 50% chance of transmitting the disease-causing mutation to their male children. Genotyping is recommended in this situation. Each request for carrier testing should be considered on its merit as there are ethical considerations to take into account when testing young females who have not yet reached the age of consent for carrier status.

X-linked retinoschisis gene therapy studies in humans are underway, intending to restore protein function and to explore possible improvement of visual function. Successful preclinical studies in a mouse model were observed and the safety and tolerability of an *RS1* adeno-associated virus (AAV8) administered in humans by intravitreal injection have been demonstrated.[11]

References

1. Molday RS, Kellner U, Weber BHF. X-linked juvenile retinoschisis: clinical diagnosis, genetic analysis, and molecular mechanisms. *Prog Retinal Eye Res* 2012;**31**:195–212.
2. Cukras CA, Huryn LA, Jeffrey BP, Turriff A, Sieving PA. Analysis of anatomic and functional measures in X-linked retinoschisis. *Investig Ophthalmol Vis Sci* 2018;**59**:2841–7.
3. Grigg JR, Hooper CY, Fraser CL, Cornish EE, McCluskey PJ, Jamieson RV. Outcome measures in juvenile X-linked retinoschisis: a systematic review. *Eye (Lond)* 2020;**34**(10):1760–9. https://doi.org/10.1038/s41433-020-0848-6.
4. Chen C, Xie Y, Sun T, Tian L, Xu K, Zhang X, et al. Clinical findings and RS1 genotype in 90 Chinese families with X-linked retinoschisis. *Mol Vis* 2020;**26**:291–8.
5. Fahim AT, Ali N, Blachley T, Michaelides M. Peripheral fundus findings in X-linked retinoschisis. *Br J Ophthalmol* 2017;**101**(11):1555–9.
6. Khan KN, Robson A, Mahroo OAR, Arno G, Inglehearn CF, Armengol M, et al. A clinical and molecular characterisation of CRB1-associated maculopathy. *Eur J Hum Genet* 2018;**26**(5):687–94.
7. Audo I, Robson AG, Holder GE, Moore AT. The negative ERG: clinical phenotypes and disease mechanisms of inner retinal dysfunction. *Surv Ophthalmol* 2008;**53**(1):16–40.
8. Sergeev YV, Caruso RC, Meltzer MR, Smaoui N, MacDonald IM, Sieving PA. Molecular modeling of retinoschisin with functional analysis of pathogenic mutations from human X-linked retinoschisis. *Hum Mol Genet* 2010;**19**:1302–13.
9. Sergeev YV, Vitale S, Sieving PA, Vincent A, Robson AG, Moore AT, et al. Molecular modeling indicates distinct classes of missense variants with mild and severe XLRS phenotypes. *Hum Mol Genet* 2013;**22**(23):4756–67.
10. Apushkin MA, Fishman GA. Use of dorzolamide for patients with X-linked retinoschisis. *Retina* 2006;**26**:741–5.
11. Cukras C, Wiley HE, Jeffrey BG, et al. Retinal AAV8-RS1 gene therapy for X-linked retinoschisis: initial findings from a phase I/IIa trial by intravitreal delivery. *Mol Ther* 2018;**26**:2282–94.

13A.12

Occult macular dystrophy

Yu Fujinami-Yokokawa, Anthony G. Robson, Panagiotis I. Sergouniotis, and Kaoru Fujinami

Occult macular dystrophy (OMD or OCMD) is a non-syndromic macular condition that was first described in 1989. Affected individuals typically report progressive worsening of visual acuity and have evidence of macular dysfunction on pattern or multifocal electroretinogram (ERG). Characteristically, fundus examination is unremarkable and full-field ERGs are typically within normal limits.[1-3]

A subset of patients with OMD is found to carry heterozygous pathogenic variants in the *RP1L1* gene.[4,5] These mutations are associated with an autosomal dominant mode of inheritance. Despite this, affected individuals often report having no relevant family history. Notably, examining first-degree relatives of patients may be informative even if these relatives are visually asymptomatic.[5,6]

OMD appears to be genetically heterogeneous and many people who are clinically diagnosed with the condition do not carry *RP1L1* mutations. It is expected that some of these cases will be due to high-effect genetic alterations in specific genes and will show familial clustering (non-*RP1L1* OMD). Other cases will be caused by the complex interplay between multiple genetic and environmental/lifestyle factors (multifactorial disease). Several *RP1L1* OMD phenocopies have also been described (Table 13A.7). A proposed classification is outlined in Fig. 13A.56.

TABLE 13A.7 Main differential diagnoses for *RP1L1* occult macular dystrophy (Miyake disease).

- Non-organic or functional visual loss.
- Solar retinopathy
- Laser-induced retinal injury
- Alkyl nitrite ('poppers') related maculopathy
- Macular dystrophies (including *ABCA4*-retinopathy)
- Cone dysfunction disorders (including achromatopsia)
- Cone/cone-rod dystrophies
- Congenital stationary night-blindness (including fundus albipunctatus)
- Age-related macular degeneration
- Autoimmune retinopathies
- Acute macular neuroretinopathy
- Optic neuropathies (including retrobulbar optic neuritis and dominant optic atrophy)

Clinical characteristics

RP1L1 occult macular dystrophy (Miyake disease)

The clinical presentation of this disorder is highly variable and the age of onset ranges from childhood to late adulthood (median 24 years).[5,7,8] A large study of affected individuals with East Asian ancestries highlighted that the most common symptoms at presentation are reduced visual acuity (~75%) and/or photophobia (~35%).[5] It is not unusual to diagnose the condition in an asymptomatic individual as part of a routine optometric/ophthalmic examination or through family work-up, prompted by a diagnosis in a family member.[7-9] A myopic refractive error is often noted although this is likely to be non-specific.[7,9]

The best-corrected visual acuity is variable, with reported median values ranging from 0.4 to 0.7 LogMAR.[3,6-8] In general, over three-quarters of cases retain vision better than 1.0 logMAR.[7,9] There is a significant correlation between LogMAR vision and age at presentation suggesting that early-onset disease tends to be associated with a more severe phenotype.[7,9]

Fundus examination is unremarkable, which can make the diagnosis challenging.[1-3] Fundus autofluorescence imaging is also within normal limits in many patients, but some affected individuals (~50%) have subtle abnormalities; these include a circular area of hyperautofluorescence at the fovea[3,7,10] or non-congruent parafoveal arcs of mildly increased signal. These fundus autofluorescence findings differ from the typical features of other cone-mediated retinal dystrophies (e.g. presence of a low autofluorescence signal surrounded by a ring of increased autofluorescence or multiple foci of increased and decreased signal).

Characteristic microstructural changes at the level of the photoreceptors help make a clinical diagnosis.[6-8,11] OCT imaging reveals irregularity ('blurring') of the ellipsoid zone (corresponding to photoreceptor outer segment abnormalities) in most patients (~80% of cases) (Fig. 13A.57); focal ellipsoid zone disruption at the foveola and/or absence of the interdigitation zone (representing the interdigitation of the apical processes of the RPE with the cone outer segments) at the macula may also be present.[5,6,11] A minority of patients have only very subtle photoreceptor

242 SECTION | III Genetic disorders affecting the posterior segment

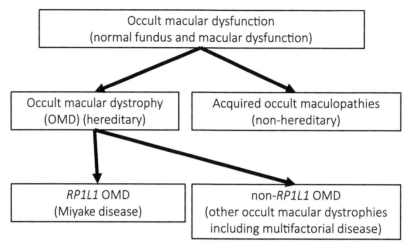

FIG. 13A.56 Classification of occult macular dysfunction disorders. Conditions associated with normal fundus and localised macular dysfunction can be broadly divided into hereditary and non-hereditary forms. Hereditary forms (i.e. occult macular dystrophy (OMD)) can be further divided into *RP1L1* OMD (*RP1L1*-associated occult macular dystrophy or Miyake disease) and non-*RP1L1* OMDs. Acquired occult maculopathies include non-hereditary phenocopies of *RP1L1* OMD.

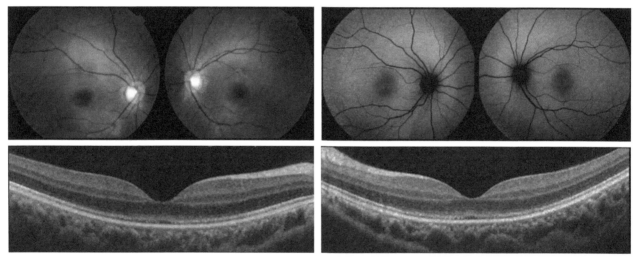

FIG. 13A.57 Fundus imaging in an individual with occult macular dystrophy and a heterozygous pathogenic variant in *RP1L1*, c.133C>T (p.Arg45Trp). Fundus photographs and autofluorescence imaging are within normal limits. On OCT, there is irregularity/blurring of the ellipsoid zone and loss of the interdigitation zone of the photoreceptors.

architectural changes; the ellipsoid zone appears preserved and central vision can be within normal limits in these cases.[7] Furthermore, in a subset of affected individuals, there is a degree of asymmetry between eyes in terms of OCT findings; this makes diagnosis more challenging.

Visual field testing by kinetic and static perimetry detects central loss of retinal sensitivity in most patients but this is not a universal finding.[7] Full-field ERG findings are within normal limits in the majority of cases but mild generalised cone system dysfunction (on light-adapted ERGs) has been noted in a small number of patients.[5] A key to diagnosis is the presence of macular dysfunction, as detected by multifocal ERG or pattern ERG (Fig. 13A.58).[1,2,5,12,13] It is highlighted that multifocal ERGs provide higher spatial resolution than standard pattern ERG methods, and may be more sensitive to highly localised foveal dysfunction.[5] Patients may be classified into three functional groups based on multifocal ERG findings:

- Group 1, paracentral dysfunction with relatively preserved central/peripheral function;
- Group 2, homogeneous central dysfunction with preserved peripheral function; and
- Group 3, widespread dysfunction over the recorded area.[13]

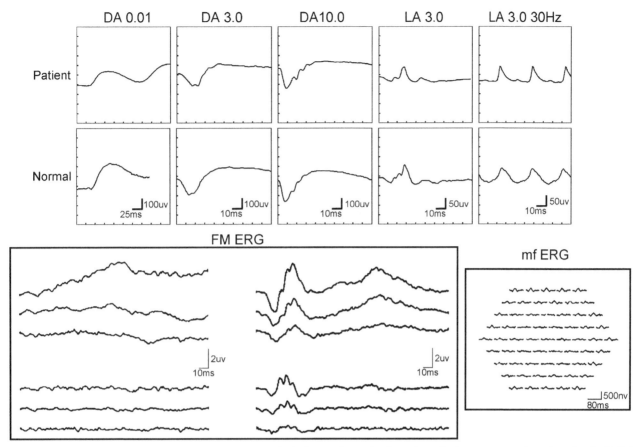

FIG. 13A.58 Electroretinographic (ERG) findings in an individual with occult macular dystrophy and a heterozygous pathogenic variant in *RP1L1*, c.133C>T (p.Arg45Trp). Traces from a representative normal subject are also shown for comparison. The full-field ERGs are normal in both dark-adapted (DA) and light-adapted (LA) conditions. Focal macular (FM) and multifocal (mf) ERGs demonstrate widespread amplitude reduction in the central retinal responses.

Non-*RP1L1* occult macular dystrophy

A proportion of people with OMD do not have pathogenic variants in *RP1L1*.[6] Efforts to clinically and genetically characterise this group of patients are ongoing. It is becoming evident though that there are differences in the retinal microstructural abnormalities compared to those seen in *RP1L1* OMD. Notably, OCT imaging highlights that the majority of patients with non-*RP1L1* OMD lack the classical findings of ellipsoid zone blurring and interdigitation zone loss.

Molecular pathology

Heterozygous variants in the *RP1L1* gene are an important cause of OMD. The *RP1L1* gene encodes RP1-like1, a 2401 amino-acid photoreceptor-specific protein originally identified based on its sequence similarity to the *RP1* gene.[14,15] The RP1 and RP1L1 proteins are thought to biochemically interact with each other and to have synergistic roles.[14,16] They exhibit identical temporal expression patterns and are likely to co-localise to the photoreceptor ciliary axoneme, a structure that has a role in the formation and maintenance of outer segment discs.[4,5,14,15] A heterozygous *RP1L1* knock-out mouse model was reported to have normal retinal morphology, while homozygous knock-out mice developed retinal degeneration (albeit, milder compared to their *RP1* homozygous knock-out counterparts).[14]

Multiple multigenerational families with *RP1L1* OMD from different parts of the world have been identified. Two mutational hotspots have emerged: one around amino-acid 45 and another between amino-acids 1196 and 1201. Most OMD-associated variants are missense changes and the most prevalent mutation is *RP1L1* c.133C > T (p.Arg45Trp).[6,17]

Several cases with biallelic truncating *RP1L1* variants have been reported.[5,8] Affected individuals with these genotypes develop adult-onset retinitis pigmentosa. The markedly different phenotypes between *RP1L1*-associated autosomal dominant OMD (predominantly affecting central macular cones) and *RP1L1*-associated autosomal recessive retinitis pigmentosa (predominantly involving rod photoreceptors) highlights the diverse molecular disease mechanisms of *RP1L1*-related disorders (e.g. a gain of function/dominant negative effect for OMD versus a loss of function mechanism in retinitis pigmentosa).[7] It is highlighted that the related gene *RP1* is also associated with several markedly different retinal phenotypes that may be inherited as autosomal dominant or autosomal recessive traits.[18]

Molecular investigations should be considered in all individuals with suspected OMD. Genetic testing can point to an aetiological diagnosis and inform clinical management, avoiding superfluous investigations and allowing better guidance concerning risks to family members. Generally, misdiagnosis or delayed diagnosis is common in patients with OMD who are often thought to have unexplained visual loss or suspected optic nerve disease (Table 13A.7). Given this, there should be a low threshold for genetic screening.

Targeted (single gene or panel-based analysis) or comprehensive (exome or genome sequencing) genetic testing can be offered. These approaches have high analytical validity in terms of detecting missense variants in the two hot spots of the *RP1L1* gene.[6–8] Variants outside of the two hot spots may be difficult to interpret or even to technically detect due to the presence of highly repetitive regions in the gene (especially between amino-acids 1312 and 2160). Analysis of genes that may be associated with non-*RP1L1* OMD such as *CRX* and *GUCY2D* should be considered.[19,20]

Clinical management

There are presently no approved curative treatments for OMD. Routine ophthalmic examination with refraction, visual field testing, fundus autofluorescence imaging, and OCT should be offered. Multifocal or pattern ERG[21,22] may be used to detect and further classify macular dysfunction, with relevance to changes in retinal structure and visual acuity. Full-field ERGs[23] can be used to exclude or assess generalised retinal involvement.

When heterozygous mutations are detected (e.g. in *RP1L1*, *CRX* or *GUCY2D*), ophthalmic examination and, if appropriate, OCT, electrodiagnostic assessment and/or genetic testing should be offered to first-degree relatives of affected individuals.

References

1. Miyake Y, Ichikawa K, Shiose Y, Kawase Y. Hereditary macular dystrophy without visible fundus abnormality. *Am J Ophthalmol* 1989;**108**(3):292–9. https://doi.org/10.1016/0002-9394(89)90120-7.
2. Miyake Y, Horiguchi M, Tomita N, Kondo M, Tanikawa A, Takahashi H, et al. Occult macular dystrophy. *Am J Ophthalmol* 1996;**122**(5): 644–53. https://doi.org/10.1016/s0002-9394(14)70482-9.
3. Miyake Y, Tsunoda K. Occult macular dystrophy. *Jpn J Ophthalmol* 2015;**59**(2):71–80. https://doi.org/10.1007/s10384-015-0371-7.
4. Akahori M, Tsunoda K, Miyake Y, Fukuda Y, Ishiura H, Tsuji S, et al. Dominant mutations in RP1L1 are responsible for occult macular dystrophy. *Am J Hum Genet* 2010;**87**(3):424–9. https://doi.org/10.1016/j.ajhg.2010.08.009.
5. Davidson AE, Sergouniotis PI, Mackay DS, Wright GA, Waseem NH, Michaelides M, et al. RP1L1 variants are associated with a spectrum of inherited retinal diseases including retinitis pigmentosa and occult macular dystrophy. *Hum Mutat* 2013;**34**(3):506–14. https://doi.org/10.1002/humu.22264.
6. Fujinami K, Kameya S, Kikuchi S, Ueno S, Kondo M, Hayashi T, et al. Novel RP1L1 variants and genotype-photoreceptor microstructural phenotype associations in cohort of Japanese patients with occult macular dystrophy. *Invest Ophthalmol Vis Sci* 2016;**57**(11): 4837–46. https://doi.org/10.1167/iovs.16-19670.
7. Fujinami K, Yang L, Joo K, Tsunoda K, Kameya S, Hanazono G, et al. Clinical and genetic characteristics of East Asian patients with occult macular dystrophy (Miyake disease): East Asia occult macular dystrophy studies report number 1. *Ophthalmology* 2019;**126**(10): 1432–44. https://doi.org/10.1016/j.ophtha.2019.04.032.
8. Zobor D, Zobor G, Hipp S, Baumann B, Weisschuh N, Biskup S, et al. Phenotype variations caused by mutations in the RP1L1 gene in a large mainly German cohort. *Invest Ophthalmol Vis Sci* 2018;**59**(7): 3041–52. https://doi.org/10.1167/iovs.18-24033.
9. Kato Y, Hanazono G, Fujinami K, Hatase T, Kawamura Y, Iwata T, et al. Parafoveal photoreceptor abnormalities in asymptomatic patients with RP1L1 mutations in families with occult macular dystrophy. *Invest Ophthalmol Vis Sci* 2017;**58**(14):6020–9. https://doi.org/10.1167/iovs.17-21969.
10. Fujinami K, Tsunoda K, Hanazono G, Shinoda K, Ohde H, Miyake Y. Fundus autofluorescence in autosomal dominant occult macular dystrophy. *Arch Ophthalmol* 2011;**129**(5):597–602. https://doi.org/10.1001/archophthalmol.2011.96.
11. Nakamura N, Tsunoda K, Mizuno Y, Usui T, Hatase T, Ueno S, et al. Clinical stages of occult macular dystrophy based on optical coherence tomographic findings. *Invest Ophthalmol Vis Sci* 2019;**60**(14): 4691–700. https://doi.org/10.1167/iovs.19-27486.
12. Piao CH, Kondo M, Tanikawa A, Terasaki H, Miyake Y. Multifocal electroretinogram in occult macular dystrophy. *Invest Ophthalmol Vis Sci* 2000;**41**(2):513–7.
13. Yang L, Joo K, Tsunoda K, Kondo M, Fujinami-Yokokawa Y, Arno G, et al. Spatial functional characteristics of East Asian patients with occult macular dystrophy (Miyake disease); EAOMD report no. 2. *Am J Ophthalmol* 2021;**221**:169–80. https://doi.org/10.1016/j.ajo.2020.07.025.
14. Conte I, Lestingi M, den Hollander A, Alfano G, Ziviello C, Pugliese M, et al. Identification and characterisation of the retinitis pigmentosa 1-like1 gene (RP1L1): a novel candidate for retinal degenerations. *Eur J Hum Genet* 2003;**11**(2):155–62. https://doi.org/10.1038/sj.ejhg.5200942.

15. Yamashita T, Liu J, Gao J, LeNoue S, Wang C, Kaminoh J, et al. Essential and synergistic roles of RP1 and RP1L1 in rod photoreceptor axoneme and retinitis pigmentosa. *J Neurosci* 2009;**29**(31):9748–60. https://doi.org/10.1523/JNEUROSCI.5854-08.2009.
16. Bowne SJ, Daiger SP, Malone KA, Heckenlively JR, Kennan A, Humphries P, et al. Characterization of RP1L1, a highly polymorphic paralog of the retinitis pigmentosa 1 (RP1) gene. *Mol Vis* 2003;**9**:129–37.
17. Tsunoda K, Usui T, Hatase T, Yamai S, Fujinami K, Hanazono G, et al. Clinical characteristics of occult macular dystrophy in family with mutation of RP1l1 gene. *Retina* 2012;**32**(6):1135–47. https://doi.org/10.1097/IAE.0b013e318232c32e.
18. Huckfeldt RM, Grigorian F, Place E, Comander JI, Vavvas D, Young LH, et al. Biallelic RP1-associated retinal dystrophies: expanding the mutational and clinical spectrum. *Mol Vis* 2020;**26**:423–33.
19. Fujinami-Yokokawa Y, Fujinami K, Kuniyoshi K, Hayashi T, Ueno S, Mizota A, et al. Clinical and genetic characteristics of 18 patients from 13 Japanese families with CRX-associated retinal disorder: identification of genotype-phenotype association. *Sci Rep* 2020;**10**(1):9531. https://doi.org/10.1038/s41598-020-65737-z.
20. Liu X, Fujinami K, Kuniyoshi K, Kondo M, Ueno S, Hayashi T, et al. Clinical and genetic characteristics of 15 affected patients from 12 Japanese families with GUCY2D-associated retinal disorder. *Transl Vis Sci Technol* 2020;**9**(6):2. https://doi.org/10.1167/tvst.9.6.2.
21. Bach M, Brigell MG, Hawlina M, et al. ISCEV standard for clinical pattern electroretinography (PERG): 2012 update. *Doc Ophthalmol* 2013;**126**:1–7. https://doi.org/10.1007/s10633-012-9353-y.
22. Hoffmann MB, Bach M, Kondo M, Li S, Walker S, Holopigian K, et al. ISCEV standard for clinical multifocal electroretinography (mfERG) (2021 update). *Doc Ophthalmol* 2021;**142**(1):5–16. https://doi.org/10.1007/s10633-020-09812-w.
23. McCulloch DL, Marmor MF, Brigell MG, et al. ISCEV standard for full-field clinical electroretinography (2015 update). *Doc Ophthalmol* 2015;**130**:1–12. https://doi.org/10.1007/s10633-014-9473-7.

13A.13

North Carolina macular dystrophy

Kent W. Small, Jingyan Yang, and Fadi Shaya

North Carolina macular dystrophy is an autosomal dominant condition associated with congenital, non-progressive macular abnormalities. It is rare and has been found worldwide, in less than 100 families (from the United States, Europe, Australia, Korea, China and others). It was first reported in a large family of Irish origin from North Carolina in 1971.[1,2] The initial cross-sectional studies suggested that the disease was progressive. However, Small and co-workers re-examined the original family almost 20 years later and realised that North Carolina macular dystrophy is a stationary developmental disorder with highly variable expressivity.[3,4] The differential diagnosis of this condition is broad ranging from congenital toxoplasmosis in severe cases[5] to early age-related macular degeneration in mild cases. In general, the clinical diagnosis of North Carolina macular dystrophy cannot be made in a single, isolated individual without genotype analysis or without examining other family members.

Clinical characteristics

North Carolina macular dystrophy presents with a variable macular phenotype and a non-progressive disease course. As it is a developmental disorder, macular lesions are thought to be present at birth. A striking feature is a relatively good (or at least better than expected from clinical examination) vision even in cases with severe-appearing macular malformations. Colour vision and full-field electroretinograms (ERG) are typically preserved, while pattern and multifocal ERG recordings can reveal significant amplitude reductions in the central retina.

Three grades of severity have been described in terms of fundus appearance[5]:

- Grade 1: bilaterally symmetrical, small (<50 μm) yellow-white, drusen-like specks in the central macula. Visual acuity of 0.0 to 0.2 LogMAR.
- Grade 2: bilaterally symmetrical, confluent yellow-white specks in the central macula. Visual acuity of 0.1 to 1.0 (Fig. 13A.59)
- Grade 3: bilaterally symmetrical, coloboma-like lesion in the central macula. Visual acuity of 0.5 to 1.0 LogMAR (Figs 13A.60 and 13A.61)

In general, patients are born with their disease at a particular grade and do not progress from one grade to another.

A macular coloboma-like excavation (grade 3) is seen in roughly one-third of affected individuals. This corresponds to a well-demarcated area of absence of the RPE and choriocapillaris causing posterior bowing into the choroid and deformation of the sclera with surrounding subretinal fibrosis (more so on the sharply shelving temporal edge than the more smoothly sloping nasal edge). The patient's fixation is typically on the nasal edge of the lesion rather than in the central area, where the fovea would be expected.

A small subset of patients experiences moderate to severe visual loss due to the formation of choroidal neovascularisation with a disciform scar. This complication is usually seen in patients with coloboma-like lesions (grade 3 fundus appearance) and is unlikely to occur in individuals with mild phenotypes (grade 1 fundus appearance).

Molecular pathology

North Carolina macular dystrophy is genetically heterogeneous. Most individuals are found to carry genetic variants in an intragenic region in chromosome 6 located over 10,000 base-pairs from the neighbouring gene. More than five single nucleotide changes and several copy number variants in this region have been implicated in North Carolina macular dystrophy. These mutations impact sequences that are likely to have a role in regulating the spatio-temporal expression of a retinal transcription factor, *PRDM13*. *PRDM13* is expressed in the fetal retina and dorsal spinal column and is not found in adult tissues.[6] Overexpression of this gene has been shown to impair retinal development in animal models.[7] Intriguingly, a few changes near *PRDM13* that have been linked to North Carolina macular dystrophy have also been implicated in a

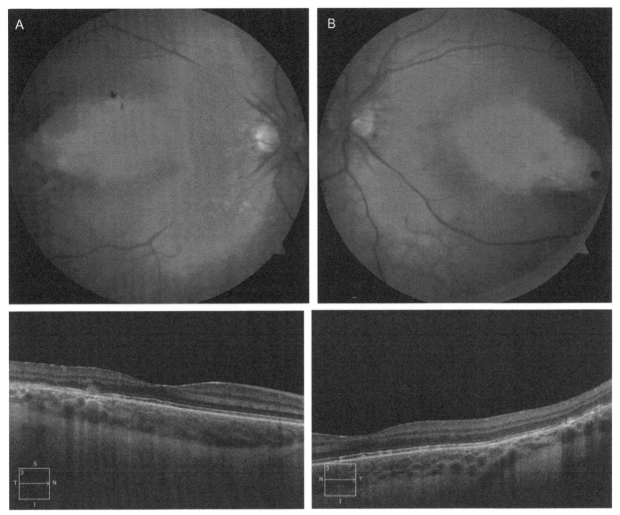

FIG. 13A.59 Fundus photography (A) and macular OCT (B) in a 60-year-old patient with North Carolina macular dystrophy (grade 2 fundus appearance).

different condition, progressive bifocal chorioretinal atrophy (PBCRA).[8]

Several patients with a phenotype resembling North Carolina macular dystrophy have been found to have copy number variants in a different chromosomal location in chromosome 5.[6,8]

Clinical management

As North Carolina macular dystrophy seems to develop *in utero* and is congenital, there is currently no treatment for preventing the disease. However, choroidal neovascularisation, a key complication, is treatable with standard anti-VEGF intravitreal injections (e.g. intravitreal bevacizumab).[9]

Family members of patients with North Carolina macular dystrophy should be offered ophthalmic examination and, if appropriate, genetic testing. Each affected individual has a 50% chance of transmitting the mutated allele to their children. However, predicting whether the children will have a mild or a more severe phenotype is not possible. It is worth noting though that most (at least 50%) of affected individuals have a lifetime of relatively good vision.

248 SECTION | III Genetic disorders affecting the posterior segment

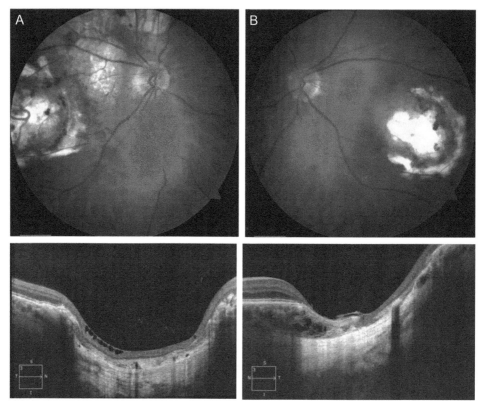

FIG. 13A.60 Fundus photography (A) and macular OCT (B) in a patient with North Carolina macular dystrophy (grade 3 fundus appearance).

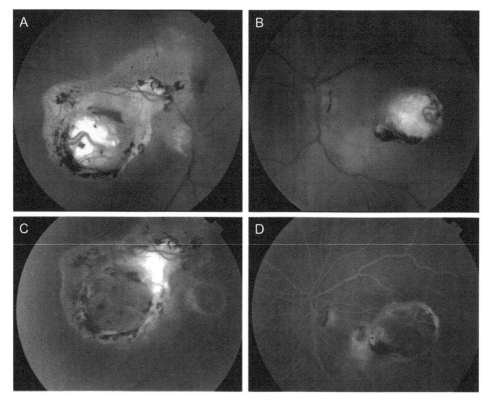

FIG. 13A.61 Fundus photography (A and B) and late frames of fluorescein angiogram (C and D) from a 63-year-old patient with North Carolina macular dystrophy (grade 3 fundus appearance).

References

1. Lefler WH, Wadsworth JA, Sidbury JB. Hereditary macular degeneration and amino-aciduria. *Am J Ophthalmol* 1971;**71**:224–30.
2. Frank HR, Landers MB, Williams RJ, Sidbury JB. A new dominant progressive foveal dystrophy. *Am J Ophthalmol* 1974;**78**:903–16.
3. Small KW. North Carolina macular dystrophy, revisited. *Ophthalmology* 1989;**96**:1747–54.
4. Kw S, Killian J, McLean WL. North Carolina's dominant progressive foveal dystrophy: how progressive is it? *Br J Ophthalmol* 1991;**75**:401–6.
5. Small KW, Tran EM, Small L, Rao RC, Shaya F. Multimodal imaging and functional testing in a North Carolina macular disease family: toxoplasmosis, fovea plana, and torpedo maculopathy are phenocopies. *Ophthalmol Retina* 2019;**3**(7):607–14.
6. Small KW, DeLuca AP, Whitmore SS, et al. North Carolina macular dystrophy is caused by dysregulation of the retinal transcription factor PRDM13. *Ophthalmology* 2016;**123**(1):9–18.
7. Manes G, Joly W, Guignard T, Smirnov V, Berthemy S, Bocquet B, et al. A novel duplication of PRMD13 causes North Carolina macular dystrophy: overexpression of PRDM13 orthologue in drosophila eye reproduces the human phenotype. *Hum Mol Genet* 2017;**26**(22):4367–74.
8. Silva RS, Arno G, Cipriani V, Pontikos N, Defoort-Dhellemmes S, Kalhoro A, et al. Unique noncoding variants upstream of PRDM13 are associated with a spectrum of developmental retinal dystrophies including progressive bifocal chorioretinal atrophy. *Hum Mutat* 2019;**40**(5):578–87.
9. Bakall B, Bryan JS, Stone EM, et al. Choroidal neovascularization in North Carolina macular dystrophy responsive to anti–vascular endothelial growth factor therapy. *Retin Cases Brief Rep* 2018;**15**(5):509–13.

13A.14

Genetic disorders mimicking age-related macular disease

Veronika Vaclavik, Panagiotis I. Sergouniotis, Graeme C.M. Black, and Francis L. Munier

Drusen are accumulations of extracellular debris that are located immediately beneath the RPE. On fundoscopy, they are typically visualised as small, discrete, yellow–white lesions, usually in the macular region. They are a major clinical sign of age-related macular degeneration (AMD) (discussed in Section 13A.15), but they may also just be part of the normal ageing process. Notably, when drusen are detected at an early age (e.g. <50 years), a range of rare disorders should be considered. These conditions are mainly genetic, and most of them affect only the eye (i.e. they are non-syndromic). It is however important to be aware of multisystemic (i.e. syndromic) causes (e.g. C3 glomerulopathy; Table 13A.8).

When assessing patients with early-onset drusen, obtaining a comprehensive medical and family history and examining relatives is key. Performing careful fundoscopy and requesting multimodal imaging are also crucial as, for some of these disorders, the distribution and nature of the drusen is characteristic and gives an indication of the underlying condition (Table 13A.9). Early diagnosis allows precise management, ongoing surveillance, and prompt treatment (e.g. for choroidal neovascularisation) for affected individuals, as well as accurate risk estimation for family members.[1,2]

TABLE 13A.8 Classification of isolated and syndromic disorders characterised by drusen.

Category	Aetiology/associations
Non-syndromic	Age-related macular degeneration (AMD) (Section 13A.15)
	Sorsby fundus dystrophy Autosomal dominant drusen Late onset retinal degeneration (L-ORD) North Carolina macular dystrophy (Section 13A.13)
Syndromic	C3 glomerulopathy Pseudoxanthoma elasticum (Section 13B.2.3) Vitamin A deficiency

Clinical characteristics

Early in the disease process, people with familial drusen-associated retinopathies are asymptomatic; thus, affected individuals are occasionally identified through routine optometric or ophthalmic examination. It is also not unusual to diagnose these conditions in asymptomatic individuals through family workup, triggered by a diagnosis in a family member. As with AMD, patients may develop central visual loss and/or distortion. Where symptoms result from choroidal neovascularisation, they may be of sudden onset and, if left untreated, lead to severe visual impairment. In general, this group of conditions exhibits marked interfamilial and intrafamilial variability in terms of retinal appearance, severity, and progression. A few selected disorders are discussed below; for a more comprehensive list, see Ref. 1.

Autosomal dominant drusen (also known as familial dominant drusen, Doyne honeycomb retinal dystrophy, or malattia leventinese)

Individuals with autosomal dominant drusen (also known as familial dominant drusen, Doyne honeycomb retinal dystrophy, or malattia leventinese[3]) generally become symptomatic in the fourth or fifth decade of life. Early symptoms include reduced central vision, distortion, and/or paracentral scotomas. Visual loss occurs predominantly as a result of chorioretinal atrophy and scarring at the macula although choroidal neovascularisation occurs in a minority of cases.

The characteristic sign of this disorder is early-onset drusen at the posterior pole and in the peripapillary area, with a radial distribution in some cases. Notably, drusen abutting the optic nerve head have been shown to be suggestive of this condition. Macular drusen tend to become confluent with age, and Bruch's membrane becomes increasingly separated from the RPE.

Multimodal imaging can be helpful in individuals with autosomal dominant drusen. Fundus autofluorescence imaging reveals that drusen in this condition are mostly

TABLE 13A.9 Selected retinal phenotypes associated with drusen formation.

Syndrome	Gene	Chromosomal location	Protein product	Main mode of inheritance	Clinical features	Drusen
Autosomal dominant drusen	EFEMP1	2p16	EGF-containing fibulin-like extracellular matrix protein 1	Autosomal dominant	Central visual loss associated with atrophy	Drusen that may be visualised in the peripapillary region (including temporal to the optic disc) and tend to become confluent later in the disease process
Sorsby fundus dystrophy	TIMP3	22q12.3	Tissue inhibitor of metalloproteinases-3	Autosomal dominant	Early-onset bilateral neovascularisation; macular atrophy in the absence of neovascularisation	Typical drusen and reticular pseudodrusen, often situated outside the arcades or nasal to the disc
Late onset retinal degeneration (L-ORD)	C1QTNF5	11q23.3	C1q- and tumour necrosis factor-related protein 5	Autosomal dominant	Progressive macular atrophy	Sub-RPE deposits and reticular pseudodrusen
North Carolina macular dystrophy	PRDM13	6q16	PR domain-containing protein 13	Autosomal dominant	Stationary; symptomatic due to extensive macular atrophy	Fine drusen that hyperautofluorescence on fundus autofluorescence imaging; drusen-like lesions in the far periphery
Early-onset macular drusen	CFH	1q31.1	Complement factor H	Autosomal dominant	Progressive macular atrophy; neovascularisation	Cuticular drusen (basal laminar drusen)

hyperautofluorescent (in contrast to the more variable AMD drusen). On OCT, there is hyper-reflective thickening of the RPE/Bruch's membrane complex corresponding to drusen deposition; the photoreceptor layer's integrity is often preserved although abnormalities may be noted (Figs 13A.62–13A.64). OCT angiography (OCT-A) is able to capture early neovascularisation without the need for fundus fluorescein angiography.

As dysfunction in autosomal dominant drusen is usually confined to the macula, the full-field electroretinograms (ERGs) tend to be within normal limits.[1–3]

Sorsby fundus dystrophy

Individuals with Sorsby fundus dystrophy typically develop symptoms after the third decade of life. The most common presentation involves central visual loss and/or distortion; this is often preceded by a period of disturbance of dark adaptation and night vision.[1,4] Although the phenotypic expressivity of Sorsby fundus dystrophy is variable, the penetrance of the condition later in life is high.[5]

Early-onset drusen are present in most affected individuals. They preferentially accumulate along the temporal arcades and progress over time to include the central macula. They may resemble typical drusen (i.e. lesions located between the basal lamina of the RPE and Bruch's membrane) or reticular drusen/pseudodrusen (i.e. lesions located between the apical surface of the RPE and photoreceptor outer segments). Visual loss can occur due to gradual chorioretinal degeneration at the macula, which can also extend peripherally (often in a multilobulated pattern). Choroidal neovascularisation is a common complication that generally develops from the fourth to the sixth decade of life; it is usually bilateral and often results in severe visual loss.

Fundus autofluorescence imaging and infrared reflectance imaging may identify irregular signals in the peripheral macula in early disease; the central fovea is often spared early on. OCT may identify drusen-like deposits and highlight Bruch's membrane thickening, an important early

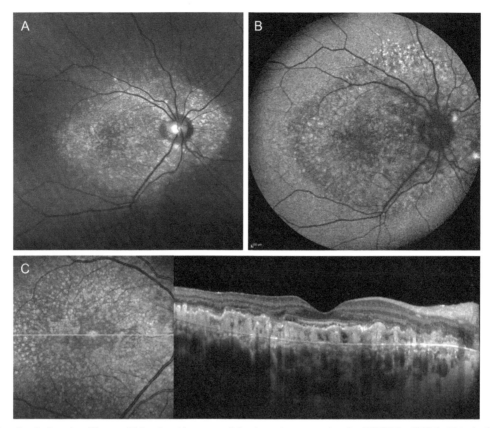

FIG. 13A.62 Imaging findings in a 25-year-old female with autosomal dominant drusen carrying the *EFEMP1* c.1033C > T (p.Arg345Trp) variant in heterozygous state. (A) Fundus photograph of the right posterior pole: large confluent drusen and small radial drusen are noted at the posterior pole and nasally to the optic disc. (B) Short wavelength autofluorescence image of the right eye: hyperautofluorescent lesions (corresponding to drusen) are noted. (C) Infrared reflectance and OCT scan of the right eye: diffuse deposition of the hyper-reflective material underneath the RPE is noted. The inner retina appears preserved.

sign of the condition (Figs 13A.65–13A.67). OCT and OCT-A are helpful for detecting neovascularisation.

Variable electroretinographic findings have been reported, but, in most cases, the full-field ERGs remain normal until extensive and peripheral chorioretinal atrophy becomes apparent.[1,2]

Late onset retinal degeneration (L-ORD)

Individuals with late onset retinal degeneration (L-ORD) develop rod photoreceptor dysfunction, generally in the fifth to sixth decade of life. Central visual loss may subsequently occur as a result of chorioretinal atrophy or neovascularisation.

Drusen-like sub-RPE deposits and reticular pseudo-drusen-like changes are typically noted from the fifth decade of life onwards. Later on, widespread, well-demarcated areas of chorioretinal atrophy are observed in the mid-periphery and at the posterior pole.

Full-field ERGs are generally in keeping with a rod-cone pattern of dysfunction with a significant rod system involvement.[1,6–9]

The presence of anteriorly placed zonular insertions into the anterior lens is a characteristic sign of L-ORD (best visualised on retroillumination)[10] (Fig. 13A.68).

North Carolina macular dystrophy

Individuals with North Carolina macular dystrophy (NCMD) may have macular changes that resemble fine, confluent drusen. These lesions are congenital and typically non-progressive. It is noted that drusen-like retinal deposits may be observed in a linear configuration in the far periphery, a feature that can aid diagnosis via widefield imaging.[11] For further information on this condition, see Section 13A.13.

CFH-related disorders

Variants in the complement factor H (*CFH*) gene have been associated with a number of clinical entities including AMD, cuticular drusen, atypical haemolytic uremic syndrome, and C3 glomerulopathies. For example, the *CFH* c.1204C > T (p.His402Tyr) common variant imparts a high

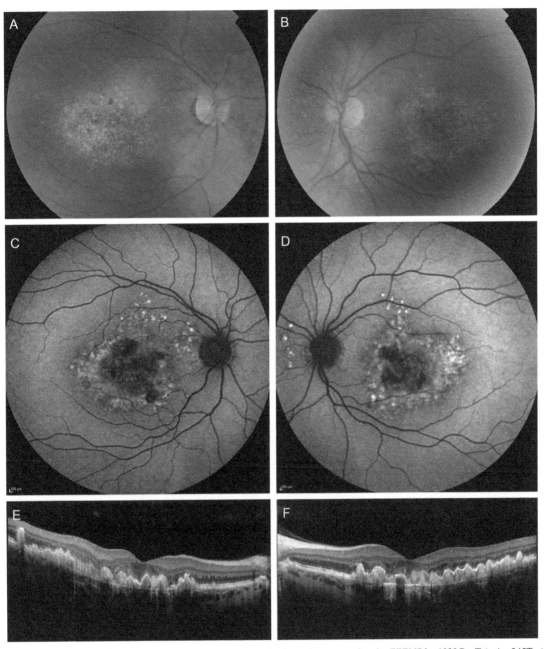

FIG. 13A.63 Imaging findings in a 51-year-old female with autosomal dominant drusen carrying the *EFEMP1* c.1033C>T (p.Arg345Trp) variant in heterozygous state. (A and B) Fundus photographs of the right and left posterior poles: large, confluent drusen, hyperpigmented areas of RPE changes, and areas of macular fibrosis are noted. Drusen nasal to the optic disc are seen in the left eye. (C and D) Short wavelength autofluorescence images of the right and left eye: hyperautofluorescent lesions (corresponding to drusen) and central areas of hypoautofluorescence (corresponding to a atrophy/fibrosis) are noted. (E and F) Fovea-centred OCT scans of the right and left eye: diffuse and subfoveal deposition of the hyper-reflective material underneath the RPE is noted. Large drusen correspond to dome-shaped elevations of the RPE band, while smaller drusen correspond to hyper-reflective saw-tooth elevations.

risk for AMD (see Section 13A.15). Notably, rare variants in this gene have been associated with early-onset drusen; such genetic alterations in *CFH* should be suspected in patients who have extensive drusen deposition, drusen with crystalline appearance, and/or drusen nasal to the optic disc in combination with other patient characteristics, such as early age at disease onset, cuticular drusen on fluorescein angiography ('stars-in-the-sky' appearance[13]), and a positive family history for AMD.[14]

Rare pathogenic variants in *CFH* can also lead to extraocular phenotypes and have been associated with atypical haemolytic uremic syndrome (AHUS) and C3 glomerulopathy.

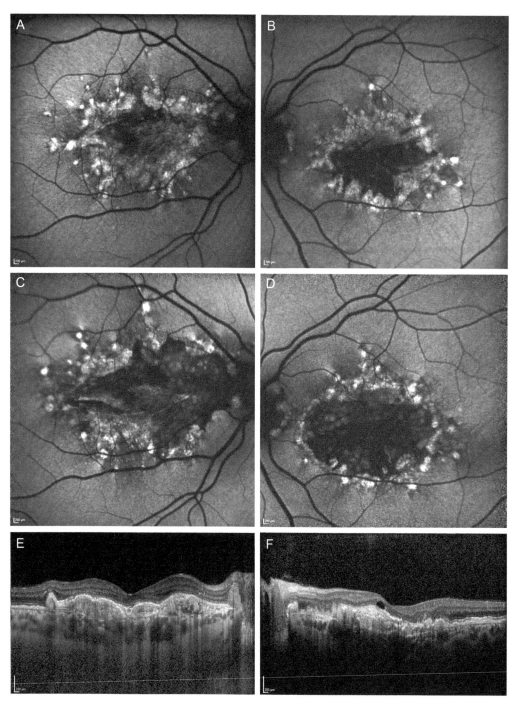

FIG. 13A.64 Evolution of imaging findings in a 35-year-old female with autosomal dominant drusen. (A and B) Short wavelength autofluorescence images of the right and left eye: hyperautofluorescent lesions (corresponding to drusen) and central areas of hypoautofluorescence (corresponding to atrophy and fibrosis following long-standing drusen deposition) are noted. (C and D) Short wavelength autofluorescence images of the right and left eye 6 years later: concentric extension of the radial drusen and enlargement of the central hypoautofluorescence areas are noted. (E and F) Fovea-centred OCT scans of the right and left eye: diffuse and subfoveal deposition of the hyper-reflective material underneath the RPE is noted; the appearance is similar to that of a fibrovascular pigment epithelial detachment.

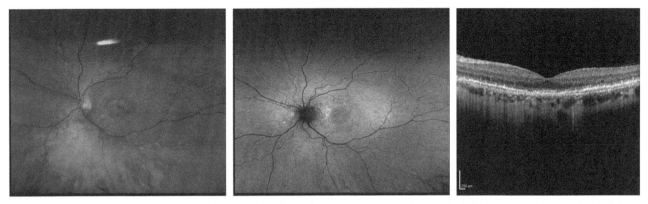

FIG. 13A.65 Imaging findings in a 50-year-old male with Sorsby fundus dystrophy carrying the *TIMP3* c.610A>T (p.Ser204Cys) variant in heterozygous stage. Widefield fundus imaging demonstrating drusen outside vascular arcades and nasal to the disc. These show minimal hyperautofluorescence on fundus autofluorescence imaging, but they correspond to hyper-reflective lesions on OCT.

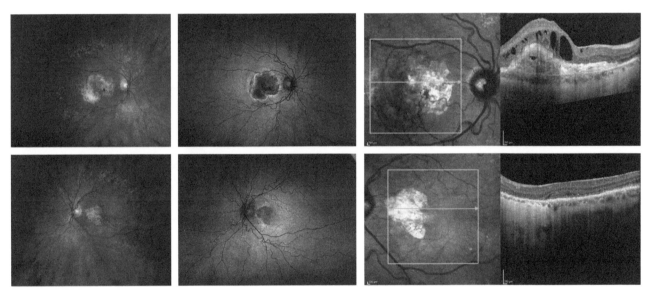

FIG. 13A.66 Imaging findings in a 51-year-old male with Sorsby fundus dystrophy (brother of the patient shown in Fig. 13A.65). Imaging findings in the right (top row) and left (bottom row) eye are shown. The right eye shows classically distributed drusen deposits and macular scarring secondary to a choroidal neovascular membrane. The left eye manifests drusen and macular atrophy in the absence of neovascularisation.

Although drusen are seldom seen in AHUS, they may develop in association with C3 glomerulopathies.

C3 glomerulopathies are a group of autoinflammatory renal disorders resulting in glomerular complement deposition. Their presentation ranges from asymptomatic haematuria and proteinuria to acute manifestations associated with severe hypertension and fluid retention.[15]

Drusen in individuals with C3 glomerulopathies are likely to be present from an early age; they are generally seen at the posterior pole and tend to extend to the equator and beyond. In one disease subgroup (formerly classified as 'dense deposit disease' or 'membranoproliferative glomerulonephritis type II'), drusen may be noted from around 20 years of age (Fig. 13A.69). Choroidal neovascularisation and central serous retinopathy may occur.[12]

Given the association between drusen and kidney disease, it is recommended that the presence of renal dysfunction is excluded in people who have drusen and are under the age of 40 years.

Molecular pathology

Retinopathies that feature drusen are genetically heterogeneous and most of them segregate as autosomal dominant traits (see Table 13A.9). Interestingly, autosomal dominant

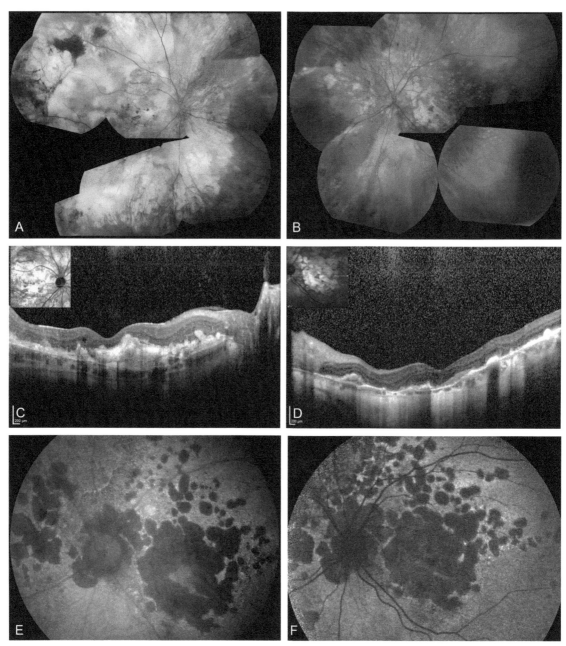

FIG. 13A.67 Imaging findings in a 73-year-old female with Sorsby fundus dystrophy. The right retina shows scarring, atrophy, and hyperpigmentation (A). The left retina shows multiple areas of atrophy, with haemorrhages nasal to the disc corresponding to a documented choroidal neovascular membrane (B). Drusen are visible temporally in both eyes. OCT reveals thickening and a subfoveal hyper-reflective area corresponding to fibrosis in the right eye (C); retinal thinning (with an absent photoreceptor layer) and areas of thickened Bruch's membrane (temporally to the fovea) are noted in the left eye (D). Fundus autofluorescence imaging of the left eye shows a small preserved island of retina in the middle of a large hypoautofluorescent area corresponding to atrophy. Multiple smaller nummular areas of reduced signal are visible beyond the arcades, around the optic nerve and nasally to the disc (E). Autofluorescence imaging of the left eye of the same patient 3 years later, at age 74 years, revealed enlargement of existing atrophy and some new areas of hypo-autofluorescence, mainly superonasally (F).

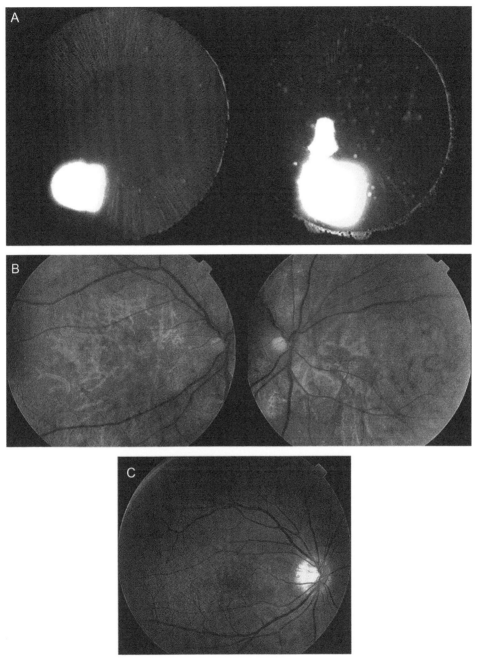

FIG. 13A.68 Findings in late onset retinal degeneration (L-ORD). (A) Peripupillary iris atrophy and long, anteriorly inserting zonules in two siblings with L-ORD. (B) Widespread chorioretinal atrophic changes throughout the nasal and temporal fundus of a patient with L-ORD. (C) Drusen-like deposits in the right fundus of an asymptomatic individual who has molecularly confirmed L-ORD and is in the fourth decade of life. *(Adapted from Ref. 10.)*

drusen, Sorsby fundus dystrophy, and L-ORD all result from heterozygous mutations in extracellular matrix proteins.

All cases of autosomal dominant drusen are caused by a single heterozygous missense substitution in *EFEMP1*, c.1033C > T (p.Arg345Trp).[1–3] Haplotype analysis suggests that this mutation has occurred recurrently in different geographies.[16] The gene product, fibulin-3, is a secreted glycoprotein that is broadly expressed and interacts with tissue inhibitor of metalloproteinase-3 (TIMP3, see below), collagen XVIII, and tropoelastin. These interactions likely contribute to basement membrane integrity.[17]

Sorsby fundus dystrophy is caused by heterozygous mutations in the *TIMP3* gene.[18] Disease-associated variants are found almost exclusively in the final exon of the gene (exon 5), and most of them are missense changes.

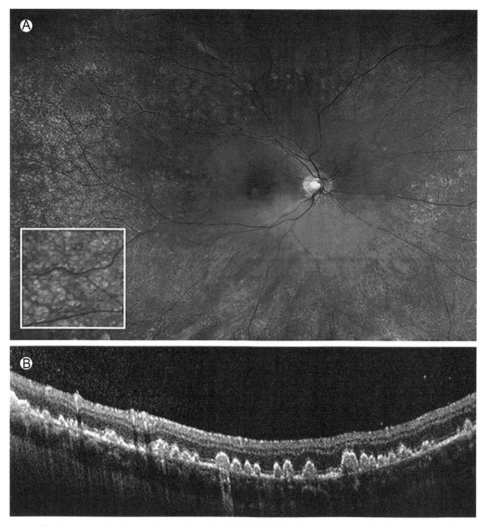

FIG. 13A.69 Drusen in G3 glomerulopathy (dense deposit disease). (A) Widefield fundus photograph showing numerous drusen-like deposits in the macula and peripheral retina of the right eye; insert shows an area with magnification. (B) OCT shows subretinal dome-shaped lesions mostly in the peripheral retina. *(Adapted from Ref. 12.)*

These substitutions usually create a new cysteine residue that allows intermolecular disulphide bond formation. *TIMP3* encodes an inhibitor of matrix metalloproteinases, which play an important role in the regulation of extracellular matrix turnover. Mutant TIMP3 is believed to form dimers and to accumulate within Bruch's membrane, leading to widespread thickening of the membrane. This interferes with the normal critical functions of Bruch's membrane, RPE, and choroid.[2,4,5,19,20]

The majority of cases of L-ORD are caused by a single heterozygous missense substitution in the *C1QTNF5* gene, c.489C>G (*p*.Ser163Arg) (although other pathogenic missense variants in this gene have also been described). *C1QTNF5* encodes a secreted and membrane-associated protein that is highly expressed in the RPE and plays a role in RPE cellular adhesion. Mutant proteins are believed to exert dominant negative effects as they appear to tie up wild-type protein in structurally suboptimal and unstable multimers.[21,22]

Disruption of the complement cascade is a key aetiological factor underlying AMD, as discussed in Section 13A.15. The rare heterozygous variants in *CFH* that have been described in cases with early-onset drusen phenotypes are generally loss of function changes (i.e. truncating variants) that alter the transcripts encoding both factor H and factor H-like-1 (FHL1) proteins.[23,24] Genotype–phenotype correlations have been described with respect to *CFH* variants: factor H contains 20 complement control protein (CCP) domains and variants affecting CCP domains 1–4 and 6–8 result in retinal phenotypes, while those associated with renal disease alter CCP domains 19 and 20. Since the *CFH* c.3628C>T (p.Arg1210Cys) variant, which is known to be associated with AMD, may also cause AHUS, this correlation is not absolute.[23–26]

Clinical management

Currently, there are no specific treatments although choroidal neovascularisation can be managed successfully with anti-VEGF intravitreal injections (e.g. bevacizumab or ranibizumab[27–30]). It is worth noting that the risk of early neovascularisation is greatest for Sorsby fundus dystrophy (compared to autosomal dominant drusen, L-ORD, and North Carolina macular dystrophy). Consequently, presymptomatic genetic testing for Sorsby fundus dystrophy can be considered to enable affected individuals to self-monitor and to identify visual acuity reduction or metamorphopsia as early as possible.[28] Presymptomatic testing should be undertaken in cooperation with clinical genetics departments (see Chapter 4).

Given the phenotypic similarity between autosomal dominant drusen and AMD, retinal laser has been used to attempt drusen reabsorption as a prophylactic treatment for dominant drusen. Significant changes in retinal morphology were observed although there was no clear impact on short-term visual function.[31,32]

Where syndromic manifestations are suspected (e.g. complement-related disorders; pseudoxanthoma elasticum), referral to other specialties including clinical genetics should be considered.

References

1. Khan KN, Mahroo OA, Khan RS, Mohamed MD, McKibbin M, Bird A, Michaelides M, Tufail A, Moore AT. Differentiating drusen: drusen and drusen-like appearances associated with ageing, age-related macular degeneration, inherited eye disease and other pathological processes. *Prog Retin Eye Res* 2016;**53**:70–106.
2. Rahman N, Georgiou M, Khan KN, Michaelides M. Macular dystrophies: clinical and imaging features, molecular genetics and therapeutic options. *Br J Ophthalmol* 2020;**104**(4):451–60.
3. Stone EM, et al. A single EFEMP1 mutation associated with both Malattia Leventinese and Doyne honeycomb retinal dystrophy. *Nat Genet* 1999;**22**(2):199–202.
4. Christensen DRG, Brown FE, Cree AJ, Ratnayaka JA, Lotery AJ. Sorsby fundus dystrophy—a review of pathology and disease mechanisms. *Exp Eye Res* 2017;**165**:35–46.
5. Gliem M, Müller PL, Mangold E, Holz FG, Bolz HJ, Stöhr H, Weber BH, Charbel IP. Sorsby fundus dystrophy: novel mutations, novel phanotypic characteristics and treatment outcomes. *Investig Ophthalmology Vis Sci* 2015;**56**(4):2664–76.
6. Borooah S, Papastavrou VT, Lando L, Moghimi S, Lin T, Dans K, Motevasseli T, Cameron JR, Freeman WR, Dhillon B, Browning AC. Characterizing the natural history of foveal-sparing atrophic late-onset retinal degeneration. *Retina* 2021;**41**(6):1329–37.
7. Borooah S, Papastavrou V, Lando L, Han J, Lin JH, Ayyagari R, Dhillon B, Browning AC. Reticular pseudodrusen in late-onset retinal degeneration. *Ophthalmol Retina* 2020; S2468-6530(20)30493-0.
8. Borooah S, Collins C, Wright A, Dhillon B. Late-onset retinal macular degeneration: clinical insights into an inherited retinal degeneration. *Br J Ophthalmol* 2009;**93**(3):284–9.
9. De Zaeytijd J, Coppieters F, De Bruyne M, Van Royen J, Roels D, Six R, Van Cauwenbergh C, De Baere E, Leroy BP. Longitudinal phenotypic study of late-onset retinal degeneration due to a founder variant c.562C>A p.(Pro188Thr) in the *C1QTNF5* gene. *Ophthalmic Genet* 2021;1–12.
10. Subrayan V, Morris B, Armbrecht AM, Wright AF, Dhillon B. Long anterior lens zonules in late-onset retinal degeneration (L-ORD). *Am J Ophthalmol* 2005;**140**(6):1127–9.
11. Green DJ, Lenassi E, Manning CS, McGaughey D, Sharma V, Black GC, Ellingford JM, Sergouniotis PI. North Carolina macular dystrophy: phenotypic variability and computational analysis of disease-associated noncoding variants. *Invest Ophthalmol Vis Sci* 2021;**62**(7):16.
12. Lent-Schochet D, et al. Drusen in dense deposit disease: not just age-related macular degeneration. *Lancet* 2020;**395**(10238):1726.
13. Boon CJ, van de Ven JP, Hoyng CB, den Hollander AI, Klevering BJ. Cuticular drusen: stars in the sky. *Prog Retin Eye Res* 2013;**37**:90–113.
14. Kersten E, Geerlings MJ, den Hollander AI, de Jong EK, Fauser S, Peto T, Hoyng CB. Phenotype characteristics of patients with age-related macular degeneration carrying a rare variant in the complement factor H gene. *JAMA Ophthalmol* 2017;**135**(10):1037–44.
15. Smith RJH, Appel GB, Blom AM, Cook HT, D'Agati VD, Fakhouri F, Fremeaux-Bacchi V, Józsi M, Kavanagh D, Lambris JD, Noris M, Pickering MC, Remuzzi G, de Córdoba SR, Sethi S, Van der Vlag J, Zipfel PF, Nester CM. C3 glomerulopathy—understanding a rare complement-driven renal disease. *Nat Rev Nephrol* 2019;**15**(3):129–43.
16. Takeuchi T, Hayashi T, Bedell M, Zhang K, Yamada H, Tsuneoka H. A novel haplotype with the R345W mutation in the EFEMP1 gene associated with autosomal dominant drusen in a Japanese family. *Invest Ophthalmol Vis Sci* 2010;**51**(3):1643–50. https://doi.org/10.1167/iovs.09-4497 Epub 2009 Oct 22. PMID: 19850834 Free PMC article.
17. Zayas-Santiago A, Cross SD, Stanton JB, Marmorstein AD, Marmorstein LY. Mutant Fibulin-3 causes proteoglycan accumulation and impaired diffusion across Bruch's membrane. *Invest Ophthalmol Vis Sci* 2017;**58**(7):3046–54.
18. Weber BH, Vogt G, Pruett RC, Stöhr H, Felbor U. Mutations in the tissue inhibitor of metalloproteinases-3 (TIMP3) in patients with Sorsby's fundus dystrophy. *Nature Genet* 1994;**8**:352–6.
19. Anand-Apte B, Chao JR, Singh R, Stöhr H. Sorsby fundus dystrophy: insights from the past and looking to the future. *J Neurosci Res* 2019;**97**(1):88–97.
20. Schoenberger SD, Agrawal A. A novel mutation at the N-terminal domain of the TIMP3 gene in Sorsby fundus dystrophy. *Retina* 2013;**33**:429–35.
21. Hayward C, Shu X, Cideciyan AV, Lennon A, Barran P, Zareparsi S, Sawyer L, Hendry G, Dhillon B, Milam AH, Luthert PJ, Swaroop A, Hastie ND, Jacobson SG, Wright AF. Mutation in a short-chain collagen gene, CTRP5, results in extracellular deposit formation in late-onset retinal degeneration: a genetic model for age-related macular degeneration. *Hum Mol Genet* 2003;**12**(20):2657–67.
22. Stanton CM, Borooah S, Drake C, Marsh JA, Campbell S, Lennon A, Soares DC, Vallabh NA, Sahni J, Cideciyan AV, Dhillon B, Vitart V, Jacobson SG, Wright AF, Hayward C. Novel pathogenic mutations in C1QTNF5 support a dominant negative disease mechanism in late-onset retinal degeneration. *Sci Rep* 2017;**7**(1):12147.

23. Taylor RL, Poulter JA, Downes SM, McKibbin M, Khan KN, Inglehearn CF, Webster AR, Hardcastle AJ, Michaelides M, Bishop PN, Clark SJ, Black GC. United Kingdom inherited retinal dystrophy consortium. Loss-of-function mutations in the CFH gene affecting alternatively encoded factor H-like 1 protein cause dominant early-onset macular drusen. *Ophthalmology* 2019;**126**(10):1410–21.
24. Tzoumas N, Hallam D, Harris CL, Lako M, Kavanagh D, Steel DHW. Revisiting the role of factor H in age-related macular degeneration: insights from complement-mediated renal disease and rare genetic variants. *Surv Ophthalmol* 2021;**66**(2):378–401.
25. Martinez-Barricarte R, Pianetti G, Gautard R, Misselwitz J, Strain L, Fremeaux-Bacchi V, Skerka C, Zipfel PF, Goodship T, Noris M, Remuzzi G, de Cordoba SR. European working party on the genetics of HUS. The complement factor H R1210C mutation is associated with atypical hemolytic uremic syndrome. *J Am Soc Nephrol* 2008;**19**(3):639–46.
26. Raychaudhuri S, Iartchouk O, Chin K, Tan PL, Tai AK, Ripke S, Gowrisankar S, Vemuri S, Montgomery K, Yu Y, Reynolds R, Zack DJ, Campochiaro B, Campochiaro P, Katsanis N, Daly MJ, Seddon JM. A rare penetrant mutation in CFH confers high risk of age-related macular degeneration. *Nat Genet* 2011;**43**(12):1232–6.
27. Fung AT, Stöhr H, Weber BHF, Holz FG, Yannuzzi LA. Atypical Sorsby fundus dystrophy with a novel tyr159cys timp-3 mutation. *Retin Cases Brief Rep* 2013;**7**:71–4.
28. Menassa N, Burgula S, Empeslidis T, Tsaousis KT. Bilateral choroidal neovascular membrane in a young patient with a Sorsby fundus dystrophy: the value of promt treatment. *BMJ Case Rep* 2017;**2017**: bcr2017220488.
29. Kaye R, Lotery AJ. Long term outcome of bevacizumab therapy in Sorsby fundus dystrophy. A case series. ARVO E-abstract B0260. *Investig Ophthalmology Vis Sci* 2017;**58**(8):229.
30. Sohn EH, et al. Responsiveness of choroidal neovascular membranes in patients with R345W mutation in fibulin 3 (Doyne honeycomb retinal dystrophy) to anti-vascular endothelial growth factor therapy. *Arch Ophthalmol* 2011;**129**(12):1626–8.
31. Lenassi E, et al. Laser clearance of drusen deposit in patients with autosomal dominant drusen (p.Arg345Trp in EFEMP1). *Am J Ophthalmol* 2013;**155**(1):190–8.
32. Cusumano A, Falsini B, Giardina E, Cascella R, Sebastiani J, Marshall J. Doyne honeycomb retinal dystrophy—functional improvement following subthreshold nanopulse laser treatment: a case report. *J Med Case Rep* 2019;**13**(1):5.

13A.15

Genetic architecture of age-related macular degeneration

Johanna M. Colijn and Caroline C.W. Klaver

Age-related macular degeneration (AMD) is a degenerative retinal condition that predominantly affects the macula. It is typically seen in individuals who are ≥50 years old and its prevalence increases with age. It can be broadly classified as early stage (including medium/large-sized drusen and RPE abnormalities) and late stage (including atrophy and neovascularisation). Late-stage AMD results in loss of central vision, often leading to severe visual impairment.

AMD is a multifactorial disorder, caused by a complex interplay of multiple genetic and environmental/lifestyle factors. Large-scale genetic studies have identified variants in more than 30 genomic locations (loci), of which the most important are in chromosome 1 (*CFH* [complement factor H], locus) and chromosome 10 (*ARMS2* [age-related maculopathy susceptibility 2] locus).[1] Major non-genetic risk factors include age, smoking and low dietary intake of antioxidants (e.g. carotenoids, zinc). In general, dysregulation in the complement, lipid metabolism, angiogenesis, inflammation, and extracellular matrix pathways have been implicated in AMD pathogenesis.[2]

Clinical characteristics

The early stages of AMD are often asymptomatic. Some patients experience difficulty seeing in low contrast situations, in changing light conditions or in low luminance settings, especially while reading. When evaluating vision in the clinic these issues are often not highlighted as visual acuity tests are conducted in optimal light and contrast conditions. The late stages of AMD are often symptomatic, giving distorted vision or scotoma. When patients are unilaterally affected in the non-dominant eye, symptoms can remain unnoticed for a long time until the other eye experiences a visual decline. Atrophy usually progresses slowly but disability can increase substantially once it reaches the fovea. In contrast, when neovascularisation occurs, the symptoms can progress quickly due to macular oedema.

AMD is diagnosed on fundus examination and/or through retinal imaging. Early-stage features include drusen and RPE abnormalities. Drusen are yellow-white lesions corresponding to lipid-rich depositions between the RPE and Bruch's membrane (see also Section 13A.14). In general, small (<63 μm in diameter), well-defined drusen are considered part of the normal ageing process and are of limited significance. Conversely, larger drusen have been closely associated with progression to late-stage AMD. RPE abnormalities, typically corresponding to focal hypo- and hyperpigmentation in the macular region, have also been linked with an increased risk of developing advanced disease.[2,3]

Geographic atrophy is a manifestation of late-stage AMD. It is characterised by RPE degeneration followed by loss of photoreceptors and choriocapillaris. This leads to large (≥175 μm in diameter), well-defined areas of retinal atrophy. Progression of geographic atrophy is slow (months to years), and in some people, the fovea is spared for a long time providing reasonable visual acuity. A risk factor for the development or progression of geographic atrophy is the presence of reticular pseudo-drusen. Unlike conventional drusen, these drusen-like deposits are located above the RPE (in the subretinal space and the extracellular matrix between the photoreceptors), hence the term pseudo-drusen.[2]

Choroidal neovascularisation is the other late-stage feature of AMD. It is associated with the formation of new blood vessels originating from the choroid. These vessels cross through Bruch's membrane and proliferate in the sub-RPE and/or subretinal space. They often leak and bleed causing the presence of fluid within or below the retina. Over time, neovascularisation can result in the formation of fibrous scars.

Two atypical neovascular AMD variants have been described: retinal angiomatous proliferation (associated with intraretinal vascular complexes that arise from both the deep retinal capillaries and the choroid) and polypoidal choroidal vasculopathy (associated with polyp-like aneurysmal dilations of choroidal vessels leading to serosanguineous RPE detachments, exudative changes and, often subretinal fibrosis). The latter appears to be particularly prevalent in individuals of Asian and African ancestries.[4,5]

Advances in multimodal imaging technologies have significantly enriched our understanding of the natural history and pathogenesis of AMD. Also, imaging modalities such as OCT, OCT angiography, fluorescein angiography, and fundus autofluorescence imaging play a major role in the effective diagnosis and management of this disorder.

Molecular pathology

It is estimated that 50%–70% of AMD risk is attributable to genetic factors and, according to family and twin studies, people with a family history of AMD have a higher risk for developing the disease.[6] The first genes associated with AMD, including *CFH* and *ARMS2*, were identified through linkage studies. However, the biggest advances in unravelling the genetics of AMD came from large genome-wide association studies (GWAS). Over 50 independently associated variants in over 30 genomic loci have been identified, although the risk driven by *CFH* and *ARMS2* remains the most prominent.[1]

Both common (minor allele frequency ≥ 5%) and rare (minor allele frequency < 5%) variants have been implicated in AMD.[7] The high-risk version ('allele') of some common AMD-associated variants can be found in up to 50% of the general population but the effect sizes of these changes are generally small. In contrast, the high-risk allele of rare variants is identified in a much smaller proportion of the general population but the effect sizes are often larger. For example, an important risk allele of the *CFH* variant rs121913059 is present in 0.01% of the population but it is associated with 20 times higher risk of late-stage AMD.[1] It is worth highlighting that not all variants associated with AMD increase the risk of disease; many protective variants have also been described (Fig. 13A.70).[8] Sometimes protective and risk variants can even occur in the same gene.

One intriguing finding of AMD genetic studies is the association of a variant in the *MMP9* gene with choroidal neovascularisation but not with geographic atrophy. Having distinctive associations between the two advanced stages of AMD is uncommon. Notably, the protein encoded by this gene is involved in the breakdown of the extracellular matrix and interacts with VEGF signalling.[9,10]

In general, the variants and genes implicated in AMD have revealed three major pathways involved in the pathogenesis of the condition. These are the complement pathway, the extracellular matrix remodelling pathway and the lipid metabolism pathway.

The complement pathway is involved in the innate immune system and has a major role in the recognition and destruction of pathogens. It has three distinct sub-pathways; the classical pathway (responding to antigen–antibody complexes), the lectin pathway (responding to bacterial mannose groups) and the alternative pathway (active in a steady state but ready to intensify quickly). It is the alternative pathway that is primarily involved in AMD. Ultimately, all complement pathways generate a membrane attack complex which can penetrate a cell membrane and cause cell lysis. Complement is mostly produced in the liver but the retina can also produce some of its complement components. There is evidence suggesting that *ARMS2* encodes one of these components and that it acts as a surface complement regulator.[11] AMD-related variants in complement genes including *CFH* are thought to deregulate and overactivate the complement system. Some studies have found higher concentrations of complement in patients with AMD suggesting that inflammation may have a prominent role in disease pathogenesis. This is further supported by the observation that complement proteins are an important component of drusen.[12]

Extracellular matrix remodelling plays a key role in the maintenance of Bruch's membrane, which consists mainly of collagen and elastin fibres. There is a balance between the synthesis and the breakdown of the extracellular matrix, keeping this layer in its proper form. The RPE plays a major role in the regulation of this process through the expression of extracellular matrix related genes. In AMD there is an increase in collagen and the Bruch's membrane thickens.[13] This diminishes the outflow of the retina on one side and decreases the blood supply on the other side.[14] These pathophysiological changes likely facilitate the formation of drusen.[15]

The third pathway involves lipid metabolism. Although it is not fully understood how this pathway contributes to AMD, population studies suggest that high-density lipoprotein (HDL)-cholesterol levels are elevated and triglycerides are decreased in AMD patients. In addition, the composition of HDL seems to be different in AMD patients compared to controls individuals. Genes involved in this pathway exert their effect through lipid transport from the systemic circulation to the retina and, within the retina, between the RPE cells and photoreceptors.[16] It is noted that diet has an impact on the development of AMD with a high intake of products containing omega-3 lowering the risk of acquiring the condition.[17]

Overall, significant progress has been made in elucidating the molecular pathology of AMD. Importantly, the mechanism of interaction between genetic, environmental/lifestyle and epigenetic factors (such as methylation, microRNAs and long non-coding RNAs) are now beginning to emerge. At present, genetic testing for AMD is not routinely performed in most clinical settings. However, several studies have highlighted the value of genomic investigations in this group of patients. For example, genotyping assays testing common and rare AMD-associated variants have been shown to identify people at high risk of developing the advanced disease.[18] This is done through the estimation of the cumulative impact of multiple disease-implicated genetic variants. Although individually these variants generally confer small amounts of risk, when combined they can be used to create a meaningful risk score, known as polygenic score. Polygenic scores can be in turn combined with clinical and/or environmental risk factors to predict which individuals have a

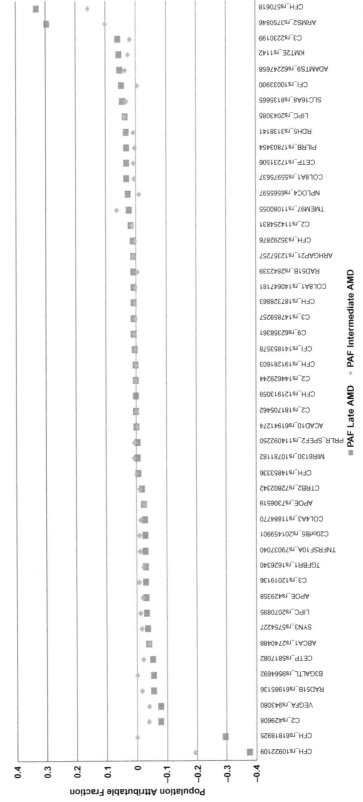

FIG. 13A.70 Bar graph showing the contribution (population-attributable fraction, PAF) of 49 AMD-associated genetic variants to intermediate (*light blue*) and late (*green*) AMD. Negative values suggest a protective effect. The high-effect *CFH* rs121913059 variant is not included for intermediate AMD because it was too rare to make meaningful calculations. (*Adapted from Ref. 8.*)

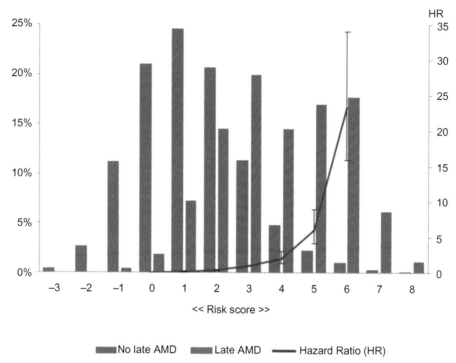

FIG. 13A.71 Likelihood of developing late AMD for people with different cumulative genetic risk score (polygenic score). The x-axis represents the polygenic score category, the left y-axis represents the frequency as a percentage and the right y-axis represents the hazard ratio (HR) of incident late AMD. The *black line* also represents the hazard ratio of incident late AMD. Category 3 is the reference category and has a hazard ratio of 1.0. Error bars indicate the 95% confidence interval. *(Adapted from Ref. 19.)*

higher chance of developing AMD and progressing to more severe forms (Fig. 13A.71).[20]

Clinical management

Studies have shown that patients with early-stage AMD can decrease their risk of progression towards late-stage disease by adjusting to a beneficial lifestyle. This includes quitting smoking, increasing physical activity and eating fruits, vegetables and fatty fish. It is thought that antioxidants and polyunsaturated fatty acids in these food products reduce the oxidative stress in the retina.[2] A large multicentre clinical trial showed that supplements containing vitamin C, vitamin E, copper, zinc, lutein and zeaxanthin reduce the risk of progression to late-stage AMD by 25% after 6 years of follow-up.[21]

In the clinic, patients with early-stage AMD should be given an Amsler grid for self-monitoring any distortion of vision. Currently, there is no treatment available for geographic atrophy. In the case of choroidal neovascularisation, intravitreal anti-VEGF injections can be administered in a monthly routine for three months. It has been shown that if the treatment is started directly after the first symptoms/signs of neovascularisation, more visual acuity is likely to be preserved. After the initial series of three injections, the effectiveness of the treatment should be evaluated. There are several treatment regimens available and factors that should be taken into account include the efficacy of the delivered therapeutic agent, the baseline visual acuity of the patient, cost considerations, and the risk of complications from the injections. If one type of anti-VEGF agent is not effective, a switch can be made to another. Usually, injections are continued until the retina is 'dry', and the time between the injections is extended gradually to avoid recurrence of macula oedema.

As AMD is a heritable disease, patients should be made aware of the risk to their family members. Siblings and children of the patient should also be informed about the risk of developing the disease. The effect of a healthy lifestyle on the prevention of the late stages of AMD should be highlighted. The symptoms that should prompt patients to seek ophthalmic review and the importance of starting treatment as soon as the first suggestive signs are noted should be discussed. At this moment there are multiple algorithms available for estimating an individual's risk for developing visually significant complications; these are based on current phenotype, age, environmental/lifestyle factors and, in some cases, also genetics. With genetic testing becoming cheaper and prediction models becoming more accurate, personalised counselling for AMD is on its way.

There are many ongoing trials for improving treatment regimens for choroidal neovascularisation,[4] as well as for finding a treatment for geographic atrophy[5] by, for example, intervening in the complement pathway. However, to find a cure for AMD, basic science focusing on the pathophysiological mechanisms is needed. Together with studies on genetics and epidemiology more and more of the disordered processes in AMD are elucidated providing new targets for therapy.

At the same time, major advancements are taking place in retinal imaging. These new techniques together with artificial intelligence facilitate the diagnostic process and sometimes help in earlier detection of AMD.[22] Once these technologies become more widely available, intervention can start at an earlier stage in the disease course, hopefully reducing the risk of blindness in the future.

References

1. Fritsche LG, et al. A large genome-wide association study of age-related macular degeneration highlights contributions of rare and common variants. *Nat Genet* 2016;**48**(2):134–43. https://doi.org/10.1038/ng.3448.
2. Mitchell P, Liew G, Gopinath B, Wong TY. Age-related macular degeneration. *Lancet* 2018;**392**(10153):1147–59.
3. Chakravarthy U, Peto T. Current perspective on age-related macular degeneration. *JAMA* 2020;**324**(8):794–5.
4. TAC G, Georgiou M, JWB B, Michaelides M. Gene therapy for neovascular age-related macular degeneration: rationale, clinical trials and future directions. *Br J Ophthalmol* 2021;**105**(2):151–7.
5. Cabral de Guimaraes TA, Daich Varela M, Georgiou M, Michaelides M. Treatments for dry age-related macular degeneration: therapeutic avenues, clinical trials and future directions. *Br J Ophthalmol* 2021. https://doi.org/10.1136/bjophthalmol-2020-318452, bjophthalmol-2020-318452.
6. Seddon JM, et al. The US twin study of age-related macular degeneration: relative roles of genetic and environmental influences. *Arch Ophthalmol* 2005;**23**(3):321–7.
7. Ratnapriya R, Acar İE, Geerlings MJ, Branham K, Kwong A, Saksens NTM, et al. Family-based exome sequencing identifies rare coding variants in age-related macular degeneration. *Hum Mol Genet* 2020;**29**(12):2022–34.
8. Colijn JM, Meester-Smoor M, Verzijden T, et al. Genetic risk, lifestyle, and age-related macular degeneration in Europe: the EYE-RISK consortium. *Ophthalmology* 2020;**128**(7):1039–49. https://doi.org/10.1016/j.ophtha.2020.11.024.
9. Di Y, Nie QZ, Chen XL. Matrix metalloproteinase-9 and vascular endothelial growth factor expression change in experimental retinal neovascularization. *Int J Ophthalmol* 2016;**9**(6):804–8. https://doi.org/10.18240/ijo.2016.06.02.
10. Hollborn M, Stathopoulos C, Steffen A, Wiedemann P, Kohen L, Bringmann A. Positive feedback regulation between MMP-9 and VEGF in human RPE cells. *Invest Ophthalmol Vis Sci* 2007;**48**(9):4360–7. https://doi.org/10.1167/iovs.06-1234.
11. Micklisch S, Lin Y, Jacob S, et al. Age-related macular degeneration associated polymorphism rs10490924 in ARMS2 results in deficiency of a complement activator. *J Neuroinflammation* 2017;**14**(1):4.
12. Wang L, Clark ME, Crossman DK, Kojima K, Messinger JD, Mobley JA, et al. Abundant lipid and protein components of drusen. *PLoS One* 2010;**5**(4):e10329. https://doi.org/10.1371/journal.pone.0010329.
13. Ko F, Foster PJ, Strouthidis NG, Shweikh Y, Yang Q, Reisman CA, et al. Associations with retinal pigment epithelium thickness measures in a large cohort: results from the UK biobank. *Ophthalmology* 2017;**124**(1):105–17. https://doi.org/10.1016/j.ophtha.2016.07.033.
14. Moore DJ, Hussain AA, Marshall J. Age-related variation in the hydraulic conductivity of Bruch's membrane. *Invest Ophthalmol Vis Sci* 1995;**36**(7):1290–7.
15. Curcio CA. Soft drusen in age-related macular degeneration: biology and targeting via the oil spill strategies. *Invest Ophthalmol Vis Sci* 2018;**59**(4). https://doi.org/10.1167/iovs.18-24882.
16. van Leeuwen EM, Emri E, Merle BMJ, Colijn JM, Kersten E, Cougnard-Gregoire A, et al. A new perspective on lipid research in age-related macular degeneration. *Prog Retin Eye Res* 2018;**67**:56–86. https://doi.org/10.1016/j.preteyeres.2018.04.006.
17. Zhong Y, Wang K, Jiang L, Wang J, Zhang X, Xu J, et al. Dietary fatty acid intake, plasma fatty acid levels, and the risk of age-related macular degeneration (AMD): a dose-response meta-analysis of prospective cohort studies. *Eur J Nutr* 2021. https://doi.org/10.1007/s00394-020-02445-4.
18. de Breuk A, Acar IE, Kersten E, MMVAP S, Colijn JM, Haer-Wigman L, et al. EYE-RISK consortium. Development of a genotype assay for age-related macular degeneration: the EYE-RISK Consortium. *Ophthalmology* 2020. https://doi.org/10.1016/j.ophtha.2020.07.037.
19. Buitendijk GHS, Rochtchina E, Myers C, van Duijn CM, Lee KE, Klein BEK, et al. Prediction of age-related macular degeneration in the general population: the three continent AMD consortium. *Ophthalmology* 2013;**120**(12):2644–55. https://doi.org/10.1016/j.ophtha.2013.07.053.
20. Ajana S, Cougnard-Gregoire A, Colijn JM, BMJ M, Verzijden T, de Jong PTVM, et al. Predicting progression to advanced age-related macular degeneration from clinical, genetic, and lifestyle factors using machine learning. *Ophthalmology* 2020;**128**(4):587–97.
21. Age-Related Eye Disease Study 2 Research Group. Lutein + zeaxanthin and omega-3 fatty acids for age-related macular degeneration: the Age-Related Eye Disease Study 2 (AREDS2) randomized clinical trial. *JAMA* 2013;**309**:2005–15.
22. Gong D, Kras A, Miller JB. Application of deep learning for diagnosing, classifying, and treating age-related macular degeneration. *Semin Ophthalmol* 2021;**22**:1–7. https://doi.org/10.1080/08820538.2021.1889617.

Chapter 13B

Syndromic retinal disease

Chapter outline

13B.1 Ciliopathies — 268
    13B.1.1 Bardet-Biedl syndrome — 270
    13B.1.2 Joubert syndrome — 274
    13B.1.3 Alström syndrome — 277
    13B.1.4 Usher syndrome — 281
13B.2 Retinopathy in inborn errors of metabolism — 285
    13B.2.1 Gyrate atrophy of the choroid and retina — 288
    13B.2.2 Bietti corneoretinal crystalline dystrophy — 292
    13B.2.3 Pseudoxanthoma elasticum — 295
    13B.2.4 MIDD (maternally inherited diabetes, deafness and maculopathy) — 302
    13B.2.5 Long-chain L-3 hydroxyacyl-CoA dehydrogenase (LCHAD) deficiency — 306
    13B.2.6 Neuronal ceroid lipofuscinosis (Batten disease) — 309
    13B.2.7 Cobalamin C deficiency — 313
    13B.2.8 Cohen syndrome — 317

13B.1

Ciliopathies

Cilia are evolutionarily conserved hair-like organelles that are a feature of almost all human cells. Based on the ability to beat rhythmically or lack thereof, cilia can be classified as motile and non-motile.[1] Motile cilia are mainly present in specialised cells, such as spermatozoa and the epithelial lining of the respiratory tract.[2] In contrast, a single non-motile cilium, also known as primary cilium, can be found in almost every human cell. While motile cilia predominantly serve to move fluid across membrane surfaces, primary cilia form 'antennas' that allow transduction of sensory information from the extracellular environment.[3] Disruption of cilia leads to a wide range of human disorders, known as ciliopathies. Defects in more than 300 genes have been found to be associated with this group of conditions.[4]

The photoreceptor outer segment is a highly specialised primary cilium and defects in it may lead to retinal degeneration, a feature of many non-motile ciliopathies.[1] Retinal involvement can occur either as part of an isolated ophthalmic condition (e.g. retinal dystrophy) or as part of a multisystemic disorder.[5] Other organ systems which are commonly affected in ciliopathies include the kidney, the central nervous system, the liver and the skeleton. There is significant clinical variability, both in terms of disease severity and the number of affected organs, even among individuals with the same mutation(s). Prevalent ciliopathies are represented schematically in Fig. 13B.1. A selected subset of conditions is discussed in this chapter which does not intend to be exhaustive but outlines the key signs and symptoms of these disorders, and highlights the importance of systematically evaluating children with retinal dystrophies.

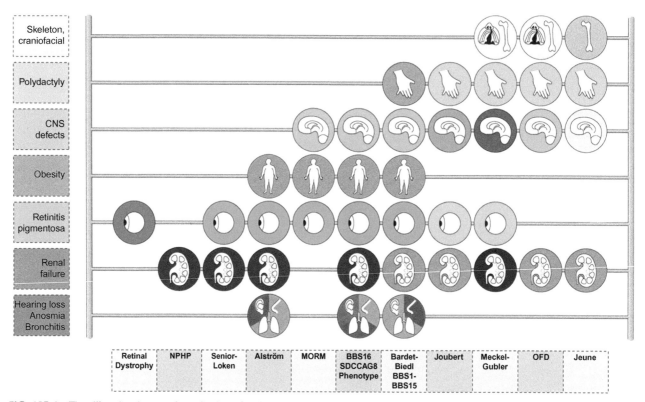

FIG. 13B.1 The ciliopathy abacus: schematic view of various ciliopathies represented as an abacus with affected organs. *NPHP*, nephronophthisis; *MORM*, mental retardation, truncal obesity, retinal dystrophy, and micropenis syndrome; *OFD*, oral-facial-digital syndrome.

References

1. Chen HY, Kelley RA, Li T, Swaroop A. Primary cilia biogenesis and associated retinal ciliopathies. *Semin Cell Dev Biol* 2020;**S1084-9521** (19), 30167-3.
2. Wallmeier J, Nielsen KG, Kuehni CE, Lucas JS, Leigh MW, Zariwala MA, et al. Motile ciliopathies. *Nat Rev Dis Primers* 2020;**6**(1):77.
3. May-Simera H, Nagel-Wolfrum K, Wolfrum U. Cilia—the sensory antennae in the eye. *Prog Retin Eye Res* 2017;**60**:144-80.
4. TJP vD, Kennedy J, van der Lee R, de Vrieze E, Wunderlich KA, Rix S, et al. CiliaCarta: an integrated and validated compendium of ciliary genes. *PLoS One* 2019;**14**(5), e0216705.
5. Bujakowska KM, Liu Q, Pierce EA. Photoreceptor cilia and retinal ciliopathies. *Cold Spring Harb Perspect Biol* 2017;**9**(10):a028274.

13B.1.1

Bardet-Biedl syndrome

Hélène Dollfus

Bardet-Biedl syndrome (BBS) is an emblematic ciliopathy primarily characterised by retinal degeneration, obesity, polydactyly, renal dysfunction, hypogonadism and occasional cognitive impairment. It is typically inherited as an autosomal recessive trait and it has been associated with at least 24 genes.[1]

Clinical characteristics

The diagnosis of Bardet-Biedl syndrome is based on a set of cardinal features that emerge throughout infancy and childhood. These include:

- retinal dystrophy,
- truncal obesity,
- postaxial polydactyly,
- renal malformations or dysfunction,
- hypogonadotropic hypogonadism,
- intellectual disability.

Additional features may include olfactory dysfunction (anosmia or hyposmia), diabetes, cardiac anomalies (including laterality defects such as situs inversus), liver abnormalities and Hirschprung disease. Not all symptoms are always present and there can be significant intra- and inter-familial variability.[1,2]

Retinal dysfunction is one of the commonest features of Bardet-Biedl syndrome (>95% of cases[2]). Visual symptoms, including abnormal visual behaviour and photophobia, are often what brings affected individuals to medical attention, typically in the first decade of life. Profound electroretinographic abnormalities can be detected as early as 3 years of age. Typically, there is panretinal degeneration with equal rod and cone photoreceptor involvement. A rod-cone pattern similar to retinitis pigmentosa (with or without early macular involvement) or, less commonly, a cone-rod pattern may also be seen (Fig. 13B.2).[3] There is progressive degeneration classically leading to severe visual handicap before adulthood. However, significant variability in the ophthalmic phenotype has been described including milder and/or later onset presentations.

Obesity, another major feature, is present in around 90% of affected individuals. It typically develops in the first year of life and worsens with age.[1,2] The mean body mass index is $36 kg/m^2$[1] and the origin of obesity is both central (hypothalamic eating control) and peripheral (adipose tissue).

Limb anomalies are common; they are present at birth and have significant diagnostic value. Polydactyly (additional digits at the side of the hand and/or foot) is noted in around 80% of cases but can manifest itself very subtly.[2] Other frequently reported limb malformations include brachydactyly (short digits) or syndactyly (fused digits) (Fig. 13B.3).

Renal structural anomalies are present in about 50% of cases and may lead to renal dysfunction and kidney failure in late childhood. Hyperechogenic or cystic kidneys may be observed in prenatal ultrasound evaluation and this finding combined with polydactyly in an infant is highly suggestive of Bardet-Biedl syndrome. Importantly, chronic kidney disease is a major contributor to morbidity and mortality in this group of patients and it has been reported that 6% of children and 8% of adults develop end-stage renal disease requiring dialysis and/or transplantation.[4]

  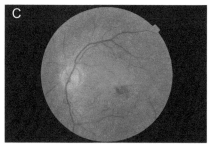

FIG. 13B.2 Fundus photography from: (A) an adult patient of African ancestry with a homozygous *BBS1* mutation and significant retinal pigment migration; (B) an adult patient with biallelic *BBS10* mutations and advanced signs of retinitis pigmentosa; (C) a 20 year old patient with biallelic BBS12 mutations and early signs of retinal degeneration.

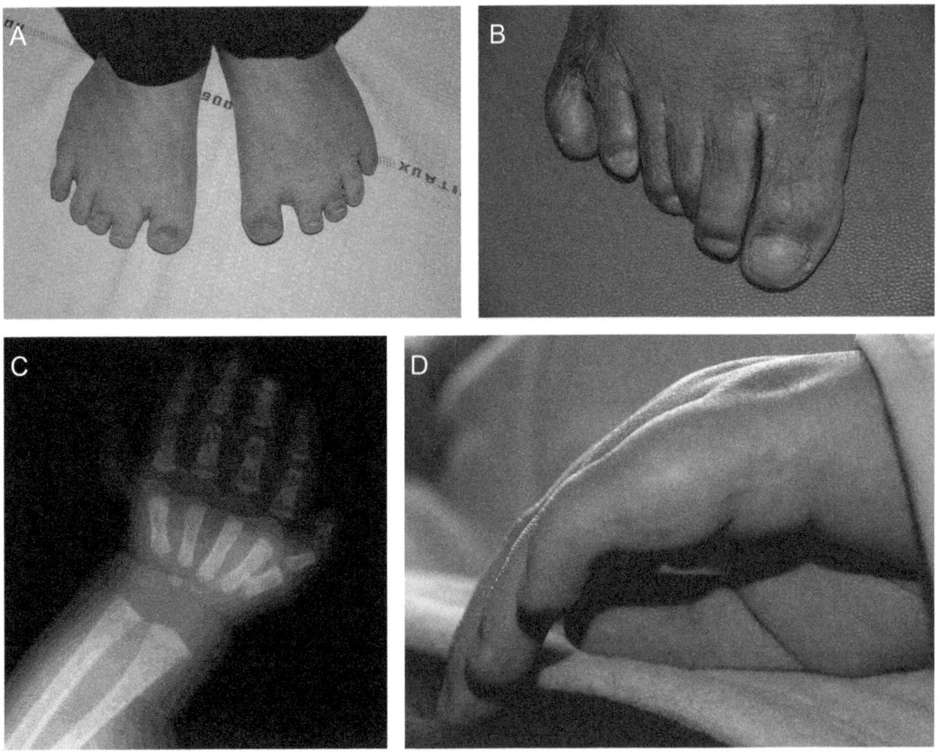

FIG. 13B.3 Limb anomalies in Bardet-Biedl syndrome. (A) Brachydactyly of both feet in an affected adult. (B) Postaxial polydactyly in an adult patient. (C) X-ray of the hand of a child with postaxial polydactyly. (D) Lateral scar showing the site of removal of postaxial polydactyly.

Hypogonadism and genitourinary malformations can occur including micropenis and/or small-volume testes in males and hydrometrocolpos (a neonatal vaginal malformation leading to a large abdominal tumour) in females. Infertility is common but reproduction has been observed for female and, more rarely, for male patients.

Neuropsychiatric symptoms can include developmental delay, intellectual disability, learning difficulties, speech deficit, and behavioural problems. Intellectual function ranges from severe intellectual deficiency to average intelligence.[5] Importantly, a subset of people with Bardet-Biedl syndrome has accomplished high academic levels stressing that the level of intellectual performance is highly variable. Slow ideation and hyperemotive status are common.

Molecular pathology

Bardet-Biedl syndrome is an autosomal recessive condition. Although there is emerging evidence suggesting that genetic modifiers influence the severity of the clinical presentation,[6] genetic counselling is presently based on classic Mendelian concepts (e.g. 25% recurrence risk for parents with an affected child).

Bardet-Biedl syndrome, like many other ciliopathies, is genetically highly heterogeneous. More than 24 genes have been implicated, all encoding proteins required for primary cilium function.[1] Many of these genes have also been associated with other diseases including non-syndromic retinitis pigmentosa (*BBS1*, *BBS2*, *ARL6*, *BBS8*, *C8orf37*), Joubert syndrome (*MKS1*, *CEP290*), Senior-Løken syndrome (*CEP290*, *SDCCAG8*), nephronophthisis (*CEP290*, *IFT172*), and Jeune thoracic dysplasia (*IFT172*).

It is noteworthy that mutations in two genes, *BBS1* and *BBS10*, account for more than 40% of cases. Several relatively prevalent, recurrent mutations have been reported including the c.1169T>G (p.Met390Arg) change in *BBS1* and the c.271dupT (p.Cys91fs) variant in *BBS10*. Genotype-phenotype correlations remain to be ascertained in large cohorts but meso-axial polydactyly is a characteristic feature of *LZTFL1* (BBS17) variants.

Genetic investigations are typically performed by way of high-throughput sequencing mainly through multigene panels (including ciliopathy gene panels and retinal dystrophy gene panels). These have proven to be very effective, detecting mutations in over 80% of cases and enabling differential diagnosis from overlapping conditions such as Alström syndrome (see Section 13B.1.3) (Fig. 13B.4).

Clinical management

Multidisciplinary management is required and special attention should be given to kidney function and the

272 SECTION | III Genetic disorders affecting the posterior segment

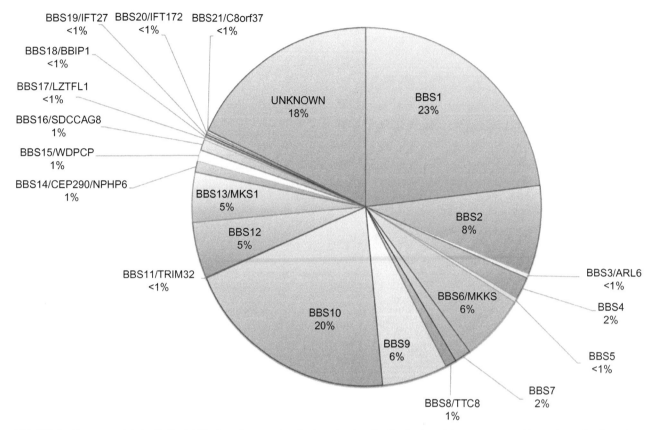

FIG. 13B.4 Genes associated with Bardet-Biedl syndrome and a pie chart showing the distribution of the genes mutated in a cohort of patients from France. *(Courtesy of Dr. Jean Muller.)*

prevention of obesity. Regular medical follow-up visits are advisable including paediatric assessment, internal medicine input (e.g. nephrology, endocrinology), medical genetics review and assessment by other specialities on an ad hoc basis (e.g. orthopaedics for removal of extra digits).

At the time of diagnosis, basic investigations have to be undertaken including:

- examination for polydactyly (feet and hands) and other developmental anomalies (cardiac malformations including situs inversus, genital malformations);
- kidney evaluation including imaging (kidney ultrasound including the urogenital tract), blood biochemistry (urea, creatinine clearance, cystatin C, blood cell count) and urine biology (pH, electrolytes, urea, creatinine, albumin and β2-microglobulin).

On follow-up visits, symptoms of polyuria, polydipsia, respiratory dysfunction, sleep apnea, constipation, auditory and olfactory problems should be enquired about. Physical examination (including body mass index assessment) and kidney biochemical evaluation are warranted at least once a year. A general work-up including testing for diabetes and metabolic syndrome is advocated in cases with severe obesity. Mood/psychological/psychiatric problems may necessitate appropriate follow-up.

Surgical intervention may be required for end-stage kidney failure, severe obesity or polydactyly. It is of paramount importance that anaesthetic expert evaluation is undertaken before attempting general anaesthesia as advanced procedures are often required for individuals with Bardet-Biedl syndrome (e.g. video-laryngoscopy or intubation under fibroscopy).

The ophthalmic management is similar to that of individuals with non-syndromic retinitis pigmentosa. Slit-lamp examination including screening for early lens opacification and retinal dystrophy features (such as pigment epithelial irregularity, 'salt and pepper' appearance, intraretinal pigment clumping and/or retinal vessel narrowing) should be performed. Ophthalmic investigations typically include electroretinographic assessment (particularly useful for the initial diagnosis of retinal dysfunction) and imaging (including OCT, which is useful to detect macular oedema).

References

1. Forsyth RL, Gunay-Aygun M. Bardet-Biedl syndrome overview. 2003 Jul 14 [updated 2020 Jul 23]. In: Adam MP, Ardinger HH, Pagon RA, et al., editors. *GeneReviews® [internet]*. Seattle, WA: University of Washington, Seattle; 1993–2020. Available from: https://www.ncbi.nlm.nih.gov/books/NBK1363/.

2. Niederlova V, Modrak M, Tsyklauri O, Huranova M, Stepanek O. Meta-analysis of genotype-phenotype associations in Bardet-Biedl syndrome uncovers differences among causative genes. *Hum Mutat* 2019;**40**:2068–87.
3. Scheidecker S, Hull S, Perdomo Y, Studer F, Pelletier V, Muller J, et al. Predominantly cone-system dysfunction as rare form of retinal fegeneration in patients with molecularly confirmed Bardet-Biedl syndrome. *Am J Ophthalmol* 2015;**160**(2):364–372.e1.
4. Forsythe E, Sparks K, Best S, Borrows S, Hoskins B, Sabir A, et al. Risk factors for severe renal disease in Bardet-Biedl syndrome. *J Am Soc Nephrol* 2017;**28**:963–70.
5. Kerr EN, Bhan A, Héon E. Exploration of the cognitive, adaptive and behavioral functioning of patients affected with Bardet-Biedl syndrome. *Clin Genet* 2016;**89**(4):426–33.
6. Kousi M, Söylemez O, Ozanturk A, Mourtzi N, Akle S, Jungreis I, et al. Evidence for secondary-variant genetic burden and non-random distribution across biological modules in a recessive ciliopathy. *Nat Genet* 2020;**52**:1145–50. https://doi.org/10.1038/s41588-020-0707-1.

13B.1.2

Joubert syndrome

Frauke Coppieters, Elise Héon, and Monika K. Grudzinska Pechhacker

Joubert syndrome is a group of congenital neurodevelopmental disorders characterised by the triad of developmental delay, hypotonia and a distinct cerebellar/brainstem malformation; the latter is recognised on neuroimaging as the 'molar tooth sign' and is a diagnostic hallmark of the condition.[1] Some patients may also have kidney abnormalities, liver disease, skeletal anomalies and/or additional brain changes.[2] Ophthalmic manifestations (including retinal degeneration and eye movement abnormalities) may be present; these vary between the different genetic subtypes and can range from mild to severe.[3,4]

Mutations in over 35 genes have been implicated in Joubert syndrome. Because most of these genes encode proteins that localise at the basal body or the ciliary transition zone of non-motile (primary) cilia,[5] Joubert syndrome is considered a ciliopathy. In addition to genetic heterogeneity, Joubert syndrome exhibits significant clinical heterogeneity. One consistent aspect though is the need for multidisciplinary clinical management.[5]

Clinical characteristics

Affected individuals typically present in infancy with hypotonia, abnormal eye movements and disturbed respiratory control; ataxia and/or intellectual disability are often noted in childhood. In addition to these core features, many patients also have other impaired organ systems including the eye, the kidney, the liver, and the skeleton. Significant variability in disease severity has been described both within and among affected families.[6] The general prognosis typically depends on the severity of breathing dysregulation (which can be life threatening soon after birth) and the extent of renal and hepatic complications.

The typical congenital respiratory control issues of Joubert syndrome include short alternate episodes of apnea and hyperpnea, and episodic hyperpnea alone; these generally improve with age. Abnormal respiratory patterns and sleep behaviours, sleep-disordered breathing and recurrent respiratory infections have also been described.[7,8]

The most common neurological features are hypotonia (evolving to ataxia with broad-based gait), congenital feeding problems (dysphagia), and developmental delay.

Cognition ranges from normal (in a minority of cases) to severe intellectual disability. Due to oral motor apraxia, expressive speech is affected more than comprehension. Seizures occur in at least 10% of affected individuals. Additional neurological features include encephalocele, hydrocephalus, dystonic episodes, temperature regulation problems, and behavioural and psychiatric problems.[6]

The diagnostic hallmark of Joubert syndrome is the 'molar tooth' sign which is visible on axial brain magnetic resonance imaging (MRI) at the junction of the midbrain and hindbrain. This appearance results from characteristic malformations such as cerebellar vermis aplasia/hypoplasia, an abnormally deep interpeduncular fossa (at the level of the isthmus and upper pons), and thick and horizontally oriented superior cerebellar peduncles (Fig. 13B.5).

Ophthalmic manifestations are varied and are not present in all genetic subtypes of Joubert syndrome.[3] Although the visual acuity is not always quantifiable due to developmental/neurological issues, very severe visual impairment affects only a small subset of cases (<5%). The combination of variability in ophthalmic features and inability to reliably assess vision (before 9 years of age on average) makes predicting visual function before school age particularly challenging.[3] Among Joubert syndrome patients with ophthalmic phenotypes, the majority have abnormal eye movements; these include (congenital) oculomotor apraxia (~80%), nystagmus (~70%) and strabismus (~70%). Oculomotor apraxia is typically seen as head thrusting to shift gaze and it is associated with the inability to voluntarily initiate saccades. This feature usually improves with time. It is highlighted that children found to have oculomotor apraxia should have a brain MRI (to specifically look for the molar tooth sign) even in the absence of other syndromic features.

Ptosis (unilateral or bilateral) is relatively common (~40%) but rarely obstructs vision enough to require surgical intervention. Chorioretinal coloboma may be present (~30%) (Fig. 13B.6A) and is often associated with an increased risk for liver disease. Optic nerve atrophy has been described in a subset of cases (~20%).[4]

Variable forms of retinal degeneration have been reported in up to 30% of patients with Joubert syndrome (Fig. 13B.6B and C). The retinal phenotype can vary from

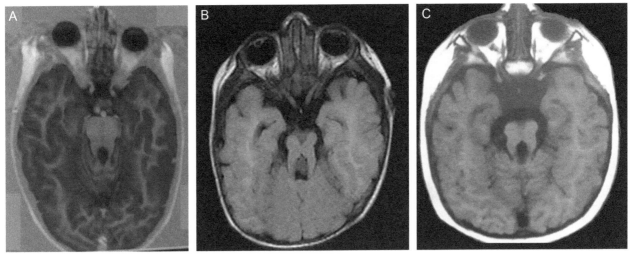

FIG. 13B.5 Brain MRI in three unrelated individuals with Joubert syndrome showing the typical midbrain-hindbrain malformation known as the "molar tooth" sign. Axial sections at the pontine or pontomesencephalic junction are displayed. *(B and C: Adapted from Ref. 9.)*

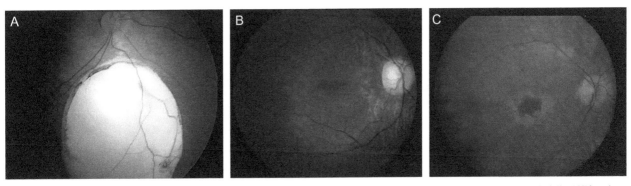

FIG. 13B.6 Fundus findings in individuals with Joubert syndrome. (A) Inferior chorioretinal coloboma in an individual with biallelic *AHI1* variants. (B) Retinal degeneration in a 12-year-old patient with Joubert syndrome; macular changes and vessel attenuation are noted. (C) Retinal degeneration in a 19-year-old patient with Joubert syndrome; significant maculopathy is noted.

an early childhood onset severe panretinal dystrophy (e.g. Leber congenital amaurosis) to adult-onset cone-rod or rod-cone dystrophy. Since fundoscopy in affected infants and young children is often unremarkable (or only reveals subtle changes), further assessment of retinal function with electroretinography is important.[3,10] Notably, affected individuals with retinal involvement have been shown to have a higher incidence of kidney disease[3] (which affects their general prognosis).

About a quarter of Joubert syndrome patients have renal manifestations, which include fibrocystic kidney disease, nephronophthisis and cystic dysplasia. Kidney failure is the leading cause of death in affected individuals after the first year of life. Juvenile nephronophthisis is one of the commonest renal features of Joubert syndrome; this generally manifests as acute or chronic renal insufficiency in the late first or early second decade of life. It eventually evolves to end-stage renal failure requiring dialysis or kidney transplantation. Infantile nephronophthisis is more severe and generally presents in the first few years of life. Early-onset anaemia often accompanies nephronophthisis.

About 15% of patients with Joubert syndrome present with congenital hepatic fibrosis, characterised by raised serum liver enzymes, hepato(spleno)megaly or abnormal liver echogenicity. Severe, less frequent complications include portal hypertension and oesophageal varices.

Approximately 15% of Joubert syndrome patients have polydactyly and over 5% have scoliosis. Rarer clinical manifestations include skeletal dysplasia, endocrine abnormalities, oral manifestations, hearing loss, Hirschsprung disease, and congenital heart disease.[6,8]

Joubert syndrome shows considerable clinical overlap with other related ciliopathies such as the Senior-Løken syndrome, Bardet-Biedl syndrome, Meckel syndrome, Mainzer-Saldino syndrome, acrocallosal syndrome, COACH syndrome, and hydrolethalus syndrome. Other differential diagnoses include cerebellar and brainstem congenital defects such as the Dandy-Walker malformation and Poretti-Boltshauser syndrome.

Molecular pathology

Joubert syndrome is associated with pathogenic variants in over 35 genes. It is generally inherited as an autosomal recessive trait except for one subtype, *OFD1*-related disease, which is inherited in an X-linked recessive manner.

Given the substantial genetic heterogeneity of this condition, mutational analysis of affected individuals is usually performed using either gene panel testing or exome/genome sequencing. Pathogenic variants are identified in 60–90% of Joubert syndrome patients depending on the cohort and testing strategy. Mutations in six genes (*AHI1, CC2D2A, CEP290, CPLANE1, KIAA0586,* and *TMEM67*) account for most cases. A founder effect has been described for *TMEM216* c.218G>T (p.Arg73Leu) and *TMEM237* c.52C>T (p.Arg18Ter) in individuals of Ashkenazi Jewish and Canadian Hutterite ancestries respectively.

Broad genotype–phenotype correlations have been established. These highlight an increased risk of developing specific organ dysfunction in specific Joubert syndrome subtypes including: *CEP290* (retinal dystrophy and kidney disease), *AHI1* (retinal dystrophy), *TMEM67* (coloboma, liver fibrosis, kidney disease), and *RPGRIP1L/NPHP1/TMEM237* (kidney disease).[3]

Clinical management

A detailed workup at diagnosis is important to document the full extent of the disease. Patient management requires a multidisciplinary approach. Many children will present in early life as a result of neurological manifestations including abnormal breathing, hypotonia and oculomotor apraxia. Close working between paediatric neurology, clinical genetics and ophthalmology are key to optimising surveillance of potential multisystemic complications (e.g. progressive renal, hepatic and retinal disease). In addition to biomedical literature on patient management,[7] detailed healthcare recommendations for Joubert syndrome have been published.[8]

In cases where the underlying molecular defect is known, prenatal diagnosis and preimplantation genetic testing can be offered. In cases where the molecular defect is unknown, fetal MRI and ultrasound may reveal aberrant clinical features in at risk pregnancies such as posterior fossa malformations and cerebellar vermis hypoplasia. However, the absence of fetal abnormalities cannot exclude Joubert syndrome.

References

1. Romani M, Micalizzi A, Valente EM. Joubert syndrome: congenital cerebellar ataxia with the molar tooth. *Lancet Neurol* 2013;**12**(9):894–905.
2. Parisi MA. The molecular genetics of Joubert syndrome and related ciliopathies: the challenges of genetic and phenotypic heterogeneity. *Transl Sci Rare Dis* 2019;**4**(1–2):25–49.
3. Brooks BP, Zein WM, Thompson AH, Mokhtarzadeh M, Doherty DA, Parisi M, et al. Joubert syndrome: ophthalmological findings in correlation with genotype and hepatorenal disease in 99 patients prospectively evaluated at a single center. *Ophthalmology* 2018;**125**(12):1937–52.
4. Wang SF, Kowal TJ, Ning K, Koo EB, Wu AY, Mahajan VB, et al. Review of ocular manifestations of joubert syndrome. *Genes (Basel)* 2018;**9**(12).
5. Reiter JF, Leroux MR. Genes and molecular pathways underpinning ciliopathies. *Nat Rev Mol Cell Biol* 2017;**18**(9):533–47.
6. Bachmann-Gagescu R, Dempsey JC, Phelps IG, O'Roak BJ, Knutzen DM, Rue TC, et al. Joubert syndrome: a model for untangling recessive disorders with extreme genetic heterogeneity. *J Med Genet* 2015;**52**(8):514–22.
7. Vilboux T, Doherty DA, Glass IA, Parisi MA, Phelps IG, Cullinane AR, et al. Molecular genetic findings and clinical correlations in 100 patients with Joubert syndrome and related disorders prospectively evaluated at a single center. *Genet Med* 2017;**19**(8):875–82.
8. Bachmann-Gagescu R, Dempsey JC, Bulgheroni S, Chen ML, D'Arrigo S, Glass IA, et al. Healthcare recommendations for Joubert syndrome. *Am J Med Genet A* 2020;**182**(1):229–49.
9. Brancati F, Barrano G, Silhavy JL, Marsh SE, Travaglini L, Bielas SL, et al. CEP290 mutations are frequently identified in the oculo-renal form of Joubert syndrome-related disorders. *Am J Hum Genet* 2007;**81**(1):104–13.
10. Ruberto G, Parisi V, Bertone C, Signorini S, Antonini M, Valente EM, et al. Electroretinographic assessment in Joubert syndrome: a suggested objective method to evaluate the effectiveness of future targeted treatment. *Adv Ther* 2020;**37**(9):3827–38.

13B.1.3

Alström syndrome

Isabelle Meunier, Hélène Dollfus, Vasiliki Kalatzis, Béatrice Bocquet, and Catherine Blanchet

Alström syndrome is a rare autosomal recessive multisystemic condition. Key manifestations include cone-rod dystrophy, hearing impairment and metabolic disturbance leading to diabetes and obesity in childhood. Dilated cardiomyopathy can occur in infancy or later in life and renal dysfunction is reported in half of the cases. Wide inter- and intra-familial variability have been described and delayed diagnosis is common, often due to the gradual onset of the classic symptoms during childhood.[1] It is important to consider Alström syndrome in all infants with cone-rod dystrophy, even in the absence of apparent systemic features.

The vast majority of affected individuals carry biallelic mutations in *ALMS1*, a gene with a role in the formation, function and maintenance of primary cilia. Alström syndrome is therefore considered a ciliopathy.

Clinical characteristics

Individuals with Alström syndrome often present in the first few months of life either with nystagmus or, less commonly, with infantile-onset cardiomyopathy.

Nystagmus is classically combined with marked photophobia and abnormal visual behaviour. These ophthalmic findings are due to a severe, infantile-onset cone-rod dystrophy that sometimes mimics achromatopsia or Leber congenital amaurosis.[2,3] Despite the significant visual impairment, fundus examination often only reveals very subtle macular changes including white dots and/or a 'peau d'orange' pattern. There is typically no or minimal pigment migration in the peripheral retina. With time, the optic nerve head becomes pale and the vessels narrow. Characteristic imaging findings are shown in Figs. 13B.7 and 13B.8 (see also[2]).

Congestive heart failure develops in over 60% of patients at some point in their lives as a result of infantile-, adolescent- or adult-onset cardiomyopathy.[4] The onset, progression and clinical outcome of heart disease vary, even among individuals with the same mutation. The spectrum of possible cardiac manifestations includes a transient but severe dilated cardiomyopathy in the first weeks of life and progressive restrictive cardiomyopathy identified between the second and fourth decade of life.[3] Hypertension is frequently reported, even during childhood, and is potentially intensified by renal dysfunction.

Progressive bilateral hearing loss is also a feature of Alström syndrome. This occurs in at least 70% of cases[5] and is usually diagnosed between age 1 and 10 years.[3] Hearing impairment succeeds visual impairment and is mostly moderate but can range from mild to severe. The audiometric pattern is variable and there is a progressive deterioration of 10–15 degrees per decade. The typical mechanism of hearing loss is sensorineural although mixed mechanisms with a conductive component, due to fluid in the middle ear, are also common. Most children with Alström syndrome report frequent episodes of otitis media which can lead to transient worsening or fluctuations in hearing.[5] Associated sinusitis, bronchitis and pneumonia may occur.[6]

With time, and after visual and hearing impairment, endocrinologic and metabolic dysfunction develops. These include insulin resistance and early-onset diabetes mellitus, diabetes insipidus, hypertriglyceridemia, infertility, hypothyroidism and growth hormone deficiency. The result is obesity in early childhood and a short adult stature with scoliosis later in life. In females, hyperandrogenism with hirsutism (excessive growth of dark or coarse hair in a male-like pattern) and alopecia have been reported; in males, hypergonadotropic hypogonadism has been described.[3]

Alström syndrome is commonly associated with chronic kidney disease that is slowly progressive and has onset in mid-childhood or adulthood. Recurrent urinary tract infections with or without urological malformations may occur in the first decade of life. Hepatic manifestations are common and range from elevated transaminases to non-alcoholic fatty liver disease, cirrhosis and hepatic failure in the second to third decade of life.[3] Other features of Alström syndrome include acanthosis nigricans, soft dysmorphic features (rounded face, deep-set eyes, thick ears), dental anomalies (gingivitis, discoloured teeth), flat feet, short fingers and seizures.

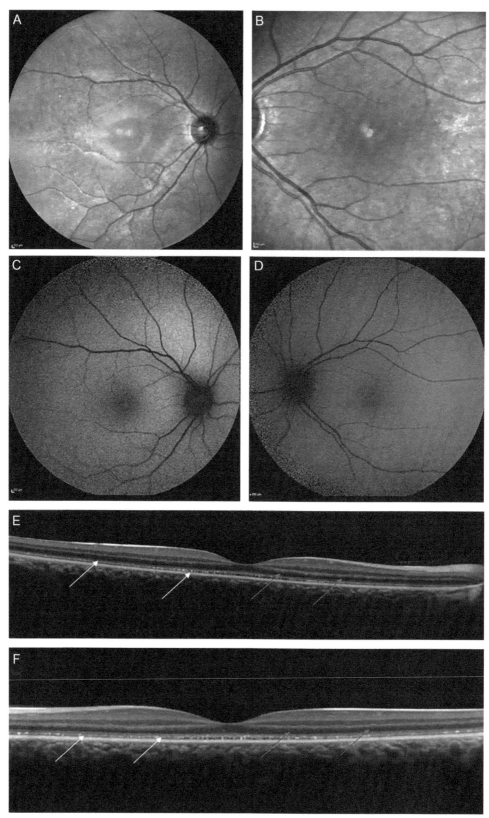

FIG. 13B.7 Fundus images from a female child with genetically confirmed Alström syndrome. The patient originally presented with nystagmus and marked photophobia at age 4 months. There was also a history of transient dilated cardiomyopathy that appeared at one month of age, and of recurrent otitis media. Truncal obesity started becoming evident from 3 years of age. At the age of 13 years, the patient developed insulin resistance with an early-onset diabetes mellitus and severe hearing impairment. The visual acuity was 1.3 LogMAR in each eye. Infrared reflectance (A and B) and fundus autofluorescence (C and D) imaging are within normal limits. OCT (E and F) reveals a granular pattern of the ellipsoid zone *(white arrows)* and mild thinning of the outer nuclear layer *(red arrows)*.

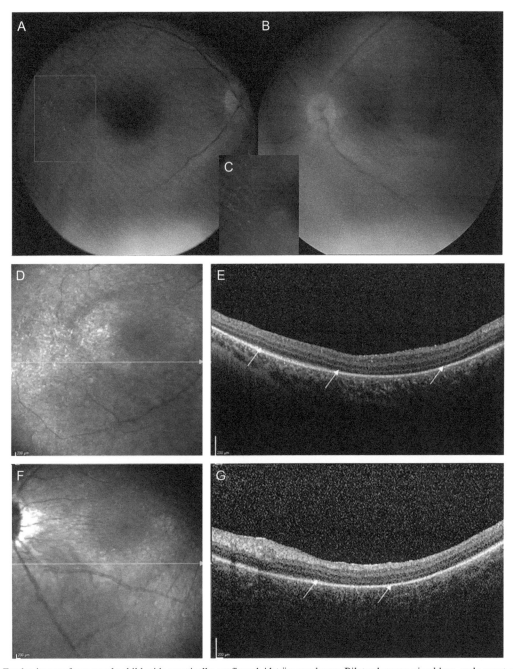

FIG. 13B.8 Fundus images from a male child with genetically confirmed Alström syndrome. Bilateral severe visual loss, early-onset nystagmus and major photophobia were reported. There were no extraocular signs and no cardiac history. The visual acuity was 1.0 LogMAR in each eye. On colour photographs (A–C), white dots are visible temporally to the fovea (C magnified image of the temporal part, *blue square*). On infrared reflectance photographs (D and F), a subtle 'bull's eye' pattern can be observed. On OCT (E and G), the outer nuclear layers appear disorganised and there is photoreceptor loss *(white arrows)*.

Molecular pathology

Biallelic mutations in the *ALMS1* gene should be identified to make the diagnosis of Alström syndrome. *ALMS1* encodes a large protein that localises at the base of primary cilia and has multiple functions. Over 300 mutations have been described in this gene; most of them are nonsense or frameshift variants associated with a complete lack of protein expression. There is a higher frequency of some specific *ALMS1* mutations in certain ethnic populations, for example, the c.10538_10539ins19 (p.Lys3513fs) variant in French Acadians and the c.10775del (p.Thr3592fs) change in patients with English ancestries.[3]

Clinical management

The role of the ophthalmologist is critical in considering the possibility of Alström syndrome in the differential diagnosis, ordering appropriate genetic testing and initiating referral to other specialists. Affected individuals should be referred to an internist or a paediatrician as dysfunction of many organs will occur over time. Low-fat, low-sugar diets and an exercise regimen should be implemented to prevent hyperinsulinemia.[7] Ventilation tubes are often required to improve recurrent otitis media. Hearing aids or, more rarely, cochlear implants, may be needed. For more information on the clinical management of Alström syndrome see 1.

In terms of ophthalmic interventions, photophobia should be reduced using filtering or tinted glasses (or lenses), and by wearing a cap or a hat. Low vision rehabilitation should be offered and any amblyopia and/or strabismus should be managed.

References

1. Tahani N, Maffei P, Dollfus H, Paisey R, Valverde D, Milan G, et al. Consensus clinical management guidelines for Alström syndrome. *Orphanet J Rare Dis* 2020;**15**(1):253.
2. Nasser F, Weisschuh N, Maffei P, Milan G, Heller C, Zrenner E, et al. Ophthalmic features of cone-rod dystrophy caused by pathogenic variants in the ALMS1 gene. *Acta Ophthalmol* 2018;**96**(4):e445–54.
3. Paisey RB, Steeds R, Barrett T, et al. Alström syndrome. 2003 Feb 7 [updated 2019 Jun 13]. In: Adam MP, Ardinger HH, Pagon RA, et al., editors. *GeneReviews® [internet]*. Seattle, WA: University of Washington, Seattle; 1993–2020. Available from: https://www.ncbi.nlm.nih.gov/books/NBK1267/.
4. Brofferio A, Sachdev V, Hannoush H, Marshall JD, Naggert JK, Sidenko S, et al. Characteristics of cardiomyopathy in Alström syndrome: prospective single-center data on 38 patients. *Mol Genet Metab* 2017;**121**:336–43. https://doi.org/10.1016/j.ymgme.2017.05.017.
5. Lindsey S, Brewer C, Stakhovskaya O, Kim HJ, Zalewski C, Bryant J, et al. Auditory and otologic profile of Alström syndrome: comprehensive single center data on 38 patients. *Am J Med Genet A* 2017;**173**:2210–8. https://doi.org/10.1002/ajmg.a.38316.
6. Boerwinkle C, Marshall JD, Bryant J, Gahl WA, Olivier KN, Gunay-Aygun M. Respiratory manifestations in 38 patients with Alström syndrome. *Pediatr Pulmonol* 2017;**52**:487–93. https://doi.org/10.1002/ppul.23607.
7. Lee N-C, Marshall JD, Collin GB, Naggert JK, Chien Y-H, Tsai W-Y, et al. Caloric restriction in Alström syndrome prevents hyperinsulinemia. *Am J Med Genet A* 2009;**149A**:666–8. https://doi.org/10.1002/ajmg.a.32730.

13B.1.4

Usher syndrome

Francesco Testa and Francesca Simonelli

Usher syndrome is a group of clinically and genetically heterogeneous autosomal recessive disorders characterised by sensorineural hearing loss and progressive retinal degeneration (retinitis pigmentosa); vestibular dysfunction is also present in some cases. Usher syndrome is the most common syndromic cause of deafness and the most prevalent cause of dual sensory impairment. Historically, it has been divided into three clinical subtypes that differ in severity and progression of audiovestibular symptoms (Table 13B.1).

Mutations in at least 10 genes have been implicated in Usher syndrome (Table 13B.1).[1] These genes encode proteins from different protein families and most of these molecules localise at or around ciliary structures in the photoreceptors and olfactory epithelia.

Clinical characteristics

Individuals with Usher syndrome have a hearing impairment that is generally present at birth, and visual loss that is typically noted in early or late childhood (although it can be diagnosed later in life as vision problems may be overlooked in a deaf child). The ophthalmic phenotype is in keeping with retinitis pigmentosa; early findings include night vision problems and peripheral visual field constriction while central visual loss occurs later in the disease process.

The electroretinographic responses are abnormal (with the rod system being more severely affected than the cone system) or undetectable. Notably, electrodiagnostic testing can help confirm the presence of retinal dysfunction even in very young children in whom fundus examination is unremarkable. Typical fundoscopic findings in Usher syndrome include arteriolar narrowing, optic nerve head pallor and bone spicule-like pigmentary changes in the mid-peripheral retina; atypical presentations have also been reported (Fig. 13B.9). Fundus autofluorescence imaging reveals a perimacular concentric hyperautofluorescent ring in many patients; this is considered the transition zone between degenerated and relatively spared retina (see also Section 13A.1 and Fig. 13B.10. OCT can help detect and monitor photoreceptor degeneration and foveal lesions such as an epiretinal membrane or cystoid macular oedema.

Some differences in the onset and progression of visual impairment have been identified among the three Usher syndrome subtypes (although these differences are much less evident than those in vestibular and auditory dysfunction—see below). Individuals with Usher syndrome type 1 almost invariably report symptom onset in the first decade of life, whereas people with Usher syndrome type 2 or 3 report onset over a greater age range, up to the fourth decade of life. In general, visual field constriction starts earlier and is more rapid in Usher syndrome type 1.[2]

TABLE 13B.1 Clinical features and genes associated with Usher syndrome.

	Associated genes	Hearing impairment	Speech	Vestibular function	Retinitis pigmentosa[a]
Usher syndrome type 1	MYO7A, USH1C, CDH23, PCDH15, USH1G, CIB2	Severe-to-profound; pre-lingual	Delayed	Poor balance; delayed motor development	Symptom onset in first decade of life; legal blindness typically in fourth decade of life
Usher syndrome type 2	USH2A, ADGRV1, WHRN	Mild-to-severe; typically pre-lingual; non-progressive	Generally normal	Normal balance	Symptom onset in second decade of life; legal blindness typically in sixth decade of life
Usher syndrome type 3	CLRN1	Mild-to-severe; post-lingual; progressive	Generally normal	Variable; mild dysfunction in ~50% of cases	Variable symptom onset, typically in the second decade of life

[a]Retinal dysfunction can be detected by electroretinography in the first years of life.

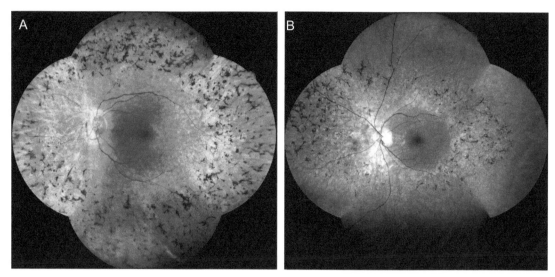

FIG. 13B.9 Fundus photography from two adult patients with Usher syndrome. (A) The typical fundoscopic findings of the condition are noted including attenuation of retinal arterioles, bone-spicule pigmentation in the retinal periphery. (B) Some atypical findings are shown: although there is mid-peripheral bone spicule pigmentation, the disc and macula appear normal, there is minimal retinal arteriolar narrowing, and the far periphery appears preserved.

The audiovestibular phenotype in Usher syndrome is variable. Individuals with Usher syndrome type 1, the most severe form, typically have bilateral, prelingual/congenital, severe-to-profound sensorineural hearing loss (Fig. 13B.11A). Cochlear implants are highly recommended in this subset of patients as they can help with speech development. Vestibular areflexia and abnormal vestibular function are common and can lead to motor delay; affected children are often slow to gain head control, sit unsupported or walk.[3]

Usher syndrome type 2, the most common form of the disorder, is clinically distinct from type 1. It is associated with bilateral, moderate-to-severe sensorineural hearing loss with normal vestibular function. Hearing impairment is typically prelingual/congenital and non-progressive. The higher frequencies are more severely affected (Fig. 13B.11B) and hearing aids generally help to normalise speech. Environmental trauma (e.g. noise) or a genetic susceptibility (e.g. presbycusis) in addition to the congenital, stable deficit of Usher syndrome type 2 may combine to produce a severe-to-profound hearing loss in older patients; in these cases, cochlear implantation may be warranted.[4]

Usher syndrome type 3 is the least common form (except in the Finnish and Ashkenazi Jewish populations). In this subtype, hearing loss is often post-lingual but it can evolve progressively to profound deafness (typically after the fourth decade of life). Vestibular dysfunction of variable intensity occurs in about 50% of affected individuals.

Molecular pathology

Usher syndrome is inherited in an autosomal recessive manner and at least 10 disease-associated genes have been identified, including six implicated in type 1, three in type 2 genes and one in type 3.[1,3,4] Interestingly, mild/hypomorphic mutations in some of these genes can cause either non-syndromic deafness or non-syndromic retinitis pigmentosa (e.g. the c.2276G>T (p.Cys759Phe) variant in USH2A[4]).

For Usher syndrome type 1, the most commonly mutated gene is MYO7A which accounts for about 50% of cases in most populations studied; mutations in CDH23 and PCDH15 are typically the two next most common causes of this disease subtype.[3] About 80% of Usher syndrome type 2 cases are caused by abnormalities in the USH2A gene including truncating variants like the common c.2299delG (p.Glu767fs) mutation and/or severe missense changes; around 10% or less of cases are caused by ADGRV1 mutations.[4] Mutations in the CLRN1 gene have been implicated in Usher syndrome type 3 with two prevalent mutations, c.528T>G (p.Tyr176Ter) and c.144T>G (p.Asn48Lys), accounting for most cases of Finnish and Ashkenazi Jewish ancestries, respectively.[1]

The proteins encoded by all these genes are part of a dynamic protein complex that is present in hair cells of the inner ear and in photoreceptor cells of the retina.[6] MYO7A is also an essential RPE protein and CLRN1 appears to be predominantly expressed in the retinal Müller glia.[1]

Multigene panel testing has a high diagnostic yield in patients with Usher syndrome (over 90%) and facilitates early diagnosis and optimal management.[5] Notably, genomic rearrangements and intronic variants have been identified in a subset of patients[3–5]; this should be taken into account when requesting or designing a gene panel test.

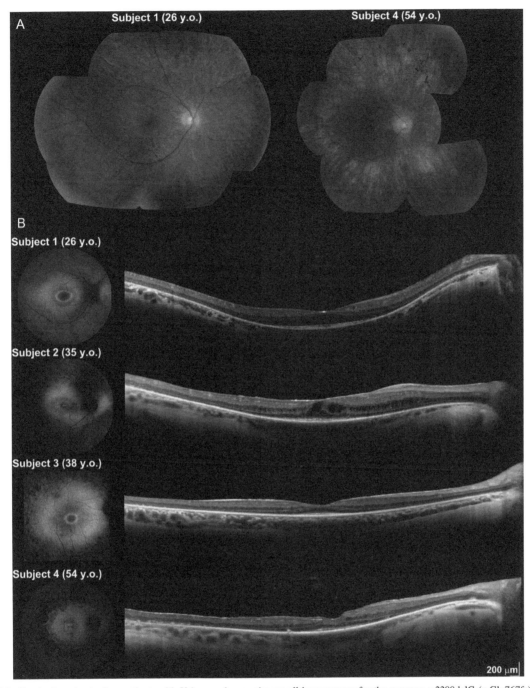

FIG. 13B.10 Fundus imaging in four patients with Usher syndrome who are all homozygous for the common c.2299delG (p.Glu767fs) mutation in *USH2A*. (A) Colour fundus photographs of the right eye of subject 1 and subject 4 reveal changes suggestive of retinitis pigmentosa (B) Fundus autofluorescence imaging and foveal OCT scans of the right eye of the four subjects. A ring of the increased signal is noted on fundus autofluorescence imaging in all four cases. OCT reveals that the photoreceptor inner segment ellipsoid zone is relatively preserved in the area within the hyperautofluorescent ring. Macular oedema is present in subject 2 and, to a lesser extent subject 1 and 4. *(Adapted from Ref. 7.)*

Clinical management

The possibility of Usher syndrome should be carefully considered in infants and young children with sensorineural hearing loss including those detected through the newborn hearing screen. Additionally, Usher syndrome type 1 must be high on the list of differential diagnoses in cases with severe, congenital hearing impairment, especially when there is a co-existing motor delay.

Cochlear implantation in children with severe, congenital hearing impairment is increasingly common. Early

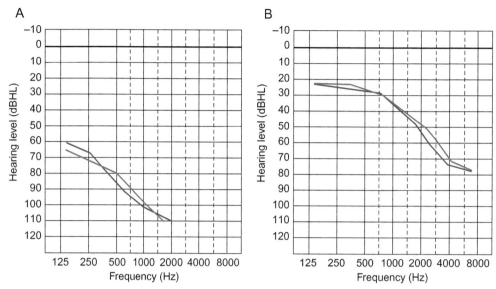

FIG. 13B.11 Audiograms showing bilateral profound hearing impairment (A) and bilateral mild-to-severe hearing loss (B) in two patients with Usher syndrome. *Redline*, right ear; *blue line*, left ear.

diagnosis of Usher syndrome can ensure that affected children are offered timely cochlear implantation and speech therapy and are managed in the light of eventual potential dual sensory impairment.[3]

Electroretinography may be of use early in the disease process when clinical signs of pigmentary retinopathy are less obvious. Yearly ophthalmic review is advised to monitor the disease and to look for potentially treatable complications such as cataracts or cystoid macular oedema.

While there is currently no available cure for retinal degeneration associated with Usher syndrome, there are numerous therapeutic strategies under development. These include gene replacement, gene editing, nonsense suppression and antisense oligonucleotide-based approaches.[1,3]

References

1. Toms M, Pagarkar W, Moosajee M. Usher syndrome: clinical features, molecular genetics and advancing therapeutics. *Ther Adv Ophthalmol* 2020;**12**, 2515841420952194.
2. Stingl K, Kurtenbach A, Hahn G, Kernstock C, Hipp S, Zobor D, et al. Full-field electroretinography, visual acuity and visual fields in Usher syndrome: a multicentre European study. *Doc Ophthalmol* 2019; **139**(2):151–60.
3. Koenekoop RK, Arriaga MA, Trzupek KM, Lentz JJ. Usher syndrome type I. [updated 2020 Oct 8]. In: Adam MP, Ardinger HH, Pagon RA, Wallace SE, LJH B, Stephens K, Amemiya A, editors. *GeneReviews®*. Seattle, WA: University of Washington, Seattle; 1993–2020.
4. Lentz J, Keats B. Usher syndrome type II. [updated 2016 Jul 21]. In: Adam MP, Ardinger HH, Pagon RA, Wallace SE, LJH B, Stephens K, Amemiya A, editors. *GeneReviews®*. Seattle, WA: University of Washington, Seattle; 1993–2020.
5. Bonnet C, Riahi Z, Chantot-Bastaraud S, Smagghe L, Letexier M, Marcaillou C, et al. An innovative strategy for the molecular diagnosis of Usher syndrome identifies causal biallelic mutations in 93% of European patients. *Eur J Hum Genet* 2016;**24**(12):1730–8.
6. Kremer H, van Wijk E, Märker T, Wolfrum U, Roepman R. Usher syndrome: molecular links of pathogenesis. *Hum Mol Genet* 2006. https://doi.org/10.1093/hmg/ddl205.
7. Lenassi E, Saihan Z, Bitner-Glindzicz M, Webster AR. The effect of the common c.2299delG mutation in USH2A on RNA splicing. *Exp Eye Res* 2014;**122**:9–12.

13B.2

Retinopathy in inborn errors of metabolism

Inborn errors of metabolism are a diverse group of genetic conditions whose pathophysiology is intrinsically linked to the impairment of a biochemical pathway. Over 1500 different types have been recognised. Based on their molecular aetiology, these conditions have been classified into several groups including mitochondrial disorders of energy metabolism, congenital disorders of glycosylation, disorders of lipids etc.[1] The ubiquitous nature of the affected metabolic processes leads to multisystem manifestations in many of these disorders. Tissues with high metabolic requirements, such as the retina, are particularly vulnerable (Table 13B.2).

Inborn errors of metabolism usually present in infancy or early childhood, although they can occasionally manifest in adolescence or adulthood. There is wide variability in severity: some conditions are manifest from birth and result in premature death, while others are not apparent until adult life. Notably, patients may occasionally have a very mild phenotype or remain completely asymptomatic. Hypotonia, failure to thrive, jaundice and seizures are common features of many infantile-onset subtypes. Later systemic manifestations may include developmental delay, intellectual impairment, dysmorphism, seizures, skeletal, cardiovascular, neurological, renal and liver disease. Ophthalmic features are common and the ophthalmologist has an important role both in the diagnostic process and in the ongoing management of affected individuals (Box 13B.1).

Although a small number of inborn errors of metabolism are diagnosed through newborn screening (e.g. phenylketonuria and galactosaemia), diagnostic metabolic profiling is required to reach a precise diagnosis in most patients. This relies on the application of several relatively rapid tests that are generally guided by the patient's range of clinical symptoms and signs. These tests remain a key foundation for diagnosis and guide the introduction of treatment regimens. Molecular genetic testing can be confirmatory of

TABLE 13B.2 Examples of inborn errors of metabolism that may have retinopathy as a clinical feature.

High order category	Name of condition	Retinopathy at presentation	Retinopathy as late complication
Disorders of nitrogen-containing compounds	Gyrate atrophy of the choroid and retina	Yes	
Disorders of vitamins, cofactors & minerals	Cobalamin C deficiency		Yes
Disorders of lipids	Bietti corneoretinal crystalline dystrophy	Yes	
	LCHAD		Yes
Storage disorders	Mucopolysaccharidoses (types I, II, III, IV, VII)		Yes
	Neuronal ceroid lipofuscinosis	Yes	Yes
Disorders of peroxisomes and oxalate	Zellweger disease		Yes
	Refsum disease	Yes	
Mitochondrial disorders of energy metabolism	Kearns-Sayre syndrome (Chapter 21)		Yes
	MELAS /MIDD		Yes

LCHAD, long-chain L-3 hydroxy-acyl-CoA dehydrogenase deficiency; *MELAS*, mitochondrial encephalopathy, lactic acidosis, and stroke-like episodes; *MIDD*, maternally inherited diabetes and deafness.

BOX 13B.1 Ophthalmic manifestations of inborn errors of metabolism

Individuals with inborn errors of metabolism can have a range of ophthalmic manifestations. These may involve the anterior segment (e.g. corneal clouding, angle-closure glaucoma, cataract, lens sub-luxation) or the optic nerve (e.g. optic neuropathy due to the direct effect of the metabolic disorder or secondary to glaucoma or raised intracranial pressure). Abnormalities of ocular motility such as vertical or horizontal gaze palsy, abnormal saccadic movements, nystagmus, oculomotor apraxia and strabismus may be an early feature of certain subtypes.[2] A cherry-red spot can be the initial finding in other.[3]

Retinal degeneration is common in a wide range of inborn errors of metabolism. This can occur in several scenarios.

1. *Retinal abnormalities recognised in a patient already diagnosed (or in the process of being diagnosed) with an inborn error of metabolism.*

    In these cases, the ophthalmologist is required to monitor the ophthalmic complications and to evaluate the impact of any visual impairment in the overall context of the condition. This is exemplified by *mucopolysaccharidoses* (MPS) (see also Section 9.4).

    The mucopolysaccharidoses are a group of autosomal recessive disorders caused by defects in enzymes that are responsible for the degradation of glycosaminoglycans. The lifespan and quality of life of many affected individuals have improved due to the wider availability of enzyme replacement therapy and haematopoietic stem cell treatment.

    There is significant heterogeneity in the ophthalmic phenotype which may include corneal clouding, glaucoma, optic nerve involvement and retinopathy. Retinopathy typically occurs in MPS I (Hurler, Hurler-Scheie and Scheie), MPSII (Hunter), MPSIII (Sanfilippo), and MPSIV (Morquio).[4] The onset and severity are variable, but night vision problems are usually noticed in the second or third decade of life; there is progressive visual field constriction and the central vision is typically affected later in the disease process. RPE mottling is an early fundoscopic sign while pigment clumping, retinal vessel attenuation, disc pallor and maculopathy may be observed later (Fig. 13B.12). OCT shows photoreceptor degeneration, thickening of the external limiting membrane and, in some cases, maculopathy. Treatment of the underlying condition does not appear to significantly influence the severity or progression of retinopathy in this group of patients.

2. *The identification of retinal abnormalities facilitates the diagnosis of an inborn error of metabolism.*

    In these cases, it is critical that the ophthalmologist considers the possibility of a metabolic disorder in the differential diagnosis, orders appropriate investigations and initiates referral(s) to other specialists. Investigations include metabolic testing (in association with metabolic medicine, paediatrics and/or clinical genetics) and, potentially, genomic testing. This is particularly important for certain treatable subtypes. This is exemplified by *classic/adult Refsum disease*.

    Refsum disease is an autosomal recessive disorder caused by defects in peroxisomal enzymes that break down phytanic acid, a type of fat found in certain foods. This heterogeneous condition is classified into two subgroups based on differences in clinical presentation and molecular pathology: infantile Refsum disease and classic/adult Refsum disease. Most cases of the latter have mutations in the *PHYH* gene and typically present in late childhood or adolescence with gradual deterioration of night vision due to retinitis pigmentosa. When the disease progresses, additional features including anosmia, progressive sensorineural hearing loss, peripheral polyneuropathy with ataxia (unsteady gate) and cardiac arrhythmias may be noted. The primary treatment for adult Refsum disease is to restrict or avoid foods that contain phytanic acid including dairy products, beef, lamb and certain fish.[5]

3. *Genetic testing in individuals with apparently isolated retinal disease highlights mutations in genes causing inborn errors of metabolism.*

    A significant number of severe metabolic conditions are now being recognised in an attenuated form; mild/hypomorphic genetic variants in the associated disease-causing genes can lead to retinal degeneration without significant systemic abnormalities. This is exemplified by conditions caused by mutations in the *PCYT1A* gene.

    PCYT1A encodes the rate-limiting enzyme in phosphatidylcholine biosynthesis and has been linked to two different phenotypes: spondylometaphyseal dysplasia with cone-rod dystrophy and also a complex phenotype with congenital lipodystrophy, severe fatty liver disease and reduced HDL cholesterol without any retinal or skeletal involvement. Notably, several individuals with isolated retinal dystrophy have been found to carry 'mild' mutations in *PCYT1A*,[6] highlighting that defects in this enzyme can result in non-syndromic retinal phenotypes.

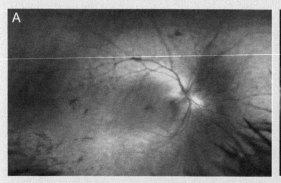
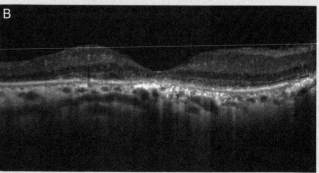

FIG. 13B.12 Fundus imaging of the right eye of a 43-year-old patient with mucopolysaccharidosis type I (MPS I Scheie). Mild corneal clouding was present impacting image quality. Bone spicule pigmentation and RPE mottling are noted on multicolour imaging (A). Foveal OCT reveals disruption of the outer nuclear layers (B).

the indicated diagnosis and, particularly in cases with milder symptoms (e.g. apparently isolated ophthalmic disease), can also be used as a firstline diagnostic investigation.

Patients with inborn errors of metabolism are optimally managed in a multidisciplinary setting due to the diverse disease manifestations. Treatment can be preventative, with a dietary restriction or supplementation or regular medications (e.g. substrate reduction therapy or enzyme replacement therapy). Supportive treatment may be required during periods of metabolic crisis or intercurrent infection. Specific symptoms may be treated with surgery, physiotherapy and/or psychological support.

References

1. Ferreira CR, van Karnebeek CDM, Vockley J, Blau N. A proposed nosology of inborn errors of metabolism. *Genet Med* 2019; **21**(1):102–6.
2. Koens LH, Tijssen MAJ, Lange F, Wolffenbuttel BHR, Rufa A, Zee DS, et al. Eye movement disorders and neurological symptoms in late-onset inborn errors of metabolism. *Mov Disord* 2018;**33**(12):1844–56.
3. Poll-The BT, Maillette de Buy Wenniger-Prick CJ. The eye in metabolic diseases: clues to diagnosis. *Eur J Paediatr Neurol* 2011; **15**(3):197–204.
4. Ashworth JL, Biswas S, Wraith E, Lloyd IC. The ocular features of the mucopolysaccharidoses. *Eye (Lond)* 2006 May;**20**(5):553–63.
5. Kumar R, De Jesus O. Refsum disease. [Updated 2020 Aug 1]. In: *StatPearls [Internet]*. Treasure Island (FL): StatPearls Publishing; 2020 Jan. Available from: https://www.ncbi.nlm.nih.gov/books/NBK560618/.
6. Testa F, Filippelli M, Brunetti-Pierri R, Di Fruscio G, Di Iorio V, Pizzo M, et al. Mutations in the PCYT1A gene are responsible for isolated forms of retinal dystrophy. *Eur J Hum Genet* 2017;**25**(5):651–5.

13B.2.1

Gyrate atrophy of the choroid and retina

Karolina M. Stepien, Panagiotis I. Sergouniotis, and Graeme C.M. Black

Gyrate atrophy of the choroid and retina, also known as ornithine aminotransferase deficiency, is an autosomal recessive disorder that primarily affects the eye. Affected individuals typically present in childhood with night vision problems and have characteristic fundus abnormalities (well-demarcated circumferential patches of chorioretinal atrophy). The condition is diagnosed based on the retinal findings, the presence of high ornithine levels in the blood, and the detection of mutations in the *OAT* gene. Treatment mainly involves dietary modifications and management of complications.

Clinical characteristics

Affected individuals usually present in the first or second decade of life with nyctalopia and/or progressive myopia.[1] Following this, concentric visual field loss is typically noted during the second decade of life. This slowly progresses to complete loss of vision at around the fourth to the fifth decade of life.[2,3] Loss of visual acuity may be secondary to macular oedema and/or photoreceptor loss. Early lens opacities are common, with most patients requiring surgery by the third or fourth decade of life.[4]

The retinal manifestations are highly characteristic and strongly point towards the diagnosis. Fundoscopy shows sharply-demarcated, scalloped areas of chorioretinal atrophy arising peripherally in the retina (Figs. 13B.13 and 13B.14). These are initially patchy but gradually enlarge, become confluent and spread centrally towards the macula. The atrophic areas are visible as areas of absent signal on fundus autofluorescence imaging. OCT

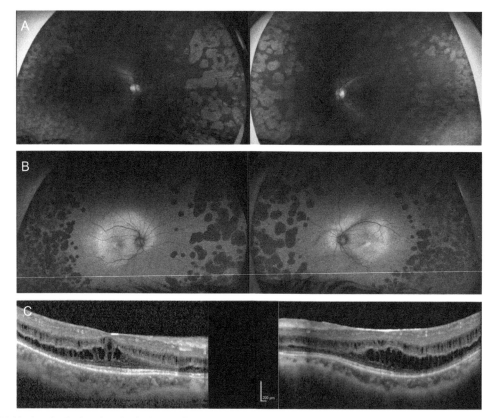

FIG. 13B.13 Multicolour fundus imaging (A), fundus autofluorescence imaging (B) and OCT scans (C) from a 10-year-old male who has gyrate atrophy and carries the c.1192C>T (p.Arg398Ter) mutation in the *OAT* gene in the homozygous state. Minimal bone spicule pigmentation, relatively preserved retinal vasculature and circular areas of chorioretinal atrophy are evident on multicolour imaging. Sharply demarcated peripheral areas of reduced signal are observed on fundus autofluorescence imaging. Increased signal is noted at the posterior pole bilaterally. OCT at the level of the fovea reveals macular oedema, a common finding in this condition.

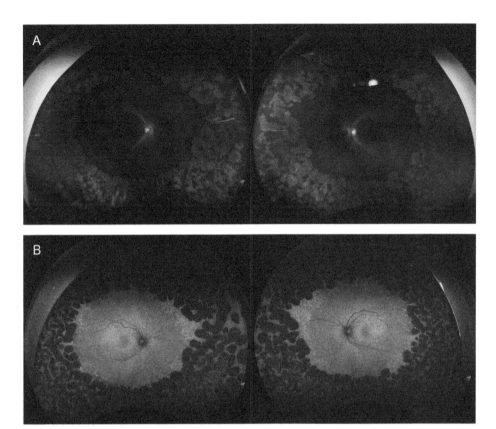

FIG. 13B.14 Multicolour fundus imaging (A) and fundus autofluorescence imaging (B) from a 12-year-old female who has gyrate atrophy and carries the c.722C>T (p.Pro241Leu) mutation in the *OAT* gene in the homozygous state. She presented with reduced visual acuities (0.2 LogMAR right and 0.3 LogMAR left), progressive myopia and mild bilateral posterior subcapsular lens opacities. Fundoscopy revealed peripheral chorioretinal degeneration with a scalloped edge. The electroretinograms were grossly attenuated in the light-adapted state and extinguished in the dark-adapted state. Gyrate atrophy was suspected and plasma ornithine levels were measured. These were significantly elevated at 670 μmoL/L (normal 41–129 μmoL/L). The patient subsequently commenced on a low-protein diet with supplemental essential amino-acids, vitamins and minerals, but found this very difficult adhere to and eventually abandoned it.

demonstrates outer retinal thinning over the areas of atrophy. Notably, at the border between the preserved and atrophic retina, the inner segment ellipsoid zone disappears with a steep or near-vertical slope. This is similar to choroideremia, an important differential diagnosis, and in contrast to retinitis pigmentosa where the disappearance of the ellipsoid zone is gradual with a swallower slope. Retinal tubulations, i.e. round or ovoid hyporeflective spaces with hyper-reflective borders located at the outer nuclear layer, are often seen on OCT. This non-specific sign is associated with photoreceptor rearrangement after outer retinal loss. Hyper-reflective deposits may be present in the ganglion cell layer. Early in the disease process, there is some thickening of the retina even when it appears anatomically preserved. Macular oedema is common and may be present from an early age (Fig. 13B.13). Electrodiagnostic testing reveals rod-cone dysfunction in the early stages which progresses to an extinguished full-field electroretinogram.[1]

Most gyrate atrophy patients have no extraocular features but a minority of cases have broader symptoms. Occasionally, newborns with gyrate atrophy develop hyperammonemia which may lead to poor feeding, vomiting and diarrhoea.[5] Furthermore, a small number of patients have systemic features which may include muscle weakness, thin hair and central nervous system manifestations including diffuse brain atrophy.[2]

Although it is uncommon for affected individuals to have muscle symptoms,[6] abnormalities on muscle biopsy have been described.[7,8] The significance of these changes is unknown and they usually emerge before puberty and progress with age. In the small subset of cases who do have muscle symptoms, these are usually due to secondary creatine deficiency. High ornithine concentrations inhibit the activity of arginine-glycine transamidinase, the rate-limiting enzyme in creatine biosynthesis.[8] This leads to deficiency of high-energy creatine which in turn may result in selective destruction of cells that are especially dependent on their energy reservoirs.[8] Body composition assessment and monitoring is required in this group of patients. Decreased brain creatine may also be noted in individuals with gyrate atrophy.[9] While creatine supplement dosing (equivalent to the normal requirement) may be sufficient enough to correct the creatine deficiency of the myocyte, it does not correct it in the neurons and the retina, due to their less efficient uptake mechanisms.[10] The decreased brain creatine concentrations have been thought to lead to learning difficulties which, despite being a finding occasionally described in the literature, it is often not clinically apparent in affected individuals.

Molecular pathology

OAT is the only gene known to be mutated in gyrate atrophy. It encodes ornithine aminotransferase, an enzyme that catalyses the conversion of the amino-acid ornithine to pyrroline 5-carboxylate with the assistance of vitamin B6 (which acts as a co-factor).[3] Over 80 mutations have been described in OAT including nonsense, frameshift and, most commonly, missense variants. These changes lead to a significant reduction in enzyme activity and result in the accumulation of ornithine in the blood, urine, cerebrospinal fluid, and aqueous humour of affected individuals.

Although OAT is expressed in most tissues, its deficiency is toxic mainly to RPE (where it is highly expressed) and potentially to the outer retina and choroid.[11] Although the biochemical defect in gyrate atrophy is known, our understanding of the pathogenesis of the condition is limited. In particular, it is unclear how systemic metabolic abnormalities result in chorioretinal degeneration with only minimal involvement of other tissues.

Clinical management

In those suspected of having gyrate atrophy, the diagnosis can be rapidly confirmed by measuring plasma ornithine levels. These are often 10–20-fold above normal and may result in concomitant overflow ornithinuria.[12] Patients also have a modest but significant reduction in plasma lysine (reduced to ~50% of normal), glutamate, glutamine, and creatine.[3] Genetic confirmation of the diagnosis in the presence of relevant eye signs and elevated ornithine levels is not necessary although this may support family management and decision making.

In two mouse models, severe dietary restriction of arginine has been shown to dramatically alter the disease course, slowing deterioration of the electroretinogram and resulting in histologic improvement when compared to controls.[11,13] In people with gyrate atrophy in whom ornithine levels are controlled through consumption of an arginine-restricted diet, retinal degeneration also appears to be slower.[14] This suggests that plasma ornithine plays an essential role in the disease process and emphasises the importance of early diagnosis and early correction of ornithine accumulation to prevent irreversible retinal degeneration.[12]

There is presently no cure for gyrate atrophy and treatment options remain limited, in the form of a low-protein diet and synthetic amino-acid supplements. Generally, management requires support from a paediatric and, later in life, an adult metabolic specialist and metabolic dietician. It is highlighted, compliance with diet is often suboptimal, particularly in children and adolescents, as a result of the highly restrictive nature of the diet. Achieving the plasma ornithine target of 400 µmol/L is often not straightforward and this affects outcomes and quality of life.

A small proportion of patients (<5%) exhibit a reduction in plasma ornithine in response to pharmacologic supplementation with pyridoxine hydrochloride (i.e. vitamin B6) at a dose of 200–400 mg per day. Pyridoxine responsiveness in humans is mediated by an unidentified mechanism and appears not to depend on the specific OAT genotype.[15] About one-third of patients have a small amount of residual OAT activity and this could partly explain why these individuals are more likely to benefit from pyridoxine administration. Importantly, pyridoxine treatment should be considered in all patients, and it should be continued based on its efficacy in individual cases.

Lysine supplementation (dose of 4 g per day) is recommended for competitive inhibition of ornithine. Notably, hypolysinaemia, resulting from the effect of high ornithine levels on lysine catabolism and distribution,[3] may cause osteoporosis in affected individuals.[16]

Patients with gyrate atrophy have secondary creatine deficiency in both muscle and brain. Creatine monohydrate (starting dose of 1.5 g, four times per day) helps reduce muscle weakness and improves muscle mass.[17] The clinical response can be monitored by body composition and physiotherapy assessment.

In addition to dietary management, ophthalmic management requires regular refraction (especially during childhood), monitoring of cataracts and low vision aid support.

References

1. Sergouniotis PI, Davidson AE, Lenassi E, Devery SR, Moore AT, Webster AR. Retinal structure, function, and molecular pathologic features in gyrate atrophy. *Ophthalmology* 2012;**119**(3):596–605. https://doi.org/10.1016/j.ophtha.2011.09.017.
2. Kaiser-Kupfer MI, Kuwabara T, Askanas V, et al. Systemic manifestations of gyrate atrophy of the choroid and retina. *Ophthalmology* 1981;**88**:302–6.
3. Valle D, Simell O. The hyperornithinaemias. In: Scriver C, Beaudet A, Sly W, Valle D, editors. *The metabolic and molecular bases of inherited diseases.* New York: McGraw-Hill; 1995. p. 1147–85.

4. Kaiser-Kupfer MI, Kuwabara T, Uga S, et al. Cataract in gyrate atrophy: clinical and morphologic studies. *Invest Ophthalmol Vis Sci* 1983;**24**:432–6.
5. Cleary MA, Dorland L, de Koning TJ, Poll-The BT, Duran M, Mandell R, et al. Ornithine aminotransferase deficiency: diagnostic difficulties in neonatal presentation. *J Inherit Metab Dis* 2005;**28**(5):673–9.
6. Valtonen M, Nanto-Salonen K, Heinanen K, Alanen A, Kalimo H, Simell O. Skeletal muscle of patients with gyrate atrophy of the choroid and retina and hyperornithinaemia in ultralow-field magnetic resonance imaging and computed tomography. *J Inherit Metab Dis* 1996;**19**:729–34.
7. Fleury M, Barbier R, Ziegler F, Mohr M, Caron O, Dollfus H. Tranchant, Warter JM. Myopathy with tubular aggregates and gyrate atrophy of the choroid and retina due to hyperornithinaemia. *J Neurol Neurosurg Psychiatry* 2007;**78**(6):656–7.
8. Heinanen K, Nanto-Salonen K, Komu M, Erkintalo M, Heinonen OJ, Pulkki K, et al. Muscle creatine phosphate in gyrate atrophy of the choroid and retina with hyperornithinaemia-clues to pathogenesis. *Eur J Clin Invest* 1999;**29**(5):426–31.
9. Nanto-Salonen K, Komu M, Lundbom N, Heinanen K, Alanen A, Sipila I, et al. Reduced brain creatine in gyrate atrophy of the choroid and retina with hyperornithinemia. *Neurology* 1999;**53**(2):303–97.
10. O'Gorman E, Beutner G, Walliman T, Brdiczka D. Differential effects of creatine depletion on the regulation of enzyme activities and on creatine-stimulated mitochondrial respiration in skeletal muscle, heart and brain. *Biochim Biophys Acta* 1996;**1276**:161–70.
11. Wang T, Lawler AM, Steel G, Sipila I, Milam AH, Valle D. Mice lacking ornithine-delta-amino transferase have paradoxical neonatal hypoornithinaemia and retinal degeneration. *Nat Genet* 1995;**11**:185–90.
12. Wang T, Steel G, Milam AH, Valle D. Correction of ornithine accumulation prevents retinal degeneration in a mouse model of gyrate atrophy of the choroid and retina. *PNAS* 2000;**97**(3):1224–9.
13. Bisaillon JJ, Radden LA, Szabo ET, Hughes SR, Feliciano AM, Nesta AV, et al. The retarded hair growth (rhg) mutation in mice is an allele of ornithine aminotransferase (oat). *Mol Genet Metab Rep* 2014;**1**:378–90.
14. Kaiser-Kupfer MI, Caruso RC, Valle D. Gyrate atrophy of the choroid and retina. Long-term reduction of ornithine slows retinal degeneration. *Arch Ophthalmol* 1991;**109**:1539–48.
15. Doimo M, Desbats MA, Baldoin MC, Lenzini E, Basso G, Murphy E, et al. Functional analysis of missense mutations of OAT, causing gyrate atrophy of choroid and retina. *Hum Mutat* 2013;**34**(1):229–36.
16. Ahmet I, Serdar KS. Osteoporosis associated with gyrate atrophy: a case report. *Doc Ophthalmol* 2006;**113**:61–4.
17. Vannas-Sulonen K, Sipila I, Vannas A, Simell O, Rapola J. Gyrate atrophy of the choroid and retina. A five-year follow-up of creatine supplementation. *Ophthalmology* 1985;**92**(12):1719–27.

13B.2.2

Bietti corneoretinal crystalline dystrophy

Erin C. O'Neil and Tomas S. Aleman

Bietti crystalline dystrophy is a rare autosomal recessive degeneration characterised by yellow-white crystalline lipid deposits in the retina and, occasionally, in the cornea.[1-3] Affected individuals usually present with night vision problems, decreased visual acuity and/or visual field defects between the second and fourth decade of life.[4] Progression to legal blindness typically occurs within the next two decades.[5,6]

Bietti crystalline dystrophy is caused by biallelic mutations in *CYP4V2*, a gene that encodes an enzyme involved in the selective hydrolysis of fatty acids.[6-9] Notably, the disease is more prevalent in people with East Asian ancestries.

Clinical characteristics

The frequently reported variability in disease expression of Bietti crystalline dystrophy to a large extent represents variation in the severity of a remarkably consistent phenotype. The disease is characterised by an early, predominantly central retinal degeneration with yellow-white refractile crystals and associated RPE atrophy that eventually leads to overt chorioretinal atrophy (Fig. 13B.15A).[10-12] Crystals are easily visualised as bright spots on near-infrared fundus reflectance imaging (Fig. 13B.15A). Visual fields typically show pericentral and mid-peripheral scotomas; there is significant rod sensitivity loss co-localising to these areas, suggestive of a regional predilection for rod dysfunction at least early in the disease process.[10] Retina-wide rod dysfunction becomes evident on full-field electroretinography in patients with more advanced disease. Localised cone dysfunction is less severe and explains the initial relative preservation of visual acuity (until the fovea is involved by the expansion of the areas of chorioretinal atrophy). There are no major dark adaptation defects despite abnormal rod-mediated sensitivities, arguing against a primary visual cycle abnormality.[10] Occasionally, Bietti crystalline dystrophy may mimic retinitis pigmentosa with bone spicules and vascular attenuation, though without optic nerve pallor. The resemblance is more notable in advanced disease as the crystalline deposits seen in Bietti dystrophy often diminish or even disappear in areas of significant chorioretinal atrophy.

A consistent topography of disease is observed with a predilection for the central and peripapillary retina (Fig. 13B.15A). The peripapillary changes expand to involve the retina nasal to the fovea.[10,13] Generally, this regional predilection suggests vulnerability of areas with high rod photoreceptor cell density, and greater photoreceptor-to-RPE cell ratios.

Near-infrared fundus autofluorescence imaging can highlight early abnormalities, including parafoveal and perifoveal hypoautofluorescence extending nasally and superiorly to the eccentricity of the optic nerve. Extensive loss of near-infrared autofluorescence signal in areas of relatively preserved short-wavelength (blue light) fundus autofluorescence signal suggests that RPE depigmentation and overaccumulation of lipofuscin within the RPE cells may precede photoreceptor outer segment loss. OCT reveals that retinal crystals are primarily located in the apical aspect of the RPE (Fig. 13B.15B). Only rarely are the crystals observed within the inner retina or deeper to the apical RPE and, when this occurs, it is typically in regions with severe RPE abnormalities, suggesting displacement from their original location.[10] The observed localised rod photoreceptor dysfunction in the context of extensive outer segment loss and a deceivingly preserved outer nuclear layer is a pattern uncommon in most primary photoreceptor diseases but somewhat frequent in Bietti crystalline dystrophy.[10] Choroidal neovascularisation has been reported in several affected individuals[4,14-18]; occasional cystoid macular oedema in the absence of neovascularisation has also been described.[5] Outer retinal tubulations are frequently seen on OCT, as is the case in most conditions with predominant RPE defects.[13] It is noted that, a small proportion of patients show very small crystals in the subepithelial and anterior stroma of the peripheral cornea.[19]

Molecular pathology

CYP4V2 (cytochrome P450 family 4 subfamily V polypeptide 2) is the only known disease-associated gene for Bietti crystalline dystrophy. Abnormal function of the encoded enzyme results in deranged fatty acid metabolism, elevated serum triglycerides, abnormal cholesterol storage, and reduced conversion of fatty acid precursors to ω-3

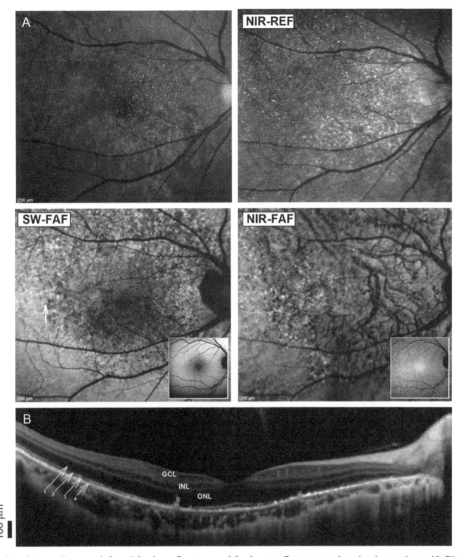

FIG. 13B.15 (A) Fundus photography, near-infrared fundus reflectance and fundus autofluorescence imaging in a patient with Bietti crystalline dystrophy. The normal appearance of short-wavelength fundus autofluorescence imaging (SW-AF) and near-infrared wavelength fundus autofluorescence imaging (NIR-AF) are shown as insets. (B) OCT imaging in a patient with Bietti crystalline dystrophy. The nuclear layers are labelled: *ONL*, outer nuclear layer; *INL*, inner nuclear layer; *GCL*, ganglion cell layer. *(Adapted from Ref. 10.)*

polyunsaturated fatty acids and docosahexaenoic acid (DHA), a known component of ophthalmic membranes. Although *CYP4V2* is expressed in the brain, placenta, lung, liver and kidney, clinical manifestations of disease only affect serum fatty acid levels and ocular tissues.[20] Immunohistochemical analysis demonstrates that the CYP4V2 protein localises to the endoplasmic reticulum of the RPE, although it can also be seen in the nuclear layers of the retina, ganglion cells, and corneal epithelium.[9]

Over 110 disease-causing variants have been described. Several regionally-specific, founder mutations have been identified. For example, the most commonly encountered mutation, c.802-8_810delinsGC, is particularly prevalent in affected individuals with Chinese ancestries.[3] The c.197T>G (p.Met66Arg) change is seen frequently in South Asia, while the c.332T>C (p.Ile111Thr) mutation is commonly noted in patients from Lebanon, Germany and/or Italy. At this time, there is no definite genotype–phenotype correlation around disease severity or onset.[20]

Clinical management

Although genetic testing can ultimately confirm the diagnosis and help with counselling, other causes of retinal crystals and chorioretinal atrophy should be considered, including primary hyperoxaluria, cystinosis, Sjögren-Larsson syndrome and drug toxicity.[19,21]

Regular, at least annual, follow-up should be offered with particular attention to the emergence of choroidal neovascularisation. Self-monitoring of distortion with Amsler grids should be instructed. Generally, intravitreal anti-VEGF injections are effective in treating neovascularisation in Bietti crystalline dystrophy.[10] Although no curative therapies are available for the retinal degeneration seen in this condition, recent work has raised hope that reducing intracellular free cholesterol within RPE cells with the use of cyclodextrins or d-tocopherol may modify the phenotype.[22]

References

1. Kaiser-Kupfer MI, Chan CC, Markello TC, Crawford MA, Caruso RC, Csaky KG, et al. Clinical biochemical and pathologic correlations in Bietti's crystalline dystrophy. *Am J Ophthalmol* 1994;**118**:569–82.
2. Lee J, Jiao X, Hejtmancik JF, Kaiser-Kupfer M, Chader GJ. Identification, isolation, and characterization of a 32-kDa fatty acid-binding protein missing from lymphocytes in humans with Bietti crystalline dystrophy (BCD). *Mol Genet Metab* 1998;**65**:143–54.
3. Zhang X, Xu K, Dong B, Peng X, Li Q, Jiang F, et al. Comprehensive screening of CYP4V2 in a cohort of Chinese patients with Bietti crystalline dystrophy. *Mol Vis* 2018;**24**:700–11.
4. Atmaca LS, Muftuoglu O, Atmaca-Sonmez P. Peripapillary choroidal neovascularization in Bietti crystalline retinopathy. *Eye (Lond)* 2007;**21**:839–42.
5. Lockhart CM, Smith TB, Yang P, Naidu M, Rettie AE, Nath A, et al. Longitudinal characterisation of function and structure of Bietti crystalline dystrophy: report on a novel homozygous mutation in CYP4V2. *Br J Ophthalmol* 2018;**102**:187–94.
6. Mataftsi A, Zografos L, Milla E, Secretan M, Munier FL. Bietti's crystalline corneoretinal dystrophy: a cross-sectional study. *Retina* 2004;**24**:416–26.
7. Garcia-Garcia GP, Martinez-Rubio M, Moya-Moya MA, Perez-Santonja JJ, Escribano J. Current perspectives in Bietti crystalline dystrophy. *Clin Ophthalmol* 2019;**13**:1379–99.
8. Li A, Jiao X, Munier FL, Schorderet DF, Yao W, Iwata F, et al. Bietti crystalline corneoretinal dystrophy is caused by mutations in the novel gene CYP4V2. *Am J Hum Genet* 2004;**74**:817–26.
9. Nakano M, Kelly EJ, Wiek C, Hanenberg H, Rettie AE. CYP4V2 in Bietti's crystalline dystrophy: ocular localization, metabolism of omega-3-polyunsaturated fatty acids, and functional deficit of the p.H331P variant. *Mol Pharmacol* 2012;**82**:679–86.
10. Fuerst NM, Serrano L, Han G, Morgan JI, Maguire AM, Leroy BP, et al. Detailed functional and structural phenotype of Bietti crystalline dystrophy associated with mutations in CYP4V2 complicated by choroidal neovascularization. *Ophthalmic Genet* 2016;**37**:445–52.
11. Miyata M, Oishi A, Hasegawa T, Ishihara K, Oishi M, Ogino K, et al. Choriocapillaris flow deficit in Bietti crystalline dystrophy detected using optical coherence tomography angiography. *Br J Ophthalmol* 2018;**102**:1208–12.
12. Oishi A, Oishi M, Miyata M, Hirashima T, Hasegawa T, Numa S, et al. Multimodal imaging for differential diagnosis of Bietti crystalline dystrophy. *Ophthalmol Retina* 2018;**2**:1071–7.
13. Aleman TS, Han G, Serrano LW, Fuerst NM, Charlson ES, Pearson DJ, et al. Natural history of the central structural abnormalities in choroideremia: a prospective cross-sectional study. *Ophthalmology* 2017;**124**:359–73.
14. Battaglia Parodi M, Casalino G, Iacono P, Introini U, Adamyan T, Bandello F. The expanding clinical spectrum of choroidal excavation in macular dystrophies. *Retina* 2018;**38**:2030–4.
15. Gupta B, Parvizi S, Mohamed MD. Bietti crystalline dystrophy and choroidal neovascularisation. *Int Ophthalmol* 2011;**31**:59–61.
16. Kobat SG, Gul FC, Yusufoglu E. Bietti crystalline dystrophy and choroidal neovascularization in childhood. *Int J Ophthalmol* 2019;**12**:1514–6.
17. Li Q, Li Y, Zhang X, Xu Z, Zhu X, Ma K, et al. Utilization of fundus autofluorescence, spectral domain optical coherence tomography, and enhanced depth imaging in the characterization of Bietti crystalline dystrophy in different stages. *Retina* 2015;**35**:2074–84.
18. Mamatha G, Umashankar V, Kasinathan N, Krishnan T, Sathyabaarathi R, Karthiyayini T, et al. Molecular screening of the CYP4V2 gene in Bietti crystalline dystrophy that is associated with choroidal neovascularization. *Mol Vis* 2011;**17**:1970–7.
19. Song WK, Clouston P, MacLaren RE. Presence of corneal crystals confirms an unusual presentation of Bietti's retinal dystrophy. *Ophthalmic Genet* 2019;1–5.
20. Ng D, Lai T, Ng T, Pang C. Genetics of Bietti crystalline dystrophy. *Asia Pac J Ophthalmol* 2016;**5**:245–52.
21. Kovach JL, Isildak H, Sarraf D. Crystalline retinopathy: unifying pathogenic pathways of disease. *Surv Ophthalmol* 2019;**64**:1–29.
22. Hata M, Ikeda HO, Iwai S, Iida Y, Gotoh N, Asaka I, et al. Reduction of lipid accumulation rescues Bietti's crystalline dystrophy phenotypes. *Proc Natl Acad Sci U S A* 2018;**115**:3936–41.

13B.2.3

Pseudoxanthoma elasticum

Julie De Zaeytijd

Pseudoxanthoma elasticum (PXE) is an autosomal recessive multisystemic disorder characterised by the gradual accumulation of deposits of calcium and other minerals in elastin fibres. This aberrant calcification/mineralisation most prominently affects the eye, the skin and the blood vessels. In the eye, it typically results in breaks in Bruch's membrane, leading to a characteristic (but nondiagnostic) sign known as angioid streaks. The skin is often lax and redundant, and distinctive changes are typically noted in the skin of the neck and the axillae (armpits). Vascular calcification is often present and this can result in accelerated atherosclerosis and an increased risk of cerebrovascular incidents and peripheral arterial disease.

PXE is an important differential diagnosis for people with angioid streaks. Early identification allows monitoring for choroidal neovascularisation and cardiovascular complications. Precise and timely diagnosis can be achieved through genetic testing. PXE is primarily caused by biallelic mutations in *ABCC6*, a gene that is most highly expressed in the liver and kidney and encodes a transmembrane protein with a role in calcium metabolism.[1-4]

Clinical features

Affected individuals develop skin changes in the first or second decade of life but these are often overlooked or not recognised as a sign of PXE. Most patients are visually asymptomatic early on. On many occasions, investigations are initiated following the identification of fundoscopic abnormalities on routine ophthalmic/optometric examination. The characteristic retinal signs of the condition are typically present before the third decade of life and include diffuse mottling of the fundus (known as 'peau d'orange') and angioid streaks.[2,5]

Peau d'orange (orange-peel), the earliest ophthalmic feature of PXE, first appears in the papillomacular area and spreads centrifugally as a circumferential ring, giving the fundus a speckled appearance (with darker spots against a retinal background with a more yellow-white reflex). Areas of peau d'orange represent the transition between the more posterior calcified Bruch's membrane (which has a yellowish reflex) and the anterior non-calcified normal retinal periphery. The darker spots actually represent the remaining healthy retina (Fig. 13B.16).[5,6]

Angioid streaks are the most characteristic ophthalmic feature of PXE. They represent breaks in the calcified Bruch's membrane and appear as jagged irregular lines approximately of the calibre of a retinal vessel. They are typically bilateral (although they can be asymmetrical) and are most prominent close to the disc. They tend to radiate from a peripapillary ring towards the equator of the eye and it is thought that their orientation may be related to force lines due to the pull of intrinsic and extrinsic ocular muscles. There is significant variability; in some patients, angioid streaks may be limited to a few almost imperceptible lines, whereas in others they present as a spidery web (Fig. 13B.17). In older patients, RPE atrophy is seen as a paler line adjacent to angioid streaks. Individuals with angioid streaks that cross the posterior pole are prone to developing recurrent subretinal choroidal neovascularisation which, when untreated, rapidly evolves into a fibrovascular scar, often with consequent central visual loss (Fig. 13B.18).[5]

Pattern dystrophy-like changes may also be a feature of PXE. In many cases, these progress to chorioretinal atrophy that occurs independently to the development of choroidal neovascularisation.[7-9] Such changes are associated with material deposited below the neurosensory retina and are best visualised on autofluorescence imaging (Fig. 13B.19).[10] Reticular drusen are also noted in a subset of patients. OCT occasionally reveals the presence of subretinal fluid occurring without apparent neovascularisation; this often appears as a shallow serous detachment and it is thought to be linked to impaired RPE pump function. When unrelated to choroidal neovascularisation, this abnormality rarely affects visual acuity and tends to be unresponsive to treatment (Fig. 13B.20).[9]

'Comet'-like lesions are a highly suggestive or even pathognomonic feature for PXE. These solitary, nodular, yellow-white changes are typically located in the mid-peripheral fundus.[7] A comet 'tail' may or may not be present. These tails are generally visible as triangles of chorioretinal atrophy adjacent to the comet-like lesion and always points towards the posterior pole. Occasionally a 'comet rain' can be observed[8] (Fig. 13B.21). It is noted that optic nerve head drusen are thought to be more common in patients with PXE than in the general population (Fig. 13B.22).[5]

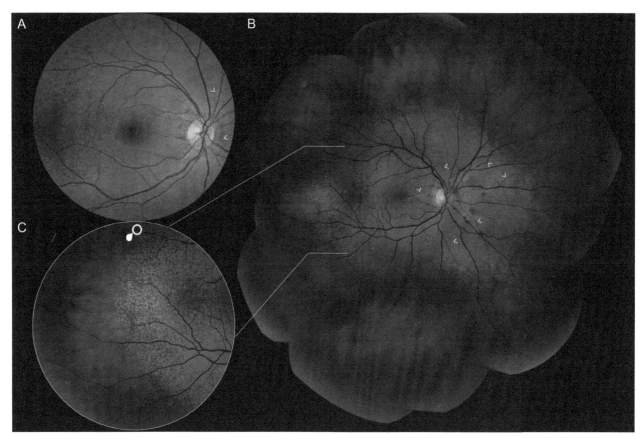

FIG. 13B.16 Peau d'orange appearance of the fundus in pseudoxanthoma elasticum. (A) Typical early image of peau d'orange, visible in the temporal area of the posterior pole. (B) Composite fundus image highlighting a uniform yellow reflex at the posterior pole and an associated circumferential area of peau d'orange outside the vascular arcades. The latter corresponds to the transition zone between the affected central and unaffected peripheral retina. (C) Detail of the transition zone of B showing a speckled appearance: darker dots of unaffected retina are present against a more yellow background. The *arrowheads* highlight the location of angioid streaks.

Dermatological abnormalities are present in over 95% of patients. They manifest as plaques of small, yellowish papules ('xanthoma-like' lesions) most frequently in the neck and axillae as well as in other flexural areas (antecubital/popliteal fossae, groin). Loss of skin elasticity may lead to increased skin laxity with excessive skin folding again primarily seen in the axillae and the groin.[3,11] (Fig. 13B.23). Histopathological analysis of affected areas reveals clumping and fragmentation of elastic fibres with associated calcification.[12]

Cardiovascular complications consequent upon increased atherosclerosis (such as myocardial infarction, angina pectoris, cardiac valve anomalies) were historically reported in patients with PXE but without precise comparison to control populations. However, comprehensive studies using modern cardiac echography and SPECT (single-photon emission computed tomography) suggest that the risk of cardiac involvement in people with PXE is similar to that of the general population.[13] On the other hand, patients with PXE (and, less commonly, their carrier relatives) have been shown to have an increased risk of cerebrovascular incidents as well as accelerated atherosclerosis in small and medium-sized arteries leading to peripheral artery disease with intermittent claudication.[14]

Other recognised complications include gastric haemorrhage and organ asymptomatic calcification (breast, kidney, liver spleen) seen on ultrasound.

A set of diagnostic criteria for PXE were published in 1994 and updated in 2010.[13,15,16] The authors recommend that a definitive diagnosis requires the presence of two (or more) major criteria not belonging to the same (skin, eye, genetic) category. Major diagnostic criteria include:

1. *Skin* (i) Yellowish papules and/or plaques on the lateral side of the neck and/or flexural areas of the body. (ii) Increase of morphologically altered elastin with fragmentation, clumping and calcification of elastic fibres in a skin biopsy taken from clinically affected skin.

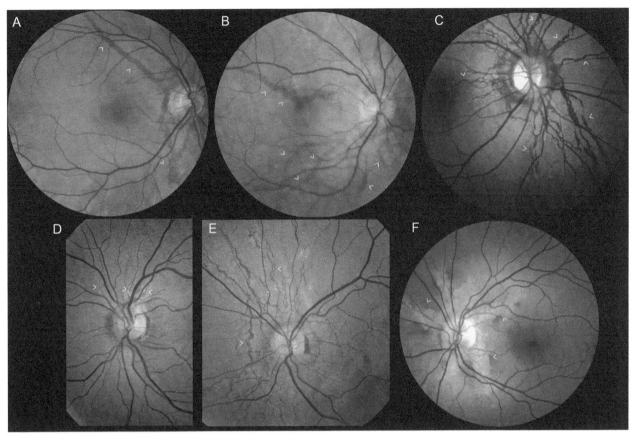

FIG. 13B.17 Angioid streaks in pseudoxanthoma elasticum. (A) Broad, red angioid streaks extending from the optic disc inferiorly and towards the supero-temporal arcade. (B) Broad angioid streaks, almost indistinguishable from retinal blood vessels, crisscrossing the macular area. (C) Darker, red-brown angioid streaks surrounding the optic disc and tapering towards the periphery. (D) Few, almost imperceptible angioid streaks hiding among the blood vessels. (E) 'Spidery' web of angioid streaks. (F) RPE atrophy adjacent to angioid streaks giving a feathery appearance. The *arrowheads* highlight the location of angioid streaks.

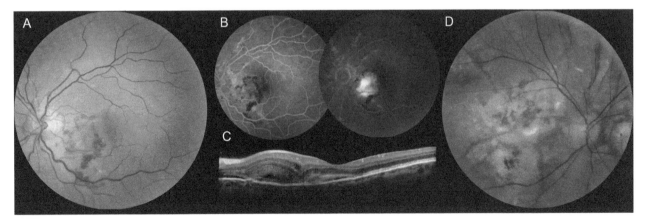

FIG. 13B.18 Neovascularisation in pseudoxanthoma elasticum. (A) Left eye of a 44-year-old female showing an active subretinal neovascular membrane with an intraretinal haemorrhage juxtafoveally. (B) Early and late phase of fluorescein angiography confirming the presence of choroidal neovascularisation. (C) OCT revealing the presence of intra- and subretinal fluid and pointing to the presence of active choroidal neovascularisation. (D) Right eye of a 73-year-old male showing fibrovascular scarring at the macular area and beyond the vascular arcades subsequent upon choroidal neovascular membrane formation before the introduction of anti-VEGF treatment.

298 SECTION | III Genetic disorders affecting the posterior segment

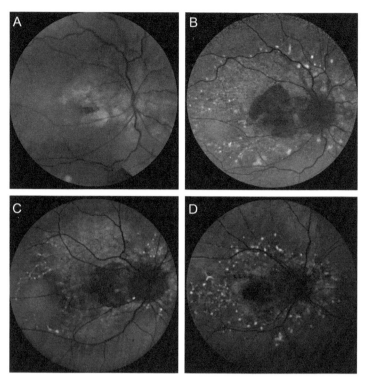

FIG. 13B.19 Pattern dystrophy in pseudoxanthoma elasticum. Fundus photography (A) and short-wavelength autofluorescence imaging (B) from a 56-year-old affected female reveal macular atrophy with a degree of foveal sparing and without underlying neovascularisation. A pattern dystrophy phenotype is unmasked in (B). Fundus autofluorescence images from two other patients who have pseudoxanthoma elasticum and a pattern dystrophy appearance are shown in (C) and (D).

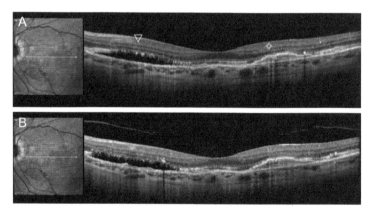

FIG. 13B.20 OCT imaging from a patient with pseudoxanthoma elasticum at baseline (A) and 2 years later (B). Persistent subretinal fluid occurring independently to the development of choroidal neovascularisation is noted. There was no response to initial intravitreal anti-VEGF treatment. Two years later, no major change is observed; the patient's best-corrected visual acuity remains stable despite discontinuing treatment with intravitreal injections. The *arrow* highlights a shallow neurosensory detachment. The star highlights a small fibrotic scar temporal to the fovea.

2. *Eye* (i) Peau d'orange fundal appearance. (ii) One or more angioid streaks, each at least as long as one disc diameter.
3. *Genetic* (i) Pathogenic variant(s) in both alleles of the *ABCC6* gene. (ii) First-degree relative who independently meets diagnostic criteria for PXE.

Molecular pathology

PXE results from biallelic mutations in the *ABCC6* gene. Considerable inter- and intra-familial variability have been reported.[18–20] The condition is thought to be fully penetrant and the differences in severity of phenotype and age of onset are likely modulated by modifying genetic and environmental factors.[21]

It is possible to identify biallelic *ABCC6* mutations in around 90% of affected individuals.[15] Around 40%–45% of pathogenic variants are missense mutations; arginine substitutions in specific protein domains (e.g. the intracellular loop 8 and the nucleotide-binding fold-1 region) are an important subset of these changes. Around 40%–45% of mutations are classic loss of functions alterations

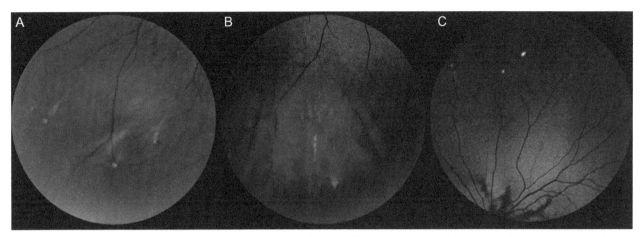

FIG. 13B.21 'Comet'-like lesions in patients with pseudoxanthoma elasticum. (A) Mid-peripheral nodular lesions with comet tails of chorioretinal atrophy pointing towards the posterior pole. (B) Cluster of comet-like lesions in the inferior mid-periphery resembling a 'comet rain'. (C) Short wavelength (blue light) autofluorescence imaging highlighting that comet-like lesions correspond to hyperautofluorescent dots. Note that angioid streaks are optimally visualised using this imaging modality (as hypoautofluorescent broad irregular lines between the blood vessels).

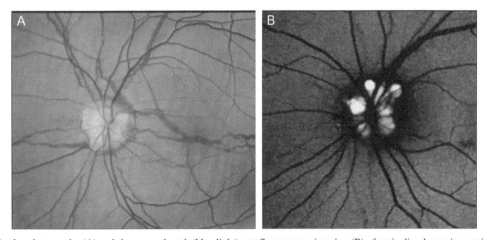

FIG. 13B.22 Fundus photography (A) and short-wavelength (blue light) autofluorescence imaging (B) of optic disc drusen in a patient with pseudoxanthoma elasticum.

(nonsense, frameshift and splice site mutations) including two frequently observed variants: c.3421C>T (p.Arg1141Ter) and c.1132C>T (p.Gln378Ter). Larger deletions have also been reported including a recurrent multiexon deletion involving exons 23 to 29.[18–20]

Although no precise genotype–phenotype correlations have been established, it is generally thought that patients with two loss of function mutations have a more severe phenotype (ophthalmic and cardiovascular) than those carrying missense mutations.[20]

ABCC6 encodes an ATP-dependent binding cassette transmembrane transporter that is primarily (and probably exclusively) expressed in the liver and the kidneys and is not expressed at all in PXE affected tissues. It has been shown that liver cells excrete large amounts of adenosine triphosphate (ATP) into the blood via this ABCC6 transporter. ATP is then hydrolysed into adenosine monophosphate (AMP) and inorganic pyrophosphate (PPi). PPi is a potent inhibitor of mineralisation and calcification.[22] If, in PXE, the transport of ATP outside the liver is impaired due to the defective ABCC6 transporter, circulating levels of PPi drop and accelerated calcification is induced. For this reason, PXE has been labelled as an inborn error of metabolism.

Several systemic and dermatological disorders may manifest features resembling PXE. Generalised arterial calcification of infancy (GACI) is characterised by severe congestive heart failure, arterial hypertension and myocardial infarction in the neonatal period. Children may show yellow papules of the skin and angioid streaks in the fundi. GACI is caused by biallelic mutations in the *ENPP1* gene but it has also been linked in a minority of cases to mutations in the *ABCC6* gene. GACI and PXE can be therefore considered

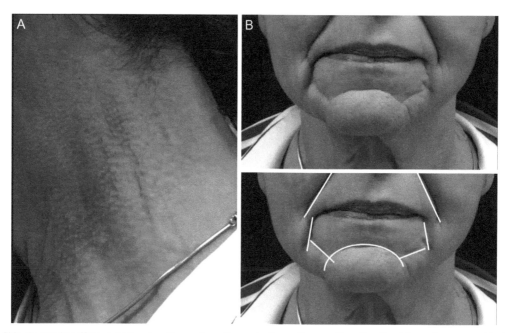

FIG. 13B.23 Skin changes in a 63-year-old female with pseudoxanthoma elasticum. Plaques of small, yellowish papules are noted in the neck (A) and deep mental creases are noted in the face (B).

allelic disorders and represent parts of the clinical spectrum of ectopic calcification.[23]

A PXE-like syndrome associated with mutations in the *GGCX* gene shows all the skin findings of PXE. These tend to be much more pronounced and widespread. In contrast, the ophthalmic phenotype in this PXE-like syndrome is rather benign with only a few, short, peripapillary angioid streaks and limited risk of neovascular membrane formation. It is worth noting that patients with *GGCX*-associated PXE-like syndrome suffer from a deficiency in vitamin K-dependent clotting factors; this is not a feature of PXE.[2]

Fundoscopic findings in β-thalassemia can resemble those of PXE. The association with the presence of angioid streaks is the strongest but other shared ophthalmic signs have also been described.[24] More broadly, angioid streaks have been reported in several conditions including high myopia, familial hyperphosphatemia, acromegaly, Paget disease, Marfan syndrome and Ehlers-Danlos syndrome.[3]

Clinical management

PXE management requires multidisciplinary workup and follow-up by a specialised team. The most significant causes of impaired vision in affected individuals are choroidal neovascularisation and the progression of macular atrophy with foveal involvement. Given the risk of developing subretinal haemorrhage or neovascularisation following ophthalmic or blunt head trauma, tailored advice regarding engagement in sports or other high-risk activities may be provided (e.g. use of protective eyewear). Affected individuals should be educated to monitor themselves to detect the earliest signs of choroidal neovascularisation (including sudden-onset visual loss, scotomata, metamorphopsia). In general, the visual prognosis in patients with PXE has improved dramatically with the introduction of intravitreal anti-VEGF treatments.[25]

Dietary and lifestyle modifications are mainly focused on atherosclerosis risk factor reduction including addressing hypertension, lipid disorders and issues with high body mass index. Notably, smoking is known to accelerate the development of complications. The use of aspirin and nonsteroidal anti-inflammatory drugs should be avoided due to the risk of gastrointestinal haemorrhage.[5]

References

1. Germain DP. Pseudoxanthoma elasticum. *Orphanet J Rare Dis* 2017;**12**(1):85.
2. Li Q, van de Wetering K, Uitto J. Pseudoxanthoma elasticum as a paradigm of heritable ectopic mineralization disorders: pathomechanisms and treatment development. *Am J Pathol* 2019;**189**(2):216–25. https://doi.org/10.1016/j.ajpath.2018.09.014.
3. Finger RP, Charbel Issa P, Ladewig MS, Gotting C, Szliska C, Scholl HP, et al. Pseudoxanthoma elasticum: genetics, clinical manifestations and therapeutic approaches. *Surv Ophthalmol* 2009;**54**(2):272–85.
4. Kranenburg G, Baas AF, de Jong PA, Asselbergs FW, Visseren FLJ, Spiering W, et al. The prevalence of pseudoxanthoma elasticum:

revised estimations based on genotyping in a high vascular risk cohort. *Eur J Med Genet* 2018;**62**(2):90–2.
5. Gliem M, Zaeytijd JD, Finger RP, Holz FG, Leroy BP, Charbel IP. An update on the ocular phenotype in patients with pseudoxanthoma elasticum. *Front Genet* 2013;**4**:14.
6. Charbel Issa P, Finger RP, Gotting C, Hendig D, Holz FG, Scholl HP. Centrifugal fundus abnormalities in pseudoxanthoma elasticum. *Ophthalmology* 2010;**117**(7):1406–14.
7. Gliem M, Muller PL, Birtel J, Hendig D, Holz FG, Charbel IP. Frequency, phenotypic characteristics and progression of atrophy associated with a diseased Bruch's membrane in pseudoxanthoma elasticum. *Invest Ophthalmol Vis Sci* 2016;**57**(7):3323–30.
8. Schoenberger SD, Agarwal A. Geographic chorioretinal atrophy in pseudoxanthoma elasticum. *Am J Ophthalmol* 2013;**156**(4):715–23.
9. Agarwal A, Patel P, Adkins T, Gass JD. Spectrum of pattern dystrophy in pseudoxanthoma elasticum. *Arch Ophthalmol* 2005;**123**(7):923–8.
10. De Zaeytijd J, Vanakker OM, Coucke PJ, De Paepe A, De Laey JJ, Leroy BP. Added value of infrared, red-free and autofluorescence fundus imaging in pseudoxanthoma elasticum. *Br J Ophthalmol* 2010;**94**(4):479–86.
11. Barteselli G, Viola F. Comet lesions in pseudoxanthoma elasticum: a spectral domain optical coherence tomography analysis. *Retina* 2015;**35**(5):1051–3.
12. Hendig D, Knabbe C, Gotting C. New insights into the pathogenesis of pseudoxanthoma elasticum and related soft tissue calcification disorders by identifying genetic interactions and modifiers. *Front Genet* 2013;**4**:114.
13. Plomp AS, Toonstra J, Bergen AA, van Dijk MR, de Jong PT. Proposal for updating the pseudoxanthoma elasticum classification system and a review of the clinical findings. *Am J Med Genet A* 2010;**152A**(4):1049–58.
14. Campens L, Vanakker OM, Trachet B, Segers P, Leroy BP, De Zaeytijd J, et al. Characterization of cardiovascular involvement in pseudoxanthoma elasticum families. *Arterioscler Thromb Vasc Biol* 2013;**33**(11):2646–52.
15. Vanakker OM, et al. Novel clinico-molecular insights in pseudoxanthoma elasticum provide an efficient molecular screening method and a comprehensive diagnostic flowchart. *Hum Mutat* 2008;**29**(1):205.
16. Lebwohl M, et al. Classification of pseudoxanthoma elasticum: report of a consensus conference. *J Am Acad Dermatol* 1994;**30**:103–7.
17. Bergen AA, Plomp AS, Schuurman EJ, Terry S, Breuning M, Dauwerse H, et al. Mutations in ABCC6 cause pseudoxanthoma elasticum. *Nat Genet* 2000;**25**(2):228–31.
18. Le Saux O, Beck K, Sachsinger C, Silvestri C, Treiber C, Goring HH, et al. A spectrum of ABCC6 mutations is responsible for pseudoxanthoma elasticum. *Am J Hum Genet* 2001;**69**(4):749–64.
19. Ringpfeil F, McGuigan K, Fuchsel L, Kozic H, Larralde M, Lebwohl M, et al. Pseudoxanthoma elasticum is a recessive disease characterized by compound heterozygosity. *J Invest Dermatol* 2006;**126**(4):782–6.
20. Legrand A, et al. Mutation spectrum in the *ABCC6* gene and genotype–phenotype correlations in a French cohort with pseudoxanthoma elasticum. *Genet Med* 2017;**19**(8):909–17.
21. Schon S, Schulz V, Prante C, Hendig D, Szliska C, Kuhn J, et al. Polymorphisms in the xylosyltransferase genes cause higher serum XT-I activity in patients with pseudoxanthoma elasticum (PXE) and are involved in a severe disease course. *J Med Genet* 2006;**43**(9):745–9.
22. Jansen RS, Duijst S, Mahakena S, Sommer D, Szeri F, Varadi A, et al. ABCC6-mediated ATP secretion by the liver is the main source of the mineralization inhibitor inorganic pyrophosphate in the systemic circulation-brief report. *Arterioscler Thromb Vasc Biol* 2014;**34**(9):1985–9.
23. Nitschke Y, Baujat G, Botschen U, Wittkampf T, du Moulin M, Stella J, et al. Generalized arterial calcification of infancy and pseudoxanthoma elasticum can be caused by mutations in either ENPP1 or ABCC6. *Am J Hum Genet* 2012;**90**(1):25–39.
24. Barteselli G, Dell'arti L, Finger RP, Charbel Issa P, Marcon A, Vezzola D, et al. The spectrum of ocular alterations in patients with beta-thalassemia syndromes suggests a pathology similar to pseudoxanthoma elasticum. *Ophthalmology* 2014;**121**(3):709–18.
25. Finger RP, Charbel Issa P, Hendig D, Scholl HP, Holz FG. Monthly ranibizumab for choroidal neovascularizations secondary to angioid streaks in pseudoxanthoma elasticum: a one-year prospective study. *Am J Ophthalmol* 2011;**152**(4):695–703.

13B.2.4

MIDD (maternally inherited diabetes, deafness and maculopathy)

Neruban Kumaran and Michel Michaelides

Mitochondrial disorders of energy metabolism are the largest subcategory of inborn errors of metabolism.[1] This clinically heterogeneous group includes several conditions with prominent ophthalmic features such as Leber hereditary optic neuropathy (Section 15.1), Kearns-Sayre syndrome (chronic progressive external ophthalmoplegia, pigmentary retinopathy and cardiac conduction block; Chapter 21) and MIDD (maternally inherited diabetes and deafness).

MIDD is inherited as a mitochondrial trait and is characterised by the presence of diabetes mellitus/impaired fasting glucose; sensorineural hearing impairment; maculopathy; and a normal or low body mass index. Patients with MIDD may also have short stature, proximal limb myopathy, left ventricular hypertrophy, renal failure, and gastrointestinal disease. Furthermore, affected female are at a greater risk of miscarriage.[2] The vast majority of affected individuals carry the m.3243A>G mutation. This genetic variant is one of the most frequent causes of mitochondrial dysfunction and is implicated not only in MIDD but also in several other disorders.[3] Notably, it is the commonest cause of MELAS, a severe condition associated with mitochondrial encephalomyopathy, lactic acidosis, and stroke-like episodes.[4]

Genetic testing (including evaluating heteroplasmy levels, i.e. the percentage of affected mitochondria, Box 13B.2) is key for diagnosis. Clinical management requires a proactive multidisciplinary approach.[3]

Clinical characteristics

MIDD is a highly variable systemic disorder that frequently has far broader manifestations than the acronym suggests. Diabetes associated with decreased insulin secretion usually develops in the fourth decade of life but the age of onset ranges from 10 to 70 years. Up to one in five affected individuals presents acutely (e.g. with diabetic ketoacidosis). Sensorineural hearing impairment is common and usually precedes the onset of diabetes. It is bilateral, predominantly affects the high-frequency range, has a gradual onset and can become severe over time. Generally, hearing loss is more common in men than in women.[5] Other extraocular features of MIDD reflect the overlap of this syndrome with MELAS and may include elevated serum lactate, neuromuscular and cardiac problems, and nephropathy with proteinuria.

A characteristic maculopathy is noted in the majority of patients (~80%) and it is often the recognition of suggestive features by the ophthalmologist that prompts appropriate genetic testing and counselling. The age of visual symptom onset is variable but tends to generally be after the fourth decade of life. The visual prognosis in MIDD is relatively good, with the majority of patients having vision better than 0.3 LogMAR. The main complaints include blurred vision and difficulty in both dim light and adapting to bright conditions.[5–7] Fundoscopic features include pattern dystrophy-like alterations (Fig. 13B.24) and RPE atrophic changes (Figs. 13B.25 and 13B.26). These progressive retinal abnormalities are typically localised at the posterior pole and frequently involve the retina surrounding the optic disc. The fovea is often relatively preserved and, as a result, the visual prognosis tends to be favourable with the majority of patients maintaining good central vision in at least one eye. It is highlighted that the presentation and rate of expansion of RPE atrophy in MIDD are comparable to age-related macular degeneration, an important differential diagnosis.[7] Cystoid macular oedema has been reported in MIDD, and a good response to intravitreal triamcinolone has been reported.[8] Furthermore, it has been suggested that diabetic retinopathy is less prevalent in individuals with MIDD, in comparison to patients with isolated diabetes, possibly due to a lower incidence of associated hypertension and a reduced metabolism that results in decreased production of vasoproliferative disease.[2] Multimodal fundus imaging is particularly useful in diagnosing and monitoring MIDD-related retinopathy.

Fundus autofluorescence imaging reveals the diffuse nature of the macular abnormalities. Decreased autofluorescence is present in regions of atrophy while areas of

> **BOX 13B.2 Heteroplasmy**
>
> Heteroplasmy is the presence either in a cell, or more pertinently in a single individual of more than one type of mitochondrial genome. In the context of human ophthalmic disease, the term is used to refer to the presence in an individual patient of both mutant and unaffected mitochondrial genomes. The ratio of mutant to unaffected mitochondria can vary between the different tissues of the same individual and, due to a bottleneck during the production of oocytes, between individuals within the same family.

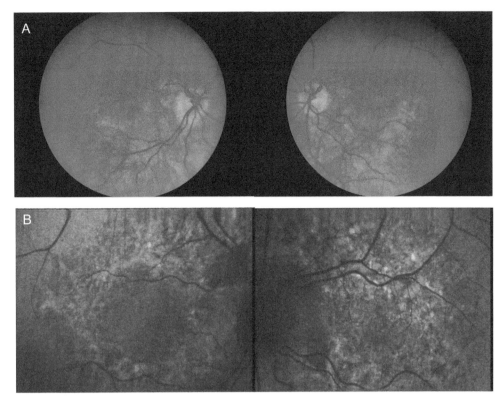

FIG. 13B.24 Fundus photography and autofluorescence imaging in a patient with MIDD and a pattern dystrophy-like appearance. (A) Colour fundus photographs showing bilateral yellow-white spots in the macula. (B) Fundus autofluorescence imaging showing diffuse speckled macular autofluorescence which is greater than what would be expected on the basis of the corresponding fundus photographs.

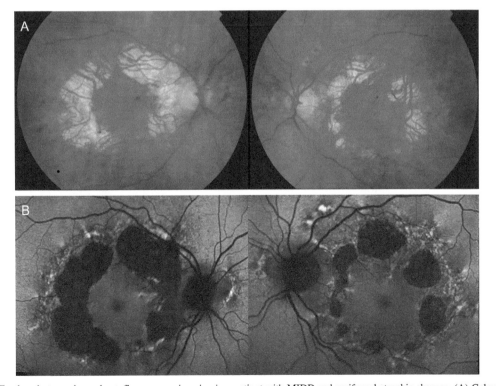

FIG. 13B.25 Fundus photography and autofluorescence imaging in a patient with MIDD and perifoveal atrophic changes. (A) Colour fundus photographs showing bilateral, symmetrical, discontinuous, circumferentially distributed/oriented perifoveal RPE atrophy. Adjacent to the areas of atrophy, pale deposits at the level of the RPE, granularity of the RPE and subretinal pigment clumping are also seen. Over time, these atrophic areas may coalesce into a ring, with the central fovea being affected at a late stage in a minority of subjects. (B) Fundus autofluorescence imaging showing decreased signal in the areas corresponding to chorioretinal atrophy; a speckled autofluorescence pattern surrounding the areas of atrophy can also be seen.

304 SECTION | III Genetic disorders affecting the posterior segment

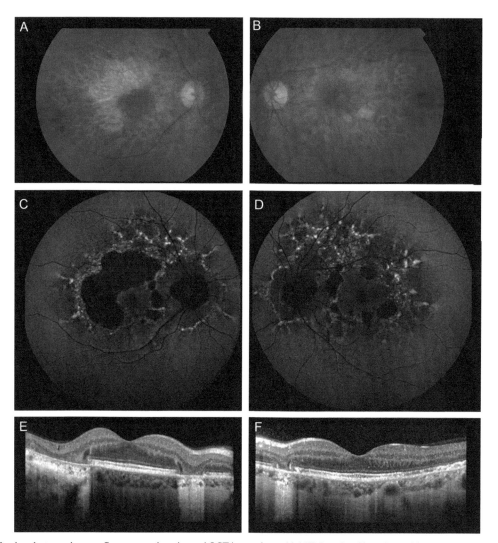

FIG. 13B.26 Fundus photography, autofluorescence imaging and OCT in a patient with MIDD and perifoveal atrophic changes. This 53-year-old female presented to the retina clinic with an undiagnosed maculopathy. Her central vision was within normal limits and her past medical history included hearing loss and migraines. She reported a family history of diabetes and hearing loss. Colour fundus photographs revealed bilateral parafoveal RPE atrophy (A,B). Fundus autofluorescence imaging revealed atrophic areas and speckled spots of both increased and decreased signal (C,D). OCT revealed preservation of central retinal structures (D,E). The m.3243A>G mutation was identified on genetic testing confirming the diagnosis of MIDD.

increased signal correspond to the subretinal deposits observed clinically. The retina surrounding the atrophic areas and/or the pattern dystrophy-like lesions demonstrates a typical speckled autofluorescence pattern, with spots of both increased and decreased signal. The diffuse nature of these abnormality is not evident on biomicroscopy (Figs. 13B.24–13B.26).[6,7,9,10]

OCT imaging in patients with MIDD reveals a consistent sequence of features that is similar to findings in individuals with age-related macular degeneration. This includes ellipsoid zone loss and subretinal deposits early on and loss of both RPE and outer nuclear layers at later disease stages. Outer retinal tubulations are occasionally observed.[9]

Full-field electroretinography reveals no evidence of generalised retinal dysfunction in patients with MIDD. In contrast, pattern and multifocal electroretinograms are abnormal in keeping with macular dysfunction.[11]

Other ophthalmic manifestations of the condition include early cataracts[7] and, rarely, progressive external ophthalmoplegia.

Molecular pathology

MIDD has been described in association with several mitochondrial DNA mutations. However, the m.3243A>G change accounts for the vast majority of cases. This mutation alters one of the 22 transfer RNA molecules that are encoded by mitochondrial DNA (more specifically, the one that carries the amino-acid leucine). These transfer RNAs are essential for the incorporation of amino-acids in nascent proteins and, without them, various proteins encoded by the mitochondrial genome do not form. This leads to dysfunction of the mitochondrial respiratory chain which decreases energy production and results in the production of high amounts of toxic reactive oxygen species.[4,5] It has been suggested that macular RPE are more significantly affected than the peripheral retina, because of the higher metabolic demand in this region.

The m.3243A>G mutation is considered to be the most common pathogenic mitochondrial variant.[3] It is associated with diverse clinical manifestations that collectively constitute a wide spectrum ranging from MELAS at the severe end to asymptomatic carriers at the mild end. More severe phenotypes may be the result of a higher abundance of the pathogenic variant in affected organs.[4] However, the relationship between mutational load (heteroplasmy) and clinical status is imprecise and making predictions about the phenotypic expression of a m.3243A>G mutation carrier is challenging.

Clinical management

Early diagnosis of MIDD allows optimisation of diabetes management, informs prognosis and enables genetic counselling. Obtaining a timely molecular diagnosis is therefore key. In the first instance, mitochondrial DNA mutations are usually screened for in blood leucocytes. It is common to evaluate mutation (heteroplasmy) levels in other tissues as well, especially if the pathogenic variant is undetectable in blood. Examples include buccal mucosa (saliva sample), urinary epithelial cells (urine sample), hair follicles and (most reliably) skeletal muscle. It is noted that the mitochondrial mutation load in the blood significantly reduces over time and calculating age-related levels enhances precision.[3]

MIDD is most often associated with a decrease in insulin secretion rather than with impaired insulin sensitivity. Therefore, the most appropriate oral glucose lowering agents are insulin secretagogues such as sulfonylureas. Metformin is generally avoided in MIDD patients as it increases the risk of lactic acidosis.[5] Regarding hearing, early referral to audiology is recommended and auditory intervention (e.g. early fitting of hearing aids) should be considered. Ototoxic medications and prolonged noise exposure should be avoided.[5]

Many patients diagnosed with MIDD, as well as those with MELAS, have multiple morbidities which are frequently overlooked or undiagnosed. Referral for specialist physician care should be sought wherever possible, considering the risk of neurological, muscle or gastrointestinal complications. Furthermore, all patients require a full cardiac workup to evaluate the risk of heart failure and cardiomyopathy, and to identify potential conduction disorders. Renal function should also be assessed (e.g. to look for proteinuria) as early treatment (e.g. with angiotensin-converting enzyme inhibitors) can prevent kidney disease progression.[5] No proven intervention is available for MIDD or MELAS associated maculopathy but regular screening for traditional diabetes-related microvascular complications is advisable. Discussing the management of MELAS is beyond the scope of this chapter and further information can be found in Ref. 4.

Coenzyme Q10 (CoQ10) is a potential therapeutic option for patients with MIDD and MELAS. This molecule is thought to play a role in the respiratory chain in mitochondria and may improve the mutation-associated dysfunction in MIDD. As such, one randomised study suggested that 150mg of coenzyme Q10 may be effective in preventing hearing loss and delaying progression in diabetes.[12]

References

1. Ferreira CR, van Karnebeek CDM, Vockley J, Blau N. A proposed nosology of inborn errors of metabolism. *Genet Med* 2019;**21**(1):102–6.
2. Murphy R, Turnbull DM, Walker M, Hattersley AT. Clinical features, diagnosis and management of maternally inherited diabetes and deafness (MIDD) associated with the 3243A>G mitochondrial point mutation. *Diabet Med* 2008;**25**:383–99.
3. de Laat P, Rodenburg RR, Roeleveld N, Koene S, Smeitink JA, Janssen MC. Six-year prospective follow-up study in 151 carriers of the mitochondrial DNA 3243 A>G variant. *J Med Genet* 2020. jmedgenet-2019-106800.
4. El-Hattab AW, Almannai M, Scaglia F. MELAS. 2001 Feb 27 [updated 2018 Nov 29]. In: Adam MP, Ardinger HH, Pagon RA, et al., editors. *GeneReviews® [internet]*. Seattle, WA: University of Washington, Seattle; 1993–2020. Available from: https://www.ncbi.nlm.nih.gov/books/NBK1233/.
5. Robinson KN, Terrazas S, Giordano-Mooga S, Xavier NA. The role of heteroplasmy in the diagnosis and management of maternally inherited diabetes and deafness. *Endocr Pract* 2020;**26**(2):241–6.
6. Michaelides M, Jenkins SA, Bamiou DE, Sweeney MG, Davis MB, Luxon L, et al. Macular dystrophy associated with the A3243G mitochondrial DNA mutation. Distinct retinal and associated features, disease variability, and characterization of asymptomatic family members. *Arch Ophthalmol* 2008;**126**(3):320–8.
7. Müller PL, Treis T, Pfau M, Esposti SD, Alsaedi A, Maloca P, et al. Progression of retinopathy secondary to maternally inherited diabetes and deafness—evaluation of predicting parameters. *Am J Ophthalmol* 2020;**213**:134–44.
8. Qian CX, Branham K, Khan N, Lundy SK, Heckenlively JR, Jayasundera T. Cystoid macular changes on optical coherence tomography in a patient with maternally inherited diabetes and deafness (MIDD)-associated macular dystrophy. *Ophthalmic Genet* 2017;**38**(5):467–72.
9. Müller PL, Maloca P, Webster A, Egan C, Tufail A. Structural features associated with the development and progression of RORA secondary to maternally inherited diabetes and deafness. *Am J Ophthalmol* 2020;**218**:136–47.
10. Rath PP, Jenkins S, Michaelides M, Smith A, Sweeney MG, Davis MB, et al. Characterisation of the macular dystrophy in patients with the A3243G mitochondrial DNA point mutation with fundus autofluorescence. *Br J Ophthalmol* 2008;**92**(5):623–9.
11. Bellmann C, Neveu MM, Scholl HP, Hogg CR, Rath PP, Jenkins S, et al. Localized retinal electrophysiological and fundus autofluorescence imaging abnormalities in maternal inherited diabetes and deafness. *Invest Ophthalmol Vis Sci* 2004;**45**(7):2355–60.
12. Suzuki S, Hinokio Y, Ohtomo M, Hirai M, Hirai A, Chiba M, et al. The effects of coenzyme Q10 treatment on maternally inherited diabetes mellitus and deafness, and mitochondrial DNA 3243 (A to G) mutation. *Diabetologia* 1998;**41**(5):584–8.

13B.2.5

Long-chain L-3 hydroxyacyl-CoA dehydrogenase (LCHAD) deficiency

Kristina Teär Fahnehjelm, Anna Nordenström, and Jane L. Ashworth

Long-chain 3-hydroxyacyl CoA dehydrogenase (LCHAD) deficiency is an autosomal recessive disorder of long-chain fatty acid oxidation. Affected individuals often present in the first months of life and common symptoms and signs include fatigue, episodes of hypoglycaemia and muscular problems. Sudden death from cardiomyopathy or organ failure can occur. Early diagnosis (often through newborn screening) and initiation of treatment (including strict dietary management) have considerably improved survival, and many patients are now living into adulthood. Despite these advances, many affected individuals develop visual loss due to a progressive chorioretinopathy.[1]

LCHAD deficiency is caused by mutations in *HADHA*. This gene encodes the alpha subunit of the mitochondrial trifunctional protein. Notably, there is significant phenotypic variability, even between individuals with the same underlying mutation.

Clinical characteristics

Before neonatal screening, the diagnosis of LCHAD deficiency was usually established in the first year of life following metabolic decompensation.[2] Fatty acids are used for energy production in the body and LCHAD deficiency can lead to symptoms during catabolic situations such as fasting, infections with fever or exercise; these may result in hypoketotic hypoglycaemia, rhabdomyolysis with muscular pain, metabolic acidosis, elevated transaminases and creatine kinases, cardiomyopathy, and ultimately in severe cases, sudden death.[2–4] In contrast to individuals with other fatty acid oxidation deficiencies, patients with LCHAD deficiency are at risk of developing visual impairment due to progressive retinal dysfunction.[5] Other long-term complications include peripheral neuropathy, hepatomegaly and learning difficulties.

Visual symptoms such as light hypersensitivity and decreased vision may gradually develop, often at around 2–4 years of age. There is a characteristic sequence of fundoscopic changes. Early on, there is irregular macular pigmentation (Fig. 13B.27) followed by a progressive patchy chorioretinal atrophy that starts in the peripapillary region or at the macula, with relative foveal sparing, and extends peripherally (Fig. 13B.28). Rarely, affected individuals develop hyperpigmented foveal scars, sometimes with retinal fibrosis. Patchy areas of the decreased signal may be seen on fundus autofluorescence imaging while OCT reveals abrupt transition zones and outer retinal tubulations (similar to gyrate atrophy or choroideremia). Interestingly, despite the early involvement of the posterior pole, affected individuals tend to retain good central vision in at least one eye.[1] Electroretinographic responses are attenuated and delayed from an early age. Both cone and rod photoreceptors appear to be affected but the primary defect is thought to be at the level of the RPE.[6–9]

In general, LCHAD-chorioretinopathy progresses rapidly in patients with dramatic neonatal symptoms but tends to be less severe in individuals receiving early and strict dietary management.[5,10] In terms of systemic status, patients with advanced disease are becoming increasingly rare, probably due to early detection and treatment that helps prevent hypoglycaemia and thereby improve outcomes.[11] Newborn screening is one method by which early diagnosis can be achieved and LCHAD deficiency is now included in many newborn screening programmes in Europe and North America.[12]

Molecular pathology

LCHAD deficiency is a recessive disorder caused by homozygous or compound heterozygous mutations in the *HADHA* gene. *HADHA* encodes one of the subunits of an enzyme that resides in the inner mitochondrial membrane (mitochondrial trifunctional protein). This enzyme catalyses the final steps of the mitochondrial oxidation of long-chain fatty acids.

Over 80 mutations have been described. A founder effect has been suggested for some of these changes, including the c.1528G>C (p.Glu510Gln) variant that is present in more than 80% of cases from Sweden and Finland.[8,11]

The pathophysiology of LCHAD-chorioretinopathy is thought to be due to toxic effects of fatty acids and intermediate metabolites on the RPE, energy deficiency and affected metabolism of docosahexaenoic acid.[13] Affected RPE cells are generally found to be small with accumulated lipids and triglycerides, fewer melanosomes, deficient tight junctions and increased apoptosis.[14]

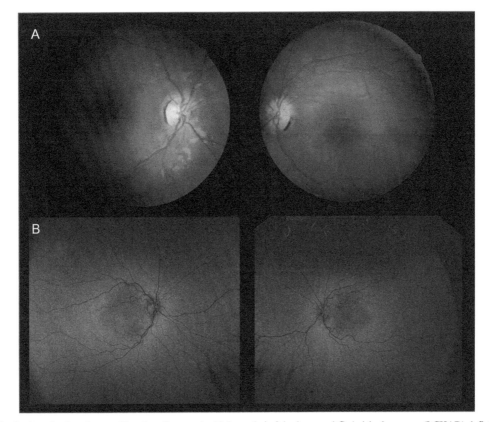

FIG. 13B.27 Fundus imaging in a 4-year-old patient diagnosed with long-chain 3-hydroxyacyl CoA dehydrogenase (LCHAD) deficiency by neonatal screening. There was no complaint of photophobia or night vision problems and the visual acuity was slightly subnormal for age (at 0.20 LogMAR right and 0.10 LogMAR left). Colour fundus photography revealed subtle pigmentary disturbances parafoveally with a mottled appearance (A). Fundus autofluorescence imaging revealed irregular autofluorescence at the macula bilaterally (B).

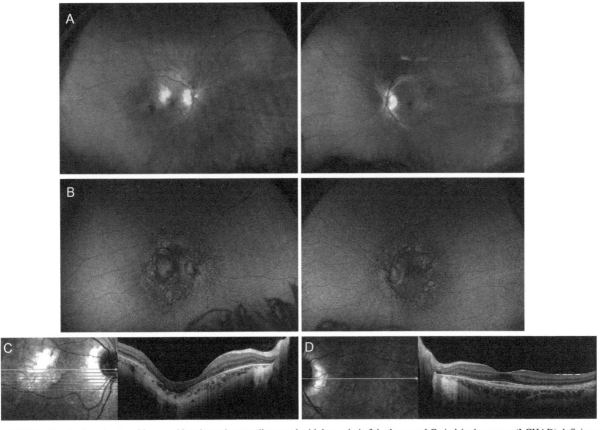

FIG. 13B.28 Fundus imaging in a 20-year-old patient who was diagnosed with long-chain 3-hydroxyacyl CoA dehydrogenase (LCHAD) deficiency at an early age. The best-corrected visual acuity was 0.20 LogMAR in the right and 0.0 LogMAR in the left eye. Electroretinography had been performed several times and showed progressively subnormal responses especially in the light-adapted state. Fundoscopy revealed peripapillary atrophy, hypopigmentation at the posterior pole and juxtafoveal chorioretinal atrophy. The left eye also had peripapillary atrophy (A). Fundus autofluorescence imaging revealed macular and peripapillary hyper- and hypoautofluorescent lesions in the right more than in the left eye (B). OCT revealed outer retinal atrophic changes centrally in the right eye (C). In the left eye there was irregularity at the level of the RPE and the photoreceptor outer segments (D).

Clinical management

Treatment of LCHAD deficiency consists of a low-fat diet (with restricted intake of long-chain fatty acids), supplemented with medium-chain triglycerides and essential fatty acids. Therapy in affected children aims to prevent catabolic episodes by giving frequent feeds and, in many cases, continuous night feeds.[11,12] Further research is required to determine the optimal long-term treatment for LCHAD deficiency and to assess the necessity of adhering to a challenging and complicated low-fat diet later in life.

Regarding ophthalmic management, given the risk of visual loss, affected individuals should be offered regular annual assessments of visual function and ophthalmic status.

References

1. Boese EA, Jain N, Jia Y, Schlechter CL, Harding CO, Gao SS, et al. Characterization of chorioretinopathy associated with mitochondrial trifunctional protein disorders: long-term follow-up of 21 cases. *Ophthalmology* 2016;**123**(10):2183–95.
2. Hagenfeldt L, von Döbeln U, Holme E, Alm J, Brandberg G, Enocksson E, et al. 3-hydroxydicarboxylic aciduria—a fatty acid oxidation defect with severe prognosis. *J Pediatr* 1990;**116**(3):387–92.
3. Hagenfeldt L, Venizelos N, von Döbeln U. Clinical and biochemical presentation of long-chain 3-hydroxyacyl-CoA dehydrogenase deficiency. *J Inherit Metab Dis* 1995;**18**(2):245–8.
4. Wanders RJ, IJlst L, van Gennip AH, Jakobs C, de Jager JP, Dorland L, et al. Long-chain 3-hydroxyacyl-CoA dehydrogenase deficiency: identification of a new inborn error of mitochondrial fatty acid beta-oxidation. *J Inherit Metab Dis* 1990;**13**(3):311–4.
5. Fahnehjelm KT, Holmstrom G, Ying L, Haglind CB, Nordenstrom A, Halldin M, et al. Ocular characteristics in 10 children with long-chain 3-hydroxyacyl-CoA dehydrogenase deficiency: a cross-sectional study with long-term follow-up. *Acta Ophthalmol* 2008;**86**(3):329–37. https://doi.org/10.1111/j.1600-0420.2007.01121.x.
6. Schrijver-Wieling I, van Rens GH, Wittebol-Post D, Smeitink JA, de Jager JP, de Klerk HB, et al. Retinal dystrophy in long chain 3-hydroxy-acyl-coA dehydrogenase deficiency. *Br J Ophthalmol* 1997;**81**(4):291–4.
7. Tyni T, Pihko H, Kivela T. Ophthalmic pathology in long-chain 3-hydroxyacyl-CoA dehydrogenase deficiency caused by the G1528C mutation. *Curr Eye Res* 1998;**17**(6):551–9.
8. Tyni T, Kivela T, Lappi M, Summanen P, Nikoskelainen E, Pihko H. Ophthalmologic findings in long-chain 3-hydroxyacyl-CoA dehydrogenase deficiency caused by the G1528C mutation: a new type of hereditary metabolic chorioretinopathy. *Ophthalmology* 1998;**105**(5):810–24. https://doi.org/10.1016/s0161-6420(98)95019-9.
9. Tyni T, Immonen T, Lindahl P, Majander A, Kivela T. Refined staging for chorioretinopathy in long-chain 3-hydroxyacyl coenzyme A dehydrogenase deficiency. *Ophthalmic Res* 2012;**48**(2):75–81. https://doi.org/10.1159/000334874.
10. Immonen T, Turanlahti M, Paganus A, Keskinen P, Tyni T, Lapatto R. Earlier diagnosis and strict diets improve the survival rate and clinical course of long-chain 3-hydroxyacyl-CoA dehydrogenase deficiency. *Acta Paediatr* 2016;**105**(5):549–54.
11. Haglind CB, Nordenstrom A, Ask S, von Dobeln U, Gustafsson J, Stenlid MH. Increased and early lipolysis in children with long-chain 3-hydroxyacyl-CoA dehydrogenase (LCHAD) deficiency during fast. *J Inherit Metab Dis* 2015;**38**(2):315–22. https://doi.org/10.1007/s10545-014-9750-3.
12. Fraser H, Geppert J, Johnson R, Johnson S, Connock M, Clarke A, et al. Evaluation of earlier versus later dietary management in long-chain 3-hydroxyacyl-CoA dehydrogenase or mitochondrial trifunctional protein deficiency: a systematic review. *Orphanet J Rare Dis* 2019;**14**(1):258.
13. Harding CO, Gillingham MB, van Calcar SC, Wolff JA, Verhoeve JN, Mills MD. Docosahexaenoic acid and retinal function in children with long-chain 3-hydroxyacyl-CoA dehydrogenase deficiency. *J Inherit Metab Dis* 1999;**22**(3):276–80.
14. Polinati PP, Ilmarinen T, Trokovic R, Hyotylainen T, Otonkoski T, Suomalainen A, et al. Patient-specific induced pluripotent stem cell-derived RPE cells: understanding the pathogenesis of retinopathy in long-chain 3-hydroxyacyl-CoA dehydrogenase deficiency. *Invest Ophthalmol Vis Sci* 2015;**56**(5):3371–82.

13B.2.6

Neuronal ceroid lipofuscinosis (Batten disease)

Dipak Ram and Jane L. Ashworth

Neuronal ceroid lipofuscinoses are a group of lysosomal storage disorders that are associated with neurodegeneration and intracellular accumulation of lipofuscin (autofluorescent material consisting of two-thirds protein, one-third lipid, and trace amounts of metals such as iron). This group of genetic conditions is also known as Batten disease. Notably, in the past, the term Batten disease was used to refer only to a specific, juvenile-onset form of neuronal ceroid lipofuscinosis.

There are various disease subtypes that may present at different ages with varying outcomes. Common features include progressive psychomotor and cognitive decline, epilepsy, retinal degeneration and reduced lifespan.[1-3]

Neuronal ceroid lipofuscinoses are typically inherited as autosomal recessive traits although a rare, adult-onset, autosomal dominant form has also been described. There is significant genetic heterogeneity and at least 13 genes have been linked to these conditions.[4,5]

Clinical characteristics

Neuronal ceroid lipofuscinoses typically present with neurodegeneration, leading to progressive motor and cognitive decline. Seizures and progressive visual loss are also common early features. The most common presentations include:

1. *Infantile-onset (CLN1 disease)*: Children with this phenotype usually present between the ages of 6 months and 2 years with developmental delay. Seizures are common, particularly myoclonic jerks. Retinal degeneration is evident by 2 years of age. Affected children typically develop regression of previously acquired skills and death occurs within the first decade of life.
2. *Late infantile-onset (CLN2 disease)*: Children usually present between the ages of 2 and 4 years with seizures. Myoclonic jerks are common and, following the onset of seizures, there is usually loss of previously acquired skills. Retinal involvement occurs early and leads to progressive visual impairment relatively rapidly. Most children do not survive beyond the second decade of life.
3. *Juvenile-onset (CLN3 disease)*: In this phenotype, children usually present with progressive visual loss, between the ages of 4 and 8 years. The ophthalmologist, therefore, plays a crucial role in early diagnosis, particularly as the associated behavioural changes may initially be very subtle. Seizures tend to present later on, in adolescence and can be associated with varying levels of developmental regression. Life expectancy varies but death most commonly occurs in the second or third decade of life.
4. *Adult-onset*: Individuals usually present in their 30s with epilepsy alongside motor and cognitive decline. Retinal involvement is less common in this group of patients.

Although these are typical presentations of neuronal ceroid lipofuscinosis, it is highlighted that some affected individuals would not fall in any of the above categories. However, patients presenting with developmental regression, especially when this is associated with visual difficulties or seizures, should always alert a clinician to suspect neuronal ceroid lipofuscinosis.[1-4]

The rate of decline in vision and progression of retinopathy is rapid compared to other retinal degenerations. Ophthalmic features include central visual loss, colour vision abnormalities, photophobia, night vision problems and peripheral visual loss. A subset of affected children may be noted to be 'overlooking', i.e. holding their eyes in a raised position when attempting to fixate; this is thought to be due to the relative preservation of the inferior retina. Rotary nystagmus may occasionally be present. Fundoscopy may initially be normal, but often shows granularity of the macular RPE or a 'bull's eye' maculopathy pattern. Importantly, at these early stages, neuronal ceroid lipofuscinosis is often misdiagnosed as non-syndromic retinal dystrophy (e.g. Stargardt disease or cone/cone-rod dystrophy) or medically unexplained visual loss (Fig. 13B.29). At later stages, vascular attenuation, disc pallor and peripheral retinal changes occur. In general, intraretinal pigment migration is not a prominent finding.

The full-field electroretinogram is usually severely attenuated in both light- and dark-adapted states at diagnosis. In very early cases there may be an electronegative bright flash electroretinogram (i.e. loss of b-wave amplitude in the presence of a relatively preserved a-wave) in keeping with inner retinal dysfunction. OCT imaging

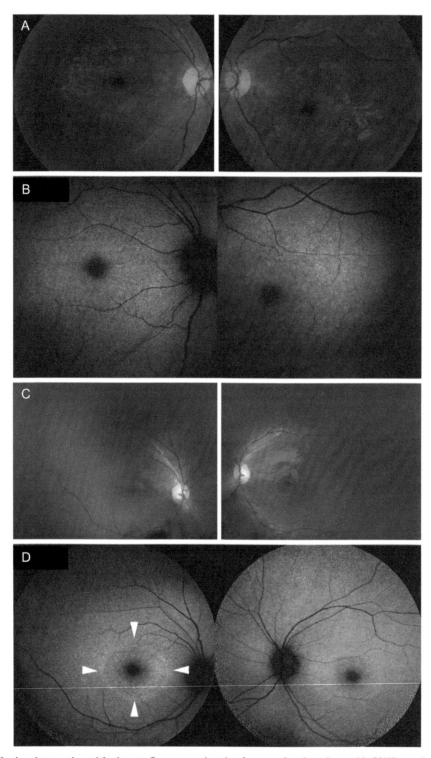

FIG. 13B.29 Colour fundus photographs and fundus autofluorescence imaging from two female patients with *CLN3*-associated neuronal ceroid lipofuscinosis. (A and B) Disease onset in the first case was at age 5 years. Initial clinical findings included nystagmus, reduced central vision and poor night vision. There were no neurological signs. A subtle bull's eye maculopathy pattern was noted with a degree of retinal vascular attenuation. (C and D) Disease onset in this second case was at age 3 years. Initial clinical findings included reduced central vision and poor night vision. There was a degree of clumsiness and the possibility of an autism spectrum disorder was raised. Initially, the diagnosis of Stargardt disease was made. A subtle bull's eye maculopathy pattern was noted and a ring of increased autofluorescence signal was present *(white arrowheads)*. *(Adapted from Ref. 6.)*

FIG. 13B.30 Multicolour fundus imaging (A) and OCT (B) from the right eye of a 10-year-old patient with *CLN3*-associated neuronal ceroid lipofuscinosis. Pigment mottling and vascular attenuation are noted. OCT reveals near-complete loss of outer retinal layers and atrophy of the nerve fibre and ganglion cell layers at the central macula.

reveals photoreceptor loss and may demonstrate macular oedema. OCT can be particularly informative in *CLN3* disease as, unlike many other retinal degenerations, abnormalities are often noted in all retinal layers from an early stage (Fig. 13B.30).[6–8]

Molecular pathology

In neuronal ceroid lipofuscinosis and other lysosomal storage disorders, there is abnormal recycling of several molecules. Over time, this leads to intracellular accumulation of cytotoxic material (including lipofuscin) which builds up particularly in the brain and retina.

CLN1, *CLN2*, and *CLN3* were the first genes implicated in these disorders; traditionally, they have been associated with infantile-onset, late infantile-onset and juvenile-onset forms of neuronal ceroid lipofuscinosis respectively. At least 10 more disease-associated genes have been described in recent decades. Although a few genotype–phenotype correlations have been proposed, certain genetic defects can lead to a wide range of disease severity and outcomes. For example, mutations in *CLN1* (also known as *PPT1*) are not only associated with infantile-onset but also with juvenile-onset forms of the condition. Given the observed clinical heterogeneity, an appropriate way to classify this group of disorders would involve using both genotypic and phenotypic terms (e.g. '*CLN1* disease—infantile-onset' or '*CLN1* disease—juvenile-onset').[1–4]

Mutations in *CLN3* are the commonest cause of neuronal ceroid lipofuscinosis. More than 80 pathogenic variants have been described in this gene and one of the most prevalent mutations is a ~1000 base-pair deletion leading to loss of exons 7 and 8. It is highlighted that a small number of patients with *CLN3* are found to have a non-syndromic retinal disease with unremarkable neuroimaging and normal electron microscopic assessment of their peripheral lymphocytes. These affected individuals often have a rod-cone dystrophy pattern with early macular involvement and generally fall into two phenotypic groups: a late onset one (with the onset of visual symptoms between the second and the fourth decade of life) and an early-onset one (with the onset of visual problems before the second decade of life). In general, the onset of retinal degeneration in both these groups is still later than that seen in patients with *CLN3* mutations who have multisystemic involvement.[9–11]

Clinical management

In patients who present with developmental regression, brain MRI imaging is indicated. The most common neuroimaging features seen in individuals with neuronal ceroid lipofuscinoses are cerebral and cerebellar atrophy. Electroencephalograms may show encephalopathy or epileptiform activity but these changes are not specific to this group of disorders.

Technological advances have transformed the way neuronal ceroid lipofuscinoses are diagnosed. In the past, electron microscopy of skin tissue was used to demonstrate the accumulation of lipofuscin. Electron microscopy of peripheral blood samples (buffy coat studies) has also been used as a less invasive method of demonstrating ultrastructural abnormalities suggestive of these disorders. In recent years, rapid enzyme testing has been typically used when a patient is suspected to have neuronal ceroid lipofuscinosis. These tests can highlight a deficiency in the activity of certain lysosomal enzymes. Notably, *PPT1* deficiency is associated with *CLN1/PPT1* gene mutations and *TPP1* deficiency is associated with *CLN2/TPP1* gene mutations. Enzymatic assays for other disease forms are currently unavailable. Genetic testing is the gold standard method for confirming the diagnosis of neuronal ceroid lipofuscinosis. Liaison with paediatric neurologists and metabolic physicians is recommended to facilitate these enzymatic and genetic studies.[1–4]

At present, there is no curative treatment available to people with neuronal ceroid lipofuscinosis. Management strategies mainly revolve around supportive therapies. Anti-epileptic drugs are frequently required to manage seizures, which can be refractory. Medications may also be used to manage spasticity and its associated discomfort. Physiotherapy, occupational therapy, speech and language therapy and dietician input are integral in maintaining a patient's wellbeing. Psychological support for the affected individual and their family should also be offered at an early stage given the progressive debilitating nature of the condition. Support groups can be helpful not just for patients, but also for the wider family. Towards the terminal aspect of a patient's journey, it is important to adopt a palliative care approach. This should involve a thorough discussion with the family, and include a limitation of treatment agreement. The main goal is to ensure that the patient is kept as comfortable as possible.

Once a genetic diagnosis of neuronal ceroid lipofuscinosis is confirmed in a patient, it is important to offer appropriate genetic counselling to the family. As outlined above, most cases are inherited in an autosomal recessive fashion. The families should be made aware of the 1 in 4 risk of future children being affected. If there are any concerns raised about family members or if the family wants to explore pre-implantation genetic testing, a referral to clinical genetics is warranted.

Extensive research is ongoing in many aspects of neuronal ceroid lipofuscinosis, including enzyme replacement therapy, gene therapy and bone marrow transplantation.

Cerliponase alfa has been approved for use as an intraventricular enzyme replacement therapy in *CLN2* disease as it has been shown to either stop or slow down the progression of the disease. Early diagnosis of *CLN2* disease is therefore vital to achieve the best potential outcome. Presently, no similar treatment options are available for other forms of neuronal ceroid lipofuscinosis.[4,12,13]

References

1. Schulz A, Kohlschütter A, Mink J, Simonati A, Williams R. NCL diseases—clinical perspectives. *Biochim Biophys Acta* 2013;**1832**: 1801–6.
2. Mole SE, Cotman SL. Genetics of the neuronal ceroid lipofuscinoses (Batten disease). *Biochim Biophys Acta* 2015;**1852**:2237–41.
3. Williams RE, Mole SE. New nomenclature and classification scheme for the neuronal ceroid lipofuscinoses. *Neurology* 2012;**79**:183–91.
4. Mole SE, Anderson G, Band HA, Berkovic SF, Cooper JD, Kleine Holthaus SM, et al. Clinical challenges and future therapeutic approaches for neuronal ceroid lipofuscinosis. *Lancet Neurol* 2019;**18**(1):107–16.
5. Naseri N, Sharma M, Velinov M. Autosomal dominant neuronal ceroid lipofuscinosis: clinical features and molecular basis. *Clin Genet* 2020;**12**.
6. Wright GA, Georgiou M, Robson AG, Ali N, Kalhoro A, Holthaus SK, et al. Juvenile batten disease (CLN3): detailed ocular phenotype, novel observations, delayed diagnosis, masquerades, and prospects for therapy. *Ophthalmol Retina* 2020;**4**(4):433–45.
7. Bozorg S, Ramirez-Montealegre D, Chung M, Pearce D. Juvenile neuronal ceroid lipofuscinosis (JNCL) and the eye. *Surv Ophthalmol* 2009;**54**(4):463–71.
8. Preising MN, Abura M, Jäger M, Wassill KH, Lorenz B. Ocular morphology and function in juvenile neuronal ceroid lipofuscinosis (CLN3) in the first decade of life. *Ophthalmic Genet* 2017;**38**(3):252–9.
9. Ku CA, Hull S, Arno G, Vincent A, Carss K, Kayton R, et al. Detailed clinical phenotype and molecular genetic findings in CLN3-associated isolated retinal degeneration. *JAMA Ophthalmol* 2017;**135**(7):749–60.
10. Chen FK, Zhang X, Eintracht J, Zhang D, Arunachalam S, Thompson JA, et al. Clinical and molecular characterization of non-syndromic retinal dystrophy due to c.175G > A mutation in ceroid lipofuscinosis neuronal 3 (CLN3). *Doc Ophthalmol* 2019;**138**(1):55–70.
11. Wang F, Wang H, Tuan HF, Nguyen DH, Sun V, Keser V, et al. Next generation sequencing-based molecular diagnosis of retinitis pigmentosa: identification of a novel genotype-phenotype correlation and clinical refinements. *Hum Genet* 2014;**133**(3):331–45.
12. Johnson TB, Cain JT, White KA, Ramirez-Montealegre D, Pearce DA, Weimer JM. Therapeutic landscape for batten disease: current treatments and future prospects. *Nat Rev Neurol* 2019;**15**(3):161–78.
13. Schulz A, Ajayi T, Specchio N, et al. Study of intraventricular Cerliponase Alfa for CLN2 disease. *N Engl J Med* 2018;**378**:1898–907.

13B.2.7

Cobalamin C deficiency

Tomas S. Aleman, Bart P. Leroy, and Giacomo M. Bacci

Cobalamin deficiency is a group of multisystemic disorders associated with abnormal intake of vitamin B12 (cobalamin) in cells. Multiple disease subtypes have been described based on the primary metabolic site of dysfunction. Cobalamin C deficiency is the most common form. It is associated with developmental and neurological complications, haematological abnormalities and renal failure. Various degrees of ophthalmic involvement have been described ranging from normal vision to severely impaired visual function due to maculopathy, retinopathy and/or optic neuropathy.

Cobalamin C deficiency is caused by biallelic mutations in the *MMACHC* gene.[1] Although the disease-implicated gene is known, our understanding of disease pathophysiology remains incomplete. Notably, interventions focused on improving biochemical parameters are generally not sufficient to prevent organ damage.[2]

Clinical characteristics

Cobalamin C deficiency can be challenging to diagnose clinically due to its heterogeneity. The severity of presentation can vary considerably, ranging from severe, infantile-onset forms to late-presenting, relatively mild subtypes. In general, patients presenting in the first year of life are defined as early-onset, whereas those exhibiting clinical signs in childhood or later are defined as late onset. The latter form is less prevalent and less likely to be associated with ophthalmic abnormalities.[2]

Infants with the early-onset cobalamin C deficiency have a multisystemic disorder with ophthalmological, neurological, haematological, renal, gastrointestinal and cardiopulmonary manifestations. Eye involvement is variable and may include nystagmus, strabismus, retinopathy with early macular affection and optic atrophy.[3] As mentioned above, the ophthalmic disease tends to be mild, subclinical or absent in individuals with late onset forms.[4]

Retinal abnormalities can be observed even a few days after birth.[5] Initially, only subtle macular RPE changes may be noted, but by 6–8 months of life, the diagnosis of maculopathy is evident and a 'bull's eye' pattern and/or central atrophic lesions (with associated central photoreceptor and ganglion cell loss) can be seen (Fig. 13B.31).[6] In the peripheral retina, pigmentary changes are usually not observed early on. They may however be noticed later in the course of disease even if good metabolic control is achieved (Fig. 13B.32). Electroretinographic studies are in keeping with rod-cone dysfunction but the traces can be remarkably preserved in some patients despite severe degeneration.[3,4]

Neurological features are often severe and include hypotonia, developmental delay, microcephaly and epilepsy. Megaloblastic anaemia, hypersegmented neutrophils, thrombocytopenia and/or pancytopenia are the most common haematological findings. Haemolytic uremic syndrome (progressive renal failure associated with microangiopathic anaemia and thrombocytopenia) may occur.[2]

Molecular pathology

Cobalamin C deficiency is inherited as an autosomal recessive trait and is caused by mutations in the *MMACHC* gene. The encoded protein plays various roles in cells including acting as a trafficking chaperone for cobalamin intermediates and as a catalytic enzyme for B12 metabolism.[7]

Cobalamin is generally converted into two active forms: methylcobalamin and adenosylcobalamin. The former is required as a coenzyme for methionine synthase to convert homocysteine to methionine. The latter is necessary for the activity of the mitochondrial enzyme methylmalonyl-CoA mutase that converts L-methylmalonyl-CoA to succinyl-CoA. As a result, abnormalities in *MMACHC* cause decreased methionine synthesis and accumulation of homocysteine and methylmalonic acid. Hypomethioninemia leads to depletion of glutathione in the cellular environment. This could contribute to oxidative stress injury to the RPE. It is well known that glutathione is one of the principal antioxidant agents in cellular protection against reactive oxygen species and patients with cobalamin C deficiency have been shown to have severely abnormal glutathione levels.[8] It has been speculated that changes in oxidative status may be linked to worsening of clinical signs despite appropriate systemic treatment. This could be particularly relevant in developing and specialised tissues like the macula during the

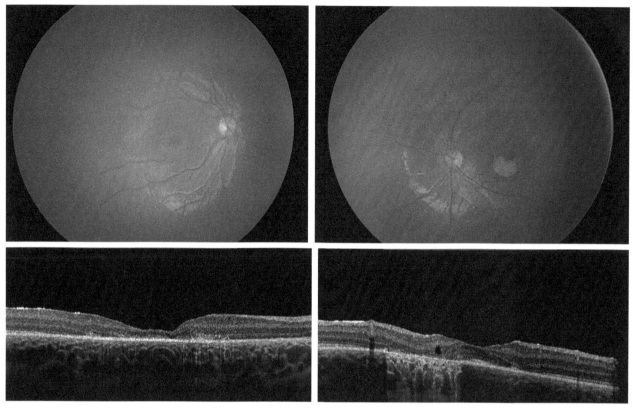

FIG. 13B.31 Widefield fundus photographs and foveal OCT scans in an individual with cobalamin C deficiency. A 'bull's eye' pattern of atrophy is noted and OCT reveals thinning of inner and outer retinal layers, intraretinal cystic spaces and irregularities of the ellipsoid zone.

first months of life. Despite appropriate methionine supplementation and adequate methionine plasma levels, patients with early-onset cobalamin C deficiency continue to have progressive maculopathy demonstrating that the mechanism is probably more complex. Intriguingly, individuals with cobalamin F deficiency, another disorder of intracellular intake of cobalamin, do not show signs of ophthalmic involvement despite hypomethioninemia early in life.[9]

Over 100 mutations have been reported in *MMACHC*. The c.271dupA (p.Arg91fs) and c.331C>T (p.Arg111Ter) variants are among the most common causes of the early-onset cobalamin C deficiency phenotype. Other, less prevalent mutations such as c.394C>T (p.Arg132Ter) are associated with the late onset presentation with milder manifestations and only subtle or no retinal alterations.[10] Although the exact molecular pathology has not yet been elucidated, variability in the ophthalmic phenotype may be related to the biochemical consequences of the different mutations.

Clinical management

The diagnosis of cobalamin C deficiency is typically made by paediatricians with the help of metabolic specialists. Clinical examination, biochemical analysis (including total and free plasma homocysteine as well as plasma and urine levels of methylmalonic acid, methionine, folate and vitamin B12) and molecular genetic analysis (screening of the *MMACHC* gene) are essential in the diagnostic pathway.[11]

Management of cobalamin C deficiency, like other metabolic diseases, requires a multidisciplinary approach. This is key to obtaining the best outcome in terms of function and quality of life. Although ophthalmic manifestations lack strict correlation to metabolic status, they may be mitigated by prenatal or early treatment.[3] There is wide variation in treatment regimens between different centres but these typically include parenteral (e.g. intramuscular) hydroxocobalamin and, variably, oral betaine and oral methionine.[11]

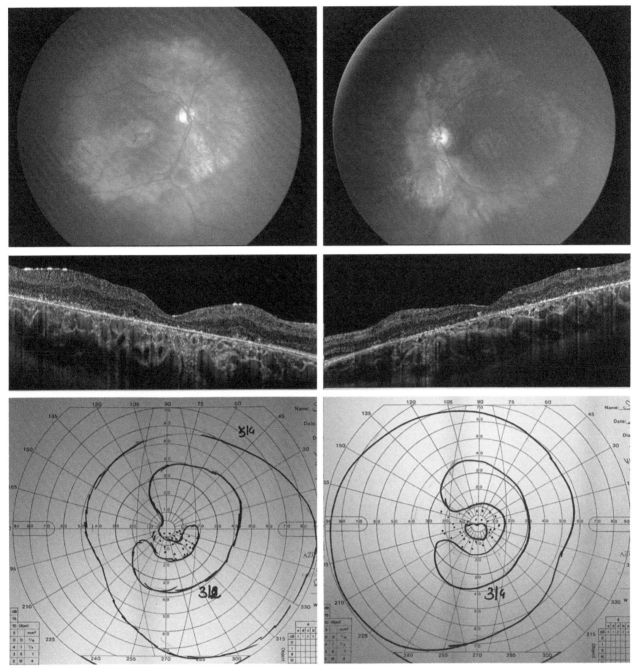

FIG. 13B.32 Widefield fundus photographs, foveal OCT scans and visual field test result in an individual with cobalamin C deficiency. Diffuse retinal pigmentary changes involving the macular area are noted. There is annular bone-spicule pigmentation along the main vascular arcades and relative sparing of the retinal periphery. OCT reveals loss of the ellipsoid zone and diffuse thinning of both the inner and outer retinal layers. Perimetry testing reveals scotomata corresponding to the atrophic areas on fundoscopy.

References

1. Lerner-Ellis JP, Tirone JC, Pawelek PD, Doré C, Atkinson JL, Watkins D, et al. Identification of the gene responsible for methylmalonic aciduria and homocystinuria, cblC type. *Nat Genet* 2006;**38**(1): 93–100. https://doi.org/10.1038/ng1683.
2. Martinelli D, Deodato F, Dionisi-Vici C. Cobalamin C defect: natural history, pathophysiology, and treatment. *J Inherit Metab Dis* 2011; **34**(1):127–35. https://doi.org/10.1007/s10545-010-9161-z.
3. Brooks BP, Thompson AH, Sloan JL, Manoli I, Carrillo-Carrasco N, Zein WM, et al. Ophthalmic manifestations and long-term visual

outcomes in patients with cobalamin C deficiency. *Ophthalmology* 2016;**123**(3):571–82.
4. Weisfeld-Adams JD, McCourt EA, Diaz GA, Oliver SC. Ocular disease in the cobalamin C defect: a review of the literature and a suggested framework for clinical surveillance. *Mol Genet Metab* 2015;**114**(4):537–46.
5. Bacci GM, Donati MA, Pasquini E, Munier F, Cavicchi C, Morrone A, et al. Optical coherence tomography morphology and evolution in cblC disease-related maculopathy in a case series of very young patients. *Acta Ophthalmol* 2017;**95**(8):e776–82.
6. Bonafede L, Ficicioglu CH, Serrano L, Han G, Morgan JI, Mills MD, et al. Cobalamin C deficiency shows a rapidly progressing maculopathy with severe photoreceptor and ganglion cell loss. *Invest Ophthalmol Vis Sci* 2015;**56**:7875–87.
7. Kim J, Hannibal L, Gherasim C, Jacobsen DW, Banerjee R. A human vitamin B12 trafficking protein uses glutathione transferase activity for processing alkylcobalamins. *J Biol Chem* 2009;**284**(48):33418–24.
8. Pastore A, Martinelli D, Piemonte F, Tozzi G, Boenzi S, Di Giovamberardino G, et al. Glutathione metabolism in cobalamin deficiency type C (cblC). *J Inherit Metab Dis* 2014;**37**(1):125–9.
9. Alfadhel M, Lillquist YP, Davis C, Junker AK, Stockler-Ipsiroglu S. -Eighteen-year follow-up of a patient with cobalamin F disease (cblF): report and review. *Am J Med Genet A* 2011;**155A**(10):2571–7.
10. Morel CF, Lerner-Ellis JP, Rosenblatt DS. Combined methylmalonic aciduria and homocystinuria (cblC): phenotype-genotype correlations and ethnic-specific observations. *Mol Genet Metab* 2006;**88**(4): 315–21.
11. Huemer M, Diodato D, Schwahn B, Schiff M, Bandeira A, Benoist JF, et al. Guidelines for diagnosis and management of the cobalamin-related remethylation disorders cblC, cblD, cblE, cblF, cblG, cblJ and MTHFR deficiency. *J Inherit Metab Dis* 2017;**40**(1):21–48.

13B.2.8

Cohen syndrome

Kate E. Chandler

Cohen syndrome is an autosomal recessive multisystemic disorder characterised by intellectual disability in conjunction with two of the following three key clinical features: characteristic facial dysmorphism, progressive retinal degeneration, and intermittent neutropenia. Less specific clinical features include microcephaly, short stature, truncal obesity with slender extremities and joint hyperextensibility. The degree of intellectual disability varies from moderate to severe, but special educational schooling is usually required. An overfriendly disposition is typically described and autistic spectrum behavioural traits are frequently reported. Visual problems are significant in Cohen syndrome and include high myopia and progressive retinopathy; by adult life, most individuals are registered as visually impaired. There is an increased risk of diabetes in adult patients but general health is otherwise good and lifespan is normal. Cohen syndrome is caused by biallelic variants in the VPS13B gene.

Clinical characteristics

It is important to consider the diagnosis of Cohen syndrome in children with microcephaly who present with hypotonia, neutropenia, and global developmental delay. Ophthalmologists should consider this diagnosis in young children with developmental delay, high myopia, night vision problems, and pigmentary retinopathy. The distinctive facial dysmorphism is an important clue.[1]

Ophthalmic abnormalities are a prominent feature of Cohen syndrome. Congenital anomalies of the eye are a rare manifestation and include microphthalmia, microcornea, and colobomata.[2–4] Most affected individuals present with early-onset myopia and progressive retinopathy.[2] Myopia starts in the preschool years and progresses during childhood to high-grade myopia, with a median refractive error of −11.0 diopters. Retinal pigmentary changes (peripheral and/or central in a 'bull's eye' maculopathy pattern) can be identified in young children.[2] Rod-cone dysfunction is evident on electroretinography by 5 years of age in most affected individuals.[5] By 10 years of age, many patients have widespread pigmentary retinopathy with symptoms of nyctalopia and visual field loss (Fig. 13B.33). The severity and rate of progression of the visual handicap are variable, even within the same family.[2]

Ophthalmic complications resulting in severe visual impairment include macular atrophy, macular oedema, retinal detachment, optic nerve involvement,[6] early-onset cataracts,[7] lens sub-luxation (Fig. 13B.34)[2] and keratoconus.[8,9] By mid-adult life, the visual handicap is usually marked and affected individuals are registered visually impaired.[5]

High myopia in Cohen syndrome is attributed to high corneal and lenticular refractive power in the presence of a normal axial length.[10] Keratometry and biometry measurements often indicate shallow anterior chambers and thick crystalline lenses.[10] Reduced corneal thickness has also been reported.[8]

Children with Cohen syndrome have global developmental delay that is typically evident in the first year of life. Learning difficulties requiring special educational provision are a consistent feature.[11] When formally assessed, intelligence quotients range between 30 and 70. Profound cognitive impairment is reported in up to 20% of cases.[12]

Individuals with Cohen syndrome are generally described as having a sociable and cheerful disposition.[3] However, hyperactivity, attention deficit, pervasive developmental disorder and autistic spectrum behaviour have all been reported.[7,11]

The typical facial appearance in Cohen syndrome is characterised by:

- thick hair, eyebrows, and eyelashes
- wave-shaped downward-slanting palpebral fissures
- prominent convex nose and
- short upturned philtrum (vertical groove between the nose and upper lip) (Figs. 13B.34 and 13B.35).

The facial dysmorphism is less distinct in preschool children making a clinical gestalt diagnosis difficult.[15]

Neutropenia may be detected at birth but tends to be intermittent.[16] It can be associated with a normal total white cell count and it can therefore go undetected if a differential count is not requested. Neutropenia is an isolated finding and no other blood cell lineages are affected. Bone marrow analysis shows normal cellularity[17] and there is no evidence for associated malignant development. Individuals with Cohen syndrome frequently suffer from minor skin and oral infections but life-threatening infections are uncommon.[6,17]

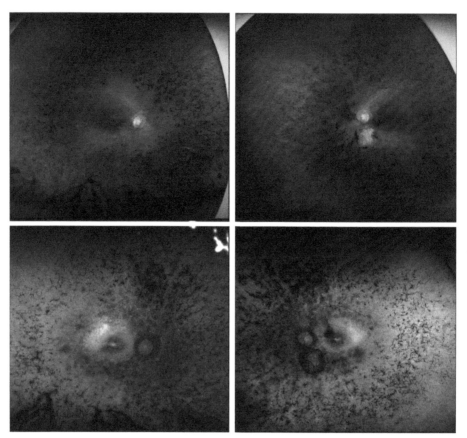

FIG. 13B.33 Widefield retinal imaging in a young adult with Cohen syndrome. There is both central and peripheral retinal involvement with extensive pigmentary changes peripherally, attenuation of retinal vessels and evidence of atrophy and oedema in both maculae.

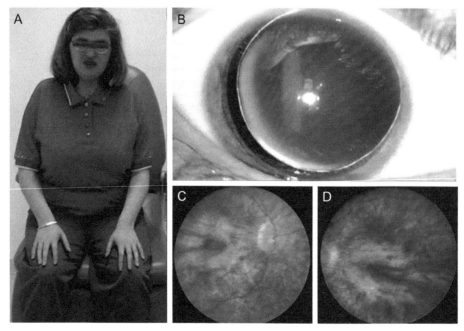

FIG. 13B.34 Findings in a female with Cohen syndrome. (A) typical facial gestalt, truncal obesity, and small/narrow hands with long tapering fingers. (B) Slit-lamp photograph of the same patient showing lens sub-luxation due to high myopia. (C and D) Fundus photographs of the posterior pole of both eyes illustrating atrophic macular changes attenuated retinal vessels and pale optic discs. *(Adapted from Ref. 4.)*

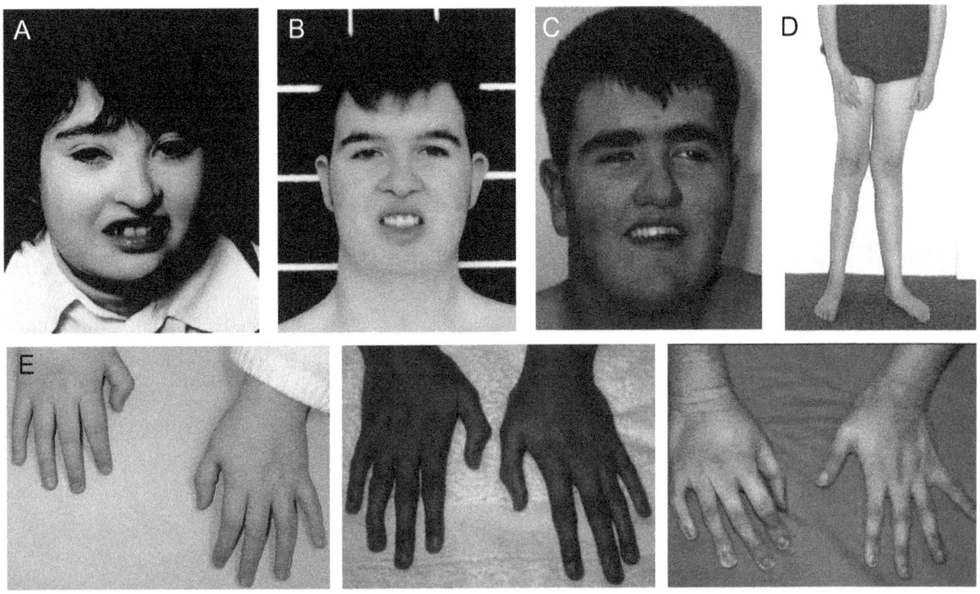

FIG. 13B.35 Extraocular findings in individuals with Cohen syndrome. (A–C) Characteristic facial appearance including thick eyebrows in three unrelated individuals. (D) Slender extremities with truncal obesity. (E) Hands showing long tapering fingers. *(A and B: Adapted from Ref. 13; C–E: Adapted from Ref. 14.)*

Birth weight is often low for gestation (3rd centile on average) and failure-to-thrive may occur in infancy because of feeding difficulties.[18] This usually improves over the first months of life, but some patients require nasogastric tube or gastrostomy feeding. Typically, Cohen syndrome children show excess weight gain from the age of 8 years. Although they may display a large appetite and obsession with food, hyperphagia is not usually present. Classically they develop truncal obesity but with slender arms and legs; generalised obesity is uncommon.[19,20] Most individuals with Cohen syndrome are of short stature, with a height less than 0.4 centiles, especially after puberty.[21] Pubertal development is often delayed but progresses normally.

Impaired glucose tolerance and type 2 diabetes are reported in several adults with Cohen syndrome. These appear to be linked to insulin response dysregulation and abnormal patterns of fat distribution (in keeping with truncal obesity with a large waist circumference).[20]

Cardiac abnormalities have been reported in a minority of individuals with Cohen syndrome. Reduced left ventricular function was described in Finnish patients over 40 years of age[19] but has not been reported in other studies. Low levels of high-density lipoprotein (HDL) values with normal low-density lipoprotein levels (LDL), triglyceride, and total cholesterol suggest a possible increased risk of heart disease and metabolic syndrome in adults with Cohen syndrome.[20]

Generalised joint laxity is common and joint dislocation may occur.[21] Kyphoscoliosis and pectus carinatum are reported in older patients.[7,19]

In association with neutropenia, individuals with Cohen syndrome have altered periodontal bacterial flora resulting in increased susceptibility to periodontitis and premature tooth loss.[22]

In summary, individuals with molecularly confirmed Cohen syndrome typically have 6 of the following 8 diagnostic criteria[14]:

- developmental delay
- microcephaly
- typical facial gestalt
- truncal obesity and slender extremities
- overly sociable behaviour
- joint hypermobility
- high myopia and/or retinal dystrophy (by age 5 years)
- intermittent neutropenia

Molecular pathology

Cohen syndrome is an autosomal recessive disorder caused by mutations in the *VPS13B* gene.[13] The exact function of *VPS13B* is unknown but it is thought to play a role in the intracellular transport of proteins.[23] It may also be involved in post-translational modification by glycosylation of newly synthesised proteins. Notably, Cohen syndrome patients have been shown to have an unusual pattern of glycosylation of serum proteins.[24] More than 200 *VPS13B* mutations have been reported, predominantly nonsense or frameshift variants.[6,13] Missense mutations occur, often in association with a milder phenotype.[25] There does not appear to be a mutational hotspot in *VPS13B*,[26] and many mutations are family-specific. Founder mutations have been identified in Finnish, Amish, Irish communities and in Greek islands.[13,27–29] Intragenic copy number variants are identified in up to a third of cases and comprehensive genetic analysis should include targeted testing for these changes (including appropriate informatics analysis of high-throughput sequencing data and/or microarray analysis).[30] Overall, *VPS13B* mutations and deletions are detected in around 90% of affected individuals.[30]

Clinical management

Diagnostic testing: When considering the diagnosis of Cohen syndrome, the following should be arranged:

- detailed assessment by a clinical dysmorphologist to delineate the typical dysmorphic features, body habitus, growth parameters, and behavioural pattern,
- differential white cell count to look for neutropenia. As this is an intermittent feature of the condition, it may not be present at the time of testing. Repeated blood count evaluation is not indicated but a review of past results may reveal a previously low neutrophil count.
- full ophthalmic assessment by a paediatric ophthalmologist, including full-field electroretinography to look for evidence of a panretinal dysfunction. Electrophysiological assessment should be repeated if the diagnosis is suspected but the examination of an infant or young child is unremarkable.

Definitive diagnosis of Cohen syndrome can be made by identification of biallelic mutations and/or copy number variants in *VPS13B* on molecular genetic testing.

Visual impairment: Early detection and assessment of a child's visual problems play an important role in preparing the family for their child's visual prognosis, and provide invaluable information around which they can plan for the child's special needs. Regular monitoring of vision is needed to identify correctable complications, such as glaucoma, cataracts, and lens sub-luxation, as well as to advise about visual deterioration. When appropriate, the individual should be registered as visually impaired and information should be given on low vision aids.

Intellectual disability: All children with Cohen syndrome require ongoing assessment for developmental delay and learning difficulties by a multidisciplinary team. This should include a community paediatrician, a speech and language therapist, a physiotherapist, an occupational therapist, and an educational psychologist. Neuropsychological assessments can help clarify the level of learning difficulties and highlight the presence of specific behavioural traits. Individual educational programs focusing on the child's specific learning needs and independence skills are recommended.

Neutropenia: Treatment interventions to improve the neutrophil count are rarely required, as individuals with Cohen syndrome are usually well, despite their neutropenia. In individuals who have recurrent and severe infections, repeated infusions of granulocyte colony-stimulating factor (G-CSF) in combination with prophylactic antibiotics have been used to good effect. Referral to a haematologist for management supervision is recommended for these individuals.

Endocrine issues: Annual screening for diabetes and thyroid disorders should be carried out from puberty.

Cardiovascular issues: Cardiac assessment including an echocardiogram to assess left ventricular function is recommended in adults over the age of 40 years. Annual monitoring of blood pressure and lipid levels, in particular HDL, is recommended in adults.

Musculoskeletal issues: Assessment of the degree of joint hyperextensibility should be carried out and those with joint sub-luxation or dislocation should be referred for specialist assessment to a rheumatologist or orthopaedic surgeon. Physiotherapy and exercises, such as swimming, to promote joint stability are advisable and weight control should be encouraged.

Dental issues: Good oral hygiene and regular dental assessment are recommended to prevent gingivitis and premature tooth loss.

Family management and counselling: Cohen syndrome follows an autosomal recessive inheritance pattern. Parents of an affected child are heterozygote mutation carriers and are asymptomatic. They have a 25% risk of having an affected child with each subsequent pregnancy. Carrier testing and prenatal diagnosis are possible where pathogenic *VPS13B* mutations have been identified.

References

1. Rodrigues JM, Fernandes HD, Caruthers C, Braddock SR, Knutsen AP. Cohen syndrome: review of the literature. *Cureus* 2018;**10**(9): e3330. https://doi.org/10.7759/cureus.3330.
2. Chandler KE, Biswas S, Lloyd IC, Parry N, Clayton-Smith J, Black GCM. The ophthalmic findings in Cohen syndrome. *Br J Ophthalmol* 2002;**86**:1395–8.
3. Cohen Jr MM, Hall BD, Smith DW, Graham CB, Lampert KJ. A new syndrome with hypotonia, obesity, mental deficiency and facial, oral, ocular and limb anomalies. *J Pediatr* 1973;**83**:280–4.
4. Taban M, Memoracion-Peralta DS, Wang H, Al-Gazali LI, Traboulsi EI. Cohen syndrome: report of nine cases and review of the literature, with emphasis on ophthalmic features. *J AAPOS* 2007;**11**(5):431–7.
5. Kivitie-Kallio S, Summanen P, Raitta C, Norio R. Ophthalmologic findings in Cohen syndrome—a long term follow up. *Ophthalmology* 2000;**107**:1737–45.
6. Hennies HC, Rauch A, Seifert W, Schumi C, Moser E, Al-Taji E, et al. Allelic heterogeneity in the *COH1* gene explains clinical variability in Cohen syndrome. *Am J Hum Genet* 2004;**75**:138–45.
7. Douzgou S, Petersen MB. Clinical variability of genetic isolates of Cohen syndrome. *Clin Genet* 2011;**79**(6):501–6.
8. Douzgou S, Samples JR, Georgoudi N, Petersen MB. Ophthalmic findings in the Greek isolate of Cohen syndrome. *Am J Med Genet A* 2011;**155A**(3):534–9.

9. Khan A, Chandler K, Pimenides D, Black GC, Manson FD. Corneal ectasia associated with Cohen syndrome: a role for COH1 in corneal development and maintenance? *Br J Ophthalmol* 2006;**90**:390–1.
10. Summanen P, Kivitie-Kallio S, Norio R, Raitta C, Kivela T. Mechanism of myopia in Cohen syndrome mapped to chromosome 8q22. *Invest Ophthalmol Vis Sci* 2002;**43**:1686–93.
11. Chandler KE, Moffett M, Clayton-Smith J, Baker GA. Neuropsychological assessment of a group of UK patients with Cohen syndrome. *Neuropediatrics* 2003;**34**:7–13.
12. Kivitie-Kallio S, Larsen A, Kajasto K, Norio R. Neurological and psychological findings in patients with Cohen syndrome: a study of 18 patients aged 11 months to 57 years. *Neuropediatrics* 1999;**30**:181–9.
13. Kolehmainen J, Black GC, Saarinen A, Chandler K, Clayton-Smith J, Traskelin AL, et al. Cohen syndrome is caused by mutations in a novel gene, *COH1*, encoding a transmembrane protein with a presumed role in vesicle-mediated sorting and intracellular protein transport. *Am J Hum Genet* 2003;**72**:1359–69.
14. Kolehmainen J, Wilkinson R, Lehesjoki AE, Chandler K, Kivitie-Kallio S, Clayton-Smith J, et al. Delineation of Cohen syndrome following a large-scale genotype-phenotype screen. *Am J Hum Genet* 2004;**75**(1):122–7.
15. El Chehadeh-Djebbar S, Blair E, Holder-Espinasse M, Moncla A, Frances AM, Rio M, et al. Changing facial phenotype in Cohen syndrome: towards clues for an earlier diagnosis. *Eur J Hum Genet* 2013;**21**(7):736–42.
16. Fryns JP, Legius E, Devriendt K, Meire F, Standaert L, Baten E, et al. Cohen syndrome: the clinical symptoms and stigmata at a young age. *Clin Genet* 1996;**49**:237–41.
17. Olivieri O, Lombardi S, Russo C, Corrocher R. Increased neutrophil adhesive capability in Cohen syndrome, an autosomal recessive disorder associated with granulocytopenia. *Haematologica* 1998;**83**:778–82.
18. Kivitie-Kallio S, Norio R. Cohen syndrome: essential features, natural history, and heterogeneity. *Am J Med Genet* 2001;**102**:125–35.
19. Kivitie-Kallio S, Eronen M, Lipsanen-Nyman M, Marttinen E, Norio R. Cohen syndrome: evaluation of its cardiac, endocrine and radiological features. *Clin Genet* 1999;**56**:41–50.
20. Limoge F, Faivre L, Gautier T, Petit JM, Gautier E, Masson D, et al. Insulin response dysregulation explains abnormal fat storage and increased risk of diabetes mellitus type 2 in Cohen syndrome. *Hum Mol Genet* 2015;**24**(23):6603–13.
21. Chandler KE, Kidd A, Al-Gazali L, Black GCM, Clayton-Smith J. Diagnostic criteria, clinical characteristics and natural history of Cohen syndrome. *J Med Genet* 2003;**40**:233–41.
22. Alaluusua S, Kivitie-Kallio S, Wolf J, Haavio M-L, Asikainen PS. Periodontal findings in Cohen syndrome with chronic neutropenia. *J Periodontol* 1997;**68**:473–8.
23. Velayos-Baeza A, Vettori A, Copley RR, Dobson-Stone C, Monaco AP. Analysis of the human VPS13 gene family. *Genomics* 2004;**84**(3):536–49.
24. Duplomb L, Duvet S, Picot D, Jego G, El Chehadeh-Djebbar S, Marle N, et al. Cohen syndrome is associated with major glycosylation defects. *Hum Mol Genet* 2014;**23**(9):2391–9.
25. Gueneau L, Duplomb L, Sarda P, Hamel C, Aral B, Chehadeh SE, et al. Congenital neutropenia with retinopathy, a new phenotype without intellectual deficiency or obesity secondary to VPS13B mutations. *Am J Med Genet A* 2014;**164A**(2):522–7.
26. Seifert W, Holder-Espinasse M, Kühnisch J, Kahrizi K, Tzschach A, Garshasbi M, et al. Expanded mutational spectrum in Cohen syndrome, tissue expression, and transcript variants of COH1. *Hum Mutat* 2009;**30**(2):E404–20.
27. Bugiani M, Gyftodimou Y, Tsimpouka P, Lamantea E, Katzaki E, d'Adamo P, et al. Cohen syndrome resulting from a novel large intragenic COH1 deletion segregating in an isolated Greek island population. *Am J Med Genet A* 2008;**146A**(17):2221–6.
28. Falk MJ, Feiler HS, Neilson DE, Maxwell K, Lee JV, Segall SK, et al. Cohen syndrome in the Ohio Amish. *Am J Med Genet A* 2004;**128**:23–8.
29. Murphy AM, Flanagan O, Dunne K, Lynch SA. High prevalence of Cohen syndrome amongst Irish travellers. *Clin Dysmorphol* 2007;**16**:257–9.
30. Balikova I, Lehesjoki AE, de Ravel TJ, Thienpont B, Chandler KE, Clayton-Smith J, et al. Deletions in the *VPS13B* (*COH1*) gene as a cause of Cohen syndrome. *Hum Mutat* 2009;**30**:E845–54.

Chapter 14

Familial vitreoretinopathies

Chapter outline

14.1 Familial exudative vitreoretinopathy
    spectrum   324
    14.1.1 Familial exudative
            vitreoretinopathy   324
    14.1.2 Norrie Disease   329
    14.1.3 *KIF11*-related disorders   331

14.2 Incontinentia pigmenti   334
14.3 Stickler syndrome and allied collagen
    vitreoretinopathies   339
14.4 Wagner disease   347
14.5 Knobloch syndrome   351

The development of the vitreous can be divided into three stages: primary, secondary and tertiary. The primary vitreous is formed by the hyaloid vascular system, a vascular network of blood vessels supporting the lens and the avascular inner retina of the developing eye. This system spontaneously regresses at around 14 weeks gestation once the lens no longer requires oxygen and around the appearance of the first retinal blood vessels. Failure of this regression results in persistent fetal vasculature, which is generally a sporadic and unilateral condition. The secondary vitreous starts developing at around 9 weeks of gestation; it displaces the primary vitreous and becomes the adult vitreous gel. The tertiary vitreous forms from surface neuroectoderm and develops into the lens zonules.

The process of normal retinal vascular development occurs from 18 to 38 weeks gestation. It is controlled by key developmental molecules (e.g. VEGF) and pathways (including Wnt signalling). Defective peripheral vascular development and retinal avascularity are seen classically in retinopathy of prematurity where the resultant pre-retinal neovascularization, vitreoretinal fibrosis and traction can lead to retinal detachment and visual impairment. Developmental peripheral retinal non-perfusion is also characteristic of a range of genetic conditions including familial exudative vitreoretinopathy (FEVR) spectrum (Section 14.1) and incontinentia pigmenti (Section 14.2). The former group of conditions includes classic FEVR (Section 14.1.1), severe vitreoretinal dysplasia phenotypes such as Norrie disease (Section 14.1.2), *KIF11*-related disorders (Section 14.1.3), osteoporosis-pseudoglioma (OPPG) syndrome (Section 14.1.1), and a subset of the complex microphthalmia disorders (Chapter 16).

The adult vitreous cortex is virtually acellular and is composed of a highly hydrated extracellular matrix. Its network of macromolecules includes glycosaminoglycans (GAGs; predominantly hyaluronan) and fibrillar proteins, of which the main group is collagen. Nearly all of the collagen in the vitreous is found in heterotypic fibrils containing collagen types II, V, IX and XI, the proteins mutated in Stickler syndrome (Section 14.3). Another vitreous glycosaminoglycan, chondroitin sulphate, forms the proteoglycan versican, the molecule mutated in Wagner disease (Section 14.4). At the vitreoretinal interface, the extracellular composition of the vitreous changes and the inner limiting membrane comprises a different range of collagen subtypes that includes type XVIII, or endostatin, encoded by the gene mutated in Knobloch syndrome (Section 14.5).

14.1

Familial exudative vitreoretinopathy spectrum

14.1.1

Familial exudative vitreoretinopathy

Johane Robitaille

Familial exudative vitreoretinopathy (FEVR) is a group of developmental disorders characterised by abnormal retinal angiogenesis leading to incomplete vascularisation of the peripheral retina. FEVR was initially described in the 1960s as a familial entity that resembled retinopathy of prematurity in individuals who were born full-term; it has since been shown to follow autosomal dominant, X-linked recessive or autosomal recessive inheritance patterns.[1,2] FEVR can be part of a multisystemic disorder, for example in Norrie disease (a FEVR subtype associated with defects in the NDP gene; discussed in Section 14.1.2) or in osteoporosis-pseudoglioma syndrome (a condition associated with biallelic LRP5 gene mutations).

FEVR exhibits significant variability in phenotypic expression (Table 14.1) but penetrance approaches 100% when intravenous fluorescein angiography is used to visualise the peripheral vascular deficits in cases that appear normal on fundoscopy. Complications frequently result in vision loss, and early detection is key in optimising visual outcomes. Genetic testing often facilitates timely diagnosis although it is worth noting that genomic analysis fails to identify the molecular defect in many clinically diagnosed FEVR cases.[4–6]

Clinical characteristics

FEVR has highly heterogeneous clinical presentation and exhibits significant variability even amongst affected members of the same family. An asymmetric appearance between the eyes of a proband is also common.[4]

At the most severe end of the spectrum, a small number of children present with bilateral severe congenital blindness resulting from failure of retinal vascularisation and congenital retinal detachments; this has variously been described as 'retinal dysplasia', 'pseudoglioma', and 'congenital retinal non-attachment'. A subset of patients may present with leukocoria, associated with a white fibrous retrolental mass and a shallow anterior chamber.

On the milder end of the spectrum, affected individuals may be asymptomatic with good visual function and only subtle signs of peripheral retinal avascularity on widefield fluorescein angiography; it is thought that patients in this category represent at least half of FEVR cases. Although many of these individuals remain stable over time, complications leading to vision loss may arise from secondary subretinal/intraretinal exudation and fibrovascular proliferation that result in temporal dragging of the retina and tractional folds/detachment (Fig. 14.1). Furthermore, FEVR patients are at increased risk of developing rhegmatogenous retinal detachments. These tend to occur in the second or third decade of life, in contrast to tractional detachments which are typically present in the first years of life. Overall, retinal detachment is thought to occur in 20%–30% of eyes.

TABLE 14.1 Familial exudative vitreoretinopathy clinical staging system as per Ref. 3

Stage	Clinical feature
1	Avascular peripheral retina
1A	without exudate or leakage
1B	with exudate or leakage
2	Retinal neovascularisation
2A	without exudate or leakage
2B	with exudate or leakage
3	Extra-macular retinal detachment
3A	without exudate or leakage
3B	with exudate or leakage
4	Macula-involving retinal detachment
4A	without exudate or leakage
4B	with exudate or leakage
5	Total retinal detachment

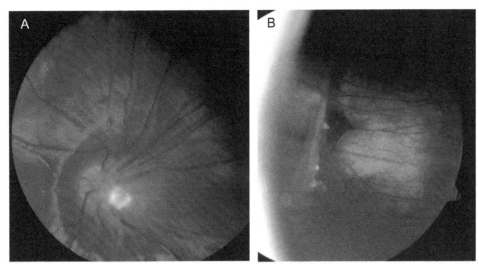

FIG. 14.1 Fundus imaging in two siblings who have familial exudative vitreoretinopathy (FEVR) and carry compound heterozygous mutations in *LRP5*. (A) Images from the older, more severely affected sibling reveal macular dragging with narrowing of the temporal arcades and glial tissue overlying the optic nerve. (B) The younger sibling had areas of fibrous proliferation temporally with abnormal retinal appearance beyond the proliferation. *(Courtesy of Dr. Paul Shuckett and Dr. Ravi Dookeran.)*

FEVR may be isolated or in association with a range of systemic manifestations that are important to identify when present. The most frequent of these is osteoporosis which may occur in individuals carrying heterozygous or biallelic *LRP5* variants.[7] Notably, biallelic mutations in this gene can also cause osteoporosis-pseudoglioma syndrome,[8] a condition associated with severe FEVR and infantile- or childhood-onset skeletal abnormalities. Affected individuals often have severe osteoporosis, recurrent fractures (including vertebrae), and craniotabes (skull softening). Microcephaly may be present and around 25% of cases have cognitive impairment.

Molecular pathology

FEVR is genetically heterogeneous and pathogenic variants in over 11 genes have been implicated in this condition (including *NDP, LRP5, TSPAN12, FZD4, KIF11, ZNF408, CTNNB1, JAG1, RCBTB1, ATOH7, ILK*). The majority of molecularly confirmed cases carry mutations in *NDP, LRP5, FZD4* or *TSPAN12* (Table 14.2). These four genes are involved in a biological pathway that is highly conserved through evolution and has a role in Wnt signalling (Fig. 14.2). Wnt signalling is one of the key cascades regulating development. It is critical for retinal

TABLE 14.2 Genes associated with familial exudative vitreoretinopathy.

Gene	Mode of inheritance	Frequency	Syndromic associations
FZD4	AD	8%–25%	
LRP5	AD, AR	8%–20%	Osteoporosis-pseudoglioma syndrome; microcephaly
NDP	XLR	5%–6%	Norrie disease
TSPAN12	AD, AR	2%–8%	
KIF11	AD	5%–8%	Microcephaly with or without chorioretinopathy, lymphedema, and/or mental retardation
ZNF408	AD	1%–3%	
ILK	AD	1%–2%	
ATOH7	AR	<1%	Optic nerve aplasia/hypoplasia
CTNNB1	AD	<1%	
JAG1	AD	Rare	
RCBTB1	AD	Rare	

AD, autosomal dominant; *AR*, autosomal recessive; *XLR*, X-linked recessive.

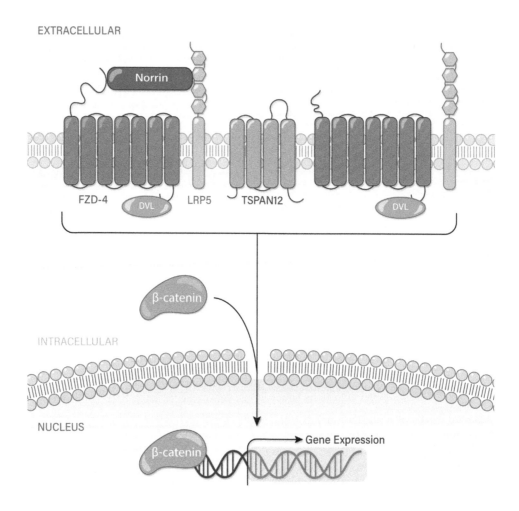

FIG. 14.2 Schematic showing the position of the NDP (norrin), LRP5, FZD4 and TSPAN12 proteins within the Wnt signalling pathway (see text for further details).

angiogenesis via FZD4–NDP signalling. Briefly, *NDP* encodes Norrin, a protein secreted by Müller glia cells in the retina that binds to the transmembrane protein FZD4 and its co-receptor LRP5, both of which are expressed by vascular endothelial cells. TSPAN12 acts as an additional co-receptor that amplifies FZD4-mediated intracellular signalling which ultimately leads to the regulation of retinal angiogenesis.[11]

Despite the successful identification of numerous disease-associated genes, approximately 50% of FEVR pedigrees remain undiagnosed at the molecular level. Furthermore, no clear genotype-phenotype correlation has emerged for the ophthalmic features. For example, FEVR that presents with bilateral retinal detachments in infancy may be due to defects in several genes including *ATOH7*, *CTNNB1*, *NDP*, *FZD4* or *LRP5*.

LRP5 and *FZD4* are the most commonly mutated genes in FEVR cases in most populations. LRP5 encodes a single-pass transmembrane protein that is a member of the low-density lipoprotein receptor family. As mentioned above, mutations in *LRP5* have been associated with a spectrum of phenotypes ranging from osteoporosis-pseudoglioma syndrome (biallelic variants) to isolated forms of FEVR (biallelic or heterozygous variants). In general:

- Homozygous or compound heterozygous *LRP5* mutations that typically result in complete loss of protein function cause osteoporosis-pseudoglioma syndrome.
- Autosomal recessive forms of FEVR are typically associated with missense variants that are likely to be hypomorphic; these generally cluster in the extracellular epidermal growth factor-like domains 1–3.
- A range of *LRP5* mutations have been associated with autosomal dominant FEVR; these include presumed loss of function alleles, suggesting that FEVR represents a phenotype associated with haploinsufficiency (i.e. the presence of a single functional copy of a gene is insufficient to maintain normal function). Thus, careful examination of carrier parents of children with autosomal recessive FEVR is necessary.

FZD4, the transmembrane protein receptor for Norrin, has a highly conserved N-terminal cysteine-rich domain, and a C-terminal intracellular domain. Mutations cause autosomal dominant forms of FEVR and many are clustered in these C- and N-terminal domains. Overall, more than 50% of the mutations are missense variants, with the remainder being nonsense and insertion/deletion mutations.[12]

Clinical management

Careful dilated fundoscopy and widefield fluorescein angiography are key. Examination under anaesthetic is often required in young children for diagnosis and imaging acquisition.[9] Complications resulting in visual loss can evolve over a period of weeks at any time, particularly in the first decade. The high risk of retinal detachment warrants consideration to early preventive treatment. There are no specific surveillance guidelines and the frequency of examinations generally depends on patient age and clinical severity/stage (Table 14.1).

Infants at risk who await confirmation of their genetic status (or those from pedigrees in which no FEVR gene mutation could be identified in a first-degree affected relative) should be examined as soon as possible after birth and then monthly in the first year of life. This should continue until there is either clarification of the diagnosis through genetic testing or examination under anaesthetic with fluorescein angiography. Intravenous fluorescein angiography is necessary to detect and delineate areas of the avascular retina. Even the experienced eye can easily miss such areas on fundoscopy alone. Strong consideration of prophylactic laser treatment in infants with areas of avascular retina outlined by angiography is warranted given the unacceptably high risk of vision-threatening complications that may occur rapidly at any time. Intravitreal anti-VEGF injections may also have a role in the treatment of FEVR, particularly as an adjunct therapy to either laser or surgical management.[9]

Diagnostically, individuals with suspected FEVR remain challenging and the list of differentials is long (See Box 14.1). In new cases, past medical history should include information about gestational age at birth, general development and bone fractures. Family history should enquire about relatives with vision loss, retinal detachment, bone fractures and delayed general development. Depending on the appearance and age at presentation, investigations may need to include Toxocara titres and an ultrasound to rule out retinoblastoma.

Genetic testing and/or referral to a clinical geneticist is recommended in new and atypical cases of FEVR as well as

---

BOX 14.1 Differential diagnosis and phenocopies of familial exudative vitreoretinopathy (FEVR)

FEVR has a wide differential diagnosis[9] and there is significant overlap with several conditions including
- Retinopathy of prematurity: a condition associated with premature birth; in contrast to FEVR, subretinal/intraretinal exudation is a rare feature.
- Coats disease: a sporadic, typically unilateral condition predominantly affecting males; in contrast to FEVR, affected individuals rarely have tractional changes or neovascularisation.
- Persistent fetal vasculature: a sporadic, typically unilateral condition associated with retinal folds; unilateral persistent fetal vasculature may underlie asymmetric FEVR and fluorescein angiography may detect avascular areas in the contralateral eye; individuals with bilateral persistent fetal vasculature should be considered as FEVR cases with or without an associated syndromic condition.[10]
- Norrie disease: an X-linked recessive variant of FEVR, often associated with learning difficulties and hearing impairment (Section 14.1.2).
- Incontinentia pigmenti: an X-linked dominant condition with characteristic skin findings (see Section 14.2).
- KIF11-related disorders: a group of autosomal dominant conditions including 'microcephaly with or without chorioretinopathy, lymphedema, or mental retardation' (see Section 14.1.3).
- Adams–Oliver syndrome: an autosomal dominant condition associated with mutations in the *ARHGAP31* gene and characterised by malformations of the limbs and abnormal skin development, particularly on the scalp.
- Cerebroretinal microangiopathy with calcification and cysts, also known as Coats plus syndrome: an autosomal recessive condition associated with mutations in the *CTC1* gene.
- Facioscapulohumeral muscular dystrophy: an autosomal dominant disorder associated with deletions of tandem repeats (termed D4Z4) located on chromosome 4q35 and characterised by muscle weakness and wasting.
- Dystroglycanopathy spectrum disorders caused by biallelic mutations in a number of genes including *POMGNT1* (previously known as muscle–eye–brain disease or Walker–Warburg syndrome).

Since these conditions may have variable phenotypes, a strong index of suspicion of a syndromic diagnosis is recommended for all new cases of FEVR.

in individuals with bilateral persistent fetal vasculature. The significant variability of isolated and syndromic FEVR should be taken into account during genetic counseling. It is noted that several syndromes presenting as FEVR are associated with cognitive delay that may not be apparent on initial diagnosis. Importantly, asymptomatic relatives of affected individuals, especially young children who are at risk of retinal detachment or osteoporosis should be identified and managed accordingly. Families at risk of having affected children should be counselled to alert their treating physician during pregnancy so that an ophthalmologist will examine the newborn as soon as possible after birth is identified.

References

1. Criswick VG, Schepens CL. Familial exudative vitreoretinopathy. *Am J Ophthalmol* 1969;**68**(4):578–94.
2. Gilmour DF. Familial exudative vitreoretinopathy and related retinopathies. *Eye (Lond)* 2015;**29**(1):1–14.
3. Kashani AH, Learned D, Nudleman E, Drenser KA, Capone A, Trese MT. High prevalence of peripheral retinal vascular anomalies in family members of patients with familial exudative vitreoretinopathy. *Ophthalmology* 2014;**121**(1):262–8.
4. Chen C, Sun L, Li S, Huang L, Zhang T, Wang Z, et al. The spectrum of genetic mutations in patients with asymptomatic mild familial exudative vitreoretinopathy. *Exp Eye Res* 2020;**192**:107941.
5. Chen C, Wang Z, Sun L, Huang S, Li S, Zhang A, et al. Next-generation sequencing in the familial exudative vitreoretinopathy-associated rhegmatogenous retinal detachment. *Invest Ophthalmol Vis Sci* 2019;**60**(7):2659–66.
6. Li JK, Li Y, Zhang X, Chen CL, Rao YQ, Fei P, et al. Spectrum of variants in 389 Chinese probands with familial exudative vitreoretinopathy. *Invest Ophthalmol Vis Sci* 2018;**59**(13):5368–81.
7. Downey LM, Bottomley HM, Sheridan E, Ahmed M, Gilmour GF, Inglehearn CF, et al. Reduced bone mineral density and hyaloid vasculature remnants in a consanguineous recessive FEVR family with a mutation in LRP5. *Br J Ophthalmol* 2006;**90**:1163–7.
8. Ai M, Heeger S, Bartels CF, Schelling DK, Osteoporosis-Pseudoglioma Collaborative Group. Clinical and molecular findings in osteoporosis-pseudoglioma syndrome. *Am J Hum Genet* 2005;**77**(5):741–53.
9. Boyce T, et al. *Familial exudative vitreoretinopathy.* American Academy of Ophthalmology, EyeWiki; December 2020. Accessed December 2020 https://eyewiki.aao.org/Familial_Exudative_Vitreoretinopathy_(FEVR.
10. Kartchner JZ, Hartnett ME. Familial exudative vitreoretinopathy presentation as persistent fetal vasculature. *Am J Ophthalmol Case Rep* 2017;**6**:15–7.
11. Wang Z, Liu CH, Huang S, Chen J. Wnt signaling in vascular eye diseases. *Prog Retin Eye Res* 2019;**70**:110–33.
12. Robitaille JM, Zheng B, Wallace K, Beis MJ, Tatlidil C, Yang J, et al. The role of frizzled-4 (*FZD4*) mutations in familial exudative vitreoretinopathy (FEVR) and coats disease. *Br J Ophthalmol* 2011;**95**(4):574–9.

14.1.2

Norrie disease

Johane Robitaille and Graeme C.M. Black

Norrie disease is a rare syndromic form of familial exudative vitreoretinopathy (FEVR) associated with mutations in the *NDP* gene. The condition is inherited as an X-linked recessive trait, making diagnosis and identification of carrier females important. Congenital visual impairment in Norrie disease is combined with progressive hearing loss and, in some cases, developmental delay and/or intellectual disability.

Clinical characteristics

Norrie disease often presents with ophthalmic manifestations in infancy before the development of systemic features (Fig. 14.3). Congenital blindness with bilateral retinal detachments or retinal dysplasia is found in many patients whilst other cases manifest low vision in childhood that tends to evolve to no light perception by adulthood; the term Norrie disease has been classically used to describe these severe disease subtypes. In contrast, a small number of male *NDP* mutation carriers have a milder disease course and can occasionally be asymptomatic and may rarely have normal retinal vasculature on fluorescein angiography.

Female carriers of *NDP* mutations may also have phenotypic manifestations resulting from uneven X-chromosome inactivation patterns. Such individuals should also be carefully examined and managed accordingly. While most female *NDP* carriers have no phenotypic manifestations, abnormalities can range from subtle peripheral avascular areas to retinal folds/detachments. The presence of peripheral retinal vascular changes may not be useful in identifying carrier status due to lack of specificity of these changes; confirmation of carrier status with genetic testing is therefore recommended.[1,2]

Approximately 50% of male patients with Norrie disease manifest progressive loss of cognitive function and psychosis with onset in the first decade of life.[3] Progressive sensorineural hearing loss in the second and third decades occurs in approximately one-third of patients, although in some pedigrees, especially with formal audiological testing, this proportion may be much higher. Extraocular manifestations in carriers have not been systematically studied.

Molecular pathology

Norrie disease is caused by mutations in *NDP*, a gene located on the short arm of the X chromosome.[4] *NDP* encodes norrin, a secreted protein with a cysteine-knot

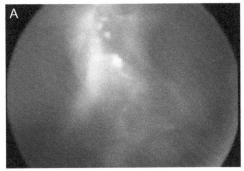 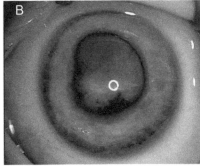

FIG. 14.3 Findings in a child with Norrie disease. (A) Fundus photograph of the right eye showing a falciform fold that incorporates close to 95% of the retina leaving bare choroid over 360 degrees. Genetic testing identified a missense mutation in the *NDP* gene in both the child and his mother. A different missense mutation in the same codon has been reported in a patient with a milder form of FEVR. This illustrates the variability of the Norrie disease spectrum and highlights the prognostic dilemma for other features of the condition that can emerge with time. (B) Anterior segment image of the left eye revealing a fibrotic, totally detached retina that is visible immediately behind a clear lens. *(Courtesy of Dr. Rajeev Muni.)*

motif that activates the Wnt/β-catenin pathway through FZD4 and LRP5 (Fig. 14.2). Nonsense mutations and some missense variants have been linked to Norrie disease whilst milder phenotypes are commonly associated with missense changes.[5,6] A small number of whole gene deletions have been described and these are often associated with severe ophthalmic disease. Deletions encompassing the neighbouring monoamine oxidase genes (*MAOA* and *MOAB*) have been associated with more severe intellectual disability.

A single individual diagnosed with unilateral Coats disease was found to carry an *NDP* mutation in retinal tissue but not in blood.[7] As Coats disease is generally not associated with germline mutations, affected males with unilateral findings do not require molecular investigation for *NDP* mutations. Bilateral Coats disease is less common and, in a subset of these cases, the phenotype may be part of a multisystemic disorder such as cerebroretinal microangiopathy with calcification and cysts or facioscapulohumeral muscular dystrophy (see Box 14.1).

Clinical management

The ophthalmic management of Norrie disease is similar to that of other forms of FEVR (discussed in Section 14.1). For affected males, careful monitoring of hearing loss, and consideration for cochlear implantation in adult life requires multidisciplinary care.

In families of severely affected boys, a molecular diagnosis in a proband may lead to further testing of at-risk family members. Both prenatal and pre-implantation genetic testing may be considered. *In utero* diagnosis and planned preterm delivery with treatment of retinal neovascularisation before retinal detachment may also be offered. Reduction of vision loss can be achieved in affected males when treatment is initiated before 40 weeks corrected gestational age.[8]

References

1. Khan AO, Aldahmesh MA, Meyer B. Correlation of ophthalmic examination with carrier status in females potentially harboring a severe Norrie disease gene mutation. *Ophthalmology* 2008;**115**(4):730–3.
2. Shastry BS, Hiraoka M, Trese DC, Trese MT. Norrie disease and exudative vitreoretinopathy in families with affected female carriers. *Eur J Ophthalmol* 1999;**9**(3):238–42.
3. Smith SE, Mullen TE, Graham D, Sims KB, Rehm HL. Norrie disease: extraocular clinical manifestations in 56 patients. *Am J Med Genet A* 2012;**158A**(8):1909–17.
4. Chen ZY, Hendriks RW, Jobling MA, Powell JF, Breakefield XO, Sims KB, et al. Isolation and characterization of a candidate gene for Norrie disease. *Nat Genet* 1992;**1**(3):204–8.
5. Chen ZY, Battinelli EM, Fielder A, Bundey S, Sims K, Breakefield XO, et al. A mutation in the Norrie disease gene (NDP) associated with X-linked familial exudative vitreoretinopathy. *Nat Genet* 1993;**5**(2):180–3.
6. Meindl A, Lorenz B, Achatz H, Hellebrand H, Schmitz-Valckenberg P, Meitinger T. Missense mutations in the NDP gene in patients with a less severe course of Norrie disease. *Hum Mol Genet* 1995;**4**(3):489–90.
7. Black GC, Perveen R, Bonshek R, Cahill M, Clayton-Smith J, Lloyd IC, et al. Coats' disease of the retina (unilateral retinal telangiectasis) caused by somatic mutation in the NDP gene: a role for norrin in retinal angiogenesis. *Hum Mol Genet* 1999;**8**(11):2031–5.
8. Esmer AC, Sivrikoz TS, Gulec EY, Sezer S, Kalelioglu I, Has R, et al. Prenatal diagnosis of persistent hyperplastic primary vitreous: report of 2 cases and review of the literature. *J Ultrasound Med* 2016;**35**(10):2285–91.

14.1.3

KIF11-related disorders

Johane Robitaille and Graeme C.M. Black

Heterozygous variants in the *KIF11* gene have been reported to cause a range of ophthalmic phenotypes including familial exudative vitreoretinopathy (FEVR) and a characteristic chorioretinopathy with focal atrophic patches. Systemic manifestations are common in this condition and include microcephaly and developmental delay. A characteristic facial phenotype may also be present.

Clinical characteristics

Individuals with *KIF11*-related disease typically present in infancy or early childhood with microcephaly, developmental delay and abnormal vision. However, the phenotype is highly variable and later presentations have been described.

Ophthalmic manifestations are thought to fall into one of two major categories: chorioretinal atrophy and FEVR (with some manifesting either or both features) (Fig. 14.4). Notably, there are examples of phenotypic overlap either within families or within a single individual (e.g. there are descriptions of cases who presented with FEVR in one eye and chorioretinal changes in the other).[1,2]

In general, chorioretinopathy with sharply demarcated atrophic patches is the most common eye abnormality; optic atrophy and retinal vessel attenuation may also be present. These features may progress with time and individuals generally have reduced cone- and rod-mediated full-field electroretinograms.[1,3] Fundoscopic features in keeping with FEVR spectrum may also be noted including avascularity of the peripheral retina and tractional retinal fold/detachments (see also Section 14.1.1). Other associated

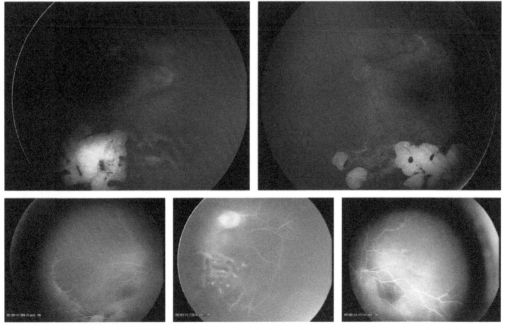

FIG. 14.4 Fundus imaging in a patient with a *KIF11*-related disease, showing bilateral macular dragging and narrowing of the vascular arcades (upper panel). Fluorescein angiography (lower panel) demonstrates hypovascularisation of the retina in the posterior pole with vascular malformations and peripheral retinal non-perfusion. There is also leakage at the optic disc in the right eye.

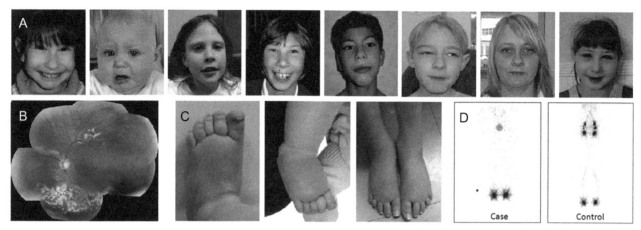

FIG. 14.5 Findings in patients with *KIF11*-related disease. (A) Facial features of affected individuals; characteristic findings in this syndrome include upward-slanting palpebral fissures, a broad nose with a rounded tip, a long philtrum with a thin upper lip, a prominent chin, and prominent ears. (B) Composite colour photograph of the left fundus of an affected individual showing focal areas of peripheral chorioretinal atrophy. (C) Bilateral, congenital lower-limb primary lymphedema in affected individuals; the dorsa of the feet are involved, showing pitting oedema, deep interphalangeal creases, small dysplastic nails, and wide-calibre veins. (D) Comparison of lower-limb lymphoscintigraphy (imaging taken 2 h after injection of radioactive isotope [technetium 99] into the spaces between the toes) between an affected individual and an unaffected control. The patient's lymphoscintigraphy shows no significant main-tract filling suggesting initial lymphatic vessel dysfunction. *(Reproduced with permission from Ref. 4.)*

ophthalmic abnormalities include anterior segment developmental anomalies, cataracts, microcornea, and microphthalmia.[2] A small subset of mutation carriers has no overt ophthalmic phenotype.[4]

Extraocular features are also variable. A group of affected children are now recognised to have a syndrome that encompasses microcephaly, lower limb lymphedema and dysmorphic facial features. Microcephaly ranges from mild to severe and is often associated with mild to moderate developmental delay; neuroimaging often reveals simplified gyri.[5] The lower limb lymphedema is variable and may be absent or may resolve through childhood. A characteristic facial phenotype with upward-slanting palpebral fissures, broad nose with rounded tip, long philtrum (vertical groove between the nose and upper lip) with thin upper lip, prominent chin, and prominent ears has been described in some cases (Fig. 14.5).[4]

Molecular pathology

The *KIF11* gene encodes a microtubule motor protein involved in mitosis.[6] Non-mitotic functions have also been described although their exact role in retinal development and maintenance has not been fully elucidated.[3]

A large proportion of *KIF11* mutations are likely to result in loss of protein function (i.e. nonsense, frameshift, splicing variants). Germline *de novo* mutations are common although variants may also be inherited, usually from a parent who has some features of the condition.[7]

Biallelic mutations in *TUBGCP6*, *TUBGCP4* and *CDK19* can cause a similar phenotype to *KIF11*-related disease ('microcephaly with or without chorioretinopathy, lymphedema, or mental retardation'). Whilst the majority of *TUBGCP4* and *TUBGCP6* related cases have chorioretinopathy rather than a FEVR-like presentation,[1] very few cases of *CDK19*-related disease have been reported with at least one proband has been found to manifest bilateral retinal folds and an avascular peripheral retina.[8]

Patient management

The ophthalmic management of *KIF11*-related FEVR is similar to that in other forms of the condition discussed earlier in Section 14.1. The input of a multidisciplinary team is otherwise necessary, usually involving paediatrics, ophthalmology, and clinical genetics.

References

1. Hull S, Arno G, Ostergaard P, Pontikos N, Robson AG, Webster AR, et al. Clinical and molecular characterization of familial exudative vitreoretinopathy associated with microcephaly. *Am J Ophthalmol* 2019;**207**:87–98.
2. Shurygina MF, Simonett JM, Parker MA, Mitchell A, Grigorian F, Lifton J, et al. Genotype phenotype correlation and variability in microcephaly associated with chorioretinopathy or familial exudative vitreoretinopathy. *Invest Ophthalmol Vis Sci* 2020;**61**(13):2.
3. Birtel J, Gliem M, Mangold E, Tebbe L, Spier I, Müller PL, Holz FG, Neuhaus C, Wolfrum U, Bolz HJ, Charbel Issa P. Novel Insights Into the Phenotypical Spectrum of KIF11-Associated Retinopathy, Including a New Form of Retinal Ciliopathy. *Invest Ophthalmol Vis Sci* 2017;**58**(10):3950–9.
4. Ostergaard P, Simpson MA, Mendola A, Vasudevan P, Connell FC, van Impel A, et al. Mutations in KIF11 cause autosomal-dominant

microcephaly variably associated with congenital lymphedema and chorioretinopathy. *Am J Hum Genet* 2012;**90**(2):356–62.
5. Mirzaa GM, Enyedi L, Parsons G, Collins S, Medne L, Adams C, et al. Congenital microcephaly and chorioretinopathy due to de novo heterozygous KIF11 mutations: five novel mutations and review of the literature. *Am J Med Genet A* 2014;**164A**(11):2879–86.
6. Valentine MT, Fordyce PM, Krzysiak TC, Gilbert SP, Block SM. Individual dimers of the mitotic kinesin motor Eg5 step processively and support substantial loads in vitro. *Nat Cell Biol* 2006;**8**(5):470–6.
7. Chen C, Sun L, Li S, Huang L, Zhang T, Wang Z, et al. Novel variants in familial exudative vitreoretinopathy patients with KIF11 mutations and the genotype-phenotype correlation. *Exp Eye Res* 2020;**199**:108165.
8. Mukhopadhyay A, Kramer JM, Merkx G, Lugtenberg D, Smeets DF, Oortveld MAW, et al. *CDK19* is disrupted in a female patient with bilateral congenital retinal folds, microcephaly and mild mental retardation. *Hum Genet* 2010;**128**(3):281–91.

14.2

Incontinentia pigmenti

Arundhati Dev Borman and Robert H. Henderson

Incontinentia pigmenti, also known as Bloch–Sulzberger syndrome, is an X-linked dominant disorder affecting several organ systems that have ectodermal origin including the skin, the teeth, the eyes and the central nervous system. The name of the condition derives from the histological finding of loose melanin in the dermis, in which the cells appear to be 'incontinent' of their pigment.

Most cases occur sporadically[1] although familial incontinentia pigmenti is noted in 10%–25% of probands. Since this X-linked dominant condition is generally associated with male lethality, approximately 95% of affected individuals are female.[2] Incontinentia pigmenti generally demonstrates high penetrance but variable expressivity, even within families; the latter is often associated with variability in X-chromosome inactivation patterns.

Clinical characteristics

Affected individuals typically present in the neonatal period with characteristic skin lesions. Clinical diagnostic criteria have been proposed and a key consideration is whether the suspected individual has a family history of the condition. In cases with negative family history, the presence of at least one typical dermatological feature (i.e. one 'major criterion') is required whereas the presence of additional features ('minor criteria') such as dental anomalies, retinal disease, central nervous system abnormalities, and abnormal hair or nails, supports the diagnosis (Table 14.3).[3,4] When there is at least one affected first-degree relative, the presence of any of the following key criteria strongly supports the diagnosis: (i) presence/history of typical skin lesions; (ii) atrophic, hairless linear skin streaks; (iii) abnormal dentition, (iv) retinal disease, and (v) alopecia or coarse wiry hair. Notably, it is often the ophthalmic and neurological features that have the most devastating functional and life-threatening sequelae.

The dermatological findings in incontinentia pigmenti are striking and almost all affected individuals display cutaneous features in the first few months of life. The lesions progress through four stages which can persist into adulthood. They begin as vesicles (marked erythema with blistering)—stage 1, become verrucous (wart-like papules and keratotic patches)—stage 2, then appear hyperpigmented (patchy streaks and whorls)—stage 3, and eventually become hypopigmented—stage 4 (Fig. 14.6). These lesions appear along the lines of Blaschko (embryonic lines of normal cell development in the skin that are invisible under normal conditions) and are most visible on the trunk and limbs.

TABLE 14.3 Landy and Donnai incontinentia pigmenti diagnostic criteria, with modifications.

Major criteria	Minor criteria (supportive evidence)
Skin changes that are distributed along Blaschko's lines and classically evolve sequentially through the following four stages: • vesiculobullous stage • verrucous stage • hyperpigmented stage • hypopigmented/atrophic stage	Dental anomalies (hypodontia, anodontia, microdontia, abnormally shaped teeth); Retinal disease; Central nervous system abnormalities; Alopecia or abnormal hair (coarse wiry hair, anomalies of eyebrows, and eyelashes); Abnormal nails (pitting ridging, hypertrophic curved nails)

Conditions for confirming the diagnosis:
• If there is no evidence of incontinentia pigmenti in a first-degree relative and genetic testing data are lacking, at least one major and one minor criterion are required to make the diagnosis of sporadic incontinentia pigmenti;
• If there is an affected first-degree relative and/or if there is a positive genetic test result, the presence of any single major or minor criterion is sufficient for the diagnosis of incontinentia pigmenti.

Adapted from Refs. 3,4.

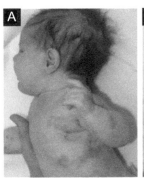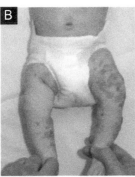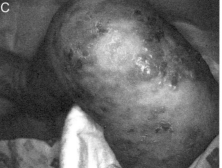

FIG. 14.6 Cutaneous features of incontinentia pigmenti. (A) Stage 1—red, blister-like vesicular lesions affecting the trunk. (B) Stage 1—red, blister-like vesicular lesions affecting the lower limbs. (C) Stage 2—lesions on the right lower limb including papules and keratotic patches with some healing residual stage 1 blisters. (D) Stage 3—marbled hyperpigmentation along the lines of Blashko on the trunk.

Both retinal and non-retinal ophthalmic manifestations occur in incontinentia pigmenti.[5] Retinal findings are typically noted in the first year of life, the most notable being retinal detachment. However, the most common finding is peripheral retinal non-perfusion/avascularity, similar to that seen in retinopathy of prematurity. This can be visible in the neonatal period, but may not be detected on fundoscopy alone. In general, affected infants require fundus fluorescein angiography to delineate the integrity of the retinal vasculature (Fig. 14.7A). The timing, schedule and type of intervention following the identification of angiographic abnormalities are still a matter of discussion (see 'Clinical management' section). Retinal manifestations of incontinentia pigmenti may halt at any stage and may spontaneously regress, often leaving several sequelae (Fig. 14.8). Of those with sight-threatening complications, around 70% have retinal anomalies and around 10% have a retinal detachment. There is an increased risk of tractional retinal detachment, seen as early as age 2 weeks, especially when there is peripheral retinal neovascularisation and ischaemic optic neuropathy at presentation.[6] Other retinal findings include neovascularisation at the edges of the avascular retina with arborised and anastomotic vessels; haemorrhage; persistent fetal vasculature; foveal atrophy and macular vascular aneurysms.[1]

Around 30% of patients with incontinentia pigmenti have neurological abnormalities, which include motor retardation, learning difficulties, convulsive disorders, spastic paralysis and cortical blindness.[1] Over 90% of central nervous system symptoms occur by age 2 years; seizures (around 70%), followed by motor (40%) and cognitive (30%) delay are the most prevalent features. Neuroimaging in early childhood is important to identify cerebral ischaemia in the setting of vasoocclusive disease; other findings include cerebral and cerebellar atrophy, ventricular dilatation, hydrocephalus, agenesis of the corpus callosum and cystic lesions.

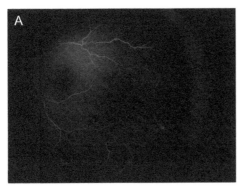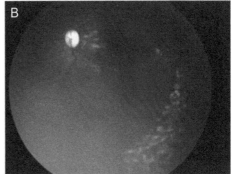

FIG. 14.7 Retinal findings in incontinentia pigmenti. (A) Fundus fluorescein angiogram demonstrating an area of avascular peripheral retina. (B) Colour fundus photograph following retinal laser photocoagulation to the avascular peripheral retina.

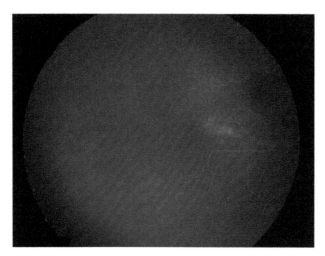

FIG. 14.8 Fundus fluorescein angiogram in a 16-year-old patient with incontinentia pigmenti and untreated, non-perfused peripheral retina.

Between 60% and 80% of patients with incontinentia pigmenti have dental anomalies including anodontia and abnormally shaped teeth (e.g. conical). Alopecia, nail abnormalities (pitting and ridging), and breast abnormalities may also be seen.

Molecular pathology

Incontinentia pigmenti is caused by pathogenic variants in the *IKBKG* gene (previously known as *NEMO*) located on the long arm of the X chromosome (Xq28). *IKBKG* encodes a regulatory subunit of an inhibitory complex which is required for nuclear factor kappa beta (NFκβ) function. This transcription factor complex regulates immune response and cellular apoptosis.

An ~11,700 base-pair deletion removing exons 4–10 from the *IKBKG* gene is the most common disease-associated variant. It is present in up to 80% of cases and causes loss of NFκβ function. Generally, most other causative variants are frameshift or nonsense mutations that also lead to loss of functional protein, although insertion/deletion and missense mutations have also been described.[7,8] In those with some residual protein function, an alternative phenotype of anhidrotic ectodermal dysplasia with immunodeficiency, that is not discussed here, maybe observed.[9]

In the absence of the protective activity against inflammation and apoptosis that the NFκβ pathway provides, endothelial and other cells throughout the body can overexpress factors that induce cell migration such as eotaxin (specific for eosinophils). This may explain the serum eosinophilia and extensive involvement of epidermal eosinophils seen in affected individuals.[1] Combined with other factors, eosinophils cause excessive inflammation, leading to vaso-occlusion and ischaemia. This is the likely mechanism by which the dermatological, retinal and neurological features manifest in incontinentia pigmenti. Retinal avascularity and under-perfusion occur as a result of retinal arteriolar occlusion, precipitating ischaemia, and leading to neovascularisation and retinal detachment in the most severe cases.

Molecular investigation in incontinentia pigmenti often begins with screening for the common exon 4–10 deletion. In probands without this variant, sequencing of the *IKBKG* gene is typically undertaken.[7] Serum eosinophil analysis is no longer routinely undertaken as molecular analysis has surpassed this investigation.

Clinical management

Screening and follow-up recommendations to assess retinal involvement and progression have been made, and efforts to develop formal guidelines are ongoing. It has been suggested that a detailed ophthalmic examination with fluorescein angiography under general anaesthetic should be performed as soon as possible after diagnosis (ideally within the first few months of life). Serial re-examinations with angiography are recommended throughout the first year, at 3 monthly intervals, and then 6 monthly until 3 years of age.[6,10,11] If no retinal manifestations are identified by the age of 1 year, they are unlikely to occur later. The frequency of anaesthesia needs to be balanced against the risk of associated neuro-cognitive deficit. It is also noted there is controversy as to whether any treatment prevents further deterioration.

As the retinal features of incontinentia pigmenti bear similarities to those seen in retinopathy of prematurity, prophylactic ablation (usually laser photocoagulation, occasionally cryotherapy) to the ischaemic retina has been recommended and has been widely implemented, with the intent of halting fibrovascular proliferation and preventing retinal detachment (Fig. 14.7B). However, the natural history in some untreated patients is that of stability or vascular remodelling with any angiogenic drive being potentially helpful. Which patients will therefore progress, and which will remain stable without adverse sequelae, remains unclear (Fig. 14.9). Tractional retinal detachments tend to occur early (even within the first few weeks of life, perhaps reflecting that the vaso-occlusive processes began *in utero*), and have a poor surgical response, whilst rhegmatogenous retinal detachments tend to occur later in life.[6] Retinal laser and cryotherapy themselves carry a risk of inducing cicatricial retinal changes and breaks. In some patients treated with retinal laser ablation, subsequent fluorescein angiography has demonstrated new areas of non-perfusion in previously vascularised areas (Fig. 14.10). Whether treatment is inducing these changes is a matter

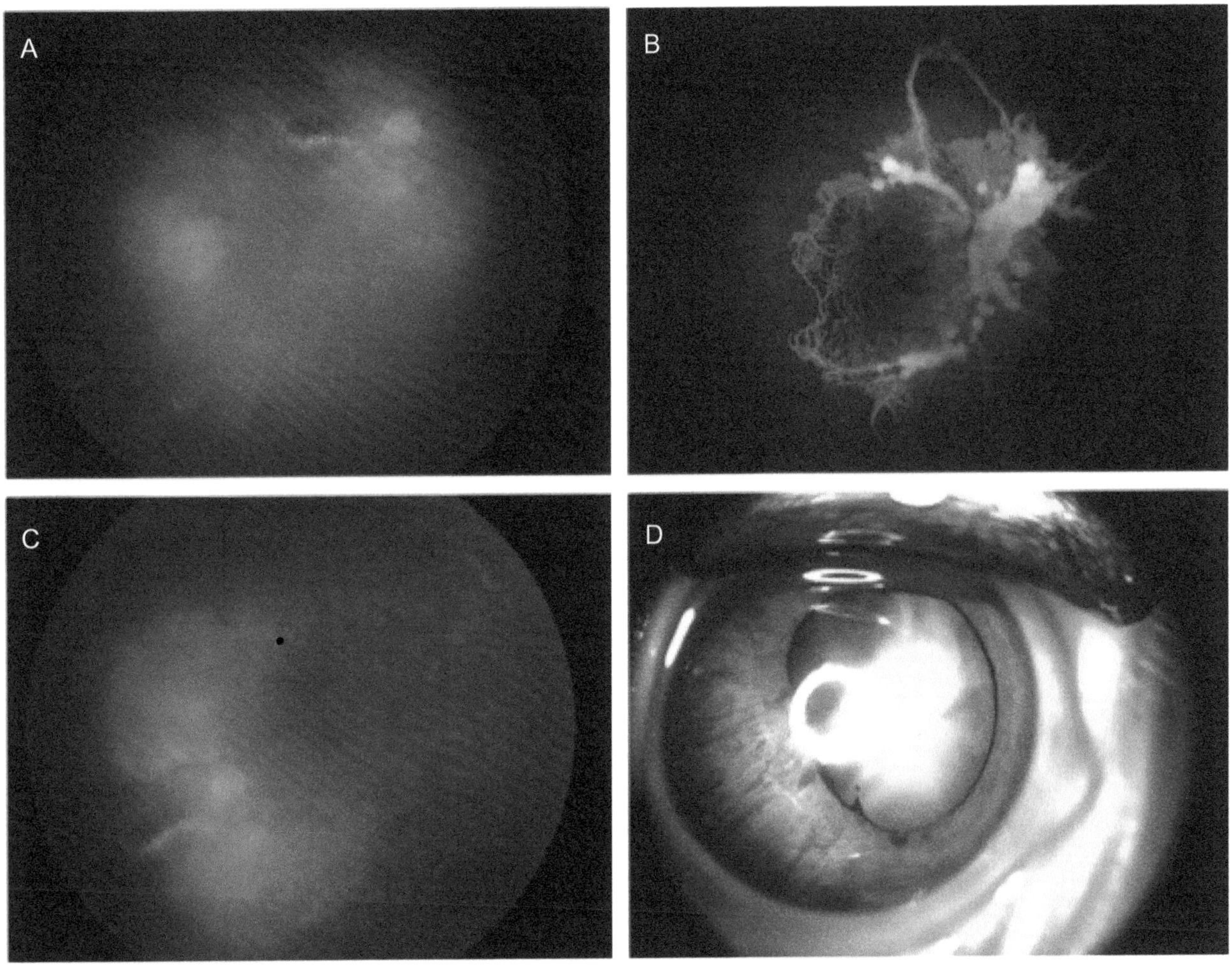

FIG. 14.9 Retinal sequelae of incontinentia pigmenti in an affected individual over the first 4 months of life. (A) Colour fundus photograph demonstrating avascular peripheral retina. (B) Fundus fluorescein angiogram demonstrating peripheral retinal non-perfusion and primitive vascular architecture with widespread capillary drop-out inside the arcades. (C) Colour fundus photograph following retinal laser photocoagulation. (D) Subsequent retinal detachment following retinal laser photocoagulation.

of active investigation. A comprehensive study into the long-term retinal outcomes in incontinentia pigmenti suggested that laser photocoagulation and cryotherapy should be limited to those patients who show documented progression of neovascularisation, vitreous traction or haemorrhage.[6] There is a paucity of evidence for use of anti-VEGF agents but the retinal non-perfusion is often widespread with minimal neovascularisation.

Incontinentia pigmenti requires multidisciplinary care, and the dermatological, neurological and dental abnormalities are managed by respective specialists from as early an age as possible. The X-linked dominant inheritance pattern in this condition presents further challenges to the management of these families, particularly in relation to genetic counselling. Approximately 65% of affected individuals carry *de novo* pathogenic variants in *IKBKG*.[12] Particular emphasis should be given to pregnancy history, especially recurrent miscarriages, due to the increased risk of male lethality *in utero*. The expected ratio amongst liveborn children of affected females is approximately 33% unaffected females, 33% affected females and 33% unaffected males.

Surviving affected males are likely either to have a 47, XXY karyotype (Klinefelter syndrome, in which there is an extra X chromosome) or to display somatic mosaicism (i.e. there are two genetically distinct populations of cells within the same individual). A male with somatic and germline mosaicism may transmit the pathogenic *IKBKG* variant to his daughters, but not to his sons. Females with the pathogenic variant will be affected. It is possible to undertake prenatal testing if the pathogenic variant in the family is known.

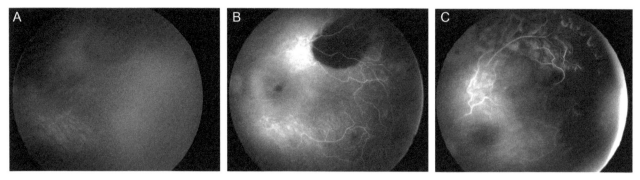

FIG. 14.10 Retinal findings in a 2-month-old female with incontinentia pigmenti. (A) Left fundus photograph showing supero-temporal intraretinal haemorrhage. (B) Fluorescein angiogram demonstrating widespread temporal capillary dropout, masking from the haemorrhage, and vascular remodelling with a supero-temporal arteriole crossing the horizontal midline. (C) Fluorescein angiogram performed 3 months later revealing new abnormalities temporally to the original area of retinal non-perfusion and leak, posterior to previous laser scars.

References

1. Swinney CC, Han DP, Karth PA. Incontinentia pigmenti: a comprehensive review and update. *Ophthalmic Surg Lasers Imaging Retina* 2015;**46**(6):650–7.
2. Wang R, Lara-Corrales I, Kannu P, Pope E. Unraveling incontinentia pigmenti: a comparison of phenotype and genotype variants. *J Am Acad Dermatol* 2019;**81**(5):1142–9.
3. Landy SJ, Donnai D. Incontinentia pigmenti (Bloch-Sulzberger syndrome). *J Med Genet* 1993;**30**(1):53–9.
4. Minic S, Trpinac D, Obradovic M. Incontinentia pigmenti diagnostic criteria update. *Clin Genet* 2014;**85**(6):536–42.
5. Goldberg MF, Custis PH. Retinal and other manifestations of incontinentia pigmenti (Bloch-Sulzberger syndrome). *Ophthalmology* 1993;**100**(11):1645–54.
6. Chen CJ, Han IC, Tian J, Munoz B, Goldberg MF. Extended follow-up of treated and untreated retinopathy in incontinentia pigmenti: analysis of peripheral vascular changes and incidence of retinal detachment. *JAMA Ophthalmol* 2015;**133**(5):542–8.
7. Conte MI, Pescatore A, Paciolla M, Esposito E, Miano MG, Lioi MB, et al. Insight into IKBKG/NEMO locus: report of new mutations and complex genomic rearrangements leading to incontinentia pigmenti disease. *Hum Mutat* 2014;**35**(2):165–77.
8. Fusco F, Paciolla M, Napolitano F, Pescatore A, D'Addario I, Bal E, et al. Genomic architecture at the Incontinentia Pigmenti locus favours de novo pathological alleles through different mechanisms. *Hum Mol Genet* 2012;**21**(6):1260–71.
9. Doffinger R, Smahi A, Bessia C, Geissmann F, Feinberg J, Durandy A, et al. X-linked anhidrotic ectodermal dysplasia with immunodeficiency is caused by impaired NF-kappaB signaling. *Nat Genet* 2001;**27**(3):277–85.
10. Holmstrom G, Thoren K. Ocular manifestations of incontinentia pigmenti. *Acta Ophthalmol Scand* 2000;**78**(3):348–53.
11. Balaratnasingam C, Lam GC. Retinal sequelae of incontinentia pigmenti. *Pediatr Int* 2009;**51**(1):141–3.
12. Scheuerle AE, Ursini MV. Incontinentia Pigmenti. 1999 Jun 8 [Updated 2017 Dec 21], In: Adam MP, Ardinger HH, Pagon RA, et al., editors. *GeneReviews® [internet]*. Seattle, WA: University of Washington, Seattle; 1993–2020. Available from: https://www.ncbi.nlm.nih.gov/books/NBK1472/.

14.3

Stickler syndrome and allied collagen vitreoretinopathies

Martin P. Snead

Stickler syndrome is the most common cause of retinal detachment in childhood, the most common cause of familial retinal detachment and a recognised cause of dual sensory impairment. It is clinically and genetically heterogeneous and at least 10 distinct disease subtypes have been identified (Table 14.4).[1] Stickler syndrome usually results from abnormalities in type II, IX or XI collagen molecules, the major structural components of the extracellular matrix in vitreous and cartilage. Heterozygous mutations in *COL2A1* (the gene encoding type II collagen) and *COL11A1* (the gene encoding type XI collagen) underlie the two most prevalent forms of the condition. Several associated disorders, including Czech dysplasia, spondyloepiphyseal dysplasia congenita (SEDC) and Kniest dysplasia have an identical ophthalmic phenotype and are also caused by heterozygous mutations in *COL2A1*; they may therefore be considered as part of the same spectrum in terms of differential diagnosis.

The pathognomonic hallmark of all but one form of Stickler syndrome is a congenital abnormality of vitreous embryogenesis (the only exception is type 3 Stickler syndrome where there is no eye involvement; Table 14.4).[1] Other features include congenital myopia, retinal detachment, hearing loss, cleft palate and premature arthropathy. Using a combination of accurate phenotyping and molecular genetic analysis, it is possible to identify the underlying genetic mutation in over 95% of cases so that prophylactic retinopexy can be offered to the high-risk subgroups to reduce the risk of blindness.

Clinical characteristics

The clinical features of Stickler syndrome are highly variable. The possibility of this diagnosis should, in particular, be considered in the following scenarios:

1. Neonates with a cleft palate or Pierre-Robin sequence (i.e. combination of micrognathia, cleft palate and displacement of the tongue towards the back of the oral cavity—glossoptosis) in association with myopia. Myopia is unusual in a neonate and may sometimes be mistaken for congenital glaucoma (buphthalmos).
2. Infants with joint dysplasia (especially bilateral 'Perthes' disease) in association with myopia and/or deafness.
3. Individuals with a family history of retinal detachment ('ocular-only' variants of Stickler syndrome[2]).
4. Sporadic cases of retinal detachment associated with joint hypermobility, midline clefting, or deafness.
5. Infants with rhizomelic (proximal) limb shortening with a midline cleft or a high arch palate.

The ophthalmic hallmark of Stickler syndrome results from embryological defects in normal vitreous development that can be visualised with dilated slit-lamp examination. The identification of, and differentiation between, the membranous vitreous phenotype and the beaded vitreous phenotype is valuable for the diagnosis of specific disease subtypes and can facilitate familial analysis and genetic variant interpretation. This pathognomonic feature is important for the clinical diagnosis of the ocular-only subgroups which have no systemic features to suggest the diagnosis (Figs 14.11–14.13).[1,2]

The association of this group of conditions with congenital quadrantic lamellar cataracts (Fig. 14.14) is well recognised and can be a useful diagnostic marker. However, this sign can be present in both type 1 and type 2 Stickler syndrome and does not therefore distinguish between subgroups in the way that the differing vitreous phenotypes allow. Congenital myopia is common and is typically combined with high astigmatic error. It is however important to recognise that approximately 15% of patients exhibit no significant refractive error.

The greatest ophthalmic risk of Stickler syndrome is that of rhegmatogenous retinal detachment due to giant retinal tears; these occur in the pars plana due to an abnormally anterior separation of the posterior hyaloid membrane (Fig. 14.15). Up to 60% of individuals with type 1 Stickler syndrome develop retinal detachment and approximately half of these patients suffer bilateral retinal detachment.[4] The risk of retinal detachment in type 2 Stickler syndrome is somewhat lower, affecting around 40% of patients.[5]

Assessment of these features requires refraction under cycloplegia, slit-lamp examination (for lens and vitreous

TABLE 14.4 Stickler syndrome and allied collagenopathies.

Condition		Gene	Cytogenetic location	Distinguishing features
Stickler syndrome				
	Type 1	COL2A1	12q13.11	Membranous (type 1) congenital vitreous anomaly, retinal detachment, congenital megalophthalmos, deafness, arthropathy, cleft palate **High risk of blindness**
	Ocular only	COL2A1	12q13.11	Membranous (type 1) congenital vitreous anomaly, retinal detachment congenital megalophthalmos; no systemic features **High risk of blindness**
	Type 2	COL11A1	1p21.1	Beaded (type 2) congenital vitreous anomaly, retinal detachment, congenital megalophthalmos, deafness, arthropathy, cleft palate
	Type 2 recessive	COL11A1	1p21.1	Autosomal recessive; beaded (type 2) congenital vitreous anomaly, retinal detachment, congenital megalophthalmos, cleft palate, profound severe congenital deafness
	Type 3	COL11A2	6p21.32	**Non-ocular Stickler syndrome** Normal vitreous and ophthalmic examination; deafness, arthropathy, cleft palate
	Type 4	COL9A1	6q13	Autosomal recessive; deafness, myopia, vitreoretinopathy, retinal detachment, epiphyseal dysplasia
	Type 5	COL9A2	1p34.2	Autosomal recessive; deafness, myopia, vitreoretinopathy, retinal detachment, epiphyseal dysplasia
	Type 6	COL9A3	20q13.33	Autosomal recessive; deafness, myopia, vitreoretinopathy, retinal detachment, epiphyseal dysplasia
	Type 7	BMP4	14q22.2	Hypoplastic vitreous, retinal detachment, deafness, arthropathy, palate abnormality, renal dysplasia
	Type 8	LOXL3	2p13.1	Autosomal recessive; myopia, hypoplastic vitreous, arthropathy, palate abnormality; normal facies; normal hearing
Kniest dysplasia		COL2A1	12q13.11	(usually) Membranous (type 1) congenital vitreous anomaly, retinal detachment, congenital megalophthalmos, severe arthropathy, short stature, phalangeal dysplasia
Spondyloepiphyseal dysplasia congenita (SEDC)		COL2A1	12q13.11	(usually) Membranous (type 1) congenital vitreous anomaly, retinal detachment, congenital megalophthalmos, severe short stature, rhizomelic limb shortening, barrel chest
Wagner syndrome		VCAN	5q14.2-q14.3	Early cataract, pseudoexotropia, erosive vitreoretinopathy
Czech dysplasia		COL2A1	12q13.11	Hypoplastic vitreous, retinal detachment, cleft palate, normal stature, spondyloarthropathy, short postaxial toes

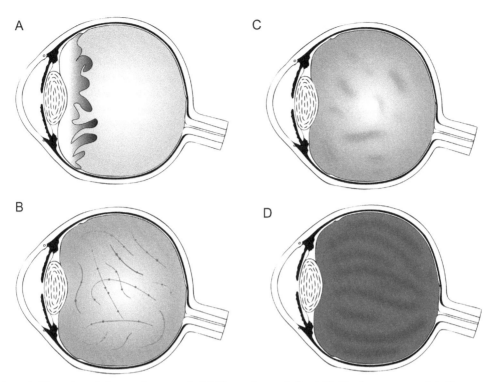

FIG. 14.11 Schematic illustration of vitreous phenotypes in Stickler syndrome. A (top left): Membranous congenital vitreous anomaly [mutations causing *COL2A1* haploinsufficiency]. B (bottom left): Beaded congenital vitreous anomaly [mutations with a dominant negative effect on *COL11A1*]. C (top right): Hypoplastic congenital vitreous anomaly [splicing mutations in *COL2A1*]. D (bottom right): Normal vitreous architecture: compact lamellar array [mutations in *COL11A2*]. *(Reproduced with permission from Ref. 2.)*

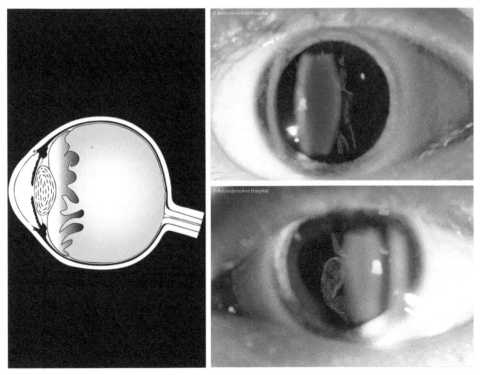

FIG. 14.12 The slit-lamp appearance of membranous congenital vitreous anomaly [due to a mutation causing *COL2A1* haploinsufficiency]. *(Reproduced with permission from Ref. 3.)*

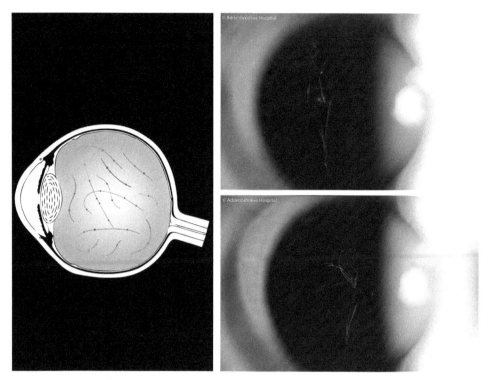

FIG. 14.13 The slit-lamp appearance of beaded congenital vitreous anomaly [due to a mutation with a dominant negative effect on COL11A1]. *(Reproduced with permission from Ref. 3.)*

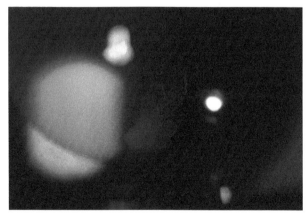

FIG. 14.14 Congenital quadrantic lamellar cataract in an individual with Stickler syndrome. This well-recognised sign is a useful diagnostic marker. However, it can be present in both type 1 and type 2 Stickler syndrome and does not therefore distinguish between subgroups in the way that the differing vitreous phenotypes allow. *(Reproduced with permission from Ref. 2.)*

phenotyping), fundus examination with indirect ophthalmoscopy and (if possible) scleral depression. In very young patients this may require examination under general anaesthesia. Retinal breaks at the ora serrata are frequently identified.

Sensorineural hearing loss is common in Stickler syndrome and combined conductive and sensorineural hearing impairment is often seen in individuals with palate abnormalities. It is important to recognise subclinical hearing loss as it contributes to the risk of dual sensory loss. Referral for otoscopy, pure tone audiogram and tympanometry may be required to investigate and distinguish conductive from sensorineural hearing loss and tympanic membrane dysfunction. Notably, it has been shown that biallelic variants in *COL11A1*, the gene associated with type 2 Stickler syndrome, can result in profound or even total sensorineural hearing loss.[6–8]

The craniofacial features of Stickler syndrome result largely from delayed development of the lower jaw and midface, leading to failed or incomplete fusion of the palatal processes. Cleft palate is a feature in approximately 40% of type 1 Stickler syndrome patients and 25% of type 2 Stickler syndrome patients; Pierre-Robin sequence is present in 13% and 1.5% of type 1 and type 2 patients respectively.[9] Examination of the palate by inspection and palpation is important as some patients will be unaware of their high palate or subclinical soft cleft. Other facial features include malar hypoplasia (i.e. abnormally flat cheekbones), flattened or broad nasal bridge, anteriorly-facing nostrils and a long philtrum (vertical groove between the nose and upper lip). It is noted that these features are highly variable and reliance on facial phenotype for diagnosis is unwise. Slit-lamp examination for the pathognomonic vitreous phenotypes (Figs 14.11–14.13) is far more dependable

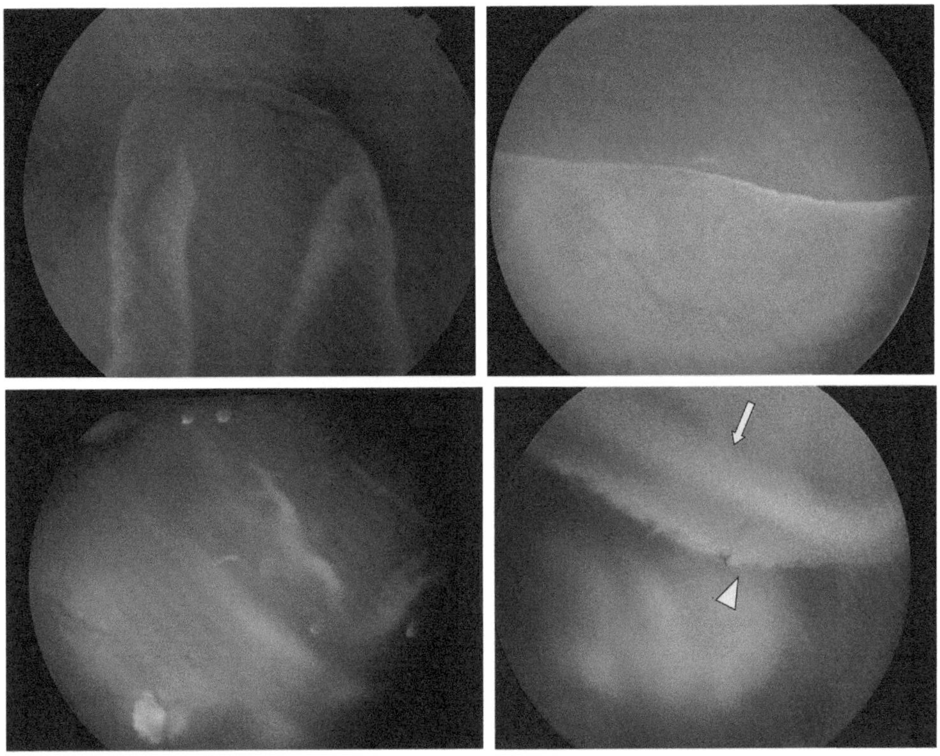

FIG. 14.15 Four examples of giant retinal tear detachments in type 1 Stickler syndrome. *(Reproduced with permission from Ref. 2.)*

and indicates Stickler syndrome sub-groups in clinically challenging situations, such as cases complicated by double heterozygosity (e.g. for *COL2A1* and *COL11A1*), as well as cases with 'ocular only' forms with minimal or no systemic involvement.[2]

Joint hypermobility (Fig. 14.16) is common in childhood and can be assessed and compared against age and sex-matched normative data using the Beighton score. Any joint(s) may be affected but those most commonly involved are the hips, knees and lumbar spine.[10] Radiological imaging may be normal even in the presence of obvious joint hypermobility but in a subset of cases, it reveals evidence of epiphyseal dysplasia, particularly of the hips and knees. If present, the radiological changes may simulate Perthes disease (i.e. idiopathic avascular necrosis of the femoral epiphysis seen in children), although they probably represent dysplastic development of the femoral head rather than true avascular necrosis.[1]

Adult patients are likely to exhibit a degenerative arthropathy typically but not exclusively affecting the hips, knees and lumbar spine.[10] Of note, the musculoskeletal features of Kniest dysplasia (Fig. 14.17), spondyloepiphyseal dysplasia congenita (Fig. 14.18) and Czech dysplasia are more severe than those of Stickler syndrome and include rhizomelic (proximal) limb shortening, kyphosis and scoliosis. Odontoid hypoplasia may also be present, predisposing to cervicomedullary instability and imaging of the cervical spine should be considered.[3]

Molecular pathology

Stickler syndrome is genetically heterogeneous and molecular analysis typically involves evaluating the coding sequence of at least 9 disease-implicated genes (*COL2A1, COL11A1, COL11A2, COL9A1, COL9A2, COL9A3, LRP2, LOXL3; BMP4*, Table 14.4). Dosage analysis to look for deletions or duplications is important; where this is not possible using high-throughput sequencing strategies, multiplex ligation-dependent probe amplification (MLPA) can be utilised.[11] Analysis of the complete sequence (exons and introns) of specific genes is required for patients with phenotypes strongly suggestive of Stickler syndrome but in whom no variant is identified by conventional strategies as many of these individuals will carry deep intronic mutations.[12,13] Confirmation of the pathogenicity of deep intronic changes may require a combination of approaches including comparison of variant frequency on DNA sequence variation databases, co-segregation of a variant with phenotype, *in silico* prediction software and

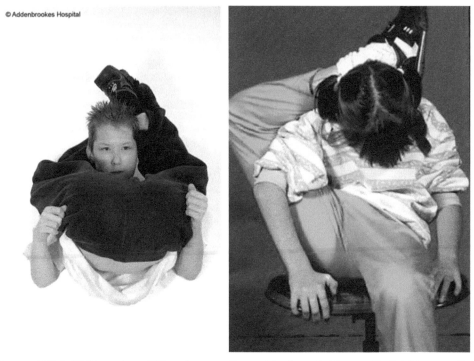

FIG. 14.16 Joint hypermobility in Stickler syndrome. This can be assessed in a variety of ways, but the use of the Beighton score allows comparison with age- and sex-matched normative data. *(Reproduced with permission from Ref. 3.)*

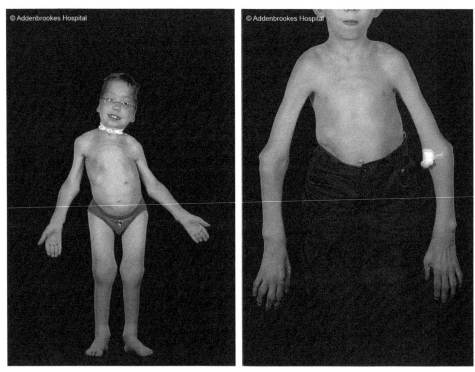

FIG. 14.17 Findings in Kniest dysplasia. *(Reproduced with permission from Ref. 3.)*

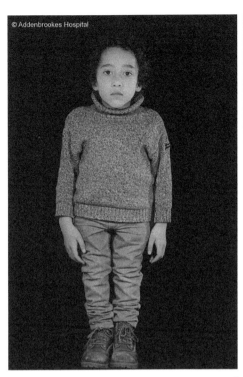
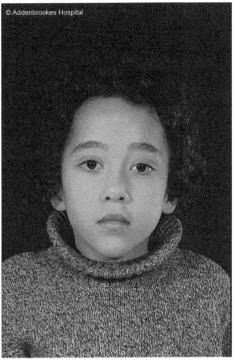

FIG. 14.18 Findings in spondyloepiphyseal dysplasia congenital, SEDC. *(Reproduced with permission from Ref. 3.)*

'minigene' functional analysis to assess the impact on exon splicing.[12]

Most cases of Stickler syndrome (type 1) are inherited in an autosomal dominant fashion and will harbour heterozygous *COL2A1* loss of function variants resulting in haploinsufficiency of type II collagen—the major structural protein of the vitreous body.[1] Type II procollagen exists in two alternatively spliced forms. The short form (which is expressed in cartilage but not in the eye) has exon 2 spliced out and so loss of function variants in exon 2 result in ocular-only Stickler syndrome without systemic involvement.[2] Spondylo-epiphyseal dysplasia congenita (SEDC), Kniest dysplasia, vitreoretinopathy with phalangeal epiphyseal dysplasia (VPED) and Czech dysplasia are also caused by mutations in *COL2A1* but usually result from variants with dominant negative effects (rather than mutations causing haploinsufficiency) and lead to more severe skeletal dysplastic changes.[3]

Type 2 Stickler syndrome is caused by variants in *COL11A1* encoding the α1 chain of type XI collagen. Affected individuals exhibit a characteristic vitreous phenotype (Fig 14.13). Typical molecular changes result in either substitution of an obligate glycine within the Gly-Xaa-Yaa amino-acid sequence repeat region of the molecule or mRNA mis-splicing, or are deletions/duplications leading to in-frame sequence alterations.[13,14] *COL11A1*, like the other main genes implicated in Stickler syndrome *COL2A1* and *COL11A2*, is subject to alternative splicing.

The natural alternative splicing of exon 9 in *COL11A1* modifies the effect of mutations occurring at that location reducing the severity of the associated skeletal dysplasia.[6]

Autosomal recessive Stickler syndrome is much less common and has been reported in association with biallelic variants in genes for Type IX collagen, *COL9A1*, *COL9A2*, and *COL9A3* (see Table 14.4).[15] Individuals carrying compound heterozygous *COL11A1* variants have also been reported; in these cases, a variant leading to alternative splicing can modify the effect of another mutation to result in recessive type 2 Stickler syndrome with unusually profound hearing loss.[6,7]

Heterozygous or biallelic variants in *COL11A2* cause phenotypes overlapping with Stickler syndrome but without ophthalmic manifestations.[3] They are therefore not associated with an increased risk of retinal detachment (Table 14.4).

Wagner vitreoretinopathy (see Section 14.4) was historically considered to be a variant of Stickler syndrome. It is now clear that these two disorders are genetically and phenotypically separate and although Wagner disease should be considered in the differential diagnosis of ocular-only Stickler syndrome, it exhibits a different vitreous phenotype.[16]

Other, non-collagen-encoding, genes may result in phenotypes similar to Stickler syndrome. Biallelic mutations in *LRP2* cause the overlapping ultra-rare recessive conditions Donnai–Barrow and facio-oculo-acustico-renal (FOAR) syndromes; a minority of patients with these disorders have milder phenotypes that overlap with Stickler syndrome, but the absence of vitreous anomaly and hearing loss, and the presence of renal abnormalities should aid in differentiating these disorders. In a small number of cases diagnosed with Stickler syndrome, biallelic mutations in *LOXL3* have been identified (Table 14.4).

Clinical management

Stickler syndrome is an important diagnosis not to miss—it is the most common cause of familial retinal detachment and the most common cause of rhegmatogenous retinal detachment in children. Half of all patients with genetically confirmed type 1 Stickler syndrome who experience retinal detachment suffer detachment in their second eye within 4 years of the first eye.[4] Molecular genetic analysis has a key role to play in risk stratification for the prevention of blindness and dual sensory impairment. Bilateral 360° prophylactic retinopexy has been shown to substantially reduce the risk of giant retinal tear detachment in the high-risk subgroup[4] and can be expected to reduce the risk in the rarer but allelic conditions of spondyloepiphyseal dysplasia congenita and Kniest dysplasia.[3] For these reasons, it is important to ensure that probands and affected relatives with Stickler syndrome are reviewed and prospectively managed in a tertiary vitreoretinal service with the necessary support.

Once a proband is diagnosed with Stickler syndrome, management requires a multidisciplinary approach. Patients should be referred for formal audiology assessment, and to rheumatology and podiatry for the specialist management of the associated arthropathy and soft tissue complications. The majority of cases of Stickler syndrome are inherited in an autosomal dominant fashion and therefore prospective family management is important. This requires input from genetic counsellors/clinical genetics and may involve exploration of the wider familial implications. Examination of first-degree relatives can be undertaken and, in families where a causative mutation has been identified, cascade predictive genetic testing can be offered. The examination of parents and siblings is important and assists in the assessment of the status of a neonate where vitreous phenotyping may be impossible without examination under anaesthesia. Genetic counselling is also valuable to those patients seeking family planning advice; in these situations, pre-implantation genetic testing can be considered.

References

1. Snead MP, Martin H, Bale P, Shenker N, Baguley DM, Alexander P, et al. Therapeutic and diagnostic advances in Stickler syndrome. *Ther Adv Rare Dis* 2020. https://doi.org/10.1177/2633004020978661.
2. Snead MP, McNinch AM, Poulson AV, Bearcroft P, Silverman B, Gomersall P, et al. Stickler syndrome, ocular only variants and a key diagnostic role for the ophthalmologist. *Eye* 2011;**25**:1389–400.
3. Snead, MP. Retinal detachment in childhood. In: Lyons, C, Hoyt, C (eds) *Paediatric ophthalmology and strabismus*. 6th ed. Philadelphia: Elsevier Saunders, 2020.
4. Fincham GS, Pasea L, Carroll C, McNinch AM, Poulson AV, Richards AJ, et al. Prevention of retinal detachment in Stickler syndrome: the Cambridge Prophylactic Cryotherapy protocol. *Ophthalmology* 2014;**121**(8):1588–97.
5. Poulson AV, Hooymans J, Richards AJ, Bearcroft P, Murthy R, Baguley D, et al. Clinical features of type 2 Stickler syndrome. *J Med Genet* 2004;**41**(8):e107. http://www.jmedgenet.com/cgi/content/full/41/8/e107.
6. Richards AJ, Fincham G, McNinch A, Hill D, Castle B, Lees M, et al. Alternative splicing modifies the effect of mutations in COL11A1 and results in recessive type 2 Stickler syndrome with profound hearing loss. *J Med Genet* 2013;**50**(11):765–71.
7. Nixon TRW, Richards AJ, Lomas A, Abbs S, Vasudevan P, McNinch AM, et al. Inherited and de novo bi-allelic pathogenic variants in COL11A1 result in type 2 Stickler syndrome with severe hearing loss. *Mol Genet Genomic Med* 2020;e1354. https://doi.org/10.1002/mgg3.1354.
8. Alexander P, Gomersall P, Stancel-Lewis J, Fincham GS, Poulson AV, Richards AJ, et al. Auditory dysfunction in type 2 Stickler syndrome. *European Archives of Oto-Rhino-Laryngology* 2021 278:2261–68.
9. Zimmermann J, Stubbs DJ, Richards AJ, Alexander P, McNinch AM, Matta B, et al. Stickler syndrome: airway complications in a case series of 502 patients. *Anesth Analg* 2020;(December 16). https://doi.org/10.1213/ANE.0000000000004582.
10. McArthur N, Rehm A, Shenker N, Richards AJ, McNinch AM, Poulson AV, et al. Stickler syndrome in children: a radiological review. *Clin Radiol* 2018;**73**(7):678.e13–8.
11. Vijzelaar R, Waller S, Errami A, Donaldson A, Lourenco T, Rodrigues M, et al. Deletions within COL11A1 in type 2 Stickler syndrome detected by multiplex ligation-dependent probe amplification (MLPA). *BMC Med Genet* 2013;**26**(14):48.
12. Richards AJ, McNinch A, Whittaker J, Treacy B, Oakhill K, Poulson A, et al. Splicing analysis of unclassified variants in COL2A1 and COL11A1 identifies deep intronic pathogenic mutations. *Eur J Hum Genet* 2012;**20**(5):552–8.
13. Richards AJ, McNinch A, Martin H, Oakhill K, Rai H, Waller S, et al. Stickler syndrome and the vitreous phenotype: mutations in COL2A1 and COL11A1. *Hum Mutat* 2010;**31**(6):E1461–71.
14. Richards AJ, Pope FM, Yates JRW, Scott JD, Snead MP. A family with Stickler syndrome type 2 has a mutation in the COL11A1 gene resulting in the substitution of glycine 97 by valine in α1(XI) collagen. *Hum Mol Genet* 1996;**5**:1339–43.
15. Nixon TRW, Alexander P, Richards A, McNinch A, Bearcroft PWB, Cobben J, et al. Homozygous type IX collagen variants (COL9A1, COL9A2 and COL9A3) causing recessive Stickler syndrome—expanding the phenotype. *Am J Med Genet* 2019;**179A**:1498–506.
16. Richards AJ, Martin S, Yates JRW, Baguley DM, Pope FM, Scott JD, et al. COL2A1 exon 2 mutations: relevance to the Stickler and Wagner syndromes. *Br J Ophthalmol* 2000;**84**:364–71.

14.4

Wagner disease

Cyril Burin des Roziers, Pierre-Raphaël Rothschild, Antoine Brézin, and Sophie Valleix

Wagner disease (previously known as Wagner syndrome) is an autosomal dominant non-syndromic vitreoretinopathy that was first described in 1938 by H. Wagner in a family of Swiss ancestry.[1] Clinical signs and symptoms usually manifest in childhood and can lead to severe visual loss. Wagner disease has a recognisable ophthalmic phenotype but this appears to be broader than initially appreciated.[2]

This familial vitreoretinopathy is caused by specific mutations altering the normal splicing of exon 8 of the VCAN gene. VCAN encodes a large chondroitin sulphate proteoglycan, named versican. Versican is one of the main structural macromolecules of the vitreous; it interacts with numerous other proteins and it is required to establish the gel-like consistency of the vitreous during development and adulthood. Versican is also expressed in the extracellular matrix of a large number of tissues. It is involved in cell adhesion, proliferation, migration and angiogenesis and it is thought to play a central role in tissue morphogenesis and maintenance.

The most typical feature of Wagner disease is an optically empty vitreous cavity with avascular strands and veils, resulting from vitreous liquefaction and structural modifications of the vitreoretinal interface. The presentation is variable and typically includes myopia, presenile cataracts, and chorioretinal atrophy. Retinal detachment is a major cause of visual loss and can present at a very young age. Molecular genetic analysis can detect VCAN mutations in most cases allowing the distinction of Wagner disease from other vitreoretinal disorders with overlapping clinical features.

Clinical features

Despite the ubiquitous expression of VCAN throughout the body, no extraocular manifestations have been robustly associated with Wagner disease to date. The ophthalmic features are often subtle or non-specific initially and many patients will present for evaluation of Wagner disease due to a family history of the disorder. Myopia associated to an increased axial length is a common finding. It is usually mild (< 6 diopters), but can be severe and present early in life.[3] An optically empty vitreous is seen in almost all patients with genetically confirmed Wagner disease. This vitreous abnormality is usually noted in early adolescence although it may be detected as early as age 2 years. The core vitreous appears fibrillary suggesting that strands are present within an empty core. The vitreous cortex (posterior hyaloid) in Wagner disease is partially detached and folded at the equator creating a circumferential avascular 360° veil that is virtually pathognomonic of the disease (Fig. 14.19A). The vitreous cortex is also detached at the fovea where it creates a localised posterior vitreous detachment without perifoveal separation.[4] Visualising this abnormality on OCT can be helpful and assist the diagnostic process (Fig. 14.19B). A tractional retinal schisis can be found peripherally at the insertion of the veil whereas at the fovea, the posterior hyaloid can be confused with an idiopathic epiretinal membrane (Fig. 14.19C). The typical picture is that of a dense posterior hyaloid forming a bridge over the foveal pit.

Retinal detachment is a common feature and can be tractional, rhegmatogenous, or, occasionally, exudative.[4,5] The tractional subtype is the most typical form of detachment associated with Wagner disease and it is due to abnormal peripheral vitreoretinal adhesions. Whilst retinal detachment has long been thought to occur late in the course of the disease (this was considered by some authors a key sign to distinguish Wagner disease from Stickler syndrome), it is now clear that almost half of the patients with Wagner vitreoretinopathy present with retinal detachment at a young age.[4]

Chorioretinal degeneration is a constant feature of Wagner disease and is usually progressive. The severity of chorioretinal involvement can vary greatly from mild nyctalopia without obvious fundus findings to severe night-blindness with visual field constriction and wipeout of the RPE and choroid—so-called 'erosive vitreoretinopathy'.[6] Electroretinography (full-field ERG) can be normal in the early stages, but both rod- and cone-mediated responses are usually affected later in the course of the disorder. Together with visual field testing, ERG parameters can be used to monitor disease progression. Pigment clumping is commonly observed on fundoscopy. In general, there is no topographic predilection, but Wagner disease can be confused with retinitis pigmentosa or even enhanced S-cone syndrome when pigmentary changes are prominent

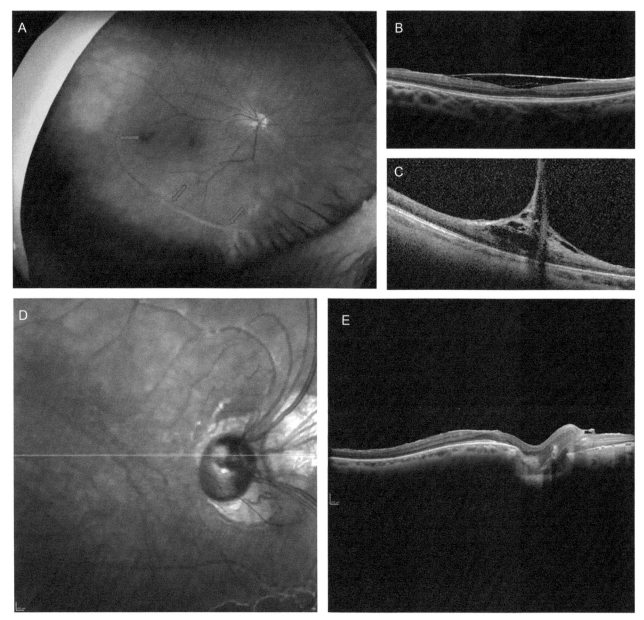

FIG. 14.19 Fundus imaging in patients with Wagner disease. (A) Wide-field fundus imaging showing a typical avascular vitreous veil *(blue arrows)* at the equator; this corresponds to the folding of the vitreous cortex within an optically empty vitreous. (B) OCT showing the peculiar relationship between the retina and the vitreous cortex in patients with Wagner disease. The upper panel shows a posterior vitreous detachment that is localised at the fovea without perifoveal separation; this is in contrast to the usual findings in age-related posterior vitreous detachment. Also, note the presence of an abnormal thickened and double-layered posterior hyaloid. (C) A vitreous cortex fold that is responsible for the avascular veil is seen on OCT imaging. (D,E) Infrared fundus photograph (D) and OCT of the macula (E) demonstrating how retinal vessels exiting the optic nerve are directed to the nasal aspect of the retina leading to a 'situs inversus' configuration.

and diffuse. It is noted that retinal vessels often present sheathing or appear dragged temporally in keeping with an ectopic fovea; this can lead to pseudostrabismus due to an increased angle kappa (Fig. 14.19A). This vascular misplacement is congenital but can mimic familial exudative vitreoretinopathy. Rarely, the retinal vessels exiting the optic nerve are directed to the nasal aspect of the retina and apear to have a 'situs inversus' configuration (Fig. 14.19D and E).

Early-onset cataract is another common finding; a peculiar aspect of the lens opacities has been reported in the form of tiny dots within the crystalline lens called 'dot-shaped cataract'. Surgery for nuclear lens opacities is usually required by the age of 40 years.[7] Postoperative but also spontaneous anterior uveitis have been recognised.[4,8] Congenital glaucoma due to anterior segment malformation but also early-onset glaucoma from various mechanisms can occasionally be encountered during the course of the disease.[9]

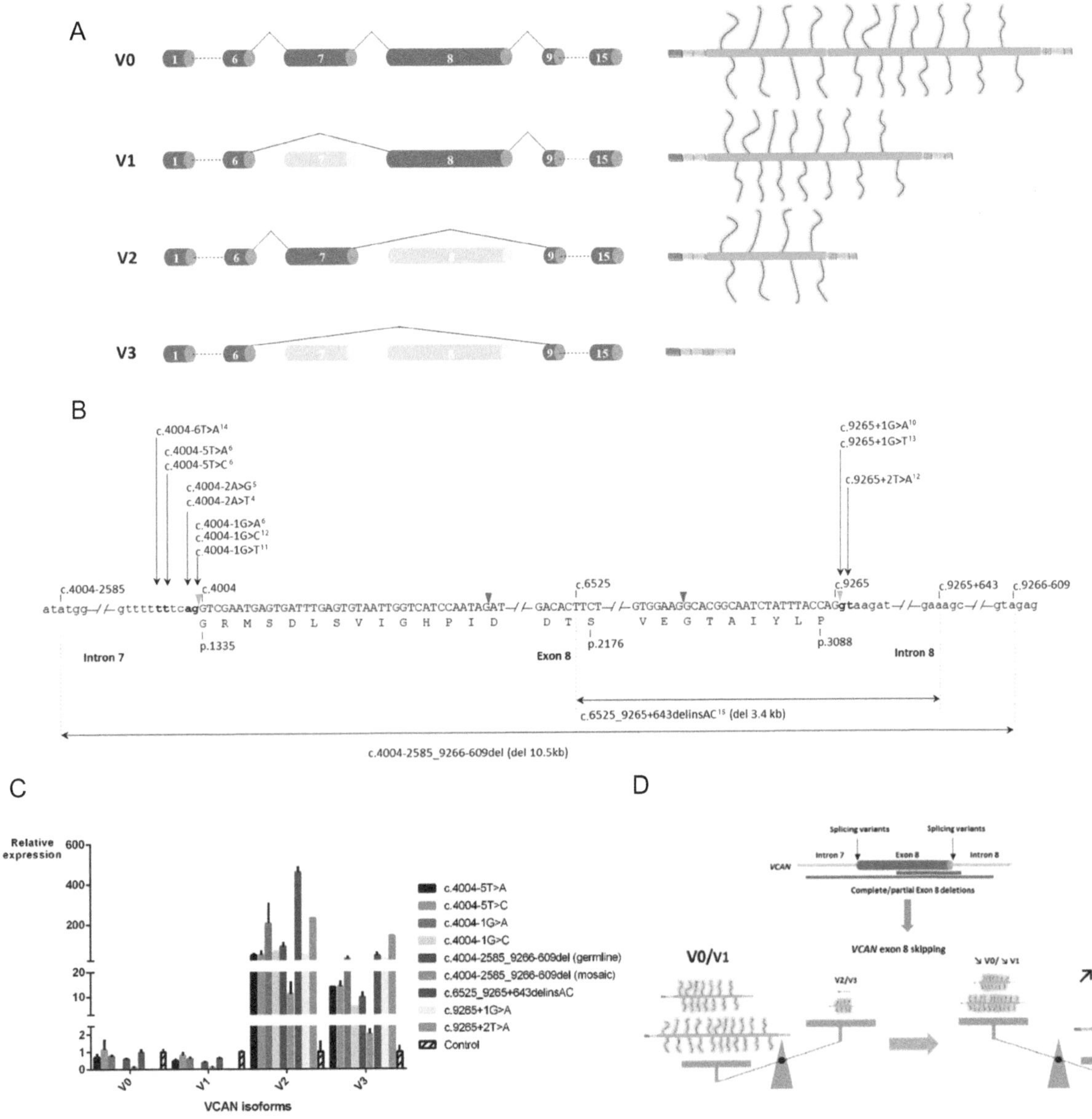

FIG. 14.20 The role of *VCAN* gene mis-splicing in the molecular pathology of Wagner disease. (A) *VCAN* exon 7 and 8 physiological alternative splicing results in the production of four different transcript isoforms. (B) Review of various *VCAN* mutations associated with Wagner disease. The *green arrows* indicate the original splice sites and the *red arrows* indicate cryptic splice sites activated due to loss of intron 7 acceptor or intron 8 donor original splice site. (C) Relative expression analysis by quantitative reverse transcription PCR (RT-PCR) of *VCAN* V0, V1, V2, and V3 mRNA isoforms from members of different family members with *VCAN* mutations. (D) Nucleotide substitution affecting intron 7 acceptor or intron 8 donor splice sites as well as complete or partial *VCAN* exon 8 deletions leading to *VCAN* 8 exon skipping and imbalanced quantitative ratios of all VCAN isoforms. The specific increase of the V2 isoform is considered a key event underlying the molecular pathogenesis of Wagner disease.

Molecular pathology

VCAN is a large gene (110,000 base-pairs) composed of 15 exons, with exons 7 and 8 spanning approximately 3000 and 5000 base-pairs respectively. These exons encode the attachment sites of chondroitin sulphate chains and are subject to alternative splicing, producing at least four physiological isoforms of the encoded versican protein. These include V0 (with both exons 7 and 8), V1 (with exon 8 only), V2 (with exon 7 only), and V3 (without exons 7 and 8), whose expression is temporally and spatially regulated in a specific manner (Fig. 14.20A).

Most pathogenic variants associated with Wagner disease are splice mutations flanking exon 8 (Fig. 14.20B).[4–6,10–14] It has been shown in patients' blood or lymphoblastoid cell lines that specific mutations localised to the splice donor or acceptor sites around exon 8, alter splicing and result in the skipping of this exon, which in turn, lead to ectopic/increased level of VCAN transcripts lacking exon 8 (V2 and V3 isoforms) with a decreased ratio of the V0 and V1 transcript isoforms (Fig. 14.20C).

Some patients with clinical features of Wagner disease and characteristically imbalanced quantitative ratios of VCAN transcript isoforms remained with no detectable disease-associated splice mutations in VCAN for many years.[6] In-depth molecular analysis of such individuals revealed that they were heterozygous for deletions overlapping VCAN exon 8.[15] This finding further supports the notion that the characteristic degenerative liquefaction of the vitreous observed in Wagner disease patients is the result of an inappropriate balance between splice isoform of versican in the eye (Fig. 14.20B–D).

Molecular diagnosis of Wagner disease can be enhanced by using a gene panel approach including VCAN and several other genes of interest (whose mutations are known to be associated with familial vitreoretinopathies with overlapping features). This approach can allow the detection of all types of Wagner disease-related mutations. For example, short-read sequencing technologies have successfully detected a VCAN exon 8 deletion that was *de novo* and in a mosaic state.[15]

Clinical management

Given the heterogeneity of the clinical manifestations and the progressive course of the disorder, genetic testing should be performed to confirm the diagnosis of Wagner disease. Screening for congenital myopia that requires prompt treatment to prevent amblyopia is recommended for siblings of patients with Wagner disease. Early-onset glaucoma and cataract should also be promptly diagnosed and treated. Postoperative severe inflammation (uveitis or vitreoretinal proliferation) should be anticipated and managed accordingly. Abnormal vitreoretinal adherence should be taken into account when planning vitreoretinal surgery. Currently, as a limited number of families with genetic confirmation have been reported, there is no consensus on the treatment of Wagner disease-associated retinal detachments. However, retinal detachment surgery by scleral buckling alone without vitrectomy is the preferred technique when applicable to avoid postoperative inflammation (uveitis or vitreoretinal proliferation), and to overcome the tractional component that cannot be relieved by dissection owing to the abnormal vitreoretinal interface. Electroretinography and visual field testing performed at regular intervals help monitor chorioretinal degeneration. Finally, OCT is useful for non-invasive, detailed clinical evaluation of the vitreoretinal interface, the macula area, and the peripheral retina providing key information for disease monitoring and vitreoretinal surgery planning.

References

1. Wagner H. Ein bisher unbekanntes Erbleiden des Auges (degeneratiohyaloideo-retinalis hereditaria), beobachtet im Kanton Zurich. *Klin Mbl Augenheilk* 1938;**100**:840–57.
2. Rothschild P-R, Burin-des-Roziers C, Audo I, Nedelec B, Valleix S, Brézin AP. Spectral-domain optical coherence tomography in Wagner syndrome: characterization of vitreoretinal interface and foveal changes. *Am J Ophthalmol* 2015;**160**:1065–1072.e1. https://doi.org/10.1016/j.ajo.2015.08.012.
3. Fryer AE, Upadhyaya M, Littler M, Bacon P, Watkins D, Tsipouras P, et al. Exclusion of COL2A1 as a candidate gene in a family with Wagner-Stickler syndrome. *J Med Genet* 1990;**27**:91–3.
4. Brézin AP, Nedelec B, Barjol A, Rothschild P-R, Delpech M, Valleix S. A new VCAN/versican splice acceptor site mutation in a French Wagner family associated with vascular and inflammatory ocular features. *Mol Vis* 2011;**17**:1669–78.
5. Miyamoto T, Inoue H, Sakamoto Y, Kudo E, Naito T, Mikawa T, et al. Identification of a novel splice site mutation of the CSPG2 gene in a Japanese family with Wagner syndrome. *Invest Ophthalmol Vis Sci* 2005;**46**:2726–35. https://doi.org/10.1167/iovs.05-0057.
6. Mukhopadhyay A, Nikopoulos K, Maugeri A, de Brouwer APM, van Nouhuys CE, Boon CJF, et al. Erosive vitreoretinopathy and Wagner disease are caused by intronic mutations in CSPG2/Versican that result in an imbalance of splice variants. *Invest Ophthalmol Vis Sci* 2006;**47**:3565–72. https://doi.org/10.1167/iovs.06-0141.
7. Graemiger RA, Niemeyer G, Schneeberger SA, Messmer EP. Wagner vitreoretinal degeneration. Follow-up of the original pedigree. *Ophthalmology* 1995;**102**:1830–9.
8. Meredith SP, Richards AJ, Flanagan DW, Scott JD, Poulson AV, Snead MP. Clinical characterisation and molecular analysis of Wagner syndrome. *Br J Ophthalmol* 2007;**91**:655–9. https://doi.org/10.1136/bjo.2006.104406.
9. Jewsbury H, Fry AE, Watts P, Nas V, Morgan J. Congenital glaucoma in Wagner syndrome. *J AAPOS* 2014;**18**:291–3. https://doi.org/10.1016/j.jaapos.2013.12.014.
10. Kloeckener-Gruissem B, Bartholdi D, Abdou MT, Zimmermann DR, Berger W. Identification of the genetic defect in the original Wagner syndrome family. *Mol Vis*. 2006;**12**:350–5.
11. Chen X, Zhao K, Sheng X, Li Y, Gao X, Zhang X, et al. Targeted sequencing of 179 genes associated with hereditary retinal dystrophies and 10 candidate genes identifies novel and known mutations in patients with various retinal diseases. *Invest Ophthalmol Vis Sci* 2013;**54**:2186–97. https://doi.org/10.1167/iovs.12-10967.
12. Kloeckener-Gruissem B, Neidhardt J, Magyar I, Plauchu H, Zech J-C, Morlé L, et al. Novel VCAN mutations and evidence for unbalanced alternative splicing in the pathogenesis of Wagner syndrome. *Eur J Hum Genet* 2013;**21**:352–6. https://doi.org/10.1038/ejhg.2012.137.
13. Ronan SM, Tran-Viet K-N, Burner EL, Metlapally R, Toth CA, Young TL. Mutational hot spot potential of a novel base pair mutation of the CSPG2 gene in a family with Wagner syndrome. *Arch Ophthalmol* 2009;**127**:1511–9. https://doi.org/10.1001/archophthalmol.2009.273.
14. Rothschild P-R, Audo I, Nedelec B, Ghiotti T, Brézin AP, Monin C, et al. De novo splice mutation in the versican gene in a family with Wagner syndrome. *JAMA Ophthalmol* 2013;**131**:805–7. https://doi.org/10.1001/jamaophthalmol.2013.681.
15. Burin-des-Roziers C, Rothschild P-R, Layet V, Chen J-M, Ghiotti T, Leroux C, et al. Deletions overlapping VCAN exon 8 are new molecular defects for Wagner disease. *Hum Mutat* 2017;**38**:43–7. https://doi.org/10.1002/humu.23124.

14.5

Knobloch syndrome

Irina Balikova

Knobloch syndrome is an autosomal recessive disorder first described in 1971 by Knobloch and Layer.[1] It is characterised by the triad of occipital skull abnormalities (ranging from an isolated scalp defect to meningoencephalocele), high myopia and vitreoretinal degeneration (often with retinal detachment). There is significant clinical heterogeneity and several additional ocular (e.g. nystagmus, iris abnormalities, macular atrophy, cone-rod dysfunction) and extraocular (e.g. epilepsy, developmental delay, renal abnormalities) features have been described.[2-4]

Timely diagnosis of Knobloch syndrome is important as it allows informed counselling and highlights the risk of potential complications including retinal detachment, lens sub-luxation and glaucoma (angle-closure or pigmentary). Ophthalmologists have a key role in the recognition of this condition as the ocular manifestations can be characteristic whilst the extraocular features may be subtle or absent.

Knobloch syndrome is caused by biallelic mutations in the *COL18A1* gene, encoding collagen XVIII. This collagen subtype is widely expressed throughout the body and is found in association with various basement membranes. It has a role in multiple biological processes including brain and eye development. Cleavage of this matrix molecule produces endostatin, a potent anti-angiogenic protein.[5]

Clinical features

A subset of individuals with Knobloch syndrome present with nystagmus in infancy. Typically, high myopia (ranging from −6 to −20D) is noted in the first years of life. This is associated with increased axial length and progresses with age.[6] Strabismus is seen in a subset of patients.[2,3,7]

Anterior segment abnormalities are a common finding in people with Knobloch syndrome. These include a smooth/featureless iris with flat crypts (Fig. 14.21), persistent pupillary membrane and iris transillumination. Lens sub-luxation is a recognised complication and early-onset cataracts, especially cortical opacities, are common.[2,3,7,8]

Vitreoretinal degeneration is considered a cardinal feature of Knobloch syndrome. Affected individuals typically have a collapsed/synertic vitreous with abnormal vitreous condensations and strands. These are generally more prominent than what would be expected in cases with non-syndromic high myopia. OCT imaging often reveals an abnormal vitreoretinal interface and poor foveal differentiation.[9] The presence of the latter together with infantile nystagmus and iris transillumination in a patient with Knobloch syndrome may lead to an erroneous clinical diagnosis of albinism.[7,10]

Fundus examination reveals myopic fundi with a markedly tessellated appearance, prominent choroidal vessels and peripapillary atrophy. Macular atrophic lesions are very common. These changes can be associated with white fibrillar vitreous condensations and may or may not have a 'punched out' appearance (Fig. 14.22). In some cases, the macular defects are mild, and there are only subtle RPE irregularities. It is worth highlighting though

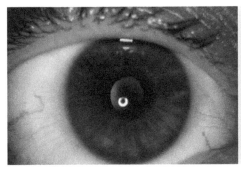 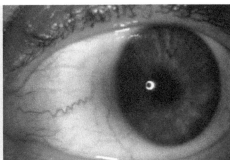

FIG. 14.21 Iris anomalies in a patient with Knobloch syndrome. Smooth iris with hypoplastic iris crypts and hypopigmented appearance. Ectopia of the pupil is also noted on the left side.

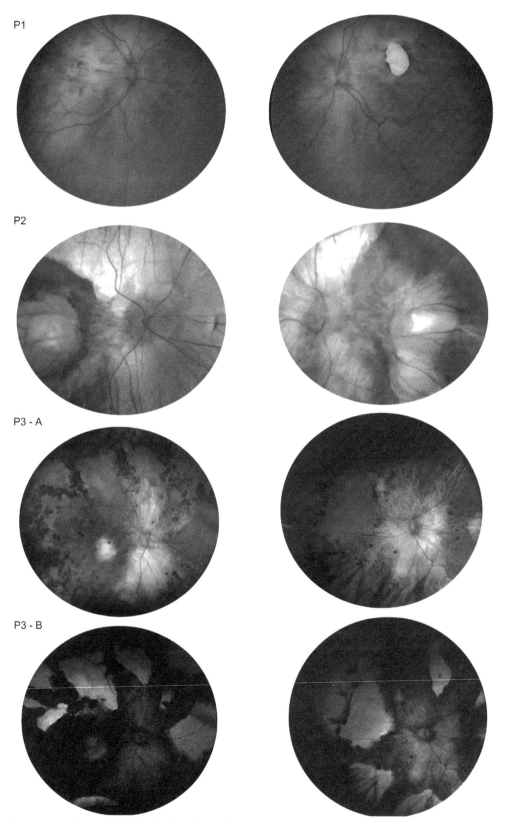

FIG 14.22 Fundus appearance in patients with Knobloch syndrome. (P1) Patient 1 has pigment alterations on the right side and a macular atrophic lesion on the left side. (P2) Patient 2 has extensive chorioretinal atrophy and punched-out macular lesions bilaterally. (P3A) Patient 3 has chorioretinal atrophy, most pronounced peripapillary and extending superiorly and inferiorly along the vascular arcades. The maculae show punched-out lesions. Dense zones of pigment proliferation are present in the mid-periphery. (P3B) Short wavelength (blue light) fundus autofluorescence imaging in patient 3 showing zones of reduced signal corresponding to the atrophic lesions and areas of pigmentations. The findings in P1 and P2 have also been reported in Ref. 8.

that these changes can progress to atrophic lesions in middle life. Other common complications that can threaten vision include retinal detachment[2,3,7,8] and acute angle-closure.[11]

Electroretinography (full-field ERG) in children with Knobloch syndrome often reveals both cone and rod system dysfunction. It is unclear to what extent there is progression; this is likely to be slow and some deterioration of the ERG parameters over time has been documented in a small number of patients.[2,10]

Occipital skull abnormalities are common, but not universally present in patients with Knobloch syndrome. These changes can be very variable ranging from subtle focal defects in the scalp/hair of the occipital region (cutis aplasia) to purely bone lesions to extensive meningoencephaloceles (i.e. sac-like protrusions of intracranial tissue through a defect in the skull) (Fig 14.23). Careful examination of the occiput for cutaneous abnormalities, palpable swelling, alopecia or patches of white hair is recommended in all Knobloch syndrome suspects.

Neuroimaging is typically requested and can be used to help support the diagnosis. In addition to visualising the occipital defects, it can help identify additional abnormalities of the central neural system. A variety of developmental brain anomalies has been associated with Knobloch syndrome including occipital skull defects (with or without encephalocele), polymicrogyria, pachygyria, heterotopic grey matter, agenesis of septum pellucidum, cerebellar and cerebral hypoplasia, hamartomatous lesions and sub-ependymal nodules. Notably, a subset of patients has normal neuroimaging, emphasising that some people with *COL18A1* mutations have non-syndromic eye disease.[2,4] Most affected individuals have normal intelligence, although some have developmental delay. Other systemic problems described in individuals with Knobloch syndrome include epilepsy and renal abnormalities such as congenital hydronephrosis, duplex kidney and bifid ureter.[2–4]

Molecular pathology

Most patients with Knobloch syndrome carry biallelic mutations in the *COL18A1* gene. This gene encodes three different versions (isoforms) of the α1 subunit of collagen XVIII. Mutations in parts of the gene that would lead to a deficit in all three isoforms are thought to cause a more severe ophthalmic phenotype and to predispose to epilepsy. The vast majority of pathogenic variants in *COL18A1* are premature terminations, splice site mutations or whole exon deletions; the abundance of such loss of function mutations suggests that Knobloch syndrome is probably the 'null' *COL18A1* phenotype. One of the most common *COL18A1* mutations is the c.4063_4064delCT variant; this frameshift change has been described in people from various ancestries.[2]

COL18A1 is expressed in multiple organs, with the highest levels detected in the kidney and lung. The associated protein is found throughout the eye although it is not seen in the photoreceptors.[12]

Molecular genetic testing of patients suspected of having Knobloch syndrome is likely to be undertaken either using gene panels for retinal disease or by exome/genome analysis. The presence of large deletions (involving single or multiple exons) underlines the importance of incorporating copy number analysis in these investigations. Notably, in many cases, the condition will be identified as part of workup in a child with an apparently non-syndromic ophthalmic phenotype.

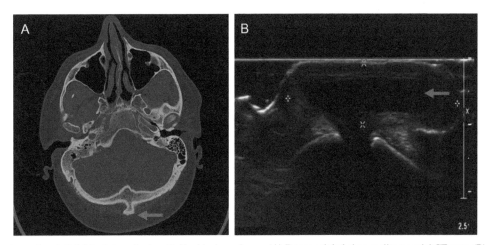

FIG. 14.23 Spectrum of occipital defects in patients with Knobloch syndrome. (A) Bone occipital abnormality on axial CT scan. (B) Occipital meningocele ultrasound image. The abnormalities are highlighted with *arrows*. *(Adapted from Ref. 8.)*

Clinical management

Establishing the extent of the disease is an important initial step in the management of individuals with Knobloch syndrome. A multidisciplinary team approach is recommended and ophthalmologists should work alongside paediatricians and/or clinical geneticists to ensure effective systemic evaluation. Neuroimaging is typically requested to document the characteristic signs of the condition.[4] Electrodiagnostic testing and OCT allow precise characterisation of the ophthalmic phenotype.

Affected children should be regularly followed up to ensure that adequate refraction is performed and that appropriate glasses are issued; amblyopia management should be initiated where appropriate. Surveillance for lens subluxation, angle-closure, and retinal detachment are particularly important

At present, there are no available gene-specific therapies for Knobloch syndrome. Also, there are limited treatment options for neurological manifestations. Whilst most retinal detachments can be repaired, such surgery is usually challenging in Knobloch syndrome and insertion of silicon oil long-term may be required.

References

1. Knobloch WH, Layer JM. Retinal detachment and encephalocele. *J Pediatr Ophthalmol* 1971;**8**:181e4.
2. Hull S, et al. Molecular and clinical findings in patients with Knobloch syndrome. *JAMA Ophthalmol* 2016;**134**(7):753–62.
3. Khan AO, et al. The distinct ophthalmic phenotype of Knobloch syndrome in children. *Br J Ophthalmol* 2012;**96**(6):890–5.
4. Caglayan AO, et al. Brain malformations associated with Knobloch syndrome—review of literature, expanding clinical spectrum, and identification of novel mutations. *Pediatr Neurol* 2014;**51**(6): 806–13. e8.
5. Heljasvaara R, Aikio M, Ruotsalainen H, Pihlajaniemi T. Collagen XVIII in tissue homeostasis and dysregulation—lessons learned from model organisms and human patients. *Matrix Biol* 2017;**57-58**: 55–75. https://doi.org/10.1016/j.matbio.2016.10.002.
6. AlBakri A, Ghazi NG, Khan AO. Biometry, optical coherence tomography, and further clinical observations in Knobloch syndrome. *Ophthalmic Genet* 2017;**38**(2):138–42.
7. Levinger N, Hendler K, Banin E, Hanany M, Kimchi A, Mechoulam H, et al. Variable phenotype of Knobloch syndrome due to biallelic COL18A1 mutations in children. *Eur J Ophthalmol* 2020;**25**. 1120672120977343.
8. Balikova I, Sanak NS, Fanny D, Smits G, Soblet J, de Baere E, et al. Three cases of molecularly confirmed Knobloch syndrome. *Ophthalmic Genet* 2020;**41**(1):83–7.
9. Thau A, Tsukikawa M, Wangtiraumnuay N, Capasso J, Affel E, Alnabi WA, et al. Optical coherence tomography in Knobloch syndrome. *Ophthalmic Surg Lasers Imaging Retina* 2019;**50**(8):e203–10.
10. Gradstein L, Hansen RM, Cox GF, Altschwager P, Fulton AB. Progressive retinal degeneration in a girl with Knobloch syndrome who presented with signs of ocular albinism. *Doc Ophthalmol* 2017;**134** (2):135–40.
11. Wawrzynski J, Than J, Gillam M, Foster PJ. Acute angle closure in Knobloch syndrome. *J Glaucoma* 2021. https://doi.org/10.1097/IJG.0000000000001781. January 13.
12. Määttä M, Heljasvaara R, Pihlajaniemi T, Uusitalo M. Collagen XVIII/endostatin shows a ubiquitous distribution in human ocular tissues and endostatin-containing fragments accumulate in ocular fluid samples. *Graefes Arch Clin Exp Ophthalmol* 2007;**245**(1):74–81.

Chapter 15

Genetic disorders affecting the optic nerve

Chapter outline

15.1 Leber hereditary optic neuropathy 356
15.2 Autosomal dominant optic neuropathy 362
15.3 Autosomal recessive optic neuropathy 368
15.4 Wolfram syndrome spectrum 371

Amongst the wide range of aetiologies underlying bilateral optic neuropathies, genetic conditions represent an important subgroup. The recognition of individuals affected by this broad and expanding group of disorders has historically been challenging and many patients remained undiagnosed. The advent of high-throughput DNA sequencing technologies has enabled a unified approach facilitating prompt diagnosis.

Generally, the genes underlying most cases of hereditary optic neuropathy encode proteins that are localised in mitochondria. These include components of the five multimeric protein complexes that form the mitochondrial respiratory chain, namely complex I [NADH dehydrogenase], complex II [succinate dehydrogenase], complex III [ubiquinol cytochrome *c* oxidoreductase], complex IV [cytochrome *c* oxidase], and complex V [adenosine triphosphate (ATP) synthase]. These complexes are embedded in the inner membrane of mitochondria and coordinate oxidative phosphorylation, a key metabolic pathway that leads to the formation of ATP and is the main source of energy in eukaryotic cells. It is worth highlighting that the structure, function and maintenance of mitochondria depend upon the expression of genes encoded by both the nuclear and the mitochondrial genome. For example, amongst the mitochondrial respiratory chain complexes, complex I is the largest and consists of 7 subunits encoded by the mitochondrial genome and 38 subunits encoded by the nuclear genome. As a result, optic neuropathies can result from both maternally inherited, mitochondrial DNA mutations (e.g. Leber hereditary optic neuropathy, Section 15.1) or from mutations in nuclear-encoded genes (e.g. dominant optic atrophy, Section 15.2; Wolfram syndrome, Section 15.4).

Pathophysiological studies on cellular models of hereditary optic neuropathies often highlight alterations of mitochondrial physiology and structure. In some cases, the enzymatic activity of the mitochondrial respiratory complexes I to IV is abnormal; in other cases, such as dominant optic atrophy, the most discriminative cellular feature concerns the structure of the mitochondrial network. In general, it is unclear why mutations in this wide range of ubiquitously expressed and universally important molecules preferentially/primarily affect the retinal ganglion cells. It is noted though that the significant amount of energy required to repolarise a ganglion cell membrane after an action potential, makes the small calibre and unmyelinated axons of the papillomacular bundle a particularly vulnerable site.

In the past decade, few therapeutic advances have been made for this group of conditions. Idebenone has shown benefit for Leber hereditary optic neuropathy and whilst it has also been studied in related conditions (e.g. dominant optic atrophy) robust evidence is lacking. Gene therapy is proposed as a promising therapeutic strategy for a number of these disorders and clinical trials are ongoing.

15.1

Leber hereditary optic neuropathy

Ungsoo S. Kim and Patrick Yu-Wai-Man

Leber hereditary optic neuropathy (LHON) is a mitochondrially inherited optic neuropathy characterised by bilateral, usually sequential, visual loss. LHON was the first human condition reported to be caused by point mutations in the mitochondrial DNA (mtDNA). Three so-called 'primary', mtDNA mutations are recognised, accounting for around 90% of cases globally (Fig. 15.1).[1–5]

Clinical characteristics

LHON is characterised by subacute, bilateral, painless visual loss. The peak age of onset is between 20 and 35 years, although visual impairment can occur anytime from early childhood to late adulthood. About 12% of LHON carriers lose vision before the age of 12 years; in contrast, about

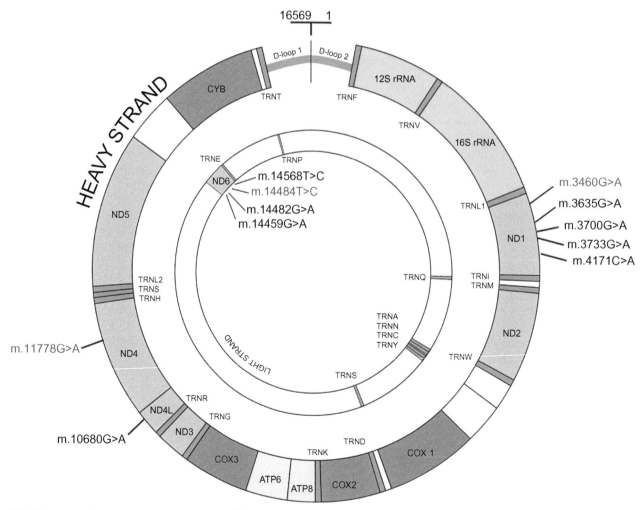

FIG. 15.1 Map of the mitochondrial genome. *ND1–ND6*, subunits of complex I; *CYB*, a subunit of complex III; *COX1-3*, subunits of complex IV; *ATP6 and 8*, subunits of complex V; *COX*, cytochrome *c* oxidase; *CYB*, cytochrome *B*; *TRN*, transfer RNA. The primary mitochondrial DNA mutations causing Leber hereditary optic neuropathy are highlighted with red font.

8% of carriers lose vision when they are older than 50 years of age.[6,7] In about 20%–30% of cases, both eyes are affected simultaneously. In those patients that present with unilateral optic nerve involvement, the fellow eye is usually affected within 2–4 months.[8] If the second eye remains unaffected, another underlying cause for the optic neuropathy should be considered given that 97% of patients with LHON develop vision loss in the second eye within 1 year of the first eye being affected.[9]

The visual impairment is severe and the majority of patients will meet the criteria for legal blindness with visual acuities worse than 1.0 LogMAR. The visual prognosis is partly dependent on the underlying mtDNA mutation with the m.14484T>C variant carrying the highest likelihood of spontaneous visual recovery. This tends to occur 12–24 months after disease onset although later recovery has occasionally been described.[10] With the caveat that the reported rates vary depending on the criteria used to define recovery, 37%–58% of patients with the m.14484T>C sequence change will experience visual improvement compared to 4%–25% for those with the m.11778G>A variant.[4,11]

The clinical stages of LHON have been classified based on the disease status and the time since the onset of visual loss into (1) asymptomatic (mutation carriers); (2) sub-acute (from the onset of visual loss to 6 months after); (3) dynamic (6–12 months); and (4) chronic (>12 months).[12]

In the acute stage, the optic disc classically shows peripapillary telangiectatic microangiopathy, swelling of the retinal nerve fibre layer and vascular tortuosity (Fig. 15.2). However, the fundus can look entirely normal, which can lead to an incorrect diagnosis of functional visual loss, especially in children. In the right clinical context, other features that point towards LHON include the relative preservation of the pupillary light reflexes and the lack of vascular leakage at the optic disc on fluorescein angiography. Early visual field changes before disease conversion have been described in some LHON mutation carriers.[13,14]

OCT imaging has been used to document changes in retinal nerve fibre layer thickness in LHON (Fig. 15.3). In the acute stage, there is increased thickness before gradual thinning is observed as the disease progresses into the chronic stage.[15,16] In contrast, thinning of the ganglion cell—inner plexiform layer within the central macula has been reported as an early sign of disease conversion, occurring before a decrease in visual acuity becomes evident.[17] Studies using OCT-angiography have confirmed the presence of peripapillary telangiectasia, predominantly in the temporal area.[18,19] Although visual electrophysiology is not required when investigating a typical case of LHON, attenuation of the pattern visual evoked potentials and a decrease in the N95 component of the pattern electroretinogram are seen in affected individuals, in keeping with a primary retinal ganglion cell pathology.

The majority of affected individuals with LHON will only develop vision problems. However, a wide range of extraocular features has been reported in association with LHON-associated mutations. The strongest causal associations are with dystonia, myoclonus and multiple sclerosis-

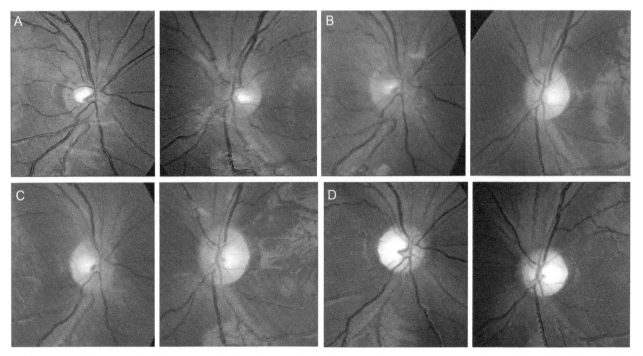

FIG. 15.2 Optic disc changes in a 14-year-old male who carries the m.11778G>A mitochondrial DNA mutation and presented with visual loss in the left eye, followed by the right eye 3 months later. (A) Mild peripapillary retinal nerve fibre layer (RNFL) swelling with segmental RNFL loss along the papillomacular bundle was observed in the left eye 3 months after the onset of visual loss. Visual acuity was 0.0 LogMAR in the right eye and counting fingers in the left eye. (B) The right optic disc was still hyperaemic 2 months after the right eye became affected. Temporal pallor of the left optic disc was now apparent. Visual acuity was 1.3 LogMAR in the right eye and counting fingers in the left eye. (C) and (D) Pallor of the neuroretinal rim and more prominent peripapillary RNFL loss with disease progression; panels C and D refer to 8 and 13 months from first disease onset, respectively.

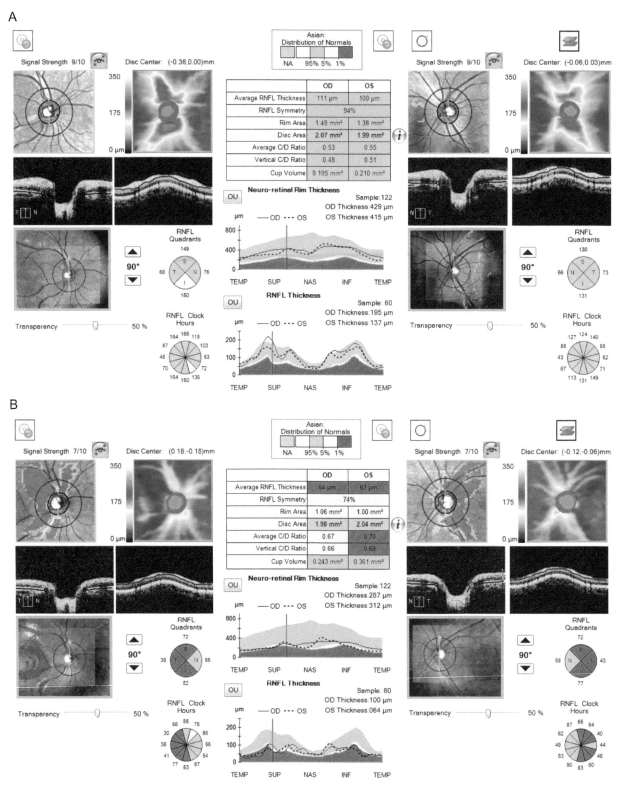

FIG. 15.3 OCT imaging documenting changes in peripapillary retinal nerve fibre layer (RNFL) thickness in Leber hereditary optic neuropathy (LHON). Findings from an 18-year-old man who carries the m.14484T>C mtDNA mutation and presented with a 3-week history of visual disturbance in his left eye are shown. His visual acuities were 1.0 LogMAR in the right and 0.0 LogMAR in the left eye. (A) At the baseline assessment, OCT imaging showed slightly increased RNFL thickness in the right eye and a normal RNFL profile in the left eye. (B) The left eye became involved 2 months after disease onset in the right eye. Four years later, there had been no significant visual recovery and the patient's visual acuities were counting fingers in both eyes. OCT imaging showed marked RNFL thinning with some relative nasal sparing.

like illness (Harding disease).[20] The latter was originally described in female LHON carriers of the m.11778G>A mutation, but it was subsequently reported to be also associated with the m.3460G>A and m.14484T>C mutations. Notably, although classic LHON is less common in female carriers compared to males, women are more likely to develop the LHON—multiple sclerosis overlap syndrome.[21]

Molecular pathology

In 1988, Wallace and colleagues reported the first mtDNA point mutation causing human disease with the identification of the m.11778G>A variant in families with LHON.[2] The so-called 'primary' mtDNA mutations are m.3460G>A, m.11778G>A and m.14484T>C. They account for around 90% of LHON cases worldwide[22,23]; the m.11778G>A change accounts for around 70% of cases in Northern Europe and for up to 90% of cases in Asia.[24] The m.14484T>C variant is the most common pathogenic variant in individuals of French Canadian ancestries due to a mutational founder event.[25] Other rarer mtDNA mutations have been reported in affected individuals that were initially found to be negative for the three primary LHON mutations.[26] These include variants (e.g. m.3890G>A, m.13094T>C) which can have variable multisystemic phenotypic impacts such as Leigh syndrome.[27,28]

The m.3460G>A, m.11778G>A and m.14484T>C mutations give rise to alterations in the MTND1, MTND4 and MTND6 core subunits of the mitochondrial respiratory chain complex I. These abnormalities ultimately lead to compromised oxidative phosphorylation (Fig. 15.1). As a result, there is decreased production of ATP and increased levels of reactive oxygen species, both of which contribute to retinal ganglion cell loss.

Where present, a mtDNA mutation is not in itself sufficient to cause visual loss, with the overall penetrance usually quoted at 50% for a male carrier and 10% for a female carrier (with variation both within and between families). Male patients are predominantly affected and comprise 80%–90% of all cases.[10] Whilst the mitochondrial genetic basis of LHON accounts for the transmission down the maternal line, it does not explain the predilection for male carriers to lose vision.[29] Putative reasons for the incomplete penetrance of LHON include: (1) the complex interplay between the mtDNA background (haplogroups) and the nuclear genome; (2) environmental factors, in particular, smoking[6,30]; (3) the capacity to mount a compensatory response with increased mitochondrial biogenesis[31]; and (4) hormonal factors resulting in female carriers being relatively protected by higher levels of circulating oestrogen levels.[32] In addition, some families demonstrate *heteroplasmy*—that is, the presence of both normal and mutation-positive mitochondrial genomes in the same individual; where heteroplasmy is present, this can also contribute to phenotypic variability (see also Section 13B.2.4, Box 13B.2).

Clinical management

Several possible aetiologies and investigations should be considered when evaluating individuals with suspected LHON. Demyelinating optic neuritis is an important differential diagnosis, especially in individuals with acute presentation who are in the early stages, and in whom the fellow eye remains unaffected. Importantly, visual loss in LHON is painless compared with demyelinating optic neuritis where periorbital or retrobulbar pain with eye movement is a typical sign reported by over 90% of patients.[33] Of note, bilateral demyelinating optic neuritis is uncommon and genetic testing for LHON should be requested when the fellow eye becomes involved.[34] Testing for anti-aquaporin 4 (AQP4) and anti-myelin oligodendrocyte glycoprotein (MOG) antibodies as well as brain MRI evaluation are needed to explore the possibility of neuromyelitis optica (NMO) spectrum disorder and MOG-IgG associated optic neuritis.[35,36] Other inflammatory optic neuropathies need to be excluded and appropriate investigations to rule out systemic disorders such as granulomatosis with polyangiitis (previously known as Wegener granulomatosis) and sarcoidosis should be considered. Nutritional (including those caused by poor diet and chronic malabsorption states) and toxic (including those secondary to drugs such as ethambutol and linezolid) optic neuropathies should also be included in the differential diagnosis.[37]

Genetic testing is the gold standard to confirm the clinical diagnosis of LHON. It is however expected that this will be negative for the three commonest mutations in around 10% of cases. In this scenario, other rarer mtDNA mutations should be sought. It is noted that a small number of individuals with an LHON-like phenotype are found to have biallelic mutations in nuclear genes and therefore are affected by autosomal recessive conditions (see also Section 15.3). This suggests that a sequencing approach that encompasses first the mitochondrial genome and then, where appropriate, the nuclear genome should be utilised.

The management of LHON remains largely supportive with visual rehabilitation for the patient, and genetic counselling that can be extended to the wider family depending on individual circumstances.

Treatment options for LHON remain limited. Notably, idebenone has been approved by the European Medicines Agency for use in LHON. Idebenone is an analogue of ubiquinone, a fat-soluble molecule located within the inner membrane of mitochondria that increases the efficiency of electron transport along the mitochondrial respiratory chain.[38] The cumulative evidence indicates that idebenone has a visual benefit in a subgroup of treated patients and the likelihood of a positive response is higher when the treatment is initiated early.[39] An expert panel recommended that idebenone should be started as soon as possible at a dose of 900 mg per day in patients with a disease duration of less than 1 year and that treatment should be provided for at least 1 year before discontinuing in non-responders.[12]

Besides idebenone, EPI-743 and MTP-131 are two other compounds with purported mitochondrial neuroprotective properties that are being considered in LHON.[40,41] Adequately powered and well-designed randomised clinical trials will be essential to objectively investigate the safety and efficacy of these experimental drugs. Gene therapy is an attractive strategy for LHON as replacement of the missing/defective mitochondrial complex subunit should promote retinal ganglion cell survival by restoring mitochondrial function. Several clinical trials are underway[42] and promising results have been reported for patients treated with a recombinant replication-defective adeno-associated virus, serotype 2, carrying a modified DNA encoding the human mitochondrial ND4 protein (rAAV2/2-*ND4*) and a specific mitochondrial targeting sequence (MTS). Unilateral intravitreal injection of the rAAV2/2-*ND4* gene therapy vector in patients who carried the m.11778G>A mutation, and had a visual loss of up to 1 year resulted in bilateral visual improvement.[43,44]

Environmental triggers are thought to contribute to the overall risk of losing vision in LHON. LHON carriers should be strongly advised not to smoke and to moderate their alcohol consumption.[30] If there are other suitable alternatives, it is also advisable to avoid exposure to medications that have been reported to have a detrimental effect on mitochondrial function.

References

1. Mascialino B, Leinonen M, Meier T. Meta-analysis of the prevalence of Leber hereditary optic neuropathy mtDNA mutations in Europe. *Eur J Ophthalmol* 2012;**22**:461–5.
2. Wallace DC, Singh G, Lott MT, et al. Mitochondrial DNA mutation associated with Leber's hereditary optic neuropathy. *Science* 1988;**242**:1427–30.
3. Puomila A, Hamalainen P, Kivioja S, et al. Epidemiology and penetrance of Leber hereditary optic neuropathy in Finland. *Eur J Hum Genet* 2007;**15**:1079–89.
4. Yu-Wai-Man P, Griffiths PG, Brown DT, Howell N, Turnbull DM, Chinnery PF. The epidemiology of Leber hereditary optic neuropathy in the North East of England. *Am J Hum Genet* 2003;**72**:333–9.
5. Yu-Wai-Man P, Griffiths PG, Chinnery PF. Mitochondrial optic neuropathies—disease mechanisms and therapeutic strategies. *Prog Retin Eye Res* 2011;**30**:81–114.
6. Dimitriadis K, Leonhardt M, Yu-Wai-Man P, et al. Leber's hereditary optic neuropathy with late disease onset: clinical and molecular characteristics of 20 patients. *Orphanet J Rare Dis* 2014;**9**:158.
7. Majander A, Bowman R, Poulton J, et al. Childhood-onset Leber hereditary optic neuropathy. *Br J Ophthalmol* 2017;**101**:1505–9.
8. Harding AE, Sweeney MG, Govan GG, Riordan-Eva P. Pedigree analysis in Leber hereditary optic neuropathy families with a pathogenic mtDNA mutation. *Am J Hum Genet* 1995;**57**:77–86.
9. Riordan-Eva P, Sanders MD, Govan GG, Sweeney MG, Da Costa J, Harding AE. The clinical features of Leber's hereditary optic neuropathy defined by the presence of a pathogenic mitochondrial DNA mutation. *Brain* 1995;**118**(Pt 2):319–37.
10. Newman NJ. Hereditary optic neuropathies: from the mitochondria to the optic nerve. *Am J Ophthalmol* 2005;**140**:517–23.
11. Johns DR, Heher KL, Miller NR, Smith KH. Leber's hereditary optic neuropathy. Clinical manifestations of the 14484 mutation. *Arch Ophthalmol* 1993;**111**:495–8.
12. Carelli V, Carbonelli M, de Coo IF, et al. International consensus statement on the clinical and therapeutic management of leber hereditary optic neuropathy. *J Neuroophthalmol* 2017;**37**:371–81.
13. Hwang TJ, Karanjia R, Moraes-Filho MN, et al. Natural history of conversion of Leber's hereditary optic neuropathy: a prospective case series. *Ophthalmology* 2017;**124**:843–50.
14. Sadun AA, Salomao SR, Berezovsky A, et al. Subclinical carriers and conversions in Leber hereditary optic neuropathy: a prospective psychophysical study. *Trans Am Ophthalmol Soc* 2006;**104**:51–61.
15. Zhang Y, Huang H, Wei S, et al. Characterization of retinal nerve fiber layer thickness changes associated with Leber's hereditary optic neuropathy by optical coherence tomography. *Exp Ther Med* 2014;**7**:483–7.
16. Barboni P, Carbonelli M, Savini G, et al. Natural history of Leber's hereditary optic neuropathy: longitudinal analysis of the retinal nerve fiber layer by optical coherence tomography. *Ophthalmology* 2010;**117**:623–7.
17. Balducci N, Savini G, Cascavilla ML, et al. Macular nerve fibre and ganglion cell layer changes in acute Leber's hereditary optic neuropathy. *Br J Ophthalmol* 2016;**100**:1232–7.
18. Gaier ED, Gittinger JW, Cestari DM, Miller JB. Peripapillary capillary dilation in Leber hereditary optic neuropathy revealed by optical coherence tomographic angiography. *JAMA Ophthalmol* 2016;**134**:1332–4.
19. Takayama K, Ito Y, Kaneko H, Kataoka K, Ra E, Terasaki H. Optical coherence tomography angiography in Leber hereditary optic neuropathy. *Acta Ophthalmol* 2017;**95**:e344–5.
20. Yu-Wai-Man P, Votruba M, Burte F, La Morgia C, Barboni P, Carelli V. A neurodegenerative perspective on mitochondrial optic neuropathies. *Acta Neuropathol* 2016;**132**:789–806.
21. Pfeffer G, Burke A, Yu-Wai-Man P, Compston DA, Chinnery PF. Clinical features of MS associated with Leber hereditary optic neuropathy mtDNA mutations. *Neurology* 2013;**81**:2073–81.
22. Jia X, Li S, Xiao X, Guo X, Zhang Q. Molecular epidemiology of mtDNA mutations in 903 Chinese families suspected with Leber hereditary optic neuropathy. *J Hum Genet* 2006;**51**:851–6.
23. Mackey DA, Oostra RJ, Rosenberg T, et al. Primary pathogenic mtDNA mutations in multigeneration pedigrees with Leber hereditary optic neuropathy. *Am J Hum Genet* 1996;**59**:481–5.
24. Ueda K, Morizane Y, Shiraga F, et al. Nationwide epidemiological survey of Leber hereditary optic neuropathy in Japan. *J Epidemiol* 2017;**27**:447–50.
25. Macmillan C, Johns TA, Fu K, Shoubridge EA. Predominance of the T14484C mutation in French-Canadian families with Leber hereditary optic neuropathy is due to a founder effect. *Am J Hum Genet* 2000;**66**:332–5.

26. Jurkute N, Majander A, Bowman R, et al. Clinical utility gene card for: inherited optic neuropathies including next-generation sequencing-based approaches. *Eur J Hum Genet* 2019;**27**:494–502.
27. Ng YS, Lax NZ, Maddison P, Alston CL, Blakely EL, Hepplewhite PD, et al. MT-ND5 Mutation Exhibits Highly Variable Neurological Manifestations at Low Mutant Load *EBioMedicine*. 2018;**30**:86–93.
28. Vacchiano V, Caporali L, La Morgia C, et al. The m.3890G>A/MT-ND1 mtDNA rare pathogenic variant: Expanding clinical and MRI phenotypes. *Mitochondrion*. 2021;**60**:142–9.
29. Seedorff T. The inheritance of Leber's disease. A genealogical follow-up study. *Acta Ophthalmol* 1985;**63**:135–45.
30. Kirkman MA, Yu-Wai-Man P, Korsten A, et al. Gene-environment interactions in Leber hereditary optic neuropathy. *Brain* 2009;**132**:2317–26.
31. Caporali L, Maresca A, Capristo M, et al. Incomplete penetrance in mitochondrial optic neuropathies. *Mitochondrion* 2017;**36**:130–7.
32. Giordano C, Montopoli M, Perli E, et al. Oestrogens ameliorate mitochondrial dysfunction in Leber's hereditary optic neuropathy. *Brain* 2011;**134**:220–34.
33. Optic Neuritis Study G. Visual function 15 years after optic neuritis: a final follow-up report from the Optic Neuritis Treatment Trial. *Ophthalmology* 2008;**115**:1079–1082.e5.
34. de la Cruz J, Kupersmith MJ. Clinical profile of simultaneous bilateral optic neuritis in adults. *Br J Ophthalmol* 2006;**90**:551–4.
35. Wingerchuk DM, Banwell B, Bennett JL, et al. International consensus diagnostic criteria for neuromyelitis optica spectrum disorders. *Neurology* 2015;**85**:177–89.
36. Cobo-Calvo A, Ruiz A, D'Indy H, et al. MOG antibody-related disorders: common features and uncommon presentations. *J Neurol* 2017;**264**:1945–55.
37. Chamberlain PD, Sadaka A, Berry S, Lee AG. Ethambutol optic neuropathy. *Curr Opin Ophthalmol* 2017;**28**:545–51.
38. Lenaz G, Fato R, Genova ML, Bergamini C, Bianchi C, Biondi A. Mitochondrial complex I: structural and functional aspects. *Biochim Biophys Acta* 2006;**1757**:1406–20.
39. Klopstock T, Yu-Wai-Man P, Dimitriadis K, et al. A randomized placebo-controlled trial of idebenone in Leber's hereditary optic neuropathy. *Brain* 2011;**134**:2677–86.
40. Enns GM, Kinsman SL, Perlman SL, et al. Initial experience in the treatment of inherited mitochondrial disease with EPI-743. *Mol Genet Metab* 2012;**105**:91–102.
41. Imai T, Mishiro K, Takagi T, et al. Protective effect of bendavia (SS-31) against oxygen/glucose-deprivation stress-induced mitochondrial damage in human brain microvascular endothelial cells. *Curr Neurovasc Res* 2017;**14**:53–9.
42. Kim US, Jurkute N, Yu-Wai-Man P. Leber hereditary optic neuropathy-light at the end of the tunnel? *Asia Pac J Ophthalmol (Phila)* 2018;**7**:242–5.
43. Yu-Wai-Man P, Newman NJ, Carelli V, et al. Bilateral visual improvement with unilateral gene therapy injection for Leber hereditary optic neuropathy. *Sci Transl Med* 2020;**12**.
44. Newman NJ, Yu-Wai-Man P, Carelli V, et al. Efficacy and safety of intravitreal gene therapy for Leber hereditary optic neuropathy treated within 6 months of disease onset. *Ophthalmology* 2021;(January 12). S0161-6420(20)31187-8.

15.2

Autosomal dominant optic neuropathy

Guy Lenaers, Patrizia Amati-Bonneau, Alvaro J. Mejia-Vergara, and Alfredo A. Sadun

Autosomal dominant optic neuropathy, also known as dominant optic atrophy, is a heterogeneous group of genetic conditions associated with insidious visual loss and caused by the selective degeneration of the retinal ganglion cells and their descending axons that constitute the optic nerve.[1] Similar to Leber hereditary optic neuropathy (Section 15.1), it is the result of mitochondrial dysfunction. The typical onset of visual decline is in the first or second decade of life, although identifying a precise onset can be challenging due to the slow and gradual nature of the disorder.[2] Visual loss in dominant optic atrophy is painless, typically bilateral and relatively symmetrical. Family history of a similar presentation is common but may be absent due to subclinical disease expression or reduced penetrance.[1]

Dominant optic atrophy is commonly associated with heterozygous mutations in the OPA1 gene.[3,4] Most OPA1 mutation carriers have an isolated, ophthalmic phenotype but some patients present with a syndromic phenotype ('dominant optic atrophy plus') that includes multisystemic features such as hearing loss, myopathy, and peripheral neuropathy.

A small number of optic neuropathy families carry heterozygous variants in genes other than OPA1 (e.g. ACO2, WFS1, OPA3). However, as OPA1-associated optic atrophy is by far the most common subtype, it is the main focus of this chapter.

Clinical characteristics

Dominant optic atrophy may be subtle and difficult to detect. The onset of symptoms is typically in mid-childhood but the diagnosis is usually made several years later. This delay is due to the mild, slow and insidious course of this disorder; difficulty in communicating/appreciating visual loss by young patients may also play a role.[5] Bilateral presentation is the norm, although the disease may initially be asymmetrical. Most often, patients present with moderate loss of visual acuity in the range of 0.2 to 0.6 LogMAR.[6,7] Generally, around 45% of patients will stabilise with a vision between 0.6 and 1.0 LogMAR, whilst 40% of patients will have a visual acuity better than 0.6 LogMAR.[5,8] Dyschromatopsia is variable but frequently present. Although the condition was initially thought to be most closely associated with blue–yellow spectrum deficits, more recent studies suggested that mixed colour deficits are more prevalent.[7,9] The visual field defects are classically central and bilateral; the peripheral visual fields remain unaffected in most cases. The pupillary reflexes and circadian rhythms are normal, probably because melanopsin retinal ganglion cells are largely spared.

Fundus examination reveals optic disc pallor, predominantly temporal, in both eyes (Fig. 15.4A). This is one of the defining features of the condition. About half of the affected individuals have pallor that is limited to the temporal side of the disc whilst the remaining patients have diffuse pallor. A minority of patients have unremarkable optic disc examination at presentation but most of these individuals will develop pallor over time.[7,8] Generally, the degree of optic disc atrophy tends to correlate with the severity of visual loss.[10]

OCT imaging reveals thinning of the retinal nerve fibre layer, especially in the area of the temporal papillomacular bundle (Fig. 15.4B).[11] The macular ganglion cell layer is also thinned (Fig. 15.4C) and OCT-angiography of the peripapillary retina may reveal reduced blood flow in the temporal aspect of the optic disc.[12–14]

OPA1-related dominant optic atrophy can be associated with extra-optic-nerve manifestations. Around 25% of affected individuals are thought to develop syndromic features including hearing impairment, myopathy, sensory and/or motor neuropathy, ataxia, cerebellar/cortical atrophy, and non-specific white matter lesions; ptosis and progressive external ophthalmoplegia may also be present. The most common feature is sensorineural hearing loss; this typically develops during adolescence or young adulthood and it is often followed by muscle weakness (myopathy) and pain and tingling in the arms and legs (peripheral neuropathy).[15]

Molecular pathology

Approximately 65%–90% of dominant optic atrophy cases are found to have pathogenic variants in the OPA1 gene.[16] OPA1 is a nuclear gene that encodes a multifunctional mitochondrial protein that is, expressed in 8 isoforms.[1] This molecule is anchored to the inner membrane of the mitochondrion and has a role in the regulation of the fusion (merging) and fission (splitting) of this inner membrane.[17] Furthermore, it shapes the structure of mitochondrial cristae (inner membrane folds) and contributes to the respiratory chain complex assembly.[18]

Over 400 *OPA1* disease-implicated variants have been reported in over 1200 patients with dominant optic atrophy. Most types of genetic variation have been described, of which around a quarter are missense changes.[19] A few genotype/phenotype correlations have emerged:

- For non-syndromic dominant optic atrophy, haploinsufficiency is generally thought to be the underlying molecular mechanism; consequently, loss of function alleles (e.g. frameshift, nonsense, splicing variants) are particularly prevalent in this group, and missense variants are present in less than a third of such cases.
- Around a quarter of cases present with syndromic, 'dominant optic atrophy plus' phenotypes. Such individuals often carry missense variants within the *OPA1* GTPase domain, in particular in exons 14 and 15. This catalytic domain of the OPA1 protein appears to play a critical role in mitochondrial homeostasis and it is thought that missense changes altering its sequence increase the risk of extraocular manifestations, potentially through a dominant negative mechanism.
- Biallelic *OPA1* mutations are associated with an uncommon syndromic optic neuropathy (Behr syndrome, see also Section 15.3) featuring early-onset optic atrophy, spinocerebellar ataxia, spasticity, peripheral neuropathy, and gastrointestinal dysmotility. Affected individuals generally carry a loss of function allele *in trans* to a hypomorphic allele.[20]

The different clinical presentations associated with missense *OPA1* mutations make an accurate interpretation of these variants critical. A locus-specific database (www.lovd.nl/OPA1) has been developed; this valuable resource can help with the assessment of previously reported sequence alterations, including recurrent variants such as the c.1499G>A (p.Arg500His) mutation, which commonly leads to optic atrophy and sensorineural hearing loss.[21]

ACO2 and other genes associated with dominant optic atrophy: Apart from *OPA1*, other genes have been implicated in dominantly inherited optic neuropathies including *OPA3*, *OPA4*, *OPA5*, *OPA8*, *SPG7*, *AFG3L2* and *SSBP1*.[1] Although mutations in each of these loci are found in less than 1% of dominant optic atrophy patients, sequencing these genes allows a more robust approach to genomic analysis and diagnosis. *ACO2* is thought to be the most commonly mutated gene after *OPA1*. It encodes a mitochondrial protein that has a role in the mitochondrial Krebs (citric acid) cycle. Over 50 families carrying heterozygous *ACO2* variants have been described and mutation carriers present with optic atrophy that is indistinguishable from

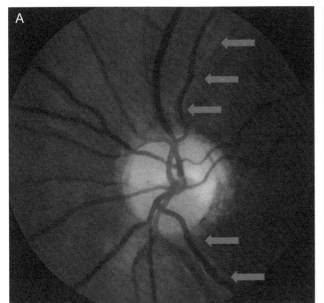
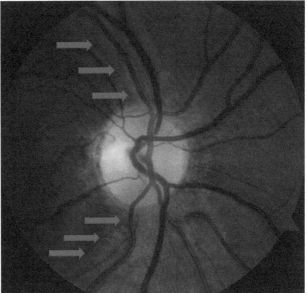

FIG. 15.4 Fundus imaging in a 12-year-old female with dominant optic atrophy. She first presented at age 8 years with loss of central vision. Four years later her visual acuities were 0.9 LogMAR in each eye. Both her father and older sister have a diagnosis of dominant optic atrophy and were found to carry a pathogenic variant in *OPA1*. (A) Colour fundus photography reveals a wedge-like, pale area with defined borders supero- and infero-temporally to each optic disc (the demarcation line is highlighted with *blue arrows*). This finding is in keeping with retinal nerve fibre layer (RNFL) loss predominantly affecting the papillomacular bundle.

(Continued)

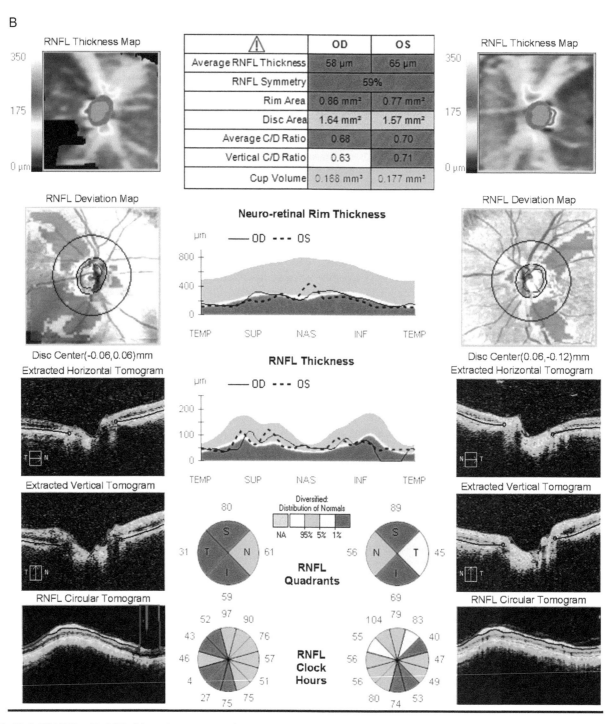

FIG. 15.4, CONT'D (B) OCT of the optic nerve and peripapillary RNFL. The abnormal areas seen on fundus photography corresponds to a region of significant RNFL thinning.

(Continued)

FIG. 15.4, CONT'D (C) OCT of the macular ganglion cell layer- inner plexiform layer (GCL + IPL) complex. Significant, generalised thinning of this complex is noted centrally. This corresponds to the temporal RNFL loss as these fibres are a projection of the ganglion cells in the macular region.

that induced by *OPA1* variants. A small number of cases with biallelic *ACO2* changes has been reported with affected individuals having a more severe clinical presentation, including infantile cerebellar and retinal degeneration.[22] Heterozygous variants in *WFS1* and *OPA3* can cause dominant optic atrophy associated with deafness and cataract, respectively. Notably, biallelic mutations in these genes are seen in association with the distinct multisystemic conditions Wolfram syndrome (Section 15.4) and 3-methylglutaconic aciduria (Costeff syndrome).

Clinical management

The differential diagnosis of dominant optic atrophy is broad and includes Leber hereditary optic neuropathy, toxic or nutritional optic neuropathy and normal tension glaucoma. Clinical suspicion of the condition should prompt genetic testing. Establishing that there is at least one affected member in two consecutive generations, and distinguishing an autosomal dominant from a mitochondrial maternal transmission, typically orientates genetic investigations. If a mutation is identified, segregation analysis should be considered. Genetic counselling for affected families conveys the fact that the disease is inherited as an autosomal dominant trait; prenatal diagnosis is feasible but remains uncommon.

Individuals with syndromic dominant atrophy should be managed by a multidisciplinary team including geneticists, neuro-ophthalmologists, neurologists, and otorhinolaryngologists. Mitochondrial respiratory chain studies to evaluate the severity of the energetic deficiency are often recommended in multisystemic disorders affecting mitochondrial function (including 'dominant optic atrophy plus'). Either a skin biopsy to generate fibroblasts or a skeletal muscle biopsy can be used to measure the enzymatic activity of the five mitochondrial respiratory chain complexes.

It is currently not possible to reverse the visual loss associated with dominant optic atrophy. However, many patients may be helped significantly by low vision aids. Affected individuals are advised not to smoke and to moderate their alcohol consumption. If there are other suitable alternatives, it is recommended to avoid exposure to drugs that have been reported to have a detrimental effect on mitochondrial function.

Patients with syndromic dominant optic atrophy often experience dual sensory impairment as the visual defect is combined with hearing loss. For this group of patients, cochlear implantation should be considered.

Several experimental approaches are under investigation although robust clinical trials in humans are lacking. These include oral idebenone (see Section 15.1)[23–25] and gene therapy using adeno-associated viral vectors.[26]

References

1. Lenaers G, Neutzner A, Le Dantec Y, Jüschke C, Xiao T, Decembrini S, et al. Dominant optic atrophy: culprit mitochondria in the optic nerve. *Prog Retin Eye Res* 2020;(December 16):100935.
2. Newman NJ. Hereditary optic neuropathies: from the mitochondria to the optic nerve. *Am J Ophthalmol* 2005;**140**(3):517–23.
3. Delettre C, Lenaers G, Griffoin JM, et al. Nuclear gene OPA1, encoding a mitochondrial dynamin-related protein, is mutated in dominant optic atrophy. *Nat Genet* 2000;**26**:207–10. https://doi.org/10.1038/79936.
4. Alexander C, Votruba M, Pesch UE, et al. OPA1, encoding a dynamin-related GTPase, is mutated in autosomal dominant optic atrophy linked to chromosome 3q28. *Nat Genet* 2000;**26**:211–5. https://doi.org/10.1038/79944.
5. Miller NR, Subramanian PS, Patel VR. In: Williams & Wilkins, editor. *Walsh & Hoyt's clinical neuro-ophthalmology: the essentials*. 3rd ed. Wolters Kluwer; 2016.
6. Kjer B, Eiberg H, Kjer P, Rosenberg T. Dominant optic atrophy mapped to chromosome 3q region. II Clinical and epidemiological aspects. *Acta Ophthal Scand* 1996;**74**:3–7.
7. Votruba M, Fitzke FW, Holder GE. Clinical features in affected individuals from 21 pedigrees with dominant optic atrophy. *Arch Ophthalmol* 1998;**116**(3):351. https://doi.org/10.1001/archopht.116.3.351.
8. Pineles SL, Balcer LJ. 5. Visual loss: optic neuropathies. In: *Liu, Volpe, and Galetta's neuro-ophthalmology*. 3rd ed. vol. 1. Elsevier; 2019. p. 101–96. https://doi.org/10.3399/bjgp10x539335.
9. Kline LB, Glaser JS. Dominant optic atrophy. The clinical profile. *Arch Ophthalmol* 1979;**97**(9):1680–6.
10. Newman NJ. 8. Hereditary optic neuropathies. In: *Neuro-ophthalmology*. vol. 1; 2008. p. 191–205.
11. Park SW, Hwang JM. Optical coherence tomography shows early loss of the inferior temporal quadrant retinal nerve fiber layer in autosomal dominant optic atrophy. *Graefes Arch Clin Exp Ophthalmol* 2014;**253**(1):135–41. https://doi.org/10.1007/s00417-014-2852-7.
12. Balducci N, Ciardella A, Gattegna R, et al. Optical coherence tomography angiography of the peripapillary retina and optic nerve head in dominant optic atrophy. *Mitochondrion* 2017;**36**:60–5. https://doi.org/10.1016/j.mito.2017.03.002.
13. Himori N, Kunikata H, Inoue M, Takeshita T, Nishiguchi KM, Nakazawa T. Optic nerve head microcirculation in autosomal dominant optic atrophy and normal-tension glaucoma. *Acta Ophthalmol* 2017;**95**(8):e799–800. https://doi.org/10.1111/aos.13353.
14. Inoue M, Himori N, Kunikata H, et al. The reduction of temporal optic nerve head microcirculation in autosomal dominant optic atrophy. *Acta Ophthalmol* 2016;**94**(7):e580–5. https://doi.org/10.1111/aos.12999.
15. Chao de la Barca JM, Prunier-Mirebeau D, Amati-Bonneau P, et al. OPA1-related disorders: diversity of clinical expression, modes of inheritance and pathophysiology. *Neurobiol Dis* 2016;**90**:20–6. https://doi.org/10.1016/j.nbd.2015.08.015.
16. Chun BY, Rizzo III JF. Dominant optic atrophy: updates on the pathophysiology and clinical manifestations of optic atrophy 1 mutation. *Curr Opin Ophthalmol* 2016;**27**:475–80.
17. MacVicar T, Langer T. OPA1 processing in cell death and disease—the long and short of it. *J Cell Sci* 2016;**129**:2297–306.
18. Cogliati S, Frezza C, Soriano ME, et al. Mitochondrial cristae shape determines respiratory chain supercomplexes assembly and respiratory efficiency. *Cell* 2013;**155**:160–71.
19. Le Roux B, Lenaers G, Zanlonghi X, et al. OPA1: 516 unique variants and 831 patients registered in an updated centralized Variome database. *Orphanet J Rare Dis* 2019;**14**:214. https://doi.org/10.1186/s13023-019-1187-1.
20. Bonneau D, Colin E, Oca F, et al. Early-onset Behr syndrome due to compound heterozygous mutations in OPA1. *Brain* 2014;**137**:e301.
21. Yu-Wai-Man P, Trenell MI, Hollingsworth KG, Griffiths PG, Chinnery PF. OPA1 mutations impair mitochondrial function in both pure and complicated dominant optic atrophy. *Brain J Neurol* 2011;**134**:e164.
22. Spiegel R, Pines O, Ta-Shma A, et al. Infantile cerebellar-retinal degeneration associated with a mutation in mitochondrial aconitase, ACO2. *Am J Hum Genet* 2012;**90**:518–23.
23. Del Dotto V, Fogazza M, Lenaers G, Rugolo M, Carelli V, Zanna C. OPA1: how much do we know to approach therapy? *Pharmacol Res* 2018;**131**:199–210. https://doi.org/10.1016/j.phrs.2018.02.018.

24. Smith TG, Seto S, Ganne P, Votruba M. A randomized, placebo-controlled trial of the benzoquinone idebenone in a mouse model of OPA1-related dominant optic atrophy reveals a limited therapeutic effect on retinal ganglion cell dendropathy and visual function. *Neuroscience* 2016;**319**:92–106. https://doi.org/10.1016/j.neuroscience.2016.01.042.

25. Barboni P, Valentino ML, La Morgia C, et al. Idebenone treatment in patients with OPA1-mutant dominant optic atrophy. *Brain J Neurol* 2013;**136**:e231. https://doi.org/10.1093/brain/aws280.

26. Sarzi E, Seveno M, Piro-Mégy C, et al. OPA1 gene therapy prevents retinal ganglion cell loss in a Dominant Optic Atrophy mouse model. *Sci Rep* 2018;**8**(1):1–6. https://doi.org/10.1038/s41598-018-20838-8.

15.3

Autosomal recessive optic neuropathy

Valerio Carelli, Chiara La Morgia, and Piero Barboni

Autosomal recessive optic neuropathy, also known as recessive optic atrophy, is a clinically and genetically heterogeneous group of rare genetic conditions associated with prominent optic nerve involvement.[1–4] At least 10 disease-associated genes have been identified. By definition, these are encoded by the nuclear genome but most of their protein products are subsequently imported and function in the mitochondria. This is in keeping with the general observation that retinal ganglion cells and their optic nerve-forming axons are selectively vulnerable to mitochondrial dysfunction.[5,6] It also highlights the common pathophysiological mechanisms between recessive optic atrophy and the much more prevalent Leber hereditary optic neuropathy (Section 15.1) and dominant optic atrophy (Section 15.2).

Recessive optic atrophy is often isolated but it can also be a feature of a multisystemic disorder. For example, it is one of the hallmarks of Wolfram syndrome (discussed in Section 15.4) and Behr syndrome (discussed below and briefly in Section 15.2). Notably, individuals with non-syndromic phenotypes suggestive of Leber hereditary optic neuropathy may rarely be found to have mutations in genes causing recessive optic neuropathies. Comprehensive genetic testing approaches facilitate prompt diagnosis of these rare disorders enabling delineation of prognosis and a better understanding of familial risk.

Clinical characteristics

Many genetic subtypes of recessive optic atrophy have been described. There is significant variability in clinical presentation although, to an extent, there is an overlap in terms of ophthalmic features; bilateral visual acuity loss, optic disc pallor and central scotomata are common findings. The clinical characteristics of selected optic atrophy subtypes are discussed below (except for Wolfram syndrome which is discussed in Section 15.4)

TMEM126A: Individuals with biallelic mutations in the *TMEM126A* gene typically have a childhood-onset, slowly progressive optic neuropathy. Sudden onset visual loss occurring in adulthood (with some similarities to Leber hereditary optic neuropathy) has been described in a subset of patients. Extraocular features tend to be absent or relatively mild, and may include sensorineural hearing loss (auditory neuropathy), peripheral polyneuropathy (sensorimotor axonal neuropathy with electrophysiological abnormalities suggestive of focal demyelination), and mild hypertrophic cardiomyopathy.[7–9]

ACO2: Biallelic mutations in *ACO2* are associated with a wide range of phenotypes. A significant number of affected individuals carry presumed loss of function variants and present in early infancy with severe truncal hypotonia, ataxia, seizures, microcephaly, and ophthalmological abnormalities including optic atrophy and, later in the disease process, retinal degeneration.[10–12] In contrast, a subset of patients presents with a later onset, milder phenotype; this includes a small number of cases presenting with isolated recessive optic neuropathy that may resemble dominant *OPA1*-related optic atrophy.[13,14]

RTN4IP1: Similar to *ACO2*-related disease, the clinical presentation in individuals with biallelic mutations in *RTN4IP1* ranges from severe mitochondrial encephalopathy to isolated optic atrophy.[15,16] Extraocular features include seizures, intellectual disability, developmental delay, growth retardation, sensorineural hearing loss, stridor, elevated lactate levels, abnormal neuroimaging and electroencephalographic patterns, and premature death. Retinal degeneration in a rod-cone dystrophy pattern may be present.[17]

NDUFS2: A similar pattern to the above disorders has been noted in patients with biallelic variants in *NDUFS2*.[18] Affected individuals typically have a constellation of symptoms of variable severity consistent with Leigh syndrome, an early-onset progressive encephalomyopathy associated with central nervous system degeneration, optic neuropathy, and premature death. It is noted though that three affected individuals from a single family were found to carry compound heterozygous *NDUFS2* variants and to have a phenotype similar to Leber hereditary optic neuropathy (subacute, painless, simultaneous, or sequential visual loss in the first few decades of life without extraocular features).

SLC25A46: Biallelic mutations in *SLC25A6* are also associated with a spectrum of neurodegenerative conditions that have optic atrophy as a core feature.[19,20] Mutations that have a severe effect on protein lead to Leigh syndrome and premature death. In contrast, variants with a less severe impact on protein function have been associated with childhood-onset optic neuropathy with no significant extraocular features in the first decades of life.[21]

YME1L1: A homozygous *YME1L1* mutation has been identified in four affected siblings from a single family of Saudi Arabian origin. All patients had intellectual disability, developmental delay, speech delay, hearing impairment, and optic atrophy. Other features included microcephaly or macrocephaly, ataxia, hyperkinesia, and athetotic/stereotypic movements. Neuroimaging was remarkable for leukoencephalopathy in all affected siblings.[22]

OPA1: As described in Section 15.2, heterozygous *OPA1* mutations are the commonest cause of dominant optic atrophy.[5] Biallelic *OPA1* variants are associated with more severe, clearly recessive phenotypes that always include optic atrophy. These may range from cases classified as Behr syndrome (early-onset optic neuropathy with neurological manifestations including ataxia, spasticity, peripheral neuropathy and, in some cases, developmental delay)[23,24] to encephalopathy with cardiomyopathy[25] and to Leigh-like syndrome.[26,27]

OPA3: Heterozygous *OPA3* mutations are a rare cause of dominant optic atrophy. Autosomal recessive phenotypes resulting from biallelic *OPA3* variants have been described under the name 'Costeff syndrome'; individuals with this condition typically have a multisystemic clinical phenotype with optic nerve involvement.[28]

Optic atrophy is a frequent feature in several syndromic disorders resulting from iron deposition in the brain. The term 'neurodegeneration with brain iron accumulation' (NBIA) has been used to describe this heterogeneous group of genetic conditions that are associated with dystonia, spasticity, parkinsonism, and dysarthria.[29] Biallelic loss of function variants in the *C19orf12* gene, encoding a mitochondrial membrane protein, account for up to 30% of NBIA cases and result in a condition termed 'mitochondrial membrane protein-associated neurodegeneration' (MPAN). MPAN is characterised initially by gait changes followed by spasticity, neuropsychiatric abnormalities, and cognitive decline; optic atrophy can be an early feature of this disorder.[30]

Many other rare forms of syndromic optic neuropathy have been reported. These are usually associated with encephalopathy, seizures, ataxia, developmental delay and many other features. Examples of such recessive conditions include disorders due to mutations in *FDXR* and *FDX2*.[31]

Molecular pathology

The molecular defects in a few selected genetic subtypes of recessive optic atrophy are discussed below.

TMEM126A encodes a protein localised in the inner mitochondrial membrane.[9] It has been recently clarified that TMEM126A is an assembly factor of the ND4 distal membrane module of complex I, explaining why *TMEM126A* mutations lead to complex I deficiency.[32] Notably, multiple affected families of north African ancestries have been found to carry the c.163C>T (p.Arg55Ter) variant.[7,8]

ACO2 encodes mitochondrial aconitase (ACO2), a ubiquitously expressed enzyme that catalyses the conversion of citrate into isocitrate and has an important role in mitochondrial DNA maintenance.[10–14] Metabolite testing does not reveal a diagnostic pattern in affected individuals on traditional biochemical testing but a putative metabolic signature has been described. Enzymatic activity in patient tissues or variant-specific assays *in vitro* has been utilised to evaluate variant pathogenicity; the findings of these assays are thought to correlate with disease severity.

RTN4IP1 encodes another mitochondria-targeted protein. It is localised in the mitochondrial outer membrane and is predicted to interact with RTN4, a molecule localised in the endoplasmic reticulum. *RTN4IP1* mutations do not seem to affect mitochondrial network morphology and generally result in a deficit in the enzymatic activity of mitochondrial complexes I and IV.[15,16] Initial reports identified a recurrent c.308G>A (p.Arg103His) mutation but since then further pathogenic missense and protein-truncating variants have been reported.

YME1L1 is an ATP-dependent enzyme that is embedded in the inner mitochondrial membrane and has a role in the maintenance of mitochondrial morphology. Notably, it enables the processing of the long forms of OPA1 (the protein that is mutated in dominant optic atrophy) to short ones. The balance of long and short OPA1 forms is key for mitochondrial dynamics and for balancing fission and fusion.[33]

Clinical management

Defining precise or unified care pathways for patients with optic neuropathies is challenging. Affected individuals will likely be referred from a variety of sources depending on whether the phenotype suggests an isolated or a multisystemic condition. Since genetic forms of optic neuropathy can be diagnosed using a single diagnostic tool (genomic sequencing), care is likely to involve collaboration between the ophthalmologist and clinical genetics. Furthermore, patients affected by syndromic disease are likely to require management that includes input from paediatrics, neurology and/or metabolic medicine.

Since many of the genes implicated in optic neuropathies can cause both dominant/recessive and isolated/syndromic disorders, genomic variant interpretation is complex and requires multidisciplinary input.

Idebenone, currently approved for Leber hereditary optic neuropathy, may be useful in some forms of recessive optic atrophy, in particular when complex I deficiency is documented.[34]

References

1. Waardenburg PJ. Different types of hereditary optic atrophy. *Acta Genet Statist Med* 1957;7:287–90.
2. Møller HU. Recessively inherited, simple optic atrophy—does it exist? *Ophthalmic Paediatr Genet* 1992;13:31–2.
3. Phillips CI, Mackintosh GI, Howe JW, Mitchell KW. Autosomal recessive 'optic atrophy' with late onset and evidence of ganglion cell dysfunction: a sibship of two females. *Ophthalmologica* 1993;206:89–93.
4. Barbet F, Gerber S, Hakiki S, Perrault I, Hanein S, Ducroq D, et al. A first locus for isolated autosomal recessive optic atrophy (ROA1) maps to chromosome 8q. *Eur J Hum Genet* 2003;11:966–71.
5. Yu-Wai-Man P, Votruba M, Burté F, La Morgia C, Barboni P, Carelli V. A neurodegenerative perspective on mitochondrial optic neuropathies. *Acta Neuropathol* 2016;132:789–806.
6. Carelli V, La Morgia C, Ross-Cisneros FN, Sadun AA. Optic neuropathies: the tip of the neurodegeneration iceberg. *Hum Mol Genet* 2017;26:R139–50.
7. Hanein S, Perrault I, Roche O, Gerber S, Khadom N, Rio M, et al. TMEM126A, encoding a mitochondrial protein, is mutated in autosomal-recessive nonsyndromic optic atrophy. *Am J Hum Genet* 2009;84:493–8.
8. La Morgia C, Caporali L, Tagliavini F, Palombo F, Carbonelli M, Liguori R, et al. First TMEM126A missense mutation in an Italian proband with optic atrophy and deafness. *Neurol Genet* 2019;5:e329.
9. Hanein S, Garcia M, Fares-Taie L, Serre V, De Keyzer Y, Delaveau T, et al. TMEM126A is a mitochondrial located mRNA (MLR) protein of the mitochondrial inner membrane. *Biochim Biophys Acta* 1830;2013:3719–33.
10. Spiegel R, Pines O, Ta-Shma A, Burak E, Shaag A, Halvardson J, et al. Infantile cerebellar-retinal degeneration associated with a mutation in mitochondrial aconitase, ACO2. *Am J Hum Genet* 2012;90:518–23.
11. Sharkia R, Wierenga KJ, Kessel A, Azem A, Bertini E, Carrozzo R, et al. Clinical, radiological, and genetic characteristics of 16 patients with ACO2 gene defects: delineation of an emerging neurometabolic syndrome. *J Inherit Metab Dis* 2019;42(2):264–75.
12. Marelli C, Hamel C, Quiles M, Carlander B, Larrieu L, Delettre C, et al. ACO2 mutations: a novel phenotype associating severe optic atrophy and spastic paraplegia. *Neurol Genet* 2018;4:e225.
13. Kelman JC, Kamien BA, Murray NC, Goel H, Fraser CL, Grigg JR. A sibling study of isolated optic neuropathy associated with novel variants in the ACO2 gene. *Ophthalmic Genet* 2018;1–4.
14. Gibson S, Azamian MS, Lalani SR, Yen KG, Sutton VR, Scott DA. Recessive ACO2 variants as a cause of isolated ophthalmologic phenotypes. *Am J Med Genet A* 2020;182(8):1960–6.
15. Angebault C, Guichet PO, Talmat-Amar Y, Charif M, Gerber S, Fares-Taie L, et al. Recessive mutations in RTN4IP1 cause isolated and syndromic optic neuropathies. *Am J Hum Genet* 2015;97:754–60.
16. Charif M, Nasca A, Thompson K, Gerber S, Makowski C, Mazaheri N, et al. Neurologic phenotypes associated with mutations in RTN4IP1 (OPA10) in children and young adults. *JAMA Neurol* 2018;75:105–13.
17. Rajabian F, Manitto MP, Palombo F, Caporali L, Grazioli A, Starace V, et al. Combined optic atrophy and rod-cone dystrophy expands the RTN4IP1 (optic atrophy 10) phenotype. *J Neuroophthalmol* 2021;41(3):e290–2.
18. Gerber S, Ding MG, Gérard X, et al. Compound heterozygosity for severe and hypomorphic NDUFS2 mutations cause non-syndromic LHON-like optic neuropathy. *J Med Genet* 2017;54:346–56.
19. Abrams AJ, Hufnagel RB, Rebelo A, Zanna C, Patel N, Gonzalez MA, et al. Mutations in SLC25A46, encoding a UGO1-like protein, cause an optic atrophy spectrum disorder. *Nat Genet* 2015;47:926–32.
20. Janer A, Prudent J, Paupe V, Fahiminiya S, Majewski J, Sgarioto N, et al. SLC25A46 is required for mitochondrial lipid homeostasis and cristae maintenance and is responsible for Leigh syndrome. *EMBO Mol Med* 2016;8:1019–38.
21. Abrams AJ, Fontanesi F, Tan NBL, Buglo E, Campeanu IJ, Rebelo AP, et al. Insights into the genotype-phenotype correlation and molecular function of SLC25A46. *Hum Mutat* 2018;39(12):1995–2007.
22. Hartmann B, Wai T, Hu H, MacVicar T, Musante L, Fischer-Zirnsak B, et al. Homozygous YME1L1 mutation causes mitochondriopathy with optic atrophy and mitochondrial network fragmentation. *Elife* 2016;5:e16078.
23. Bonneau D, Colin E, Oca F, Layet V, Layet A, Stevanin G, et al. Early-onset Behr syndrome due to compound heterozygous mutations in OPA1. *Brain* 2014;137:e301.
24. Carelli V, Sabatelli M, Carrozzo R, et al. 'Behr syndrome' with OPA1 compound heterozygote mutations. *Brain* 2015;138:e321.
25. Spiegel R, Saada A, Flannery PJ, Burté F, Soiferman D, Khayat M, et al. Fatal infantile mitochondrial encephalomyopathy, hypertrophic cardiomyopathy and optic atrophy associated with a homozygous OPA1 mutation. *J Med Genet* 2016;53:127–31.
26. Nasca A, Rizza T, Doimo M, Legati A, Ciolfi A, Diodato D, et al. Not only dominant, not only optic atrophy: expanding the clinical spectrum associated with OPA1 mutations. *Orphanet J Rare Dis* 2017;12:89.
27. Rubegni A, Pisano T, Bacci G, Tessa A, Battini R, Procopio E, et al. Leigh-like neuroimaging features associated with new biallelic mutations in OPA1. *Eur J Paediatr Neurol* 2017;21:671–7.
28. Anikster Y, Kleta R, Shaag A, Gahl WA, Elpeleg O. Type III 3-methylglutaconic aciduria (optic atrophy plus syndrome, or Costeff optic atrophy syndrome): identification of the OPA3 gene and its founder mutation in Iraqi Jews. *Am J Hum Genet* 2001;69:1218–24.
29. Wiethoff S, Houlden H. Neurodegeneration with brain iron accumulation. *Handb Clin Neurol* 2017;145:157–66.
30. Hartig MB, Iuso A, Haack T, Kmiec T, Jurkiewicz E, Heim K, et al. Absence of an orphan mitochondrial protein, c19orf12, causes a distinct clinical subtype of neurodegeneration with brain iron accumulation. *Am J Hum Genet* 2011;89:543–50.
31. Peng Y, Shinde DN, Valencia CA, Mo JS, Rosenfeld J, Truitt Cho M, et al. Biallelic mutations in the ferredoxin reductase gene cause novel mitochondriopathy with optic atrophy. *Hum Mol Genet* 2017;26:4937–50.
32. D'Angelo L, Astro E, De Luise M, Kurelac I, Umesh-Ganesh N, Ding S, Fearnley IM, Gasparre G, Zeviani M, Porcelli AM, Fernandez-Vizarra E, Iommarini L. NDUFS3 depletion permits complex I maturation and reveals TMEM126A/OPA7 as an assembly factor binding the ND4-module intermediate. *Cell Rep.* 2021;35(3):109002.
33. Anand R, Wai T, Baker MJ, Kladt N, Schauss AC, Rugarli E, et al. The i-AAA protease YME1L and OMA1 cleave OPA1 to balance mitochondrial fusion and fission. *J Cell Biol* 2014;204:919–29.
34. Amore G, Romagnoli M, Carbonelli M, Barboni P, Carelli V, La Morgia C. Therapeutic options in hereditary optic neuropathies. *Drugs.* 2021;81(1):57–86.

15.4

Wolfram syndrome spectrum

Nicole Balducci, Chiara La Morgia, Michele Carbonelli, Piero Barboni, and Valerio Carelli

Wolfram syndrome is a highly variable, multisystemic disorder also known as diabetes insipidus, diabetes mellitus, optic atrophy, and deafness (DIDMOAD). This syndromic optic neuropathy is an important cause of infantile/juvenile-onset diabetes and it is usually associated with biallelic variants in the *WFS1* gene or, less commonly, in the *CISD2* gene. A milder presentation, often described as 'Wolfram-like' syndrome has been reported; this rare subtype is caused by heterozygous mutations in *WFS1* and it is characterised by one or more of: diabetes mellitus, progressive hearing loss, and optic atrophy.[1–4]

Clinical characteristics

The diagnostic criteria for DIDMOAD were initially described in 1995.[5] Insulin-dependent diabetes mellitus and bilateral progressive optic atrophy are the hallmark features. Although these can emerge at different ages, most affected individuals present with diabetes mellitus in the first decade of life (typically around age 6 years); optic atrophy generally follows in the second decade (typically around age 11 years). Some patients also develop diabetes insipidus and/or deafness later in childhood or early adulthood. Urinary tract problems and neurological symptoms may present after the third decade of life.[6]

In terms of ophthalmic features, optic nerve involvement leads to gradual, progressive visual acuity loss and may result in severe visual impairment within a decade after the onset of disease (the best-corrected visual acuity is often worse than 1.0 LogMAR in the less affected eye). Colour vision is often significantly affected and pupillary responses are attenuated once optic nerve damage has commenced. Fundoscopy reveals diffuse optic disc pallor or atrophy (Fig. 15.5). Affected individuals may show central scotomas, but most frequently there is diffuse reduction of retinal sensitivity with variable involvement of the peripheral fields (Fig. 15.6).[7–10] As expected in retinal ganglion cell disease, visual evoked potentials show reduced amplitudes and electroretinography tends to be within normal limits. However, abnormal dark- and light-adapted full-field electroretinographic responses have been reported in a small number of cases.[11] Intriguingly, clinical and topographical changes in corneal morphology compatible with the early-stage keratoconus have been described in a significant proportion of affected individuals.[12]

Progressive retinal nerve fibre and macular ganglion cell layer thinning are detectable on OCT (Fig. 15.7), with structural parameters potentially preceding visual acuity loss.[13] A characteristic lamination at the level of the outer plexiform layer has been reported to occur exclusively in patients with optic atrophy secondary to heterozygous missense *WFS1* mutations in 'Wolfram-like' syndrome.[14] In these affected individuals, the lamination of the outer plexiform layer is characterised by three distinct sub-layers:

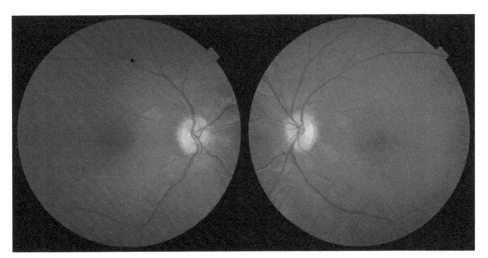

FIG. 15.5 Fundus photography of a 15-year-old girl affected by Wolfram syndrome showing bilateral diffusely pale optic discs.

372 SECTION | III Genetic disorders affecting the posterior segment

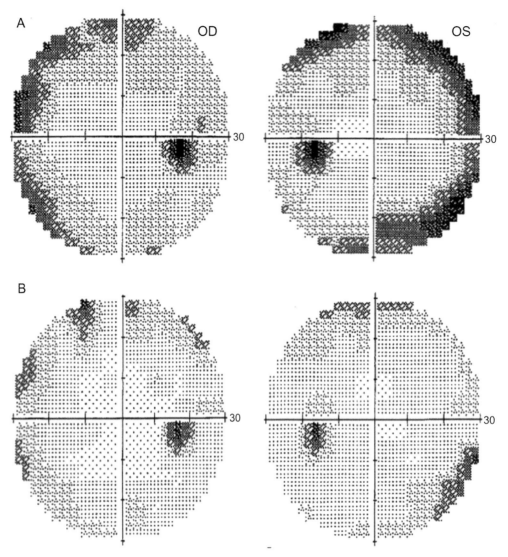

FIG. 15.6 Static perimetry (30° Humphrey visual field) in a 46-year-old woman affected by Wolfram syndrome. A progressive diffuse reduction of retinal sensitivity is noted (A). Deterioration is observed compared to a test performed 7 years before (B).

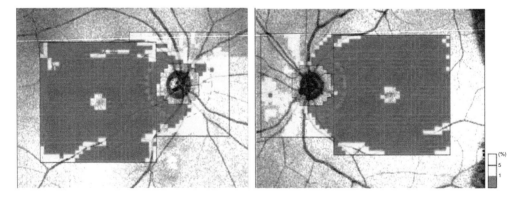

FIG. 15.7 OCT deviation map in a 46-year-old woman affected by Wolfram syndrome (patient in Fig. 15.6). There is thinning of the temporal peripapillary retinal nerve fibre layer and of the macular ganglion cell layer.

(1) an innermost highly reflective lamina, similar to what is frequently identified as the outer plexiform layer in normal eyes; (2) a non-reflective cleft-like middle lamina, located between the synaptic and pedicle sublayers of the outer plexiform layer; and (3) an outermost highly reflective lamina with characteristics of Henle's fibre layer. In the majority of patients, the middle outer plexiform lamina forms a nearly confluent ring, centered at the fovea and extending 1.0 to 1.5 mm from the foveolar centre up to the optic disc[14] (Fig. 15.8).

In terms of systemic features, in addition to childhood-onset insulin-dependent diabetes mellitus, many individuals with DIDMOAD also develop diabetes insipidus (40%–70% of cases); this is thought to be due to hypothalamic dysfunction and it often occurs late in the second decade of life. These abnormalities lead to symptoms of excessive thirst and urination; other symptoms may include dehydration, weakness and mouth dryness. Sensorineural hearing loss is noted in about 30%–60% of cases and can range from deafness beginning at birth to mild hearing loss during adolescence that worsens over time. The hearing impairment usually affects the high frequencies and then progresses slowly.[3,5,15]

Urinary tract abnormalities develop in at least 30% of cases, usually in the third decade of life. These vary widely and may be complicated by diabetes insipidus. They typically include bladder instability/atony (leading to incontinence) and structural abnormalities (ureteric obstruction, hydroureter, severe hydronephrosis) predisposing to recurrent urinary tract infections.[3]

Neurological manifestations are also common; they occur in at least 30% of cases, typically after the third decade of life, and may include ataxia, peripheral or autonomic neuropathy, anosmia, myoclonus and/or cognitive impairment. Brainstem atrophy is almost always present and, in the most severe cases, it can result in premature death secondary to central apnea. Psychiatric and behavioural problems including depression and dementia may develop later in the course of the disorder.[3,5,16]

The life expectancy is significantly reduced for most DIDMOAD patients. On average this tends to be 30–40 years although some affected individuals live well into middle age. The most common causes of death are respiratory failure, autonomic dysfunction and dysphagia.[3,5]

The differential diagnosis of DIDMOAD includes mitochondrial disorders and, in particular, mitochondrial encephalopathy lactic acidosis and stroke-like episodes (MELAS),[17,18] autosomal dominant optic atrophy, *WFS1*-associated deafness, Friedreich ataxia, Bardet–Biedl syndrome, Alström syndrome, CAPOS syndrome (cerebellar ataxia—areflexia—pes cavus—optic atrophy—sensorineural hearing loss), congenital rubella syndrome and thiamine-responsive anaemia with diabetes mellitus and deafness.[19]

The term 'Wolfram-like' syndrome has been used to describe a heterogeneous group of autosomal dominantly inherited disorders associated with heterozygous mutations in the *WFS1* gene. They include a milder range of phenotypes such as isolated optic atrophy, optic atrophy

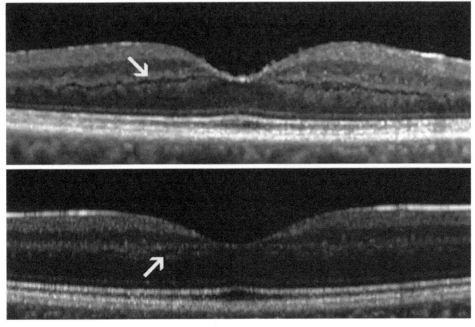

FIG. 15.8 Macular OCT from a 69-(upper panel) and a 23-year-old patient with 'Wolfram-like' syndrome and heterozygous mutations in *WFS1*. Outer plexiform layer (OPL) lamination is noted. The *yellow arrows* indicate the cleft-like middle structure of the OPL. *(Adapted from Ref. 14.)*

associated with low-frequency hearing loss, isolated low-frequency sensorineural hearing loss, diabetes alone or associated with hearing impairment.[4,20–23]

Molecular pathology

Two genes have been associated with DIDMOAD: *WFS1* and *CISD2*. *WFS1* is responsible for about 90% of cases and encodes wolframin, a transmembrane glycoprotein, localised primarily in the endoplasmatic reticulum. Wolframin is highly expressed in the pancreas (β-cells), the brain (neurons) and the retina (including in retinal ganglion cells, photoreceptors and Müller cells). It has been shown that *WFS1* mutations lead to elevated endoplasmatic reticulum stress levels and aggregation of misfolded proteins. This can trigger apoptosis especially in cell types that rely heavily on the endoplasmatic reticulum for protein folding (e.g. pancreatic β-cells, retinal ganglion cells). As discussed above, DIDMOAD is typically caused by biallelic mutations in *WFS1*; however, heterozygous mutations in the same gene have also been associated with disease, albeit with milder phenotypes ('Wolfram-like' syndrome). There appears to be no clear genotype–phenotype correlation in *WFS1*-associated disorders and wide heterogeneity may be observed even within the same family.[1,2,6,24,25]

Biallelic mutations in the *CISD2* gene are a rare cause of DIDMOAD. Individuals affected by this disease subtype often present with optic atrophy; diabetes mellitus and hearing impairment typically develop later and diabetes insipidus is generally absent. Upper intestinal ulcers and defective platelet aggregation have been described in some individuals with *CIDS2*-associated DIDMOAD. The *CISD2* gene encodes an endoplasmatic reticulum intermembrane small protein that localises to mitochondrial and endoplasmic reticulum membranes.[1,26,27]

Notably, in a small number of patients who have phenotypes overlapping with DIDMOAD, but who lack mutations in *WFS1* or *CIDS2*, mitochondrial DNA variants have been identified (including known LHON-associated mutations).[28,29] Conversely, individuals with genetic forms of optic neuropathy who test negative for *OPA1*, *OPA3* and mitochondrial DNA mutations may occasionally carry pathogenic variants in *WFS1*.[30]

It has been previously proposed that mitochondrial dysfunction plays a role in the pathogenic mechanism of DIDMOAD.[31] It has now been shown that the WFS1 protein is enriched in cellular membranes, where the endoplasmatic reticulum comes into close proximity to mitochondria. This confirms the central role of the endoplasmatic reticulum in disease pathophysiology and suggests that mitochondrial dysfunction is mostly a secondary event in DIDMOAD.[17,18,32]

Clinical management

There are no effective therapies that can delay the progression of the underlying disease. Specialist multidisciplinary care is required and careful clinical monitoring and supportive therapy can alleviate symptoms and improve quality of life. Particular emphasis should be placed on the optimisation of glycaemic control with insulin therapy. The use of idebenone[33] or docosahexaenoic acid may delay optic nerve atrophy, although the efficacy of these agents has not been proven in large-scale studies. Low vision aids can be valuable in daily life.

Careful evaluation of the underlying inheritance pattern is important for counselling and reproductive advice. It can also help with confirming the diagnosis and it has the potential to inform multidisciplinary patient management.[2,3]

References

1. Pallotta MT, Tascini G, Crispoldi R, Orabona C, Mondanelli G, Grohmann U, et al. Wolfram syndrome, a rare neurodegenerative disease: from pathogenesis to future treatment perspectives. *J Transl Med* 2019;**17**(1):238.
2. Tranebjærg L, Barrett T, Rendtorff ND. WFS1 Wolfram syndrome spectrum disorder. In: Adam MP, Ardinger HH, Pagon RA, et al., editors. *GeneReviews® [internet]*. Seattle, WA: University of Washington, Seattle; 1993–2020. 2009 February 24 [updated 2020 April 9]. Available from: https://www.ncbi.nlm.nih.gov/books/NBK4144/.
3. Urano F. Wolfram Syndrome: diagnosis, management, and treatment. *Curr Diab Rep* 2016;**16**:6.
4. Valero R, Bannwarth S, Roman S, Pasquis-Flucklinger V, Vialettes B. Autosomal dominant transmission of diabetes and congenital hearing impairment secondary to a missense mutation in the WFS1 gene. *Diabet Med* 2008;**25**:657–61.
5. Barrett TG, Bundey SE, Macleod AF. Neurodegeneration and diabetes: UK nationwide study of Wolfram (DIDMOAD) syndrome. *Lancet* 1995;**346**(8988):1458–63.
6. de Heredia ML, Clèries R, Nunes V. Genotypic classification of patients with Wolfram syndrome: insights into the natural history of the disease and correlation with phenotype. *Genet Med* 2013;**15**(7):497–506. https://doi.org/10.1038/gim.2012.180.
7. Barrett TG, Bundey SE, Fielder AR, Good PA. Optic atrophy in Wolfram (DIDMOAD) syndrome. *Eye (Lond)* 1997;**11**(Pt 6):882–8.
8. Hoekel J, Chisholm SA, Al-Lozi A, Hershey T, Tychsen L, Washington University Wolfram Study Group. Ophthalmologic correlates of disease severity in children and adolescents with Wolfram syndrome. *J AAPOS* 2014;**18**(5):461–465.e1.
9. Soares A, Mota Á, Fonseca S, Faria O, Brandão E, Falcão Dos Reis F, et al. Ophthalmologic manifestations of Wolfram syndrome: report of 14 cases. *Ophthalmologica* 2019;**241**(2):116–9.
10. Ustaoglu M, Onder F, Karapapak M, Taslidere H, Guven D. Ophthalmic, systemic, and genetic characteristics of patients with Wolfram syndrome. *Eur J Ophthalmol* 2020;**30**(5):1099–105.
11. Scaramuzzi M, Kumar P, Peachey N, Nucci P, Traboulsi EI. Evidence of retinal degeneration in Wolfram syndrome. *Ophthalmic Genet* 2019;**40**(1):34–8.

12. Waszczykowska A, Zmysłowska A, Braun M, Zielonka E, Ivask M, Koks S, et al. Corneal abnormalities are novel clinical feature in Wolfram syndrome. *Am J Ophthalmol* 2020;**217**:140–51.
13. Zmyslowska A, Fendler W, Waszczykowska A, Niwald A, Borowiec M, Jurowsk P, et al. Retinal thickness as a marker of disease progression in longitudinal observation of patients with Wolfram syndrome. *Acta Diabetol* 2017;**54**:1019–24.
14. Majander A, Bitner-Glindzicz M, Chan CM, Duncan HJ, Chinnery PF, Subash M, et al. Lamination of the outer plexiform layer in optic atrophy caused by dominant WFS1 mutations. *Ophthalmology* 2016;**123**(7):1624–6.
15. Karzon R, Narayanan A, Chen L, Lieu JEC, Hershey T. Longitudinal hearing loss in Wolfram syndrome. *Orphanet J Rare Dis* 2018;**13**(1):102. https://doi.org/10.1186/s13023-018-0852-0. 29945639.
16. Chaussenot A, Bannwarth S, Rouzier C, Vialettes B, Mkadem SA, Chabrol B, et al. Neurologic features and genotype-phenotype correlation in Wolfram syndrome. *Ann Neurol* 2011;**69**:501–8.
17. La Morgia C, Maresca A, Amore G, Gramegna LL, Carbonelli M, Scimonelli E, et al. Calcium mishandling in absence of primary mitochondrial dysfunction drives cellular pathology in Wolfram Syndrome. *Sci Rep* 2020;**10**:4758.
18. La Morgia C, Maresca A, Caporali L, Valentino ML, Carelli V. Mitochondrial diseases in adults. *J Intern Med* 2020;**287**:592–608.
19. Astuti D, Sabir A, Fulton P, Zatyka M, Williams D, Hardy C, et al. Monogenic diabetes syndromes: locus-specific databases for Alström, Wolfram, and Thiamine-responsive megaloblastic anemia. *Hum Mutat* 2017;**38**(7):764–77.
20. Bonnycastle LL, Chines PS, Hara T, et al. Autosomal dominant diabetes arising from a Wolfram syndrome 1 mutation. *Diabetes* 2013;**62**(11):3943–50.
21. Eiberg H, Hansen L, Kjer B, Hansen T, Pedersen O, Bille M, et al. Autosomal dominant optic atrophy associated with hearing impairment and impaired glucose regulation caused by a missense mutation in the WFS1 gene. *J Med Genet* 2006;**43**:435–40.
22. Grenier J, Meunier I, Daien V, Baudoin C, Halloy F, Bocquet B, et al. WFS1 in optic neuropathies: mutation findings in nonsyndromic optic atrophy and assessment of clinical severity. *Ophthalmology* 2016;**123**:1989–98.
23. Lesperance MM, Hall JW, San Agustin TB, Leal SM. Mutations in the Wolfram Syndrome Type 1 Gene (WFS1) define a clinical entity of dominant low-frequency sensorineural hearing loss. *J Clin Res Pediatr Endocrinol* 2016;**8**(4):482–3.
24. Riachi M, Yilmaz S, Kurnaz E, Aycan Z, Çetinkaya S, Tranebjærg L, et al. Functional assessment of variants associated with Wolfram syndrome. *Hum Mol Genet* 2019;**28**(22):3815–24.
25. Takeda K, Inoue K, Tanizawa Y, Marsuzaki Y, Oba J, Watanabe Y, et al. WFS1 (Wolfram syndrome 1) gene product: predominant subcellular localization to endoplasmatic reticulum in cultured cells and neuronal expression in rat brain. *Hum Mol Genet* 2001;**10**:477–84.
26. Amr S, Heisey C, Zhang M, Xia X-J, Shows KH, Ajlouni K, et al. A homozygous mutation in a novel zinc-finger protein, ERIS, is responsible for Wolfram syndrome 2. *Am J Hum Genet* 2007;**81**:673–83.
27. Mozzillo E, Delvecchio M, Carella M, et al. A novel CISD2 intragenic deletion, optic neuropathy and platelet aggregation defect in Wolfram syndrome type 2. *BMC Med Genet* 2014;**15**:88.
28. Hofmann S, Bezold R, Jaksch M, Obermaier-Kusser B, Mertens S, Kaufhold P, et al. Wolfram (DIDMOAD) syndrome and Leber hereditary optic neuropathy (LHON) are associated with distinct mitochondrial DNA haplotypes. *Genomics* 1997;**39**:8–18.
29. Rötig A, Cormier V, Chatelain P, Francois R, Saudubray JM, Rustin P, et al. Deletion of mitochondrial DNA in a case of early-onset diabetes mellitus, optic atrophy, and deafness (Wolfram syndrome, MIM 222300). *J Clin Invest* 1993;**91**(3):1095–8.
30. Galvez-Ruiz A, Galindo-Ferreiro A, Schatz P. Genetic testing for wolfram syndrome mutations in a sample of 71 patients with hereditary optic neuropathy and negative genetic test results for OPA1/OPA3/LHON. *Neuroophthalmology* 2017;**42**(2):73–82.
31. Bu X, Rotter JI. Wolfram syndrome: a mitochondrial-mediated disorder? *Lancet* 1993;**342**:598–600.
32. Angebault C, Fauconnier J, Petergnani S, Rieusset J, Danese A, Affortit J, et al. ER-mitochondria cross-talk is regulated by Ca^{2+} sensor NCS1 and is impaired in Wolfram Syndrome. *Sci Signal* 2018;**11**:eaaq1380.
33. Bababeygy SR, Wang MY, Khaderi KR, Sadun AA. Visual improvement with the use of idebenone in the treatment of Wolfram syndrome. *J Neuroophthalmol* 2012;**32**:386–9.

Section IV

Genetic disorders affecting both the anterior and posterior segment

Chapter 16

Developmental eye disorders

Chapter outline
16.1 Microphthalmia–anophthalmia–coloboma spectrum   378
16.2 Nanophthalmia and posterior microphthalmia   385

16.1

Microphthalmia–anophthalmia–coloboma spectrum

Mariya Moosajee and Graeme C.M. Black

Microphthalmia, anophthalmia, and ocular coloboma, collectively considered together under the acronym MAC, represent a vast clinical spectrum that is a common cause of childhood visual impairment.[1-4] MAC conditions carry profound lifelong impact for both child and family, with significant social, emotional, and health implications including elevated early mortality rates in syndromic cases.

Most MAC cases are sporadic; around two-thirds are syndromic with up to one-third displaying craniofacial anomalies. Inheritance can be autosomal or X-linked, and in both cases, it can be either dominant or recessive. The occurrence of *de novo* mutations, mosaicism and incomplete penetrance makes clinical prediction of inheritance patterns challenging.

This group of conditions is highly genetically heterogenous with over 90 disease-causing genes already identified.[4] Genomic testing of MAC patients using high-throughput sequencing is increasingly successful in identifying the underlying cause. Given that *de novo* mutations are common in apparently sporadic cases (i.e. the majority of individuals with MAC), testing of parent-offspring trios is generally the most effective strategy in this group of patients.

Clinical characteristics

From an ophthalmic viewpoint, the diagnosis of a MAC condition can often be made relatively easily. Since these are congenital eye anomalies, they are present at birth and many cases are being detected in the newborn period—some cases of anophthalmia may even be detected antenatally. However, more subtle forms of microphthalmia will not be noted at birth. Also, depending on size and position, chorioretinal or optic disc colobomata may be asymptomatic and remain undetected even into adulthood.

Where a MAC anomaly is identified, the delineation of any associated eye abnormalities and the subclassification of the developmental ocular defect is important. This will help to guide ophthalmic management, assist interpretation of genomic results and facilitate estimation of recurrence risks. Ophthalmologists must assess both anterior and posterior segments as affected eyes often have more than one ocular defect. Visual electrophysiology is valuable in assessing the level of vision as well as retinal function (as several MAC-implicated genes have also been associated with retinal degeneration).

Anophthalmia: True anophthalmia occurs when eye development is aborted at around 3–4 weeks gestation when the optic vesicle is developing, leading to the absence of the eye, optic nerve, and chiasm. Often, a small cystic remnant is detectable, a situation termed 'clinical anophthalmia'; this suggests that the optic vesicle has formed but subsequently degenerated leaving an extremely hypoplastic optic nerve, chiasm or tract. In these instances, the phenotype may overlap with the most severe forms of microphthalmia (Fig. 16.1). The precise anatomy can be confirmed by an

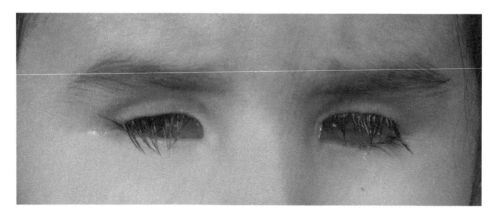

FIG. 16.1 External photograph from a child with bilateral anophthalmia. MRI scan of the orbits highlighted that the intraorbital structures were small. There is no cornea, lens, or sclera detectable. The axial length was 5 mm for the right globe and 4 mm for the left globe. The optic nerves were hypoplastic in the intraconal orbital space and were undetectable at the distal optic canal. MRI scan of the brain did not show any accompanying central nervous system abnormalities. This patient was therefore classified as having severe microphthalmia or clinical anophthalmia.

MRI scan of the brain and orbits although this is in many cases somewhat academic. The absence of the eye has a profound impact on the growth of the orbit, especially during the first 3 years of life.

Microphthalmia: Microphthalmia is a clinically and functionally heterogeneous group of conditions. Most bilateral cases have associated multisystemic features that form part of a syndrome; this tends to be more variable in unilateral cases and careful systemic investigation is required.[5] A microphthalmic eye has reduced overall volume; it may be associated with reduced overall size or with a reduction in the growth of either the anterior or the posterior segment. The mean axial length of a full-term neonate is 17 mm; for an adult, this is 23.8 mm. An axial length of <19 mm at 1 year of age or <21 mm in an adult using ultrasound B-scan constitutes a diagnosis of microphthalmia, representing ≥2 standard deviations below normal. Assessment of corneal diameter is also required: microcornea is defined as horizontal diameter <9 mm in a newborn and/or <10 mm for children >2 years of age. Posterior microphthalmia is a rare subset of microphthalmia in which the total axial length of the eyeball is reduced, although the anterior segment dimensions including corneal diameter, anterior chamber depth, and anteroposterior length of the lens are within normal limits (see Section 16.2).[6]

The impact on vision is dependent on severity and eye size, as well as on whether the microphthalmia is 'simple' (i.e. without other ocular defects) or 'complex' and therefore associated with other ocular malformations such as iris/chorioretinal coloboma, anterior segment dysgenesis, developmental cataract, vitreoretinal dysplasia, retinal dystrophy, and optic nerve hypoplasia (see Box 16.1).

Ocular coloboma: Ocular coloboma is the commonest of the MAC conditions. It represents incomplete fusion of the optic fissure, a process that occurs between 5 and 7 weeks gestation, and can affect one or more structures involved in the process including the iris, lens, ciliary body, retina, choroid, and optic nerve (Fig. 16.2). True colobomata are found in the inferonasal quadrant of the eye and may be strikingly asymmetrical. Iris coloboma for example typically manifests as a wedge-shaped defect inferonasally (so-called 'keyhole' pupil). Where incomplete optic fissure closure affects the ciliary body, there may be a concomitant defect in the lens shape (e.g. a notch can be seen) as a result of zonular deficiency. On fundoscopy, a chorioretinal coloboma manifests as an area of choroidal deficiency that reveals bare white sclera with areas of hyperpigmentation at the margins of the defect (Fig. 16.3). A large coloboma may be associated with a superior visual field defect and, where a large chorioretinal coloboma encroaches upon the macula, there may be a significant impact on central vision. Up to a third of patients with chorioretinal coloboma, in particular, those with high myopia, can be at increased risk of retinal detachment and may require monitoring. Failure of closure of the most posterior portion of the optic fissure results in an optic nerve coloboma, leaving a defect in the inferior portion of the optic nerve.

Syndromic MAC: All children with ocular developmental defects should be investigated within a multidisciplinary team, including paediatricians and clinical geneticists, to identify syndromic features and thereafter manage co-morbid complications. There is a considerable number of recognisable syndromes associated with MAC

BOX 16.1 Complex microphthalmia

Where microphthalmic eyes have associated structural ocular defects, the characterisation of such defects is important for understanding functional consequences for vision, and for determining clinical management options. Such phenotypes are linked to genes known to be mutated in association with a range of ocular abnormalities:

Anterior segment (including cornea and lens): A normal lens is critical to normal ocular development and microphthalmia/microcornea are well-recognised associations of congenital cataracts. Mutations in genes that cause isolated (e.g. crystallins, connexins, transcription factors such as *MAF*) as well as syndromic congenital cataract (e.g. *BCOR*, *NHS*, and *EPHA2*), have been associated with MAC. Interestingly, biallelic variants in several genes that are typically associated with dominant isolated cataracts have been implicated in syndromic microphthalmia (e.g. *PITX3*).[16] Two genes associated with characteristic anterior segment phenotypes are *FOXE3* and *PAX6*. *FOXE3* controls very early lens development and biallelic mutations underlie primary congenital aphakia associated with microphthalmia.[17] *PAX6* is a key transcription factor and heterozygous missense mutations are an uncommon cause of complex microphthalmia which may be associated with iris abnormalities and anterior segment dysgenesis.[18]

Posterior segment (including retina and vitreous): Three genes that are associated with MAC and posterior segment abnormalities are *ATOH7*, *RBP4*, and *MIR-204*. *ATOH7* is expressed in retinal progenitor cells and is required for normal retinal ganglion cell development. Biallelic mutations in this gene have been described in severe vitreoretinal phenotypes including persistent fetal vasculature and 'congenital retinal non-attachment'. Such patients may have a consequent reduction in ocular size and carry a diagnosis of microphthalmia.[19] Heterozygous and biallelic mutations in *RBP4*, which binds circulating retinol, can cause vitamin A deficiency and colobomatous microphthalmia with retinal dystrophy.[20] *MIR-204* encodes a microRNA, a molecule involved in the regulation of gene expression; a single family with bilateral colobomata and retinal dystrophy has been shown to result from a heterozygous mutation in *MIR-204*.[21]

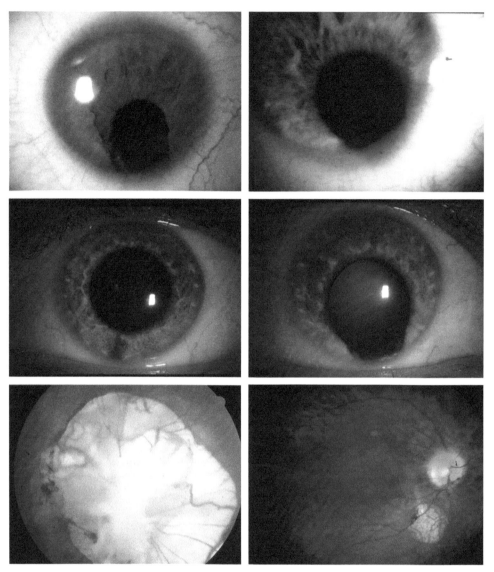

FIG. 16.2 Optic fissure closure defects (colobomata). (Top panel) Iris coloboma associated with microcornea. (Middle panel) Bilateral iris coloboma showing variable expressivity. (Bottom panel) Large optic nerve coloboma and small asymptomatic chorioretinal coloboma.

defects (e.g. CHARGE syndrome; anophthalmia–oesophageal–genital (AEG) syndrome, Warburg Micro syndrome, Matthew Wood syndrome; see Table 16.1). Several of these conditions result from perturbations of specific biochemical and metabolic pathways (e.g. transcription factors, retinoic acid signalling, BMP growth factor signalling), and there is often considerable overlap within and among the different conditions. For this reason, a systematic clinical approach (see Chapter 5) that is aligned to an agnostic genomic testing strategy is the most efficient method for achieving a diagnosis. In particular, this should aim to identify treatable or preventable systemic sequelae including associated cardiovascular, renal, and neurological pathologies.

Molecular pathology

Genetic testing of MAC patients using high-throughput sequencing is increasingly available and, as more and more genes are described, increasingly successful in identifying the underlying cause. There are relatively few published case series that are large enough and sufficiently unbiased in their case selection to judge true diagnostic rates.[7–9] Often, published cohorts do not include prediagnosed patients with chromosomal abnormalities and large-scale genomic alterations that may be diagnosed with conventional array CGH (comparative genomic hybridisation; see Chapter 2); aneuploidies, translocations, large deletions

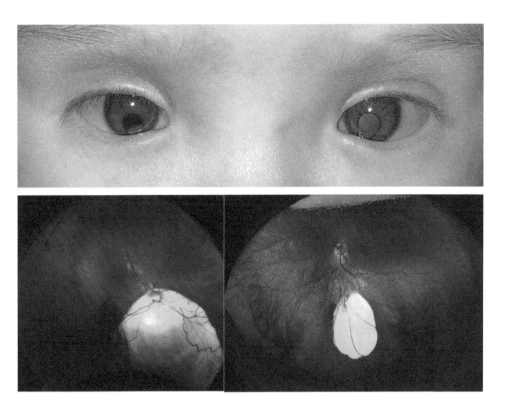

FIG. 16.3 External photograph (top) and widefield fundus imaging in a 5-month-old girl with bilateral iris and chorioretinal coloboma. Ultrasound B-scan revealed an axial length of 20mm in the right globe and 21mm in the left globe, indicating that the colobomatous changes are not associated with microphthalmia.

and duplications account for 10%–25% of MAC cases.[5] Furthermore, patients with classically diagnosed syndromes (renal coloboma syndrome, CHARGE) may also not be included in the construction of 'undiagnosed patient' cohorts. In general, the diagnostic rate for severe anophthalmia/microphthalmia is reported to be 30%–75%; in broader cohorts with bilateral MAC it is ~20%–30% and, for unilateral microphthalmia, it is ~10%.[7–9]

Given the large number of conditions caused by *de novo* mutational events, for apparently sporadic cases (i.e. the majority of cases) it is advisable, where possible, to test parent-offspring trios.

MAC phenotypes resulting from heterozygous mutations

Several MAC phenotypes result from heterozygous gene mutations. Classical examples include *SOX2* and *OTX2*, two transcription factors that are critical to early ocular development (see Box 16.2). Indeed, *SOX2* and *OTX2* mutations are the most frequently identified causes of bilateral anophthalmia and severe microphthalmia. Heterozygous pathogenic *SOX2* variants, including whole gene deletions, account for 15%–40% of such cases and 10%–15% of all anophthalmia/microphthalmia patients. Heterozygous *OTX2* variants underlie 2%–10% of anophthalmia and microphthalmia cases (see Box 16.1).[4] More rare causes include

- *MAB21L2*: Heterozygous mutations in *MAB21L2*, a gene encoding a BMP4 signalling antagonist, are a rare cause of MAC phenotypes. Like *SOX2* and *OTX2*, similar/identical *MAB21L2* genetic variants may be associated with autosomal dominant isolated or syndromic colobomatous microphthalmia.[10]
- *YAP1*: Heterozygous mutations in *YAP1*, a gene encoding a transcriptional coactivator, have been seen in patients with isolated bilateral coloboma as well as in a family with autosomal dominant syndromic colobomatous microphthalmia.[11]

MAC phenotypes resulting from biallelic mutations

A significant number of genes that are mutated in both isolated and syndromic severe MAC cases carry biallelic pathogenic mutations. These include:

- *VSX2*, another ocular transcription factor expressed in early retinal progenitor cells. Associated with isolated microphthalmia.[12]
- *ALDH1A3*, *STRA6*, and *RARB*, genes encoding enzymes involved in retinoic acid metabolism. Biallelic

TABLE 16.1 Genes commonly implicated in microphthalmia–anophthalmia–coloboma.

Gene	Syndromic	Isolated	Mode of inheritance	Named syndrome	Common ocular abnormality	Common systemic abnormality
SOX2	+	+	Autosomal dominant	AEG syndrome	A/M C Retinal dystrophy	Central nervous system abnormalities, microcephaly, intellectual disability, seizures, hypopituitarism, growth retardation, cardiac defects
OTX2	+	+	Autosomal dominant		A/M C Retinal dystrophy	Central nervous system abnormalities, microcephaly, intellectual disability, seizures, hypopituitarism, growth retardation, cardiac defects
RARB	+	−	Autosomal dominant / recessive		A/M	Diaphragmatic hernia, intellectual disability, cardiac/pulmonary/genital anomalies
YAP1	+	+	Autosomal dominant		A/M C	Hearing loss, intellectual disability, orofacial clefting
CHD7	+	+	Autosomal dominant	CHARGE syndrome	C	Choanal atresia, congenital heart defects, intellectual disability, genital abnormalities, deafness
PAX2	+	−	Autosomal dominant	Renal-coloboma syndrome	C	Progressive renal dysfunction and renal hypoplasia
MAB21L2	+	+	Autosomal dominant		A/M C	Macrocephaly, intellectual disability, skeletal dysplasia
RAX	+	+	Autosomal dominant		A/M C	Intellectual disability
TFAP1	+	+	Autosomal dominant	BOFS	A/M C Congenital cataract	Orofacial clefting (cleft lip, cleft palate), branchial skin anomalies
PTCH1	+	−	Autosomal dominant	BCNS	A/M C ASOD, cataract	Basal cell carcinoma, macrocephaly, skeletal abnormalities
VSX2	−	+	Autosomal recessive		A/M	
TENM3	−	+	Autosomal recessive		C	
C12orf57	+	+	Autosomal recessive	Tentamy syndrome	C	Structural brain abnormalities (agenesis corpus callosum), intellectual disability, skeletal abnormalities
ALDH1A3	+	+	Autosomal recessive		A/M C	
STRA6	+	+	Autosomal recessive	PDAC syndrome	A/M C	Pulmonary agenesis or hypoplasia, diaphragmatic hernia, cardiac defects
SMOC1	+	−	Autosomal recessive	Waardenburg anophthalmia	A/M	Limb abnormalities including postaxial oligosyndactyly, facial anomalies and intellectual disability
HCCS	+	−	X-linked dominant	MIDAS syndrome	A/M	Congenital skin lesions (head and neck), structural brain abnormalities (agenesis corpus callosum), intellectual disability, congenital heart disease
BCOR	+	−	X-linked dominant / recessive	Lenz OFCD syndrome	A/M Sclerocornea	Intellectual disability, structural brain abnormalities with microcephaly

A/M, anophthalmia/microphthalmia; C, Coloboma; AEG, anophthalmia–esophageal–genital; CHARGE, coloboma, heart defect, atresia choanae, retarded growth and development, genital hypoplasia, ear anomalies/deafness; BOFS, branchiooculofacial syndrome; BCNS, basal cell nevus syndrome; PDAC, pulmonary hypoplasia/agenesis, diaphragmatic hernia/eventration, anophthalmia/microphthalmia, and cardiac defect; MIDAS, microphthalmia, dermal aplasia, and sclerocornea (also known as LSDMCA1); OFCD, oculofaciocardiodental.

BOX 16.2 The role of *SOX2* and *OTX2* in microphthalmia–anophthalmia–coloboma (MAC)

SOX2: Pathogenic variants in *SOX2* have been identified in patients with isolated MAC that can be either bilateral (most frequently) or unilateral. They are, however, most commonly identified in individuals with syndromic microphthalmia. When present, extraocular features may include structural brain anomalies (ventriculomegaly, hypoplasia of the corpus callosum, periventricular heterotopia, pituitary abnormalities) that may be associated with microcephaly, neuro-cognitive delay, and seizures. Sensorineural hearing loss, oesophageal atresia, growth retardation, and genital anomalies are also recognised; indeed, *SOX2* mutations are known to cause anophthalmia–oesophageal–genital (AEG) syndrome.

A large number of pathogenic variants has been described including truncating and missense changes; whole gene deletions are common and represent up to a third of cases. Missense substitutions in the DNA-binding or transactivation domain of SOX2 have been suggested to produce a milder phenotype involving ocular coloboma. Both parental mosaicism and non-penetrance have been described; familial cases where affected individuals have unilateral disease have also been reported. Consequently, careful examination and genetic analysis of parents is important for accurate genetic counselling.[4]

OTX2: Mutations in *OTX2* are the second most frequently identified cause of bilateral anophthalmia and severe microphthalmia. Pathogenic variants have been identified in patients with an isolated developmental eye disease that ranges from the MAC spectrum (again, this can be either bilateral or unilateral) to Leber congenital amaurosis. Like *SOX2*, *OTX2* mutations are frequently identified in patients with syndromic microphthalmia. When present, extraocular features typically include structural brain malformations and learning disability as well as other abnormalities including sensorineural hearing loss, pituitary hypoplasia, growth retardation, and male genital anomalies.

A large number of heterozygous pathogenic variants has been described, most associated with loss of protein function. Like *SOX2*, a significant proportion of cases carrying whole gene deletions has been described. *De novo OTX2* mutations are commonly encountered, and both complete non-penetrance, as well as parental mosaicism, are recognised. Again, careful examination and genetic analysis of parents are required.[4]

ALDH1A3 mutations have been seen in up to 10% of cases in some consanguineous populations and may lead to isolated or syndromic disease. *STRA6* mutations also cause both syndromic and isolated MAC phenotypes and underlie PDAC (pulmonary agenesis or hypoplasia, diaphragmatic hernia, anophthalmia, and cardiac defects). *RARB* has only been described in a small number of patients with syndromic disease.[4]

- A wide range of genetic defects have been identified during the investigation of patients with MAC phenotypes that are at the milder end of the spectrum. These include mutations in a large number of genes that have each been seen in relatively small numbers of cases (e.g. *TENM3*, *C12orf57*, and *ABCB6*).[5]

MAC phenotypes resulting from mutations in the X chromosome

True X-linked MAC is rare and has been associated with mutations in several genes, including:

- *BCOR*. A small number of missense variants in *BCOR* have been associated with X-linked recessive syndromic MAC, in which severe microphthalmia is associated with significant intellectual disability.[13] Notably, X-linked dominant loss of function mutations in *BCOR* cause oculofaciocardiodental (OFCD) syndrome in females (Section 11.2.3) which is generally associated with congenital cataracts but in some cases may cause microphthalmia.

- *NAA10*. Mutations in *NAA10* can cause a range of phenotypes including X-linked syndromic MAC. Typically, there is anophthalmia/microphthalmia, which is often severe and associated with intellectual disability and skeletal abnormalities (including digital abnormalities such as syndactyly). Splicing mutations, and variants in the 3' untranslated region altering the consensus polyadenylation sequence have been described.[14]

- *HCCS*. The *HCCS* gene encodes holocytochrome *c*-type synthase, an enzyme that localises to the inner mitochondrial membrane and catalyses the covalent attachment of heme to cytochrome *c*. Loss of function variants in this gene leads to the X-linked dominant disorder 'linear skin defects with multiple congenital anomalies' (LSDMCA1) formerly known as MIDAS (microphthalmia, dermal aplasia, and sclerocornea). Affected females have unilateral or bilateral microphthalmia associated with linear skin defects that are found on the head and neck.[15]

Clinical management

Management of individuals with MAC is broadly focused in three areas: diagnosis based on detailed phenotyping and early genomic analysis; management of ocular abnormalities; identification and management of extraocular abnormalities.

Patients with anophthalmia and significant microphthalmia should be referred to an adnexal specialist for socket expansion using enlarging conformers. Ocular prosthetics may be required. If a child has a non-seeing eye, cosmesis can be addressed by fitting cosmetic shells or contact lenses. There are no specific treatment options to improve visual function in microphthalmia patients, hence current management focuses on maximising existing vision by correcting refractive error where appropriate and preventing amblyopia. Those with poor vision can be supported by low visual aids and training. Management of associated defects such as cataracts or glaucoma may be necessary. Individuals with chorioretinal colobomata, in particular those with very high myopia, require surveillance for retinal detachments.

MAC patients with both bilateral and unilateral defects should be referred to a clinical geneticist/paediatrician for investigation of systemic features; this may include renal ultrasound, cardiac assessment, and/or MRI imaging of the brain to identify associated structural abnormalities. Where appropriate, referral to neurology and endocrinology may be indicated. Generally, this will be dictated either by clinical signs/symptoms or by the molecular findings. Early medical management will aim to reduce morbidity and maintain as much function as possible. The wider medical and social teams should provide support to the child and family within the community.

References

1. Llorente-Gonzalez S, Peralta-Calvo J, Abelairas-Gomez JM. Congenital anophthalmia and microphthalmia: epidemiology and orbitofacial rehabilitation. *Clin Ophthalmol* 2011;**5**:1759–65.
2. Verma AS, Fitzpatrick DR. Anophthalmia and microphthalmia. *Orphanet J Rare Dis* 2007;**2**:47.
3. Mitry D, Bunce C, Wormald R, et al. Causes of certifications for severe sight impairment (blind) and sight impairment (partial sight) in children in England and Wales. *Br J Ophthalmol* 2013;**97**(11):1431–6.
4. Williamson KA, Fitzpatrick DR. The genetic architecture of microphthalmia, anophthalmia and coloboma. *Eur J Med Genet* 2014;**57**(8):369–80.
5. Plaisancié J, Ceroni F, Holt R, Zazo SC, Calvas P, Chassaing N, et al. Genetics of anophthalmia and microphthalmia. Part 1: *Hum Genet* 2019 Sep;**138**(8-9):799–830.
6. Harding P, Moosajee M. The molecular basis of human anophthalmia and microphthalmia. *J Dev Biol* 2019;**7**(3):16.
7. Chassaing N, Causse A, Vigouroux A, Delahaye A, Alessandri JL, Boespflug-Tanguy O, et al. Molecular findings and clinical data in a cohort of 150 patients with anophthalmia/microphthalmia. *Clin Genet* 2014;**86**:326–34.
8. Gerth-Kahlert C, Williamson K, Ansari M, Rainger JK, Hingst V, Zim-mermann T, et al. Clinical and mutation analysis of 51 probands with anophthalmia and/or severe microphthalmia from a single center. *Mol Genet Genomic Med* 2013;**1**:15–31.
9. Gonzalez-Rodriguez J, Pelcastre EL, Tovilla-Canales JL, Garcia-Ortiz JE, Amato-Almanza M, Villanueva-Mendoza C, et al. Mutational screening of CHX10, GDF6, OTX2, RAX and SOX2 genes in 50 unrelated microphthalmia–anophthalmia–coloboma (MAC) spectrum cases. *Br J Ophthalmol* 2010;**94**:1100–4.
10. Rainger J, Pehlivan D, Johansson S, Bengani H, Sanchez-Pulido L, Williamson KA, et al. Monoallelic and biallelic mutations in MAB21L2 cause a spectrum of major eye malformations. *Am J Hum Genet* 2014;**94**(6):915–23.
11. Williamson KA, Rainger J, Floyd JA, Ansari M, Meynert A, Aldridge KV, et al. Heterozygous loss-of-function mutations in YAP1 cause both isolated and syndromic optic fissure closure defects. *Am J Hum Genet* 2014;**94**(2):295–302.
12. Ferda Percin E, Ploder LA, Yu JJ, Arici K, Horsford DJ, Rutherford A, et al. Human microphthalmia associated with mutations in the retinal homeobox gene CHX10. *Nat Genet* 2000;**25**(4):397–401.
13. Ng D, Thakker N, Corcoran CM, Donnai D, Perveen R, Schneider A, et al. Oculofaciocardiodental and Lenz microphthalmia syndromes result from distinct classes of mutations in BCOR. *Nat Genet* 2004;**36**(4):411–6.
14. Johnston JJ, Williamson KA, Chou CM, Sapp JC, Ansari M, Chapman HM, et al. *NAA10* polyadenylation signal variants cause syndromic microphthalmia. *J Med Genet* 2019;**56**(7):444–52.
15. Slavotinek A. Genetics of anophthalmia and microphthalmia. Part 2: Syndromes associated with anophthalmia-microphthalmia. *Hum Genet* 2019;**138**(8–9):831–46.
16. Bidinost C, Matsumoto M, Chung D, Salem N, Zhang K, Stockton DW, et al. Heterozygous and homozygous mutations in PITX3 in a large Lebanese family with posterior polar cataracts and neurodevelopmental abnormalities. *Invest Ophthalmol Vis Sci* 2006;**47**(4):1274–80.
17. Anand D, Agrawal SA, Slavotinek A, Lachke SA. Mutation update of transcription factor genes FOXE3, HSF4, MAF, and PITX3 causing cataracts and other developmental ocular defects. *Hum Mutat* 2018;**39**(4):471–94.
18. Hall HN, Williamson KA, FitzPatrick DR. The genetic architecture of aniridia and Gillespie syndrome. *Hum Genet* 2019;**138**(8–9):881–98.
19. Khan K, Logan CV, McKibbin M, Sheridan E, Elçioglu NH, Yenice O, et al. Next generation sequencing identifies mutations in Atonal homolog 7 (ATOH7) in families with global eye developmental defects. *Hum Mol Genet* 2012;**21**(4):776–83.
20. Khan KN, Carss K, Raymond FL, Islam F, Nihr BioResource-Rare Diseases Consortium, Moore AT, et al. Vitamin A deficiency due to bi-allelic mutation of RBP4: there's more to it than meets the eye. *Ophthalmic Genet* 2017;**38**(5):465–6.
21. Conte I, Hadfield KD, Barbato S, Carrella S, Pizzo M, Bhat RS, et al. MiR-204 is responsible for inherited retinal dystrophy associated with ocular coloboma. *Proc Natl Acad Sci U S A* 2015;**112**(25):E3236–45.

16.2

Nanophthalmia and posterior microphthalmia

Arif O. Khan

Nanophthalmia is a disorder of ocular axial elongation that results in very high hypermetropia and thickening of the posterior sclera and choroid. Associated secondary complications include characteristic macular horizontal wrinkles or folds. There is a significant risk of angle-closure glaucoma over time. If anterior chamber dimensions are grossly within normal limits, the phenotype is termed posterior microphthalmia.

Clinical characteristics

Affected children present with decreased vision resulting either from uncorrected high hypermetropia or from secondary retinal changes. Associated amblyopia is common, and refractive accommodative esotropia can occur. The corneas are steep with keratometry values that are inversely correlated to axial length (unlike common paediatric hypermetropia, see below). Macular findings range from fine horizontal wrinkles that may not be visible without careful slit-lamp biomicroscopy to grossly apparent papillomacular folds. These secondary macular changes do not occur in other 'small eye' phenotypes such as microphthalmia. Retinal dystrophy with associated cystoid macular degeneration can occur when the underlying cause is mutations in the *MFRP* gene (see below). There is a high risk for later angle-closure glaucoma and predisposition to uveal effusion, particularly after intraocular surgery.[1,2]

Nanophthalmia and posterior microphthalmia should be distinguished from common paediatric hypermetropia. Typical clinical parameters for affected individuals who are older than 2 years of age are as follows: axial length of 14.50–17.50mm (mean 16mm), corneal power of 47–51 D (mean 48.5 D), horizontal corneal diameter of 10–12mm (mean 11.20mm), and hypermetropia of 12–18 D (mean 15 D). Key distinguishing features include lack of anisometropia, corneal steepening that is inversely correlated to axial length, retinal findings and high risk of uveal effusion.[3]

Molecular pathology

Nanophthalmia and posterior microphthalmia represent a genetically heterogenous disease spectrum.

Biallelic loss of function mutations in *PRSS56*, encoding a serine protease expressed in the neurosensory retina, cornea, sclera, and optic nerve, have been found in patients with posterior microphthalmia with essentially normal anterior segment development[4,5] (Fig. 16.4). Studies in mouse models have shown that this gene is essential for axial length development and iridocorneal angle configuration.[6]

Biallelic mutations in *MFRP* have been associated with both nanophthalmia and posterior microphthalmia with or without retinal degeneration[7,8] (Figs. 16.4–16.6). Pathogenic variants include loss of function and missense changes, with exon 5 being a mutation hotspot.[5] The encoded protein is a member of the frizzled-related protein family and it is predominantly expressed in the RPE. The relevant knockout mouse model develops a retinal degeneration phenotype, often with scattered yellow dots, while the zebrafish model has a phenotype of short axial length with macular folds.[9–11]

Rare forms of nanophthalmia are associated with heterozygous mutations in *TMEM98* (transmembrane protein 98) or *MYRF* (myelin regulatory factor). In *TMEM98*, pathogenic variants include mainly missense mutations as well as a C-terminal frameshift change.[12,13] In *MYRF*, mutations are generally loss of suggesting a haploinsufficiency mechanism. *TMEM98* encodes a transmembrane protein that is known to interact with *MYRF* to regulate self-cleavage of the MYRF protein which acts as a transcription factor.[14] Notably, *MYRF* loss of function mutations may be associated with non-ocular phenotypic features including congenital heart defects.

Rare heterozygous mutations in *BEST1*, which act to alter splicing, have been described in autosomal dominant vitreoretinochoroidopathy (ADVIRC), a complex ocular phenotype that can be associated with high hypermetropia (Section 13.1.9).[15]

Clinical management

Affected children need full refractive correction to prevent amblyopia and strabismus. Macular involvement can limit visual potential, depending upon the extent of the wrinkles or fold. It is important for patients to be informed regarding the risk for angle-closure glaucoma and to be followed up for this possibility.[16,17] Angle-closure glaucoma in nanophthalmia often necessitates lens removal rather than peripheral iridotomies. If intraocular surgery is needed, prophylactic sclerostomies should be considered because of the high risk for postoperative serous retinal detachment.[18]

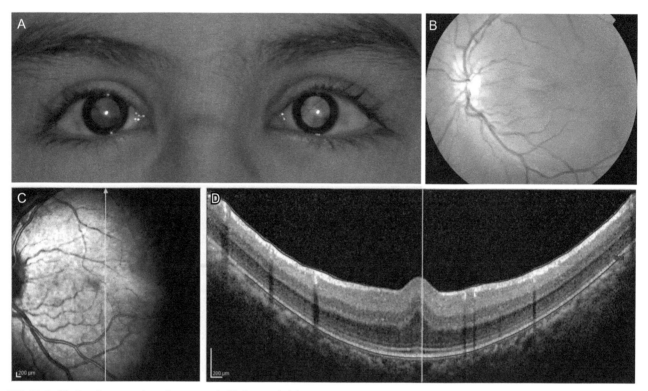

FIG. 16.4 Findings in a young boy who presented with decreased vision and was found to have *PRSS56*-related posterior microphthalmia. For both eyes, the cycloplegic refraction was +14.50, the anterior chamber depth was 3.15 mm, the axial length was 16.20 mm, and the average keratometry was 52 D. (A) The retinal vessels can be visualised on a conventional anterior segment photograph, indicative of very high hyperopia. (B) Left fundus photograph showing a congested disc and a papillomacular fold. (C, D) Left eye vertical OCT scan demonstrating a papillomacular fold.

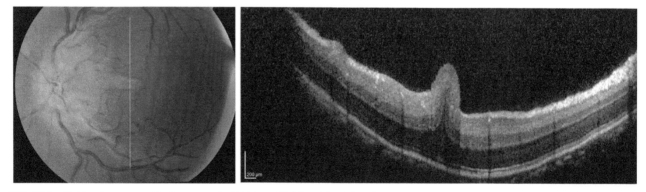

FIG. 16.5 Findings in a young girl who presented with decreased vision and was found to have *MFRP*-related posterior microphthalmia. For both eyes, the cycloplegic refraction was +13.00, the anterior chamber depth was 3.10 mm, the axial length was 16.25 mm, and the average keratometry was 50 D. (A) Left fundus photograph showing a congested disc and a papillomacular fold. The right eye was similar (not shown). (B) Left eye vertical OCT scan demonstrating the papillomacular fold.

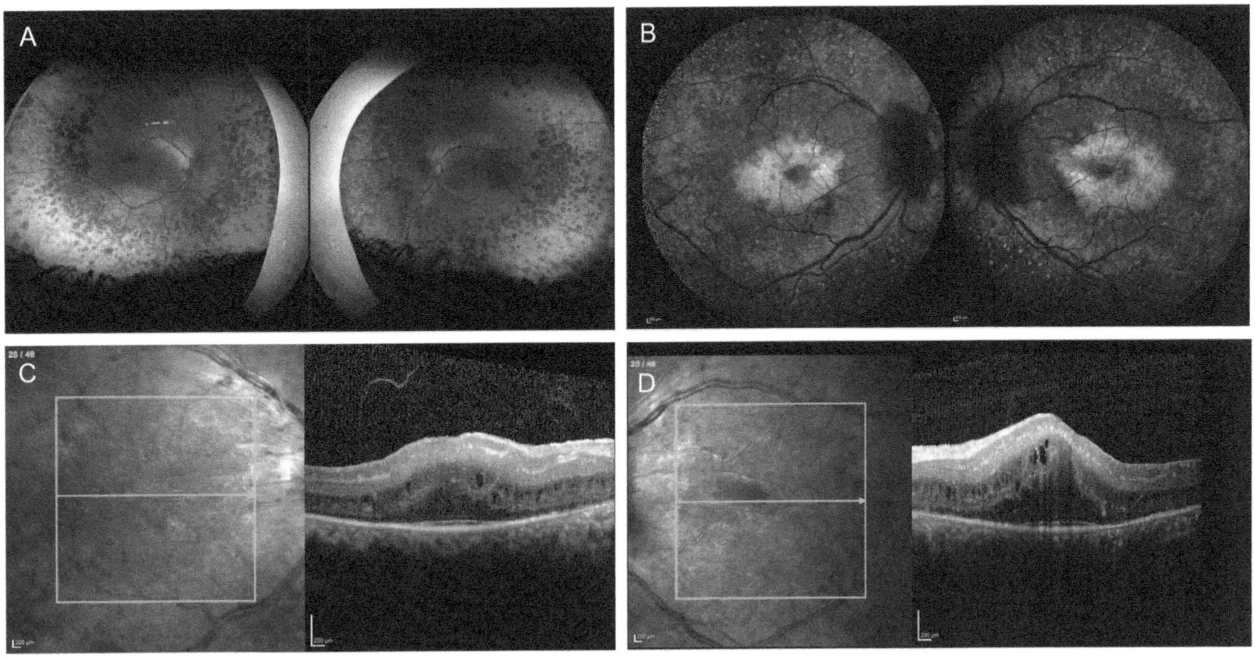

FIG. 16.6 Widefield fundus imaging (A), fundus autofluorescence imaging (B) and OCT (C, D) in a young boy who is homozygous for an *MFRP* mutation, c.662insT (p.Thr223HisfsTer16). The boy was first seen at age 13 years. His visual acuities were 2.0 LogMAR in the right and 1.0 LogMAR in the left eye. Full-field electroretinograms were grossly attenuated, down to noise levels in all but the light-adapted single flash recording. Fundus imaging revealed widespread thinning of the RPE with some pigment clumps in the periphery. There was a retinal fold extending from the disc to the macula in the left eye and a vitreous opacity overlying the disc in each eye.

References

1. Khan AO. Posterior microphthalmos versus nanophthalmos. *Ophthalmic Genet* 2008;**29**(4):189.
2. Khan AO. Nanophthalmos in children: morphometric and clinical characterization. *J AAPOS* 2020;**24**(4):259.
3. Nowilaty SR, Khan AO, Aldahmesh MA, Tabbara KF, Al-Amri A, Alkuraya FS. Biometric and molecular characterization of clinically diagnosed posterior microphthalmos. *Am J Ophthalmol* 2013;**155**(2):361–72. e7.
4. Gal A, Rau I, El Matri L, Kreienkamp HJ, Fehr S, Baklouti K, et al. Autosomal-recessive posterior microphthalmos is caused by mutations in PRSS56, a gene encoding a trypsin-like serine protease. *Am J Hum Genet* 2011;**88**(3):382–90.
5. Almoallem B, Arno G, De Zaeytijd J, Verdin H, Balikova I, Casteels I, et al. The majority of autosomal recessive nanophthalmos and posterior microphthalmia can be attributed to biallelic sequence and structural variants in MFRP and PRSS56. *Sci Rep* 2020;**10**(1):1289.
6. Nair KS, Hmani-Aifa M, Ali Z, Kearney AL, Ben Salem S, Macalinao DG, et al. Alteration of the serine protease PRSS56 causes angle-closure glaucoma in mice and posterior microphthalmia in humans and mice. *Nat Genet* 2011;**43**(6):579–84.
7. Sundin OH, Leppert GS, Silva ED, Yang JM, Dharmaraj S, Maumenee IH, et al. Extreme hyperopia is the result of null mutations in MFRP, which encodes a Frizzled-related protein. *Proc Natl Acad Sci U S A* 2005;**102**(27):9553–8.
8. Mukhopadhyay R, Sergouniotis PI, Mackay DS, Day AC, Wright G, Devery S, et al. A detailed phenotypic assessment of individuals affected by MFRP-related oculopathy. *Mol Vis* 2010;**16**:540–8.
9. Katoh M. Molecular cloning and characterization of MFRP, a novel gene encoding a membrane-type Frizzled-related protein. *Biochem Biophys Res Commun* 2001;**282**:116–23.
10. Hawes NL, Chang B, Hageman GS, Nusinowitz S, Nishina PM, Schneider BS, et al. Retinal degeneration 6 (rd6): a new mouse model for human retinitis punctata albescens. *Invest Ophthalmol Vis Sci* 2000;**41**(10):3149–57.
11. Collery RF, Volberding PJ, Bostrom JR, Link BA, Besharse JC. Loss of zebrafish Mfrp causes nanophthalmia, hyperopia, and accumulation of subretinal macrophages. *Invest Ophthalmol Vis Sci* 2016;**57**(15):6805–14.
12. Garnai SJ, Brinkmeier ML, Emery B, Aleman TS, Pyle LC, Veleva-Rotse B, et al. Variants in myelin regulatory factor (MYRF) cause autosomal dominant and syndromic nanophthalmos in humans and retinal degeneration in mice. *PLoS Genet* 2019;**15**(5), e1008130.
13. Prasov L, Guan B, Ullah E, Archer SM, Ayres BM, Besirli CG, et al. Novel TMEM98, MFRP, PRSS56 variants in a large United States high hyperopia and nanophthalmos cohort. *Sci Rep* 2020;**10**(1):19986.
14. Cross SH, Mckie L, Hurd TW, Riley S, Wills J, Barnard AR, et al. The nanophthalmos protein TMEM98 inhibits MYRF self-cleavage and is required for eye size specification. *PLoS Genet* 2020;**16**(4), e1008583.

15. Yardley J, Leroy BP, Hart-Holden N, Lafaut BA, Loeys B, Messiaen LM, et al. Mutations of VMD2 splicing regulators cause nanophthalmos and autosomal dominant vitreoretinochoroidopathy (ADVIRC). *Invest Ophthalmol Vis Sci* 2004;**45**(10):3683–9.
16. Khan AO. The relationships among cycloplegic refraction, keratometry, and axial length in children with refractive accommodative esotropia. *J AAPOS* 2011;**15**(3):241–4.
17. Hazin R, Khan AO. Isolated microcornea: case report and relation to other "small eye" phenotypes. *Middle East Afr J Ophthalmol* 2008;**15**(2):87–9.
18. Rajendrababu S, Babu N, Sinha S, Balakrishnan V, Vardhan A, Puthuran GV, et al. A randomized controlled trial comparing outcomes of cataract surgery in nanophthalmos with and without prophylactic sclerostomy. *Am J Ophthalmol* 2017;**183**:125–33.

Chapter 17

Aniridia

Graeme C.M. Black and Mariya Moosajee

Aniridia is a pan-ocular bilateral congenital eye anomaly characterised by complete or partial iris hypoplasia, foveal, hypoplasia, and infantile-onset nystagmus. There are several frequently-associated ocular abnormalities, often of later onset, including cataract, glaucoma, corneal opacification and vascularisation. Occasionally, retinal involvement may include exudative retinopathy. Microphthalmia and ocular coloboma (iris, chorioretinal, and/or optic disc) have also been reported in a subset of cases.

Isolated aniridia is mainly caused by genetic variants in the *PAX6* gene.[1,2] Whilst familial cases are inherited as an autosomal dominant trait of very high penetrance, sporadic cases (those where neither parent is affected) should be investigated carefully to exclude deletions of *WT1*, a gene associated with childhood renal tumours.

Clinical characteristics

The hallmark feature of aniridia is a developmental abnormality of the anterior segment that results in complete, partial, or segmental iris hypoplasia (Figs 17.1 and 17.2). The phenotype is variable between and within families; however, there is typically little variability between the two eyes of the same affected individual. The majority of patients will develop other anterior segment abnormalities including cataract (cortical/anterior polar which is present in the second and third decade) and glaucoma (for which there is a lifelong risk) (Fig. 17.2).[3,4] Congenital glaucoma rarely occurs in aniridia but when it is present, a large corneal diameter and corneal oedema may be the presenting findings. 'Aniridia-related keratopathy' due to limbal stem cell deficiency may cause ocular surface disease, discomfort, and worsening of vision in adulthood.[5]

Individuals with aniridia characteristically show nystagmus and foveal hypoplasia which is present in >90% of patients and which contributes to the early visual impact of the condition (Fig. 17.1). Usually, there is a significant reduction in visual acuity (0.9 LogMAR or worse).[6] A small number of cases may also have optic nerve hypoplasia. Despite their many ocular problems, most individuals with aniridia can retain useful vision with appropriate management.

Extraocular manifestations are now emerging; these include diabetes and sleep and behavioural disorders that may require further clinical investigations. *PAX6* is widely expressed in the central nervous system and agenesis of the corpus callosum, absence of the anterior commissure and pineal gland, and/or anosmia have been described in a small number of affected individuals.[7]

Molecular pathology

The majority of aniridia cases result from heterozygous variants in *PAX6*, a gene that is located on chromosome 11p13 and encodes a highly conserved transcription factor regulating a multitude of genes involved in eye development. Mechanistically, the condition is generally caused by *PAX6* haploinsufficiency, and the majority of detected mutations are nonsense, frameshift or splice site variants. Whole gene deletions and regulatory mutations have also been described.[1,2,8–10] A small number of frameshift duplications and deletions resulting in C-terminal extensions have been reported in association with phenotypes that encompass exudative retinopathy.[11]

Missense variants in *PAX6* represent 10%–15% of mutations described and may also result in a classic aniridia phenotype.[1,2] However, some pathogenic missense changes (in particular those within the paired domain) have also been shown to be responsible for a range of ocular abnormalities including Peters anomaly (Section 10.4), isolated foveal hypoplasia, and microphthalmia/coloboma (Chapter 16).[8,12]

A small number of individuals with an aniridia-like phenotypes have been found to carry mutations in either *PITX2* or *FOXC1*. In these cases, there is no associated foveal hypoplasia and the visual acuity is likely to be considerably better compared to that in individuals with defects in *PAX6*.[13]

Sporadic aniridia: The chromosomal location of *PAX6* on chromosome 11p13 was identified through detection of visible cytogenetic rearrangements. Some of these affected individuals had a complex and recognisable phenotype associated with Wilms tumour, Aniridia, Genitourinary abnormalities, and mental Retardation ('WAGR' syndrome).[14] Indeed, this

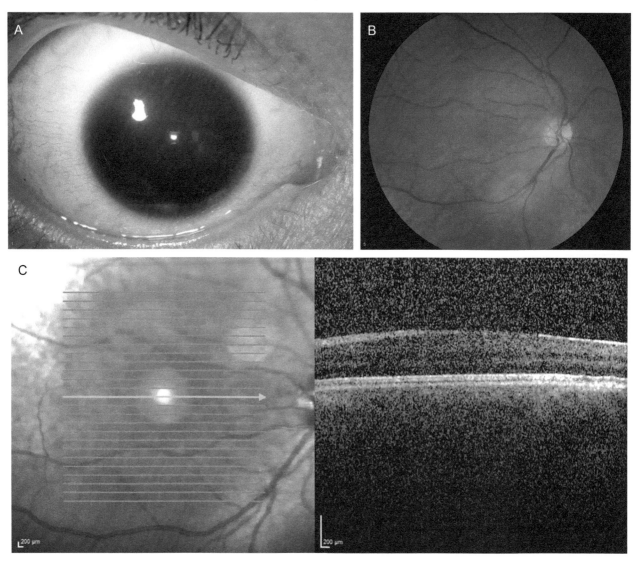

FIG. 17.1 Imaging from a 2-year-old child who was referred with nystagmus and poor vision. Bilateral aniridia with iris hypoplasia were noted (A), fundus examination lacked a macular reflex (B), and foveal hypoplasia was detected on OCT (C).

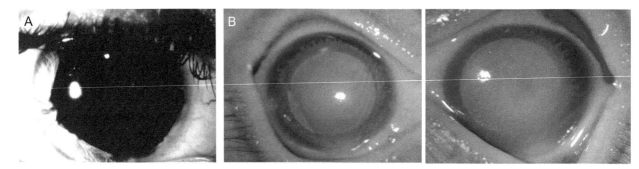

FIG. 17.2 Partial aniridia (A) and bilateral aniridia with corneal clouding secondary to developmental glaucoma (B).

association led to the identification of *WT1*, the tumour suppressor gene that lies close to *PAX6* and which, when mutated or deleted, gives rise to a predisposition to childhood renal tumours (known as Wilms tumour or nephroblastoma). Approximately 90% of children with Wilms tumour develop a mass by 4 years of age; 98% develop a mass by 7 years of age. Children with no family history of aniridia—that is those where neither parent is affected—must be treated as having a possible *WT1* deletion. In these scenaria, the identification of the mutational mechanism giving rise to aniridia is mandatory and, where this is not possible and the affected individual is under 5-7 years, childhood screening for Wilms tumour should be instituted. Screening for *WT1* deletions may be undertaken cytogenetically (e.g. using FISH analysis) or by exclusion through the identification of a *PAX6* mutation. Knowledge of the turnaround times for genomic analysis will define which methodology is utilised and whether a renal investigation is required in the interim.

Gillespie syndrome: Gillespie syndrome is a distinct, ultrarare condition associated with a variant of aniridia. Abnormal development of the iris sphincter results in 'congenital mydriasis' (apparently dilated pupil) with an iris remnant that shows a scalloped edge and iridolenticular strands. The condition is not associated with foveal hypoplasia. Gillespie syndrome is otherwise a multisystemic disorder characterised by cerebellar hypoplasia, ataxia, intellectual disability, and hypotonia. There may be associated cardiac and skeletal abnormalities. The condition is caused by mutations in *ITPR1* and may be dominant, most frequently associated with *de novo* loss of function mutations. Homozygous or compound heterozygous hypomorphic variants in *ITPR1* may give rise to an identical but recessively inherited trait.[8]

Clinical management

Fundus imaging including OCT should be attempted in individuals with suspected aniridia to document foveal hypoplasia; this can support a clinical diagnosis especially where iris defects are subtle. Anterior segment OCT can help to delineate the anterior segment structures even in the presence of a corneal opacity. Ultrasound B-Scan can be performed to assess axial length due to the association with microphthalmia. High-frequency ultrasound biomicroscopy can be performed, usually under anaesthesia, in infants with corneal opacification (including severe corneal oedema resulting from associated congenital glaucoma) to assess iris hypoplasia.

Ophthalmic management of aniridia relies upon screening for treatable secondary complications such as cataracts and glaucoma. In middle life, ocular surface abnormalities secondary to keratopathy and limbal stem cell failure represent a significant cause of morbidity. Conservative support includes review of refractive error, management of glare/photophobia, and visual impairment support.[6]

Aniridia is a highly penetrant autosomal dominant trait and, in the majority of cases, one parent is affected. As described above, understanding the underlying molecular pathology in sporadic cases is crucial as these individuals may also have non-ocular aspects to the phenotype, and so may require a paediatric assessment. Therefore, when a child presents with aniridia *without family history* they should be under joint care with a paediatrician and have genetic testing immediately (either by a clinical geneticist or an ophthalmologist specialising in genetic eye disease) to rule out an associated *WT1* deletion.[9,10]

Novel therapeutic modalities, such as drugs (e.g. Ataluren) which suppress premature termination codons, have shown some promise in animal models.[15]

References

1. Prosser J, van Heyningen V. PAX6 mutations reviewed. *Hum Mutat* 1998;**11**(2):93–108.
2. Lima Cunha D, Arno G, Corton M, Moosajee M. The spectrum of *PAX6* mutations and genotype-phenotype correlations in the eye. *Genes (Basel)* 2019;**10**(12):1050.
3. D'Oria F, Barraquer R, Alio JL. Crystalline lens alterations in congenital aniridia. *Arch Soc Esp Oftalmol* 2021;(Feb 18). S0365-6691(21)00028-9.
4. Bajwa A, Burstein E, Grainger RM, Netland PA. Anterior chamber angle in aniridia with and without glaucoma. *Clin Ophthalmol* 2019;**13**:1469–73.
5. Yazdanpanah G, Bohm KJ, Hassan OM, Karas FI, Elhusseiny AM, Nonpassopon M, et al. Management of congenital aniridia-associated keratopathy: long-term outcomes from a tertiary referral center. *Am J Ophthalmol* 2020;**210**:8–18. https://doi.org/10.1016/j.ajo.2019.11.003.
6. Landsend ECS, Lagali N, Utheim TP. Congenital aniridia—a comprehensive review of clinical features and therapeutic approach. *Surv Ophthalmol* 2021;(March 3). S0039-6257(21)00065-5.
7. Yogarajah M, Matarin M, Vollmar C, Thompson PJ, Duncan JS, Symms M, et al. PAX6, brain structure and function in human adults: advanced MRI in aniridia. *Ann Clin Transl Neurol* 2016;**3**(5):314–30.
8. Hall HN, Williamson KA, FitzPatrick DR. The genetic architecture of aniridia and gillespie syndrome. *Hum Genet* 2019;**138**(8–9):881–98.
9. Clericuzio C, Hingorani M, Crolla JA, van Heyningen V, Verloes A. Clinical utility gene card for: WAGR syndrome. *Eur J Hum Genet* 2011;**19**(4). https://doi.org/10.1038/ejhg.2010.220.
10. Moosajee M, Hingorani M, Moore A. Aniridia. In: Adam MP, Ardinger HH, Pagon RA, Wallace SE, Bean LJH, Stephens K, Amemiya A, editors. *GeneReviews*®. Seattle (WA): University of Washington, Seattle; 2018 [Internet].
11. Hingorani M, Williamson KA, Moore AT, van Heyningen V. Detailed ophthalmologic evaluation of 43 individuals with PAX6 mutations. *Invest Ophthalmol Vis Sci* 2009;**50**(6):2581–90.
12. Williamson KA, Hall HN, Owen LJ, Livesey BJ, Hanson IM, Adams GGW, et al. Recurrent heterozygous PAX6 missense variants cause severe bilateral microphthalmia via predictable effects on DNA-protein interaction. *Genet Med* 2020;**22**(3):598–609.
13. Perveen R, Lloyd IC, Clayton-Smith J, Churchill A, van Heyningen V, Hanson I, et al. Phenotypic variability and asymmetry of Rieger syndrome associated with PITX2 mutations. *Invest Ophthalmol Vis Sci* 2000;**41**(9):2456–60.
14. Richardson R, Hingorani M, Van Heyningen V, Gregory-Evans C, Moosajee M. Clinical utility gene card for: aniridia. *Eur J Hum Genet* 2016;**24**(11). https://doi.org/10.1038/ejhg.2016.73.
15. Gregory-Evans CY, Wang X, Wasan KM, Zhao J, Metcalfe AL, Gregory-Evans K. Postnatal manipulation of Pax6 dosage reverses congenital tissue malformation defects. *J Clin Invest* 2014;**124**(1):111–6. https://doi.org/10.1172/JCI70462.

Chapter 18

Albinism

Eulalie Lasseaux, Magella M. Neveu, Mathieu Fiore, Fanny Morice-Picard, and Benoît Arveiler

Albinism is a heterogeneous group of non-progressive conditions characterised by decreased or absent ocular pigmentation and variable skin pigmentation. Historically, albinism has been split into three main types:

- oculocutaneous albinism (OCA1 to OCA8)
- X-linked ocular albinism (OA1)
- syndromic albinism including Hermansky-Pudlak syndrome (HPS-1 to HPS-11) and Chediak-Higashi syndrome (CHS-1) (Table 18.1).

A distinct condition known as isolated foveal hypoplasia or FHONDA (Foveal Hypoplasia, Optic Nerve Decussation defects and Anterior segment dysgenesis) features some of the characteristic ocular abnormalities that are present in albinism: nystagmus, foveal hypoplasia and optic nerve misrouting.[1] Although pigmentation defects are not encountered in this disorder, some consider it to be a subtype of albinism.

The clinical diagnosis of albinism can be obvious in individuals where most of the clinical signs are present and marked. However, there is considerable variability and reaching a clinical diagnosis may be challenging in certain cases, for example in individuals where the skin and hair do not appear hypopigmented. Notably, a range of molecular defects has been shown to underlie this presentation including variants in *GPR143* (associated with X-linked ocular albinism), variants in *SLC38A8* (associated with isolated foveal hypoplasia), hypomorphic variants in genes implicated in oculocutaneous albinism (e.g. *TYR* and *OCA2*), and variants in genes involved in certain syndromic forms of albinism (e.g. HPS-3, HPS-5, HPS-6, HPS-8). This overlap suggests that a biological distinction between oculocutaneous, ocular, and syndromic forms of the condition is somewhat artificial.[2–4]

A comprehensive ophthalmic examination is key for the diagnosis of albinism. Variable degrees of nystagmus and foveal hypoplasia are almost always present; these often result in reduced visual acuity. Whenever one of these two features is noted, even in a subtle way (e.g. nystagmus that is not present at birth or tends to disappear with age; low grade foveal hypoplasia), albinism should be considered as a differential diagnosis and relevant investigations should be initiated.

Genetic analysis has a high diagnostic yield in albinism suspects and is increasingly being used to guide diagnosis. Early testing is recommended in all individuals who have nystagmus and one or more features of albinism. A key motivation for this is to promptly identify the potentially life-threatening complications that can be linked to syndromic forms of the condition.[2,5]

Clinical characteristics

Ophthalmological aspects

The ocular signs of albinism include:

- nystagmus, most frequently of the pendular or jerk type;
- iris transillumination due to lack of melanin pigment in the posterior epithelial layer of the iris;
- fundal hypopigmentation associated with lack of pigmentation in the RPE (Fig. 18.1);
- foveal hypoplasia associated with complete or partial failure of foveal pit formation and specialisation; the fovea appears featureless, the foveal reflex is lacking and there is a continuation of the inner retinal layers at the foveola on OCT (Fig. 18.2).

The visual acuity in individuals with albinism is typically outside normal limits and generally ranges between 0.3 and 1.0 LogMAR. Other features include reduced or absent stereopsis, strabismus, and refractive errors (usually hypermetropic astigmatism).[6]

An attempt to perform slit-lamp examination should be made in family members of affected individuals. Notably, obligate female carriers of X-linked ocular albinism often have pigmentary retinal changes, resulting in a radial tapetal reflex and a 'mud splatter' appearance on fundus autofluorescence imaging (Fig. 18.3); iris transillumination may also be present.[7] Very rarely, female carriers can be affected and may present with the hallmarks of the condition: nystagmus, foveal hypoplasia and reduced visual acuity.

Individuals with albinism typically have electrophysiological[8] and functional MRI[9] evidence of retino-cortical misrouting, i.e. the majority of nerve fibres from each eye cross at the chiasm and project to the contralateral hemisphere of the visual cortex (Figs 18.4–18.6). This

TABLE 18.1 List of human genes associated with albinism.

Gene	Classification	Albinism type
TYR	OCA1	Oculocutaneous albinism Type 1
OCA2	OCA2	Oculocutaneous albinism Type 2
TYRP1	OCA3	Oculocutaneous albinism Type 3
SLC45A2	OCA4	Oculocutaneous albinism Type 4
Unknown	OCA5	Oculocutaneous albinism Type 5
SLC24A5	OCA6	Oculocutaneous albinism Type 6
LRMDA/C10orf11	OCA7	Oculocutaneous albinism Type 7
DCT	OCA8	Oculocutaneous albinism Type 8
GPR143	OA1	Ocular albinism Type 1 (X-linked)
LYST	CHS-1	Chediak–Higashi Syndrome Type 1
BLOC3S1/HPS1	HPS-1	Hermansky–Pudlak Syndrome Type 1
AP3B1	HPS-2	Hermansky–Pudlak Syndrome Type 2
BLOC2S1/HPS3	HPS-3	Hermansky–Pudlak Syndrome Type 3
BLOC3S2/HPS4	HPS-4	Hermansky–Pudlak Syndrome Type 4
BLOC2S2/HPS5	HPS-5	Hermansky–Pudlak Syndrome Type 5
BLOC2S3/HPS6	HPS-6	Hermansky–Pudlak Syndrome Type 6
BLOC1S8/DTNBP1	HPS-7	Hermansky–Pudlak Syndrome Type 7
BLOC1S3	HPS-8	Hermansky–Pudlak Syndrome Type 8
BLOC1S6	HPS-9	Hermansky–Pudlak Syndrome Type 9
AP3D1	HPS-10	Hermansky–Pudlak Syndrome Type 10
BLOC1S5	HPS-11	Hermansky–Pudlak Syndrome Type 11
SLC38A8	FHONDA	Foveal Hypoplasia, Optic Nerve Decussation defects, and Anterior segment dysgenesis

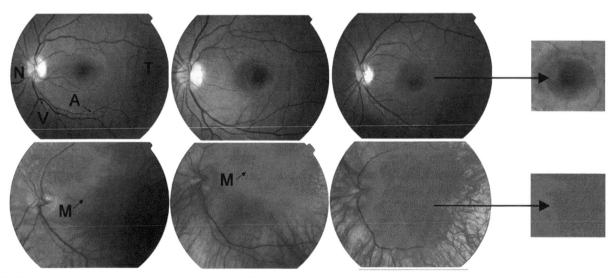

FIG. 18.1 Fundus photographs from the left eye of three unaffected individuals (top row) and three patients with albinism (bottom row). Each unaffected subject appears to have a formed fovea (darkened central area), a pigmented fundus, and two main temporal arteries that exit the optic nerve head and arc around the fovea. Each arterial branch (A) has a vein associated with it (V). Smaller vessels branch from the main arteries and project towards the fovea but do not pass through it. The macular region is magnified on the right, highlighting the fovea. Individuals with albinism lack a fully formed fovea. The fundus is hypopigmented and the choroidal vessels are visible through the RPE. The main arteries are present, but exit the optic nerve head at a wider angle than in the unaffected subjects. The presumptive foveal region (M) is magnified on the right; this appears featureless and vessels pass through it. N corresponds to nasal and T corresponds to temporal retina.

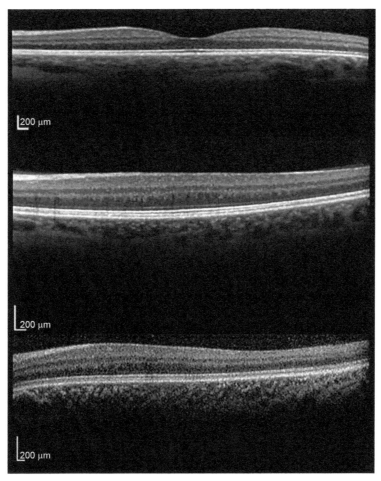

FIG. 18.2 OCT imaging showing foveal hypoplasia in albinism. A normal fovea with a formed foveal pit is shown in the top row. Hypoplastic foveae from two individuals with albinism are shown in the middle and bottom rows; note the absence of a foveal pit in these horizontal scans that include the fixation point (i.e. the retinal centre). In general, that most salient feature of foveal hypoplasia is the incomplete excavation of the inner retinal layers, including the ganglion cell layer and the inner plexiform layer.

finding is in contrast to normal routing, where the nerve fibres from each eye decussate equally at the chiasm and project to both the ipsilateral and contralateral hemispheres. Misrouting can be detected using visual evoked potential (VEP) recordings and the interhemispheric difference can be quantified using a variety of metrics including the asymmetry index and the chiasm coefficient.[10] VEP findings can be variable, particularly in younger children, and technical factors should be taken into account (including electrode placement and the use of a pattern or a flash stimulus) (see also Chapter 6). Current evidence suggests that misrouting can be optimally detected using a high-frequency hand-held flash VEP in children <3 years of age, a standard flash VEP in 3–6-year-old subjects, and a pattern VEP in individuals ≥6 years of age.[10,11]

Dermatological aspects

Differences in skin and hair colour are principally the result of variation in the melanin content of the skin. Melanin is a molecule that is predominantly synthesised by melanocytes. These cells migrate from the neural crest into the epidermis during the first 2 months of gestation and produce melanin within specialised vesicles known as melanosomes. Melanosomes are subsequently extruded from melanocytes and are transferred into neighbouring keratinocytes where they protect the nucleus from ultraviolet mutagenic damage. In general, differences in skin colour are related to variation in the number, size, composition and distribution of melanosomes in keratinocytes.[12,13]

Albinism is associated with hypopigmentation of the skin and hair that varies from total absence of pigment to normal pigmentation. In the most 'complete' forms of the disorder

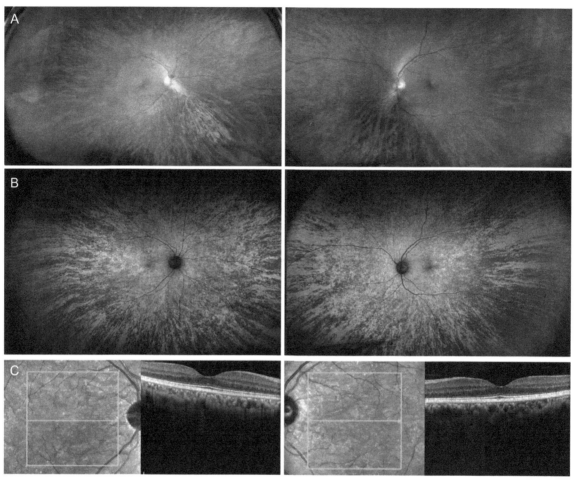

FIG. 18.3 Widefield fundus photography (A), fundus autofluorescence imaging (B) and OCT scans (C) from a visually asymptomatic 52-year-old female whose son was diagnosed with X-linked ocular albinism. The *GPR143* c.887G>A (p.Gly296Glu) change was detected in heterozygous state. Pigmentary changes arranged in radial streaks are noted in the peripheral retina. These correspond to radial patches of relative hypoautofluorescence. A formed foveal pit is noted bilaterally. *(Courtesy of Dr Panagiotis I. Sergouniotis.)*

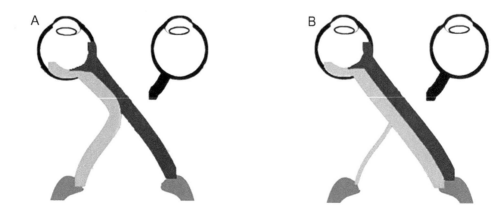

FIG. 18.4 Schematic of the visual pathways in unaffected individuals and people with albinism. (A) Normal visual pathways. Nerve fibres originating from the temporal retina project to the ipsilateral hemisphere *(light grey line)*. Nerve fibres originating from the nasal retina, cross at the chiasm, and project to the contralateral hemisphere *(dark grey line)*. (B) Misrouting in albinism. The majority of optic nerve fibres decussate to the contralateral hemisphere *(light and dark grey lines)*.

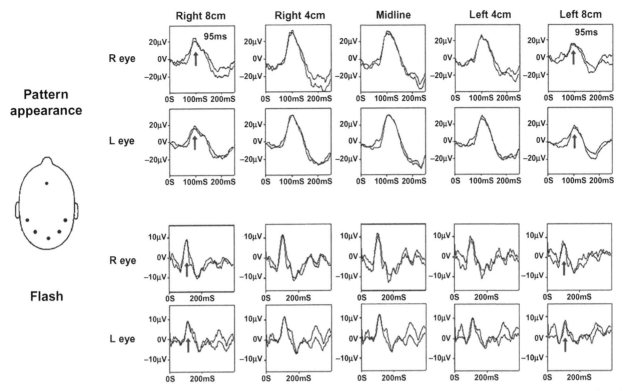

FIG. 18.5 Pattern appearance (rows 1 and 2) and flash (rows 3 and 4) VEPs recorded from a 7-year-old visually normal child using five electrode positions across the visual cortex (schematic on the left showing electrode positions) to monocular stimulation. Two electrodes over the right hemisphere (right 8 cm and 4 cm), two electrodes over the left hemisphere (left 8 cm and 4 cm) and one electrode over the midline were utilised. VEP responses were symmetrical and of similar amplitude and peak time from the left and right hemispheres.

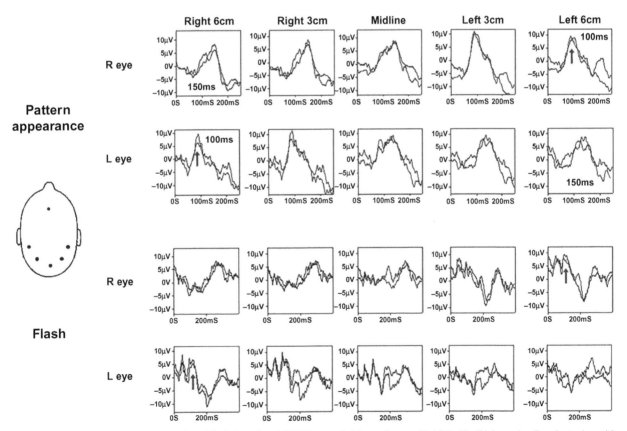

FIG. 18.6 Pattern appearance (rows 1 and 2) and flash (rows 3 and 4) VEPs recorded from a 7-year-old child with albinism using five electrode positions across the visual cortex (schematic on left showing electrode positions) to monocular stimulation. Right eye (rows 1 and 3): VEP responses are of shorter peak time and/or larger amplitude from the left hemisphere compared with the right hemisphere. Left eye (rows 2 and 4): VEP responses are of shorter peak time and/or larger amplitude from the right hemisphere compared with the left hemisphere.

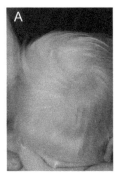  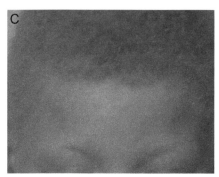

FIG. 18.7 Different hair phenotypes in people with albinism. (A) White hair in a patient with OCA1. (B) Light brown hair in a patient with OCA6. (C) Yellow hair in a patient with OCA2.

the hair, eyelashes and eyebrows are white and the skin is pale and does not tan (Fig. 18.7A). At birth, the hair may be described as light blonde and an abnormality may not be evident (especially given that newborn hair are often sparse and short). It is highlighted that pigmentation can develop with age: skin that was initially very pale may become creamy, irides that were light blue or almost pink can become pigmented and white hair may become yellow or even light brown. Notably, the degree of hair hypopigmentation appears to be independent of the severity of the ophthalmologic impairment (Fig. 18.7B and C).[14,15]

Skin hypopigmentation is associated with an increased risk of sunburns and non-melanoma skin cancers. In the absence of photoprotection, sun-exposed body sites are prone to developing dark-brown freckles, thickened, coarse and rough skin (pachydermia), premalignant skin changes (e.g. actinic keratosis) and malignant skin lesions. The relative risk of both squamous and basal cell carcinoma is much higher in people with albinism compared to the general population and these malignancies tend to occur at a younger age than in unaffected individuals (often in the third to fourth decade of life).[16] Melanoma is rare and only a few cases have been reported in the biomedical literature.[17,18] Cumulative sun exposure is an important risk factor, and albinism-associated skin cancers are overall less frequent in European countries compared to Africa.[19,20]

Haematological aspects: Hermansky-Pudlak and Chediak-Higashi syndromes

Albinism may infrequently be associated with platelet dysfunction due to δ-granule defects. Platelets are small anucleate cell fragments that circulate in the blood, playing a crucial role in managing vascular integrity and regulating haemostasis.[21] They contain dense (δ-) granules, specialised secretory organelles that have a role in haemostasis; δ-granules are a subtype of lysosome-related organelles (LROs), a group that also includes melanosomes.[22,23]

The diagnosis of platelet δ-granule abnormalities is not straightforward. These cannot be detected with a simple whole blood count test, or with peripheral blood smears as δ-granules are too small to be observed using light microscopy. However, as these granules are rich in calcium which makes them inherently electron-opaque and easily detectable by electron microscopy (Fig. 18.8).

Screening for platelet defects may involve quantitative assays that measure granule components or indirect tests that assess platelet granule release through functional assays[25] (Table 18.2). Generally, the adoption of these investigations into clinical practice remains inconsistent[26]; one key issue is the need for many of these tests to be performed within a few hours of sample collection. Typically, cases with relevant symptoms and signs have to undergo two different tests, for example, the adenosine triphosphate (ATP) release test plus a flow cytometric analysis of platelet mepacrine uptake.[27]

A decreased number of platelet δ-granules leads to a defective haemostatic response. Patients suffer from haemorrhagic diathesis mainly characterised by prolonged bleeding and a tendency to bleed or bruise easily. Ecchymoses, petechiae, epistaxis and gingival bleeding or menorrhagia in women, are classic symptoms and should be enquired about in all albinism suspects. Individuals affected by syndromic forms of albinism generally demonstrate mild to moderate bleeding, but in a subset of cases, this can be severe or even life-threatening.[28]

The two conditions that are associated with both albinism and platelet δ-granule defects are Hermansky-Pudlak syndrome and Chediak-Higashi syndrome. These disorders are caused by deficiency or alteration of lysosome-related organelle (LRO) biogenesis and their key features are discussed below.

Hermansky-Pudlak syndrome (HPS): Hermansky-Pudlak syndrome is a group of autosomal recessive hypopigmentation disorders associated with bleeding

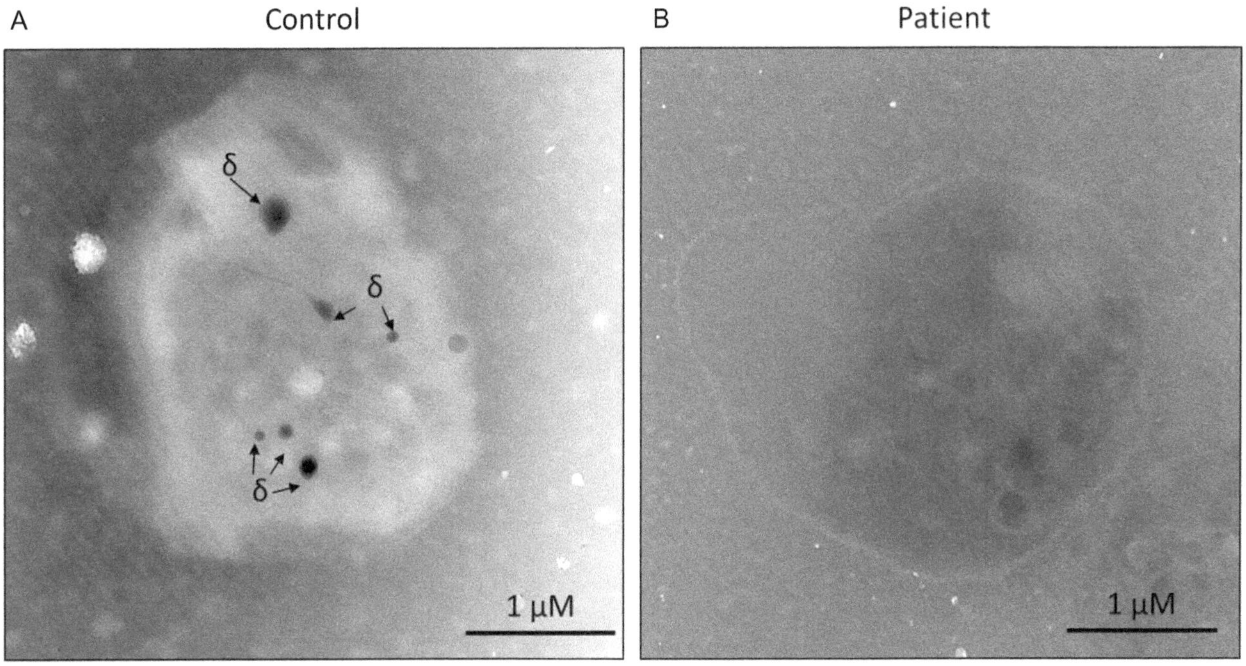

FIG. 18.8 Detection and enumeration of δ-granules using a rapid ultrastructural technique on fresh preparations of platelets (electron microscopy). (A) Platelet appearance in a healthy subject. (B) Platelet appearance in patient with a severe defect of δ-granules. *(Adapted from Ref. 24.)*

TABLE 18.2 List of selected laboratory tests that can be used to evaluate platelet granule defects.

Quantitative and semi-quantitative assays	Indirect tests
Electron microscopy	Serotonin (^{14}C-5HT) secretion assay
Flow cytometry (mepacrine)	Adenosine triphosphate (ATP) release test
HPLC (high-performance liquid chromatography)/ELISA (enzyme-linked immunosorbent assay) to measure platelet serotonin or nucleotide levels	CD63 exposure assay

diathesis.[29] To date, 11 subtypes have been described. Some of these are associated with milder phenotypes (e.g. HPS-5, HPS-8) while the most severe forms (e.g. HPS-1, HPS-4) develop additional complications including pulmonary fibrosis and granulomatous colitis. Pulmonary fibrosis in Hermansky-Pudlak syndrome is characterised by progressive and irreversible fibrosis of the lungs, ultimately requiring lung transplantation and often leading to premature death from respiratory failure. Pulmonary fibrosis is typically detected in the fourth or fifth decade of life using high resolution CT imaging. Lung biopsy is contraindicated in this group of patients due to the bleeding tendency associated with the underlying syndrome. Granulomatous colitis resembling Crohn's disease may be present, with wide variability.[30]

Chediak-Higashi Syndrome (CHS): Chediak-Higashi syndrome is an autosomal recessive disorder that is associated with variants in the *LYST* gene.[31] In addition to hypopigmentation and platelet dysfunction, Chediak-Higashi syndrome is characterised by immunodeficiency and neurologic symptoms. Myeloid cells from bone marrow aspirates show giant intracytoplasmic granules which are pathognomonic of the disease.

Molecular pathology

At least 20 genes are known to be associated with albinism (Table 18.1). Some of them encode key melanogenic enzymes (*TYR, TYRP1, DCT*) while others encode specific receptor, solute carrier or integral proteins located in the melanosomes where melanin is synthesized (*OCA2, SLC45A2, GPR143*). A large group of albinism-implicated genes have a role in the biogenesis of melanosomes and other lysosome-related organelles (LROs).

Due to the wide clinical variability observed among albinism patients, testing of all albinism-related genes using high-throughput sequencing is required to establish a diagnosis. Pathogenic variants can be single nucleotide changes or larger-scale alterations such as multiexon deletions; the latter group represents over 10% of albinism variants and comprehensive testing that targets this type of variation should be requested (e.g. high-resolution array-CGH testing or additional bioinformatic analysis of high-throughput sequencing data).[5,32] Importantly, even after in-depth testing of all known albinism-related genes, a subset of patients (~30%) remains without a molecular diagnosis.

Analyses of large-scale cohorts of people with albinism have allowed the estimation of the relative contribution of the different forms of this condition. In populations of European ancestries, OCA1 [TYR] stands out as the most frequent form and accounts for 40%–45% of individuals with molecularly diagnosed albinism. The second most prevalent form is OCA2 [OCA2] (25%–30%), followed by OCA4 [SLC45A2] (about 10%). OA1 [GPR143], the X-linked form causing 'ocular albinism' represents about 7% of cases. All other forms are much rarer, although syndromic subtypes collectively account for 6%–7% of cases. Isolated foveal hypoplasia (FHONDA) represents less than 1% of cases.[5,15,33-35]

The pathogenicity of certain variants has not been clearly established. The c.1205G > A (p.Arg402Gln) variant in *TYR* is a good example as its effect has been the subject of extensive debate in the biomedical literature.[36-40] This change is frequent in the population worldwide (17%), and even more so in individuals of European ancestries (27%). Notably, up to 50% of OCA1 [TYR] patients are found to be compound heterozygotes for the *TYR* c.1205G > A change (combined with a clearly pathogenic *TYR* variant).[5] It is highlighted that the effect/penetrance of this missense change appears to be modified by a promoter variant, *TYR* c.-301C>T.[40] In general, people with albinism carrying the *TYR* c.1205G > A change have a milder phenotype compared to individuals with two severely pathogenic *TYR* variants.[4] These milder cases present with some level of pigmentation and tend to tan with age. The mean visual acuity in this group is higher than in people with classic OCA1.

Clinical management

Misdiagnosis is common in people with albinism. To an extent, this is because many affected individuals present only with ocular features and do not have significant skin/hair hypopigmentation. The role of the ophthalmologist is therefore key both in suspecting albinism and in initiating appropriate genetic investigations (e.g. in infants with nystagmus or in adults with foveal hypoplasia).[2]

Genetic testing is a robust and reliable way of establishing an aetiological diagnosis in albinism suspects. Notably, given the limitations of routine haematological investigations, genetic tests represent a valuable means for diagnosing syndromic forms of albinism, such as Hermansky-Pudlak and Chediak-Higashi syndromes, on time.[26] Obtaining a genetic diagnosis can also facilitate genetic counselling and help refine the risk to family members and future pregnancies (for example through the identification of an X-linked form of ocular albinism).[41]

Unprotected exposure of the skin of individuals with albinism to ultraviolet radiation can lead to skin cancer. Where solar exposure is extensive and sunscreen difficult to obtain, the malignant cutaneous manifestations can be life-shortening. The importance of lifelong sun protection can therefore not be overestimated. Recommendations should be made around avoiding prolonged light exposure (sun, tanning beds, etc.), wearing clothes that cover the exposed skin and applying sunscreen before outdoor activities (SPF 30+ with frequent and liberal re-application even on cloudy days). An early dermatological consultation is recommended and lifelong, periodic skin examinations should be arranged (surveillance should start at adolescence as skin cancer may appear as early as the teenage years).[14,42]

Ophthalmolmic management should involve periodic updating of glasses prescriptions. Filtering lenses and a cap or a hat can be helpful when there is light hypersensitivity. Evaluation by a low vision specialist should be considered and educational support should be provided to affected children (including close-to-board seating and high-contrast, large-print reading material). In some cases, extraocular muscle surgery can be performed to address ocular misalignment and/or to improve head posture.

A multidisciplinary approach coordinated by a specialist centre is recommended for all patients with albinism, and in particular with those with Hermansky-Pudlak and Chediak-Higashi syndromes.

References

1. Poulter JA, Al-Araimi M, Conte I, van Genderen MM, Sheridan E, Carr IM, et al. Recessive mutations in SLC38A8 cause foveal hypoplasia and optic nerve misrouting without albinism. *Am J Hum Genet* 2013;**93**(6):1143–50.
2. Arveiler B, Michaud V, Lasseaux E. Albinism: an underdiagnosed condition. *J Invest Dermatol* 2020. https://doi.org/10.1016/j.jid.2019.12.010. pii: S0022-202X(19)33564-X.
3. Campbell P, Ellingford JM, Parry NRA, Fletcher T, Ramsden SC, Gale T, et al. Clinical and genetic variability in children with partial albinism. *Sci Rep* 2019;**9**(1):16576.
4. Monfermé S, Lasseaux E, Duncombe-Poulet C, Hamel C, Defoort-Dhellemmes S, Drumare I, et al. Mild form of oculocutaneous albinism type 1: phenotypic analysis of compound heterozygous patients with the R402Q variant of the TYR gene. *Br J Ophthalmol* 2019;**103**:1239–47.

5. Lasseaux E, Plaisant C, Michaud V, Pennamen P, Trimouille A, Gaston L, et al. Molecular characterization of a series of 990 index patients with albinism. *Pigment Cell Melanoma Res* 2018;**31**:466–74.
6. Kruijt CC, de Wit GC, Bergen AA, Florijn RJ, Schalij-Delfos NE, van Genderen MM. The phenotypic spectrum of albinism. *Ophthalmology* 2018;**125**(12):1953–60. https://doi.org/10.1016/j.ophtha.2018.08.003.
7. Acton JH, Greenberg JP, Greenstein VC, Marsiglia M, Tabacaru M, Theodore Smith R, et al. Evaluation of multimodal imaging in carriers of X-linked retinitis pigmentosa. *Exp Eye Res* 2013;**113**:41–8. https://doi.org/10.1016/j.exer.2013.05.003.
8. Dorey SE, Neveu MM, Burton LC, Sloper JJ, Holder GE. The clinical features of albinism and their correlation with visual evoked potentials. *Br J Ophthalmol* 2003;**87**:767–72.
9. Puzniak RJ, Ahmadi K, Kaufmann J, Gouws A, Morland AB, Pestilli F, et al. Quantifying nerve decussation abnormalities in the optic chiasm. *Neuroimage Clin* 2019;**24**:102055. https://doi.org/10.1016/j.nicl.2019.102055.
10. Kruijt CC, de Wit GC, Talsma HE, Schalij-Delfos NE, van Genderen MM. The detection of misrouting in albinism: evaluation of different VEP procedures in a heterogeneous cohort. *Invest Ophthalmol Vis Sci* 2019;**60**(12):3963–9.
11. Neveu MM, Jeffery G, Burton LC, Sloper JJ, Holder GE. Age-related changes in the dynamics of human albino visual pathways. *Eur J Neurosci* 2003;**18**:1939–49.
12. Lambert MW, Maddukuri S, Karanfilian KM, Elias ML, Lambert WC. The physiology of melanin deposition in health and disease. *Clin Dermatol* 2019;**37**(5):402–17.
13. Pavan WJ, Sturm RA. The genetics of human skin and hair pigmentation. *Annu Rev Genomics Hum Genet* 2019;**20**:41–72. https://doi.org/10.1146/annurev-genom-083118-015230.
14. Marti A, Lasseaux E, Ezzedine K, Léauté-Labrèze C, Boralevi F, Paya C, et al. Lessons of a day hospital: comprehensive assessment of patients with albinism in a European setting. *Pigment Cell Melanoma Res* 2018;**31**:318–29.
15. Mauri L, Manfredini E, Del Longo A, Veniani E, Scarcello M, Terrana R, et al. Clinical evaluation and molecular screening of a large consecutive series of albino patients. *J Hum Genet* 2017;**62**:277–90. https://doi.org/10.1038/jhg.2016.123.
16. Kiprono SK, Chaula BM, Beltraminelli H. Histological review of skin cancers in African albinos: a 10-year retrospective review. *BMC Cancer* 2014;**14**:157. https://doi.org/10.1186/1471-2407-14-157.
17. Perry PK, Silverberg NB. Cutaneous malignancy in albinism. *Cutis* 2001;**67**:427–30.
18. Ribero S, Carrera C, Tell-Marti G, Pastorino C, Badenas C, Garcia A, et al. Amelanotic melanoma in oculocutaneous albinism: a genetic, dermoscopic and reflectance confocal microscopy study. *Br J Dermatol* 2017;**177**:e333–5. https://doi.org/10.1111/bjd.15687.
19. Kromberg JG, Castle D, Zwane EM, Jenkins T. Albinism and skin cancer in southern Africa. *Clin Genet* 1989;**36**:43–52.
20. Yakubu A, Mabogunje OA. Skin cancer in African albinos. *Acta Oncol (Stockh)* 1993;**32**:621–2.
21. Gremmel T, Frelinger AL, Michelson AD. Platelet physiology. *Semin Thromb Hemost* 2016;**42**:191–204. https://doi.org/10.1055/s-0035-1564835.
22. Masliah-Planchon J, Darnige L, Bellucci S. Molecular determinants of platelet delta storage pool deficiencies: an update. *Br J Haematol* 2013;**160**:5–11. https://doi.org/10.1111/bjh.12064.
23. Selle F, James C, Tuffigo M, Pillois X, Viallard J-F, Alessi M-C, et al. Clinical and laboratory findings in patients with δ-storage pool disease: a case series. *Semin Thromb Hemost* 2017;**43**:48–58. https://doi.org/10.1055/s-0036-1584568.
24. Fiore M, Garcia C, Sié P, Favier R, Lavenu-Bombled C, Hurtaud MF, et al. Déficit en granules denses plaquettaires: Une cause sous-estimée de saignements inexpliqués. *Hématologie* 2017;**23**:243–54. https://doi.org/10.1684/hma.2017.128.
25. Cattaneo M, Cerletti C, Harrison P, Hayward CPM, Kenny D, Nugent D, et al. Recommendations for the standardization of light transmission Aggregometry: a consensus of the Working Party from the Platelet Physiology Subcommittee of SSC/ISTH. *J Thromb Haemost* 2013. https://doi.org/10.1111/jth.12231.
26. Gresele P, Harrison P, Bury L, Falcinelli E, Gachet C, Hayward CP, et al. Diagnosis of suspected inherited platelet function disorders: results of a worldwide survey. *J Thromb Haemost* 2014;**12**:1562–9. https://doi.org/10.1111/jth.12650.
27. Le Blanc J, Mullier F, Vayne C, Lordkipanidzé M. Advances in platelet function testing-light transmission aggregometry and beyond. *J Clin Med* 2020;**9**(8):2636. https://doi.org/10.3390/jcm9082636.
28. Lowe GC, Lordkipanidzé M, Watson SP, UK GAPP Study Group. Utility of the ISTH bleeding assessment tool in predicting platelet defects in participants with suspected inherited platelet function disorders. *J Thromb Haemost* 2013;**11**:1663–8. https://doi.org/10.1111/jth.12332.
29. El-Chemaly S, Young LR. Hermansky-Pudlak syndrome. *Clin Chest Med* 2016;**37**:505–11. https://doi.org/10.1016/j.ccm.2016.04.012.
30. Ambrosio AL, Di Pietro SM. Storage pool diseases illuminate platelet dense granule biogenesis. *Platelets* 2017;**28**:138–46. https://doi.org/10.1080/09537104.2016.1243789.
31. Nurden A, Nurden P. Advances in our understanding of the molecular basis of disorders of platelet function. *J Thromb Haemost* 2011;**9**(Suppl 1):76–91. https://doi.org/10.1111/j.1538-7836.2011.04274.x.
32. Morice-Picard F, Lasseaux E, Cailley D, Gros A, Toutain J, Plaisant C, et al. High-resolution array-CGH in patients with oculocutaneous albinism identifies new deletions of the TYR, OCA2, and SLC45A2 genes and a complex rearrangement of the OCA2 gene. *Pigment Cell Melanoma Res* 2014;**27**:59–71. https://doi.org/10.1111/pcmr.12173.
33. Hutton SM, Spritz RA. Comprehensive analysis of oculocutaneous albinism among non-Hispanic caucasians shows that OCA1 is the most prevalent OCA type. *J Invest Dermatol* 2008;**128**:2442–50. https://doi.org/10.1038/jid.2008.109.
34. Rooryck C, Morice-Picard F, Elçioglu NH, Lacombe D, Taieb A, Arveiler B. Molecular diagnosis of oculocutaneous albinism: new mutations in the OCA1–4 genes and practical aspects. *Pigment Cell Melanoma Res* 2008;**21**:583–7. https://doi.org/10.1111/j.1755-148X.2008.00496.x.
35. Simeonov DR, Wang X, Wang C, Sergeev Y, Dolinska M, Bower M, et al. DNA variations in oculocutaneous albinism: an updated mutation list and current outstanding issues in molecular diagnostics. *Hum Mutat* 2013;**34**:827–35. https://doi.org/10.1002/humu.22315.
36. Ghodsinejad Kalahroudi V, Kamalidehghan B, Arasteh Kani A, Aryani O, Tondar M, Ahmadipour F, et al. Two novel tyrosinase (TYR) gene mutations with pathogenic impact on oculocutaneous albinism type 1 (OCA1). *PLoS One* 2014;**9**. https://doi.org/10.1371/journal.pone.0106656, e106656.
37. Kubal A, Dagnelie G, Goldberg M. Ocular albinism with absent foveal pits but without nystagmus, photophobia, or severely reduced vision. *J AAPOS* 2009;**13**:610–2. https://doi.org/10.1016/j.jaapos.2009.09.015.

38. Norman CS, O'Gorman L, Gibson J, Pengelly RJ, Baralle D, Ratnayaka JA, et al. Identification of a functionally significant tri-allelic genotype in the Tyrosinase gene (TYR) causing hypomorphic oculocutaneous albinism (OCA1B). *Sci Rep* 2017;**7**:4415. https://doi.org/10.1038/s41598-017-04401-5.
39. Oetting WS, Pietsch J, Brott MJ, Savage S, Fryer JP, Summers CG, et al. The R402Q tyrosinase variant does not cause autosomal recessive ocular albinism. *Am J Med Genet A* 2009;**149A**:466–9. https://doi.org/10.1002/ajmg.a.32654.
40. Michaud V, Lasseuax E, Green DJ, Gerrard DT, Plaisant C, UK Biobank Eye and Vision Consortium, Genomics England Research Consortium, Fitzgerald T, Birney E, Arveiler B, Black GC, Sergouniotis PI. The contribution of common regulatory and protein-coding *TYR* variants in the genetic architecture of albinism. *medRxiv* 2021. https://doi.org/10.1101/2021.11.01.21265733.
41. Lenassi E, Clayton-Smith J, Douzgou S, Ramsden SC, Ingram S, Hall G, et al. Clinical utility of genetic testing in 201 preschool children with inherited eye disorders. *Genet Med* 2020;**22**(4):745–51. https://doi.org/10.1038/s41436-019-0722-8.
42. Federico JR, Krishnamurthy K. Albinism. In: *StatPearls [internet]*. Treasure Island, FL: StatPearls Publishing; 2020 January. 30085560.

Section V

Genetic disorders affecting ocular motility

Chapter 19

Infantile nystagmus

Jay E. Self and Helena Lee

Nystagmus is a clinical sign associated with rhythmic, repetitive and uncontrolled movements of the eyes. It can be differentiated from other eye movement abnormalities by the characteristic initial slow drift away from fixation. This is either followed by a corrective fast movement (jerk nystagmus) or by a corrective slow movement (pendular nystagmus). In terms of direction, nystagmus can be horizontal, vertical, rotatory or mixed.

Nystagmus can be broadly grouped into infantile and acquired forms; this chapter focuses on the former subgroup. The onset of infantile nystagmus is usually within the first 4–6 months of life. It may occur in isolation or it may be associated with a visual or a multisystemic disorder. Visual function in affected individuals ranges from severely abnormal to almost normal, and progressive visual loss can occur in a subset of patients. Given the heterogeneity of the patient population with infantile nystagmus and the inherited nature of many underlying disorders, genetic testing is a valuable early investigation. Notably, a detailed clinical assessment, often including electrodiagnostic testing and retinal imaging, is necessary to accurately interpret the genetic test results.[1]

Clinical characteristics

When infantile nystagmus is encountered in clinical practice, it can pose a significant diagnostic dilemma or it can be considered a 'secondary finding' and be largely ignored from a diagnostic standpoint (e.g. in patients with classic albinism). Many different classification systems exist for nystagmus but the key groups in terms of underlying diagnosis, include:

- *Idiopathic infantile nystagmus*: This relatively common nystagmus subtype is a diagnosis of exclusion and is encountered in individuals who have no associated afferent visual pathway disease. It is almost always bilateral and it occurs in the horizontal plane (including in upgaze and downgaze). It is often of the jerk type (although it may initially be pendular or mixed) and it beats in the direction of horizontal gaze, i.e. the fast phase has a rightward direction on right gaze and a leftward direction on left gaze. It typically dampens on convergence and it is associated with a null zone, i.e. there is a direction of gaze in which nystagmus is attenuated. Affected individuals may have an abnormal head posture.
- *Fusion maldevelopment nystagmus syndrome (FMNS, formerly known as manifest latent nystagmus)*: This nystagmus subtype is caused by early loss of binocularity and is typically seen in children with strabismus (e.g. infantile esotropia); less commonly it is seen in children with congenital cataract or other disorders associated with early-onset visual impairment. The characteristic finding is horizontal jerk nystagmus with the fast phase towards the fixing eye (i.e. reversing with alternate occlusion). It intensifies with monocular occlusion and dampens when looking towards the slow phase (i.e. in full adduction). Thus, it is best identified by alternating occlusion with the eyes in far lateral gaze (as this generally brings out the biggest change in intensity). Affected individuals may have esotropia and dissociated vertical deviation. In cases with no other clinical features, genetic testing is currently not employed for individuals with this nystagmus subtype. Although a genetic component related to familial

strabismus may be present, Mendelian disease is rare. For cases with fusion maldevelopment nystagmus and additional ophthalmic features please see below.
- *Nystagmus associated with ophthalmic disease (including albinism, foveal hypoplasia, achromatopsia, congenital stationary night-blindness, Leber congenital amaurosis, congenital cataract and optic nerve hypoplasia)*: Individuals with a wide range of early childhood onset ophthalmic disorders can present with nystagmus. The clinical features may include abnormal visual behaviour (including poor fixation and poor visual contact), photophobia and night vision problems. The signs can be subtle but may point to the underlying diagnosis (e.g. foveal hypoplasia may point to albinism, *SLC38A8*-associated foveal hypoplasia or aniridia spectrum; see also Table 19.1). The nystagmus can be horizontal or multiplanar and often has a high frequency. Variability in nystagmus intensity (which is a product of frequency and amplitude) may be noted in different positions of gaze. Occasionally, the waveform is erratic and roving eye movements may be present, often at the onset of nystagmus. Electrodiagnostic testing (ERG, VEP) is often a key for diagnosis, along with genetic testing, and retinal imaging (where possible). Notably, genetic testing is unlikely to be informative in cases of optic nerve hypoplasia as this disorder is generally multifactorial (and rarely has a significant monogenic component). Individuals with optic nerve hypoplasia should be jointly managed with the paediatric neurology and endocrinology teams, and initial investigations should typically include pituitary function tests and neuroimaging.
- *Acquired nystagmus or presence of significant oscillopsia*: When assessing infants and young children with nystagmus, it should be borne in mind that some affected individuals will have 'acquired' nystagmus with an underlying neurological cause. Patients are typically older than 6 months and present with new onset of nystagmus; associated oscillopsia can be reported in older children. Nystagmus is unlike the idiopathic infantile or fusion maldevelopment subtypes and there might be a purely vertical or a very asymmetrical pattern. Generally, such cases require early neuroimaging. They are rarely caused by true congenital/genetic disorders and in most cases, non-genetic investigations are warranted in the first instance.
- *Non-nystagmus eye movements*: Several involuntary abnormalities of fixation that are not nystagmus have been described. The key difference between these phenotypes and nystagmus is that the abnormal slow movement that is present in nystagmus is absent in these other conditions. This group includes oscillations and intrusions in which there is a fast movement away from fixation followed by a corrective saccade; this corrective saccade may be immediate (oscillation) or delayed (intrusion). Opsoclonus and ocular flutter are forms of oscillation and are associated with bursts of fast eye movements. Square-wave jerks are a form of intrusion and are associated with horizontal excursions from fixation. These phenotypes can be misdiagnosed as nystagmus but have different aetiologies and investigation pathways; genetic testing rarely forms part of the initial work-up. Cases with opsoclonus or ocular flutter will typically be urgently investigated via joint management with paediatric neurology[1]

Molecular pathology

While a detailed description of the molecular pathology of all forms of nystagmus is outside the remit of this chapter, some general principles around genetic testing in nystagmus cases are discussed below.

Idiopathic infantile nystagmus: Focused gene sequencing is not suitable for this form of nystagmus, except in rare cases

TABLE 19.1 Genetic causes of foveal hypoplasia.

Gene	Mode of inheritance	Comment on ocular phenotype	Reference/section
AHR	Autosomal recessive	Typically no other ocular abnormalities are noted	Ref. 2
PAX6	Autosomal dominant/de novo	Aniridia is common but foveal hypoplasia can be isolated in a subset of cases	Chapter 17
SLC38A8	Autosomal recessive	Occasionally anterior segment dysgenesis and optic nerve decussation defects are noted	Chapter 18
Albinism subtypes (≥19 genes)	Autosomal recessive	The signs of albinism can be subtle or absent	Chapter 18
GPR143-associated albinism	X-linked	The signs of albinism can be subtle or absent	Chapter 18

where there is a strong X-linked family history and a phenotype suggestive of mutations in *FRMD7*.[3] Thus, broad, high-throughput sequencing approaches are generally more appropriate.

FRMD7 mutations are identified in approximately 50% of cases with idiopathic infantile nystagmus and large X-linked pedigrees, and in around 5% of cases with idiopathic infantile nystagmus and no relevant family history. Although *FRMD7* is located on the X chromosome, mutations can manifest in females; this can make the identification of obvious X-linked pedigrees (i.e. those with only affected males) challenging. Penetrance in females is around 30%–50% and it is thought to be higher in individuals carrying missense variants compared to those carrying a loss of function variants.[4]

Other cases of presumed idiopathic infantile nystagmus that yield a positive genetic test result are due to initial misdiagnosis. For example, spinocerebellar ataxias (e.g. *CACNA1A*-related disease) are now known to present in some patients, as early-onset nystagmus with late-onset ataxia; in these cases, the early clinical manifestations can be easily mistaken for idiopathic infantile nystagmus.[5] Other examples include patients that are only later found to have clinical features suggestive of an underlying ophthalmic disorder. This can happen for several reasons including the age-dependent yield of certain investigations and the evolving nature of certain clinical signs. Specific examples include milder albinism phenotypes, congenital stationary night-blindness (e.g. *CACNA1F*-retinopathy) and aniridia spectrum (*PAX6*-oculopathy).[6-8] It is important to note that complex genotypes are common in some disorders (such as albinism) meaning that clinical genomic tests frequently yield inconclusive results. This also tends to be more common in cases with milder phenotypes.[7]

Nystagmus associated with ophthalmic disease: Genetic testing is clearly an effective adjunct to diagnosis in patients with nystagmus secondary to ophthalmic disorders as most of them have a strong genetic component. Early differentiation of syndromic versus non-syndromic forms (e.g. in congenital cataracts or in albinism cases) and progressive versus stationary forms (e.g. achromatopsia versus cone-rod dystrophy cases) is important for directing management, defining prognosis and determining treatment options (e.g. in patients with *RPE65*-retinopathy).[8,9]

Clinical management

The standard clinical workup for nystagmus typically includes evaluation of the child as a whole, orthoptic assessment, refraction, nystagmus phenotyping, anterior and posterior segment examination and examination of relatives; in most cases, retinal imaging, electrophysiology and/or neuroimaging are required. It is important to note that many of the features which direct clinicians to the underlying cause of nystagmus are often subtle or difficult to identify in younger children. The interpretation of genetic test results relies on the identification of these clinical features, and a detailed clinical workup is imperative in all cases.

Refractive correction is a priority when managing children with nystagmus. Contact lenses (soft or rigid gas permeable) may be superior to glasses in improving visual function, especially in individuals with high refractive errors. Otherwise, evaluation by a low vision specialist should be initiated and educational support should be provided where appropriate.

Pharmacological treatment (with baclofen, gabapentin, memantine, etc.) has been used mainly for the management of acquired nystagmus in patients over 16 years of age. Notably, many of the drugs that have been tried have limited/modest supportive evidence of efficacy and significant side effect profiles.

There is evidence supporting the use of extraocular muscle surgery to correct abnormal head postures occurring secondary to an eccentric null zone in idiopathic infantile nystagmus. The operation most commonly performed is the Kestenbaum-Anderson procedure. For more information see Ref. 1.

References

1. Self JE, et al. Management of nystagmus in children: a review of the literature and current practice in UK specialist services. *Eye (Lond)* 2020;**34**:1515–34. https://doi.org/10.1038/s41433-019-0741-3.
2. Mayer AK, Mahajnah M, Thomas MG, Cohen Y, Habib A, Schulze M, et al. Homozygous stop mutation in AHR causes autosomal recessive foveal hypoplasia and infantile nystagmus. *Brain* 2019;**142**(6):1528–34. https://doi.org/10.1093/brain/awz098.
3. Choi JH, Jung JH, Oh EH, Shin JH, Kim HS, Seo JH, et al. Genotype and phenotype spectrum of FRMD7-associated infantile nystagmus syndrome. *Invest Ophthalmol Vis Sci* 2018;**59**(7):3181–8. https://doi.org/10.1167/iovs.18-24207.
4. Michaud V, Defoort-Dhellemmes S, Drumare I, Pennamen P, Plaisant C, Lasseaux E, et al. Clinical and molecular findings of FRMD7 related congenital nystagmus as a differential diagnosis of ocular albinism. *Ophthalmic Genet* 2019;**40**(2):161–4.
5. Self J, et al. Infantile nystagmus and late onset ataxia associated with a CACNA1A mutation in the intracellular loop between s4 and s5 of domain 3. *Eye (Lond)* 2009;**23**:2251–5. https://doi.org/10.1038/eye.2008.389.
6. Vincent MC, Pujo AL, Olivier D, Calvas P. Screening for PAX6 gene mutations is consistent with haploinsufficiency as the main mechanism leading to various ocular defects. *Eur J Hum Genet* 2003;**11**:163–9. https://doi.org/10.1038/sj.ejhg.5200940.
7. Norman CS, et al. Identification of a functionally significant tri-allelic genotype in the Tyrosinase gene (TYR) causing hypomorphic oculocutaneous albinism (OCA1B). *Sci Rep* 2017;**7**:4415. https://doi.org/10.1038/s41598-017-04401-5.

8. Thomas MG, Maconachie G, Sheth V, McLean RJ. Gottlob I development and clinical utility of a novel diagnostic nystagmus gene panel using targeted next-generation sequencing. *Eur J Hum Genet* 2017;**25**(6):725–34. https://doi.org/10.1038/ejhg.2017.44.

9. Lenassi E, Clayton-Smith J, Douzgou S, Ramsden SC, Ingram S, Hall G, et al. Clinical utility of genetic testing in 201 preschool children with inherited eye disorders. *Genet Med* 2020;**22**(4):745–51. https://doi.org/10.1038/s41436-019-0722-8.

Chapter 20

Congenital cranial dysinnervation disorders

Chapter outline

20.1 Congenital fibrosis of the extraocular
     muscles   408
20.2 Duane retraction syndrome   412
20.3 Horizontal gaze palsy and progressive
     scoliosis   418
20.4 Moebius syndrome   422

Strabismus is a common group of disorders that have a significant genetic component. Within this highly heterogeneous group sits a small number of recognisable Mendelian conditions including certain congenital cranial dysinnervation disorders (CCDDs; this section) and mitochondrial cytopathies (that cause chronic external ophthalmoplegia; Chapter 21).

CCDDs are congenital, non-progressive conditions that result from abnormal development of the cranial nerves and their nuclei. They include both sporadic and familial disorders, and since they frequently involve the extraocular and facial musculature and result in abnormal eye movements, they often present to the ophthalmologist.

CCDDs can be due to errors in cranial motor neuron identity or abnormalities of cranial nerve axon growth/guidance; these can result in phenotypes such as congenital fibrosis of the extraocular muscles (Section 20.1) or Duane syndrome (Section 20.2). Notably, the phenotypic spectrum of CCDDs may include facial weakness (Moebius syndrome, Section 20.4) or a range of neurological, cardiovascular and/or skeletal abnormalities (horizontal gaze palsy and scoliosis, Section 20.3).

20.1

Congenital fibrosis of the extraocular muscles

Mary C. Whitman and Elizabeth C. Engle

Congenital fibrosis of the extraocular muscles (CFEOM) is a stationary disorder characterised by paralytic strabismus and congenital ptosis. It is genetically heterogeneous and both autosomal dominant and autosomal recessive subtypes have been described. Genotype–phenotype correlations have been reported and their recognition can now direct both patient management and family counselling.

Clinical characteristics

Affected individuals are born with varying degrees of ophthalmoplegia and ptosis. Vertical eye movement deficits, especially of upgaze, are a hallmark of the condition and patients' eyes are often stuck in infraduction. Combined with congenital ptosis, this often results in a prominent chin-up head position. Horizontal eye movement deficits vary, ranging from full horizontal motility to nearly complete ophthalmoplegia.[1] Marcus Gunn jaw winking and other evidence of misinnervation (oculomotor synkinesis) have been reported in individuals with CFEOM. The former manifests as elevation of a ptotic upper eyelid with specific movements of the jaw. It is often first noted in young infants when they are feeding and results from aberrant innervation of the levator palpebrae superioris muscle by axons intended to run in the motor branch of the trigeminal nerve (cranial nerve V) and to innervate the pterygoid muscle.

Three main forms of CFEOM have been described—CFEOM1, CFEOM2, and CFEOM3. These subtypes were initially defined clinically, however now that the genetic aetiologies have been elucidated, genetic classification is more informative.

CFEOM1, the 'classic' CFEOM phenotype, is an autosomal dominant disorder that can arise *de novo* and presents with prominent congenital bilateral ptosis, inability to elevate either eye above the midline (eyes are typically 20–30 degrees infraducted), and often restricted horizontal movements. Patients develop a compensatory chin-up head position. It is usually symmetric and can be rarely associated with additional neurologic or syndromic abnormalities (Fig. 20.1).

CFEOM2, which has been mainly reported in patients from the Middle East, is inherited as an autosomal recessive trait. Typically, there is bilateral ptosis, wide-angle exotropia, and severe limitation of horizontal and vertical movements. Affected individuals often have small, sluggish pupils, and can have sub-normal visual acuity consistent with retinal dysfunction; additional neurological or syndromic abnormalities are generally not present (Fig. 20.2).

CFEOM3 has the most variable clinical phenotype. It is inherited as an autosomal dominant trait, can arise *de novo* and ranges from relatively mild to quite severe, even within the same family. The eye movement deficits may be unilateral or asymmetric, ptosis may or may not

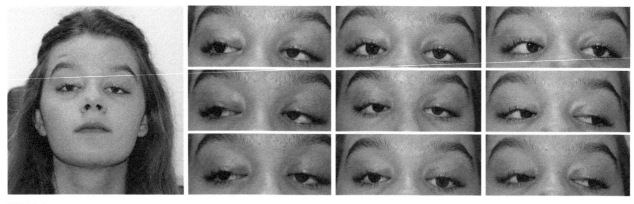

FIG. 20.1 Eye alignment and motility patterns in an individual with CFEOM1 and a heterozygous *KIF21A* mutation. Note the bilateral ptosis with brow recruitment and chin-up head position. In the top panel, showing up-right, up and up-left gaze, note the inability to elevate the eyes above the midline. In the middle panel, showing right, straight, and left gaze, note the relatively preserved horizontal motility. In the bottom panel, showing down-right, down, and down-left gaze, note that downgaze is restricted.

Congenital cranial dysinnervation disorders Chapter | 20 | 409

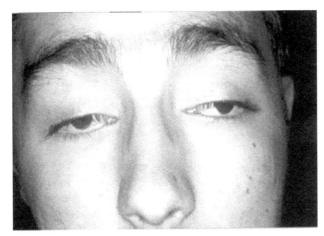

FIG. 20.2 Brother (top) and sister (bottom) with CFEOM2 and biallelic *PHOX2A* mutations. Note the bilateral ptosis and the significant exotropia. The sister has undergone ptosis surgery. *(Adapted from Ref. 2.)*

be present, and patients may retain some ability to elevate their eyes above the midline. Horizontal motility deficits are also quite variable. CFEOM3 can be isolated or associated with a variety of other neurological abnormalities (Fig. 20.3).

Molecular pathology

CFEOM was originally believed to result from an intrinsic defect of the extraocular muscles resulting in fibrosis of the muscles, hence the name 'congenital fibrosis of the extraocular muscles'. A variety of evidence, including human autopsy and MRI studies and mouse models, have now shown that CFEOM is a disorder of the development and targeting of the oculomotor nerve (cranial nerve III); the trochlear nerve (cranial nerve IV) is also involved in CFEOM2. Hypoplasia of the extraocular muscles is a secondary effect, resulting from decreased innervation.[4]

Mutations in five genes have been reported to cause CFEOM:

KIF21A: Heterozygous missense mutations in *KIF21A* cause CFEOM1.[1] This gene encodes a kinesin motor protein that interacts directly with microtubules and moves cargos in an anterograde direction from the cell body to the developing growth cone and mature synapse. It is expressed widely in the cell bodies, axons and dendrites of multiple neuronal populations from early development through maturity, as well as in extraocular and other skeletal muscles.[1] *In vitro*, the KIF21A protein acts as an inhibitor of microtubule dynamics.

CFEOM-associated *KIF21A* missense mutations are most often in the third coiled-coil domain of the stalk, and a few are in the motor domain of the protein. The third coiled-coil domain interacts with the motor to keep the protein in an auto-inhibited state, and pathogenic variants attenuate (but do not eliminate) this autoinhibition, leading to a constitutively active molecule with increased microtubule association. *In vivo*, this leads to oculomotor axon stalling.

Knock-in mice carrying a specific *KIF21A* missense mutation have ptosis and globe retraction. Axons from the superior division of the oculomotor nerve form a

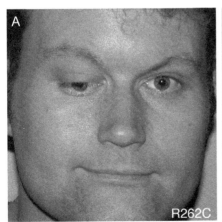 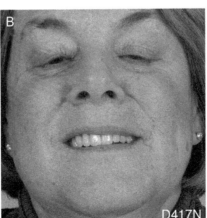 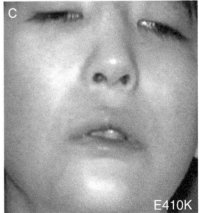

FIG. 20.3 Variability of phenotype in CFEOM3. (A) An individual with *TUBB3* c.784C > T (p.Arg262Cys) and unilateral (right-sided) ptosis and infraduction. (B) An individual with *TUBB3* c.1249G > A (p.Asp417Asn) with bilateral ptosis, infraduction and a chin-up position. (C) An individual with *TUBB3* c.1228G > A (p.Glu410Lys) with severe ptosis, exotropia, infraduction, and facial palsy. *(Adapted from Ref. 3.)*

proximal bulb consisting of stalled, wandering axons with enlarged growth cones. Distal to this bulb, the oculomotor nerve is thin. In the orbit, the superior division of the oculomotor nerve is hypoplastic, and the inferior division displays aberrant branches. This is a gain-of-function mechanism. Kif21a knock-out mice have normal oculomotor development, even though they die at birth.[5] The selective vulnerability of the superior division of the oculomotor nerve to hyperactivation of this widely expressed motor remains to be explained.

Very rarely, *KIF21A* mutations can cause CFEOM3 phenotypes, and *TUBB3* mutations can cause the classic CFEOM1 phenotype.[3] There are also a few reported families with clinical CFEOM1 who lack *KIF21A* or *TUBB3* mutations; here, the underlying genetic mutations remain to be identified.

PHOX2A: Biallelic loss of function mutations in the gene encoding the transcription factor PHOX2A cause CFEOM2.[6] This homeodomain transcription factor protein is necessary for proper differentiation of the neurons in the oculomotor and trochlear nuclei in the brainstem. Pathogenic variants result in loss of PHOX2A protein function, leading to errors of early motor neuron identity and subsequent failure of the oculomotor and trochlear nerves to develop.

TUBB3: Heterozygous missense mutations in *TUBB3* cause CFEOM3. *TUBB3* encodes the class III β-tubulin isotype. Separate genes encode multiple different α- and β-tubulin isotypes. Heterodimers of α- and β-tubulin assemble into microtubules, dynamic cytoskeletal polymers that provide structure to cells and act as highways for motor protein transport. Both microtubule dynamics and motor protein transport are critical for proper axon growth and guidance. *TUBB3*, the neuron-specific β-tubulin isotype, is expressed in all post-mitotic neurons in the developing and mature brain, although it is not the most abundant isotype in the brain.

CFEOM-associated *TUBB3* missense mutations are often but not exclusively in the C-terminal domain (i.e. towards the end) of the protein, alter binding sites for motor proteins and/or microtubule-associated proteins, and increase microtubule stability. In knock-in mice with a specific missense mutation, the developing oculomotor nerve makes an erroneous turn along its path from the brainstem to the orbit and reaches the wrong position in the orbit. The trochlear and trigeminal nerves also show growth and branching deficits.[3]

Predictable genotype–phenotype correlations occur with mutations in *TUBB3*. At least 10 distinct heterozygous *TUBB3* missense mutations have been reported to cause CFEOM3.[3,7] Amino-acid substitutions c.784C > T (p.Arg262Cys), c.904G > A (p.Ala302Thr), and c.185G > A (p.Arg62Gln) are associated with isolated CFEOM3 (without other neurological abnormalities). Other mutations cause CFEOM3 plus mutation-specific combinations of cranial and spinal peripheral neuropathies, developmental delay, intellectual/social disability, and/or brain malformations:

- c.1249G > A (p.Asp417Asn) can cause CFEOM3 followed by the development of a progressive axonal polyneuropathy during adult life.
- c.1228G > A (p.Glu410Lys), c.785G > A (p.Arg262His), and c.1249G > C (p.Asp417His) often arise *de novo* and commonly have broader phenotypes. They are associated with severe bilateral ocular phenotypes, including exotropia, infraduction, and profound limitation of eye movements. In general, people with p.Glu410Lys and p.Arg262His have facial weakness and vocal cord paralysis. Patients carrying the p.Glu410Lys substitution also have facial dysmorphism, Kallmann syndrome (anosmia with hypogonadal hypogonadism), and can develop an axonal peripheral neuropathy and cyclic vomiting.[8] The cyclic vomiting may be responsive to valproic acid. Patients with the p.Arg262His or the p.Asp417His substitution may have congenital joint contractures in addition to peripheral neuropathy with an onset in the first decade of life.[9] Most CFEOM3 patients carrying the p.Glu410Lys, p.Arg262His, or p.Asp417His variant have intellectual and social disabilities, and some meet diagnostic criteria for autism spectrum disorder. There are mutation-specific correlations with specific brain malformations including thin to absent anterior commissure and corpus callosum, dysmorphic basal ganglia, brainstem hypoplasia, and hypoplastic/absent olfactory sulci, olfactory bulbs, and facial nerves.
- c.1138C > T (p.Arg380Cys) often arises *de novo* and results in a moderate form of CFEOM

TUBB2B: A heterozygous missense mutation in *TUBB2B* segregates with CFEOM and polymicrogyria in a single-family.[10] This gene encodes another β-tubulin isotype, and the described mutation is also predicted to interfere with protein binding sites.

TUBA1A: Three heterozygous missense mutations in *TUBA1A* were identified in unrelated probands with CFEOM and brain malformations. This gene encodes an alpha tubulin isotype and the affected amino acids are at interfaces where tubulins heterodimerize or interact.[11]

Clinical management

There are currently no treatments that can restore full functionality and range of motion to the extraocular muscles. Treatment goals include management of ptosis and head position, good ocular alignment in primary position, and maximising visual outcome by prevention or treatment of amblyopia.

Strabismus surgery focuses on improving head position and alignment in the primary position while promoting the greatest range of extraocular motion possible.[12] Treatment of ptosis generally requires a frontalis suspension (sling) procedure due to poor levator function. Care must be taken not to raise the lids too high, as there is an increased risk for subsequent exposure keratopathy due to infraduction and poor to absent Bell's reflex. CFEOM patients who also have facial nerve palsy are at even greater risk of exposure keratopathy due to poor blink and ptosis repair must be very conservative and accompanied by aggressive lubrication.

People with CFEOM are at high risk for both deprivation and strabismic amblyopia and must be monitored and treated appropriately. Amblyopia is treated with patching or atropine therapy as in other children. Appropriate refractive management is also key for maximising visual outcomes.

Children with syndromic CFEOM3 require care by a team that is able to address their many developmental needs, and includes input from ENT, neurology, developmental paediatrics, endocrinology and/or orthopaedics. Patients with facial weakness and vocal cord paralysis may need tracheotomy early in life but many are successfully decannulated later in childhood.

A molecular diagnosis allows families to be counselled appropriately about the risk of recurrence in future offspring for autosomal dominant (CFEOM1 or 3) or autosomal recessive disease (CFEOM2) or where children carry a *de novo* mutation (e.g., *KIF21A* or *TUBB3*).

References

1. Yamada K, et al. Heterozygous mutations of the kinesin KIF21A in congenital fibrosis of the extraocular muscles type 1 (CFEOM1). *Nat Genet* 2003;**35**(4):318–21.
2. Yazdani A, et al. A novel PHOX2A/ARIX mutation in an Iranian family with congenital fibrosis of extraocular muscles type 2 (CFEOM2). *Am J Ophthalmol* 2003;**136**(5):861–5.
3. Tischfield MA, et al. Human TUBB3 mutations perturb microtubule dynamics, kinesin interactions, and axon guidance. *Cell* 2010;**140**(1):74–87.
4. Engle EC. The genetic basis of complex strabismus. *Pediatr Res* 2006;**59**(3):343–8.
5. Cheng L, et al. Human CFEOM1 mutations attenuate KIF21A autoinhibition and cause oculomotor axon stalling. *Neuron* 2014;**82**(2):334–49.
6. Nakano M, et al. Homozygous mutations in ARIX(PHOX2A) result in congenital fibrosis of the extraocular muscles type 2. *Nat Genet* 2001;**29**(3):315–20.
7. Whitman MC, et al. Two unique TUBB3 mutations cause both CFEOM3 and malformations of cortical development. *Am J Med Genet A* 2016;**170**(2):297–305.
8. Chew S, et al. A novel syndrome caused by the E410K amino acid substitution in the neuronal beta-tubulin isotype 3. *Brain* 2013;**136**(Pt 2):522–35.
9. Whitman MC, et al. TUBB3 Arg262His causes a recognizable syndrome including CFEOM3, facial palsy, joint contractures, and early-onset peripheral neuropathy. *Hum Genet* 2021;**140**(12):1709–31.
10. Cederquist GY, et al. An inherited TUBB2B mutation alters a kinesin-binding site and causes polymicrogyria, CFEOM and axon dysinnervation. *Hum Mol Genet* 2012;**21**(26):5484–99.
11. Jurgens JA, et al. Novel variants in TUBA1A cause congenital fibrosis of the extraocular muscles with or without malformations of cortical brain development. *Eur J Hum Genet* 2021;**29**(5):816–26.
12. Khan AO. CFEOM—pearls for management. In: De Liano RG, editor. 13th Meeting of the International Strabismological Association (18–22 March 2018), International Strabismological Association; 2019; 57–9.

20.2

Duane retraction syndrome

Arif O. Khan

Duane retraction syndrome is a congenital anomaly associated with aberrant co-innervation of the lateral rectus and medial rectus muscle by the oculomotor nerve (cranial nerve III); hypoplasia of the nucleus of the abducens nerve (cranial nerve VI) is also commonly noted. These abnormalities result in variable limitation of horizontal eye movements, and the characteristic globe retraction when the affected eye attempts to move towards the midline; the latter occurs due to co-contraction of the affected horizontal muscles on adduction.

Most cases of Duane retraction syndrome are unilateral, multifactorial, sporadic and isolated. However, rare Mendelian subtypes have been described and these can be associated with extraocular manifestations.[1,2]

Clinical characteristics

Affected individuals have a horizontal ocular misalignment that is incomitant and associated with globe retraction, palpebral aperture narrowing and, occasionally, up-shoots or down-shoots when the eye attempts to move nasally. Children with Duane retraction syndrome are often brought to medical attention because of abnormal head position, strabismus, or abnormal eye movements.

Different disease subtypes have been described with the commonest form being unilateral and associated with esotropia and deficient abduction (Huber type I, see below). In these cases, a face turn towards the affected eye is adopted to align it with the contralateral normal eye (Fig. 20.4). However, if the affected eye has low vision, the child might

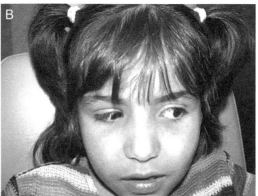
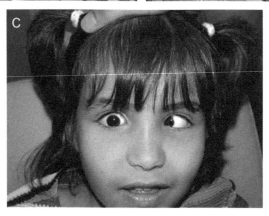

FIG. 20.4 Findings in a girl who has the common esotropic form of Duane retraction syndrome. (A) This affects her right eye and she adopts a right face-turn for binocular vision. (B) When she attempts to adduct the right eye, the globe retracts and there is a down-shoot. (C) When she tries to abduct the right eye, the duction is significantly limited.

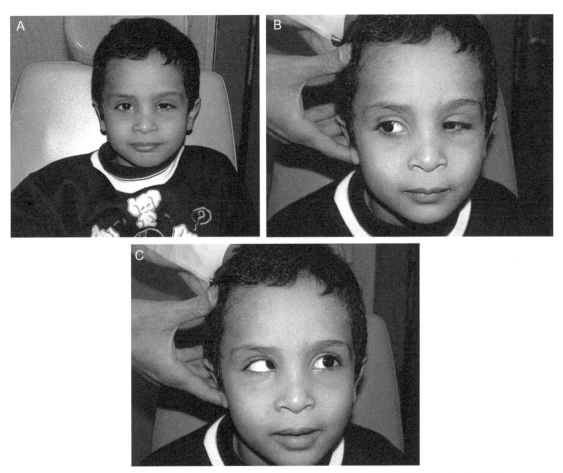

FIG. 20.5 Findings in a boy who has the common esotropic form of Duane retraction syndrome in the left eye. (A) Amblyopia is present in the affected eye and he, therefore, does not adopt a compensatory face turn for his small-angle esotropia in primary position. (A, B) Left globe retraction can be appreciated in both right gaze and primary position. (C) Left abduction is limited.

not adopt a compensatory head posture and may present with esotropia in the primary position (Fig. 20.5). Another common presenting complaint is the perception of abnormal eye movement in the unaffected eye. When a discordance is observed between the ductions of the two eyes, it may seem that the unaffected eye is abnormal with excessive movement (adduction) when in fact the actual problem is the affected eye's limited motility (abduction) (Fig. 20.6).

Although Huber defined three forms of Duane retraction syndrome, a spectrum of dysinnervation is observed rather than distinct subtypes. The Huber type I subtype, discussed above, is the most common form (Figs 20.4-20.6). Huber type II is characterised by limited adduction, typically in an exotropic eye (Fig. 20.7). Huber type III is characterised by limited adduction and abduction of the involved eye (Fig. 20.8). Rarely, an eye can be exotropic with poor adduction and paradoxically increased exotropia during attempted adduction (synergistic divergence); this is sometimes described as Duane retraction syndrome type IV (Fig. 20.9).

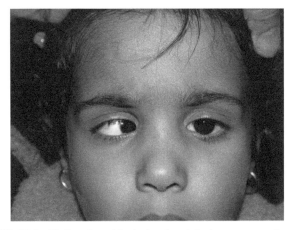

FIG. 20.6 Findings in a girl who has the relatively common esotropic form of Duane retraction syndrome in the left eye. When she fixates with her left eye, a large amount of innervation is required to abduct it to the primary position. This large amount of innervation also goes to the normal right eye, causing it to become grossly esotropic (secondary deviation). Thus, it may seem like the right eye is abnormal when in fact the abnormality is in the left eye.

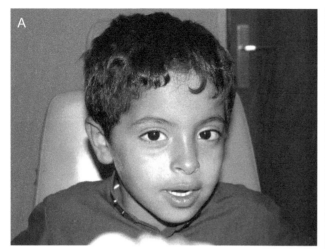
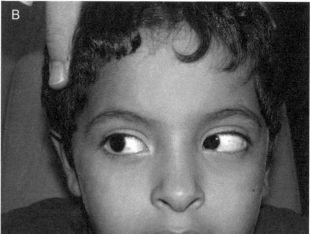
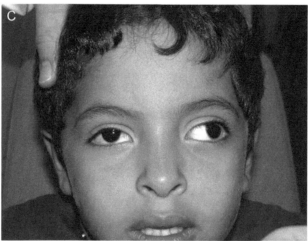

FIG. 20.7 Findings in a boy with Duane retraction syndrome (Huber type II) in the right eye. (A–C) Note that the right eye is exotropic with limited adduction.

Only a minority of cases are familial (up to 10%). Also, most affected individuals have no systemic manifestations although a variety of extraocular findings has been reported. Many of these findings are likely coincidental as Duane retraction syndrome is not an uncommon form of strabismus. However, several true associations have been described (e.g., deafness in up to ~10% of cases); these are typically congenital malformations that are directly related developmentally to the lateral rectus dysinnervation that occurs at approximately 4-6 weeks of gestation.

Molecular pathology

Typical Duane retraction syndrome does not require genetic testing or further investigations as the common forms are not associated with identifiable gene mutations, even in bilateral cases.[1,2] However, a range of genetic abnormalities has been identified in cases in which there is a family history of the condition. Certain extraocular features can provide further clues for one of these rare Mendelian subtypes. Associated genes include:

CHN1: *CHN1* encodes α1 and α2 chimaerin, which together regulate axonal guidance. Heterozygous missense mutations in this gene cause Duane retraction syndrome with or without vertical motility abnormalities. *CHN1* mutations can also cause vertical strabismus without Duane syndrome. Affected individuals do not typically have extraocular features.[3–5]

MAFB: *MAFB* encodes a transcription factor that is required for normal hindbrain development. Heterozygous (both dominantly inherited and *de novo*) loss of function mutations in this gene cause Duane retraction syndrome with or without deafness.[6]

SALL4: Heterozygous mutations in another transcription factor gene, *SALL4*, are seen in Duane-radial ray syndrome (also known as Okihiro syndrome), Holt-Oram syndrome (also known as an atriodigital syndrome) and acro-renal-ocular syndrome. *SALL4* genetic variants

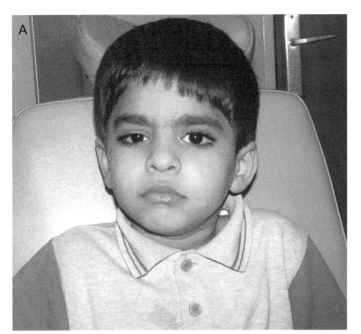
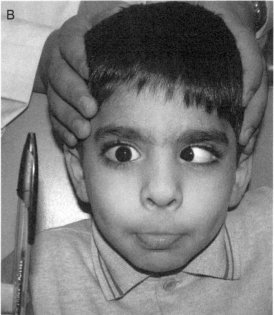
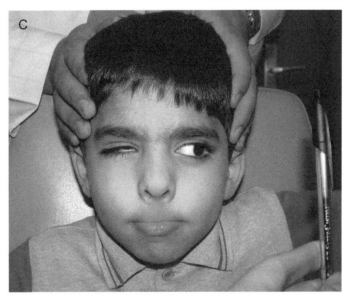

FIG. 20.8 Findings in a boy with Duane retraction syndrome (Huber type III) in the right eye. (A–C) Note that the right eye is slightly esotropic with limited adduction (with up-shoot) and limited abduction.

can also result in several syndromic conditions without ophthalmic findings.[7,8]

HOXA1: Biallelic mutations in the transcription factor gene *HOXA1* cause bilateral Duane retraction syndrome (Huber type III) in the context of bilateral deafness, cerebrovascular anomalies and autism.[1,9]

ECEL1: Biallelic mutations in the neuromuscular junction gene *ECEL1* cause Duane retraction syndrome with arthrogryposis (involving multiple joint contractures). *ECEL1* mutations can also cause other congenital cranial dysinnervation disorders (in the context of arthrogryposis) or arthrogryposis without ophthalmic findings.[10]

COL25A1: Biallelic mutations in the brain-specific membrane-bound collagen gene *COL25A1* cause Duane retraction syndrome (type IV) without extraocular features. *COL25A1* mutations are also a reported cause of isolated ptosis without Duane syndrome.[11]

Other rare syndromic conditions that may be associated with Duane retraction syndrome include[1]:

- Wildervanck syndrome (also known as a cervico-oculo-acoustic syndrome), characterised by Klippel-Feil abnormality (abnormal fusion of cervical vertebrae) and deafness; this is a sporadic, multifactorial condition that is more common in females.

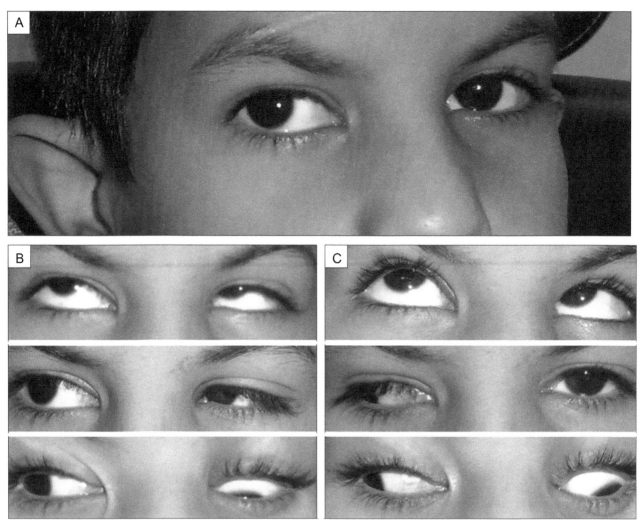

FIG. 20.9 Findings in a boy with exotropic Duane retraction syndrome in the right eye. There is synergistic divergence (i.e. paradoxically increased right abduction on attempted left gaze). Esotropic Duane retraction syndrome is also noted in the preferred fixating left eye (limited abduction and retraction during adduction), his fixating eye. (A) Preferred abnormal head position. (B) Right gaze; up, to the side, and down. (C) Left gaze; up, to the side, and down. *(Adapted from Ref. 11.)*

- Goldenhar syndrome (also known as an oculo-auriculo-vertebral syndrome) is a craniofacial microsomia that can include Duane retraction syndrome; this is a sporadic condition that is multifactorial, although there may be a genetic component in a small subset of cases.

Clinical management

Affected children should be evaluated for refractive error and amblyopia and treated appropriately. Most cases require no further intervention but prisms can be used for comfort or to improve head position. Surgery may be considered in selected cases who have a significant compensatory head posture, a significant deviation in primary position or unsightly vertical up-shoots or down-shoots. Genetic testing can be considered if there is family history or certain associated features as described above; this can enable appropriate family counselling.

References

1. Abu-Amero KK, Khan AO, Oystreck DT, Kondkar AA, Bosley TM. The genetics of non-syndromic bilateral Duane retraction syndrome. *J AAPOS* 2016;**20**(5):396–400. e2.
2. Whitman MC, Engle EC. Ocular congenital cranial dysinnervation disorders (CCDDs): insights into axon growth and guidance. *Hum Mol Genet* 2017;**26**(R1):R37–44.
3. Nugent AA, Park JG, Wei Y, Tenney AP, Gilette NM, DeLisle MM, et al. Mutant α2-chimaerin signals via bidirectional ephrin pathways in Duane retraction syndrome. *J Clin Invest* 2017;**127**(5):1664–82.
4. Miyake N, Chilton J, Psatha M, Cheng L, Andrews C, Chan WM, et al. Human CHN1 mutations hyperactivate alpha2-chimaerin and cause Duane's retraction syndrome. *Science* 2008;321(5890):839–43.

5. Angelini C, Trimouille A, Arveiler B, Espil-Taris C, Ichinose N, Lasseaux E, et al. CHN1 and Duane retraction syndrome: expanding the phenotype to cranial nerves development disease. *Eur J Med Genet* 2021;**64**(4):104188.
6. Park JG, Tischfield MA, Nugent AA, Cheng L, Di Gioia SA, Chan WM, et al. Loss of MAFB function in humans and mice causes Duane Syndrome, aberrant extraocular muscle innervation, and inner-ear defects. *Am J Hum Genet* 2016;**98**(6):1220–7.
7. Terhal P, Rösler B, Kohlhase J. A family with features overlapping Okihiro syndrome, hemifacial microsomia and isolated Duane anomaly caused by a novel SALL4 mutation. *Am J Med Genet A* 2006;**140**(3):222–6.
8. Kohlhase J, Schubert L, Liebers M, Rauch A, Becker K, Mohammed SN, et al. Mutations at the SALL4 locus on chromosome 20 result in a range of clinically overlapping phenotypes, including Okihiro syndrome, Holt-Oram syndrome, acro-renal-ocular syndrome, and patients previously reported to represent thalidomide embryopathy. *J Med Genet* 2003;**40**(7):473–8.
9. Tischfield MA, Chan WM, Grunert JF, Andrews C, Engle EC. HOXA1 mutations are not a common cause of Duane anomaly. *Am J Med Genet A* 2006;**140**(8):900–2.
10. Khan AO, Shaheen R, Alkuraya FS. The ECEL1-related strabismus phenotype is consistent with congenital cranial dysinnervation disorder. *J AAPOS* 2014;**18**(4):362–7. Erratum in: J AAPOS. 2014;18(5):517.
11. Khan AO, Al-Mesfer S. Recessive COL25A1 mutations cause isolated congenital ptosis or exotropic Duane syndrome with synergistic divergence. *J AAPOS* 2015;**19**(5):463–5.

20.3

Horizontal gaze palsy and progressive scoliosis

Arif O. Khan

Horizontal gaze palsy and progressive scoliosis (HGPPS) is an autosomal recessive disorder associated with failure of brainstem axonal decussation. It is characterised by congenital horizontal gaze palsy and juvenile scoliosis. Affected children are often not diagnosed until the development of scoliosis, which starts in the first decade of life, is progressive, and can be severe.

Clinical characteristics

Affected individuals are born with limited horizontal gaze.[1] Abduction is typically deficient but convergence is generally preserved and can be invoked to enable side cross-fixation with the adducting eye (Fig. 20.10). While this is often voluntary, there is at least one reported case where

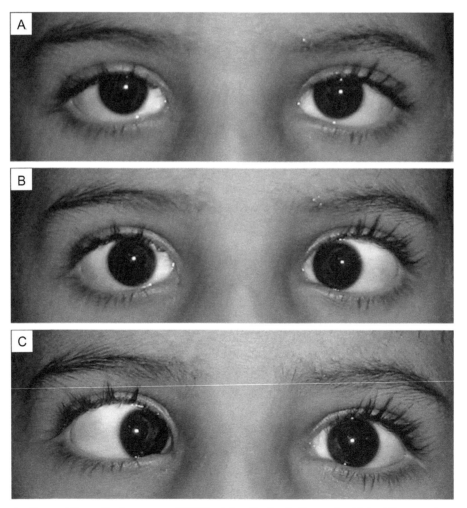

FIG. 20.10 Findings in a 5-year-old boy with a homozygous *ROBO3* mutation. Previously diagnosed as bilateral Duane syndrome type III, the child did not have primary position strabismus at distance or near fixation. The following positions of gaze are shown: (A) fixating at near with appropriate convergence; (B) attempted right gaze, using convergence to adduct the left eye; (C) attempted left gaze, using convergence to adduct the right eye. Notably, convergence cannot overcome duction limitations in Duane syndrome, and there are no signs of the abnormal synkinesis during attempted side gaze that is characteristic of Duane syndrome. *(Adapted from Ref. 1.)*

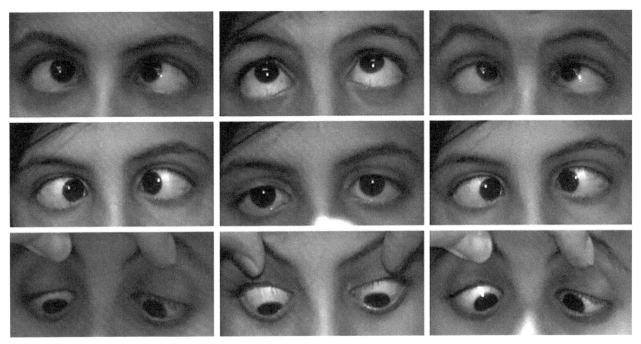

FIG. 20.11 Findings in a 9-year-old girl with a homozygous *ROBO3* mutation. Note that, on attempted side gaze, convergence substitutes horizontal gaze, while vertical versions are normal. *(Adapted from Ref. 3.)*

it was involuntary, i.e. there was synergistic convergence (Fig. 20.11). Some affected children have infantile esotropia, which facilitates side cross-fixation in the absence of horizontal gaze (Fig. 20.12). Vertical versions and ductions are preserved. Fine nystagmus and asynchronous blinking may be present. Other than these motility findings and potential refractive error, ophthalmic examination is unremarkable.

Before scoliosis is recognised, affected children are often seen by ophthalmologists because of deficient eye

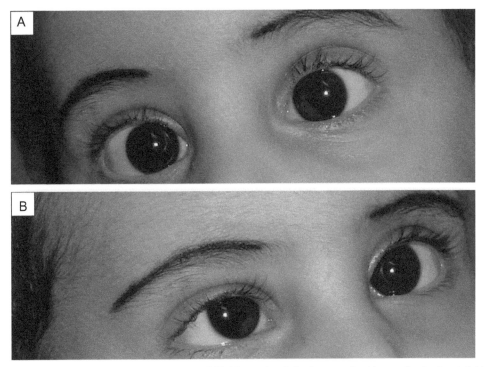

FIG. 20.12 Findings in a 10-month-old child with a homozygous ROBO3 mutation. Infantile esotropia with cross-fixation is noted; (A) right eye is used to see objects to left; (B) left eye is used to see objects to right. *(Adapted from Ref. 1.)*

movements or strabismus. Sometimes an affected infant is brought to the ophthalmologist because the inability to horizontally track (before the development of head control) is perceived as a sign of poor vision when in fact it is an eye movement abnormality (horizontal gaze palsy) rather than a problem with vision.

Scoliosis is progressive, typically severe, and can be noted as early as in the first year of life. Marked rotation of the vertebrae can cause pulmonary compromise. MRI shows characteristic brainstem features such as deep anterior and posterior medullary clefts (Fig. 20.13). There are no other systemic findings, although some affected individuals have been reported to have developmental delay.

Lack of decussation of the descending corticospinal tract and the ascending somatosensory tract in the medulla is typically noted. Because of this, an injury to the motor cortex causes ipsilateral rather than contralateral hemiparesis. Cerebellar tracts are also uncrossed or abnormally located.

Neuroimaging reveals intact abducens nerves (cranial nerve VI) bilaterally. However, deep anterior and posterior clefts are noted in the medulla and low pons. A large fourth ventricle and no decussation of the corticospinal tract, medial lemniscus and/or superior cerebellar peduncle are also noted.

Molecular pathology

The ocular motility abnormality in these patients is related to the failure of crossing of the brainstem of axons that were destined to cross—particularly those of the paramedian pontine reticular formation, medial longitudinal fasciculus, and pons. The pathogenesis of scoliosis is unclear but is likely neurogenic and related to lack of normal contralateral innervation.

HGPPS is caused by biallelic mutations in the *ROBO3* gene, which encodes an axon guidance transmembrane receptor protein predominantly expressed in the hindbrain and spinal cord commissural axons during embryonic development.[2] A range of mutations has been identified including clear loss of function mutations (around two thirds) and missense variants (around one third).[3,4]

Clinical management

Ophthalmic management involves the treatment of the associated amblyopia, refractive error and strabismus. Strabismus surgery is generally approached through conventional techniques.

The most important and challenging management issue in these patients is the early-onset, progressive and severe

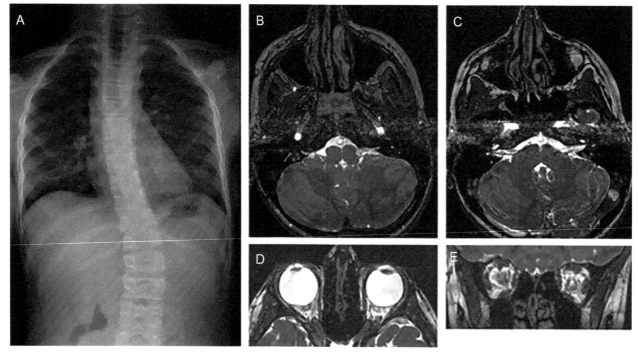

FIG. 20.13 Skeletal findings in horizontal gaze palsy and progressive scoliosis (HGPPS). (A) Spine X-ray revealing modest thoracolumbar scoliosis (concave left) that was not apparent during clothed ophthalmologic examination. Axial steady-state free precession MRI through the medulla (B) and pons (C) confirming hypoplasia with deep anterior and posterior midline clefts and a somewhat large fourth ventricle. These findings are common in typical horizontal gaze palsy and progressive scoliosis (HGPPS). Notably, T2-weighted axial (D) and coronal (E) MRI scans through the orbits demonstrate normal-appearing extraocular muscles and myopic globes. *(Adapted from Ref. 3.)*

scoliosis. Once the diagnosis is made, the patient should be promptly referred to a spinal specialist.

When HGPPS is suspected, genetic testing of the *ROBO3* gene is recommended. This can help confirm the diagnosis and enables appropriate genetic counselling for this autosomal recessive disorder.

References

1. Khan AO, Abu-Amero K. Infantile esotropia with cross-fixation, inability to v abduct, and underlying horizontal gaze palsy with progressive scoliosis. *J AAPOS* 2014;**18**(4):389–91.
2. Jen JC, Chan WM, Bosley TM, Wan J, Carr JR, Rüb U, et al. Mutations in a human ROBO gene disrupt hindbrain axon pathway crossing and morphogenesis. *Science* 2004;**304**(5676):1509–13.
3. Khan AO, Oystreck DT, Al-Tassan N, Al-Sharif L, Bosley TM. Bilateral synergistic convergence associated with homozygous ROB03 mutation (p.Pro771Leu). *Ophthalmology* 2008;**115**(12):2262–5.
4. Bosley TM, Salih MA, Jen JC, Lin DD, Oystreck D, Abu-Amero KK, et al. Neurologic features of horizontal gaze palsy and progressive scoliosis with mutations in ROBO3. *Neurology* 2005;**64**(7):1196–203.

20.4

Moebius syndrome

Mary C. Whitman and Elizabeth C. Engle

Moebius syndrome is a congenital, non-progressive anomaly that is characterised by weakness of the facial nerve (cranial nerve VII) combined with ocular abduction limitation due to abducens nerve (cranial nerve VI) weakness. It is often bilateral and affected children require multidisciplinary care. As Moebius syndrome is a multifactorial condition of uncertain aetiology, the risk of recurrence in affected families is low. However, certain Mendelian forms of congenital cranial dysinnervation disorders resemble Moebius syndrome and, in these cases, genetic testing can facilitate precise diagnosis and counselling.

Clinical characteristics

Affected individuals have bilateral (rarely unilateral) facial weakness, combined with abnormalities of horizontal eye movements, which range from isolated abduction limitation to full horizontal gaze paresis (Fig. 20.14).

Historically, studies of Moebius syndrome have used different diagnostic criteria, complicating clinical, genetic and pathophysiologic studies. In 2007, a group of clinicians and researchers defined the minimum diagnostic criteria (MDC) for classic Moebius syndrome as "congenital, uni- or bilateral, non-progressive facial weakness and limited abduction of the eye(s)". A subsequent review of patients who had received a clinical diagnosis of Moebius syndrome revealed that 19% of cases (21/112) did not meet these minimum diagnostic criteria. Of these 21 patients, 4 were found to have genetically confirmed congenital fibrosis of the extraocular muscles (CFEOM3); a pair of siblings with facial palsy and esotropia (but full ocular motility) had mutations in the *HOXB1* gene, and the remaining patients had full motility or abducens palsy without facial palsy. An additional 3 patients met the prior diagnostic criteria but also had limited vertical gaze.[1] Thus, it was proposed to refine the diagnostic criteria for Moebius syndrome to facial palsy, limited abduction, and normal vertical eye movements.[1] A subsequent study found that 50% of patients who had received a clinical diagnosis of Moebius syndrome did not meet these new criteria on detailed, multidisciplinary examination.[2]

Moebius syndrome (or sequence) can be associated with a variety of additional congenital malformations and neurological deficits. Limb malformations are common, particularly hypoplasia, transverse terminal limb defects, arthrogryposis (involving multiple joint contractures), syndactyly and brachydactyly. Poland anomaly (pectoralis muscle defect) and Klippel-Feil anomaly (abnormal fusion of cervical vertebrae) are also frequently seen. Other cranial nerve deficits and orofacial malformations including deafness, tongue weakness/malformation, swallowing or

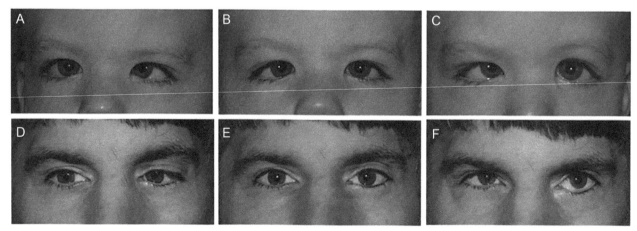

FIG. 20.14 Findings in two individuals with classic Moebius syndrome. (A–C): right, straight and left gaze, respectively, of an 18-month-old with large esotropia, marked abduction deficit bilaterally, and good adduction. (D–F): right, straight and left gaze, respectively, of an adult with orthotropia (straight eyes) and limited horizontal gaze in both directions. *(Adapted from Ref. 1.)*

respiratory difficulties, cleft palate, hypodontia, micrognathia and ear deformities can also occur. General motor disability, poor coordination, aberrant lacrimation, intellectual disability and autism have all been reported in association with Moebius syndrome.[3] Reports of the prevalence of each associated feature vary widely, as most reports are case series and use different diagnostic criteria. For example, while the prevalence of intellectual disability in Moebius syndrome is felt to be about 10% of cases, estimates vary from 0% to 75% in some series. Diagnoses of autism are particularly difficult in this population, as lack of facial expression and eye movements both affect social interaction and prevent accurate use of autism diagnostic instruments.

Although hypogonadotropic hypogonadism has been reported in association with Moebius syndrome, a careful review of those reports reveals that those patients appear to have congenital fibrosis of the extraocular muscles (CFEOM).

Molecular pathology

Despite multiple reports in the literature of familial cases of Moebius syndrome, there are no clear examples of classic Moebius syndrome (using 2014 criteria) occurring in multiple family members or showing vertical transmission. A 2014 study showed that in 5/5 families who reported multiple affected members, at least one of the affected individuals did not meet minimum diagnostic criteria for Moebius syndrome.[1] Several patients initially (mis)diagnosed with Moebius syndrome have been found to have mutations in *HOXB1* (associated with facial palsy, esotropia and full eye movements), *TUBB3* (associated with CFEOM3 with facial palsy; discussed in Section 20.1), or mutations in the myopathy genes *MYMK* or *STAC3*.[1,4,5] Many reportedly familial cases have ptosis and exotropia, rather than esotropia, or complete ophthalmoplegia and therefore are likely to have CFEOM with facial weakness or another congenital cranial dysinnervation disorder.

Isolated congenital facial palsy (with normal eye movements) can segregate as an autosomal dominant trait and, in some families, maps to chromosome 3[6] or chromosome 10.[7] These were originally reported as Moebius syndrome genetic loci, but the patients did not have abduction deficits. Deletion of chromosome 13q12.2 has been reported in one patient with Moebius syndrome, and a balanced translocation between 1p34 and 13q13 has been identified in a family with congenital facial weakness (with normal eye movements). Furthermore, *TUBB6* mutations have been reported in patients with facial nerve palsy and ptosis.

Multiple lines of evidence indicate that Moebius syndrome may not be a single disease entity. During development, the abducens and facial nerve motor neurons develop in close proximity in the hindbrain and then the facial motor neurons migrate caudally past the abducens nucleus. Vascular, genetic or teratogenic insults during early gestation could cause damage to these motor neurons or their axons, as well as cause the other associated anomalies. Radiological studies may reveal the absence of facial, abducens, and other cranial nerves, absence of the facial nerve canal in the middle ear, hypoplastic dorsal pons, and straightening or calcification of the floor of the fourth ventricle. Diffusion MRI morphometry of 21 individuals with Moebius syndrome revealed a reduction in brain volume in the pons that may correlate with limited horizontal gaze.[8] Electromyography shows neurogenic changes in almost all patients with Moebius syndrome, although more than half show heterogeneous neurogenic and myogenic changes.

Notably, Moebius sequence and related malformations are seen after early embryological exposure to a variety of teratogens. Thalidomide was prescribed for use in early pregnancy for approximately 2 years in Europe before its teratogenic properties were identified. This resulted in a cohort of exposed patients who have been followed closely. For many of these patients, the precise days of exposure were known. Facial nerve palsy, incomitant strabismus (often Duane retraction syndrome), and external ear anomalies were seen after thalidomide exposure during early gestation (days 20–25 after fertilisation).[3] Misoprostol is a synthetic prostaglandin that has been used to induce elective abortion. It is often used in Brazil, where abortion is prohibited but misoprostol is available over the counter. *In utero* exposure that does not result in abortion increases the risk of Moebius syndrome.[9,10] Several other *in utero* exposures have been suggested to cause this anomaly, including gestational hyperthermia, electric shock, benzodiazepines, cocaine, alcohol, and ergotamine.

Clinical management

Children with Moebius syndrome require care by a team that is able to address their many developmental needs, and includes input from ENT, neurology, developmental paediatrics, and/or orthopaedics. Ophthalmic care focuses on improving eye alignment in primary position, preventing or treating exposure keratopathy due to facial weakness, and maximising visual potential by prevention and treatment of amblyopia and management of refractive error.

For patients who meet the minimum diagnostic criteria for classic Moebius syndrome, the risk of recurrence in family members is very low. For patients with isolated facial palsy, which is often autosomal dominant, or other 'atypical' forms of Moebius syndrome, the risk of recurrence in family members is much higher. Genetic testing in these atypical cases (of the *HOXB1*, *TUBB3*, and *TUBB6* genes) can facilitate more precise diagnosis and counselling.

References

1. MacKinnon S, et al. Diagnostic distinctions and genetic analysis of patients diagnosed with moebius syndrome. *Ophthalmology* 2014;**121**(7):1461–8.
2. Pedersen LK, et al. Moebius sequence -a multidisciplinary clinical approach. *Orphanet J Rare Dis* 2017;**12**(1):4.
3. Miller MT, et al. The puzzle of autism: an ophthalmologic contribution. *Trans Am Ophthalmol Soc* 1998;**96**:369–85 [discussion 385-7].
4. Di Gioia SA, et al. A defect in myoblast fusion underlies Carey-Fineman-Ziter syndrome. *Nat Commun* 2017;**8**:16077.
5. Linsley JW, et al. Congenital myopathy results from misregulation of a muscle Ca2+ channel by mutant Stac3. *Proc Natl Acad Sci U S A* 2017;**114**(2):E228–36.
6. Michielse CB, et al. Refinement of the locus for hereditary congenital facial palsy on chromosome 3q21 in two unrelated families and screening of positional candidate genes. *Eur J Hum Genet* 2006;**14**(12):1306–12.
7. Verzijl HT, et al. A second gene for autosomal dominant Mobius syndrome is localized to chromosome 10q, in a Dutch family. *Am J Hum Genet* 1999;**65**(3):752–6.
8. Sadeghi N, et al. Brain phenotyping in Moebius syndrome and other congenital facial weakness disorders by diffusion MRI morphometry. *Brain Commun* 2020;**2**(1):fcaa014.
9. Cronemberger MF, et al. Ocular and clinical manifestations of Mobius' syndrome. *J Pediatr Ophthalmol Strabismus* 2001;**38**(3):156–62.
10. Pastuszak AL, et al. Use of misoprostol during pregnancy and Mobius' syndrome in infants. *N Engl J Med* 1998;**338**(26):1881–5.

Chapter 21

Progressive external ophthalmoplegia

Ungsoo S. Kim, Graeme C.M. Black, and Patrick Yu-Wai-Man

Mitochondrial disorders of energy metabolism are a diverse group of inborn errors of metabolism. They are clinically heterogeneous and include conditions with prominent ophthalmic features such as Leber hereditary optic neuropathy (Section 15.1); maternally inherited diabetes and deafness (MIDD); mitochondrial encephalopathy, lactic acidosis, and stroke-like episodes (MELAS) (Section 13B.2.4); neuropathy, ataxia, and retinitis pigmentosa (NARP); and progressive external ophthalmoplegia (PEO).

PEO is the descriptive term for a group of conditions associated with ptosis and ophthalmoparesis (i.e. defective action of the extraocular muscles leading to limitation of ocular motility). Notably, nearly half of all patients with confirmed mitochondrial disease manifest these signs.[1–3] PEO may be sporadic, inherited as a maternally-inherited (mitochondrial) trait or associated with autosomal disorders caused by mutations in nuclear genes. The spectrum of overlapping clinical phenotypes ranges from isolated PEO to 'PEO plus' syndromes with additional systemic features.[4,5] Kearns–Sayre syndrome (KSS) is an important PEO subtype defined by the triad of early-onset PEO (<20 years), pigmentary retinopathy and cardiac conduction defects; other systemic abnormalities, including additional neurological deficits, may be present.[6]

The management of individuals with PEO requires a multidisciplinary approach. Genetic investigations should be performed to identify any underlying mitochondrial or nuclear genetic defect, and a muscle biopsy is sometimes required as part of the diagnostic work-up.

Clinical characteristics

PEO arises due to the effect of mitochondrial dysfunction on the extraocular muscles.[7,8] Affected individuals present with progressive and generally symmetrical bilateral ptosis, together with more generalised weakness of the extraocular muscles. There is reduced elevation of the upper lid due to impaired function of the levator palpebrae superioris and, as the disease progresses, the visual axis becomes involved, impairing the patient's field of vision.[4] Weakness of the facial muscles may also be present and the orbicularis oculi is frequently involved. The limitation of eye movement is usually symmetrical and most affected individuals exhibit an exodeviation (Fig. 21.1).[10] However, there is marked variability in disease severity and up to half of all patients with PEO have manifest strabismus, with a proportion of those reporting transient or constant diplopia.[8]

It is important to recognise that most patients with PEO will have additional neurological and systemic deficits, some of which can be life-threatening. Common features associated with 'PEO plus' include cardiac complications (cardiac conduction defects and cardiomyopathy); endocrine abnormalities (diabetes mellitus, growth hormone insufficiency, hypogonadotrophic hypogonadism and adrenal insufficiency); and neuromuscular deficits (myopathy, peripheral neuropathy, cerebellar ataxia, sensorineural hearing loss, and cognitive impairment).[4,11]

Pigmentary retinopathy is a characteristic fundoscopic feature seen in patients with KSS (Fig. 21.2). This presents as a speckled 'salt and paper' appearance rather than the classical 'bone spicule' features typical of retinitis pigmentosa. Most patients retain reasonable visual fields and the pigmentary changes are not correlated with visual function abnormalities.[12,13]

Notably, PEO can occur in combination with optic atrophy, particularly in patients with pathogenic *OPA1* variants (Section 15.2).[14]

Patients with PEO can be variably affected by different combinations of clinical features and it is not possible to reliably predict the rate of disease progression. This variability together with the significant genetic heterogeneity makes genetic counselling particularly challenging.

Molecular pathology

PEO is genetically heterogeneous. It is caused by alterations of mitochondrial DNA (mtDNA) (either point mutations and large-scale rearrangements) or defects in nuclear genes that encode for key structural and functional components of mitochondria.[15–18]

The two most common causes of PEO are single large-scale mtDNA deletions and the m.3243A>G point mutation (see also Section 13B.2.4). Other mtDNA point mutations identified in people with PEO include m.8344A>G, m.8993T>G and m.14677T>C. The

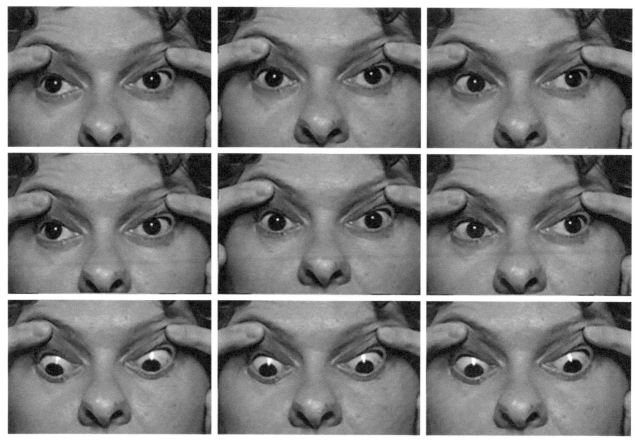

FIG. 21.1 Limited range of eye movements in a 48-year-old with progressive external ophthalmoplegia. The upper eyelids were lifted to more clearly demonstrate the range of eye moments in the nine cardinal positions of gaze. There was a significant generalised limitation of eye movements, which was worse on upgaze and mostly symmetrical. There is no manifest deviation in the primary position as these pictures were taken after the patient had undergone bilateral strabismus surgery to correct an alternating exotropia. *(Adapted from Ref. 9)*.

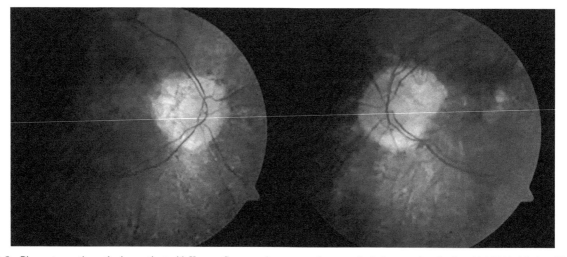

FIG. 21.2 Pigmentary retinopathy in a patient with Kearns–Sayre syndrome secondary to a single large-scale mitochondrial DNA deletion. There was generalised retinal pigment epithelial disturbance throughout the fundus, including the macula, with areas of hypo- and hyperpigmentation. A prominent ring of peripapillary atrophy was also present around both optic discs in this case. *(Adapted from Ref. 9)*.

m.8993T>G variant is typically seen in individuals with the NARP phenotype (neuropathy, ataxia, and retinitis pigmentosa).[11] Single mtDNA deletions are usually sporadic and are typically identified by long-range PCR analysis undertaken on homogenate muscle DNA from a muscle biopsy specimen.[4,18] To obviate the need for invasive muscle biopsy, urine samples are increasingly being used to detect these mtDNA deletions.[15] It is noted that the levels of the m.3243A>G variant in blood decrease with age, and determining these level in urine or a muscle biopsy is usually recommended.[19] For other mtDNA variants, a blood sample is sufficient for diagnostic purposes. Genotype–phenotype correlations have been described: in people with single large-scale mtDNA deletions, disease severity has been correlated with the size of the deletion, the deletion heteroplasmy levels in skeletal muscle, and the location of the deletion within the genome.[17,18]

An expanding list of nuclear autosomal genes has been associated with PEO spectrum phenotypes with both dominant and recessive modes of inheritance being described (Table 21.1). These disease subtypes are generally associated with defective maintenance of the mitochondrial genome; this can result either in mtDNA depletion (i.e. reduction of mtDNA copy number) or in the accumulation of multiple mtDNA deletions. These abnormalities can be detected by quantitative PCR and long-range PCR analysis of homogenate skeletal muscle DNA respectively. A muscle biopsy will also identify features pointing towards an underlying mitochondrial cytopathy; cytochrome *c* oxidase negative muscle fibres and, in some cases, ragged red fibres (representing the accumulation of mitochondria under the sarcolemmal membrane) are observed.[16] High-throughput sequencing panels are increasingly being used to screen for nuclear genes known to cause PEO, with the most commonly mutated genes being *PEO1*, *POLG1*, *OPA1* and *RRM2B*.[17]

Clinical management

Individuals with PEO have three major ophthalmological manifestations that need to be considered, namely, ptosis, ophthalmoplegia and retinal abnormalities. The most relevant is ptosis, which can be visually disabling if the visual axis is involved.[20] Ptosis surgery can be highly effective, but it needs to be undertaken by a specialist surgeon with knowledge of the possible complications to avoid the risk of corneal exposure. Although ptosis props are available, they are usually poorly tolerated by patients. For affected individuals reporting diplopia, prisms are usually effective at controlling symptoms. Strabismus surgery can be considered for correction of large deviations; this is mostly for cosmetic reasons and patients need to be advised carefully that recurrence is the norm.

Patients with PEO require input from cardiology (to assess for cardiac conduction defects or cardiomyopathy), endocrinology, audiology and neurology services to manage the multisystemic complications associated with this progressive mitochondrial disorder.

TABLE 21.1 Nuclear DNA mutations are related to progressive external ophthalmoplegia.

	Autosomal dominant	Autosomal recessive
Mitochondrial DNA replication	POLG1 C10orf2	POLG1 POLG2 RNASEH1 TFAM WFS1
mtDNA repair	DNA2	MGME1
dNTP pool maintenance	SLC25A4 RRM2B	DGUOK RRM2B TYMP TK2 ABAT SUCLA2
Mitochondrial homeostasis	OPA1	MFN2 MPV17 TYMP
Mitochondrial membrane composition		SPG7

References

1. Chinnery PF, Johnson MA, Wardell TM, Singh-Kler R, Hayes C, Brown DT, et al. The epidemiology of pathogenic mitochondrial DNA mutations. *Ann Neurol* 2000;**48**:188–93.
2. Remes AM, Majamaa-Voltti K, Karppa M, Moilanen JS, Uimonen S, Helander H, et al. Prevalence of large-scale mitochondrial DNA deletions in an adult Finnish population. *Neurology* 2005;**64**:976–81. https://doi.org/10.1212/01.WNL.0000154518.31302.ED.
3. Yu-Wai-Man P, Lai-Cheong J, Borthwick GM, He L, Taylor GA, Greaves LC, et al. Somatic mitochondrial DNA deletions accumulate to high levels in aging human extraocular muscles. *Invest Ophthalmol Vis Sci* 2010;**51**:3347–53. https://doi.org/10.1167/iovs.09-4660.
4. Orsucci D, Angelini C, Bertini E, Carelli V, Comi GP, Federico A, et al. Revisiting mitochondrial ocular myopathies: a study from the Italian network. *J Neurol* 2017;**264**:1777–84. https://doi.org/10.1007/s00415-017-8567-z.
5. Rodriguez-Lopez C, Garcia-Cardaba LM, Blazquez A, Serrano-Lorenzo P, Gutierrez-Gutierrez G, San Millan-Tejado B, et al. Clinical, pathological and genetic spectrum in 89 cases of mitochondrial progressive external ophthalmoplegia. *J Med Genet* 2020;**57**:643–6. https://doi.org/10.1136/jmedgenet-2019-106649.
6. Kisilevsky E, Freund P, Margolin E. Mitochondrial disorders and the eye. *Surv Ophthalmol* 2020;**65**:294–311. https://doi.org/10.1016/j.survophthal.2019.11.001.
7. Petty RK, Harding AE, Morgan-Hughes JA. The clinical features of mitochondrial myopathy. *Brain* 1986;**109**(Pt 5):915–38. https://doi.org/10.1093/brain/109.5.915.
8. Wabbels B, Ali N, Kunz WS, Roggenkamper P, Kornblum C. Chronic progressive external ophthalmoplegia and Kearns-Sayre syndrome: interdisciplinary diagnosis and therapy. *Ophthalmologe* 2008;**105**:550–6. https://doi.org/10.1007/s00347-007-1643-5.
9. Yu-Wai-Man P. Chapter 18—Chronic progressive external ophthalmoplegia secondary to nuclear-encoded mitochondrial genes. In: Saneto RP, Parikh S, Cohen BH, editors. *Mitochondrial case studies*. Academic Press; 2016. p. 159–69. ISBN:9780128008775. https://doi.org/10.1016/B978-0-12-800877-5.00018-8.
10. Richardson C, Smith T, Schaefer A, Turnbull D, Griffiths P. Ocular motility findings in chronic progressive external ophthalmoplegia. *Eye (Lond)* 2005;**19**:258–63. https://doi.org/10.1038/sj.eye.6701488.
11. Viscomi C, Zeviani M. MtDNA-maintenance defects: syndromes and genes. *J Inherit Metab Dis* 2017;**40**:587–99. https://doi.org/10.1007/s10545-017-0027-5.
12. McClelland C, Manousakis G, Lee MS. Progressive external ophthalmoplegia. *Curr Neurol Neurosci Rep* 2016;**16**:53. https://doi.org/10.1007/s11910-016-0652-7.
13. Ambrosio G, De Marco R, Loffredo L, Magli A. Visual dysfunction in patients with mitochondrial myopathies. I. Electrophysiologic impairments. *Doc Ophthalmol* 1995;**89**:211–8. https://doi.org/10.1007/BF01203374.
14. Hudson G, Amati-Bonneau P, Blakely EL, Stewart JD, He L, Schaefer AM, et al. Mutation of OPA1 causes dominant optic atrophy with external ophthalmoplegia, ataxia, deafness and multiple mitochondrial DNA deletions: a novel disorder of mtDNA maintenance. *Brain* 2008;**131**:329–37. https://doi.org/10.1093/brain/awm272.
15. Varhaug KN, Nido GS, de Coo I, Isohanni P, Suomalainen A, Tzoulis C, et al. Using urine to diagnose large-scale mtDNA deletions in adult patients. *Ann Clin Transl Neurol* 2020;**7**:1318–26. https://doi.org/10.1002/acn3.51119.
16. McFarland R, Taylor RW, Turnbull DM. A neurological perspective on mitochondrial disease. *Lancet Neurol* 2010;**9**:829–40. https://doi.org/10.1016/S1474-4422(10)70116-2.
17. Grady JP, Campbell G, Ratnaike T, Blakely EL, Falkous G, Nesbitt V, et al. Disease progression in patients with single, large-scale mitochondrial DNA deletions. *Brain* 2014;**137**:323–34. https://doi.org/10.1093/brain/awt321.
18. Heighton JN, Brady LI, Sadikovic B, Bulman DE, Tarnopolsky MA. Genotypes of chronic progressive external ophthalmoplegia in a large adult-onset cohort. *Mitochondrion* 2019;**49**:227–31. https://doi.org/10.1016/j.mito.2019.09.002.
19. Whittaker RG, Blackwood JK, Alston CL, Blakely EL, Elson JL, McFarland R, et al. Urine heteroplasmy is the best predictor of clinical outcome in the m.3243A>G mtDNA mutation. *Neurology* 2009;**72**:568–9. https://doi.org/10.1212/01.wnl.0000342121.91336.4d.
20. Doherty M, Winterton R, Griffiths PG. Eyelid surgery in ocular myopathies. *Orbit* 2013;**32**:12–5. https://doi.org/10.3109/01676830.2012.736599.

Section VI
Tumour predisposition syndromes

Chapter 22

Phakomatoses

Chapter outline

22.1	Neurofibromatosis type 1	430	22.3 Von Hippel–Lindau disease	439
22.2	Neurofibromatosis type 2	434	22.4 Tuberous sclerosis complex	446

The phakomatoses are a diverse group of multisystemic disorders associated with predisposition to developing tumours, usually of the hamartomatous type. Although these lesions are typically not malignant, their anatomic location and multiplicity can lead to significant morbidity and early mortality. Skin, central nervous system and eye tissues are commonly involved; internal organs and bones may also be affected. There is considerable debate about which conditions should be classified as phakomatoses; neurofibromatosis type 1 and 2 (Sections 22.1 and 22.2), Von Hippel–Lindau disease (Section 22.3) and tuberous sclerosis complex (Section 22.4) are generally considered to be the archetypes.

22.1

Neurofibromatosis type 1

D. Gareth Evans

The neurofibromatoses consist of at least three autosomal dominant conditions: neurofibromatosis type 1 (NF-1) (birth incidence 1 in 2–3000), neurofibromatosis type 2 (NF-2) (birth incidence 1 in 25–33,000) and schwannomatosis.[1] The hallmark of NF-1 is the development of heterogeneous nerve sheath tumours, neurofibromas. These result from the proliferation of various supporting elements of neural fibres including Schwann glial cells. Virtually all NF-1 patients present with at least some features in childhood but the condition may not be diagnosed until later in life. NF-1 is associated with reduced life expectancy mainly due to neoplasia (gliomas and malignant peripheral nerve sheath tumours).[2–4]

Clinical characteristics

NF-1 is characterised by the development of several key clinical features and important complications. These are featured in the relevant National Institute of Health (NIH) diagnostic criteria (Table 22.1) and are discussed below.[5] Importantly, 2 of the 6 non-familial criteria relate to the eye.

Café au lait spots: The great majority of NF-1 patients are diagnosed in childhood with multiple café au lait patches on the skin (Fig. 22.1A). These are sharply demarcated light brown spots that are generally 10 cm or less in diameter. They are typically present from birth or soon after and >98% of NF-1 cases have ≥6 patches by 5 years of age. The presence of ≥6 patches in the child of an affected parent is sufficient for diagnosis (Table 22.1). In many children without affected relatives, an eye examination can be crucial in reaching a firm diagnosis. In general, >75% of children with no affected parent and ≥6 patches will eventually be diagnosed with NF-1 (although there is overlap with Legius syndrome, an autosomal dominant condition characterised by café au lait macules and associated with mutations in the *SPRED1* gene).[6] Freckling in the axillae is another typical skin finding (Fig. 22.1B).

Neurofibromas: The clinical picture of NF-1 typically includes cutaneous neurofibromas. These peripheral nerve tumours usually start to develop around puberty but may occur earlier. The initial appearance can be of a pink-purple soft swelling that becomes a more papillomatous (warty) growth. Most cutaneous neurofibromas are nodular, well-circumscribed and have limited size (Fig. 22.1C). In contrast, plexiform neurofibromas, originating from subcutaneous or visceral peripheral nerves, are generally larger and more diffuse as they tend to extend along the course of nerves. The trigeminal and cervical nerves are frequently involved.

Cutaneous and subcutaneous nodular and plexiform neurofibromas may affect the lid and/or the orbit (Fig. 22.1D). Plexiform tumours involving multiple branches of the trigeminal nerve may cause severe disfigurement and displacement of the globe. Orbital plexiform tumours are often associated with sphenoid bone dysplasia and may lead to pulsatile proptosis.

Lisch nodules: The classical ophthalmic lesions are benign melanocytic hamartomas of the iris named after the Carl Lisch who first described them in 1937. They occur in over 90% of NF-1 patients and are detectable from early childhood (Fig. 22.2A). These lesions have been therefore part of the NIH diagnostic criteria since their inception in 1987 (Table 22.1). Lisch nodules are dome-shaped protrusions projecting from the surface of the iris. They are clear to orange-yellow or brown and have no malignant potential. They can be seen often with the naked eye but are best viewed through a slit-lamp; this can help distinguish them from flat melanocytic naevi.

Choroidal nodules: A more recent feature that is very specific to NF-1 is that of choroidal nodules or 'choroidal

TABLE 22.1 National Institute of Health (NIH) diagnostic criteria for neurofibromatosis type 1.

Neurofibromatosis type 1 is diagnosed in individuals with two or more of the following:	
1. ≥6 café au lait skin macules	>5 mm if prepubertal
	>15 mm if post-pubertal
2. ≥2 neurofibromas of any type or ≥1 plexiform neurofibroma	
3. Axillary or inguinal freckling	
4. ≥1 optic nerve glioma	
5. ≥2 Lisch iris nodules	
6. Distinctive osseous lesions such as sphenoid dysplasia or thinning of long bone cortex with/without pseudarthrosis	
7. First-degree relative affected by criteria 1–6	

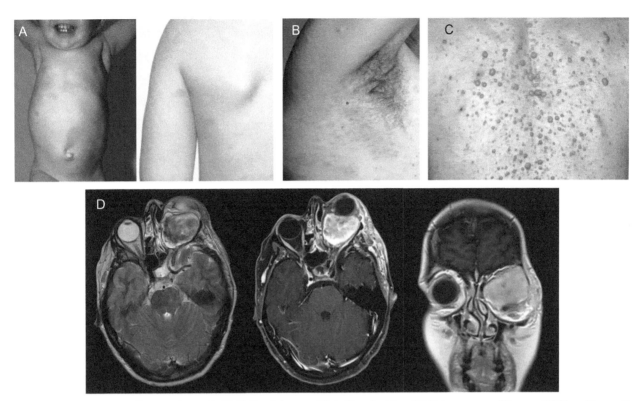

FIG. 22.1 Findings in neurofibromatosis type 1. (A) Café au lait spots. (B) Axillary freckling. (C) Cutaneous neurofibromas. (D) Neurofibroma displacing the optic nerve inferiorly seen on axial and coronal MRI scans.

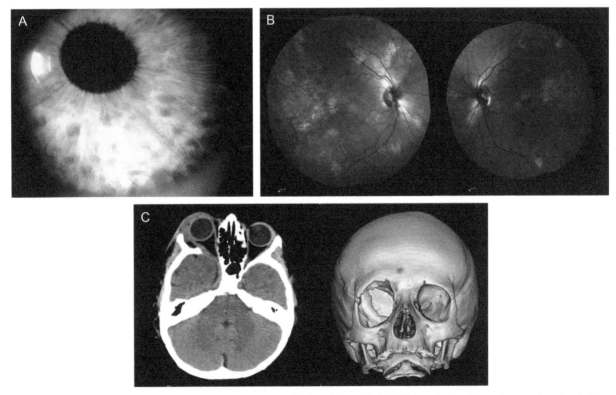

FIG. 22.2 Ophthalmic abnormalities in neurofibromatosis type 1. (A) Lisch nodules on the iris. (B) Choroidal nodules at the posterior pole of a 14-year-old boy seen on near-infrared reflectance imaging. (C) Sphenoid wing dysplasia on CT scan in a 9-year-old boy with right sphenoid wing dysplasia; this gives rise to 'bare orbit' appearance that can be diagnosed on plain X-ray. *(B: Courtesy of Mr. Vinod Sharma; C: Courtesy of Dr. Calvin Soh.)*

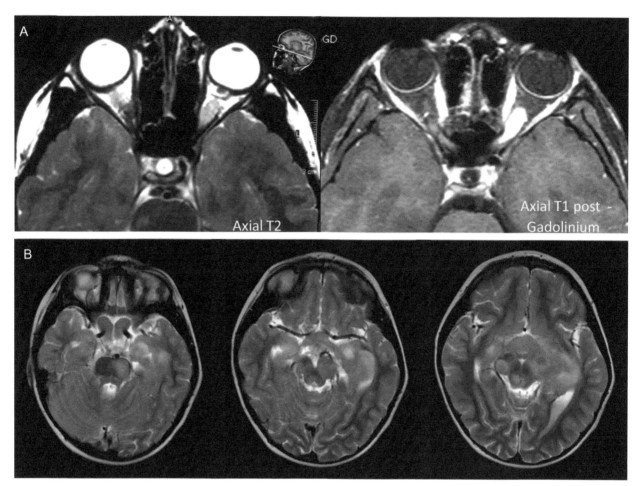

FIG. 22.3 Neuroimaging (axial MRI) findings in patients with neurofibromatosis type 1 and associated optic nerve gliomas. (A) Unilateral intraorbital optic nerve glioma, hyperintense on T2-weighted MRI (left hand-side image) and enhancing intensely with contrast (right hand-side image). (B) Optic pathway glioma involving the pre-chiasmatic optic nerves, the optic chiasm and the optic tracts. Note also the neurofibromatosis type 1 changes of myelin vacuolation in the brainstem. *(Courtesy of Dr. Calvin Soh.)*

abnormalities' (Fig. 22.2B). These are best seen on near-infrared reflectance fundus imaging.[7] They are typically found at the posterior pole and are seen in 60%–70% of NF-1 patients. They should be considered as a major criterion if Lisch nodules are not present.

Optic nerve gliomas: Pilocytic astrocytomas of the optic nerve are an important complication of NF-1. These low-grade, relatively well-defined tumours arise from glial tissue and generally grow slowly (Fig. 22.3A). Symptomatic tumours occur in about 5% of children with most presenting between birth and 6 years of age. Symptoms include visual loss, proptosis, strabismus and precocious puberty. Neuroimaging of infants identifies evidence of thickening of the optic nerves in 15%–20% of children and this finding suggests that only about one in three individuals in this group become symptomatic; indeed, many lesions appear to regress in adulthood. The tumours can occur anywhere on the optic pathway (including the optic nerve, optic chiasma, and post-chiasmatic radiations) and are not uncommonly bilateral. The typical appearance of an optic nerve glioma on an MRI scan is shown in Fig. 22.3. Because of the high proportion of lesions that do not become symptomatic, screening brain MRI scans in asymptomatic children is not recommended. It is noted that children with optic nerve gliomas are at increased risk of developing additional brain gliomas. Generally, it is rare for optic nerve lesions to become symptomatic in adulthood.

Glaucoma: Glaucoma appears to be at increased frequency in people with NF-1; this may be linked to orbital tumours.

Miscellaneous: A major complication of NF-1 is the development of malignant peripheral nerve sheath tumours. Notably, over 50% of affected children have learning difficulties.

Molecular pathology

NF-1 is caused by heterozygous, loss of function variants in the *NF1* gene. *NF1* encodes neurofibromin, a protein that has a role in controlling cellular proliferation and acts as a tumour suppressor.

Over 50% of NF-1 cases are *de novo*. Mosaicism (i.e. the presence of two or more genetically distinct cell populations in the same individual) is much less common in NF-1 than NF-2 and is usually associated with an asymmetric or segmental distribution of features. Ophthalmic specialists should be aware that segmental NF-1 can present around the eye and meet diagnostic criteria—for instance, the combination of a plexiform tumour with sphenoid wing dysplasia. In these cases, confirmation of non-mosaic NF-1 should be sought with features elsewhere and by genetic testing. The transmission risk to offspring is 50% for second generation (and beyond) affected individuals but, due to mosaicism, this may be considerably less than 50% in de novo cases.

There is no strong genotype/phenotype correlation in NF-1 although large genomic deletions affecting the NF1 locus are associated with severe disease while certain missense or in-frame deletions are linked to milder disease that is unlikely to affect the eye.

Ideally, molecular investigation in a patient with NF-1 should include RNA-level analysis. This can help detect around 96% of the underlying pathogenic variants. Mosaicism can be confirmed by biopsying café au lait patches and undertaking melanocyte culture. Identification of an identical variant in two separate biopsies confirms mosaicism.

Children of patients with NF-1 have up to a 50% chance of inheriting the condition. With molecular investigations most children can be tested with high confidence that the test will provide a definitive answer. Baseline assessment of children can occur at birth and most can be diagnosed clinically with confidence by 3 years of age.

Clinical management

A multidisciplinary approach is required to the management of individuals with NF-1.[8] Regular visual assessment from birth/diagnosis is indicated; this should be supplemented where possible with:

- near-infrared fundus imaging (to detect choroidal nodules)[9]
- OCT, including scans centred over the optic disc (to look for thinning of the circumpapillary retinal nerve fibre layer) and over the fovea (to look for thinning of the combined ganglion cell—inner plexiform layer in rectangular macula volume scans)[7]
- intraocular pressure checks (to gain insights into glaucoma risk/progression).

As discussed above, screening with MRI is not advised in asymptomatic children but should be carried out if there are any visual concerns. Whole-body MRI at transition is increasingly being used to assess internal tumour burden and to determine adult follow-up.

Surgery for isolated nodular palpebral neurofibromas is usually straightforward. This is not the case though for plexiform tumours for which determining the optimal timing of surgery is challenging and debulking, unless substantial, usually leads to rapid regrowth. MEK inhibitors, the first drug to show real promise in NF-1 tumour therapy, are gaining traction in the treatment of plexiform tumours that cause a threat to functioning. Tumour shrinkage is often achieved but chronic dosing may be required.

Treatment for optic nerve gliomas is not indicated unless there is clear evidence of progression that is either increasing symptoms or putting vision at threat. Chemotherapy can be effective and is now being supplemented with targeted therapy with MEK inhibitors. Radiotherapy should be avoided in children because of the high risk of secondary malignancy and vascular complications.

References

1. Evans DG, Howard E, Giblin C, Clancy T, Spencer H, Huson SM, et al. Birth incidence and prevalence of tumor-prone syndromes: estimates from a UK family genetic register service. *Am J Med Genet A* 2010;**152A**(2):327–32.
2. Evans DG, O'Hara C, Wilding A, Ingham SL, Howard E, Dawson J, et al. Mortality in neurofibromatosis 1: in North West England: an assessment of actuarial survival in a region of the UK since 1989. *Eur J Hum Genet* 2011;**19**(11):1187–91.
3. Uusitalo E, Leppävirta J, Koffert A, Suominen S, Vahtera J, Vahlberg T, et al. Incidence and mortality of neurofibromatosis: a total population study in Finland. *J Invest Dermatol* 2015;**135**(3):904–6.
4. Uusitalo E, Rantanen M, Kallionpää RA, Pöyhönen M, Leppävirta J, Ylä-Outinen H, et al. Distinctive cancer associations in patients with neurofibromatosis type 1. *J Clin Oncol* 2016;**34**(17):1978–86.
5. Neurofibromatosis. Conference statement. National Institutes of Health Consensus Development Conference. *Arch Neurol* 1988;**45**:575–8.
6. Denayer E, Legius E. Legius syndrome and its relationship with neurofibromatosis type 1. *Acta Derm Venereol* 2020;**100**(7):adv00093.
7. Viola F, Villani E, Natacci F, Selicorni A, Melloni G, Vezzola D, et al. Choroidal abnormalities detected by near-infrared reflectance imaging as a new diagnostic criterion for neurofibromatosis 1. *Ophthalmology* 2012;**119**(2):369–75. https://doi.org/10.1016/j.ophtha.2011.07.046.
8. Jain G, Jain VK, Sharma IK, Sharma R, Saraswat N. Neurofibromatosis type 1 presenting with ophthalmic features: a case series. *J Clin Diagn Res* 2016;**10**(11):SR01–3.
9. Avery RA, Cnaan A, Schuman JS, et al. Longitudinal change of circumpapillary retinal nerve fiber layer thickness in children with optic pathway gliomas. *Am J Ophthalmol* 2015;**160**:944–52. e1.

22.2

Neurofibromatosis type 2

D. Gareth Evans

Neurofibromatosis type 2 (NF-2) or Merlin syndrome is an autosomal dominant condition characterised by a predisposition to encapsulated nerve sheath tumours, schwannomas. Like neurofibromas (the hallmark of neurofibromatosis type 1, NF-1; Section 22.1), the primary tumour cell in schwannomas is the Schwann cell, the major glial cell type in the peripheral nervous system. There are however differences between these tumours. In schwannomas, the majority of cells are Schwann cells; these tend to grow as a mass that displaces and occasionally compresses the associated nerve. In contrast, neurofibromas include a variety of cell types, including Schwann cells, connective tissue cells, blood cells etc.; this loose mixture of cells generally surrounds the nerve, but does not displace it. Compression against a nearby rigid structure (e.g., bone) may however occur. Notably, neurofibromas are not a feature of NF-2 which is now recognised as a schwannomatosis predisposition syndrome.

NF-2 is much less common than NF-1 and most affected individuals develop symptoms in adult years (although up to 20% of patients develop symptoms during childhood).[1,2] NF-2 is associated with substantially reduced life expectancy.

Clinical characteristics

NF-2 is characterised by the development of schwannomas on the vestibular branch of cranial nerve VIII; these results in loss of hearing, tinnitus and imbalance. The average age of onset of findings in NF-2 patients is 18–24 years (onset range: birth to 70 years). In general, the condition may be under-recognised in children in whom skin tumours, mononeuropathy and/or ophthalmic findings may be the first manifestation(s) (Table 22.2).[3] The diagnostic criteria for NF-2 are listed in Table 22.3 and some of the key features are discussed below.

Schwannomas: Over two-thirds of affected adults (but only one-third of children) present initially with symptoms linked to vestibular schwannoma.[5] It is almost inevitable that affected individuals will develop bilateral vestibular lesions by the fourth decade of life. These usually lead to bilateral complete deafness. Schwannomas on other cranial and spinal nerve routes as well as on peripheral nerves are also common. They cause loss of nerve function and pain

TABLE 22.2 Suggestive findings of neurofibromatosis type 2 in children.

Neurofibromatosis type 2 should be suspected in children with two or more of the following findings:

- A schwannoma at any location including intradermal
- Skin plaques present at birth or in early childhood
- A meningioma
- A cortical wedge cataract
- A retinal hamartoma
- Mononeuropathy, particularly causing facial nerve palsy, foot or wrist drop, or third nerve palsy

Having a first-degree relative with neurofibromatosis type 2 increases the likelihood of the disorder being present.

Adapted from Ref. 3.

TABLE 22.3 Simplified version of the revised Manchester criteria for neurofibromatosis type 2.[4]

Neurofibromatosis type 2 (NF-2) is diagnosed in individuals with one or more of the following:

- Bilateral vestibular schwannomas
- A first-degree relative with neurofibromatosis type 2 AND unilateral vestibular schwannoma
- A first-degree relative with neurofibromatosis type 2 AND ANY TWO NF-2 associated lesions (meningioma, schwannoma, ependymoma or juvenile cataract)
- A unilateral vestibular schwannoma AND ANY TWO other NF-2 associated lesions (meningioma, schwannoma, ependymoma or juvenile cataract)
- Multiple meningiomas (two or more) AND ANY TWO other NF-2 associated lesions (schwannoma, ependymoma or juvenile cataract)
- Constitutional or mosaic pathogenic *NF2* gene mutation in blood or identical mutations in two distinct tumours.

and, in the case of spinal tumours, they can cause spinal cord compression. Intracutaneous, often plaque-like schwannomas only appear to occur in NF-2 and distinguish this condition from schwannomatosis, an important differential diagnosis that does not typically feature hearing loss.

Meningiomas: Meningiomas, including brain and spinal cord lesions, occur at a high frequency in NF-2 with an estimated lifetime risk of 75%. The optic nerve may be affected (see below).

Intrinsic central nervous system tumours: Intrinsic tumours originating in the intracranial compartments or the spinal cord are noted in about 30% of people with NF-2. These are nearly always low-grade indolent ependymomas, a group of glial tumours. They predominate in the upper cervical spine and most remain asymptomatic.

Ophthalmic manifestations: Visual loss can be the presenting feature in a significant proportion of children diagnosed with NF-2 (up to 20% of cases).[6,7] Generally, one-third of affected individuals have decreased vision in one or both eyes. This can be due to retinal hamartomas, tumours impacting on the optic nerve (particularly optic nerve sheath meningiomas which affect about 5% of NF-2 patients) or lens opacities. In many cases, there is no underlying cause and amblyopia is common.

Congenital hamartomas of the retina (also known as combined hamartomas of the retina and retinal pigment epithelium) occur in around 15% of children presenting with NF-2 and are less common in individuals presenting in adulthood.[8] They are often asymptomatic but can affect vision in a subset of cases. These elevated fundal lesions typically have a characteristic whitish sheen superficially (corresponding to epiretinal membrane and intraretinal gliosis) and variable deeper pigmentation. They occur most commonly in a juxtapapillary location but can also be found in the macula and/or the retinal periphery. They vary in size from small discrete lesions (Fig. 22.4A) to large masses with extensive retinal and optic disc involvement (Fig. 22.4B). They are usually unilateral and unifocal but bilateral lesions have also been described. Associated retinal vascular tortuosity is typically observed.

Lens opacification is a common feature of NF-2. Presenile posterior subcapsular cataracts are seen in 60%–80% of patients but these are usually visually insignificant and only rarely require early surgery. However, cortical wedge opacities may occasionally appear in very early childhood causing visual loss and requiring intervention.

Epiretinal membranes can affect children as young as 4 years of age. They are likely of glial origin (Fig. 22.4C).

Visual function may be affected by cranial nerve tumours. More specifically, meningiomas can occur on the optic nerve sheath, compressing cranial nerve II and leading to visual loss. Schwannomas can also affect cranial nerve III, IV and/or VI (Fig. 22.5). Overall, these occur in <10% of NF-2 patients and these tumours rarely require treatment or cause troublesome symptoms. Schwannomas more frequently affect cranial nerves V and VII but again these tumours rarely require surgical or other treatment. Occasionally cranial nerve V involvement will lead to loss of corneal sensation and cranial nerve VII weakness will lead to exposure keratopathy and dry eye.

Optic disc swelling/atrophy may develop from chronic raised intracranial pressure (Fig. 22.4D). This is often due to poor venous drainage associated with obstruction from a meningioma (obstructive hydrocephalus). Intracranial pressure monitoring may be required in these cases.

Iatrogenic cranial nerve damage: It is still common to see NF-2 patients with complete hearing loss and visual problems due to the surgical damage to cranial nerve V and VII (including the intermediate nerve) whilst removing a vestibular schwannoma. Among others, this can lead to keratopathy that can be particularly challenging to manage.

Mononeuropathy: Mononeuropathies that are often apparently unrelated to any obvious tumour formation on a nerve (on MRI) are a frequent presenting feature of NF-2 in childhood. They commonly affect cranial nerve VII with a remitting and relapsing pattern that can last over 12 months. Unlike Bells palsy, recovery is generally incomplete. Treatment with steroids may be helpful. Mononeuropathies can also affect cranial nerves III, IV and VI causing double vision and squint. The onset of all mononeuropathies is fairly rapid and some gradual regain in function commonly occurs.

Molecular pathology

NF-2 is caused by heterozygous loss of function pathogenic variants in the *NF2* gene. *NF2* encodes merlin, also known as schwannomin, a cell membrane-related protein that acts as a tumour suppressor.

Over 60% of NF-2 cases are *de novo*, and over half of all *de novo* cases are due to mosaicism where the mutational event occurs after conception and only involves a proportion of cells around the body. Mosaicism is more likely in cases with later and more asymmetric presentations. The transmission risk to offspring is 50% for second generation (and beyond) affected individuals but due to mosaicism it may be considerably less than 50% in *de novo* cases.

There is a strong genotype/phenotype correlation in NF-2 with truncating variants (nonsense and frameshift) in exons 2–13 being associated with severe disease.[9] Large rearrangements and splicing variants cause intermediate disease and missense variants generally cause mild disease.

Molecular investigation in a *de novo* NF-2 patient should involve analysis of tumour samples. Tumours will contain the initiator (inherited or mosaic) pathogenic variant and usually an identifiable molecular event that causes loss of function of the second *NF2* copy. The second event is usually loss of the entire *NF2* gene as part of a whole or partial chromosome loss. Mitotic recombination (i.e. crossing over between similar sequences during mitosis) also occurs in around 10% of cases. This results in a copy number unchanged duplication of the *NF2* pathogenic variant with loss of the normal copy. Point pathogenic variants make up the remainder. Once both

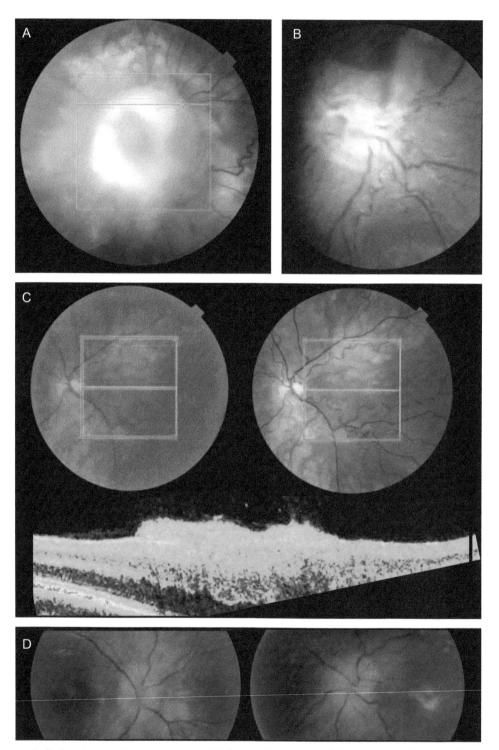

FIG. 22.4 Fundoscopic findings in neurofibromatosis type 2. (A) Juxtapapillary congenital hamartomas of the retina. (B) Extensive congenital hamartoma. (C) Epiretinal membrane causing vitreoretinal traction; associated retinal vessel tortuosity is noted. (D) Bilateral papilloedema in a 20-year-old with neurofibromatosis type 2. A hamartoma is noted in the left macula with overlying gliosis. The patient was asymptomatic but, after papilloedema was noticed, raised intracranial pressure was confirmed on lumbar puncture.

molecular events are established, the underlying *NF2* mutation can be confirmed in blood DNA. In many *de novo* cases neither variant can be confirmed (even at low allele frequency) in blood so confirmation in another tumour is required. Given that NF-2 tumours are due to loss of function of both copies of the *NF2* gene, it is thought that many of the ophthalmic features including cataract and retinal hamartoma are also caused by this mechanism.

Children of patients with NF-2 have up to a 50% chance of inheriting the condition. With molecular investigations

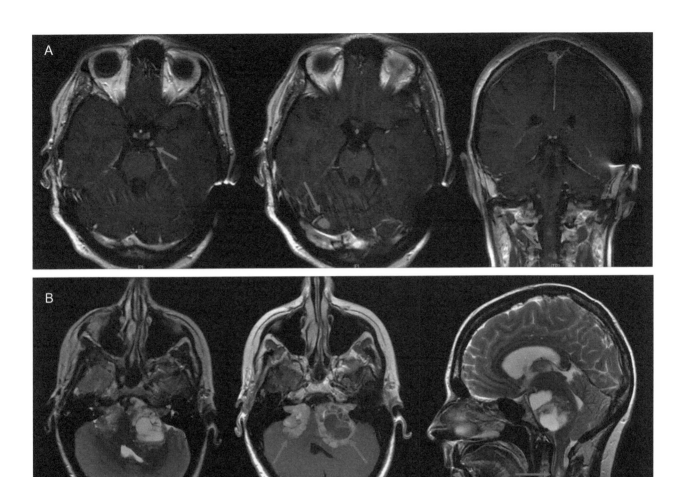

FIG. 22.5 Neuroimaging (MRI) findings in neurofibromatosis type 2. (A) Axial T2-weighted MRI scan showing bilateral cerebellopontine angle cistern tumours (left hand-side) that enhance with contrast (middle). The lesion on the left has a cystic element and is associated with significant brainstem compression. This can be visualised on coronal T2-weighted MRI (right hand-side) which also reveals an additional cervical cord ependymoma. (B) Axial and sagittal T2-weighted MRI scans revealing a schwannoma affecting cranial nerve III (left hand-side), a tentorial meningioma (middle) and multiple spinal schwannomas (right hand-side). Relevant lesions are highlighted with *red arrows*. *(Courtesy of Dr. Calvin Soh.)*

most children can be tested with high confidence that the test will give a definitive answer. If no tumour is available for sampling on the affected parent and blood analysis fails to reveal a pathogenic variant, the child can undergo molecular testing to potentially rule out a mosaic mutation inherited from the parent; baseline MRI of the brain and spine, as well as a cutaneous and eye examination, should also be performed. Baseline testing and imaging of children usually commences at around 10 years of age but can start earlier if the parent has severe NF-2.

Clinical management

NF-2 patients require multidisciplinary management by a team that includes neurosurgeons, neurologists, otolaryngologists, audiologists, geneticists and ophthalmologists. At least annual review with brain MRI and 3–5 yearly spinal MRI is required.

Until recently, the mainstay of treatment was surgical removal of tumours. Gamma knife or fractionated radiotherapy is now used for some NF-2 tumours particularly vestibular schwannomas. These therapeutic approaches should however be avoided in very young patients. The anti-VEGF agent bevacizumab has gained a substantial role in the management particularly of vestibular schwannomas but is not really effective for meningiomas. Chronic treatment is usually required and this can have effects on renal function. There is ongoing research to identify drugs to improve the treatment of NF-2 related tumours with a particular need to identify those that treat meningiomas.

The main ophthalmic management often centres around improving eyelid function and protecting the cornea in postsurgical cases. There is a strong case for regular ophthalmic review including an initial screen early in life for childhood-onset cataracts. Monitoring of the retinal and optic disc findings with OCT is recommended.

References

1. Evans DG, Howard E, Giblin C, Clancy T, Spencer H, Huson SM, et al. Birth incidence and prevalence of tumor-prone syndromes: estimates from a UK family genetic register service. *Am J Med Genet A* 2010;**152A**(2):327–32.
2. Evans DG, Bowers NL, Tobi S, Hartley C, Wallace AJ, King AT, et al. Schwannomatosis: a genetic and epidemiological study. *J Neurol Neurosurg Psychiatry* 2018;**89**(11):1215–9.
3. Evans DG. Neurofibromatosis 2. 1998 Oct 14 [Updated 2018 Mar 15]. In: Adam MP, Ardinger HH, Pagon RA, et al., editors. *GeneReviews® [Internet]*. Seattle, WA: University of Washington, Seattle; 1993–2021. Available from: https://www.ncbi.nlm.nih.gov/books/NBK1201/.
4. Evans DG, King AT, Bowers NL, Tobi S, Wallace AJ, Perry M, et al. Identifying the deficiencies of current diagnostic criteria for neurofibromatosis 2 using databases of 2777 individuals with molecular testing. *Genet Med* 2019;**21**(7):1525–33. https://doi.org/10.1038/s41436-018-0384-y.
5. Evans DG, Huson SM, Donnai D, Neary W, Blair V, Newton V, et al. A clinical study of type 2 neurofibromatosis. *Q J Med* 1992;**84**(304):603–18.
6. Anand G, Vasallo G, Spanou M, Thomas S, Pike M, Kariyawasam DS, et al. Diagnosis of sporadic neurofibromatosis type 2 in the paediatric population. *Arch Dis Child* 2018;**103**(5):463–9.
7. Gaudioso C, Listernick R, Fisher MJ, Campen CJ, Paz A, Gutmann DH. Neurofibromatosis 2 in children presenting during the first decade of life. *Neurology* 2019;**93**(10):e964–7.
8. Waisberg V, Rodrigues LO, Nehemy MB, Frasson M, de Miranda DM. Spectral-domain optical coherence tomography findings in neurofibromatosis type 2. *Invest Ophthalmol Vis Sci* 2016;**57**(9):OCT262–7.
9. Painter SL, Sipkova Z, Emmanouil B, Halliday D, Parry A, Elston JS. Neurofibromatosis type 2-related eye disease correlated with genetic severity type. *J Neuroophthalmol* 2019;**39**(1):44–9.

22.3

Von Hippel–Lindau disease

Anthony T. Moore

Von Hippel–Lindau disease (VHL) is an autosomal dominant disorder characterised by the development of vascular tumours of the central nervous system, hemangioblastomas; the cerebellum, spinal cord and retina are commonly involved. Cystic lesions of the viscera (typically of the kidney and pancreas), renal cell carcinoma and phaeochromocytoma may also be present.[1,2]

Most individuals with VHL present/manifest in adulthood and their clinical symptoms depend on the sites of the various tumours. Ophthalmologists have a key role to play in the diagnosis and management of VHL as retinal hemangioblastomas are one of the commonest and earliest features of the condition.[1,3] Prompt identification and treatment of these retinal lesions will, in many cases, prevent visual loss.

Clinical characteristics

VHL generally manifests in the third or fourth decade of life. However, in those with a known family history (~80% of cases), early molecular diagnosis and surveillance programs allow the identification of associated tumours before they become symptomatic and when treatment may be more effective.

Cerebellar and retinal hemangioblastomas are the commonest, and usually the earliest, manifestations of VHL.[1] Cerebellar lesions usually present with headache, nausea and vomiting and patients are often found to have optic disc swelling related to raised intracranial pressure. Neuroimaging reveals the typical cerebellar vascular tumours (Fig. 22.6). Less commonly, hemangioblastomas develop in the spine or brain stem and may cause symptoms related to a mass effect (Fig. 22.7).

Retinal hemangioblastomas are usually asymptomatic until they start to leak fluid. The clinical presentation typically involves unilateral visual field or central vision loss. The retinal lesions are usually seen in the mid-periphery but they can, less commonly, involve the juxtapapillary retina (Figs. 22.8 and 22.9).[3,4] They are usually multiple and tend to be asymptomatic in the early stages when they are easier to treat (Fig. 22.10).

Kidney lesions are common and can be benign (cysts) or malignant (renal cell carcinomas). Renal cell carcinomas generally present at an older age and the typical symptoms are low back pain, fatigue and haematuria (Fig. 22.11). They are a leading cause of mortality in patients with VHL and the overall survival for people with these lesions is linked to tumour size (<3 cm or ≥ 3 cm) and patient age.[5]

Other tumours and visceral cysts associated with VHL are usually identified presymptomatically once the diagnosis is made. One exception is phaeochromocytoma, an adrenal gland tumour that can occasionally be the initial reason for an affected individual to seek medical attention; the presentation is either with an abdominal mass or with cardiovascular symptoms such as hypertension or arrhythmias, related to the release of catecholamines.

Affected individuals with apparently sporadic/isolated spinal, cerebellar or retinal hemangioblastomas (or other single tumours that may be related to VHL) must have appropriate investigations including molecular genetic testing to exclude a diagnosis of VHL (Box 22.1).

Molecular pathology

VHL is caused by heterozygous variants in the *VHL* tumour suppressor gene. The associated protein has functions in multiple biological pathways including a key role in regulating the cellular response to hypoxia.[6] Notably, autosomal recessive missense mutations in the *VHL* gene are associated with a rare haematological disorder, congenital secondary polycythemia.[5]

Approximately, 80% of VHL patients have an affected parent. Mutations in a significant proportion of the remaining 20% are expected to have occurred *de novo*. Molecular genetic testing is recommended for the parents of apparently sporadic/simplex cases. If the pathogenic variant that is found in the proband cannot be detected in the blood of either parent, two possible explanations include a mutation arising in the fertilised egg during early embryogenesis (postzygotic *de novo* mutation) or a variant that is present for the first time in the proband as a result of a mutation in a germ cell (egg or sperm) of one of the parents (parental germline mosaicism). The incidence of the latter in VHL remains unknown but distinguishing between these two scenarios has important implications for estimating recurrence risk.[5]

VHL pathogenic variants are highly penetrant; almost all individuals who have a mutation in this gene will become

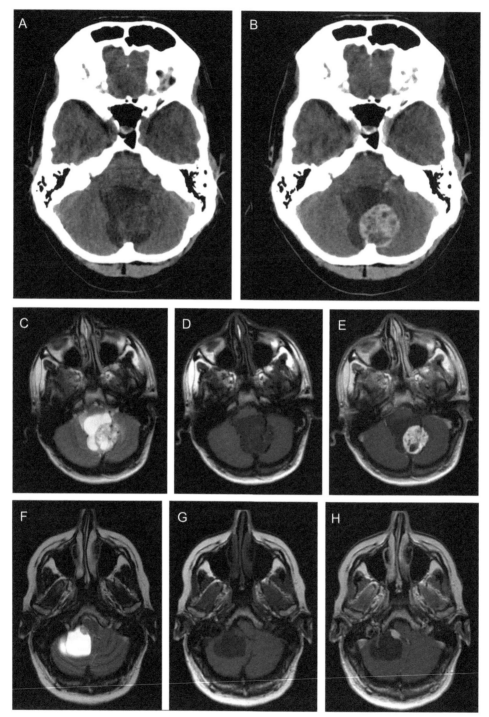

FIG. 22.6 Neuroimaging (MRI and CT) findings in two patients with Von Hippel–Lindau disease and associated solid hemangioblastomas. Images from one affected individual are shown in panels A-E and from an unrelated patient are shown in panels F–H. (A) Pre-contrast CT scan revealing a lobulated mass in the left medial cerebellum with an anteromedial cyst compressing the fourth ventricle. (B) Post-contrast CT scan highlighting that the lobulated mass enhances with contrast. (C) Axial T2-weighted MRI scan revealing a lobulated septated lesion corresponding to the mass on CT but with a larger anteromedial cystic component. (D) Axial T1-weighted MRI scan showing that the lobulated lesion is iso- to hypo-intense with the brain. A prominent draining vein extending anteriorly from the mass can be seen. (E) Axial T1-weighted post-contrast MRI scan highlighting that the mass enhances intensely with contrast. (F) Axial T2-weighted MRI scan of a cerebellar hemangioblastoma from a different patient. A cystic lesion is noted in the right cerebellum with an anteromedial mural nodule. (G) Axial T1-weighted MRI scan showing a hypointense cyst with an isointense mural nodule, (H) Axial T1-weighted post-contrast MRI revealing enhancement of the mural nodule but not of the cyst wall. *(Courtesy of Dr. Calvin Soh.)*

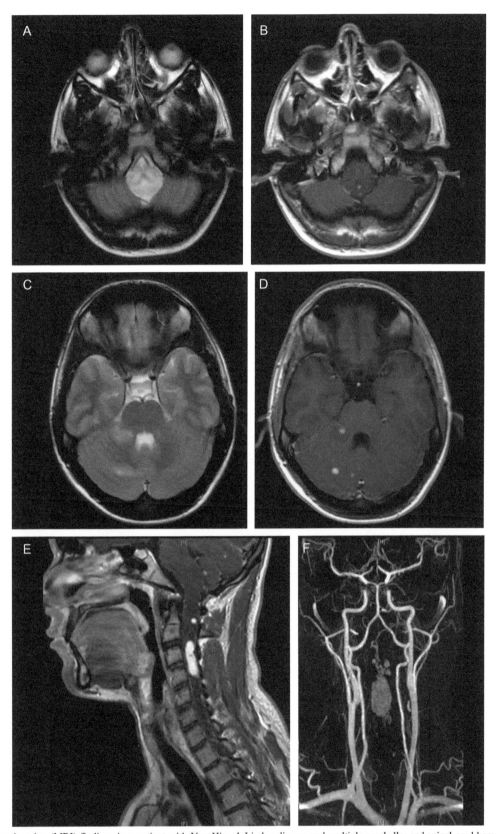

FIG. 22.7 Neuroimaging (MRI) findings in a patient with Von Hippel–Lindau disease and multiple cerebellar and spinal cord hemangioblastomas. (A) Axial T2-weighted MRI scan showing that the cervicomedullary junction is expanded by oedema. (B) Axial T1-weighted post-contrast MRI revealing two tiny enhancing hemangioblastomas on the dorsal aspect of the cervicomedullary junction. (C) Axial T1-weighted MRI scan highlighting two further areas of high signal in the right cerebellum. (D) Axial T1-weighted post-contrast MRI enhancing the visualisation of the two solid nodules that cause oedema; a further tiny dot of enhancement is noted on the right side of the vermis indicating another hemangioblastoma. (E) Sagittal T1-weighted post-contrast MRI showing two enhancing cervical cord lesions representing cord hemangioblastomas. (F) Contrast-enhanced MR angiogram revealing that these spinal cord lesions are vascular masses with tumoral blush and enlarged draining veins. *(Courtesy of Dr. Calvin Soh.)*

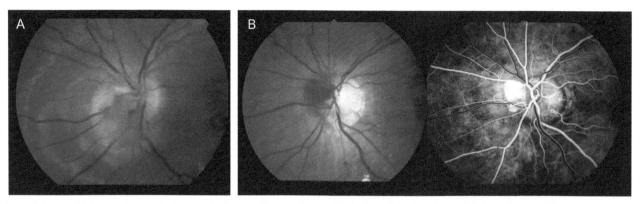

FIG. 22.8 (A) (B) Juxtapapillary retinal hemangioblastomas in two patients with Von Hippel–Lindau disease. Markedly increased fluorescence is noted on fluorescein angiography (B).

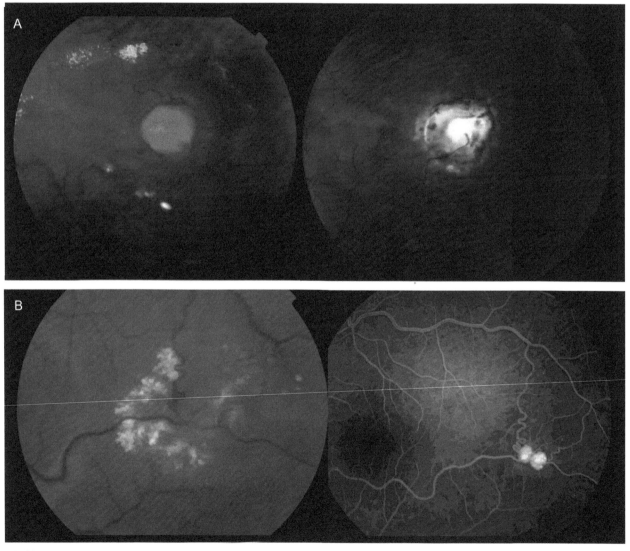

FIG. 22.9 Peripheral retinal hemangioblastomas in two patients with Von Hippel–Lindau disease. (A) Peripheral lesion before and after laser. (B) Peripheral hemangioblastoma surrounded by retinal exudate. There is markedly increased fluorescence on fluorescein angiography. *(Courtesy of Mr. Stephen Charles.)*

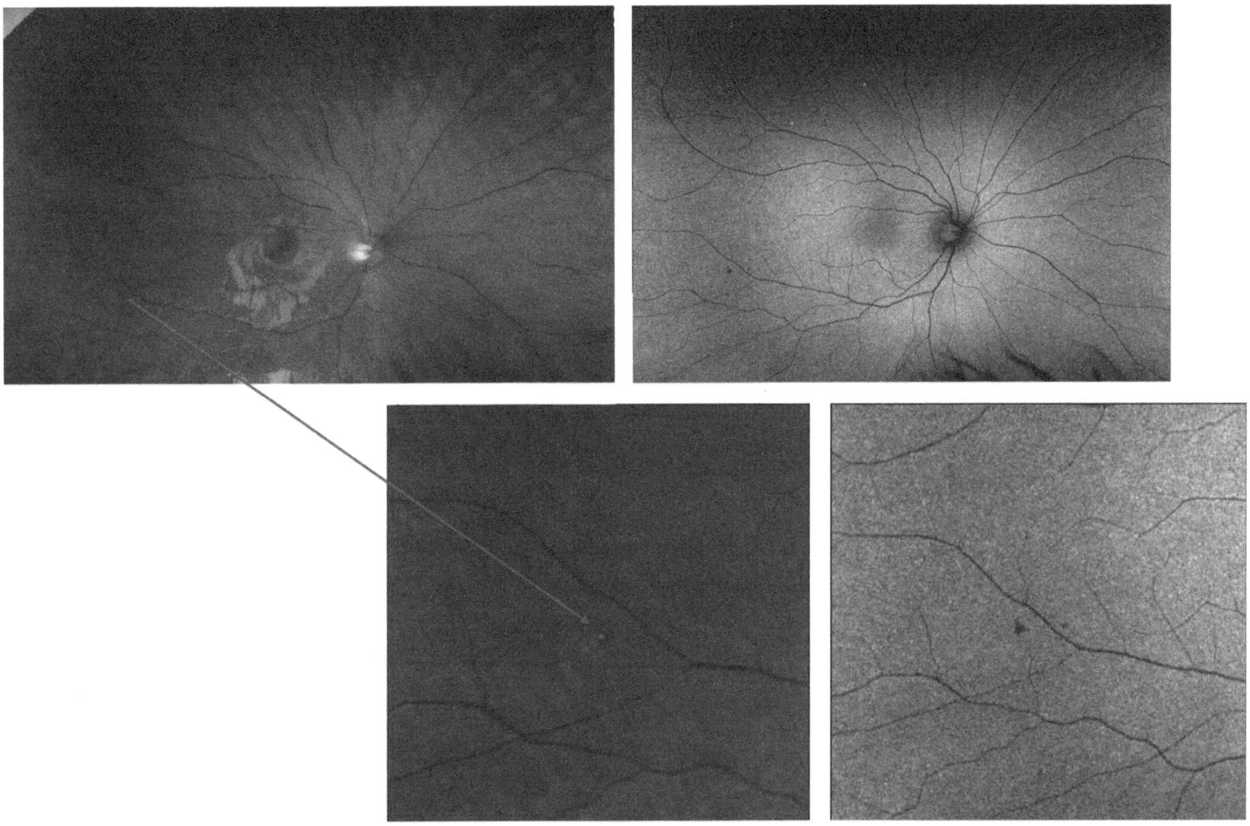

FIG. 22.10 Subtle peripheral retinal hemangioblastomas in a patient with Von Hippel–Lindau disease. Careful inspection of the wide-field multi-colour and fundus autofluorescence images may be required to identify early lesions when they are easy to treat. The location of one of the lesions is highlighted with a *red arrow* shown in an inset.

symptomatic by age 65 years. There is however significant phenotypic variability even among patients carrying identical mutations.[5]

Some evidence of genotype–phenotype correlations in VHL has emerged (reviewed in[5]); this mainly relates to the risk of developing phaeochromocytoma. There is no evidence for any relationship between genotype and severity of retinal disease.[3]

The tumours seen in VHL follow the 'two hit' hypothesis suggested by Knudsen for the development of retinoblastoma (see Chapter 25). Briefly, affected individuals are expected to carry the initiator mutation in the *VHL* gene in their cells. The second event (somatic hit) inactivates the remaining normal copy of the gene in certain cells leading to tumour development. Just as in retinoblastoma, sporadic solitary hemangioblastomas or renal cell carcinomas may be associated with biallelic somatic *VHL* mutations occurring by chance in the same cell.

Molecular genetic testing in VHL suspects is generally performed in blood samples using either Sanger sequencing of the *VHL* gene or broader high-throughput sequencing approaches.[5] In individuals who have a family history of VHL and have clinical features suggesting that they might be affected, testing is often confined to the *VHL* gene. Alternative testing strategies include panel-based testing or comprehensive genomic analysis (exome or genome sequencing). These approaches are generally more useful in individuals without a family history and/or in cases where the full criteria for the diagnosis of VHL are not met. Copy number analysis (i.e. looking for large deletions/duplications affecting the *VHL* gene) must be conducted as part of any test for this condition.

Clinical management

The management of patients with VHL involves multiple specialists including clinical geneticists, ophthalmologists, neurosurgeons, oncologists, renal physicians and general surgeons. There must be a team member who is in charge of the holistic care of the patient and acts as a liaison between the various health professionals involved; this role is usually taken by the clinical geneticist who can also provide care and advice to the wider family.

The ophthalmologist's main role is in the detection and treatment of retinal hemangioblastomas. These lesions are best detected by slit-lamp biomicroscopy or indirect

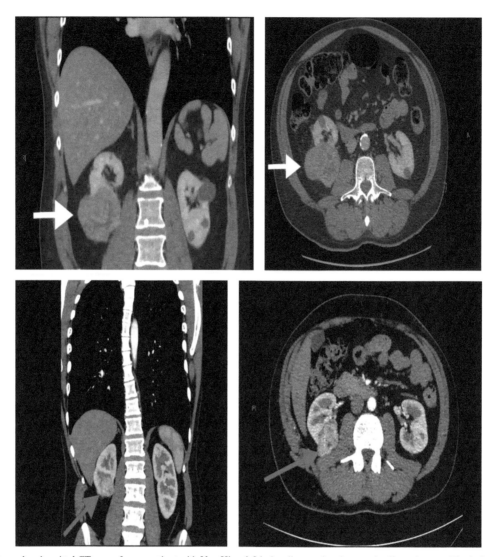

FIG. 22.11 Coronal and sagittal CT scans from a patient with Von Hippel–Lindau disease showing renal cell carcinomas. The lesions are highlighted with *white* and *red arrows*. *(Courtesy of Dr. Hosam Noweir.)*

ophthalmoscopy of the fundus after full pupil dilatation. The hemangioblastomas are usually multiple and bilateral[3,4,7]; small lesions may easily be missed (Fig. 22.10). In patients with one or more lesions, widefield fundus imaging with or without fluorescein angiography is a useful investigation that can ensure that all hemangioblastomas are identified and documented[7]; the retinal lesions show markedly increased fluorescence on fluorescein angiography. OCT imaging helps document and monitor subretinal or intraretinal fluid.

Peripheral retinal hemangioblastomas that are detected early can be usually effectively treated with laser photocoagulation, although treatment may require more than one session.[8] Larger lesions associated with subretinal fluid may be better treated with cryotherapy or, if very large, with brachytherapy or endoresection. Vitreoretinal surgery may also need to be considered if there is a tractional retinal detachment or extensive epiretinal membrane formation. Other treatment modalities such as photodynamic therapy and anti-VEGF agents have been used but the results are suboptimal. Treating optic nerve and peripapillary hemangioblastomas is generally more challenging as laser treatment may be complicated by significant loss of vision or visual field. As a result, peripapillary lesions are best observed until there is exudation. Photodynamic therapy or anti-VEGF agents may be subsequently tried in an attempt to limit exudation. However, the overall outcomes are disappointing and better treatments are needed for peripapillary tumours.[8]

Children of patients with VHL and siblings of sporadic cases should be offered presymptomatic molecular genetic diagnosis if the family mutation is known (which is usually the case); individuals carrying the mutation should undergo

> **BOX 22.1 Clinical diagnostic criteria for Von Hippel–Lindau disease (VHL).**
>
> Before the identification of the *VHL* gene, clinical criteria were used to confirm the diagnosis of Von Hippel–Lindau disease. These criteria are outlined below but it should be noted that they are less commonly used now as molecular genetic diagnosis is readily available. In general, the diagnosis of VHL should be considered:
>
> *In an individual WITHOUT relevant family history when there are TWO or more of the following:*
> - ≥2 hemangioblastomas of the retina, spine or brain.
> - a single hemangioblastoma AND a typical visceral lesion (including renal cell carcinoma, phaeochromocytoma or paraganglioma, endolymphatic sac tumour, neuroendocrine tumour of the pancreas, cystadenoma of the epididymis or broad ligament, multiple renal/pancreatic cysts)
>
> *In an individual WITH relevant family history when there is ONE or more of the following:*
> - hemangioblastoma of the retina, spine or brain
> - phaeochromocytoma or paraganglioma
> - renal cell carcinoma
> - multiple renal/pancreatic cysts
>
> Adapted from Ref. 5.

screening. If the family does not consent to molecular testing, then family members at risk of developing VHL should be offered screening after an explanation of the benefits. Screening typically includes annual eye exams starting in infancy, MRI of brain and spine, abdominal ultrasound, MRI scan of the abdomen, 24-h urine sampling for fractionated metanephrines, and audiological assessment (to exclude hearing loss from endolymphatic tumours). Various screening protocols have been published including one from the VHL Alliance (https://www.vhl.org/wp-content/uploads/2017/07/Active-Surveillance-Guidelines.pdf). Each team managing individuals with VHL must have a clear protocol in place.

References

1. Maher ER, Yates JR, Harries R, et al. Clinical features and natural history of von Hippel-Lindau disease. *Q J Med* 1990;**77**(283):1151–63.
2. Findeis-Hosey JF, McMahon KQ, Findeis SK. Von Hippel-Lindau disease. *J Pediatr Genet* 2016;**5**:116–23.
3. Webster AR, Maher ER, Moore AT. Clinical characteristics of ocular angiomatosis in Von Hippel-Lindau disease and correlation with germ line mutation. *Arch Ophthalmol.* 1999;**117**:371–8.
4. Binderup ML, Stendell A, Galanakis M, et al. Retinal hemangioblastoma: prevalance, incidence and frequency of underlying von Hippel-Lindau disease. *Br J Ophthalmol* 2018;**102**:942–7.
5. Van Leeuwaarde RS, Ahmed S, Links TP, Giles RH. Von Hippel-Lindau syndrome. In: Adam MP, Ardinger HH, Pagon RA, Wallace SE, LJH B, Stephens K, Amemiya A, editors. *GeneReviews® [Internet].* Seattle, WA: University of Washington, Seattle; 1993–2019. 2000 May 17 [updated 2018 Sep 6].
6. Gossage L, Eisen T, Maher ER. VHL, the story of a tumour suppressor gene. *Nat Rev Cancer* 2015;**15**(1):55–64.
7. Chen X, Sanfillipo CJ, Nagiel A, et al. Early detection of retinal hemangioblastomas in Von Hippel-Lindau disease using ultrawide field fluorescein angiography. *Retina* 2018;**38**:748–54.
8. Wiley HE, Krivosik V, Gaudric A, et al. Management of haemangioblastoma in Von Hippel-Lindau disease. *Retina* 2019;**39**(12):2254–63.

22.4

Tuberous sclerosis complex

Graeme C.M. Black and Elizabeth A. Jones

Tuberous sclerosis complex (TSC) is an autosomal dominant multisystem disorder associated with a variety of hamartomatous lesions that can affect many organ systems. The brain, kidney, heart, eye and skin are commonly involved. The main problems experienced by individuals with TSC are learning difficulties, epilepsy, behaviour problems, psychiatric disorders, facial angiofibromas and renal disease. Affected individuals are often diagnosed in the first few years of life. Whilst brain pathology is an important cause of morbidity and mortality in this condition, renal disease is the leading cause of mortality in adults with TSC. Clinical management requires comprehensive multidisciplinary/specialist evaluation.

TSC is caused by heterozygous pathogenic variants in either *TSC1* or *TSC2*. Genetic testing is only able to identify a causative variant in, at most, 90% of individuals with a clinical diagnosis of TSC. Therefore, the use of clinical diagnostic criteria remains critical to the correct diagnosis of patients. Notably, 30%–50% of individuals with TSC will have one or more retinal astrocytic hamartomas; the presence of these lesions is specific enough to be a major criterion for the clinical diagnosis of TSC (Table 22.4).

Around two-thirds of individuals with TSC will be affected as a result of a *de novo* variant. The remaining third will have inherited the pathogenic variant from a parent. There is significant inter- and intra-familial variability and some individuals will only become aware that they are affected when they are found to have a manifestation of TSC later in life (such as the development of renal angiomyolipomas) or when a family member presents with TSC. A family member may be asymptomatic but have significant disease (e.g. renal angiomyolipomas) and so family evaluation is important. If no pathogenic variant has been identified, it is not possible to offer family members a genetic test and so clinical screening should be offered instead.

Fundoscopy can help identify retinal astrocytic hamartomas, an important diagnostic clue either in those suspected of having TSC or when identified incidentally in a child or adult presenting for other reasons.

Clinical characteristics

TSC is highly variable in clinical presentation and findings. Affected individuals may present in infancy or early childhood with a complex neurological phenotype, later in life with renal complications or after the diagnosis is made in a family member. The diagnosis of the condition depends upon the presence of specific clinical features which form clinical diagnostic criteria (Table 22.4).[1]

In general, around 80% of patients develop seizures and around 50% have intellectual disability; autism spectrum disorder is also common. Neuroimaging frequently demonstrates the presence of cortical dysplasias (90% of cases) and sub-ependymal nodules (80% of cases); the latter affect the layer of cells just under the ependyma, the thin membrane that lines fluid-filled spaces in the brain (e.g. the lateral ventricles). Sub-ependymal giant cell astrocytomas may arise in 5%–10% of cases, most commonly in teenage years.

Dermatological manifestations are common; hypomelanotic macules, facial angiofibromas, shagreen patches and ungual fibromas are all valuable diagnostic lesions. These cause few serious issues although the facial angiofibromas may be disfiguring and ungual fibromas troublesome.

Renal manifestations are an important cause of morbidity and mortality. In addition to non-malignant lesions such as angiomyolipomas, epithelial cysts and oncocytomas, malignant lesions such as renal cell carcinomas may develop but this is uncommon.

Cardiac rhabdomyomas usually develop *in utero* and are thought to occur in around 50% of affected individuals. They usually regress spontaneously in childhood.

Individuals with TSC may have retinal astrocytic hamartomas (major diagnostic criterion) (Fig. 22.12A and B) or retinal achromic patches (minor diagnostic criterion) (Fig. 22.12C).[1] Achromic patches are sharply demarcated chorioretinal lesions similar to hypomelanotic skin lesions and are seen in around 40% of individuals with TSC. Between 30 and 50% of individuals have retinal astrocytic hamartomas which may be flat, smooth, plaque-like, translucent lesions or, more commonly, may correspond to raised multinodular 'mulberry' lesions.[2] These characteristic tumours of the neurosensory retina are generally composed of astrocytes.

TABLE 22.4 Diagnostic criteria for tuberous sclerosis complex (*2012 International TSC Consensus Conference*).[1]

A. **Genetic diagnostic criteria**
   Identification of a *TSC1* or *TSC2* pathogenic mutation in DNA from normal tissue is sufficient to make a definite diagnosis of tuberous sclerosis complex (TSC).[a]

B. **Clinical diagnostic criteria**
   - *Major features*
     Hypomelanotic macules (≥ 3, at least 5-mm diameter)
     Angiofibromas (≥ 3) or fibrous cephalic plaque
     Ungual fibromas (≥ 2)
     Shagreen patch
     Multiple retinal hamartomas
     Cortical dysplasias
     Sub-ependymal nodules
     Sub-ependymal giant cell astrocytoma
     Cardiac rhabdomyoma
     Lymphangioleiomyomatosis (LAM)
     Angiomyolipomas (≥ 2)
   - *Minor features*
     'Confetti' skin lesions
     Dental enamel pits (>3)
     Intraoral fibromas (≥ 2)
     Retinal achromic patch
     Multiple renal cysts
     Non-renal hamartomas

[a]*A pathogenic mutation should clearly inactivate TSC1 or TSC2 protein function (e.g. out-of-frame indel or nonsense mutation), prevent protein synthesis (e.g. large genomic deletion), or be a missense mutation whose pathogenicity has been established. An unremarkable genetic test does not exclude TSC.*
Definite diagnosis: Two major features or one major feature with ≥ 2 minor features.
Possible diagnosis: Either one major feature or ≥ 2 minor features.

Retinal astrocytic hamartomatous lesions can be classified into four groups based on both clinical appearance and OCT findings[3]:

- Type 1—round, relatively flat, semi-transparent retinal mass within the retinal nerve fibre layer. No disturbance of the underlying retina/RPE. Gradual transition from the normal retina to tumour is noted.
- Type 2—slightly elevated with mild traction on tumour surface; some degree of retinal disorganisation and abrupt transition from normal retina to tumour.
- Type 3—calcified mushroom lesion with internal 'moth-eaten' optically empty spaces representing intra-tumoral calcification. Gradual transition between tumour and surrounding normal retina is noted.
- Type 4—elevated, dome-shaped, optically empty tumour, with a single large cavity and optical shadowing posteriorly.

Retinal astrocytic hamartomas typically do not affect vision unless there is macular or optic disc involvement. They usually grow very slowly or are static. However, in a minority of cases, they can grow aggressively causing severe ocular complications. Vision loss can occur due to vitreous haemorrhage, retinal neovascularization, retinal detachment, subretinal haemorrhage, neovascular glaucoma, macular oedema or even, rarely, a mass effect blocking the visual axis.[2] Photodynamic therapy can be used to induce regression of vascular astrocytic hamartomatomas; mTOR inhibitors have also been used.[4]

It is worth highlighting that retinal hamartoma may also occur in isolation or may be associated with

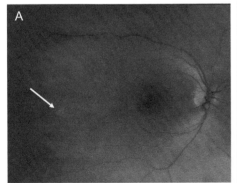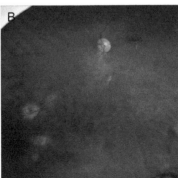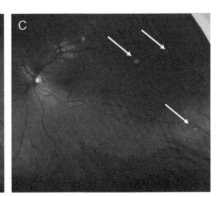

FIG. 22.12 Widefield multi-colour fundus images from three patients with tuberous sclerosis complex. (A) Raised retinal astrocytic hamartoma temporal to the macula. (B) Several retinal astrocytic hamartomas infero-temporally. (C) Punched out retinal achromic patches temporally. A selection of lesions is highlighted with *white arrows*.

neurofibromatosis type 1 (Section 22.1). The prevalence of retinal hamartomas in non-TSC populations is not known, but several case reports have been made and a series of >3500 healthy, full-term newborns identified only two cases with astrocytic hamartomas in that population.[5] The presence of a retinal hamartoma should prompt the clinician to ask about family history and any symptoms and signs that are known to occur in TSC. Further evaluations such as brain imaging may be indicated.

Molecular pathology

TSC results from heterozygous mutations in either the *TSC1* gene (encoding hamartin) or the *TSC2* gene (encoding tuberin).[6] Most pathogenic variants in *TSC1* and *TSC2* are nonsense, indel or splice-site mutations although missense mutations are more preponderant in *TSC2* (25% of such cases). A contiguous gene deletion may affect *TSC2* and *PKD1* resulting in polycystic kidney disease in addition to TSC. Disease-implicated variants are found in at most 90% of cases and *TSC2* mutations represent around two-thirds of such molecularly confirmed cases. Most affected individuals present without a family history, and, as mentioned above, around two-thirds are associated with *de novo* mutational events; the remainder are inherited from a parent.

In tuberous sclerosis hamartomas, *TSC1* or *TSC2* inactivation follows the classical Knudson two-hit hypothesis (see Section 22.3 and Chapter 25). *TSC2* pathogenic variants are statistically more likely to be associated with intellectual disability, autism, sub-ependymal nodules and retinal hamartomas compared to pathogenic variants in *TSC1*.[6] However, there is a huge overlap in the phenotypic spectrum of problems experienced by individuals with *TSC1* and *TSC2* mutations and so the distinction has little clinical utility.

TSC1 and TSC2 are interacting proteins that act together as a negative regulator of the mammalian target of rapamycin (mTOR) pathway that is critical to the regulation of protein synthesis and cellular growth. Inhibitors of this pathway (such as Sirolimus and Everolimus) are now being used to treat some of the manifestations of the disorder such as sub-ependymal giant cell astrocytomas, renal angiomyolipomas, lung involvement (lymphangioleiomyomatosis) and epilepsy.[7]

Clinical management

TSC is a multisystem condition and affected individuals require lifelong surveillance and care. Dedicated multidisciplinary clinics enable genetic diagnosis, systematic screening and expert care for the range of medical problems associated with this condition (including neurological and renal). For a detailed discussion on the management of TSC please see the relevant recommendations from the International Tuberous Sclerosis Complex Consensus Group[8] and the UK guidelines.[9]

References

1. Krueger DA, Northrup H, International Tuberous Sclerosis Complex Consensus Group. Tuberous sclerosis complex surveillance and management: recommendations of the 2012 international tuberous sclerosis complex consensus conference. *Pediatr Neurol* 2013;**49**(4):255–65.
2. Liu Y, Moore AT. Congenital focal abnormalities of the retina and retinal pigment epithelium. *Eye* 2020;**34**:1973–88.
3. Pichi F, Massaro D, Serafino M, Carrai P, Giuliari GP, Shields CL, et al. Retinal astrocytic hamartoma: optical coherence tomography classification and correlation with tuberous sclerosis complex. *Retina* 2016;**36**(6):1199–208. https://doi.org/10.1097/IAE.0000000000000829. 26618803.
4. Zhang ZQ, Shen C, Long Q, Yang ZK, Dai RP, Wang J, et al. Sirolimus for retinal astrocytic hamartoma associated with tuberous sclerosis complex. *Ophthalmology* 2015;**122**(9):1947–9. https://doi.org/10.1016/j.ophtha.2015.03.023.
5. Li L, Li N, Zhao J, et al. Findings of perinatal ocular examination performed on 3573, healthy full-term newborns. *Br J Ophthalmol* 2013;**97**:588–91.
6. Peron A, Au KS, Northrup H. Genetics, genomics, and genotype-phenotype correlations of TSC: insights for clinical practice. *Am J Med Genet C: Semin Med Genet* 2018;**178**(3):281–90.
7. Franz DN, Capal JK. mTOR inhibitors in the pharmacologic management of tuberous sclerosis complex and their potential role in other rare neurodevelopmental disorders. *Orphanet J Rare Dis* 2017;**12**(1):51. https://doi.org/10.1186/s13023-017-0596-2. 28288694. PMC5348752.
8. The International Tuberous Sclerosis Complex Consensus Group. https://www.tscinternational.org/international-tsc-consensus-guidelines/. [Accessed March 2020].
9. Amin S, Kingswood JC, Bolton PF, Elmslie F, Gale DP, Harland C, et al. The UK guidelines for management and surveillance of tuberous sclerosis complex. *QJM* 2019;**112**(3):171–82. https://doi.org/10.1093/qjmed/hcy215. 30247655.

Chapter 23

Naevoid basal cell carcinoma syndrome

D. Gareth Evans

Naevoid basal cell carcinoma syndrome (NBCCS), also known as basal cell naevus syndrome or Gorlin syndrome, is an autosomal dominant tumour-predisposing disorder characterised by two hallmark features: jaw cysts (odontogenic keratocysts) and multiple cutaneous basal cell carcinomas. Other features include skeletal anomalies, large head size and an increased risk of developing childhood-onset medulloblastoma and other types of cancer.[1,2]

Awareness among ophthalmic specialists is important as periocular basal cell carcinomas may be the presenting feature in a minority of cases. Familiarity with the diagnostic criteria (Table 23.1)[3-6] and maintaining a high index of suspicion are key to optimising the management of affected individuals. Regular dermatologic assessment is recommended and although life expectancy is not significantly altered, morbidity from complications can be substantial.[7]

Clinical characteristics

Basal cell carcinomas are the key and most problematic feature of NBCCS (Fig. 23.1A and B). Lesions are typically detected in the third decade of life although they can appear as early as 2 years of age or as late as 65 years of age. In general, there is a significant tendency for proliferation between puberty and 35 years of age.

Basal cell carcinomas of varying number, appearance and size are most commonly noted in sun-exposed areas. The most frequent initial sites are the face and the neck; the back and chest are also frequently affected but lesions below the waist are rarely encountered. Only a small fraction of these tumours becomes invasive but the invasion of deep underlying structures, especially in the face, is not uncommon (Fig. 23.1C).[6]

Other clinical features of NBCCS include:

- Jaw cysts (odontogenic keratocysts): these lesions are common (~90% of cases) and are often the initial reason for presentation. They usually present as painless swellings and can occur as early as age 5 years, although their peak occurrence is in teenage years. They have a characteristic radiological appearance on orthopantomography and untreated they can lead to major tooth disruption and jaw fractures.
- Palmar/plantar pits: these superficial lesions of the palms and feet are particularly useful diagnostically and are present in >80% of adult patients (Fig. 23.1D). These characteristic red/brown/black papules are generally 1–2 mm in size and are more clearly seen after soaking the hand/feet in warm water for up to 10 min.
- Ectopic calcification, particularly of the falx cerebri: this typical finding is visible on lateral X-rays of the skull and, although it is present in >90% of adult patients, it rarely causes symptoms (Fig. 23.2A).
- Skeletal abnormalities including bifid, wide, fused, partially missing or underdeveloped ribs: most individuals have skeletal anomalies identified on chest X-ray. These are usually asymptomatic but may give rise to a prominent or depressed sternum, downward sloping shoulders or a lump at the base of the neck. Severe skeletal defects are uncommon (Fig. 23.2B and C).
- A recognisable facial appearance with macrocephaly, bossing of the forehead, milia, coarse facial features and hypertelorism. Notably, affected children can present in infancy with very large heads that are often investigated by paediatricians.
- Medulloblastoma: this childhood brain malignancy is noted in approximately 5% of affected individuals. The peak incidence is around age 2 years.

Ophthalmologic manifestations include but are not limited to multiple periocular basal cell carcinomas (that have a high recurrence rate and may disrupt eyelid architecture); hypertelorism (due to sphenoid bone enlargement); strabismus; myelinated retinal nerve fibres (occasionally bilateral); and eyelid cysts.[8,9] Developmental defects such as developmental cataracts, anterior segment dysgenesis

TABLE 23.1 Major and minor diagnostic criteria for naevoid basal cell carcinoma syndrome.

Major criteria	Minor criteria
Basal cell carcinoma at a young age (<30 years old at diagnosis) OR excessive numbers of basal cell carcinomas out of proportion to prior sun exposure and skin type (typically >5 in a lifetime; carcinomas occurring after radiotherapy are disallowed)	Vertebral/rib anomalies (including bifid vertebrae or bifid/splayed/extra ribs)
Odontogenic keratocyst of the jaw	Macrocephaly (occipital frontal circumference >97th centile)
Palmar/plantar pits (generally ≥2)	Pre-axial/post-axial polydactyly
Lamellar (sheet-like) calcification of the falx cerebri	Cleft lip/palate
First-degree relative with naevoid basal cell carcinoma syndrome	Ovarian/cardiac fibroma
Medulloblastoma (nodular or desmoplastic)	Lymphomesenteric/pleural cysts
	Ophthalmic abnormalities (including congenital cataracts, eye developmental defects, and pigmentary changes of the retinal epithelium)

Diagnostic criteria for naevoid basal cell carcinoma syndrome have been previously proposed and refined by several groups.[3–6] The above is a combination of various classifications and intends to serve as a guide. Generally, the condition can be diagnosed in the presence of:
- two major OR
- one major and two minor criteria

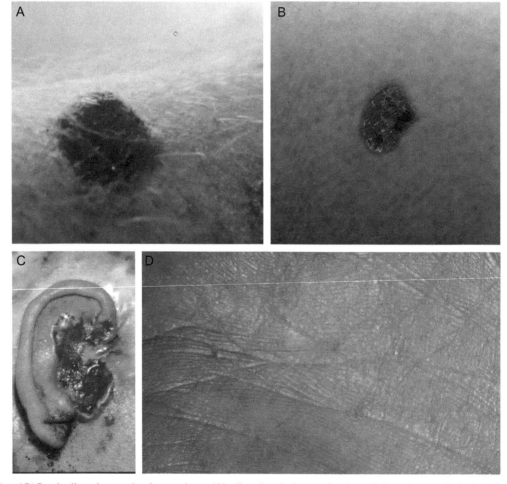

FIG. 23.1 (A and B) Basal cell carcinomas showing crusting and bleeding; these lesions can be elevated, flat or have a rolled edge with ulcer. (C) Basal cell carcinoma invading the external auditory meatus. (D) Palmar pits.

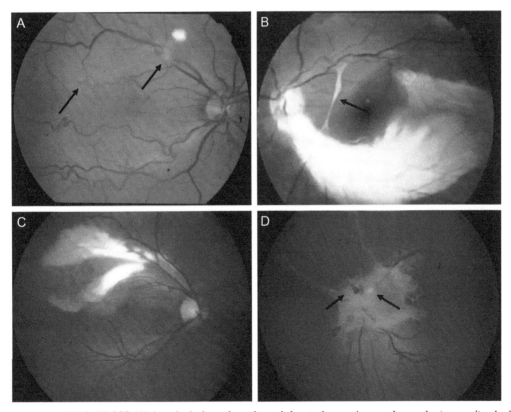

FIG. 23.2 (A) Bifid rib in the X-ray of a neonate with NBCCS (more pronounced on the left side of the image; *arrowed*). (B) Subtle bifid ribs in the X-ray of an adult with NBCCS (right side of the image; *arrowed*). (C) Central falx calcification on the skull X-ray (anteroposterior view) of a patient with NBCSS.

as well as microphthalmia and coloboma have been reported in a small subset of affected children.[10] There are several case reports of individuals with NBCCS who had notable retinal anomalies, including retinal fibrotic falciform folds leading to retinal detachment, epiretinal membranes, macular holes and combined hamartoma of the retina and retinal pigment epithelium (Fig. 23.3).[11]

Molecular pathology

NBCCS is caused by heterozygous loss of function mutations in the *PTCH1* or the *SUFU* gene.[12] About 20% of cases are *de novo* and most probands are found to have an affected parent if careful examination is performed.

FIG. 23.3 Retinal abnormalities in NBCCS. (A) An epiretinal membrane located close to the superior vascular arcades *(arrowed)* and a discrete opaque nodule (consistent with astrocytic proliferation) embedded in the superior retina. (B) Spontaneously avulsed epiretinal membrane *(arrowed)* and extensive retinal myelination. Retinal myelination is visible as an arcuate opaque region below the central macula. (C) Retinal myelination with epiretinal membrane formation resulting in wrinkling of the retinal surface. (D) Abnormal glial proliferation around the optic disc, with abnormal retinal vascular patterns.

Around 60%–70% of classically affected individuals are found to have a germline pathogenic variant in *PTCH1* with only around 5%–6% carrying mutations in *SUFU*. Jaw cysts are rare in *SUFU*-related NBCCS and the signs may be more subtle. However, meningiomas are more frequent and the risk of medulloblastoma is much higher in this disease subtype.[13]

Children of patients with NBCCS have a 50% chance of inheriting the condition. Careful clinical examination with a skeletal survey can usually diagnose NBCCS by teenage years.

Clinical management

The management of people with NBCCS involves evaluation following initial diagnosis, introduction of surveillance measures, prevention of primary manifestations and treatment of manifestations.[2]

Affected individuals should avoid more than minimal sun exposure and should cover up exposed skin by wearing long sleeves and hats; complete sunblock should be used and unnecessary radiation exposure should be avoided. Skin examination by a clinician familiar with NBCCS should be performed at least annually. Jaw X-rays (orthopantograms) are indicated every 12–18 months in patients from 8 to 30 years of age; if jaw cysts are still active at age 30 years, then further monitoring may be necessary. In infants with *SUFU*-related NBCCS, brain MRI is recommended for early detection of medulloblastoma.

Jaw cysts usually require surgical excision. A wide variety of treatment options have been described for basal cell carcinomas including surgery, topical chemotherapy (e.g. with 5% imiquimod cream), pharmacological therapy (e.g. with 5% fluorouracil cream), cryotherapy, laser and photodynamic therapy. Vismodegib, a small-molecule inhibitor of the sonic hedgehog pathway has shown promise in advanced basal cell carcinomas but side effects are unpleasant.

Mohs micrographic surgery (also known as margin-controlled excision), a technique that enables complete removal of the tumour whilst preserving as much non-involved tissue as possible, appears to be particularly effective for basal cell carcinomas. It is often the preferred management option for periocular lesions. Notably, it is not uncommon for individuals with NBCCS to lose an eye due to the invasiveness of the basal cell carcinomas in this area.[2,8]

References

1. Gorlin R. Nevoid basal cell carcinoma (Gorlin) syndrome. *Genet Med* 2004;**6**:530–9. https://doi.org/10.1097/01.GIM.0000144188.15902.C4.
2. Evans DG, Farndon PA. Nevoid basal cell carcinoma syndrome. 2002 Jun 20 [updated 2018 Mar 29]. In: Adam MP, Ardinger HH, Pagon RA, Wallace SE, Bean LJH, Mirzaa G, Amemiya A, editors. *GeneReviews® [Internet]*. Seattle, WA: University of Washington, Seattle; 1993–2021. 20301330.
3. Kimonis VE, Goldstein AM, Pastakia B, Yang ML, Kase R, DiGiovanna JJ, et al. Clinical manifestations in 105 persons with nevoid basal cell carcinoma syndrome. *Am J Med Genet* 1997;**69**(3):299–308. 9096761.
4. Bree AF, Shah MR, BCNS Colloquium Group. Consensus statement from the first international colloquium on basal cell nevus syndrome (BCNS). *Am J Med Genet A* 2011;**155A**(9):2091–7. https://doi.org/10.1002/ajmg.a.34128.
5. Foulkes WD, Kamihara J, Evans DGR, et al. Cancer surveillance in Gorlin syndrome and rhabdoid tumor predisposition syndrome. *Clin Cancer Res* 2017;**23**(12):e62–7. https://doi.org/10.1158/1078-0432.CCR-17-0595.
6. Jones EA, Sajid MI, Shenton A, Evans DG. Basal cell carcinomas in Gorlin syndrome: a review of 202 patients. *J Skin Cancer* 2011;**2011**:217378.
7. Wilding A, Ingham SL, Lalloo F, Clancy T, Huson SM, Moran A, et al. Life expectancy in hereditary cancer predisposing diseases: an observational study. *J Med Genet* 2012;**49**:264–9.
8. Wong A, Feldman BH, Nguyen Burkat C, Kim J, Yen MT, Enghelberg, et al. *Gorlin-Goltz syndrome*; 2021 https://eyewiki.aao.org/Gorlin-Goltz_syndrome.
9. Chen JJ, Sartori J, Aakalu VK, Setabutr P. Review of ocular manifestations of Nevoid Basal Cell Carcinoma Syndrome: what an Ophthalmologist needs to know. *Middle East Afr J Ophthalmol* 2015;**22**(4):421–7.
10. Ragge NK, Salt A, Collin JR, Michalski A, Farndon PA. Gorlin syndrome: the PTCH gene links ocular developmental defects and tumour formation. *Br J Ophthalmol* 2005;**89**(8):988–91.
11. Black GC, Mazerolle CJ, Wang Y, Campsall KD, Petrin D, Leonard BC, et al. Abnormalities of the vitreoretinal interface caused by dysregulated Hedgehog signaling during retinal development. *Hum Mol Genet* 2003;**12**(24):3269–76.
12. Onodera S, Nakamura Y, Azuma T. Gorlin syndrome: recent advances in genetic testing and molecular and cellular biological research. *Int J Mol Sci* 2020;**21**(20):7559. https://doi.org/10.3390/ijms21207559. 33066274.
13. Evans DG, Oudit D, Smith MJ, Rutkowski D, Allan E, Newman WG, et al. First evidence of genotype-phenotype correlations in Gorlin syndrome. *J Med Genet* 2017;**54**(8):530–6. https://doi.org/10.1136/jmedgenet-2017-104669.

Chapter 24

Congenital hypertrophy of retinal pigment epithelium (CHRPE)

Fiona Lalloo and Graeme C.M. Black

Congenital hypertrophy of the retinal pigment epithelium (CHRPE) is a common fundoscopic feature involving pigmented lesions at the level of the RPE. A prevalence of 1%–2% has been described in populations of European ancestries.[1,2] The majority of CHRPEs have neither visual nor general health implications. However, the presence of multiple (i.e. more than three),[3] bilateral, mixed pigmented/depigmented CHRPE is a specific and sensitive marker of a cancer predisposition syndrome, familial adenomatous polyposis (FAP). FAP is characterised by the presence of multiple colorectal adenomatous polyps and is caused by heterozygous mutations in the *APC* gene.

The identification of CHRPE in this rare context is an important task. However, care must be taken not to subject significant numbers of individuals with CHRPE to unnecessary psychological harm, colonoscopy or genetic testing. Therefore, it is important to recognise—and exclude—some of the commonest ophthalmic reasons to misdiagnose FAP including 'bear track' patterns (whether unilateral or bilateral), single CHRPE or even a single unilateral group of CHRPEs.

Clinical characteristics

CHRPE: CHRPE are flat or very slightly raised, round or oval, pigmented fundus lesions that are generally located in the mid-peripheral, peripapillary or macular regions (Fig. 24.1A). They may have a surrounding halo and can have depigmented lacunae within the lesion. Where it has been undertaken, OCT demonstrates a lack of overlying photoreceptors. There is hypoautofluorescence on fundus autofluorescence imaging and choroidal masking on fluorescein angiography. The lesions are largely static although, when they have been followed carefully, they have been shown to enlarge very gradually with time. A small proportion of CHRPE are not flat and contain nodules; these can very occasionally have macular oedema as a complication.

Congenital grouped pigmentation of the RPE ('bear tracks'): Congenital grouped pigmentation of the RPE is characterised by small, darkly pigmented, flat, circumscribed lesions at the level of the RPE that generally increase in size as they approach the retinal periphery. Prevalence is around 0.1% of the population. The lesions resemble animal tracks ('bear tracks') (Fig. 24.1B). They do not affect vision and electrodiagnostic and colour vision testing are normal. Mostly, lesions are unilateral and sectoral. Occasionally, the entire fundus may be involved and very rarely this pattern may be bilateral. Congenital grouped pigmentation of the RPE is usually isolated although several chance associations have been described.

Familial adenomatous polyposis (FAP): FAP is an autosomal dominant disorder that predisposes to malignancy and accounts for ~1% of all colorectal cancers.[3] CHRPEs are found in individuals with FAP both with and without extracolonic manifestation such as desmoids, osteomas, and sebaceous cysts.

Classically, affected individuals with FAP develop multiple adenomatous colorectal polyps between the second and fourth decades of life (Fig. 24.2A and B). Malignant transformation occurs in over 90% of patients by age 50 years. Tumour surveillance in mutation carriers by colonoscopy begins from around 12 years of age and leads to elective colectomy once the polyp burden is too great to manage with less invasive approaches. Polyps may also develop in the stomach, small bowel, and duodenum. Around 10% of patients develop desmoid tumours, which, whilst non-malignant, are locally invasive fibrous tissue masses that are very difficult to manage. Osteomas of the mandible and dental anomalies, such as missing or supernumerary teeth, are common. A milder form of the disease called attenuated FAP results in fewer polyps at a later age and has a later onset of malignancy.

In FAP, CHRPEs may be ovoid, pisciform, or irregular in shape and may be surrounded by a pale grey halo or an adjacent depigmented linear streak. They vary in size, from

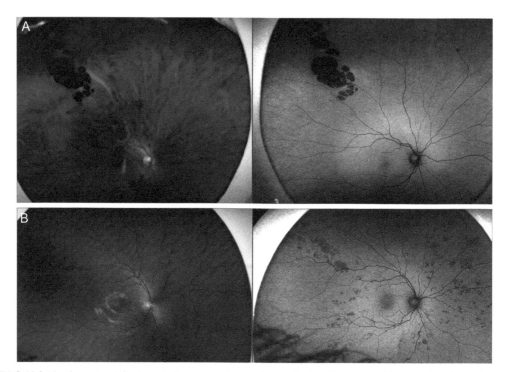

FIG. 24.1 Widefield fundus images and fundus autofluorescence images from patients with congenital hypertrophy of retinal pigment epithelium (CHRPE). (A) Single large CHRPE. (B) Congenital grouped pigmentation of the RPE ('bear tracks').

dot-like lesions to multiple disc diameters, and positions (Fig. 24.2C). In families where CHRPEs are present, they are >90% penetrant.

Molecular pathology

Familial adenomatous polyposis is caused by heterozygous mutations in *APC*. *APC* is a tumour suppressor gene and mutations follow the classical two-hit Knudson model of tumorigenesis (see Section 22.3 and Chapter 25). FAP is inherited as an autosomal dominant trait with 10%–20% of cases representing *de novo* mutations.[4] There are recognised genotype–phenotype correlations.[5] For example, where CHRPEs are seen in individuals with the FAP phenotype they suggest a mutation between, approximately, amino-acids 350 and 1400 (APC is a 2843 amino-acid protein). In former times, when sequencing technologies were more labour-intensive than currently, such observations were used to guide mutation detection; this is no longer the case.

Clinical management

The ophthalmologist is not required to be a part of the multidisciplinary care of individuals and families who carry mutations in *APC*. While the phenotypic manifestations are, in such circumstances, recognisable, they do not influence screening protocols and are no longer used either in preference to or alongside genetic testing for presymptomatic identification of mutation carriers. Consequently, the challenge for ophthalmic care is identifying potential mutation carriers, in the absence of a known family history of FAP. Given the frequency of CHRPE, and the increased availability of wide-field imaging (e.g. in the context of routine optometry), it is important not to create unnecessary anxiety in children and young adults with:

- single CHRPE
- a single unilateral cluster of CHRPE
- 'bear track' patterns are either unilateral or bilateral.

Each of these three categories is common and sadly it is not unusual to see cases investigated unnecessarily and extensively for polyposis either through colonoscopy or colonoscopy aligned to genomic investigation.

The specific and sensitive phenotypic marker of FAP is presence of multiple (i.e. more than three) CHRPEs that are bilateral and mixed pigmented/depigmented (Fig. 24.2). This is an uncommon scenario. Individuals identified with this phenotype should be referred for combined care between a cancer genetic team and a gastroenterologist.

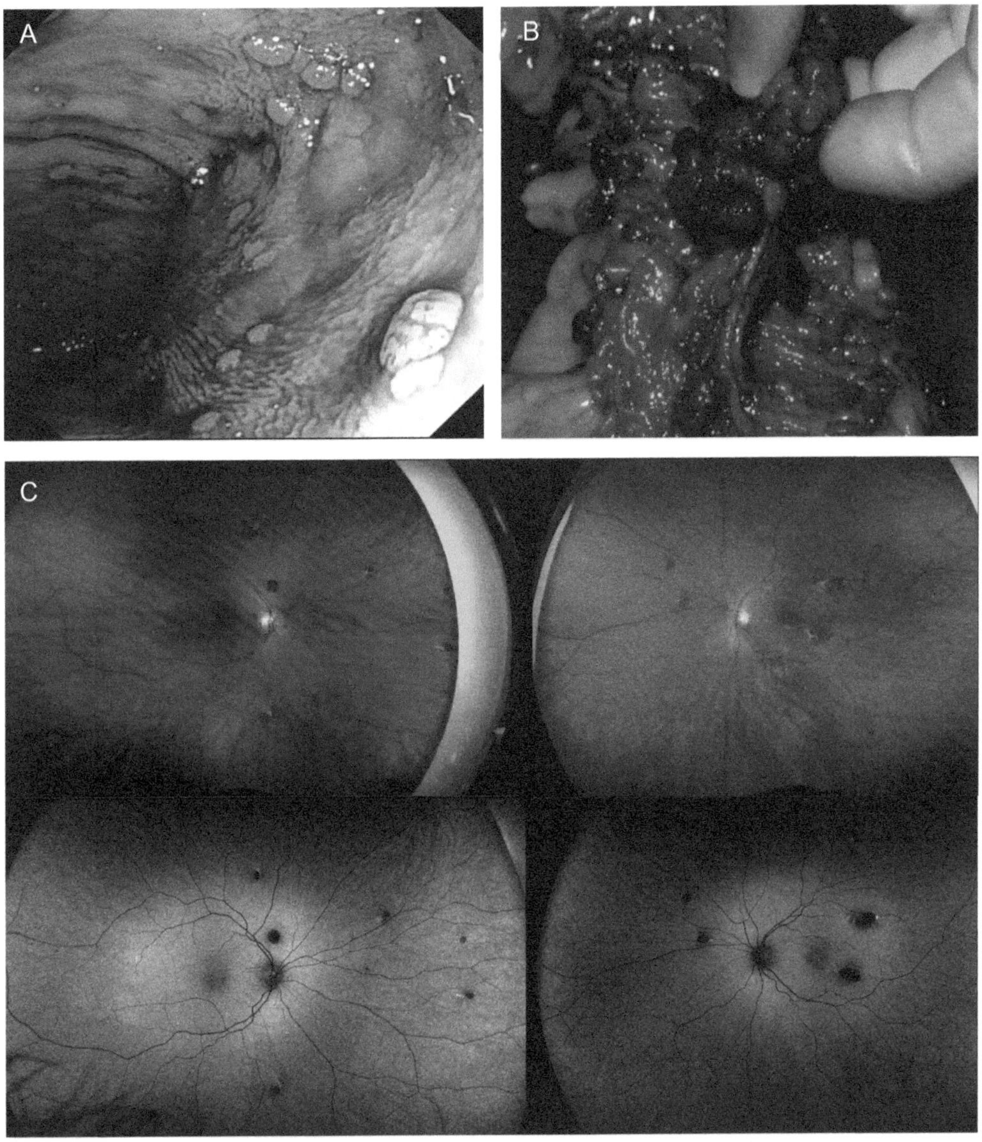

FIG. 24.2 Findings in patients with familial adenomatous polyposis (FAP). (A) Colonic polyposis in FAP with polyps highlighted with dye at colonoscopy. (B) Florid colonic polyposis in FAP *(Courtesy of Prof James Hill)*. (C) Multiple, bilateral congenital hypertrophic lesions of the retinal pigment epithelium (CHRPE) in FAP. Widefield fundus images and fundus autofluorescence images are shown.

Such a multidisciplinary team will be the most appropriate to understand the significance of any family history of bowel cancer and will employ recognised screening algorithms to identify or exclude FAP.

References

1. Shields JA, Shields CL. Tumors and related lesions of the pigmented epithelium. *Asia Pac J Ophthalmol (Phila)* 2017;**6**(2):215–23.
2. Liu Y, Moore AT. Congenital focal abnormalities of the retina and retinal pigment epithelium. *Eye* 2020;**34**:1973–88.
3. Burn J, Chapman P, Delhanty J, et al. The UK Northern Region genetic register for familial adenomatous polyposis coli: use of age of onset, congenital hypertrophy of the retinal pigment epithelium and DNA markers in risk calculations. *J Med Genet* 1991;**28**:289–96.
4. Aretz S, Uhlhaas S, Caspari R, Mangold E, Pagenstecher C, Propping P, et al. Frequency and parental origin of de novo APC mutations in familial adenomatous polyposis. *Eur J Hum Genet* 2004;**12**(1):52–8.
5. Nieuwenhuis MH, Vasen HF. Correlations between mutation site in APC and phenotype of familial adenomatous polyposis (FAP): a review of the literature. *Crit Rev Oncol Hematol* 2007;**61**(2):153–61.

Chapter 25

Retinoblastoma

Manoj V. Parulekar and Brenda L. Gallie

Retinoblastoma is the commonest primary malignant intraocular tumour of childhood accounting for ~1% of tumours in infancy. In most cases, it arises as a consequence of the silencing/loss of both copies of the tumour suppressor gene *RB1* in developing retinal cells.

The incidence of sporadic retinoblastoma is 1 in 15,000–20,000 live births, with no gender or racial predilection.[1–3] The median age at presentation is under 12 months in heritable cases (i.e. individuals with a family history of the condition), and closer to 24 months in sporadic cases. Presentation after the age of 6 years is extremely rare, although there are isolated reports of cases presenting as late as 26 years.

Heritable retinoblastoma is prototypical of a cancer susceptibility syndrome. Affected individuals carry a single pathogenic copy of the *RB1* gene that predisposes to retinoblastoma formation. All individuals with bilateral retinoblastoma carry such a pathogenic *RB1* allele. Importantly, they also have a risk of developing other cancers and lifelong surveillance is required. Nearly 20% of individuals with unilateral retinoblastoma carry a pathogenic *RB1* allele that can be transmitted to their children. For a unilaterally affected person with no detectable pathogenic allele with high sensitivity testing, the chance of having heritable retinoblastoma is <1%.

Early recognition and diagnosis, alongside effective/optimised multidisciplinary care, results in a 95% cure rate for affected individuals. By contrast, delayed diagnosis results worldwide in around 70% mortality. Understanding the genetics of retinoblastoma has enabled clinicians to develop targeted screening guidelines based on genetic risk, minimising unnecessary screening exams, and focusing resources on individuals at greatest risk.[4]

Clinical characteristics

Where there is no family history, the commonest presenting signs of retinoblastoma are leukocoria and strabismus; other well recognised presentations include red eye secondary to glaucoma, orbital cellulitis and unilateral mydriasis.[5,6] For those with a family history, screening is implemented before any such signs are observed, allowing early detection and treatment. The differential diagnosis for children with leukocoria includes Coats disease, persistent foetal vasculature, toxocariasis and congenital cataract.

When retinoblastoma is suspected, urgent referral is mandatory. Evaluation includes ultrasound biomicroscopy, examination of both eyes under anaesthetic and, where necessary, MRI scans. Retinoblastoma tumours appear creamy white and may have irregular vasculature or haemorrhagic areas on their surface (Fig. 25.1). Characteristic are tumour seeding into the vitreous and presence of calcification (on ultrasound and/or MRI, for example for tumours obscured by detachment or vitreous opacity). CT scans should be avoided in cases of suspected retinoblastoma, as exposure to radiation magnifies the risk of second cancers in individuals with germline mutations.

Occasionally, biallelic pathogenic variants in the *RB1* gene can result in a *retinoma*, a benign precursor to retinoblastoma. This is uncommon, as the genetic environment after the loss of both *RB1* alleles is unstable and usually results in multiple other mutations that promote uncontrolled proliferation and progression to retinoblastoma.[7,8] Since a retinoma is a premalignant state with characteristic clinical features that can undergo malignant transformation to retinoblastoma after many years of stability, an individual carrying such a tumour requires long-term monitoring (Fig. 25.2).

As with all cancers, tumour staging at diagnosis is important. Comprehensive staging for retinoblastoma is described in the TNM (tumour, node and metastasis) staging for cancer.[9] In addition to clinical and pathologic findings, this staging recognises the impact of genetics on outcome ('definition of heritable trait'; see Table 25.1).[7] Staging can be assigned to pathogenic allele carriers, or individuals with a positive family history even in the absence of disease manifestation.

Around 40% of children with retinoblastoma are bilaterally affected, and 60% are unilaterally affected. Over 90% of both unilateral and bilateral cases are sporadic with no family history. However, a third of the 'sporadic' cases arise from *de novo* pathogenic alleles which are present in the affected patient's germline and hence can be passed to their offspring. Of the unilateral cases, 19% are germline, and carry the same familial risks as bilateral germline cases (see Box 25.1).

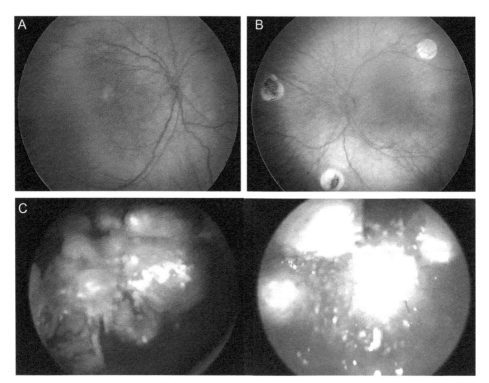

FIG. 25.1 Fundus photographs from three unrelated individuals with retinoblastoma. (A) Macular tumour detected during screening in a case with hereditary retinoblastoma. (B) Extramacular tumours in hereditary retinoblastoma treated with laser alone. (C) Advanced retinoblastoma (group D; cT2b).

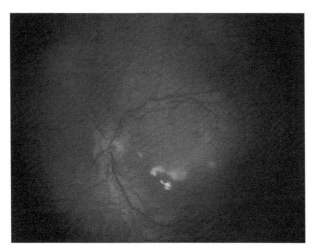

FIG. 25.2 Colour fundus photography of an individual with a retinoma.

Loss of segments of the chromosome carrying the *RB1* gene, e.g. a chromosome 13q14 deletion, can also results in retinoblastoma often associated with developmental delay and dysmorphic features.[10]

Molecular pathology

The *RB1* gene is located on the long arm of chromosome 13. It was the first tumour suppressor gene to be cloned, 15 years after Knudson predicted the inheritance pattern of retinoblastoma in 1971[11,12] (see Box 25.1). The protein product encoded by the *RB1* gene is pRB. pRB directly binds to the E2F transcription factor to regulate the cell cycle by blocking entry into the S-phase (synthesis phase; the phase of the cell cycle in which DNA is replicated) thus inhibiting cell growth.[13] Biallelic loss of pRB initiates the development of retinoma, then retinoblastoma. Inactivation of pRB can also play a role in the development of other cancer types including osteosarcoma and mesothelioma.

Pathogenic alleles in retinoblastoma are associated with loss of function, and include nonsense, indel, splice site and missense variants (the latter representing 10%–15% of the total). There are several international genetic databases that store and share information about recognised pathogenic *RB1* variants, that are used as a reference by testing laboratories. Wherever *RB1* analysis is undertaken, it is important to identify/exclude whole gene or intragenic copy number variation.

Loss of function pathogenic alleles in the *RB1* gene have approximately 90% penetrance—that is, 90% of individuals carrying an *RB1* germline pathogenic allele will develop one or more retinoblastomas. Since the remaining 10% of individuals are still at risk of second cancers, they must follow the same lifestyle guidelines and seek medical attention for any unusual symptoms. It is notable that some pathogenic alleles that are not associated with complete loss of function, including missense and some splice

TABLE 25.1 Updated classification for retinoblastoma.

Definition of the primary tumour (cT)		
cTX		Unknown evidence of intraocular tumour
cT0		No evidence of intraocular tumour
cT1		Intraocular tumour(s) with subretinal fluid ≤5 mm from the base of any tumour
	cT1a	Tumours ≤3 mm and further than 1.5 mm from the disc and fovea
	cT1b	Tumours >3 mm or closer than 1.5 mm to the disc and fovea
cT2		Intraocular tumour(s) with retinal detachment, vitreous seeding or sub-retinal seeding
	cT2a	Subretinal fluid >5 mm from the base of any tumour
	cT2b	Tumours with vitreous seeding and/or sub-retinal seeding
cT3		Advanced intraocular tumour(s)
	cT3a	Phthisis or pre-phthisis bulbi
	cT3b	Tumour invasion of the pars plana, ciliary body, lens, zonules, iris or anterior chamber
	cT3c	Raised intraocular pressure with neovascularization and/or buphthalmos
	cT3d	Hyphema and/or massive vitreous haemorrhage
	cT3e	Aseptic orbital cellulitis
cT4		Extraocular tumour(s) involving the orbit, including the optic nerve
	cT4a	Radiological evidence of retrobulbar optic nerve involvement or thickening of the optic nerve or involvement of the orbital tissues
	cT4b	Extraocular tumour clinically evident with proptosis and orbital mass
Definition of regional lymph nodes (cN)		
cNX		Regional lymph nodes cannot be assessed
cN0		No regional lymph node involvement
cN1		Evidence of preauricular, sub-mandibular, and cervical lymph node involvement
Definition of distant metastasis (M)		
cM0		No signs or symptoms of intracranial or distant metastasis
cM1		Distant metastasis without microscopic confirmation
	cM1a	Tumour(s) involving any distant site (e.g. bone marrow, liver) on clinical or radiological tests
	cM1b	Tumour involving the central nervous system (CNS) on radiological imaging (not including trilateral retinoblastoma)
pM1		Distant metastasis with microscopic confirmation
	pM1a	Histopathological confirmation of tumour at any distant site (e.g. bone marrow, liver, or other)
	pM1b	Histopathological confirmation of tumour in the cerebrospinal fluid or CNS parenchyma
Definition of heritable trait (H)		
HX		Unknown or insufficient evidence of a constitutional RB1 gene mutation
H0		Normal RB1 alleles in blood tested with demonstrated high sensitivity assays
H1		Bilateral retinoblastoma, retinoblastoma with an intracranial CNS midline embryonic tumour (i.e. trilateral retinoblastoma), patient with a family history of retinoblastoma, or molecular definition of constitutional *RB1* gene mutation

Adapted from Ref. 7.

> **BOX 25.1 The Knudson two hit hypothesis.**[11]
>
> There are two copies (alleles) of the *RB1* gene in every cell in the body, and at least one functioning allele is required to prevent the development of retinoblastoma in a retinal cell. Loss of both alleles results in retinoma or retinoblastoma.
>
> In 1971, Alfred Knudson proposed that two separate pathogenic allelic events that result in loss or inactivation of both copies of a gene are required to initiate retinoblastoma. This came to be known as the Knudson 'two hit' hypothesis.[11] Evidence suggests that tumour initiation can occur with two hits, but subsequent mutational events are necessary for the tumour to grow into retinoblastoma.[14] The two hits can occur in several broad situations and the timing of the first hit *RB1* pathogenic mutation is key to understanding familial risk which can occur:
> - prior to, or at, conception. In both these situations, all the cells in the foetus will be expected to carry the pathogenic allele, and the individual is at risk of developing one or more retinoblastoma tumours in one or both eyes (unilateral unifocal or multifocal or bilateral tumours). These individuals are also at high risk of second cancers later in life. Since they carry the pathogenic allele in their germline, there is a 50% risk of passing it on to each of their offspring;
> - after conception, when the zygote divides into two, four, eight, sixteen cells and so on. Only a portion of the cells in the body will carry the pathogenic allele; this results in mosaicism. Since the pathogenic allele may also affect a portion of gametes, offspring will be at risk, albeit lower than full germline cases[15];
> - within a retinal cell, in which case a second hit will give rise to single, unilateral retinoblastoma with no risk to offspring and no risk of second cancers.

site mutations, are recognised to be of lower penetrance. In these circumstances, unaffected members of a family carrying such an allele, may be detected as a result of cascade resulting.

Initial testing for a bilateral or unilateral multifocal retinoblastoma involves mutation analysis of blood, generally using high-throughput sequencing-based, whole gene approaches. A pathogenic allele will be identified in approximately 97% of cases.[16] Pathogenicity of deep intronic variants can be confirmed by RNA studies (reverse transcriptase PCR). Rarely, no *RB1* pathogenic allele is identified in a patient with bilateral disease. In these remaining cases genetic studies are undertaken to discover a translocation or changes on tumour DNA if available.[15,16]

For unilateral cases, tumour DNA is tested for pathogenic alleles and then blood is tested for these changes. In 15% of cases, this will identify a germline pathogenic allele, perhaps with low level mosaicism.[15] Promoter gene hypermethylation that silences *RB1* expression, a common pathogenic mechanism in tumours, is rarely present in the germline.[17]

A small number of patients carry pathogenic alleles involving large chromosome 13 rearrangements. This has historically been identified from karyotype, fluorescent in situ hybridization (FISH), or array comparative genomic hybridization (array CGH) studies. Parents of these patients may carry balanced translocations, which increase the risk for retinoblastoma in subsequent pregnancies.

A newly recognised, rare form of retinoblastoma with normal *RB1* sequences has been described. Two percent of unilateral retinoblastomas do not carry any *RB1* pathogenic allele, but carry amplification of the *MYCN* oncogene (28–121 DNA copies instead of the normal 2 copies). These *MYCNA* tumours are diagnosed at a young age (median age of 4.5 months) compared with 24 months for non-germline unilateral *RB1* mutation-negative patients. The tumours are histologically distinct with advanced features at diagnosis and poor response to chemotherapy and other eye conserving treatment.[18]

Clinical management

The treatment of retinoblastoma has evolved considerably over recent decades. Enucleation is the oldest treatment for retinoblastoma and is suitable for advanced cases. External beam radiotherapy (EBRT) was the mainstay of eye conserving treatment from the 1950s but rarely used since the introduction of chemotherapy in the 1980s. Apart from disfigurement, external beam radiotherapy significantly increases the risk of second malignancies in individuals carrying an *RB1* germline pathogenic allele.[19] Local treatment with laser and cryotherapy are now used along with chemotherapy. Laser alone may be used in early stage retinoblastoma with smaller tumours, avoiding chemotherapy, and enabling good visual outcomes. This is usually seen in children with a positive family history or a known pathogenic allele identified by testing before or soon after birth; these individuals receive early and regular screening, enabling prompt detection and treatment.[20]

Addressing the genetic aspects of retinoblastoma is an important part of the assessment and management of any new case and a detailed clinical and family history is essential. This includes a history of childhood eye cancer, removal of an eye in any family member going back at least two to three generations, and early deaths due to cancer.

RB1 mutation testing is well established in ascertaining the risk of second eye involvement, the lifelong risk of secondary cancers for the affected individual, and the risk to siblings and offspring. Importantly, successful genetic testing has been demonstrated to be clinically useful *and* cost-effective in minimising hospital visits and enabling targeted screening for those who will benefit the most.[21] The cost of screening can be offset by the benefits from reduced examinations. Consequently, every effort should be made to identify germline pathogenic alleles. Venous blood samples for genetic testing (including microarray analysis and *RB1* mutation testing) should be collected at the first available opportunity after diagnosis.

Once a pathogenic *RB1* pathogenic allele is found in an index case, while it does not alter the management for the affected individual, other family members can be specifically tested, starting with parents. This can define risk for other family members and enables targeted clinical screening (see Fig. 25.3 and Table 25.2).

- If one parent tests positive for the pathogenic allele, then other children of that parent can be offered testing. If they test negative, the risk is reduced to population risk (except for the remaining <1% that the parent is low level mosaic for that allele) and siblings can avoid screening when they test negative for the allele.
- A negative blood test reduces risk to <1% (for parents, siblings and cousins). This avoids unnecessary screening, minimises family anxiety, patient morbidity and saves resources.
- Any individuals who test positive and are under 3 years of age require screening (Table 26.2). Siblings under 3 years require screening until genetic testing has been completed.
- Older siblings and parents require eye examination to exclude retinoma or spontaneously regressed retinoblastoma.

In addition to facilitating early detection and treatment, positive mutation identification provides options around prenatal diagnosis and pre-implantation genetic diagnosis. Prenatal identification of *RB1* pathogenic allele carriers can also be used for *in utero* ultrasound monitoring since—particularly in those with highly penetrant pathogenic variants—tumours may develop *in utero*, have a predilection for the macular area with implications for vision, and may reach a considerable size at birth. Some centres offer early induction of pregnancy to enable early treatment.[20] Newborns considered at risk of retinoblastoma may have urgent genetic testing on umbilical cord blood samples at birth.

When an index case undergoes enucleation, the eye can be sent promptly, in a sterile pot (without formalin which degrades DNA) to a diagnostic laboratory in order to facilitate tumour genetic analysis. The tumour sample is studied to identify the pathogenic alleles. If neither of the two expected pathogenic alleles is detectable in tumour tissue—or if tumour DNA is not available—the risk of germline retinoblastoma cannot be excluded.

It is important to note that the introduction of modern intravitreal chemotherapy techniques has dramatically reduced enucleation rates which creates new challenges for genetic testing of unilateral cases, as tumour tissue is no longer available. However, testing of aqueous humour samples for cell free DNA can enable detection of tumour pathogenic alleles when tumour issue is not available.[22,23]

One of the big challenges in retinoblastoma management is the inequity of resources between developed and developing countries. This is particularly highlighted by the limited availability of genetic testing in countries lacking socialised healthcare systems (including some

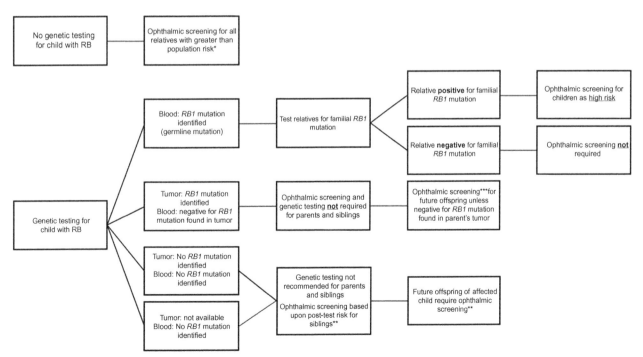

FIG. 25.3 Decision-making tree for screening siblings and relatives of children with retinoblastoma. *(Adapted from Ref. 24.)*

TABLE 25.2 Guidelines for childhood screening for retinoblastoma families.

Management Guidelines for Childhood Screening for Retinoblastoma Families

Risk Category	% risk	Eye examination schedule based upon age of unaffected child							
		Birth to 8 weeks*	>8 weeks to 12 weeks	>3 months to 12 months	>12 months to 24 months	>24 months to 36 months	>36 months to 48 months	>48 months to 60 months	5-7 years
High Risk	> 7.5	Every 2-4 weeks	Monthly		Every 2 months	Every 3 months	Every 4 months	Every 6 months	Every 6 months
Intermediate Risk	1 - 7.5	Monthly		Every 2 months	Every 3 months		Every 4-6 months		Every 6 months
Low Risk	< 1	Monthly		Every 3 months	Every 4 months		Every 6 months		Annually
General population	0.007	Screening with pediatrician							

☐ Non-sedated office examination preferred by most centers

■ Examination under anesthesia preferred by most centers

TABLE 25.3 Estimation of pre-test risk of carrying a mutant *RB1* allele for relatives of a proband with retinoblastoma.

Relative of proband	Pre-test risk for mutant allele (%)	
	Bilateral (100% carry mutation)	Unilateral (15% carry mutation)
Offspring (infant)	50	7.5
Parent	5	0.8
Sibling	2.5	0.4
Niece/nephew	1.3	0.2
Aunt/uncle	0.1	0.007
First cousin	0.05	0.007

Risk, for probands with unilateral and bilateral retinoblastoma without a family history, is shown as a percentage. Since 3rd- and 4th-degree relatives of unilateral probands have calculated risks less than the normal population risk of 0.007% (i.e. 1 in 15,000 live births), their risk is stated at 0.007%. Adapted from Ref. 24.

developed countries) where such testing is either inaccessible or is self-funded.[2] Since the cost of genetic testing and lack of genetic specialists remains a major obstacle, this can result in missed and delayed diagnosis, with adverse outcomes and increased costs in the long run, as well as increased economic difficulty for affected families. Where genetic testing is unavailable, the empiric risks listed in Table 25.3 can be used as a basis for counselling.[24]

Germline mutation carriers are at risk of second cancers later in life, including bone tumours, soft tissue sarcomas and melanoma. Such individuals need to avoid mutagenic agents or events, including minimising exposure to radiation (X-rays or CT scans), avoid radiotherapy and smoking, avoiding excess sun exposure and adopting a healthy lifestyle.[25,26] Where possible, lifelong clinical surveillance for second cancers is recommended through patient education, with advice to seek medical help if there are any suspicious signs or symptoms.

References

1. de Camargo B, de Oliveira SM, Rebelo MS, de Souza RR, Ferman S, Noronha CP, et al. Cancer incidence among children and adolescents in Brazil: first report of 14 population-based cancer registries. *Int J Cancer* 2010;**126**(3):715–20.

2. Rodriguez-Galindo C, Wilson MW, Chantada G, Fu L, Qaddoumi I, Antoneli C, et al. Retinoblastoma: one world, one vision. *Pediatrics* 2008;**122**(3):e763–7.
3. Darwich R, Ghazawi FM, Rahme E, Alghazawi N, Burnier JV, Sasseville D, et al. Retinoblastoma incidence trends in Canada: a National Comprehensive Population-Based Study. *J Pediatr Ophthalmol Strabismus* 2019;**56**(2):124–30.
4. Skalet AH, Gombos DS, Gallie BL, et al. Screening children at risk for retinoblastoma: consensus report from the american association of ophthalmic oncologists and pathologists. *Ophthalmology* 2018;**125**(3):453–8. https://doi.org/10.1016/j.ophtha.2017.09.001.
5. AlAli A, Kletke S, Gallie B, Lam WC. Retinoblastoma for paediatric ophthalmologists. *Asia Pac J Ophthalmol (Phila)* 2018;**7**(3):160–8.
6. Balmer A, Munier F. Differential diagnosis of leukocoria and strabismus, first presenting signs of retinoblastoma. *Clin Ophthalmol* 2007;**1**:431–9.
7. Gallie BL, Ellsworth RM, Abramson DH, Phillips RA. Retinoma: spontaneous regression of retinoblastoma or benign manifestation of the pathogenic allele? *Br J Cancer* 1982;**45**(4):513–21.
8. Dimaras H, Khetan V, Halliday W, et al. Loss of RB1 induces non-proliferative retinoma: increasing genomic instability correlates with progression to retinoblastoma. *Hum Mol Genet* 2008;**17**(10):1363–72. https://doi.org/10.1093/hmg/ddn024.
9. Mallipatna A, Gallie BL, Chévez-Barrios P, et al. Retinoblastoma. In: Amin MB, Edge SB, Greene FL, editors. *AJCC cancer staging manual*. Springer; 2017. p. 819–31 [chapter 68].
10. Mitter D, Ullmann R, Muradyan A, Klein-Hitpass L, Kanber D, Ounap K, et al. Genotype-phenotype correlations in patients with retinoblastoma and interstitial 13q deletions. *Eur J Hum Genet* 2011;**19**(9):947–58.
11. Knudson Jr AG. Pathogenic allele and cancer: statistical study of retinoblastoma. *Proc Natl Acad Sci U S A* 1971;**68**:820–3.
12. Friend SH, Bernards R, Rogelj S, Weinberg RA, Rapaport JM, Albert DM, et al. A human DNA segment with properties of the gene that predisposes to retinoblastoma and osteosarcoma. *Nature* 1986;**323**(6089):643–6.
13. Dimaras H, Corson TW, Cobrink D, White A, Zhao J, Munier FL, et al. *Retinoblastoma Nat Rev Dis Primers* 2015;**1**:15021.
14. Corson TW, Gallie BL. One hit, two hits, three hits, more? Genomic changes in the development of retinoblastoma. *Genes Chromosomes Cancer* 2007;**46**(7):617–34.
15. Rushlow D, Piovesan B, Zhang K, Prigoda-Lee NL, Marchong MN, Clark RD, et al. Detection of mosaic RB1 pathogenic alleles in families with retinoblastoma. *Hum Mutat* 2009;**30**(5):842–51.
16. Soliman SE, Racher H, Zhang C, MacDonald H, Gallie BL. Genetics and molecular diagnostics in retinoblastoma—an update. *Asia Pac J Ophthalmol (Phila)* 2017;**6**(2):197–207.
17. Ohtani-Fujita N, Dryja TP, Rapaport JM, Fujita T, Matsumura S, Ozasa K, et al. Hypermethylation in the retinoblastoma gene is associated with unilateral, sporadic retinoblastoma. *Cancer Genet Cytogenet* 1997;**98**(1):43–9.
18. Rushlow DE, Mol BM, Kennett JY, Yee S, Pajovic S, Thériault BL, et al. Characterisation of retinoblastomas without RB1 pathogenic alleles: genomic, gene expression, and clinical studies. *Lancet Oncol* 2013;**14**(4):327–34.
19. Schefler AC, Kleinerman RA, Abramson DH. Genes and environment: effects on the development of second malignancies in retinoblastoma survivors. *Expert Rev Ophthalmol* 2008;**3**(1):51–61.
20. Soliman SE, Dimaras H, Khetan V, et al. Prenatal versus postnatal screening for familial retinoblastoma. *Ophthalmology* 2016;**123**(12):2610–7. https://doi.org/10.1016/j.ophtha.2016.08.027.
21. Noorani HZ, Khan HN, Gallie BL, Detsky AS. Cost comparison of molecular versus conventional screening of relatives at risk for retinoblastoma. *Am J Hum Genet* 1996;**59**(2):301–7.
22. Berry JL, Xu L, Murphree AL, et al. Potential of aqueous humor as a surrogate tumor biopsy for retinoblastoma. *JAMA Ophthalmol* 2017;**135**:1221–30.
23. Gerrish A, Stone E, Clokie S, Ainsworth JR, Jenkinson H, McCalla M, et al. Non-invasive diagnosis of retinoblastoma using cell-free DNA from aqueous humour. *Br J Ophthalmol* 2019;**103**(5):721–4 [Erratum in: *Br J Ophthalmol* 2020;**104**(3):415–6].
24. Skalet AH, Gombos DS, Gallie BL, Kim JW, Shields CL, Marr BP, et al. Screening children at risk for retinoblastoma: consensus report from the American Association of Ophthalmic Oncologists and Pathologists. *Ophthalmology* 2018;**125**(3):453–8.
25. Bader JL, Meadows AT, Zimmerman LE, et al. Bilateral retinoblastoma with ectopic intracranial retinoblastoma: trilateral retinoblastoma. *Cancer Genet Cytogenet* 1982;**5**(3):203–13.
26. Brennan RC, Qaddoumi I, Billups CA, Kaluzny T, Furman WL, Wilson MW. Patients with retinoblastoma and chromosome 13q deletions have increased chemotherapy-related toxicities. *Pediatr Blood Cancer* 2016;**63**(11):1954–8.

Index

Note: Page numbers followed by *f* indicate figures, *t* indicate tables, and *b* indicate boxes.

A

ABCA4-retinopathy, 159, 207–215, 215–216*f*
Abnormal dentition, 105
aCGH. *See* Array comparative genomic hybridisation (aCGH)
Achromatopsia, 194–195, 195–196*f*, 197, 404
Achromic patch, 446, 447*f*, 447*t*
ACO2, 368–369
Acquired nystagmus, 404
Acute macular neuroretinopathy, 241*t*
Adams-Oliver syndrome, 327*b*
ADAMTSL4-related disorders, 149–150, 150*f*
Adaptive optics (AO), 59–60
Adeno-associated viral (AAV) vectors, 63
ADRP. *See* Autosomal dominant retinitis pigmentosa (ADRP)
Adult vitelliform macular dystrophy (AVMD), 219
ADVIRC. *See* Autosomal dominant vitreoretinochoroidopathy (ADVIRC)
Age-related macular degeneration (AMD), 261–265, 264*f*
AHUS. *See* Atypical hemolytic uremic syndrome (AHUS)
Albinism, 393–400, 394–398*f*, 394*t*, 399*t*, 404
Alkyl nitrite ("poppers") related maculopathy, 241*t*
Allied collagen vitreoretinopathies. *See* Stickler syndrome
Alpha (α-) mannosidosis, 131*t*
Alström syndrome, 204–205, 277, 278–279*f*, 279–280, 373
AMD. *See* Age-related macular degeneration (AMD)
American College of Medical Genetics and Genomics (ACMG), 13
Angiofibroma, 446, 447*t*
Aniridia, 389–391, 390*f*
Ankyloblepharon-ectodermal dysplasia-cleft lip/palate (AEC) syndrome, 94
Anophthalmia, 378–379, 378*f*
Anophthalmia–oesophageal–genital (AEG) syndrome, 379–380
Anterior segment developmental disorder, 97, 97*t*, 102*b*
  Axenfeld-Rieger spectrum (*see* Axenfeld-Rieger spectrum)
  Peters anomaly (*see* Peters anomaly)
  primary congenital glaucoma (*see* Primary congenital glaucoma)
  primary juvenile glaucoma (*see* Primary juvenile glaucoma)
  primary open angle glaucoma
Anti-aquaporin 4 (AQP4), 359
Anti-myelin oligodendrocyte glycoprotein (MOG), 359
ARB. *See* Autosomal recessive bestrophinopathy (ARB)
Array comparative genomic hybridisation (aCGH), 10
Arrayed primer extension (APEX), 7
ARRP. *See* Autosomal recessive retinitis pigmentosa (ARRP)
Astrocytic hamartoma, 446–448, 447*f*
Atriodigital syndrome. *See* Holt-Oram syndrome
Atypical hemolytic uremic syndrome (AHUS), 253–255, 258
Autoimmune retinopathy, 241*t*
Automated variant interpretation, 14–15, 16*t*
Autosomal dominant drusen, 250–251, 250–251*t*, 252–254*f*, 257, 259
Autosomal dominant inheritance, 1
Autosomal dominant optic atrophy, 373
Autosomal dominant optic neuropathy, 362–366, 363–365*f*
Autosomal dominant retinitis pigmentosa (ADRP), 162
Autosomal dominant vitreoretinochoroidopathy (ADVIRC), 221–222, 221*f*, 385
Autosomal recessive bestrophinopathy (ARB), 219–220, 220*f*
Autosomal recessive disease, 13
Autosomal recessive inheritance, 1–2
Autosomal recessive optic neuropathy, 368–369
Autosomal recessive retinitis pigmentosa (ARRP), 162
Axenfeld-Rieger spectrum, 105–108, 106*f*

B

Baclofen, 405
Bardet-Biedl syndrome, 25, 200, 270–272, 271*f*, 275–276, 373
Basal cell nevus syndrome (BCNS). *See* Naevoid basal cell carcinoma syndrome (NBCCS)
BBS. *See* Bardet-Biedl syndrome (BBS)
BCNS. *See* Basal cell nevus syndrome (BCNS)
Bear track patterns, 453
Behr syndrome, 363, 368–369
Benign variants, 17*t*
BEST1-related disorders, 217–223
Best disease. *See* Best vitelliform macular dystrophy (BVMD)
Bestrophinopathies. *See BEST1*-related disorders
Best vitelliform macular dystrophy (BVMD), 217–219, 217–219*f*
Bevacizumab, 247, 259, 437
Bietti crystalline dystrophy, 292–294, 293*f*
Bloch–Sulzberger syndrome. *See* Incontinentia pigmenti
Blue-cone monochromatism, 194, 196–198
BMP4 signalling, 381
Bornholm eye disease, 196, 198
Bradyopsia, 194, 196–198
Branchio-oculofacial syndrome (BOFS), 28*t*, 382*t*
Brittle cornea syndrome, 67, 70–72
BVMD. *See* Best vitelliform macular dystrophy (BVMD)

C

C3 glomerulopathy, 255
Café au lait spots, 430
Cancer predisposition syndrome, 453
Cancer susceptibility syndrome, 457
Cardiac rhabdomyomas, 446
Cardiomyopathy, 114, 115*t*, 204–205, 277, 305–306
Carrier testing, 22
CCDDs. *See* Congenital cranial dysinnervation disorders (CCDDs)
CCP. *See* Complement control proteins (CCP)
Central corneal opacification, 105
Central serous retinopathy (CSR), 255
Central visual function, 162–164
Cerebellar ataxia–areflexia–pes cavus–optic atrophy–sensorineural hearing loss (CAPOS) syndrome, 373
Cerebro-oculofacial-skeletal syndrome (COFS), 123*t*
Cerebroretinal microangiopathy with calcification and cysts, 327*b*, 330

Cerebrotendinous xanthomatosis (CTX), 129, 144–146, 144–145f
Cervico-oculoacoustic syndrome. *See* Wildervanck syndrome
CFEOM. *See* Congenital fibrosis of the extraocular muscles (CFEOM)
CHARGE syndrome, 379–380
Chediak-Higashi syndrome (CHS), 399
Chenodeoxycholic acid (CDCA), 146
Childhood lathosterolosis, 131t
Chorioretinal degeneration, 347–348
Choroidal neovascularisation, 232–235, 246, 261, 264–265, 295, 297f
Choroidal nodules/abnormalities, 430–432
Choroideremia, 172–175, 173–174f
CHRPE. *See* Congenital hypertrophy of the retinal pigment epithelium (CHRPE)
CHRRPE. *See* Combined hamartoma of the retina and RPE (CHRRPE)
CHS. *See* Chediak-Higashi syndrome (CHS)
Cilia, 268
Ciliopathies, 26, 159, 268, 268f
　Alstrom syndrome (*see* Alstrom syndrome)
　Bardet-Biedl syndrome (*see* Bardet-Biedl syndrome (BBS))
　Joubert syndrome (*see* Joubert syndrome)
　Usher syndrome (*see* Usher syndrome)
ClinVar, 4–5
COACH syndrome, 275–276
Coats disease, 327b, 330, 457
Coats plus syndrome, 327b
Cobalamin C deficiency, 313–314, 314–315f
Cockayne syndrome, 118–120
Coenzyme Q10, 305
COFS. *See* Cerebro-oculofacial-skeletal syndrome (COFS)
Cohen syndrome, 317, 318–319f, 319–320
Coloboma, 4, 26, 27–28t, 28f, 83, 189, 276, 379, 380f, 389, 449–451
Colonic polyposis, 432f
Combined hamartoma of the retina and RPE (CHRRPE), 449–451, 451f
Combined hamartomas of retina. *See* Congenital hamartoma of the retina
Complex microphthalmia, 383b
Cone/cone-rod dystrophies, 200–205, 200–205f
Cone dysfunction disorders, 194, 292
　achromatopsia, 194–195, 195–196f, 197–198
　blue-cone monochromatism, 196
　Bornholm eye disease, 196
　bradyopsia (*RGS9/R9AP*-associated retinopathy), 196–197
Congenital/childhood cataract, 113
Congenital cranial dysinnervation disorders (CCDDs), 407
　congenital fibrosis of the extraocular muscles (*see* Congenital fibrosis of the extraocular muscles (CFEOM))
　Duane retraction syndrome (*see* Duane retraction syndrome)
　horizontal gaze palsy and progressive scoliosis (*see* Horizontal gaze palsy and progressive scoliosis (HGPPS))
　Moebius syndrome (*see* Moebius syndrome)

Congenital fibrosis of the extraocular muscles (CFEOM), 408–411, 408–409f
Congenital grouped pigmentation of the RPE, 453, 454f
Congenital hamartoma of the retina, 435
Congenital heart disease, 275
Congenital hereditary endothelial dystrophy (CHED), 73, 81
Congenital hypertrophy of the retinal pigment epithelium (CHRPE), 453–455, 454f
Congenital retinal non-attachment, 324
Congenital rubella syndrome, 373
Congenital stationary night-blindness (CSNB), 181–187, 181–186f, 187t, 404
Congenital stromal dystrophy, 83
Corneal abnormalities, 67
　brittle cornea syndrome (*see* Brittle cornea syndrome)
　corneal dystrophies (*see* Corneal dystrophies)
　ectodermal dysplasias (*see* Ectodermal dysplasias, keratopathy)
　microcornea, 67
　primary megalocornea (*see* Primary megalocornea)
Corneal clouding, 67
Corneal dystrophies, 67, 73, 74t
　corneal endothelial dystrophies (*see* Corneal endothelial dystrophies)
　epithelial-stromal TGFBI corneal dystrophies (*see* Epithelial-stromal TGFBI corneal dystrophies)
　Fuchs endothelial corneal dystrophy, 73
　Meesmann epithelial corneal dystrophy (*see* Meesmann epithelial corneal dystrophy (MECD))
Corneal endothelial dystrophies, 67, 81–82
　congenital hereditary endothelial dystrophy, 81
　defects, 81
　Fuchs endothelial corneal dystrophy, 81
　Harboyan syndrome, 81
　posterior polymorphous corneal dystrophy, 81
Corneal verticillata, 89, 90f
Cortical dysplasias, 446, 447t
Costeff syndrome, 363–365
CSNB. *See* Congenital stationary night-blindness (CSNB)
CSR. *See* Central serous retinopathy (CSR)
Cuticular drusen, 251t, 252–253
Cysteamine, 87
Cystinosis, 67, 85–88
Czech dysplasia, 345

D

Dandy-Walker malformation, 275–276
Danon disease, 204–205
Deafness, 70, 227, 231–232t, 281–282, 302, 339, 340t, 371, 373, 382t, 415, 422–423, 434
Demyelinating optic neuritis, 359
Dense deposit disease, 255
Descemet membrane endothelial keratoplasty (DMEK), 82

Descemet stripping automated endothelial keratoplasty (DSAEK), 82
DHRD. *See* Doyne honeycomb retinal dystrophy (DHRD)
Diabetes insipidus, 277
Diabetes insipidus, diabetes mellitus, optic atrophy and deafness (DIDMOAD). *See* Wolfram syndrome
Dominant optic atrophy. *See* Autosomal dominant optic neuropathy
Dominant optic atrophy plus, 362–363, 366
Donnai–Barrow syndrome, 345–346
Dot-shaped cataract, 348
Doyne honeycomb retinal dystrophy (DHRD), 250–251
D-penicillamine, 89, 131t
Drusen, 207, 250, 251t, 255, 256f, 258f, 261
Duane-radial ray syndrome, 414–415
Duane retraction syndrome, 412–416, 412–416f
Dyschromatopsia, 362

E

Early-onset macular drusen (EOMD), 251t
Ectodermal dysplasias, 67, 93–95
　hypohidrotic ectodermal dysplasia, 93–94
　Keratitis-ichthyosis-deafness syndrome, 95–96
Ectopia lentis, 147, 148t
　ADAMTSL4-related disorders (*see ADAMTSL4*-related disorders)
　homocystinuria (*see* Homocystinuria)
　Marfan syndrome (*see* Marfan syndrome)
　Weill–Marchesani syndrome (*see* Weill–Marchesani syndrome)
Ectopia lentis et pupillae, 148t, 149, 150f
Ectopic calcification, 449
Ectrodactyly-ectodermal dysplasia-cleft lip/palate (EEC) syndrome, 94
Ehlers Danlos syndrome VI, 70
Electrooculogram (EOG), 34, 40–42
Electrophysiology, 33
Electroretinogram (ERG), 34
Enhanced S-cone syndrome, 176–179, 176–177f, 179f
Enucleation, 460–461
EOMD. *See* Early-onset macular drusen (EOMD)
EPI-743, 360
Epikeratoplasty, 71
Epithelial basement membrane dystrophy, 74t
Epithelial recurrent erosion dystrophy, 83
Epithelial–stromal TGFBI corneal dystrophies, 77–80
　granular corneal dystrophy type I, 77
　granular corneal dystrophy type II, 77
　lattice corneal dystrophy type I, 77
　Meretoja syndrome, 79b
　Reis–Bucklers corneal dystrophy, 77
　Thiel–Behnke corneal dystrophy, 77
Erosive vitreoretinopathy, 347–348
Everolimus, 448
Exome sequencing, 8–9
External beam radiotherapy (EBRT), 460
Eye Clinic Liaison Officer (ECLO), 23b

F

Fabry disease, 89
Facial dysmorphism, 317
Facio-oculo-acustico-renal (FOAR) syndrome, 345–346
Facioscapulohumeral muscular dystrophy, 327b, 330
Familial adenomatous polyposis (FAP), 453–454, 455f
Familial exudative vitreoretinopathy (FEVR), 324–328, 324–325t, 326f, 327b
FAP. See Familial adenomatous polyposis (FAP)
FEVR. See Familial exudative vitreoretinopathy (FEVR)
FHONDA. See Foveal hypoplasia, optic nerve decussation defects and anterior segment dysgenesis (FHONDA)
Fish-eye disease, 89–91
Flash electroretinogram, 36–38, 309–311
Fleck stromal corneal dystrophy, 83
FMNS. See Fusion maldevelopment nystagmus syndrome (FMNS)
Foetal alcohol syndrome, 26, 28t
Foetal valproate syndrome, 26
Foveal hypoplasia, 404, 404t
Foveal hypoplasia, optic nerve decussation defects and anterior segment dysgenesis (FHONDA), 393
Fraser syndrome, 27–28, 28t
Friedreich ataxia, 373
Fuchs endothelial corneal dystrophy (FECD), 81
Full-field electroretinography, 164, 292, 304
Functional visual loss, 34, 40, 43, 241t, 357
Fundus albipunctatus, 186
Fundus-related perimetry, 164
Fusion maldevelopment nystagmus syndrome (FMNS), 403

G

Gabapentin, 405
GACI. See Generalised arterial calcification of infancy (GACI)
GAGs. See Glycosaminoglycans (GAGs)
Galactokinase deficiency, 131t
Galactosaemia, 129, 140–142
Gelatinous drop-like corneal dystrophy, 83
Generalised arterial calcification of infancy (GACI), 299–300
Genetic disorders, 1–4, 2f
  autosomal dominant inheritance, 1
  autosomal recessive inheritance, 1–2
  mendelian inheritance, 1–4, 2f
  mitochondrial inheritance, 3–4
  multi-factorial disorders, 4
  X-linked recessive inheritance, 2
Genetic disorders mimicking age-related macular disease. See Drusen formation
Genetic testing, 5, 7, 13–14, 21–23
  barriers, 21
  benefits, 21
  consent, 23–24
  exome analysis, 14
  gene panel, 14
  genome analysis, 14
  massively parallel sequencing, 7–10
  Sanger sequencing, 7, 8f
  single exon analysis, 14
  single gene analysis, 14
  single variant analysis, 14
  types
    carrier testing, 22
    diagnostic tests, 22
    predictive tests, 22
    reproductive tests, 22
    research testing, 22
    segregation analysis, 22
Genetic variant, 4–5, 13–19
Genome aggregation database (gnomAD), 4–5, 15–17, 16–17t
Genome sequencing, 9–10
Genome-wide association studies (GWAS), 262
Geographic atrophy, 261
Gillespie syndrome, 391
Glaucoma, 432
Glycogen storage disease type V. See McArdle disease
Glycosaminoglycans (GAGs), 323
Goldenhar syndrome, 416
Goldmann-Favre syndrome. See Enhanced S-cone syndrome
Goldmann visual field (GVF), 167f
Gorlin syndrome. See Naevoid basal cell carcinoma syndrome (NBCCS)
Granular corneal dystrophy type I, 77
Granular corneal dystrophy type II, 77
Growth hormone deficiency, 277
GWAS. See Genome-wide association studies (GWAS)
Gyrate atrophy, choroid and retina, 288–290, 288–289f

H

Hallermann Streiff syndrome, 30
Harding disease, 357–359
Hearing loss, 275
Hemangioblastoma, 439, 440–443f, 443–444, 445b
Hermansky-Pudlak syndrome (HPS), 398–399
Heteroplasmy, 302b, 359
HGPPS. See Horizontal gaze palsy and progressive scoliosis (HGPPS)
High-throughput genomic technologies, 355
High-throughput sequencing strategies, 343–345
Holt-Oram syndrome, 414–415
Homocystinuria, 157–158
Horizontal gaze palsy and progressive scoliosis (HGPPS), 418–421
HOXA1, 415
HPS. See Hermansky-Pudlak syndrome (HPS)
Human gene mutation database (HGMD), 4–5
Hydrolethalus syndrome, 275–276
Human genome variation society (HGVS), 4–5
Hydrometrocolpos, 271
Hyperferritinemia-cataract syndrome, 129, 137–138
Hyperglycinuria, 131t
Hyperlysinemia, 148t
Hypermetropia, 176, 240
Hypertriglyceridemia, 277
Hypoplasia, 409
Hypospadias, 105
Hypothyroidism, 277

I

Idebenone, 359–360, 369
Idiopathic infantile nystagmus, 403–405
Imaging techniques, ophthalmic phenotyping
  adaptive optics, 59–60
  near-infrared fundus autofluorescence, 57
  optical coherence tomography, 53–54
  quantitative fundus autofluorescence, 57–59
  short-wavelength fundus autofluorescence, 55–57
Inborn errors of metabolism, 89–91, 129–132, 159
  cystinosis, 86–88
  Fabry disease, 89
  mucolipidosis, 85–86
  mucopolysaccharidoses, 85
  peroxisomal biogenesis disorders, 130b
  Tangier disease, 91–92
  tyrosinaemia type II, 88–89
  Wilson disease, 89
Incontinentia pigmenti, 334–337, 334t
Infertility, 277
International committee for classification of corneal dystrophies (IC3D), 73
International Society for Clinical Electrophysiology of Vision (ISCEV), 194
Intravitreal delivery, 65
Intrinsic central nervous system tumours, 435
Iron responsive element (IRE), 138
ISCEV. See International Society for Clinical Electrophysiology of Vision (ISCEV)
Isolated foveal hypoplasia, 393
Isolated nodular palpebral neurofibromas, 433

J

Jalili syndrome, 200, 204–205
Jaw cysts (odontogenic keratocysts), 449
Jeune thoracic dysplasia, 271
Joubert syndrome, 274–276

K

Kabuki syndrome, 27–28, 27–28t, 28f
Kayser–Fleischer ring, 89, 90f
KCNV2-retinopathy (cone dystrophy with supernormal rod ERG), 33, 45–46, 46f
Kearns–Sayre syndrome (KSS), 4, 302, 425, 426f
Keratitis-ichthyosis-deafness (KID) syndrome, 95–96
Kestenbaum-Anderson procedure, 405
Kjellin syndrome, 226t, 231–232t
KIF11-related disorders, 331–332, 331f
Klinefelter syndrome, 337
Klippel-Feil abnormality, 415
Kniest dysplasia, 343, 344f, 345
Knobloch syndrome, 351–354
Knudson two-hit hypothesis, 460b
KSS. See Kearns–Sayre syndrome (KSS)

L

Late-onset retinal degeneration (L-ORD), 251*t*, 252, 257*f*
Lattice corneal dystrophy type I, 77
Laser-induced retinal injury, 241*t*
LCAT deficiency (lecithin-cholesterol acyltransferase deficiency), 89–91, 91*f*
LCHAD deficiency (long-chain 3-hydroxyacyl-CoA dehydrogenase deficiency), 159, 306–307, 307*f*
Leber congenital amaurosis (LCA), 189–193, 190*f*, 404
   CRB1-associated retinal disease, 191*f*
   RPE65-associated retinal disease, 192*f*
Leber hereditary optic neuropathy (LHON), 302, 356–360, 425
Lecithin-cholesterol acyltransferase-related metabolic disease, 89–91
Legius syndrome, 430
Leigh syndrome, 368
Lens opacification, 435
Lens-related secondary glaucoma, 155
Lenticular myopia, 155
Lentivirus, 64*t*
LHON. *See* Leber hereditary optic neuropathy (LHON)
Linear skin defects with multiple congenital anomalies (LSDMCA1), 383
Lisch epithelial corneal dystrophy, 83
Lisch nodules, 430
Long-chain 3-hydroxyacyl CoA dehydrogenase (LCHAD) deficiency, 159, 306, 308
Lowe oculocelebrorenal syndrome, 129, 133–135
Lysosome-related organelles (LROs), 398

M

MAC. *See* Microphthalmia-anophthalmia-ocular coloboma (MAC)
Macrocephaly, 449
Macular corneal dystrophy, 83
Mainzer-Saldino syndrome, 275–276
Manifest latent nystagmus, 403
Marfan syndrome, 151–154, 152*t*
Margin-controlled excision. *See* Mohs micrographic surgery
Massively parallel sequencing/next-generation sequencing, 8
   exome sequencing, 8–9
   gene panels, 8
   genome sequencing, 9–10
   sequencing-by-synthesis, 7–8
   short-read approach, 7–8
   targeted sequencing, 8
Maternally-inherited diabetes and deafness (MIDD), 227, 229*f*, 302, 302*b*, 304–305, 425
Matthew Wood syndrome, 379–380
McArdle disease, 234*f*
Meckel syndrome, 275–276
Medulloblastoma, 449
Meesmann epithelial corneal dystrophy (MECD), 74–76, 75*f*
Megalocornea, 67–69, 97
Melanin, 395
Melanosomes, 395
MEK inhibitors, 433
MELAS. *See* Mitochondrial encephalopathy lactic acidosis and stroke-like episodes (MELAS)
Memantine, 405
Membranoproliferative glomerulonephritis type II, 255
Mendelian conditions, 1
Mendelian inheritance, 1–4
Mendelian retinal disorders, 159
Meningiomas, 435
Menke disease, 131*t*
Mental retardation, truncal obesity, retinal dystrophy and micropenis syndrome (MORM), 268*f*
Meretoja syndrome, 67, 79*b*, 79*f*
Merlin syndrome. *See* Neurofibromatosis type 2 (NF-2)
3-Methylglutaconic aciduria (Costeff syndrome), 363–365
Microcephaly, 332
Microcornea, 379, 380*f*
Microcornea myopia chorioretinal atrophy and telecanthus (MMCAT), 148*t*
Micropenis, 121, 121*t*, 271
Microperimetry, 164
Microphthalmia, 379, 389
Microphthalmia–anophthalmia–ocular coloboma (MAC), 378–384, 379*b*
Microphthalmia, dermal aplasia and sclerocornea (MIDAS), 383
Microspherophakia, 155
MIDD. *See* Maternally-inherited diabetes and deafness (MIDD)
Mitochondrial DNA (mtDNA), 356, 356*f*
Mitochondrial encephalopathy lactic acidosis and stroke-like episodes (MELAS), 373, 425
Mitochondrial inheritance, 3–4
Mitochondrial membrane protein-associated neurodegeneration (MPAN), 369
Miyake disease. *See RP1L1* occult macular dystrophy
Mizuo phenomenon, 236
MLPA. *See* Multiplex ligation-dependent probe amplification (MLPA)
MMCAT. *See* Microcornea myopia chorioretinal atrophy and telecanthus (MMCAT)
Moebius syndrome, 422–423
MOG-IgG associated optic neuritis, 359
Mohs micrographic surgery, 452
Mononeuropathy, 435
Mosaicism, 432–433, 435
MPAN. *See* Mitochondrial membrane protein-associated neurodegeneration (MPAN)
MPS I-H (Hurler), 86*t*
MPS I-HS (Hurler–Sheie), 86*t*
MPS II (Hunter), 86*t*
MPS III (Sanfilippo A), 86*t*
MPS IIIB (Sanfilippo B), 86*t*
MPS IIIC (Sanfilippo C), 86*t*
MPS IIID (Sanfilippo D), 86*t*
MPS I-S (Sheie), 86*t*
MPS IV (Morquio), 86*t*
MPS IX, 86*t*
MPS VI (Maroteaux–Lamy), 86*t*
MPS VII (Sly), 86*t*
MRI. *See* Magnetic resonance imaging (MRI)
MTP-131, 360
Mucolipidosis, 67, 85–86
Mucopolysaccharidoses (MPS), 67, 85
Multidisciplinary team (MDT), 13
Multifactorial disorders, 4
Multifocal Best disease, 220
Multifocal electroretinogram, 38
Multi-luminance mobility test (MLMT), 64–65
Multiplex ligation-dependent probe amplification (MLPA), 10, 343–345
Myopia, 317
Myotonic dystrophy, 233*f*

N

Naevoid basal cell carcinoma syndrome (NBCCS), 449–452, 450*t*, 451*f*
Nance–Horan syndrome, 127–128
Nanophthalmia, 385–386, 387*f*
NARP. *See* Neuropathy, ataxia and retinitis pigmentosa (NARP)
NBCCS. *See* Naevoid basal cell carcinoma syndrome (NBCCS)
NBIA. *See* Neurodegeneration with brain iron accumulation (NBIA)
NCMD. *See* North Carolina macular dystrophy (NCMD)
Near-infrared fundus autofluorescence (NIR-AF), 57
Nephroblastoma, 389–391
Nephronophthisis, 268*f*, 271, 275
Neuhauser syndrome, 69
Neurodegeneration with brain iron accumulation (NBIA), 369
Neurofibromas, 430
Neurofibromatosis type 1 (NF-1), 430–433, 430*t*
Neurofibromatosis type 2 (NF-2), 434–438, 434*t*
Neuromyelitis optica (NMO), 359
Neuronal ceroid lipofuscinoses (Batten diseases), 309–312
Neuropathy, ataxia and retinitis pigmentosa (NARP), 425
Neutropenia, 317
NF-1. *See* Neurofibromatosis type 1 (NF-1)
NF-2. *See* Neurofibromatosis type 2 (NF-2)
Night-blindness, 23
NMO. *See* Neuromyelitis optica (NMO)
Non-*RP1L1* occult macular dystrophy, 243
Nonsyndromic congenital cataract, 114–117, 115*t*
Nonsyndromic retinitis pigmentosa, 162–170, 163–164*f*, 168*t*
Normal-tension glaucoma, 104, 366
Norrie disease, 2, 329–330, 329*f*
North Carolina macular dystrophy (NCMD), 246–248, 247–248*f*

NR2E3-associated dominant retinitis pigmentosa, 178–179, 178f
Nystagmus, 403–405

O

Occipital skull abnormalities, 353
Occult macular dystrophy (OMD/OCMD), 241–244, 241t
OCT. See Optical coherence tomography (OCT)
Ocular albinism, 400
Ocular coloboma, 379, 389
Oculo-auriculo-vertebral syndrome. See Goldenhar syndrome
Oculoauriculovertebral (Goldenhaar) syndrome, 30
Oculocerebrorenal (Lowe) syndrome, 118, 131t, 133–135
Oculocutaneous albinism, 50f, 187t, 393, 394t
Oculodentodigital syndrome, 27t
Oculofaciocardiodental syndrome, 125–126
Oculomotor development, 409–410
Oculomotor synkinesis, 408
Oguchi disease, 186
Okihiro syndrome. See Duane-radial ray syndrome
OMD/OCMD. See Occult macular dystrophy (OMD/OCMD)
ON-bipolar cells, 37, 47–48, 167f, 181–183
Ophthalmic electrodiagnostic tests, 34–38
  electrooculogram (see Electrooculogram)
  electroretinogram (see Electroretinogram (ERG))
  multifocal electroretinogram (see Multifocal electroretinogram)
  pattern electroretinogram (see Pattern electroretinogram)
  visual evoked potential (see Visual evoked potential (VEP))
Ophthalmic imaging techniques
  adaptive optics, 53
  near-infrared fundus autofluorescence, 53
  optical coherence tomography, 53
  quantitative fundus autofluorescence, 53
  short-wavelength fundus autofluorescence, 53
Ophthalmic syndromes, 25–30, 27–28t
Ophthalmoparesis, 425
Ophthalmoplegia, 427
Optical coherence tomography (OCT), 53–54, 165–166
Optic nerve gliomas, 432
Optic nerve hypoplasia, 404
Oral-facial-digital syndrome (OFD), 268f
Oral manifestations, 275
Oscillopsia, 404
Osteomas, 453
Osteoporosis, 144–145, 157, 290, 325, 327–328
Osteoporosis pseudo-glioma syndrome, 325

P

Palmar/plantar pits, 449
Pattern dystrophy, 225–235, 226t
Pattern electroretinogram, 39–40

Peau d'orange retinopathy, 227, 231–232t, 295, 296f, 298
PEO. See Progressive external ophthalmoplegia (PEO)
Peripapillary telangiectasia, 357
Peroxisome biogenesis disorder 14B, 131t
Persistent fetal vasculature, 323, 327–328, 327b, 335, 457
Peters anomaly, 109–110
Phaeochromocytoma, 439, 443
Phakomatoses, 429
  neurofibromatosis type 1 (see Neurofibromatosis type 1 (NF-1))
  neurofibromatosis type 2 (see Neurofibromatosis type 2 (NF-2))
  tuberous sclerosis complex (see Tuberous sclerosis complex (TSC))
  Von Hippel–Lindau disease (see Von Hippel–Lindau disease (VHL))
Phase I studies/trials, 64
Phase II studies/trials, 64
Phase III studies/trials, 63–64
Phenylketonuria, 285–287
Phototherapeutic keratectomy, 76, 79–80
Pigmentary retinopathy, 23, 425
Pierre Robin sequence, 25, 339, 342–343
Pilocytic astrocytomas, 432
Pituitary abnormalities, 105
Polygenic risk score (PRS), 102b, 103f, 262–264, 264f
Polypoidal choroidal vasculopathy, 261
Poretti-Boltshauser syndrome, 275–276
Posterior amorphous corneal dystrophy, 83
Posterior microphthalmia. See Nanophthalmia
Posterior amorphous corneal dystrophy, 74t, 83
Posterior polymorphous corneal dystrophy (PPCD), 73, 81
Predictive testing, 22
Primary congenital glaucoma, 97t, 98–100, 99–100f, 102–103
Primary hyperoxaluria, 293
Primary juvenile glaucoma, 97t, 102–104
Primary megalocornea, 68–69
Primary open angle glaucoma, 102b
Progressive external ophthalmoplegia (PEO), 425–427, 427t
Pseudo-glioma, 324
Pseudoxanthoma elasticum (PXE), 227, 230f, 295–300, 296–300f
Ptosis, 425, 427
PXE. See Pseudoxanthoma elasticum (PXE)
Pyridoxine, 157–158, 290
Pyridoxine hydrochloride, 290

Q

Quantitative fundus autofluorescence (qAF), 57–59
Quantitative polymerase chain reaction (qPCR), 10

R

Recessive optic atrophy. See Autosomal recessive optic neuropathy

Refsum disease, 192, 285t, 286b
Reis–Bucklers corneal dystrophy, 77
Renal cell carcinoma, 439, 443, 444f, 445b, 446
Renal coloboma syndrome, 380–381
Reproductive genetic testing, 22
Retinal abnormalities, 427
Retinal angiomatous proliferation, 261
Retinal dysfunction, 270
Retinal dysplasia, 324
Retinal dystrophies, 159, 268
Retinal ganglion cells, 34
Retinal gene therapy, 63–66, 64t, 65–66f
Retinal nerve fibre layer (RNFL), 358f
Retinitis pigmentosa (RP), 33, 162, 167f
Retinoblastoma, 443, 457–462, 458f, 459t, 461f, 462t
Retino-cortical misrouting, 393–395
Retino-cortical pathway, 34
Retinoma, 457
Retinopathy of prematurity, 323–324, 327b, 335–337
Retinoschisis, 238
Retrobulbar optic neuritis, 241t
RGS9/R9AP-associated retinopathy. See Bradyopsia
Rhizomelic chondroplasia punctata, 129
Rieger syndrome, 27–29, 27t
Rod-cone dystrophy, 162
Rod monochromacy, 194
RP. See Retinitis pigmentosa (RP)
RP1L1 occult macular dystrophy, 241–242
RPE65-retinopathy, 159
Rubinstein-Taybi syndrome, 28f, 28t

S

Sanger sequencing, 7
Schnyder corneal dystrophy, 83
Schwannomas, 434
Schwannomatosis predisposition syndrome, 434
Scoliosis, 420
S-cone monochromacy, 194
SEDC. See Spondyloepiphyseal dysplasia congenital (SEDC)
Segregation testing, 22
Senior-Loken syndrome, 275–276
Sensorineural hearing loss, 342
Severe early childhood onset retinal dystrophy (SECORD). See Leber congenital amaurosis (LCA)
SFD. See Sorsby fundus dystrophy (SFD)
Short-wavelength fundus autofluorescence (SW-AF), 55–57
Sialidosis type II, 131t
Sirolimus, 448
Situs inversus configuration, 347–348
Sjögren-Larsson syndrome, 293
Skeletal abnormalities, 449
Skeletal dysplasia, 275
Skin hypopigmentation, 398
Smith-Lemli-Opitz syndrome, 123t
Solar retinopathy, 241t
Sorsby fundus dystrophy (SFD), 251–252, 251t, 255–256f
Spinocerebellar ataxia, 200

Spondyloepiphyseal dysplasia congenital (SEDC), 343, 345, 345*f*
Sporadic aniridia, 389–391
Stargardt disease, 9–10, 18, 38, 40, 44–45, 44*f*, 57, 59*f*, 60, 207, 225*f*, 226–227, 309, 310*f*
Stickler syndrome, 2–3, 25–26, 339–346, 341–342*f*
and allied collagenopathies, 340*t*
Stomatin-deficient cryohydrocytosis, 130*b*, 131*t*
Strabismus, 240, 407, 411
Structural variant analysis, 10–11, 10*f*
Subretinal gene delivery, 66
Sulphite oxidase deficiency, 148*t*
Syndromic cataract, 118
cerebrotendinous xanthomatosis (*see* Cerebrotendinous xanthomatosis (CTX))
Cockayne syndrome (*see* Cockayne syndrome)
galactosaemia (*see* Galactosaemia)
hyperferritinemia-cataract syndrome (*see* Hyperferritinemia-cataract syndrome)
Lowe oculocerebrorenal syndrome (*see* Lowe oculocerebrorenal syndrome)
Nance-Horan syndrome, 118
oculofaciocardiodental syndrome (*see* Oculofaciocardiodental syndrome)
Warburg Micro syndrome (*see* Warburg Micro syndrome)

T

Tangier disease, 91–92
Temtamy syndrome, 382*t*
Thiamine-responsive anaemia, 373
Thiel–Behnke corneal dystrophy, 77
Traboulsi syndrome, 148*t*

Tuberous sclerosis complex (TSC), 446–448
Tyrosinaemia type II, 88–89

U

Ultrasound biomicroscopy (UBM), 110*f*
Urinary tract abnormalities, 373
Usher syndrome, 14, 281–284, 281*t*, 282–284*f*

V

Variant interpretation
benign nature, 15
PM2, 15–17
PM5, 15
PP3, 17–18
PS1, 15
PS4, 17
PVS1, 15
Variants of uncertain significance (VUS), 169
VEP. *See* Visual evoked potential (VEP)
Vestibular schwannoma, 434–435, 434*t*, 437
VHL. *See* Von Hippel-Lindau disease (VHL)
Vismodegib, 452
Visual electrophysiology, 33, 378
Visual evoked potential (VEP), 34, 42–43, 393–395, 396–397*f*
Vitamin A deficiency, 187*t*, 250*t*, 379
Vitamin A supplementation, 169–170
Vitreoretinal degeneration, 351
Vitreoretinopathy with phalangeal epiphyseal dysplasia (VPED), 345
Vitreous veils, 236, 239*f*
Von Hippel–Lindau disease (VHL), 439–445, 441*f*, 445*b*
VPED. *See* Vitreoretinopathy with phalangeal epiphyseal dysplasia (VPED)
VUS. *See* Variants of uncertain significance (VUS)

W

Waardenburg anophthalmia syndrome, 382*t*
Wagner disease, 347–350, 349*f*
Wagner vitreoretinopathy, 345
WAGR syndrome. *See* Wilms tumour, aniridia, genitourinary abnormalities and mental retardation (WAGR syndrome)
Walker-Warburg syndrome, 327
Warburg Micro syndrome, 121–123, 121*t*, 123*t*, 379–380
Wegener granulomatosis, 359
Weill–Marchesani syndrome, 155–156
WFS1-associated deafness, 373
Wildervanck syndrome, 415
Wilms tumour, aniridia, genitourinary abnormalities and mental retardation (WAGR syndrome), 389–391
Wilson disease, 89
WNT signalling, 323, 325–326, 326*f*, 410
Wolfram syndrome, 363–365, 368, 371–374
Wolfram-like syndrome, 371–374, 373*f*

X

X chromosome inactivation, 329
X-linked endothelial corneal dystrophy, 84
X-linked ocular albinism, 393, 396*f*
X-linked recessive inheritance, 2
X-linked retinitis pigmentosa (XLRP), 2, 23, 165
X-linked retinoschisis, 2, 236–240, 237*f*
XLRP. *See* X-linked retinitis pigmentosa (XLRP)

Z

Zellweger syndrome, 18, 129, 131*t*
Zika virus, 26

Index of Genes

Note: Page numbers followed by *f* indicate figures, *t* indicate tables, and *b* indicate boxes.

A

AASS, 148*t*
ABAT, 427*t*
ABCA1, 91
ABCA4, 9–10, 15, 38, 40, 44–46, 44*f*, 46*f*, 49*f*, 57–60, 59*f*, 159, 168*t*, 169–170, 200–205, 207–214, 208–216*f*, 225, 225*f*, 226*t*, 229, 231*t*, 241*t*
ABCC6, 227, 231–232*t*, 295, 298–300
ACO2, 362–365, 368–369
ADAM9, 203*f*
ADAMTS10, 148*t*, 155
ADAMTS17, 107, 148*t*, 156
ADAMTS18, 148*t*
ADAMTSL4, 148*t*, 149, 150*f*, 153
ADGRV1, 281*t*, 282
AFG3L2, 363–365
AGBL1, 74*t*, 81–82
AGBL5, 168*t*
AHR, 404*t*
AIPL1, 203*f*
ALDH1A3, 381–383, 382*t*
ALMS1, 277, 279
AP3B1, 394*t*, 399
APC, 453–454
AP3D1, 394*t*, 399
ARHGAP31, 327*b*
ARHGEF18, 190*t*
ARL6, 271
ARMS2, 261–262, 263*f*
ARSB, 86*t*
ASPH, 148*t*
ATF6, 45, 197
ATOH7, 325–326, 325*t*, 383*b*
ATP7A, 131*t*
ATP7B, 131*t*

B

BBIP1, 272*f*
BBS1, 168*t*, 270*f*, 271
BBS2, 168*t*, 271
BBS4, 272*f*
BBS5, 272*f*
BBS7, 272*f*
BBS8, 271
BBS10, 270*f*, 271
BBS17, 271
BCOR, 26–29, 125–126, 382*t*, 383, 383*b*
BEST1, 44*f*, 45, 168*t*, 212, 217, 217–221*f*, 221–223, 225–227, 226*t*, 229, 231*t*, 385
BFSP1, 114
BFSP2, 114, 115*t*
B3GALTL, 110, 263*f*
BLOC1S3, 394*t*, 399
BLOC1S5, 394*t*, 399
BLOC1S6, 394*t*, 399
BLOC1S8/DTNBP1, 394*t*, 399
BLOC2S1/HPS3, 394*t*, 399
BLOC2S2/HPS5, 394*t*, 399
BLOC2S3/HPS6, 394*t*, 399
BLOC3S1/HPS1, 394*t*, 399–400
BLOC3S2/HPS4, 394*t*, 399

C

CABP4, 186
CACNA1A, 405
CACNA2D4, 203*f*
CACNA1F, 45–46, 48*f*, 182*f*, 186, 405
CBS, 148*t*, 157
CDH3, 168*t*
CDH23, 281*t*, 282
CDHR1, 45–46, 168*t*
CEP290, 9–10, 189–193, 190*f*, 190*t*, 271, 276
CERKL, 45–46, 168*t*
CFH, 251*t*, 252–255, 261–262, 263*f*
CHD7, 28*t*, 382*t*
CHM, 172–174
CHN1, 414
CHRDL1, 67–69, 68*f*
CHST6, 74*t*, 83
CIB2, 281*t*
CISD2, 371, 374
CLN1, 309, 311
CLN2, 309, 311–312
CLN3, 168*t*, 309–311, 310–311*f*
CLRN1, 281*t*, 282
CNGA1, 168*t*
CNGA3, 45, 46*f*, 63, 197–198
CNGB1, 168*t*, 186–187
CNGB3, 45, 63, 196*f*, 197–198
COL2A1, 25, 28*t*, 339, 340*t*, 341*f*, 342–345
COL4A1, 97*t*, 105–107
COL4A5, 131*t*
COL5A1, 148*t*
COL5A2, 148*t*
COL8A2, 74*t*, 81–82
COL11A1, 25, 28*t*, 339, 340*t*, 341–342*f*, 342–345
COL17A1, 74*t*
COL18A1, 148*t*, 351, 353
COL25A1, 415
Complex I [NADH dehydrogenase], 355, 356*f*, 359, 369
Complex II [succinate dehydrogenase], 355
Complex III [ubiquinol cytochrome *c* oxidoreductase], 355, 356*f*
Complex IV [cytochrome *c* oxidase], 355, 356*f*, 427
Complex V [adenosine triphosphate (ATP) synthase], 355, 356*f*
C8orf37, 168*t*, 271
C10orf2, 427*t*
C12orf57, 382*t*, 383
C19orf12, 369
CPAMD8, 97*t*, 103, 105, 107, 148*t*
C1QTNF5, 251*t*, 258
CRB1, 47, 49*f*, 168*t*, 189–191, 190*t*, 190–191*f*, 236
CREBBP, 28*t*
CRX, 44*f*, 45–46, 168*t*, 178, 179*f*, 189–191, 201–204, 202*f*, 212, 244
CRYAA, 114, 115*t*
CRYAB, 114, 115*t*
CRYBA1, 114, 115*t*
CRYBA2, 114, 115*t*
CRYBA4, 114, 115*t*
CRYBB1, 114, 115*t*
CRYBB2, 114, 115*t*
CRYBB3, 114, 115*t*
CRYGB, 114, 115*t*
CRYGC, 114, 115*t*
CRYGD, 114, 115*t*
CRYGS, 115*t*
CTC1, 327
CTNNA1, 212, 226*t*, 227, 228*f*, 231*t*
CTNNB1, 325–326, 325*t*
CTNS, 86–87
CWC27, 168*t*
CYP27A1, 131*t*, 144–146
CYP51A1, 131*t*
CYP1B1, 97*t*, 98, 103, 110
CYP4V2, 292–293

D

DCN, 74*t*
DGUOK, 427*t*

DHCR7, 123*t*
DHDDS, 168*t*
DHX38, 168*t*
DMPK, 231–232*t*, 232
DNA2, 427*t*
D4Z4, 327*b*

E

ECEL1, 415
EDA, 93–94
EDAR, 93
EDARADD, 93
EFEMP1, 14, 251*t*, 252–253*f*, 257
ELOVL4, 207, 212, 226*t*, 231*t*
EP300, 28*t*
EPHA2, 115*t*, 383*b*
ERCC1, 123*t*
ERCC2, 123*t*
ERCC5, 123*t*
ERCC6, 119–120, 123*t*
ERCC8, 119–120, 123*t*
EYS, 9–10, 47, 54*f*, 56*f*, 58*f*, 168*t*

F

FAM161A, 163–164*f*, 168*t*
FDX2, 369
FDXR, 369
FHL1, 258
FOXC1, 28–29, 97*t*, 105–107, 110, 389
FOXE3, 97*t*, 110, 113, 115*t*, 379, 383*b*
FRAS1, 28*t*
FREM2, 28*t*
FRMD7, 404–405
FSCN2, 168*t*
FTH1, 138
FTL, 138, 138*f*
FYCO1, 115*t*
FZD4, 325–327, 325*t*, 326*f*, 329–330

G

GALE, 141
GALK, 131*t*, 141
GALM, 141
GALNS, 86*t*
GALT, 131*t*, 141
GJA1, 28*t*
GJA3, 114, 115*t*
GJA8, 97*t*, 110, 114, 115*t*
GJB2, 94–95
GLA, 131*t*
GNAT1, 186–187
GNAT2, 45, 197–198
GNPTAB, 86
GNS, 86*t*
GPR143, 393, 394*t*, 399–400, 404*t*
GPR179, 47–48, 186
GRHL2, 74*t*, 81
GRIP3, 28*t*
GRK1, 187
GRM6, 47–48, 186
GUCA1A, 45–46, 54*f*, 56*f*, 58*f*, 205
GUCA1B, 168*t*
GUCYA1A, 203*f*

GUCYA1B, 203*f*
GUCY2D, 45–46, 189–191, 190*f*, 190*t*, 200–201*f*, 201–204, 212, 244
GUSB, 86*t*

H

HADHA, 306
HCCS, 28*t*, 382*t*, 383
HGSNAT, 86*t*, 168*t*
HK1, 168*t*
HOXA1, 415, 423
HSF4, 115*t*
HYAL1, 86*t*

I

IDH3A, 163–164*f*, 168*t*
IDS, 86*t*
IDUA, 86*t*
IFT27, 272*f*
IFT81, 203*f*
IFT140, 168*t*
IFT172, 168*t*, 271
ILK, 325–326, 325*t*
IMGP2, 168*t*
IMPDH1, 168*t*, 189–191
IMPG1, 45, 212, 226*t*, 229, 231*t*
IMPG2, 168*t*, 212, 226*t*, 229, 231*t*
ITPA, 122–123
ITPR1, 391

J

JAG1, 107, 325–326, 325*t*

K

KCNJ13, 168*t*
KCNV2, 33, 45–46, 46*f*, 201–204, 204*f*, 212
KDM6A, 28*t*
KERA, 74*t*, 97
KIF11, 26–27, 27*t*, 323, 325–326, 327*b*, 331–332, 331–332*f*
KIF21A, 408*f*, 409–411
KMT2D, 28*t*
KRT3, 74, 74*t*, 76
KRT12, 74, 74*t*, 76

L

LCAT, 89, 91, 91*f*
LRAT, 168*t*, 169–170, 190*t*
LRIT3, 47–48, 186
LRMDA/C10orf11, 394*t*
LRP5, 324–327, 325*t*, 325–326*f*, 329–330
LTBP2, 67, 69, 97–98, 97*t*, 103, 148*t*, 155
LYST, 394*t*, 399
LZTFL1, 271

M

MAB21L2, 381, 382*t*
MAF, 110, 113, 379, 414
MAFB, 414
m.3243A>G, 227, 302, 304–305, 304*f*, 425–427
m.8344A>G, 425–427
MAK, 168*t*

MAN2B1, 131*t*
MCOLN1, 85–86
MERTK, 168*t*
MFN2, 427*t*
MFRP, 385, 386–387*f*
m.3460G>A, 357–359
m.11778G>A, 357–360, 357*f*
MGME1, 427*t*
MIP, 115*t*
MIR-184, 115*t*
MIR-204, 383*b*
Mitochondrial respiratory chain, 304–305, 355, 359–360, 366
MKS1, 271
MMACHC, 313–314
MPV17, 427*t*
m.14484T>C, 357–359, 358*f*
m.14677T>C, 425–427
m.8993T>G, 425–427
MTND1, 359
MTND4, 359
MTND6, 359
MYCNA, 460
MYO7A, 168*t*, 281*t*, 282
MYOC, 97*t*, 102–103
MYRF, 385

N

NAA10, 383
NAGLU, 86*t*
NDP, 324–326, 325*t*, 326*f*, 329–330, 329*f*
NDUFS2, 368
NEU1, 131*t*
NF1, 432–433
NF2, 434*t*, 435–436
NHS, 127, 383*b*
NR2E3, 47, 168*t*, 176, 178–179, 178–179*f*, 236
NRL, 163–164*f*, 168*t*, 178
NYX, 47–48, 48*f*, 186

O

OAT, 131*t*, 288, 288–289*f*, 290
OCA2, 393, 394*t*, 397*f*, 399–400
OCRL, 131*t*, 133–134
OFD1, 168*t*, 276
OPA1, 362–365, 363–365*f*, 369, 425, 427, 427*t*
OPA3, 362–365, 369, 374
OPA4, 363–365
OPA5, 363–365
OPA8, 363–365
OPN1LW, 33–34, 197–198
OPN1MW, 33–34, 197–198
OPTN, 97*t*, 104
OTX2, 212, 229–232, 381, 382*t*, 383*b*
OVOL2, 74*t*, 81

P

PAX2, 382*t*
PAX6, 28–29, 97, 97*t*, 106–107, 110, 113, 115*t*, 148*t*, 379*b*, 389–391, 404*t*, 405
PCARE, 168*t*, 201
PCDH15, 281*t*, 282
PCDH21, 203*f*

PCYT1A, 286b
PDE6A, 168t
PDE6B, 168t, 186–187
PDE6C, 45, 197
PDE6G, 168t
PDE6H, 45, 197
PEX1, 131t
PEX7, 131t
PEX11B, 129, 130f, 131t
PHOX2A, 409f, 410
PHYH, 286b
PIKFYVE, 74t
PITX2, 28–29, 97t, 105–107, 110, 389
PITX3, 97t, 110, 113, 115t, 379b
POC1B, 203f
POLG1, 427, 427t
POLG2, 427t
POMGNT1, 168t, 327b
PRCD, 168t
PRDM5, 70, 71f
PRDM13, 9–10, 246–247, 251t
PROM1, 15, 45–46, 168t, 212
PRPF3, 47, 63, 168, 168t
PRPF6, 168t
PRPF8, 168
PRPF31, 47, 63, 168, 168t
PRPH2, 45, 47, 168t, 201–204, 207, 212, 225–226, 225–226f, 226t, 229, 231t
PRPS1, 168t
PRSS56, 385, 386f
PTCH1, 382t, 451–452
PXDN, 97t, 100, 107, 115t
PYGM, 231–232t

R

RAB18, 122–123
RAB28, 203f
RAB3GAP1, 122–123, 122f
RAB3GAP2, 122–123
R9AP, 194, 198
RARB, 381–383, 382t
RAX, 382t
RB1, 457–461, 459t, 460b, 461f, 462t
RBP3, 168t
RBP4, 379b
RDH5, 183f, 187
RDH12, 168t, 189–193, 190t
REEP6, 168t
RGS9, 194, 198
RHO, 15, 47, 168, 168t, 186–187
RIMS2, 186
RLBP1, 168t
RNASEH1, 427t

ROBO3, 418–419f, 420–421
RP1, 47, 168t, 243–244
RP2, 165b, 167, 168t, 229–232
RPE65, 63, 168t, 169–170, 189–193, 190t, 192f, 405
RPGR, 45–47, 57, 165b, 166f, 167–168, 168t, 201–205, 205f, 212
RPGRIP1, 168t, 276
RP1L1, 45, 241, 241t, 242–243f, 243–244
RRM2B, 427, 427t
RS1, 48, 236, 237f, 238, 240
RTN4IP1, 274, 276

S

SAG, 168t, 184f, 187
SALL4, 414–415
SC5D, 131t
SDCCAG8, 271
SEMA4A, 203f
SGSH, 86t
SLC2A1, 130b, 131t, 186–187
SLC4A11, 74t, 81–82, 97
SLC6A19, 131t
SLC6A20, 131t
SLC7A14, 168t
SLC24A1, 181f
SLC24A5, 394t, 400
SLC25A4, 427t
SLC25A46, 274
SLC36A2, 131t
SLC38A8, 50f, 51, 393, 394t, 404, 404t
SLC45A2, 394t, 399–400
SMOC1, 382t
SNRNP200, 168, 168t
SOX2, 4, 381, 382t, 383b
SPATA7, 49f, 168t
SPG7, 363–365, 427t
SPG11, 231–232t
SPRED1, 430
SSBP1, 363–365
STRA6, 381–383, 382t
SUCLA2, 427t
SUFU, 451–452
SUOX, 148t

T

TACSTD2, 74t
TBC1D20, 122–123
TBK1, 97t, 104
TCF4, 74t, 81–82
TEK, 97t, 98
TENM3, 382t, 383
TFAM, 427t

TFAP1, 382t
TFAP2, 28t
TGFBI, 73, 74t, 77–80, 78f, 79b
TIMP3, 14, 251t, 255f, 257–258
TK2, 427t
TMEM98, 385
TMEM126A, 368–369
TOPORS, 168t
TP63, 94
TRPM1, 47–48, 48f, 186
TSC1, 446, 447t, 448
TSC2, 446, 447t, 448
TSPAN12, 325–326, 325t, 326f
TTC8, 168t
TTLL5, 212
TUBB3, 409f, 410–411, 423
TULP1, 168t
TYMP, 427t
TYR, 15, 50f, 393, 394t, 399–400
TYRP1, 394t, 399–400

U

UBIAD1, 74t, 83
UNC119, 203f
USH2A, 9–10, 47, 168, 168t, 281t, 282, 283f
USH1C, 168t, 281t
USH1G, 281t

V

VHL, 439–443, 445b
VIM, 115t
VPS13B, 28t, 148t, 317, 319–320
VSX2, 381, 382t

W

WDPCP, 272f
WFS1, 362–365, 371–374, 373f, 427t
WHRN, 281t
WNT10, 93
WT1, 389–391

Y

YAP1, 381, 382t
YME1L1, 369

Z

ZEB1, 74t, 81–82
ZFYVE26, 231–232t
ZNF408, 325–326, 325t
ZNF469, 70
ZNF513, 168t